IfCoLog Journal of Logics and their Applications

Volume 4, Number 4

May 2017

Disclaimer

Statements of fact and opinion in the articles in IfCoLog Journal of Logics and their Applications are those of the respective authors and contributors and not of the IfCoLog Journal of Logics and their Applications or of College Publications. Neither College Publications nor the IfCoLog Journal of Logics and their Applications make any representation, express or implied, in respect of the accuracy of the material in this journal and cannot accept any legal responsibility or liability for any errors or omissions that may be made. The reader should make his/her own evaluation as to the appropriateness or otherwise of any experimental technique described.

© Individual authors and College Publications 2017
All rights reserved.

ISBN 978-1-84890-240-4
ISSN (E) 2055-3714
ISSN (P) 2055-3706

College Publications
Scientific Director: Dov Gabbay
Managing Director: Jane Spurr

http://www.collegepublications.co.uk

Printed by Lightning Source, Milton Keynes, UK

All rights reserved. No part of this publication may be reproduced, stored in a retrieval system or transmitted in any form, or by any means, electronic, mechanical, photocopying, recording or otherwise without prior permission, in writing, from the publisher.

Editorial Board

Editors-in-Chief
Dov M. Gabbay and Jörg Siekmann

Marcello D'Agostino	Melvin Fitting	Henri Prade
Natasha Alechina	Michael Gabbay	David Pym
Sandra Alves	Murdoch Gabbay	Ruy de Queiroz
Arnon Avron	Thomas F. Gordon	Ram Ramanujam
Jan Broersen	Wesley H. Holliday	Chrtian Retoré
Martin Caminada	Sara Kalvala	Ulrike Sattler
Balder ten Cate	Shalom Lappin	Jörg Siekmann
Agata Ciabttoni	Beishui Liao	Jane Spurr
Robin Cooper	David Makinson	Kaile Su
Luis Farinas del Cerro	George Metcalfe	Leon van der Torre
Esther David	Claudia Nalon	Yde Venema
Didier Dubois	Valeria de Paiva	Rineke Verbrugge
PM Dung	Jeff Paris	Heinrich Wansing
Amy Felty	David Pearce	Jef Wijsen
David Fernandez Duque	Brigitte Pientka	John Woods
Jan van Eijck	Elaine Pimentel	Michael Wooldridge

Scope and Submissions

This journal considers submission in all areas of pure and applied logic, including:

- pure logical systems
- proof theory
- constructive logic
- categorical logic
- modal and temporal logic
- model theory
- recursion theory
- type theory
- nominal theory
- nonclassical logics
- nonmonotonic logic
- numerical and uncertainty reasoning
- logic and AI
- foundations of logic programming
- belief revision
- systems of knowledge and belief
- logics and semantics of programming
- specification and verification
- agent theory
- databases
- dynamic logic
- quantum logic
- algebraic logic
- logic and cognition
- probabilistic logic
- logic and networks
- neuro-logical systems
- complexity
- argumentation theory
- logic and computation
- logic and language
- logic engineering
- knowledge-based systems
- automated reasoning
- knowledge representation
- logic in hardware and VLSI
- natural language
- concurrent computation
- planning

This journal will also consider papers on the application of logic in other subject areas: philosophy, cognitive science, physics etc. provided they have some formal content.

Submissions should be sent to Jane Spurr (jane.spurr@kcl.ac.uk) as a pdf file, preferably compiled in LaTeX using the IfCoLog class file.

Contents

Corrigenda

Correction note to "Proof theory for non-normal modal logics:
   The neighbourhood formalism" xi
   Sarah Negri

Reminiscences

In Memoriam: Grigori E. Mints, 1939–2014 813
   Solomon Feferman and Vladimir Lifschitz

Grigori Mints, a Proof Theorist in the USSR:
   Some Personal Recollections in a Scientific Context 817
   Sergei Soloviev

About Grisha Mints 841
   Anatoly Vershik

Articles

Kant's Logic Revisited 845
   Theodora Achourioti and Michiel van Lambalgen

An Ordinal-free Proof of the Complete Cut-elimination Theorem
   for $\Pi_1^1 - CA + BI$ with the ω-rule 867
   Ryota Akiyoshi

Model Theory of Some Local Rings 885
    Paola D'Aquino and Angus Macintyre

On the Complexity of Translations from Classical
    to Intuitionistic Proofs 901
    Matthias Baaz and Alexander Leitsch

Unification for Multi-Agent Temporal Logics
    with Universal Modality 939
    Stepan I. Bashmakov, Anna V. Kosheleva and Vladimir V. Rybakov

Implicit Dynamic Function Introduction and Ackermann-like
    Function Theory 955
    Marcos Cramer

Locology and Localistic Logic:
    Mathematical and Epistemological Aspects 967
    Michel De Glas

Secure Multiparty Computation without One-way Functions 993
    Dima Grigoriev and Vladimir Shpilrain

IF Logic and Linguistic Theory 1011
    Jaakko Hintikka

Commentary on Jaakko Hintikka's "IF Logic
    and Linguistic Theory" 1023
    Gabriel Sandu

Medieval Modalities and Modern Method: Avicenna and Buridan . 1029
    Wilfrid Hodges and Spencer Johnston

On the Existence of Alternative Skolemization Methods 1075
    Rosalie Iemhoff

The Logical Cone 1087
    Reinhard Kahle

Parsing and Generation as Datalog Query Evaluation 1103
    Makoto Kanazawa

Foundations as Superstructure
   (*Reflections of a practicing mathematician*) 1213
   Yuri I. Manin

Classical and Intuitionistic Geometric Logic 1227
   Grigori Mints

Commentary on Grigori Mints' "Classical and Intuitionistic
   Geometric Logic" 1235
   Roy Dyckhoff and Sara Negri

Proof Theory for Non-normal Modal Logics:
   The Neighbourhood Formalism and Basic Results 1241
   Sara Negri

Constructive Temporal Logic, Categorically 1287
   Valeria de Paiva and Harley Eades III

The Historical Role of Kant's Views on Logic 1311
   Andrei Patkul

The Logic for Metaphysical Conceptions of Vagueness 1333
   Francis Jeffry Pelletier

A Note on the Axiom of Countability 1351
   Graham Priest

Topologies and Sheaves Appeared as Syntax and Semantics
   of Natural Language 1357
   Oleg Prosorov

Long Sequences of Descending Theories and other Miscellanea
   on Slow Consistency 1411
   Michael Rathjen

Venus Homotopically 1427
   Andrei Rodin

On a Combination of Truth and Probability:
   Probabilistic IF Logic 1447
   Gabriel Sandu

Towards Analysis of Information Structure of Computations **1457**
Anatol Slissenko

Horizons of Scientific Pluralism: Logics, Ontology, Mathematics . . . **1477**
Vladimir L. Vasyukov

Correction note to "Proof theory for non-normal modal logics: The neighbourhood formalism" and basic results"

SARA NEGRI
University of Helsinki
Finland

The sequent calculus **G3n** for system **E** presented in my paper *Proof theory for non-normal modal logics: The neighbourhood formalism and basic results* (this Journal, vol. 4(4), pp. 1241–1286, 2017) is not cut free. This can be seen by showing that the valid sequent $x : \Box(A \& B) \Rightarrow \Box(B \& A)$ is not derivable without a cut. The reason for this problem is the form of the left rule for $\triangleleft$, with the formula $y : A$ in the antecedent of the conclusion

$$\frac{y \in a, y : A, A \triangleleft a, \Gamma \Rightarrow \Delta}{y : A, A \triangleleft a, \Gamma \Rightarrow \Delta} \, L\triangleleft$$

A similar form, used for example for $L\Box$ allows to reduce the number of premisses (from two to one): so instead of the rule

$$\frac{x : \Box A, \Gamma \Rightarrow \Delta, xRy \quad y : A, x : \Box A, \Gamma \Rightarrow \Delta}{x : \Box A, \Gamma \Rightarrow \Delta} \, L\Box$$

one can use the equivalent rule

$$\frac{y : A, xRy, x : \Box A, \Gamma \Rightarrow \Delta}{xRy, x : \Box A, \Gamma \Rightarrow \Delta} \, L\Box$$

In this way, an application of rule $L\Box$ is licenced just when we have an accessibility atom of the form xRy in the antecedent of the conclusion. The reason why a similar reduction doesn't work in the case under discussion is that the formula $y : A$ doesn't behave like a relational atom: it can be principal in a right rule and therefore a cut

---

We thank Nicola Olivetti and Tiziano Dalmonte for pointing out the problem discussed in this note.

with left premiss derived by a right rule with $y : A$ principal and left premiss derived by $L \triangleleft$ with $y : A, A \triangleleft a$ in the conclusion cannot be permuted.

We can obtain a cut-free sequent calculus for system of **E** by just avoiding the simplification step used for $L\Box$ and using the rule with two premisses

$$\frac{A \triangleleft a, \Gamma \Rightarrow \Delta, y : A \quad y \in a, A \triangleleft a, \Gamma \Rightarrow \Delta}{A \triangleleft a, \Gamma \Rightarrow \Delta} L \triangleleft$$

Here y is an arbitrary label, but it is enough—by the usual argument that shows analyticity as an application of height-preserving substitution of labels—to restrict the rule to labels in the conclusion.

All the results stated in the paper hold with the two-premiss version of the rule; obvious modifications to account for the new form of the rule are needed in Lemma 3.3, Lemma 4.2, Lemma 4.5, Theorem 4.9, Theorem 5.3, Definition 5.4, and Lemma 5.5. For completeness, these modifications are detailed below.

Lemma 3.3. *Rule RE is admissible in* **G3E**.

Proof. By the following derivation:

$$\frac{\dfrac{\dfrac{x:A \Rightarrow x:B}{a \Vdash^\forall A \Rightarrow a \Vdash^\forall B}\ 3.2 \quad \dfrac{y:B\ldots \Rightarrow \ldots y:A \quad y \in a \ldots \Rightarrow \ldots y \in a}{y:B, a \in I(x), a \Vdash^\forall A, A \triangleleft a \Rightarrow x : \Box B, y \in a}L\triangleleft}{\dfrac{a \in I(x), a \Vdash^\forall A, A \triangleleft a \Rightarrow x : \Box B, a \Vdash^\forall B \quad a \in I(x), a \Vdash^\forall A, A \triangleleft a \Rightarrow x : \Box B, B \triangleleft a}{\dfrac{a \in I(x), a \Vdash^\forall A, A \triangleleft a \Rightarrow x : \Box B}{x : \Box A \Rightarrow x : \Box B}L\Box}R\Box}R\triangleleft$$

QED

Lemma 4.2(2). Sequents of the following form are derivable in **G3n*** for arbitrary formulas A and B in the propositional modal language of **G3n***:

2. $A \triangleleft a, \Gamma \Rightarrow \Delta, A \triangleleft a$

Proof. 2. By the following derivation

$$\dfrac{\dfrac{x:A, A \triangleleft a, \Gamma \Rightarrow \Delta, x \in a, x : A \quad x \in a, x : A, A \triangleleft a, \Gamma \Rightarrow \Delta, x \in a}{x:A, A \triangleleft a, \Gamma \Rightarrow \Delta, x \in a}L\triangleleft}{A \triangleleft a, \Gamma \Rightarrow \Delta, A \triangleleft a}R\triangleleft$$

where one topsequent is derivable by inductive hypothesis and the other is initial.

QED

Lemma 4.5(13).

13. If $\vdash_n A \triangleleft a, \Gamma \Rightarrow \Delta$ then $\vdash_n A \triangleleft a, \Gamma \Rightarrow \Delta, y : A$ and $\vdash_n y \in a, A \triangleleft a, \Gamma \Rightarrow \Delta$.

Theorem 4.9. Cut is admissible in **G3n***.

Proof. 4. The cut formula is $A \triangleleft a$, principal in both premises of cut. We have:

$$\dfrac{\dfrac{\mathcal{D}}{x : A, \Gamma \Rightarrow \Delta, x \in a} R \triangleleft \quad \dfrac{A \triangleleft a, \Gamma' \Rightarrow \Delta', y : A \quad y \in a, A \triangleleft a, \Gamma' \Rightarrow \Delta'}{A \triangleleft a, \Gamma' \Rightarrow \Delta'} L \triangleleft}{\Gamma, \Gamma' \Rightarrow \Delta, \Delta'} \text{Cut}$$

The cut is converted as follows:

$$\dfrac{\dfrac{\Gamma \Rightarrow \Delta, A \triangleleft a \quad A \triangleleft a, \Gamma' \Rightarrow \Delta', y : A}{\Gamma, \Gamma' \Rightarrow \Delta, \Delta', y : A} \text{Cut} \quad \dfrac{\mathcal{D}(y/x)}{y : A, \Gamma \Rightarrow \Delta, y \in a} \quad \dfrac{\Gamma \Rightarrow \Delta, A \triangleleft a \quad y \in a, A \triangleleft a, \Gamma' \Rightarrow \Delta'}{\dfrac{y \in a, \Gamma, \Gamma' \Rightarrow \Delta, \Delta'}{y : A, \Gamma^2, \Gamma' \Rightarrow \Delta^2, \Delta'} \text{Cut}} \text{Cut}}{\dfrac{\Gamma^3, \Gamma'^2 \Rightarrow \Delta^3, \Delta'^2}{\Gamma, \Gamma' \Rightarrow \Delta, \Delta'} \text{Ctr}^*}$$

where the two upper cuts are of reduced cut height and the lower ones of reduced weight of cut formula because $\mathtt{w}(y \in a) < \mathtt{w}(A \triangleleft a)$, $\mathtt{w}(y : A) < \mathtt{w}(A \triangleleft a)$. QED

Theorem 5.3. If $\Gamma \Rightarrow \Delta$ is derivable in **G3n*** (respectively **G3nM***, **G3nC***, **G3nN***), then it is valid in the class of neighbourhood frames (respectively neighbourhood frames which are supplemented, closed under intersection, containing the unit) with the * properties.

Proof. If the last rule is $L \triangleleft$, assume that the premises $A \triangleleft a, \Gamma \Rightarrow \Delta, x : A$, $y \in a, A \triangleleft a, \Gamma \Rightarrow \Delta$ are valid, and let (ρ, σ) be an arbitrary SN-realisation with (1) $\mathcal{M} \models_{\rho,\sigma} A \triangleleft a, \Gamma \Rightarrow \Delta, x : A$ and (2) $\mathcal{M} \models_{\rho,\sigma} y \in a, A \triangleleft a, \Gamma \Rightarrow \Delta$ and assume $\mathcal{M} \models_{\rho,\sigma} A \triangleleft a, \Gamma$. If (1) gives that there is B in Δ such that $\mathcal{M} \models_{\rho,\sigma} B$ we are done. Else we have $\rho(x) \in [A]$; since by assumption $[A] \subseteq \sigma(a)$, we have $\rho(x) \in \sigma(a)$, thus $\mathcal{M} \models_{\rho,\sigma} x \in a$. From (2) and it follows that there is B in Δ such that $\mathcal{M} \models_{\rho,\sigma} B$. QED

Definition 5.4($L \triangleleft$). We say that a branch in a proof search from the endsequent up to a sequent $\Gamma \Rightarrow \Delta$ is *saturated* with respect to a rule $L \triangleleft$ if the following condition holds

($L \triangleleft$) If $A \triangleleft a$ and y are in $\downarrow\Gamma$, then $y \in a$ is in Γ or $y : A$ is in Δ.

Lemma 5.5(d). Let $\mathcal{B} \equiv \{\Gamma_i \Rightarrow \Delta_i\}$ be a saturated branch in a proof-search tree for $\Gamma \Rightarrow \Delta$. Then there exists a countermodel $\mathcal{M}$ to $\Gamma \Rightarrow \Delta$, which makes all the formulas in $\boldsymbol{\Gamma}$ true, and all the formulas in $\boldsymbol{\Delta}$ false.

Proof. (d) If $A \triangleleft a$ is in $\boldsymbol{\Gamma}$, let y be an arbitrary world in the model, that is, by definition of $\mathcal{M}$, a label in $\downarrow\boldsymbol{\Gamma}$. Then by by saturation $y \in a$ is in Γ or $y : A$ is in Δ, so by inductive hypothesis $\mathcal{M} \not\models_{\rho,\sigma} y : A$ or $\mathcal{M} \models_{\rho,\sigma} y \in a$. Overall, this means that $\mathcal{M} \models_{\rho,\sigma} A \triangleleft a$. QED

In Memoriam: Grigori E. Mints, 1939–2014

Solomon Feferman
Stanford University, Stanford CA 94305-2125, USA
s.feferman@gmail.com

Vladimir Lifschitz
*Department of Computer Science, University of Texas at Austin
2317 Speedway, Stop D9500, Austin TX 78712-0233, USA*
vl@cs.utexas.edu

On May 29, 2014, ten days before his 75th birthday, Grigori ("Grisha") Mints died at Stanford, California of cardiac arrest; he had suffered a serious stroke a month before from which he never recovered. At the time of his death Mints held the position of Professor of Philosophy at Stanford University with courtesy appointments in Mathematics and Computer Science. His death unexpectedly cut short a distinguished and highly active career marked by a prodigious output of great breadth in logic and its applications. This included three books [4, 5, 9], another ten more of which he was an editor or translator, over 200 articles and over 3000 (!) reviews. His main contributions were to proof theory, constructive mathematics, intuitionistic logic, modal logic, and automated deduction.

Mints was born on June 7, 1939 in Leningrad, USSR (currently St. Petersburg, Russia). He obtained the B.S. and M.S. in Mathematics from Leningrad State University (currently St. Petersburg State University) in 1961, with a thesis on proof search in the classical predicate calculus. Working under the direction of Nikolai A. Shanin, Mints obtained the Ph.D. in Mathematics at the Leningrad S. U. in 1965, with a thesis on predicate and operator variants for theories of constructive mathematics (cf. the translation in [1]). Finally, in 1990 he was awarded the D. Sc. in Mathematics at the Leningrad S. U. for a work on proof transformations and synthesis of programs. Mints was elected to the Estonian Academy of Sciences in 2008 and to the American Academy of Arts and Sciences in 2010.

From 1961 to 1979 Mints held the position of Research Associate at the Leningrad Branch of the Steklov Mathematics Institute. After submitting his request to emigrate from the Soviet Union in 1979, he resigned his position at the Steklov Institute

---

This article originally appeared in the Bull. of Symbolic Logic 21 (2015) 31–33. Copyright is held by the Association for Symbolic Logic and it is being reprinted here with their permission.

so as not to endanger the situation of his colleagues there by his possible association with them. In the difficult period that followed, among other things Mints supported himself by doing programming jobs and translating books and articles on logic from English into Russian. Meanwhile he was able to establish connections with the Institute of Cybernetics in Tallinn, Estonia, where he obtained a part time position as Research Associate from 1980 to 1984. This turned into a full time position as Senior Research Associate from 1985 to 1991. Mints was finally permitted to emigrate to the United States in that year, when he was appointed Professor of Philosophy at Stanford University.

The direction of Mints' early work was determined to a large extent by the main interests of Nikolai A. Shanin, who, along with Andrei A. Markov, Jr., led a research group at the Steklov Institute devoted to "Russian-style" constructive mathematics.[1] Shanin's group also worked on automated reasoning, with emphasis on generating "natural" proofs, to which Mints made a number of contributions. His work in this period was also distinguished among other things by several publications on analogues of Herbrand's theorem for intuitionistic logic. Another highlight is the famous article, "What can be done in PRA?" [2] (original Russian in 1976), whose main result was obtained independently by Charles Parsons and by Gaisi Takeuti. Mints' book [4] contains the English translations from the Russian of a selection of thirteen of his articles on proof theory from the period up to 1979.[2] These concentrate on normalization theorems for classical, intuitionistic and modal systems as well as their applications to coherence theorems in category theory.

While in Tallinn, Mints studied the mathematical principles behind the program synthesizer PRIZ, designed by a group at the Institute of Cybernetics led by Enn Tyugu. Estonian computer scientists thought that their algorithm was complete, but Mints came up with an example that PRIZ could not handle. The algorithm was then improved, and Mints established the completeness of the new version in joint work with Tyugu in 1982.

At Stanford, Mints became one of the mainstays of the interdepartmental program in logic, teaching the subject at all levels, advising students, and directing doctoral dissertations. Together with Solomon Feferman, he led the seminar in logic and the foundations of mathematics. His research work continued unabated along all the general lines given above. In addition, among other things, his work [10]

---

[1] In his article [3], Mints surveyed work in the USSR on proof theory and constructive mathematics from 1925 to 1969. See also the article [12] with Sergey I. Nikolenko.

[2] Most regrettably, the volume [4] provides no information regarding the original publication data for these articles, not even their dates. These can be reconstructed from a C.V. that Mints prepared for the Stanford Philosophy Department in 2007 that is referred to but not repeated in later expansions of it.

with Philip Kremer on dynamic topological logic initiated an interesting new direction, and he contributed to linear logic for intuitionistic and natural deduction systems [7, 8]. But, most importantly, in proof theory he was noted for almost single-handedly extending Hilbert's "epsilon substitution method" to various first-order and second-order subsystems of analysis, as in [6, 11, 13, 14], with work still in progress at the time of his death.

At the professional level Mints was a member of a number of editorial boards, and of program and organizing committees for various meetings, both national and international, in which he was also an active participant. Of special concern to him was the continued fostering of ties with colleagues in the former Soviet Union. The last conference that he helped organize and at which he spoke, entitled "Philosophy, Mathematics, Linguistics: Aspects of Interaction 2014", was held at the Steklov Institute in St. Petersburg in the month of April, 2014; cf. http://www.pdmi.ras.ru/EIMI/2014/PhML/. Sadly, it was from that meeting that he returned with an illness that led in the end to his death.

Besides his extensive and enduring contributions to our subject, Grisha Mints is remembered by his colleagues, friends and students with great affection as a very warm human being — always accessible, patient, and ready to help — and for his general intellectual enthusiasm married with a keen sense of humor illuminated by a surprising font of historical knowledge.

Acknowledgements. We are grateful to Marianna Rozenfeld, Yuri Gurevich, Vladik Kreinovich, and William Tait for their helpful comments on a draft of this piece.

Cited writings of Grigori E. Mints

[1] On predicate and operator variants of the formation of theories of constructive mathematics. *Translations of the American Mathematical Society*, 100:1–68.

[2] What can be done in PRA? *Journal of Soviet Mathematics*, 14:1487–1494, 1980.

[3] Proof theory in the USSR: 1925–1969. *The Journal of Symbolic Logic*, 56:385–424, 1991.

[4] *Selected papers in proof theory*. Bibliopolis, Naples, and North-Holland, Amsterdam, 1992.

[5] *A short introduction to modal logic*. Center for the Study of Language and Information, Stanford, 1992.

[6] Gentzen-type systems and Hilbert's epsilon substitution method. I. In D. Prawitz et al., editors, *Logic, Methodology and Philosophy of Science IX*, pages 91–122. Elsevier, Amsterdam, 1994.

[7] Linear lambda-terms and natural deduction. *Studia Logica*, 60:209–231, 1998.

[8] Normal deduction in the intuitionistic linear logic. *Archive for Mathematical Logic*, 37:415–426, 1998.

[9] *A short introduction to intuitionistic logic*. Kluwer Academic/Plenum Publishers, New York, 2000.

[10] Dynamic topological logic (with P. Kremer). *Annals of Pure and Applied Logic*, 133:231–246, 2005.

[11] Cut elimination for a simple formulation of epsilon calculus. *Annals of Pure and Applied Logic*, 152:148–160, 2008.

[12] History of the Leningrad (St. Petersburg) school of constructive mathematics and proof theory (with S. I. Nikolenko). In A. Schumann, editor, *Logic in Central Europe and Eastern Europe: History, science and discourse*, pages 381–388. Rowman & Littlefield Publishers, Latham, MD, 2012.

[13] Non-deterministic epsilon substitution for ID_1: Effective proof. In U. Berger et al., editors, *Logic, Construction, Computation*, pages 325–342. Ontos Verlag, Frankfurt, 2012.

[14] Epsilon substitution for first- and second-order predicate logic. *Annals of Pure and Applied Logic*, 164:733–739, 2013.

Grigori Mints, a Proof Theorist in the USSR: Some Personal Recollections in a Scientific Context

Sergei Soloviev[*]
IRIT, Université de Toulouse, Toulouse, France
soloviev@irit.fr

Abstract

The paper is based on my recollections of Grigori Mints (1939–2014) completed by a survey of his research work in a scientific context. I speak mostly about the Soviet period of his life and work (until 1991), and sometimes go beyond the purely scientific aspects to show the atmosphere of these times.

Keywords: Grigori Mints, Biography, Logic in the USSR, History of Logic.

1

I first met Grigori when I was a second-year undergraduate at the Faculty[1] of Mathematics and Mechanics of Leningrad State University at the end of 1975 or in the beginning of 1976[2]. In the middle of our third year, we had to choose our specialization, and I had been considering mathematical logic as an option; simultaneously, I had been working on a project on uniform contact schemas under the supervision of N. K. Kossovsky, but I was attracted to the more theoretical aspects of logic. I had an acquaintance, Michael Gelfond, who was one of my teachers at the school №30 (a high school specialized in mathematics). He also was an associate of the Group of Mathematical Logic at the Leningrad Department of Mathematical Institute of the Academy of Sciences (usually called LOMI), where he defended his PhD thesis

---

[*]Partially supported by the Government of the Russian Federation Grant 074-U01 awarded to the ITMO University, St. Petersburg, Russia (associated researcher).
[1]More or less corresponds to *School*, as in *Oxford School of English*.
[2]He was often called "Grisha", a more familiar form, but for me Grigori sounds more appropriate because during several years he was my adviser.

in 1974[3]. Gelfond advised me to go to the seminar of the Group of Mathematical Logic that was held at LOMI on Mondays, and to approach Mints.

I do not remember, whether I had to call Mints before and get an appointment. To enter LOMI I had to say that I go to the seminar because it was open to the colleagues of other institutes. To Mints I had to mention Gelfond's recommendation. In any case, when I approached Mints he suggested me to take the Russian translation of S. C. Kleene's "Mathematical Logic" [13], the so called "Red Kleene"[4], and solve all the exercises. In fact, I never solved all of them because after some time, when I solved approximately one third (taken from all chapters), we had a much more lengthy and substantial discussion, and Mints proposed me to think about some original problems that were not merely exercises.

At this time he was much interested by some applications of proof theory to the theory of categories. It was Jim Lambek who first noticed the link between categories with additional structure and deductive systems. He published a series of three papers called "Deductive systems and categories" [20–22]. Let me mention that two of these papers appeared in Springer Lecture Notes in Mathematics and were accessible in the LOMI's scientific library. Mints knew also about S. Mac Lane's works on coherence, but as far as I remember, most of all his attention was attracted by the recent paper by Mann [24] on the connection between the equivalence of proofs and the equality of morphisms in Cartesian Closed Categories probably because (in difference from Lambek) it considered natural deductions that were well known to Mints. This connection opened an interesting perspective in that certain problems of category theory, first of all the so called coherence problems (problems of commutativity of diagrams) may have nice proof-theoretical solutions.

In this essay I will try to render my impression of the style of Grigori Mints as a researcher. He was always very open, receptive towards the newest tendencies in all domains of world science related to proof theory and logic. Of course my impressions are subjective, and alone they cannot give a true idea of the whole extent and significance of his works, but I will try to complete this subjective part by a more academic survey based on publications, documents, and testimonies of G. E. Mints colleagues and friends.

Among the events that impressed me at this early period of my acquaintance with Mints was the visit (and talk) of an outstanding logician, G. Kreisel (1923–2015) to LOMI in June 1976 that Mints organized.

---

[3]The head of the Group of Mathematical Logic (and Gelfond's PhD adviser) was N. Shanin. Gelfond emigrated to the United States in 1978. He is now a professor of computer science at Texas Tech University.

[4]The translation (in red cover) was published in 1973. The book was translated by Yu. A. Gastev. Mints was the editor of the translation.

The weather was unusually cold, but the central heating was already switched off because it was June.

At this time the building that LOMI occupied today[5] was under renovation and the institute was temporarily "exiled" to a former school far from the city center. It stood in an inner courtyard surrounded by gray buildings heavily styled since they were built in Stalin's times. Understandably, the conditions were more crowded. The group of mathematical logicians used a former classroom, and the seminars were held in the same room. I remember several tables, chairs, and a large worn leather divan, an object of amused pride in the group. Kreisel had to use an ordinary school blackboard for his talk. I also recall his coat, that seemed to me to be too light for such cold weather. Later I learned that these light coats protected against cold and rain much better than those "Made in USSR"'.

At that time I hardly asked myself what role Kresiel had played in the development of Mints as a scientist. I had no idea of the intense correspondence that Mints had with western scientists, often in spite of the obstacles and complexities typical of life in the USSR. Later I have heard from Mints that he considered Kreisel as one of his teachers[6]. He corresponded with many other Western scientists as well, for example, with A. S. Troelstra (b. 1939), S. Feferman (b. 1928), S. Mac Lane (1909–2005). In the archive of A. S. Troelstra first mention of the correspondence with G. E. Mints may be found in 1970[7].

To give a better impression of "l'air du temps", it is worth to mention that the fact of correspondence with the West did not seem strange to me at all - the idea that science is indivisible, and the borders should not be an obstacle for scientific exchanges, was common among academic researchers and the university people at this time. The academic community in the USSR remembered very well that before the Revolution of 1917 and even in 1920s scientists easily published papers and exchanged letters in all main European languages (*cf.* [85]), and did not want the return of Stalin's times.

Not long after my acquaintance with Mints I was invited to visit him at home – of course in connection with the problems he wanted to propose. A modest flat in one of the many areas of recent housing development, rather far (about 30 min. by tram or bus) from underground stations. Grigori lived there with his wife and daughter. I remember an impressive mathematical library – yellow spines of Springer Lecture

---

[5]27, Fontanka river embankment in the historic center.

[6]One of the fruits of this early collaboration between Mints and Kreisel was a lengthy paper published in Springer Lecture Notes [16]. At the end of this paper there is an appendix, and the authors notice that it is based on correspondence between two of them - obviously Mints and Kreisel.

[7]See Index of the Troelstra Archive, https://www.illc.uva.nl/Research/Publications/Reports/X-2003-01.text.pdf.

Notes in Mathematics, foreign journals...

The third year at the University (in my case 1976/77) was the year of specialization, the scientific domain for the future graduation had to be chosen. I was included in the group of geometry and mathematical logics. The University administration agreed that Mints, who did not work at the University, would become my scientific adviser, and later supervisor of my graduate work. About these years, from 1977 to 1980, it is worth to speak in more detail.

The main problem that Mints proposed me to consider was the so called coherence problem for Cartesian Closed Categories. In proof-theoretical formulation, I had to prove that all logical derivations of certain classes are equivalent.

There were also lesser problems, that later turned out to be of independent interest, for example, the problem of transformations of derivations that preserved their equivalence. Mints suggested to read an old paper (1953) by G. F. Rose [87] where an interesting transformation of formulas (the decreasing of implicative depth) was considered, and to generalize it to the derivations. It required to go to the library of LOMI and make a considerable effort with English that I did not yet know well, but the paper was there and the effort within the limits of possible.

Georg Kreisel clearly distinguished what he called the "General Proof Theory" and the "Theory of Proofs" [17]:

> A working definition of Proof Theory is essentially interested in what is traditionally called the essence or, equivalently, 'defining property' of proofs, namely their being valid arguments... general proof theory develops such refinements as the distinction between different kinds of validity, for example, logical or constructive validity (and other) familiar from the foundational literature... In contrast, the Theory of Proofs questions the utility of these distinctions compared to taking for granted the validity at least of currently used principles. Instead, this theory studies such structural features as the length of proofs and especially relations between proofs and other things, so to speak, 'the role of proofs'...

As far as I know, Mints shared his views, and his own works mostly belong to the theory of proofs in Kreisel's sense. His interest in Categorical Logic, where logical derivations are seen as morphisms in appropriate categories, and equivalence relations on derivations generated by categorical semantics are studied, is in line with this approach.

In this period Mints wrote two long papers [52, 56], that considered the correspondence between certain systems of propositional logic and categories with additional structure. Main results included a solution of the "word problem" (equality of morphisms) in free categories with additional structure of several types: closed, symmetric closed, monoidal closed, symmetric monoidal closed, and cartesian closed, in all cases based on verification of the equivalence of derivations. As the main tool, the normalization of lambda-terms associated with derivations was used. Normalization at this time was relatively well explored by proof theorists, but its use for accurate and extensive study of categorical properties of proofs was new. Mints knew about a work of Mann [24] who used normalization for partial characterization of morphisms in Cartesian Closed Categories, and wanted to complete and extend his approach. Mints knew also about works of Lambek [20–22] and Kelly-Mac Lane [12], who with some success used cut-elimination[8]. Some of the systems considered by Mints correspond to what is called nowadays, after Girard's work [8], multiplicative linear logic. In his paper [44], 10 years before Girard, Mints cited several papers by Anderson and Belnap (*e.g.*, [1]), Kreisel [15], and Prawitz [86]. Some indirect influence of Lambek [19] may be possible.

One of two papers, published in Kiev [56], was hard to find, and Mints gave me the manuscript to read.

At this period, when I wrote under Mints' direction my diploma work, he had also one PhD student, Ali Babayev. His story had some flavour of mathematical romantics. I mention it, because it shows Mints as an attentive and caring supervisor. Ali was first sent from Azerbaïdjan to Moscow for an internship under supervision of a prominent algebraist and logician Sergei Adian, but it did not go very well, and Ali felt himself somewhat lost. Mints met him during a visit to Moscow and invited to LOMI, to try to do a PhD thesis there under his own supervision. One of the problems that Mints suggested to Ali was identical to my own – he had to look for a proof of the so called coherence theorem for canonical morphisms in Cartesian Closed Categories, but we had to use different methods (Ali – lambda-calculus and natural deduction, and myself – Gentzen sequent calculus). Of course, Ali, as a PhD student had to work on several other problems. He had to explore other kinds of Closed Categories, for example, the so called Biclosed Categories, and related coherence problems. In the end we proved the coherence theorem for Cartesian Closed Categories more or less simultaneously.

Main results of this period of my work under direction of G. E. Mints were published in three papers in the volume 88 of "Zapiski" (1979). A long paper on coherence theorem contained two independent proofs, one obtained by Ali and another by myself [89]. Another paper [90] considered the preservation of equivalence of derivations under reduction of formula's depth by Rose's method. The third [91], a note of 3 pages, presented an example of exponential growth of length of natural deductions that correspond in a standard way to the sequential ones.

---

[8]Cut-elimination alone does not permit to define normal forms, and so is not enough to solve the problem of equivalence.

Mints published in the same volume two papers about various normalization problems concerning the arithmetical deductions and deductions in predicate calculus [57, 58]. To me and Ali – the younger generation – it was difficult to figure out that for him a long and a very fruitful period of relatively peaceful creative work will soon come to an end.

2

All personal recollections have only limited meaning if they are not presented in a larger context, based on documents and information gathered from other people. This section is mostly devoted to an outline of such a more objective context.

Grigori Efroimovich Mints was born in Leningrad on June 7, 1939. The names of his parents were Efroim Borukhovich Mints and Lea Mendelevna Novik.

A few more biographical details. During the war, the family was evacuated and afterwards returned to Leningrad. In 1946 Grigori entered the school №241 at Oktyabrski district of the city of Leningrad. As an overwhelming majority in his generation, at the age of 14 he was enrolled to "Komsomol" (the union of communist youth). Of course, at this period of Soviet history for most of its members "Komsomol" was no more a bridge to the career in communist party, but mere formality. He finished school in 1956 and in the same year passed the exams and entered the Faculty of Mathematics and Mechanics of Leningrad State University, together with other future members of the Group of Mathematical Logic, S. Yu. Maslov (1939–1982) and G. V. Davydov. At the same time their future wives were enrolled.

Mints was taken to the section of computational mathematics[9], that had at this time a "mixed" reputation in comparison with pure mathematics. On the one hand, the students of this section were considered as an elite of a sort, one had to have the very good marks at the entrance exams, at the other there was a risk because the graduates often were send to the institutes that worked on secret military projects, the so called "postboxes", since their street addresses were not publicly known. Remember that Soviet nuclear and space programs had at this time their "golden era", and they needed enormous amount of computations. By the way, it was also a refuge for cybernetics, that was not approved by Marxist philosophers, but they had no access to projects that had military significance. For a former student go to a "postbox" meant that it will be difficult to communicate with colleagues outside, and impossible to have contacts abroad. Happily for Mints and his friends, about 1956 the situation started to relax, and this permitted Mints, Maslov, and Davydov to be

---

[9]In 1957 another future logician, V. P. Orevkov, also entered mathematical faculty.

recruited immediately after graduation by LOMI, and become first junior members of the Group of Mathematical Logic just organized there under the leadership of one of the creators of constructive approach in mathematics, N. A. Shanin (*cf.* [26]).

In the end of 1960/61 academic year Mints defended his diploma's work under the title "An Algorithm for Proof-Search in the Classical Predicate Calculus", and was awarded the diploma "with excellence" in the specialty "mathematics". He was immediately recruited by LOMI, and had to begin his work there on August 1st. His initial position was that of a research assistant, and he remained at this post a bit more than one year.

In 1962 the first two scientific papers by Mints were published in "Doklady" of the Academy of Sciences of the USSR (DAN) [30, 31].

In 1963 he was elected by the Academic Council of LOMI to the position of Junior Researcher.

It followed afterwards almost two decades of uninterrupted and very impressive progress. In 1979 the official report signed by the administration of LOMI when the candidature of G. E. Mints for the position of senior researcher was proposed to Academic Council mentions that he has 60 published research papers and 13 articles for Mathematical Encyclopedia, Encyclopedia of Cybernetics, and other editions of similar kind. Mints was a member of the Group of Mathematical Logic, and this group itself was a remarkable association of the very talented and highly motivated researchers. In particular, it was developed and programmed by this group one of the first algorithms for automated proof-search in propositional and predicate calculus. All members of the group participated in this project.

As we shall see, one may discern more or less clearly the stages when the new interests became manifest in Mints' published works. A "cumulative effect" is obvious, *i.e.*, the intensive research work helps to master new subjects faster, and on a deeper level.

During first 3–4 years at LOMI, proof theory, which is to become later the center of G. E. Mints interests, seems not yet to take a central position. In 1963 the joint paper (with V. P. Orevkov) "A generalization of the theorems of V. I. Glivenko and G. Kreisel to a class of formulae of the predicate calculus" is published in DAN [32]. The name of G. Kreisel, who played later a very important role in Mints' scientific development, first appears in this early publication. In 1964 a long (54 p.) paper "On predicate and operator variants of the formation of theories of constructive mathematics" was published in "Trudy" of the Steklov Institute of Mathematics [33]. It contained main results of Mints' PhD ("candidate of sciences") thesis, defended in 1965.

Until the end of his work at LOMI Mints remained a junior research fellow. With other members of the Group of Mathematical Logic he often got "bonuses" (*i.e.*, complements to salary) for successful research. Generally speaking, the position of a junior researcher for a "candidate of sciences" at this time was not something unusual, though if we take into account the high research activity, typical for Mints, it seems rather questionable. His promotion to the position of a senior research fellow was considered only in the last months before he resigned. I discuss this below.

In 1965 the "Nauka" editions published the joint work that partly reflected the collective efforts of Logic Group in the development of an algorithm for automated proof search [5]. According to Mints annual reports, he participated in the development of the program modules that concerned classical propositional calculus, classical predicate calculus with functional symbols, and in programming of the module "extraction" of this algorithm. The program ran on one of the first Soviet computers "Ural".

After the defense of his "candidate nauk" (PhD) thesis the scope of G. E. Mints work quickly expanded. He got into problems related to the central themes of mathematical logic in the XXth century. At the same time it became clear that its core was certainly the theory of proofs.

A personal feature of his style was an intense work on translations and surveys, and detailed comments to these translations and papers written by other researchers, that often contained the original results.

For example, in 1967 the collection of translations that included classical works in proof theory (papers by Gentzen, Gödel, Kleene and others), called "Mathematical theory of logical inference" was published [28]. Mints translated there four papers and wrote the 39 pages appendix "Herbrand Theorem" [42]. It contains, in particular, his own results about admissibility of substitution of terms for terms, used to correct an error in Herbrand's proof.

The survey [27] (a joint paper with S. Yu. Maslov and V. P. Orevkov) was first work by Mints to be published abroad.

He wrote several appendices to the Russian translation of Kleene's "Mathematical Logic" [13].

In 1974 he published a long paper on the modal logics "The Lewis System and the System T" as an appendix to the Russian translation of R. Feys' book on modal logic [45].

An important survey [48] was published in 1975.

The same year a long "educational" paper [16] (the already mentioned joint work with Kreisel and Simpson) was published in the Springer Lecture Notes.

He wrote several appendices on proof theory to the Russian translation of Barwise's "Handbook of Mathematical Logic"[10].

Back to the 60es, among other works that illustrate the rapid thematic expansion of Mints' work, let me mention his papers on modal logic [34], on Skolem's method of quantifier elimination [36], on embedding operations [35], and on admissible rules [38]. His work on Skolem's method for constructive predicate calculus was presented at the ICM in Moscow in 1966. (The collective work on machine proof search was also presented there.)

Until the end of 60s the most important works of Mints were published in the Proceedings of Steklov Mathematical Institute (MIAN), and the short announcements of important results in "Doklady" of the Academy of Sciences (DAN). In the end of 60s the requirements for the papers to be published in the LOMI's own series, "Zapiski" were changed. The longer papers that contained the full proofs and a detailed analysis of the problems under consideration could be published.

The simplification of publishing process, according to my experience, in many cases may be stimulating for research. Since 1967, when the first volume of "Zapiski" devoted to logic (vol. 4) appeared, until the end of his work at LOMI, almost all major works written by Mints were published there.

The "Zapiski" in the 60s–80s represented, to my opinion, an interesting example of a balance between creative research work and the selection process for publication. The papers were accepted for publication only after a talk at the Logic Seminar. To be presented, the talk had to be approved, usually on the basis of the short abstract, by the senior members of the Logic Group[11]. When the volume was prepared, the text was read by some colleagues who played the role of referees. It is clear, that with this method of selection the results strongly depend on the ethical and scientific level of a research collective, but if it is scientifically and ethically adequate then the efficiency may be much higher than with "blind" selection methods that are common nowadays and assume certain level of mutual distrust.

I shall not give below a detailed account of all Mints works of the years that follow, because they are too numerous to be considered in this paper, but outline the main directions of his research and speak about some of the most significant papers.

The main topics that attracted the attention of G. E. Mints when he worked at LOMI are roughly the following:

First of all, his interest to general problems of proof theory, such as cut elimination, normalization, behavior of quantifier rules (including Herbrand theorem), never

---

[10]Russian translation of "Handbook" was published in 4 volumes, v. 4, "Proof Theory" with these appendices was published in 1982.

[11]Here only science mattered, and in this sense Mints of course was one of senior members.

disappeared. It may be said, that this interest was always present as a background or at a technical side even when the main theme was different.

Other topics were:

- Modal logic

- Derived and admissible rules

- Infinite derivations and arithmetic

- Substructural and categorical logics

- Theory of Hilbert's ε-symbol

Modal logic. All Mints' works on modal logic concern certain proof-theoretical aspects of modal systems. For example, embedding operations considered in [34] are the operations that transform the derivations of one system in the derivations of another. Some other Mints' papers of this period on modal logic: [37, 39, 45, 55]. A connection with provability logics is to be noticed, *e.g.*, in the beginning of the paper [39] Mints says: "necessity ... is interpreted as provability in classical propositional calculus"[12].

Derived and admissible rules. [38, 43]. These papers may be seen as important steps towards the works of V. Rybakov and others, who obtained the criteria of admissibility of inference rules in large classes of logics (see, *e.g.*, [88]).

Infinite derivations and arithmetic. [41, 47, 49, 50, 57, 58]. Probably the most cited is [50]. The approach proposed by Mints (to consider transfinite derivations but study them using finitistic means) turned out to be very fruitful for extraction of constructive content of classical proofs (see, *e.g.*, the recent book [14]).

Substructural and categorical logics. [44, 52, 56]. As Mints himself explained in the end of [52], his cut-elimination theorem for relevant logic [44] provided the substantial part of the normalization proof for the system that he developed for symmetric monoidal closed categories in [52]. His use of proof theory in these papers is quite elegant. The reader may see three kinds of logical systems in interaction: Hilbert-style systems, Gentzen calculi, and natural deduction. They are used to

---

[12]The connection between modal logic and provability logic is known since Gödel [9], but Mints' work may be seen as one of the inspirations for future fundamental works on provability logic, for example, by Artemov [2].

represent and explore various aspects of categorical structures. It becomes clear that not just some isolated methods, but the approach of the theory of proofs "as a whole" has a deep affinity with the theory of categories with structure (closed, symmetric closed, monoidal closed, symmetric monoidal closed, cartesian closed categories *etc.*). No doubt, these works contributed greatly to the development of categorical logic in its proof-theoretical aspect. These works and their ideas are still "in circulation". Let us cite, for example, [14] and [95] (especially Ch. 8).

Theory of Hilbert's ϵ-symbol. Mints (with Smirnov [92] and Dragalin [6]) initiated the research on ε-symbol in the USSR, though before 1979 he published only one work on this subject [46]. Mints continued to work actively on the theory of the ε-operator after 1979. His last papers on the ε-operator were [82–84]. It is interesting to notice that [46] keeps its actuality, even now. Bruno Woltzenlogel Paleo who works actually on ε-operator (in collaboration with Giselle Reis) stressed its relevance in e-mail that he sent to me recently [13].

As an attentive reader would notice, Mints edited some of the volumes of "Zapiski" cited above. He was an editor of several books translated from English (*e.g.* [13]) and himself translated from English and German. He wrote many articles on mathematical logic for the Mathematical Encyclopedia, the Encyclopedia of Cybernetics, and even for the Great Soviet Encyclopedia (third edition).

I mention this to give a better idea of his "multidirectional" activity.

He was among regular participants of Sergey Maslov's seminar, also known as the seminar on the general theory of systems. According to the recollections of Inna Davydova[14] the seminar started at 1967, and initially the meetings were organized at the Faculty of Mathematics and Mechanics of the university. Later the seminar moved to S. Maslov's home because of the administrative pressure (I had myself an opportunity to attend it in the end of the 70s – beginning of the 80s).

Mints himself wrote in the foreword to the English edition of [80]:

> The intellectual influence of the Maslov family was not restricted, however, to their scientific achievements. Their home in Leningrad (now St. Petersburg) was a meeting place of a seminar where talks on social and scientific problems were presented. One has to feel the gravity of the ideological pressure of a totalitarian state to appreciate the importance of such a free forum. The emergence of such seminars seems to be characteristic of intellectual life under oppressive regimes: recall Zilsel's seminar in Vienna where Gödel presented in January

---

[13] 6 of October 2015.

[14] See http://www.mathsoc.spb.ru/pers/maslov. Gennady Davydov and his wife Inna were friends and colleagues of Mints and Maslov.

1938 his overview of possibilities for continuing Hilbert's program. Another forum for dissident thought in the USSR was provided by a samizdat (unofficially published) journal "Summa" edited by S. Maslov which was designed as a review journal for samizdat publications.

Among the speakers were, for example, the philologist Vyacheslav Ivanov, a Foreign Fellow of British Academy since 1977 and Academician of Russian Academy of Sciences since 2000, the geneticist Raissa Berg, *cf.* the Columbia University Archive[15], the literary critic and memoirist Lidiya Ginzburg (cf. [7]).

In 1982 Maslov died tragically in a car accident.

In May 1979 the administration of LOMI finally considered Mints as a candidate for promotion to the position of senior researcher. On May 3 Mints signed an official request to submit his application, and the director of LOMI, L. D. Faddeev, endorsed the request. The meeting of the Academic Council of LOMI that had to consider the candidature was prepared as usual. On May 10 a recommendation was signed by the chef of Logic Group, N. A. Shanin. On May 25 an official appreciation of Mints research activity was signed by "troika" (direction, party secretary, and trade-union secretary). On June 28 the Academic Council of LOMI voted in favor of Mints candidature: 0 "against", 17 "for" (all of the present) of 21 members.

I do not know exactly what happened afterwards, but on August 31 Mints submitted another request, to be discharged from his position from 8 October.

The reasons of this abrupt change are not completely clear. The vote of the Council of LOMI was not the last step, after all it was only the Leningrad Department of the Mathematical Institute in Moscow (MIAN). The decision had to be confirmed there, and only after that the director of LOMI might sign the appointment order. Usually the confirmation came more or less automatically, but not always.

According to V. P. Orevkov, the direction of MIAN suggested Mints to make a presentation before the Academic Counsil there, and this was unusual. The general situation in the Academy of Sciences did not look well, for example, there were some known cases of antisemitism, and in some of these cases the director of MIAN I. M. Vinogradov was involved (see, *e.g.*, the following letter of the Academician S. P. Novikov to one of his colleagues: http://www.mi.ras.ru/~snovikov/pont.pdf). Mints might learn that his appointment will be blocked at MIAN. He also might be informed about some external pressure that would make the promotion virtually impossible (for example, due to his "too extensive" international contacts not approved by authorities).

---

[15] http://www.columbia.edu/cu/lweb/archival/collections/ldpd_6761446/

He would not like to continue as a Junior Researcher in such a circumstance. At the same time he might be reasonably convinced that, due to the same international contacts, he will be able to find a good employment at one of the Western universities[16].

All colleagues who knew Mints and with whom I had an opportunity to discuss the events of 1979 (in particular, at Mints' memorial conference in August 2015) agree, that Mints asked to be discharged on his own request from LOMI because he decided to emigrate and wanted to save from blame Shanin, who, as the chef of Logic Group, would be otherwise held "administratively responsible". It seems that the decision to emigrate was taken somewhere between June and August.

It was not possible to emigrate freely from the USSR at this time, and Mints could not know that his emigration request will be refused by the authorities.

3

If Mints remained at LOMI, he would certainly become my PhD adviser after I graduated in 1979. In reality it was no more possible. He discussed this question with Shanin, and Shanin agreed to take me as his PhD student. It turned out, though, that finally the theme of my PhD thesis (defended in 1984) was essentially inspired by my graduate work under Mints supervision.

Shanin helped me a lot as far as the presentation of my results was concerned, advised on formulations that must be satisfactory from constructive point of view, but did not intervene much in the content.

I had some opportunities to discuss mathematics with Mints. I remember him to discuss the "Algebra of Proofs" by Szabo [93] and the problem that was called (I do not remember, already at this time or later) the Mac Lane's conjecture[17]. He advised me to write S. Mac Lane about my work. I did, and our correspondence continued until the mid-90es.

Among other situations, I remember a very unpleasant moment in autumn of 1981 when I was contacted by the KGB who wanted to "ask some questions". I had no courage to refuse and was met in a park by a KGB officer in civilian clothes who did just that: asked questions about correspondence with abroad, about Maslov's seminar ... I tried to tell nothing of importance, and in spite of his explicit request to

---

[16] I already mentioned M. Gelfond who emigrated in 1978. Another colleague of Mints, V. Lifschitz, who defended his thesis under Shanin's supervision in 1969, emigrated to the USA in 1976. Both very quickly found an employment. However I am not sure that I am able to list all possible reasons.

[17] The conjecture says that the category of vector spaces is a complete model w.r.t. the axiomatic theory of Symmetric Monoidal Closed Categories.

tell nobody, I informed Mints, Maslov and Shanin about this situation, but otherwise I remember nothing in my behavior to be especially proud of. Luckily for me their interest dissolved after a couple more meetings, probably they did not have anything serious in store.

The first half of the 80s, were for Mints a "time of troubles". He submitted an emigration request to the authorities an got a refusal. He had problems to find a job.

Of course his scientific research never completely stopped. Maybe it is a right place to say that one of his most impressive traits was calm, but almost religious devotion to science, and he had to find possibilities to do what he considered as his duty in a new and much less friendly environment.

At the same time there was nothing fanatical in this devotion, there certainly remained place for social life and human relations. For example after the tragic death of Sergei Maslov in 1982, Maslov's daughter Elena and his widow Nina for many years could count on his unwavering friendship[18].

He had some contracts for translation with "Mir" and "Nauka" editions and tried to keep a usual level of scientific activity due to intense work on translations in spite of all difficulties and without an appropriate institutional affiliation. In 1981 "Mir" published the translation of G. Kreisel's selected papers [18] where Mints translated about 90 percent of the book. In a short autobiographical note published in [84], for the period 1979-1985 the collaboration with "Mir" and "Nauka" publishing houses is mentioned. In 1983 the translations (with Mints as one of translators) of Barwise's "Handbook of mathematical logic" [3] and Chang and Lee's "Symbolic Logic and Mechanical Theorem Proving" [4] were published. The A. P. Ershov's archive[19] contains the correspondence between Ershov and Mints about the project to translate H. Barendregt's "λ-calculus". This project finally was accepted, not without difficulties and delays, and the translation was published by "Mir" in 1985. Still, this sort of contracts could not give any stability, and would disappear if no adequate research position would be found.

Some hope of improvement came from his new contacts with Enn Tyugu and other Estonian scientists. Due to these contacts Mints had temporary invitations to Tallinn Institute of Cybernetics. The papers [59–62] were published. In 1983 he was an editor, with Enn Tyugu, of [63]. Joint papers [64, 65] are written for this collection. He wrote also a contribution (with Enn Tyugu) [66] to the proceedings of the IIIth Conference "Application of methods of mathematical logic" in Tallinn.

However, how far from natural his situation was, is illustrated by the fact that

---

[18]It is not only part of my personal recollections, see, *e.g.*, the A. P. Ershov's archive, http://ershov-arc.iis.nsk.su/archive/eaindex.asp, Mints to Ershov, letter of 17 Sept. 1982.

[19]See http://ershov-arc.iis.nsk.su/archive/eaindex.asp

from September 1983 to April 1985 he worked as a Senior Researcher at a computing center in the institute called *Lengipromyasomolprom*, that belonged to a large "holding" of Leningrad meat-processing plants (one of economic experiments of the late Soviet period)[20].

In 1984 Mints helped to invite Saunders Mac Lane, though he of course could not be his "host" officially. Mac Lane came with his wife Dorothy, who had to use a wheelchair. As Mac Lane wrote:

> In September 1984 we made another successful trip with the wheelchair, this time to Moscow, Leningrad and Helsinki. The occasion was an international conference and analysis to celebrate the anniversary of the Steklov Institute, the mathematical institute of the Soviet Academy of Sciences. [23, p. 303]

In Leningrad Mints himself was a principal "guide" to Saunders and Dorothy. By grace of him, I had an opportunity to meet Mac Lane and discuss mathematics. I remember also how all of us visited the Alexander Nevsky Monastery and its historical necropolis, where Leonard Euler was buried.

After the death of Brezhnev in November 1982 the USSR entered the period of rapid political changes, though it was difficult to see at the beginning how far the changes will go.

In a quick succession Andropov, and after his death Chernenko, took office of the Communist Party's General Secretary. Chernenko in his turn died in March 1985.

I remember the dinner after the defence of the PhD ("candidate nauk") thesis by Valentine Shehtman. Shehtman was from Moscow, but to organize his defence there was more difficult, the reasons being far from scientific. He had his *viva* at LOMI, and booked in advance for the evening a private room at Metropole restaurant, one of the oldest and most traditional in Leningrad. It happened that at the same time the period of mourning because of Chernenko's death was declared, and the restaurant was unusually quiet.

I remember this as a kind of photograph: Mints, Shanin, Slissenko, Shehtman, Orevkov, Matiyasevich, Sochylina (the only woman), Ruvim Gurevich[21], all in rather somber costumes (pure coincidence, not related to official mourning), all without ties (not a coincidence - somebody joked then that Shanin took as his students only those who do not wear a tie). I remember also a general feeling that the times are changing. They truly did.

---

[20]This is confirmed by a document preserved at Tallinn Institute of Cybernetics.

[21]Not to be confused with Youri Gurevich. I knew Ruvim since my student years at the faculty of mathematics and mechanics. He was a gifted mathematician, his best known result concerns the so called Tarski High School Algebra Problem [10]. He emigrated in 1987 and died prematurely in 1989.

Let me quote again Mac Lane who visited the USSR again in 1987 (this time Mac Lane went first to Moscow, then to Tbilisi in Georgia, to Leningrad and finally to Estonia, where Mints now worked):

> We then made a special trip to Tbilisi, Georgia, which was then still part of Soviet Union. But discontent over the political system was in the air ... From Leningrad we continued to Estonia, where I gave a talk at the Institute of Cybernetics in Tallinn and we were again greeted warmly by colleagues, both in Tallinn and at the University of Tartu. In Estonia too, we were much aware of the limits of freedom of speech. However, only a few weeks later glasnost and big changes took place in the Soviet system. Amazing! Within a few days, Georgians, Russians, Estonians, all were now allowed to communicate without fear.([23], p. 331.)

Since April 1985 Mints was fully employed as a Senior Researcher at the Institute of Cybernetics of the Estonian Academy of Sciences in Tallinn. I have outlined in the previous section the main directions of his research in the 60s and the 70s. In the 80s his main contributions were certainly in the domain of computer science logic. He participated actively in a pioneering research on structural synthesis of programs (SSP), the proof-theoretical aspects of structural synthesis being mostly his responsibility[22].

If we look today what came out of these studies then we shall see that some research still continues (see [94], and the bibliography there) but we may have an impression that the topic remains rather limited. In fact, it would be fair to take into account the historical context and the role of SSP in this context, because for proof theoretical methods in computer science the 80s were an early "heroic" period.

The attempts to use computers for proof search and verification started in the 60s, but the 70s and the 80s had seen the first steps to implement the idea that proofs themselves may have something to do with structuration and execution of programs. For example, the *Prolog* language, created in 1972 by A. Colmerauer and Ph. Roussel, was then a "hot topic" among proof theorists interested in applications. Another "hot topic" was the Curry-Howard correspondence [11].

In the 1970s R. Milner with his group created at Edinburgh university the ML programming language, based essentially on the principles of typed λ-calculus. In his paper on LCF (logic for computable functions) Milner wrote:"The connection between programs and logic is now recognized as a leading topic of research in the theory of computing." [29], p.146.

P. Martin-Löf was developing his Type Theory, that plays a central role in many

---

[22]Essentially, it is a form of automated synthesis of programs, based on intuitionistic propositional calculus.

modern "proof-assistants". The importance of proof theory in programming was rapidly increasing.

In a narrow, strictly technical sense, the SSP may seem today a relatively limited topic but the research and development of the SSP in the 80s and the 90s contributed a lot to the much greater domain called now formal methods in programming.

The research position in Tallinn that Mints finally got did not diminish the intensity of his work, but he certainly should feel a relief finding himself again a member of a highly motivated research group, and in a more adequate status than before. One may be not particularly interested in career-making, promotions and honors, but still feel sharply that your work is not properly appreciated.

At Mints' Memorial Conference in St. Petersburg, V. Lifshitz[23] mentioned that Mints was sometimes nicknamed a "minister of information"[24]. Estonia in Soviet times was in many ways closer to the West than the rest of the Soviet Union, including better possibilities of scientific exchange, and this also should look for him as an improvement.

In 1986 Estonia was the venue of the IVth All-Union conference "Application of methods of mathematical logic". Mints was one of its organizers, and edited (with P. Lorents) the proceedings [70].

The trip to Tallinn by train from Leningrad took only 6 hours. Many Leningrad residents enjoyed the visits to Estonian capital, especially to its historical center, an almost intact medieval city. The previous, IIIrd conference "Applications of methods of mathematical logic" in 1983, happened on the mainland, we were staying at the Olympic village in the Tallinn neighbourhood called Pirita, and often visited the city center.

This time the organizers had a more exotic plan. Its mere possibility seems to be a sign of changing times. A modest cruise ship that belonged to the Estonian Maritime Rescue was somewhat contracted, and the participants went from Tallinn to the Saaremaa island (part of the Estonian SSR). We stayed on the ship, but the conference meetings were organized at the Kuressaare Castle, a former bishop's stronghold.

Since 1986, after a long pause, Mints' papers were again published in international journals, for example rapidly appeared [71, 74–77].

In 1988 he was one of the organizers of COLOG-88, an international conference on computer science logic in Tallinn. With Per Martin-Löf he edited the proceedings

---

[23]Like Mints, he defended his PhD thesis at LOMI (with Shanin as adviser). He emigrated in 1976 and is now professor at the University of Texas at Austin.

[24]By the way, Mints wrote reviews for "Zentralblatt", "Mathematical Reviews" *etc.* since 1973. The total number of his reviews in "Zentralblatt" database is now 474. About 150 were written when he worked in Estonia (and about 15 before).

of this conference [72]. He also published a long paper there [73].

The early 80s were difficult years, and had some profound personal consequences for Mints. They marked the end of his first marriage, because his first family finally decided not to emigrate. Later, when he moved to Estonia, they remained in Leningrad, that since 1991 is again called St. Petersburg.

I remember my meeting with Grigori and his second wife, Marianna, in Tallinn. It was probably during COLOG-88 or in 1989. Before the fall of the USSR, I visited the Institute of Cybernetics a few more times. One evening Mints invited me to his home, a kind of studio in some academic residence, the type doctoral students or post-docs might have. As far as I remember, it was stuff with scientific literature. We had some tea there surrounded by the bookshelves.

Mints probably still had plans to emigrate, but they could not be definite. In 1989 he defended his Dr. Sci. thesis[25] titled "Transformations of Proofs and Program Synthesis". The defense took place at the Leningrad State University on April 26, 1989. In November 1989, he was promoted to the position of leading (or principal) researcher at the Tallinn Institute of Cybernetics.

In 1987 the borders started to open, and we could now easily go to the places that would seem impossible a few years ago. In fact, in the summer 1989, I was able to attend the ASL Logic Colloquium in West Berlin, just three months before the fall of the Berlin Wall. In 1990 I visited Mac Lane at the University of Chicago, and attended the Logic Colloquium '90 in Helsinki.

Mints was one of the invited speakers at both the Logic Colloquium '89 and the Logic Colloquium '90. In Helsinki it was probably the last time we met each other as Soviet citizens.

He was now in his element, at ease as a member of the top-level international scientific community that does not think much about borders. Of course, nothing was definitely settled yet in the ordinary, more mundane aspects of a scientist's life.

Enn Tyugu remembers:

> We visited Stanford for three months in spring of 1990. He was proposed to be a lecturer of logic instead of Barwise who took his sabbatical, I guess, in the same autumn. He impressed the Stanford people so much that he got the permanent professorship there, moved to Stanford and left our institute in August 1991.[26]

It seems symbolic that one of the Mints' last papers that Mints had published when the Soviet Union still existed was a survey on proof theory in the USSR [78].

---

[25]The degree that still exists in Russia, and is considered to be higher than PhD It may be compared to *state doctorate* that existed in many European countries until recently, and to *habilitation* that exists now.

[26]E-mail to the author, March 24, 2016.

I was never able to visit Mints when he worked at Stanford there (1991–2014), though I did see him many times on other occasions. Let this period be the subject of another paper.

Acknowledgements. I am very grateful to Evgeny Dantsin, Vladimir Orevkov, Anatol Slissenko, and Enn Tyugu for their recollections, help and advice. I would also like to thank the administration of the St. Petersburg Department of Steklov Institute of Mathematics for the opportunity to see archive documents concerning Grigori Mints.

References

[1] A. R. Anderson, N. D. Belnap. The pure calculus of entailment. *J. Symb. Logic*, 27(1):19–52, 1962.

[2] S. N. Artemov. Explicit provability and constructive semantics. *The Bulletin of Symbolic Logic*, 7(1):1–36, 2001.

[3] J. Barwise, editor. *Handbook of mathematical logic*. Russian translation in 4 volumes. Vol. 4: Proof Theory. Translated by G. Davydov, G. Mints. V. Orevkov (editor of Russian translation). "Nauka", Moscow, 1982.

[4] Ch.-L. Chang, R. Ch.-T. Lee. *Symbolic Logic and Mechanical Theorem Proving*. Russian translation. Transl. by G. Davydov, G. Mints, A. Sochylina. S. Maslov (editor of Russian translation). "Nauka", Moscow, 1983.

[5] G. Davydov, S. Maslov, G. Mints, V. Orevkov, N. Shanin, A. Slissenko. *An Algorithm for a Machine Search of a Natural Logical deduction in Propositional Calculus*. (In Russian.) "Nauka", M.-L., 1965, 39 p. Engl. transl. in: *The Automation of Reasoning*, Ed. J. Sieckmann, G. Wrightson, Springer Verlag, 1983.

[6] A. Dragalin. Intuitionistic Logic and Hilbert's ε-symbol, (Russian). *Istoriia i Metodologiia Estestvennykh Nauk*, 78–84, Moscow, 1974. Republished in: A. G. Dragalin. *Konstruktivnaia Teoriia Dokazatelstv i Nestandartnyi Analiz*, 255–263. Editorial Publ., Moscow, 2003.

[7] L. Ginzburg. *Blockade Diary*. Transl. by A. Myers. Harvill Press, 1995.

[8] G.-Y. Girard. Linear Logic. *Th. Comp. Sci.*, 50(1):1–101, 1987.

[9] K. Gödel. Eine Interpretation des intuitionistishen Aussagenkalküls. *Ergebnisse Math. Colloq.*, 4:39–40, 1933. Engl. transl. in S. Feferman, editor, *Kurt Gödel Collected Works*, I:301–303. Oxford University Press, 1995.

[10] R. Gurevič. Equational theory of positive numbers with exponentiation is not finitely axiomatizable. *Annals of Pure and Applied Logic*, 49:1–30, 1990.

[11] W. Howard. The formulas–as–types notion of construction. In J. P. Seldin and J. R. Hindley, editors, *Essays on Combinatory Logic, Lambda Calculus, and Formalism*, 479–490, Academic Press, 1980.

[12] G. M. Kelly and S. Mac Lane. Coherence in Closed Categories. *Journal of Pure and Applied Algebra*, 1(1):97–140, 1971.

[13] S. C. Kleene. *Mathematical Logic*. J. Wiley and Sons. N.Y., London, Sydney, 1967. Russian translation: G. E. Mints editor, *Matematicheskaya Logika*. Transl. Yu.A. Gastev. "Mir", Moscow, 1973.

[14] U. Kohlenbach. *Applied Proof Theory: Proof Interpretations and Their Use in Mathematics*. Spinger Monographs in Mathematics, Spinger, 2008.

[15] G. Kreisel. A survey of proof theory II. In J. E. Fenstad, editor, *Proc. 2 Scand. Logic Symp.*, pages 109–170, North-Holland, Amsterdam, 1971.

[16] G. Kreisel, G. E. Mints and S. G. Simpson. The use of abstract language in elementary metamathematics: some pedagogic examples. Springer Lecture Notes, 453:38–131. Springer, 1975.

[17] G. Kreisel. Some Facts from the Theory of Proofs and some Fictions from General Proof Theory. *Proceedings of the Fourth Scandinavian Logic Symposium and the First Soviet-Finnish Conference, Jyväskulë, Finland, June 29 – July 6, 1976. Essays on mathematical and philosophical logic*. Synthese Library, vol.122, pages 3–25. D. Reidel P.C., 1979.

[18] G. Kreisel. *Investigations in Proof Theory*. Transl. from English by Yu. Gastev and G. Mints. "Mir", M , 1981.

[19] J. Lambek. On the calculus of syntactic types. In R. Jakobson editor, *Structure of Language and Its Mathematical Aspects*. Proc. Symp. Appl. Math., pages 166–178. AMS, Providence, 1961.

[20] J. Lambek. Deductive systems and categories 1. *Math. Systems Theory*, 2(4):278–318, 1968.

[21] J. Lambek. Deductive systems and categories 2. *Lect. Notes in Math.*, 86:76–122, Springer, 1969.

[22] J. Lambek. Deductive systems and categories 3. *Lect. Notes in Math.*, 274:57–82, Springer, 1972.

[23] S. Mac Lane. *A Mathematical Autobiography*. A. K. Peters, Wellesley, Mass., 2005.

[24] C. R. Mann. The connection between equivalence of proofs and cartesian closed categories. *Proc. London Math. Soc.*, 31(3):289–310, 1975.

[25] S. Ju. Maslov. The inverse method for establishing deducibility for logical calculi. *Trudy Math. Inst. Steklov*, 98:26–87, 1968. English translation *Proc. Steklov Inst. Math.*, 98:25–96, 1968.

[26] S. Yu. Maslov, Yu. V. Matiyasevich, G. E. Mints, V. P. Orevkov, A. O. Slisenko. Nikolai Aleksandrovich Shanin (on his sixtieth birthday). *Uspekhi Mat. Nauk*, 35(2):241–245, 1980.

[27] S. J. Maslov, G. E. Mints, V. P. Orevkov. Mechanical Proof-Search and the Theory of Logical Deduction in the USSR. *Revue Internationale de Philosophie*, 25 (4), 575–584 (1971).

[28] A. V. Idelson and G. E. Mints, editors. *Mathematical Theory of Logical Inference*. (A

collection of translations.) "Nauka", Moscow, 1967.

[29] R. Milner. LCF: A Way of Doing Proofs with a Machine. MFCS 1979, *Lecture Notes in Computer Science*, 74:146–159, Springer, 1979.

[30] G. E. Mints. An analogue of Herbrand theorem for constructive predicate calculus. (In Russian.)*Dokl. Akad. Nauk SSSR*, 147(4), 1962.

[31] G. E. Mints. On the differentiability predicate and the differentiation operator. (In Rissian.) *Dokl. Akad. Nauk SSSR*, 147(5), 1962.

[32] G. E. Mints, V. P. Orevkov. A generalization of the theorems of V.I. Glivenko and G. Kreisel to a class of formulae of the predicate calculus. (In Russian.)*Dokl. Akad. Nauk SSSR*, 152(3): 553–554, 1963. = *Soviet Math. Dokl.*, 4:1365–1367, 1963.

[33] G. E. Mints. On predicate and operator variants of the formation of the theories of constructive mathematics. *Problems of the constructive direction in mathematics*. Part 3, Collection of articles. To the 60th anniversary of Andrei Andreevich Markov, *Trudy Mat. Inst. Steklov.*, 72:383–436, "Nauka", Moscow-Leningrad, 1964.

[34] G. E. Mints. Embedding operations related to the S. Kripke's "semantics". Studies in constructive mathematics and mathematical logic, Part I. *Zap. Nauchn. Sem. LOMI*, 4:152–159, Moscow, 1967.

[35] G. E. Mints, V. P. Orevkov. On embedding operations. *Ibid.*, 160–168. Moscow, 1967.

[36] G. E. Mints. The Skolem method in intuitionistic calculi. (In Russian.) *Trudy Mat. Inst. Steklov.*, 121:67–99, 1972.

[37] G. E. Mints. Cut-free calculi of the S5-type. Studies in constructive mathematics and mathematical logic, Part II. *Zap. Nauchn. Sem. LOMI*, 8:166–175. "Nauka", Leningrad, 1968.

[38] G. E. Mints. Admissible and derivable rules. *Ibid.*, 189–192.

[39] G. E. Mints. On semantics of modal logic. Studies in constructive mathematics and mathematical logic, Part III. *Zap. Nauchn. Sem. LOMI*, 16:147–152. "Nauka", Leningrad, 1969.

[40] G. E. Mints. Quantifier-free and one-quantifier systems. Studies in constructive mathematics and mathematical logic, Part IV. *Zap. Nauchn. Sem. LOMI*, 20:115–134. "Nauka", Leningrad, 1971.

[41] G. E. Mints. Exact estimates for provability of the rule of transfinite induction in initial parts of arithmetic. *Ibid.*, 134–145.

[42] G. E. Mints. Herbrand Theorem (in Russian). In A. V. Idelson and G. E. Mints, editors, *Mathematical Theory of Logical Inference (collection of Russian translations of papers by G. Gentzen, S. C. Kleene, K. Gödel etc.)*, pages 311–350. "Nauka", Moscow, 1967.

[43] G. E. Mints. Derivability of admissible rules. Studies in constructive mathematics and mathematical logic, Part V. *Zap. Nauchn. Sem. LOMI*, 32:85–90. "Nauka", Leningrad, 1972. Engl. tr.: *J. of Soviet Mathematics*, 6 (4):417–421.

[44] G. E. Mints. Cut-elimination theorem for relevant logics. (In Russian.) *Ibid.*, 90–97. Engl. tr.: *J. of Soviet Mathematics*, 6(4):422–428.

[45] G. E. Mints. Lewis' systems and system T (a survey 1965–1973). In R. Feys. *Modal*

Logic (Russian translation), pages 422–509. "Nauka", Moscow, 1974. English translation in [79].

[46] G. E. Mints. Heyting Predicate Calculus with Epsilon Symbol (Russian). Studies in constructive mathematics and mathematical logic, Part VI. *Zap. Nauchn. Sem. LOMI*, 40:101–110. "Nauka", Leningrad, 1974. English Translation in [79], 97–104.

[47] G. E. Mints. On E-theorems. *Ibid.*, 110–118.

[48] G. E. Mints. Proof theory (arithmetic and analysis). (In Russian.) *Itogi Nauki i Tekhniki. Ser. Algebra. Topol. Geom.*, 13:5–49, 1975. Engl. tr.: J. of Soviet Math., 7(4):501–531.

[49] G. E. Mints. Transfinite expansions of arithmetic formulas. Theoretical application of methods of mathematical logic, Part I. *Zap. Nauchn. Sem. LOMI*, 49:51–66. "Nauka", Leningrad, 1975.

[50] G. E. Mints. Finite investigation of infinite derivations. *Ibid.*, 67–123. Engl.tr. J. Sov. Math. 10:548–596, 1978.

[51] G. E. Mints. What can be done with PRA? Studies in constructive mathematics and mathematical logic, Part VII. *Zap. Nauchn. Sem. LOMI*, 60:93–102."Nauka", Leningrad, 1976.

[52] G. E. Mints. Closed categories and the theory of proofs. Theoretical applications of methods of mathematical logic, Part II. *Zap. Nauchn. Sem. LOMI*, 68:83–115. "Nauka", Leningrad. Otdel., Leningrad, 1977, pp 83–115. = Engl. tr.: *J. of Soviet Mathematics*, 15:45–62, 1981.

[53] G. E. Mints. A new reduction sequence for arithmetic. Studies in constructive mathematics and mathematical logic, Part VIII. *Zap. Nauchn. Sem. LOMI*, 88:106–131. "Nauka", Leningrad, 1979.

[54] G. E. Mints. A primitive recursive bound of strong normalization for predicate calculus. *Ibid.*, 131–137.

[55] G. E. Mints. On Novikov's hypothesis. *Modal and intensional logics*, "Nauka", Moscow, 1978. English translation in G. Mints, Selected papers in proof theory, Bibliopolis, Napoli, 1992.

[56] G. E. Mints. The theory of categories and the theory of proofs. *Aktualnye problemy logiki i metodologii nauki*. Kiev, "Naukova Dumka", 1979, pages 252–278. Engl. tr.: G. E. Mints. *Selected Papers in Proof Theory*. Naples, Bibliopolis, 1992.

[57] G. E. Mints. A new reduction sequence for arithmetic. *Zap. Nauchn. Sem. LOMI*, 88:106–126. "Nauka", Leningrad, 1979. Engl. transl.: *J. of Soviet Mathematics*, 20(4):2322–2333, 1982.

[58] G. E. Mints. A primitive recursive bound of strong normalization for predicate calculus. *Zap. Nauchn. Sem. LOMI*, 88:131–136. "Nauka", Leningrad, 1979. Engl. transl.: *J. of Soviet Mathematics*, 20(4):2334–2336, 1982.

[59] G. E. Mints. A simplified consistency proof for arithmetic. (Russian, English summary), *Izvestia Akad. Nauk. Ehst. SSR*, Fiz., Mat. 31:376–381, 1982.

[60] G. E. Mints. Logical Foundations of Program Synthesis. Akad. Nauk. Ehst. SSR,

preprint, Tallinn, 45 pages, 1982.

[61] G. Mints, E. Tyugu. Justifications of the Structural Synthesis of Programs. *Sci. Comput. Program.*, 2(3): 215–240, 1982.

[62] B. Volozh, M. Matskin, G. Mints, E. Tyugu. Theorem proving with the aid of program synthesizer. *Cybernetics*, 6:63–70, 1982.

[63] G. Mints and E. Tyugu (eds.). Automated Synthesis of Programs. (In Russian.) Acad. Nauk Ehst. SSR, Institute of Cybernetics, Tallinn, 1983, 203 pages.

[64] G. Mints, E. Tyugu. Justification of Structural Synthesis of Programs. (In Russian.) In: [63], 5–40.

[65] G. Mints, J. Penjam, E. Tyugu, M. Harf. Structural Synthesis of Recursive Programs. (In Russian.) In: [63], 58–72.

[66] G. Mints, E. Tyugu. Structural Synthesis and Non-Classical Logics. In: *III Conference "Application of methods of mathematical logic"*, Tallinn, 1983, pages 52–60.

[67] G. E. Mints. Structural synthesis with independent subproblems and the modal logic S4. (Russian. English summary) *Izvestia Akad. Nauk Ehst. SSR, Fiz. Mat.* 33:147–151, 1984.

[68] G. E. Mints. Calculi of resolution for non-classical logics. (Russian) *Semiotika Inf.*, 25:120–135, 1985.

[69] G. E. Mints, E. K. Tyugu. Description semantics in utopist language and automatic program synthesis. (English. Russian original) *Program. Comput. Software*, 11:251–258, 1985 (translation from *Programmirovanie*, 5:3–11, 1985.

[70] G. Mints, P. Lorents editors. IVth All-Union conference "Application of methods of mathematical logic", abstracts, v. 1-2, Inst. of Cybernetics of the Estonian Acad. Sci., Tallinn, 1986.

[71] G. Mints, E. Tyugu. Semantics of a Declarative Language. *Inf. Process. Lett.*, 23(3):147–151, 1986.

[72] P. Martin-Löf, G. Mints editors. COLOG-88 : International Conference on Computer Logic, Tallinn, USSR, December 12–16, 1988 : Proceedings, vi, 338 pp. Springer LNCS, 417. Berlin, 1990.

[73] G. Mints. Gentzen-type systems and resolution rules. Part I. Propositional logic. COLOG-88, 198–231.

[74] G. Mints, E. Tyugu. The Programming System PRIZ. *J. Symb. Comput.*, 5(3): 359–375, 1988.

[75] G. Mints. The Completeness of Provable Realizability. *Notre Dame Journal of Formal Logic*, 30(3): 420–441, 1989.

[76] G. Mints, E. Tyugu. Editorial. *J. Log. Program.*, 9(2–3): 139–140, 1990.

[77] G. Mints, E. Tyugu. Propositional Logic Programming and Priz System. *J. Log. Program.*, 9(2–3): 179–193, 1990.

[78] G. Mints. Proof Theory in the USSR 1925-1969. *J. Symb. Log.*, 56(2): 385–424, 1991.

[79] G. Mints. *Selected Papers in Proof Theory*. Bibliopolis/North-Holland, 1992.

[80] G. Mints. Foreword. V. Kreinovich, G. Mints editors, *Problems of Reducing the Exhaustive Search*. AMS Translations, 1997.

[81] G. Mints. Decidability of the Class E by Maslov's Inverse Method. A. Blass, N. Dershowitz, and W. Reisig editors. *Gurevich Festschrift*. LNCS, 6300:529–537, 2010.

[82] G. Mints. Intuitionistic Existential Instantiation and Epsilon Symbol. *arXiv*:1208.0861v1 [math.LO], 3 Aug 2012.

[83] G. Mints. Epsilon substitution for first- and second-order predicate logic. *Ann. Pure Appl. Logic*, 164(6):733–739, 2013.

[84] G. Mints. Intuitionistic Existential Instantiation and Epsilon Symbol. In *Dag Prawitz on Proofs and Meaning*, v. 7 of the series *Outstanding Contributions to Logic*, 225–238, 2015.

[85] N. Kh. Orlova, S. V. Soloviev. On the History of Logic in Russia Before Revolution: Strategies of Academic Interaction. *Logical Investigations*, 22(2):123–154, 2016.

[86] D. Prawitz. *Natural Deduction*. Almquist and Wiksell, Stockholm, 1965.

[87] G. F. Rose. Propositional Calculus and Realizability. *Transactions of the American Mathematical Society*, 75:1–19, 1953.

[88] V. Rybakov. *Admissibility of Logical Inference Rules*. Elsevier, 1997.

[89] A. A. Babayev, S. V. Solov'ev. Coherence theorem for canonical maps in cartesian closed categories. (In Russian.) *Zapiski Nauchnych Seminarov LOMI*, 88:3–29, 1979. Engl. transl.: *J. of Soviet Mathematics*, 20(4):2263–2282, 1982.

[90] S. V. Solov'ev. Preservation of equivalence of derivations under reduction of depth of formulas. *Ibid.*, 197–208. Engl. transl.: *J. Sov. Math.*, 20(4):2370–2376, 1982.

[91] S. V. Solov'ev. Growth of the length of sequential derivation transformed into natural one. *Ibid.*, 192–196. Engl. transl.: *J. Sov. Math.*, 20(4):2367–2369, 1982.

[92] V. A. Smirnov. Elimination des termes ε dans la logique intuitioniste. *Revue internationale de Philosophie*, 98:512–519, 1971.

[93] M. E. Szabo. *Algebra of Proofs*. Volume 88 of *Studies in Logic and the Foundations of Mathematics*. North-Holland, 1978.

[94] E. Tyugu. Grigori Mints and Computer Science. In S. Feferman, W. Sieg, V. Kreinovich, V. Lipschitz, Ruy de Queiroz editors. *Proofs, Categories and Computations: Essays in honor of Grigori Mints*. Dov Gabbay's College Publications, 2010.

[95] A. S. Troelstra and H. Schwichtenberg. *Basic Proof Theory*. Cambridge Tracts in Theoretical Computer Science, 43. Cambridge University Press, Cambridge, 2nd edition, 2000.

About Grisha Mints

Anatoly Vershik
St. Petersburg Department of V. A. Steklov Institute of Mathematics RAS
27, nab. r. Fontanki, St. Petersburg 191023, Russia
vershik@pdmi.ras.ru

Abstract

The author remembers his meetings and discussions with a remarkable mathematician and logician Grigori (Grisha) Mints.

I remember Grisha Mints when he was still a student. He entered the Faculty of Mathematics and Mechanics ("math-mech") of Leningrad State University when I graduated (1956), and while I was a doctoral student (since 1958) I met him sometimes. In fact, his closest friend – Sergey Maslov – a future renowned logician, attended, when he was in his terminal class at school, a mathematical seminar organized at math-mech for the school children, where I was a tutor during a whole year. So when I met him at the faculty, he usually was with Grisha, they were inseparable. And they both, rather early, have selected mathematical logic as their specialty. They both were among the best students of their promotion. By this reason, and also because the Spring of 1956 was the high time of the Khrushchev liberal "thaw", their scientific adviser N. A. Shanin, himself one of the principal followers of A. A. Markov (Jr.), could persuade the administration of the Academy of Sciences that both have to be taken to the post-graduate school at the Leningrad Department of the Steklov mathematical institute (LOMI). Note, that in Soviet times, when the director of the Mathematical institute (MIAN in Moscow) was I. M. Vinogradov, a notorious antisemite, for a Jew (as Mints) or half-Jew (as Maslov) to be taken as a post-graduate was a rare exception, even more so to be taken as a staff member.[1] Their successive research, activity, openness for human

---

The author would like to thank Sergei Soloviev for the English translation.

[1] There were only a few Jews – members of the MIAN – Mathematical Institute of Russian Academy of Sciences (Moscow and Lenigrad Department as well) during the long directorship of Vinogradov. The exceptions were, of course, admitted by Vinogradov himself for some reason. On rare occasions he yielded to the lengthy appeal of some well-known mathematicians such as

contact, helped their adviser N. A. Shanin to take them first as post-graduates and later as staff members.

Together with other young colleagues, they organized the research group that they called with a provocative irony "TREPLO". In Russian, it was an abbreviation for "Teoretichestkaya Razrabotka Evristicheskogo Poiska Logicheskih Obosnovanii" ("Theoretical Development of Heuristic Search of Logical Justifications"). This abbreviation sounded almost like "windbag" in Russian.

The plans were magnificent: to develop a mathematical theory of the machine proof of mathematical statements. To my knowledge, this aim is not yet reached, but there were surely some achievements; the role of Mints in this group was very significant. In the end of the 60s, N. A. Shanin delivered at the meetings of the Leningrad mathematical society a series of talks that presented the work of the group. Later, in the 70s, I invited the logicians (S. Maslov, and Yu. Matiyasevich) to my seminar in order to give talks about their work. I remember particularly the talk by G. Mints: he spoke about the recent result of L. Harrington [2] concerning the possible abnormal growth of the lower bound in the classical Ramsey problem. Afterwards we often discussed with him this and other themes.

To describe him as a mathematician, I have to say that he had a very broad interest in mathematics, including group theory, theory of dynamic systems, and functional analysis. He strived to apply logical methods to these domains and often obtained new proofs of known results by his methods. This helped to understand better the mainsprings of the proofs, etc. But new results require more penetration in a given domain. It was of interest to discuss with him general mathematical concepts. Grisha was always interested in philosophical aspects of mathematical theories, and we found there a common ground for many discussions. I remember our discussions concerning the Burnside type problems, ergodic theorems, concepts of universality, *etc.*. For our last meeting he prepared several extracts from Hardy's book [1] as comments to my presentation [4] about the connections of mathematics and its possible applications. I know that he had discussions also with D. K. Faddeev, Yu. I. Manin, and other well known mathematicians, who were attentive to his opinions.

In the end of the 60s and the beginning of the 70s, S. Maslov organized at his home a social and political seminar that had a very large scope, and Grisha was one

---

A. A. Markov, Yu. V. Linnik, among others, who asked him to accept their successful students. This happened very rarely and in fact I know that the Director regretted later his giving in to pressure and tried to "correct" what he considered as "defect". For example, S. Maslov was discharged from LOMI in 1970s, as well as Mints who was discharged following his decision to emigrate from the country. Remember, the outstanding mathematician V. A. Rokhlin was also dismissed from Moscow MIAN. This situation changed only after the end of Vinogradov's ditrectorship of MIAN.

of its participants. When Mints decided to emigrate, he naturally had, as always at that time, the unpleasant consequences at the institute, and had to quit, because, moreover, he became a "refuznik". Soon he moved to Tallinn, where quickly, and apparently, successfully got a new position, learned Estonian, found his place and even doctoral students.

Later he moved to the USA and obtained a chair at Stanford. His predecessor there was J. Barwise, with whom he had common interests, and who moved to Indiana University. I met him more than once, and visited his home when I was in the States, that is, Berkeley and Stanford (after 1990). With him I sent to the USA in 1990 the copies of all the issues of our illegal (in Soviet times) journal of social and political surveys "Summa" that was edited by Maslov, and where I actively participated (in the end of the 70s and the beginning of the 80s). When I visited Berkeley I donated these typed copies of "Summa" to the Slavic department in Bankroft library at UC Berkeley, where they may be consulted now. All the issues where collected in one volume called "'Summa' for free thought" and published in 2002 by the "Zvezda" publishing house at St. Petersburg [3].

According to my observations, G. Mints in the States worked fruitfully and became an active participant of mathematical events. He came many times to his native St. Petersburg where he had many colleagues and friends and where he participated in organization and the work of various conferences.

In my memory, he remained as a thoughtful, modest and witty man, far from indifferent not only to science, in particular mathematics and logic, but to all aspects of the complex modern life.

His premature and sudden demise is very saddening.

References

[1] G. H. Hardy. *A Mathematician's Apology*. Cambridge University Press, London, UK, 1967.

[2] J. Paris and L. Harrington. A Mathematical Incompleteness in Peano Arithmetic. In J. Barwise, editor, *Handbook of Mathematical Logic*, volume 90 of *Studies in Logic and the Foundations of Mathematics*, pages 1133–1142. North Holland, Amsterdam, The Netherlands, 1977.

[3] A. M. Vershik, editor, *"Summa" for free thought*, 719 pages. Publishing house of the journal "Zvezda", St. Petersburg, Russia, 2002. in Russian.

[4] A. M. Vershik. Does a freedom of the choice exist in mathematics? In G. E. Mints and O. B. Prosorov, editors, *Philosophy, Mathematics, Linguistics: Aspects of Interaction 2014. Proceedings of International Interdisciplinary Conference*, pages 84–86. VVM, St. Petersburg, Russia, April 2014.

Kant's Logic Revisited

Theodora Achourioti
*ILLC, Amsterdam University College
University of Amsterdam
Science Park 113, 1098XG Amsterdam, the Netherlands*
T.Achourioti@uva.nl

Michiel van Lambalgen
*ILLC, Department of Philosophy
University of Amsterdam
Oude Turfmarkt 141, 1012GC Amsterdam, the Netherlands*
M.vanLambalgen@uva.nl

Abstract

Kant considers his *Critique of Pure Reason* to be founded on the act of judging and the different forms of judgement, hence, take pride of place in his argumentation. The consensus view is that this aspect of the *Critique of Pure Reason* is a failure because Kant's logic is far too weak to bear such a weight. Here we show that the consensus view is mistaken and that Kant's logic should be identified with geometric logic, a fragment of intuitionistic logic of great foundational significance.

1 Preview

Below the reader will find a condensed revisionist account of Kant's so-called 'general logic', usually thought to be substandard, even when compared with the traditional logic of his day [4].[1] Ultimately our interest is in the formalisation of Kant's 'transcendental logic' (for which see [1]), but since transcendental logic takes its starting

---

The paper was originally presented at the conference "Philosophy, Mathematics, Linguistics: Aspects of Interaction 2012" (PhML-2012), held on May 22–25, 2012 at the Euler International Mathematical Institute, St. Petersburg. We are grateful to the referees for insightful comments.

[1]Not to mention the scathing verdicts from the standpoint of modern logic which we take to have started with Frege and Strawson.

point in the judgement forms listed in the Table of Judgement (most of which have their origin in general logic) we must take a close look at the actual logical forms of these judgements. The result of this investigation is that Kant's general logic is not monadic, not finitary, not classical, and perhaps linear rather than intuitionistic. We will here not elaborate on the last point[2] but we will restrict ourselves to stating a completeness theorem identifying Kant's general logic with a fragment of intuitionistic logic.

2 Validity in general logic

The key to any insightful formalisation of Kant's logic is the observation that judgements in Kant's sense participate in two kinds of logics: *general* logic and *transcendental* logic. Here is how Kant introduces 'general logic' in the first *Critique* [7]:

> [G]eneral logic abstracts from all the contents of the cognition of the understanding and of the difference of its objects, and has to do with nothing (A55-6/B80) but the mere form of thinking. (A54/B78)

And later, with a slightly different emphasis:

> General logic abstracts [...] from all content of cognition, i.e. from any relation of it to the object, and considers only the logical form in the relation of cognitions to one another, i.e. the form of thinking in general. (A55/B79)

So what is the 'mere form of thinking'?

The first two paragraphs of the *Jäsche Logik* [5] marvel at the fact that all of nature, including ourselves, is bound by rules. It continues:

> Like all our powers, the understanding is bound in its actions to rules [...] Indeed, the understanding is to be regarded in general as the source and the faculty for thinking rules in general [...] [T]he understanding is the faculty for thinking, i.e. for bringing the representations of the senses under rules.

From this it derives a characterisation of logic:

> Since the understanding is the source of rules, the question is thus, according to what rules does it itself operate? [...] If we now put aside all cognition that we have to borrow from objects and merely reflect on the use just of the understanding, we discover those of its rules which are necessary without qualification, for any purpose and without regard to any particular objects, because without them we would not think at all. [...] [T]his science of the

---

[2]Grigori Mints was planning on studying the connection between Kant's disjunctive judgement and multiplicative linear logic.

necessary laws of the understanding and of reason in general, or what is one and the same, of the mere form of thought as such, we call *logic*. [5, pp. 527-8] (cf. also A52/B76)

To appreciate the real import of this passage, one must resist the temptation to consider logic as consisting of a motley set of inference rules, such as modus ponens and syllogistic inferences, even though the *Jäsche Logik* will later list these too. Two definitions are pertinent here:

§58 A rule is an assertion under a universal condition. [5, p. 615]

Here it is important to bear in mind Kant's notion of *universal representation* as 'a representation of what is common in several objects' [5, §1, p. 589]. A rule is, therefore, applicable to a domain of indefinite extension.

The second definition is that of an *inference of reason*:

§56 An inference of reason is the cognition of the necessity of a proposition through the subsumption of its condition under a given universal rule. [5, p. 614]

At this point we will not yet provide an elaborate explanation of the notion of 'condition', but the reader is invited to take modus ponens as a concrete example. We then have the following sequence of ideas: (i) the understanding operates according to rules, (ii) the understanding's operations are necessary insofar as they pertain to the formal features of rules, and (iii) the most general formal principle is rule-application (or rule composition – as we shall see the distinction was not always made in those days). Thus Kant's logic has a general and constructive definition of validity, a consequence of the meaning of 'rule'. The *Jäsche Logik* will give concrete instances of this most general principle, such as modus ponens, but the full force of the principle will only become apparent when we come to discuss the true logical form of Kant's 'judgements'. We must note here that the general inference principle limits logic to judgements that can be seen as rules. We view Kant's emphasis on rules and their structural properties as marking the 'formal' character of his general logic. The definition of validity just given should be contrasted with the Bolzano-Tarski definition of validity: 'an argument is valid if its conclusion is true whenever its premises are' – for in this part of Kant's logic (what he calls 'general logic') there is no truth yet, there are only rules. A different kind of logic, 'transcendental logic' will introduce truth.

3 Three definitions of judgement and a Table ...

Any modern logic textbook makes a strict separation between syntax, semantics and consequence relation, and makes no reference at all to psychological processes that may be involved in a concrete case of asserting a syntactically well-formed sentence. These processes are studied in psycholinguistics, and start from the assumption that there are specific syntactic and semantic binding processes at work in the brain. For logical theorising such psycholinguistic approaches are deemed to be irrelevant. For Kant they are in fact of the essence, and his definitions of judgement also contain a cognitive component.

But the reader trying to piece together Kant's views on logic may be forgiven a sense of bewilderment when she finds not one but three seemingly very different definitions of 'judgement', none of which specifies a syntactic form, together with a 'Table of Judgement' which specifies some syntactic forms (for example, categorical, hypothetical, disjunctive, with various other subdivisions), without an indication of how these forms relate to the three definitions. Lastly, there are the examples of judgements that Kant uses in various works, whose logical forms do not fit easily in the Table of Judgement. This looks unpromising material, but we shall show that Kant's logic is nevertheless coherent and surprisingly relevant to modern concerns.

Let us begin with the three definitions of judgement:

> A judgement is the representation of the unity of the consciousness of various representations, or the representation of their relation insofar as they constitute a concept. [5, p. 597]

> A judgement is nothing but the manner in which given cognitions are brought to the objective unity of apperception. That is the aim of the copula **is** in them: to distinguish the objective unity of given representations from the subjective [...] Only in this way does there arise from this relation a judgement, i.e. a relation that is objectively valid [...][3] (B141-2)

> Judgements, when considered merely as the condition of the unification of representations in a consciousness, are rules. (*Prol.* § 23; see [8])

Even for those unfamiliar with Kant's technical vocabulary it will be obvious that 'unity' plays a central role in all three definitions. These are different ways of saying that the expressions occurring in a judgement must be bound together so that they can be simultaneously present to consciousness. The first definition posits unity simply as a requirement. The second says that unity in a judgement is achieved if the judgement has 'relation to an object'. The third definition links unity to the meaning of a judgement. Just as an example: if for a hypothetical judgement $\varphi \to \psi$ there exists a rule transforming a proof of φ into a proof of ψ, then that judgement

---

[3] Where 'objectively valid' means 'having relation to an object', which is not the same as 'true of the object'.

is unified. If the hypothetical is a truth functional material implication, then antecedent and consequent are independent, hence this is not a unified representation. The presence of a notion of unity of representation raises three questions: (i) what has this got to do with formal logic?, (ii) is there a relation between the unity and the reference to objects occurring in the second definition? and (iii) what is the relation between unity and the concrete forms of judgement given in the Table of Judgement?

3.1 Objects, concepts and general logic

Categorical judgements are composed of concepts, and objects 'fall under' concepts,[4] in a sense hinted at in the following note:

> *Refl.* **3042** Judgement is a cognition of the unity of given concepts: namely, that B belongs with various other things x, y, z under the same concept A, or also: that the manifold which is under B also belongs under A, likewise that the concepts A and B can be represented through a concept B. [9, p. 58]

It appears that both concepts and objects may fall under a given concept C. The given concept is therefore *transitive* in the sense that if (concept) M belongs to C (by being a subconcept) and (object) a belongs under M, then a belongs under C. Kant uses this semantics for concepts in his 'principle for categorical inferences of reason':

> *What belongs to the mark of a thing also belongs to the thing itself.* [5, p. 617]

The next note supplies more information about these objects 'in the logical sense' (so called because they make a cameo appearance in the section 'The logical employment of the understanding' (A68-9/B93)).

> *Refl.* **4634** We know any object only through predicates that we can say or think of it. Prior to that, whatever representations are found in us are to be counted only as materials for cognition but not as cognition. Hence an object is only a something in general that we think through certain predicates that constitute its concept. In every judgment, accordingly, there are two predicates that we compare with one another, of which one, which comprises the given cognition of the object, is the logical subject, and the other, which is to be compared with the first, is called the logical predicate. If I say: a body is divisible, this means the same as: Something x, which I cognize under the predicates that together comprise the concept of a body, I also think through the predicate of divisibility. [9, p. 149]

---

[4]Kant also uses the phrases 'object a belongs under concept C' and 'C belongs to a'.

What this *Reflexion* tells us is that an object is generic (or most general) for the 'predicates that constitute its concept', and that the quantifier 'something x' ranges over such generic objects only.

The same idea is prominent in the section of CPR entitled 'On the logical use of the understanding in general':

> [T]he understanding can make no other use of concepts than that of judging by means of them. Since no representation pertains the object immediately except intuition alone, a concept is thus never immediately related to an object, but is always related to some other representation of it (whether that be an intuition or itself already a concept). Judgement is therefore the mediate cognition of an object, hence the representation of a representation of it. (A68/B93)

An object is therefore rather like what logicians call a *type*: i.e. a set[5] $p(x)$ of formulas containing at least the free variable x;[6] free variables not identical to x can be replaced by formal parameters representing objects, hence specified by a type. As an example, consider the predicate 'body' and the type 'x is a massive body which orbits star y' – which can be used to defined the predicate 'planet', by existential quantification over y or by replacing y by a formal parameter (representing the Sun, say). Let T be the theory of the relevant concepts. If M is a concept, we say that $M(x)$ belongs to $p(x)$ if $T, p(x) \vdash M(x)$. For example, if T contains

$$\forall x(A(x) \wedge \exists y B(x,y) \to M(x)),$$

then $p(x) = \{A(x), \exists y B(x,y)\}$ belongs to $M(x)$. It is technically convenient to introduce suitable constants witnessing a type: if $p(x)$ is a (consistent) type, let a_p be a new constant satisfying $p(a_p)$.[7] These constants correspond to the 'objects in general' that we encountered in *Reflexion* **4634**. One may then view $p(x)$ and a_p as determining the same object; and in this formal sense we have that M belongs to a_p.

The next question to consider is whether Kant's theory of concepts puts a bound on the complexity of concepts, i.e. the complexity of the types belonging under the concept. The $p(x)$ given in the previous paragraph can be viewed as a single *positive primitive* formula:

Definition 1. A formula is *positive primitive* if it is constructed from atomic formulas using only $\vee$, (infinite) $\bigvee, \wedge, \exists, \bot$.

---

[5]In our context a finite set.

[6]Relations enter Kant's logic especially in connection with the hypothetical judgement (see section 3.3.2; furthermore, as Hodges observed in [4], traditional logic allowed relations in syllogisms.

[7]The constant a_p implicitly depends on the parameters and free variables (x excluded) occurring in $p(x)$.

Suppose M, P are concepts all of whose subconcepts can be defined using positive primitive types (equivalently, formulas). The judgement 'all M are P' – or in the language of *Reflexion* **4634**: 'To everything x, to which M belongs, also P belongs – may then be expressed as

$$\bigwedge_{p \in M} \bigvee_{q \in P} \forall x (p(x) \to q(x)),$$

which is equivalent to

$$\forall x (\bigvee_{p \in M} p(x) \to \bigvee_{q \in P} q(x)),$$

and this formula satisfies the definition of a *geometric implication*:

Definition 2. A formula is *geometric* or a *geometric implication* if it is of the form $\forall \bar{x}(\theta(\bar{x}) \to \psi(\bar{x}))$, where θ and ψ positive primitive.

As it turns out, Kant's theories of concepts and of judgements contain the resources to restrict the complexity of $p(x)$ to positive primitive. The reason for this is that the complexity of the relation 'M(x) belongs to p(x)' is at most that of geometric implications. For the proof we must refer the reader to [1]; but a sketch will be given in section 4.

Geometric logic – the inferential relationships between geometric formulas – is therefore naturally suggested by Kant's theory of concepts. We will see that the logical form of Kant's own examples of judgements (in so far as they are 'objectively valid' (see section 3.2)) is that of geometric implications. As a consequence, we can show by means of 'dynamical proofs' of geometric implications that judgements can be viewed as rules:

> Judgements, when considered merely as the condition of the unification of representations in a consciousness, are rules. (*Prol.* §23; see [8])

3.2 Unity, objects and transcendental logic

The second characterisation of judgement maintains that if a judgement has a certain kind of unity (the 'objective unity of apperception') then it relates to an object – has 'objective validity' – and can express a truth or falsehood of that object; it is 'truth-apt', in modern terminology. This is the domain of *transcendental logic*, which Kant defines as follows:

> [...] a science of pure understanding and of the pure cognition of reason, by means of which we think objects completely a priori. Such a science, which would determine the origin, the domain, and the objective validity of such

cognitions, would have to be called transcendental logic since it has to do merely with the laws of the understanding and reason, but solely insofar as they are related to objects a priori and not, as in the case of general logic, to empirical as well as pure cognitions of reason without distinction. (A57/B81-2)

For Kant, perceiving objects about which judgements can be made is an instance of what would now be called the binding problem: objects are always given as a 'manifold' of parts and features, which have to be bound together through a process of *synthesis*. What is very distinctive about Kant's treatment here is that the binding that binds expressions in judgement together at the same time binds parts and features together with a view toward constructing an object out of sensory material that relates to the judgement. Therefore the binding process, necessary to bring separately perceived parts and features together, is in the end a complex logical operation, described by transcendental logic:

> Transcendental logic is the expansion of the elements of the pure cognition of the understanding and the principles without which no object can be thought at all (which is at the same time a logic of truth). For no cognition can contradict it without at the same time losing all content, i.e. all relation to any object, hence all truth. (A62-3/B87)

In the *Critique*, transcendental logic is not recognisably presented as a logic, and it is commonly thought that it cannot be so presented. The article [1] shows otherwise, mainly by focussing on the semantics of transcendental logic. There is a vast difference between the notion of object as it occurs in first order models, and in Kant's logic. In the former, objects are mathematical entities supplied by the metatheory, usually some version of set theory. These objects have no internal structure, at least not for the purposes of the model theory. Kant's notions of object, as they occur in the semantics furnished by transcendental logic, are very different. For instance, there are 'objects of experience', somehow constructed out of sensory material; transcendental logic deals with *a priori* and completely general principles which govern the construction of such objects, and relate judgements to objects so that we may come to speak of *true* judgements.

3.3 The Table of Judgement (A70/B95)

The three definitions describe judgement either in terms of certain cognitive operations ('unity of representations') or in terms of a function that a judgement has to perform (establishing 'relation to an object'). There is no hint of a specific form of judgement here. We find such hints in the Table of Judgement, but there we do not find a comparison with definitions of judgement; e.g. the *Critique*'s definition occurs only at (B141-2), way after the Table of Judgement is introduced. This raises the problem of how we know that the forms proposed in the Table satisfy the three definitions, and conversely, how for instance the functional characterisation given at (B141-2) leads to specific forms of judgement.

We now turn to the forms of judgement listed in the Table of Judgement, and we discuss (some of) the inferences in which these judgements participate, in part to emphasise the many differences between Kant's logic and modern logic[8] We will also comment on the relation between the Table of Judgement and the Table of Categories (A80/B106), although a full treatment is beyond the scope of this paper.

We will begin our discussion with the title 'Relation' (A70/B95), where we find

Relation
Categorical
Hypothetical
Disjunctive

3.3.1 Categorical judgements

These are judgements in subject-predicate form, combined with quantifiers and optional negation, which can occur on the copula and on the concepts occurring in the judgement. The Table of Judgement further specifies categorical judgements with regard to Quantity and Quality:

Quantity
Universal
Particular
Singular

In the Table of Categories we find a corresponding list of 'pure concepts of the understanding':

Of Quantity
Unity
Plurality
Totality

The precise correspondence between judgement forms and Categories is a matter of controversy. Here we argue on logical grounds that Kant intended a correspondence between the universal judgement and Unity, between the particular judgement and Plurality, and between the singular judgement and Totality.[9]

---
[8]See note 1.
[9]See Frede and Krüger [3] for a different correspondence linking the singular judgement and Unity.

As explained in section 3.1, the universal judgement 'all M are P', or as Kant would have it 'To everything x to which M belongs, also P belongs', should not be interpreted as the classical $\forall x(M(x) \to P(x))$, but as

$$\forall x(\bigvee_{p \in M} p(x) \to \bigvee_{q \in P} q(x));$$

and because the subject is maintained 'assertorically', not 'problematically', we require that the types in M do not contain $\bot$. These types are therefore satisfiable – meaning that the (nonempty) collection of M's is given as that which the judgement is about, and the quantifier 'To everything x' is restricted to M, not to some universe of discourse.

The association 'universality – unity' is motivated by the fact that in the universal judgement 'all M are P' the predicate P makes no distinctions among the things falling under the subject M. Relative to P, M can hence be taken as a unit.

The things falling under M form a plurality that is not a unity (with respect to the predicate P) if there are true particular judgements 'some M are P' and 'some M are not P'.

In an unpublished note about the relation between universal and singular judgement, Kant writes:

> *Refl.* **3068** In the universal concept the sphere [=extension] of a concept is entirely enclosed in the sphere of another concept; [...] in the singular judgement, a concept that has no sphere at all is consequently merely enclosed as a part under the sphere of another concept. Thus singular judgements are to be valued equally with the universal ones, and conversely, a universal judgement is to be considered a singular judgement with regard to the sphere, much as if it were only one by itself. [9, p. 62]

Now consider (B111), where we read 'Thus **allness** (totality) is nothing other than plurality considered as a unity [...]'

Taking a plurality M to be a totality involves considering M as a unity, which means that a pair of judgements 'some M are P' and 'some M are not P' is replaced by one of 'all M are P' and 'all M are not P'. M is thus totally determined with respect to the available predicates. Since M cannot be divided using a predicate, this means that the concept M is used singularly, and hence a universal judgement 'all M are P' can equivalently be regarded as the singular judgement 'M is P', whence the correspondence between the singular judgement and totality.

Quality
Affirmative
Negative
Infinite

There is no need for our present purposes to dwell extensively on this Category, except to say that Kant makes a distinction between sentence negation as in the negative particular judgement 'some A are not B' and predicate negation, represented by the infinite judgement 'some A are non-B', which is affirmative but requires infinitary logic for its formalisation: $\bigvee_{B \cap C = \emptyset}$ (some A are C). Hence Kant's logic is not finitary. The difference with classical first order logic will only increase as we go on.

3.3.2 Hypothetical judgements

It would be a mistake to identify Kant's hypothetical judgements with a propositional conditional $p \to q$, let alone material implication as defined by its truth table: a material implication need not have any rule-like connection between antecedent and consequent. Here is the definition in the *Jäsche Logik*:

> The matter of hypothetical judgements consists of two judgements that are connected to each other as ground and consequence. One of these judgements, which contains the ground, is the *antecedent*, the other, which is related to it as consequence, is the *consequent*, and the representation of this kind of connection of two judgements to one another for the unity of consciousness is called the *consequentia* which constitutes the form of hypothetical judgements. [5, p. 601, par. 59] [10]

This definition seems to say that the hypothetical is a propositional connective, and some of Kant's examples fall into this category:

> If there is perfect justice, then obstinate evil will be punished. (A73/B98)

However, other examples exhibit a more complex structure, involving relations, variables and binding. In the context of a discussion of the possible temporal relations between cause and effect Kant writes in *CPR*:

> If I consider a ball that lies on a stuffed pillow and makes a dent in it as a cause, it is simultaneous with its effect. (A203/B246)

The hypothetical that can be distilled from this passage is:

> If a ball lies on a stuffed pillow, it makes a dent in that pillow.

From this we see that (i) the antecedent and consequent need not be closed judgements but may contain variables, and (ii) antecedent and consequent may contain relations and existential quantifiers.

---

[10] Here it is of interest to observe that in the same paragraph *consequentia* is also used to refer to an inference.

We now give an extended quote from the *Prolegomena* §29 [8] which provides another example of a hypothetical judgement whose logical structure likewise exhibits the features listed in (i) and (ii) above:

> It is, however, possible that in perception a rule of relation will be found, which says this: that a certain appearance is constantly followed by another (though not the reverse); and this is a case for me to use a hypothetical judgement and, e.g., to say: If a body is illuminated by the sun for long enough, it becomes warm. Here there is of course not yet the necessity of connection, hence not yet the concept of cause. But I continue on, and say: if the above proposition, which is merely a subjective connection of perceptions, is to be a proposition of experience, then it must be regarded as necessarily and universally valid. But a proposition of this sort would be: The sun through its light is the cause of the warmth. The foregoing empirical rule is now regarded as a law, and indeed as valid not merely of appearances, but of them on behalf of a possible experience, which requires universally and therefore necessarily valid rules [...] the concept of a cause indicates a condition that in no way attaches to things, but only to experience, namely that experience can be an objectively valid cognition of appearances and their sequence in time only insofar as the antecedent appearance can be connected with the subsequent one according to the rule of hypothetical judgements. [8, p. 105]

The logical form of the first hypothetical (a 'judgement of perception') is something like:

> If x is illuminated by y between time t and time s and $s - t > d$ and the temperature of x at t is v, then there exists a $w > 0$ such that the temperature of x at s is $v + w$ and $v + w > c$,

where d is the criterion value for 'long enough' and c a criterion value for 'warm'. We find all the ingredients of polyadic logic here: relations and quantifier alterations. The causal connection which transforms the judgement into a 'judgement of experience' arises when the existential quantifiers are replaced by explicitly definable functions.

We now move on to the logical properties of the hypothetical judgement. Here it is of some importance to note that the term *consequentia*, characterising the logical form of the hypothetical, is also used to describe the inferences from the hypothetical:

> The *consequentia* from the ground to the grounded, and from the negation of the grounded to the negation of the ground, is valid. [5, p. 623]

Furthermore, the negation of a hypothetical is not defined.[11] This strongly suggests that the hypothetical judgement is really a license for inferences. Indeed, in the

---

[11] Note that the negation of a categorical judgement is defined, although its properties do seem

Jäsche Logik Kant characterises inferences such as modus ponens and modus tollens as immediate inferences and as such needing only one premise, not two premises [5, p. 623]. Modern proof systems conceive of modus ponens as a two-premise inference, p implies q and p, therefore q. But Kant does not think of it in this way. He thinks of it as an inference with premise p, conclusion q, which is governed by a license for inference. This strongly suggests that Kant does not have a single entailment relation, as in modern logic[12], but only local entailment relations defined by specific inferences. We end this discussion of the hypothetical judgement with a further twist: its logical properties change when it is considered in a causal context, i.e. in transcendental logic:

> When the cause has been posited, the effect is posited ⟨*posita causa ponitur effectus*⟩ already flows from the above. But when the cause has been cancelled, the effect is cancelled ⟨*sublata causa tollitur effectus*⟩ is just as certain; when the effect has been cancelled, the cause is cancelled ⟨*sublato effectu tollitur causa*⟩ is not certain, but rather the causality of the cause is cancelled ⟨*tollitur causalitas causae*⟩. [6, p.336-7]

3.3.3 Disjunctive judgements

These are again not what one would think, judgements of the form $p \vee q$. The *Jäsche Logik* provides the following definition:

> A judgement is disjunctive if the parts of the sphere of a given concept determine one another in the whole or toward a whole as complements [...] [A]ll disjunctive judgements represent various judgements as in the community of a sphere [...] [O]ne member determines every other here only insofar as they stand together in community as parts of a whole sphere of cognition, outside of which, in a certain relation, nothing may be thought.(*Jäsche Logik*, §27, 28) [5, pp. 602-3]

As examples Kant provides:

> Every triangle is either right-angled or not right-angled.
> A learned man is learned either historically, or in matters of reason.

Thus the logical form is something like $\forall x(C(x) \to A(x) \vee B(x))$, where C represents the whole, A, B its parts; here it is not immediately clear whether the parts can be taken to exist outside the context of the whole. But actually the situation is much

---

to be weaker than classical negation: 'some A are not B' is the negation of 'All A are B', but it is a moot point whether the negative particular judgement has existential import. (Its infinitive counterpart does have existential import.

[12] See Hodges [4] for relevant discussion.

more complicated. The *Jäsche Logik* equivocates between concepts and judgements making up the whole, and this is intentional, as we read in the *Vienna Logic*:

> The disjunctive judgment contains the relation of different judgment insofar as they are equal, as *membra dividentia*, to the *sphaera* of a *cognitio divisa*. E.g., All triangles, as to their angles, are either right-angled or acute or obtuse. I represent the different members as they are opposed to one another and as, taken together, they constitute the whole *sphaera* of the *cognitio divisa*. This is in fact nothing other than a logical division, only in the division there does not need to be a *conceptus divisus*; instead, it can be a *cognitio divisa*. E.g., If this is not the best world, then God was not able or did not want to create a better one. This is the division of the *sphaera* of the cognition that is given to me. [5, p. 374-5]

So it is not just concepts that can be divided in the familiar way, also cognitions (*Erkenntnisse*), including judgements, can be so divided. What this means for the complexity of Kant's logic can be seen if we look at the expanded example in the *Dohna-Wundlacken Logic*:

> If this world is not the best, then God either was unfamiliar with a better [one] or did not wish to create it or could not create [it], etc. Together these constitute the whole *sphaera*. [5, p. 498]

It will be instructive to formalise this example. Let w_0 be the actual world, G a constant denoting God, let $B(w_0, w)$ represent 'w is a better world than w_0', and let $Uf(G, w)$, $Uw(G, w)$, $Uc(G, w)$ represent: 'God was unfamiliar with w', 'God was unwilling to create w' and 'God was unable to create w', respectively. We then get the combined hypothetical-disjunctive judgement:

$$\exists w B(w_0, w) \to \forall w (B(w_0, w) \to (Uf(G, w) \lor Uw(G, w) \lor Uc(G, w))).$$

It is to be noted that this hypothetical-disjunctive judgement consists entirely of relations, and that the division is formulated in terms of singular judgements containing a parameter ('God') and a variable. As in the case of the hypothetical judgement, the negation for a disjunctive judgement is not defined, which suggests that it is actually a license for inferences, using quantified forms of the disjunctive syllogism, for example:

1. Starting from the premise 'God is familiar with a better world' (which is taken to imply $\exists w(B(w_0, w) \land \neg Uf(G, w)))$ now introduces the positive primitive formula $\exists w(B(w_0, w) \lor (Uw(G, w) \lor Ua(G, w)))$.

2. Similarly the premise 'God is familiar with all better worlds' yields the formula $\forall w(B(w_0, w) \to (Uw(G, w) \lor Ua(G, w)))$.

Kant evidently believes these inferences are perfectly proper cases of the disjunctive syllogism, but the present-day reader may well ask whether his general logic has the resources to break down these inferences in smaller steps. But if the hypothetical and the disjunctive judgement are licenses for inferences, this means that they can be taken as given as far as general logic is concerned (much like a **Prolog** program is taken as given and is used only to derive atomic facts). This somewhat eases the burden on general logic, in the sense that it need not have the resources to prove hypothetical and disjunctive judgements.

As we did for the hypothetical judgement, we will also look at the intended transcendental use of the disjunctive judgement:

> The same procedure of the understanding when it represents to itself the sphere of a divided concept, it also observes in thinking of a thing as divisible; and just as in the first case the members of the division exclude each other, and yet are connected in one sphere, so in the latter case the understanding represents to itself the parts of the latter as being such that existence pertains to each of them (as substances) exclusively of the others, even while they are combined together in one whole. (B113)

The disjunctive judgement is said to involve the cognitive act of dividing a thing, while keeping the resulting parts simultaneously active in one representation. Here we are concerned with the logical principles that Kant's disjunction satisfies. Kant gives as inferences valid for a disjunctive judgement $C \to A \vee B$, the two halves of the so-called disjunctive syllogism:

C and $\neg$A implies B
C and A implies $\neg$B.

These inference rules are considerably weaker than those that are valid for the classical or intuitionistic disjunction, and remind one of the multiplicative disjunction of linear logic. Can one impose stronger inference rules on the disjunction? That is doubtful. For example, the standard right disjunction rule in sequent calculus:

$$\frac{\Gamma \Rightarrow A, \Delta}{\Gamma \Rightarrow A \vee B, \Delta}$$

is invalid for Kant, because it allows the addition of an arbitrary B to A, without the guarantee that A, B constitute a whole.

An additional consideration is the connection with divisibility; here the parts must be present simultaneously, which is what the rule just given expresses. This formulation lends some credibility to Kant's association of the disjunctive judgement with the category of simultaneity in the third Analogy of Experience. However, the new formulation raises the issue of what one should say if A and B are identical. Kant makes an important distinction between two kinds of identity in 'On the amphiboly of concepts of reflection':

> If an object is presented to us several times, but always with the same inner determinations, then it is always exactly the same if it counts as an object of pure understanding, not many but only one thing; but if it is appearance, then [...] however identical everything may be in regard to [concepts], the difference of the places of these appearances at the same time is still an adequate ground for the numerical difference of the object (of the senses) itself. Thus, in the case of two drops of water one can completely abstract from all inner difference (of quality and quantity), and it is enough that they be intuited in different places at the same time for them to be held to be numerically different. (A263-4/B319-20)

Suppose one has a 'whole' that is divided into spatially distinct parts that have 'the same inner determinations'. This hypothetical situation suggests that a logic for Kant's disjunction does not include a rule for (right) contraction:

$$\frac{\Gamma \Rightarrow A, A, \Delta}{\Gamma \Rightarrow A, \Delta}$$

But in that case also the standard rule for left disjunction introduction:

$$\frac{\Gamma, A \Rightarrow \Delta \quad \Gamma, B \Rightarrow \Delta}{\Gamma, A \vee B \Rightarrow \Delta}$$

must be dropped because otherwise right contraction becomes derivable. Instead, one would have a rule like:

$$\frac{\Gamma, A \Rightarrow \Delta \quad \Gamma, B \Rightarrow \Delta'}{\Gamma, A \vee B \Rightarrow \Delta, \Delta'}$$

3.4 Logical form of judgements

Looking back at our examples we see that, with one exception (the negative particular judgement, which, as discussed in [1] was meant by Kant to be purely negative), they are all geometric judgements. Geometric logic, i.e. the logic of geometric formulas, plays an important role in several branches of mathematics, Euclidean geometry being one but not the only example. More germane to our purposes is a result in [1], which shows that all objectively valid judgements in the sense of (B141-2) must be finite conjunctions of geometric implications.

3.5 'Functions of unity in judgements': dynamical proofs

In a dynamical proof one takes a geometric theory[13] as defining a consequence relation holding between two sets of facts. An example, taken from Coquand [2], illustrates the idea. The theory is:[14]

1. $P(x) \wedge U(x) \to Q(x) \vee \exists y R(x, y)$

2. $P(x) \wedge Q(x) \to \bot$

3. $P(x) \wedge R(x, y) \to S(x)$

4. $P(x) \wedge T(x) \to U(x)$

5. $U(x) \wedge S(x) \to V(x) \vee Q(x)$

And here is an example of a derivation of $V(a_0)$ from $P(a_0), T(a_0)$:

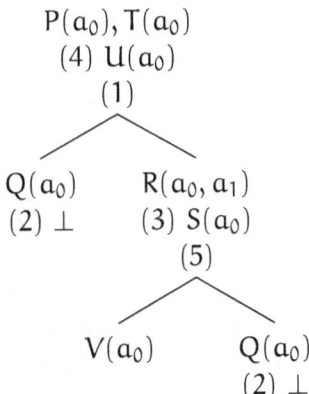

We give some comments on the derivation. The dynamical proof just given can also be taken to prove $\forall x(P(x) \wedge T(x) \to V(x))$, where the proof is the link between antecedent and consequent, hence a 'function of unity'. Furthermore, the geometric theory defines the consequence relation, hence the geometric implications occurring in it can be seen as inference rules. Disjunctions lead to branching of the tree, as we see in (1) and (5). The existential quantifier in formula (1) introduces a new term in the proof, here a_1, which appears in the right branch of (1). This constant is the 'object in general' of *Reflexion* **4634**. Lastly, a fact is derivable if it appears on

---
[13] We assume the geometric implications in the theory have antecedents consisting of conjunctions of atomic formulas only.
[14] We omit the universal quantifiers.

every branch not marked by $\bot$, which leaves $V(a_0)$. If X is a collection of facts whose terms are collected in I, F a fact with terms in I, and T a geometric theory, then there exists a dynamical proof of F from X if and only if $T, X \vdash F$ in intuitionistic logic.

It is clear how a dynamical proof of a geometric implication from a geometric theory proceeds: if T is the geometric theory and $\forall \bar{x}(\tau(\bar{x}) \to \theta(\bar{x}))$ the geometric implication (τ is a conjunction of atomic formulas, and for simplicity take θ an existentially quantified conjunction θ' of atomic formulas; we interpret θ' as a set), choose new terms not occurring in either T or $\forall \bar{x}(\tau(\bar{x}) \to \theta(\bar{x}))$, plug these terms into τ and construct a dynamical proof tree with the sets θ' at the leaves. There may occur terms in θ' not in τ; these have to be quantified existentially. Introduce any other existential quantifiers on θ' as required by θ. The result is an intuitionistic derivation of $\forall \bar{x}(\tau(\bar{x}) \to \theta(\bar{x}))$ from T. Conversely, if there is an intuitionistic derivation of $\forall \bar{x}(\tau(\bar{x}) \to \theta(\bar{x}))$ from T, then there exists a dynamical proof in the sense just sketched.

Dynamical proofs as a semantics for geometric implications can explain Kant's characterisation of judgements as rules, as well as 'a unity of the consciousness of various representations'; after all, the diagram represents 'unity' as a single spatial representation. What remains to be done is to situate a judgement's 'objective validity' relative to its other properties.

4 Completeness of the Table of Judgement

In [1] it is argued that (i) Kant's implied semantics for logic is radically different from that of classical first order logic, (ii) the implied semantics, centered around Kant's three different notions of object, can be given a precise mathematical expression, thus leading to a formalised transcendental logic, and (iii) on the proposed semantics, Kant's formal logic turns out to be geometric logic.

It is not appropriate to repeat the technical exposition here, so we will follow a different strategy starting from Kant's most fundamental characterisation of judgement:

> A judgement is nothing but the manner in which given cognitions are brought to the objective unity of apperception. (B141)

A judgement is the act of binding together mental representations; this is what the term 'unity' refers to. The aim of judgement is indicated by means of the word 'objective', which is Kant's terminology for 'having relation to an object'. But for Kant, objects are not found in experience, but they are constructed ('synthesised') from sensory matter under the guidance of the Categories, which are defined as 'concepts of an object in general, by means of which the intuition of an object is regarded as determined in respect of one of the logical functions of judgement' (B128). It is here that judgement plays an all-important role, since Kant's idea is that objects are synthesised through the act of making judgements about them.

Technically, these acts of synthesis are modelled as a kind of possible worlds structure (an 'inverse system'), where the possible worlds are finite first order models whose elements are partially synthesised objects, except for the unique top-world (the 'inverse limit') which represents (the idea of) fully synthesised objects. Bringing a (formal) judgement φ to the 'objective unity of apperception' is now characterised by the property: for any such possible worlds structure, if φ is true on all worlds, then φ is also true on the top-world. That is to say, if φ is true for all stages of synthesis of an object, then φ is true of some fully synthesised object. Kant calls judgements φ satisfying this conditional property 'objectively valid.' It turns out that the objectively valid formulas are exactly the geometric formulas. It follows that no judgement whose logical form is more complex than that allowed by the Table of Judgement can be objectively valid, i.e. this Table is complete.

It is of some interest that the key idea in the proof sheds light on Kant's logical reinterpretation of the Categories of Quantity as constraints on concepts (B113-6):

> In every cognition of an object there is, namely, **unity** of the concept, which one can call **qualitative unity** insofar as by that only the unity of the comprehension of the manifold of cognition is thought, as, say, the unity of the theme in a play, a speech, or a fable. Second, truth in respect of the consequences. The more true consequences from a given concept, the more indication of its objective reality. One could call this the **qualitative plurality** of the marks that belong to a concept as a common ground ... Third, finally, **perfection**, which consists in plurality conversely being traced back to the unity of the concept, and agreeing completely with this one and no other one, which one can call **qualitative completeness** (totality).

The phrase 'unity of the theme in a play' is probably a reference to Aristotle's 'unity of action' in tragedy, where

> the structural union of the parts [must be] such that, if any one of them is displaced or removed, the whole will be disjointed and disturbed. For a thing whose presence or absence makes no visible difference, is not an organic part of the whole (*Poetics*, VIII).

Hence we read 'qualitative unity' as the requirement that the concept under consideration is integrated with other concepts by means of a theory, and is invariant under structure-preserving mappings (homomorphisms). The latter requirement forces all subconcepts of the given concept to have the same logical complexity. We are now in a position to spell out the logical meaning of B113-6 in formal terms.

Let C be a concept which satisfies 'qualitative unity' and let T be the first order theory witnessing 'qualitative unity'. Define a 'qualitative plurality' Σ by

$$\Sigma(x) = \{\theta(x) \mid T \models \forall x(C(x) \to \theta(x)), \theta \text{ pos. prim.}\}.$$

Because we may have, for each θ, 'some θ aren't C', for all we know Σ could be a proper plurality. But 'qualitative completeness' now becomes provable:

$$\Sigma(x), T \models C(x),$$

hence by compactness there is positive primitive τ(x) such that

$$T \models \forall x(\tau(x) \leftrightarrow C(x)).$$

It follows that, as announced in section 3.1, universal judgements 'all M are P' can be expressed as geometric implications, provided the concepts M, P satisfy 'qualitative unity'.

In summary, we have shown that after formalisation, Kant's general logic turns out to be at least as rich as geometric logic, while it coincides with it when taking into account the semantics of judgements dictated by 'transcendental logic'.[15] This latter result is but one example of interesting metalogical theorems that may be proved about Kant's logic; B113-6, formally reinterpreted as a theorem about definability of concepts, is another.

References

[1] T. Achourioti and M. van Lambalgen. A formalisation of Kant's transcendental logic. *Review of Symbolic Logic*, 4(2):254–289, 2011. DOI: http://dx.doi.org/10.1017/S1755020310000341.

[2] T. Coquand. A completeness proof for geometric logic. Technical report, Computer Science and Engineering Department, University of Gothenburg, 2002. Retrieved September 29, 2010, from http://www.cse.chalmers.se/~coquand/formal.html.

[3] M. Frede and L. Krüger. Ueber die Zuordnung der Quantitaeten des Urteils und der Kategorien der Groesse bei Kant. *Kant-Studien* 61, 28–49, 1970.

[4] W. Hodges. Traditional Logic, Modern Logic and Natural Language. *Journal of Philosophical logic*, 38:589–606, 2009. DOI: http://dx.doi.org/10.1007/s10992-009-9113-y.

[5] I. Kant. *Lectures on Logic; translated from the German by J. Michael Young.* The Cambridge edition of the works of Immanuel Kant. Cambridge University Press, Cambridge, 1992.

---

[15]We simplify here and do not consider multiplicative disjunction.

[6] I. Kant. *Lectures on Metaphysics; edited by Karl Ameriks and Steve Naragon.* Cambridge University Press, Cambridge, 1997.

[7] I. Kant. *Critique of pure reason; translated from the German by Paul Guyer and Allen W. Wood.* The Cambridge edition of the works of Immanuel Kant. Cambridge University Press, Cambridge, 1998.

[8] I. Kant. *Theoretical philosophy after 1781; edited by Henry Allison and Peter Heath.* The Cambridge edition of the works of Immanuel Kant. Cambridge University Press, Cambridge, 2002.

[9] I. Kant. *Notes and fragments; edited by Paul Guyer.* The Cambridge edition of the works of Immanuel Kant. Cambridge University Press, Cambridge, 2005.

[10] M. Wolff. Vollkommene Syllogismen und reine Vernunftschluesse: Aristoteles und Kant. Eine Stellungnahme zu Theodor Eberts Gegeneinwaenden. Teil 2. *Journal for General Philosophy of Science*, 2010. DOI: http://dx.doi.org/10.1007/s10838-010-9136-7.

Received 31 July 2016

An Ordinal-Free Proof of the Complete Cut-Elimination Theorem for $\Pi_1^1\text{-}CA + BI$ with the ω-rule

Ryota Akiyoshi[*]
Waseda Institute for Advanced Study,
1-6-1 Nishi Waseda, Shinjuku-ku, Tokyo 169-8050, Japan
georg.logic@gmail.com

Abstract

This paper presents a result based on the joint project with Grisha Mints. An ordinal-free proof of the *complete* cut-elimination theorem for $\Pi_1^1\text{-}CA + BI$ with the ω-rule for (not only arithmetical) but *arbitrary* sequents is presented by iterating an extension of Buchholz' Ω-rule by the author and Mints.

1 Introduction

Takeuti showed the consistency of $\Pi_1^1\text{-}CA_0$ by proving a partial cut-elimination theorem for it in 1958 [19][1]. In 1970, Tait gave a constructive proof of the consistency of $\Sigma_2^1\text{-}DC$ (dependent choice) [18]. After these works, logicians of Feferman-Schütte's school including Buchholz, Pohlers, Jäger,... have developed proof-theoretic methods for impredicative systems in perspicuous ways. In particular, they have developed infinitary proof theory while Takeuti had worked only on finitary proof figures as Gentzen.

Here, one should mention Grisha Mints' pioneering contribution to connect these different ways of proof-theory [10]. In particular, he proposed a way of enriching infinitary derivations by finite ones. This line of investigation has been developed further by Buchholz [5–7] by showing that there is a precise correspondence between finitary proof theory and infinitary one. Also, Mints proposed a general schema of proving the normalization theorem for a finitary system using one of the corresponding infinitary system [12].

---

[*]This work was partially supported by KAKENHI 16K16690.
[1] Although he also proved the consistency of $\Pi_1^1\text{-}CA + BI$ in 1967 [20], the highlight among Takeuti's consistency results would be this earlier paper [19].

Going back to the history of proof theory, Rathjen and Arai independently obtained ordinal analysis for Π^1_2-CA [2–4, 15–17] in 1990's, which is much stronger than any iterations of Π^1_1-CA. In these results, complicated proof-theoretic ordinal notation systems have played a crucial role. When a strong impredicative theory is considered, proof-theoretic ordinals are needed even for defining the derivability relation of a suitable infinitary system.

In 2000's, Bill Tait posed a problem whether we can provide an ordinal-free proof of the complete cut-elimination theorem for Π^1_1-CA with the ω-rule[2]. Grisha Mints had had a similar direction to prove the cut-elimination theorem or the termination of ϵ-substitution method for impredicative theories[3]. At that time, I had some communication with Tait by e-mail. When I asked some questions about his work on type theory, Tait kindly introduced me to Grisha Mints. We also discussed about Tait's problem about an ordinal-free proof of the cut-elimination theorem, and then Grisha suggested me to work on the joint project to give an affirmative answer to it[4]. Since Buchholz already gave an ordinal-free proof of a partial cut-elimination for his Ω-rule, we hoped to use this method in some suitable way. Indeed, the joint paper with Grisha [1] is the first important step for this project; we extended Buchholz' Ω-rule for the lightface case to obtain the complete cut-elimination theorem for (not only arithmetical but) *arbitrary* sequents. In this paper, an ordinal-free proof of the complete cut-elimination theorem for arbitrary derivations of the full $\Pi^1_1 - CA + BI$ with the ω-rule is presented by extending the result of the joint paper [5]. This result provides another proof of Yasugi's cut-elimination theorem for the full $\Pi^1_1 - CA + BI$ with the ω-rule using an extension of Takeuti's ordinal diagrams based on arbitrary countable ordinals [22].

Finally, the author of the present paper would like to express his deepest gratitude to Grisha's kindness, advice, and support.

---

[2]We remark that Girard's proof published in 1971 of the strong normalization for second-order polymorphic calculus (System F) is not considered as a solution here. This is because, according to Tait, such a proof of the cut-elimination must involve only reasoning about well-founded trees like inductive definitions. For a modern presentation of Girard's proof, we refer to [9].

[3]Indeed, Mints gave two different ordinal-free proofs of the termination of ϵ-substitution method for the theory of non-iterated inductive definition called ID_1 [13, 14].

[4]Our joint works had been done mainly by e-mail. I met Grisha in person four times; I met him for the first time in Munich (2008). After this meeting, I visited to Stanford University twice (2008, 2010) and invited him to Keio University in Tokyo (2010).

[5]When I obtained some results concerning this paper, I asked Grisha to become the co-author of the paper. He declined my offer since much works are done by me according to him, thus I should become the sole author.

1.1 Structure of this Paper

The present paper consists of 6 sections. After introducing the basic definitions in Section 2, we define the infintary systems BI_0^Ω, BI_{n+1}^Ω and BI^Ω in Section 3.

In Section 4, we define the operators $\mathcal{R}$ (one-step reduction), $\mathcal{E}$ (reducing the cut-rank by 1), $\mathcal{E}_\omega$ (reducing the cut-rank until 0), and $\mathcal{D}_n$ (eliminating the impredicative cuts) on derivations in BI^Ω. Finally, we define the substitution operator $\mathcal{S}_T^X$.

In Section 5, we introduce BI_ω, which is Π_1^1-$CA + BI$ with the ω-rule. To take care of Takeuti's explicit/implicit distinction, we introduce another system BI_ω^e.

In Section 6, we define an embedding map g^* from derivations in BI_ω^e into the derivations in BI^Ω. By the theorems obtained so far, our main result is proved.

2 Preliminaries

First, we define a language L for second-order arithmetic and the set $PV(A)$ of free predicate variables in A which are in the scope of a second-order quantifier. We adopt Buchholz and Schütte's definition in our setting [8] and remark that the notion of $PV(A)$ is essentially introduced by Takeuti [19, 20].

0 is a term. If t is a term, then $S(t)$ is a term. If R is an n-ary predicate symbol for an n-ary primitive recursive relation and $t_1,...,t_n$ are terms, then $R(t_1,...,t_n)$ is a formula. If X is unary predicate variable and t is a term, then $X(t)$ is a formula. These are *atomic formulas*. If A is an atomic formula, then $\neg A$ is a formula. A and $\neg A$ where A is atomic are *literals*. If A is a literal, then $PV(A) := \emptyset$. If A and B are formulas, then $A \wedge B$ and $A \vee B$ are formulas. $PV(A \wedge B) = PV(A \vee B) := PV(A) \cup PV(B)$. If $A(0)$ is a formula, then $\forall x A(x)$ and $\exists x A(x)$ are formulas. $PV(\forall x A(x)) = PV(\exists x A(x)) := PV(A(0))$. If A is a formula, then $\forall X A$ and $\exists X A$ are formulas. $PV(\forall X A) = PV(\exists X A) := \{Y | Y \in FV(A) \text{ and } Y \neq X\}$.

As usual, sequents are finite sets of formulas. Moreover, if A is a formula, then $FV(A)$ is the set of free predicate variables occurring in A. Similarly, if Γ is a sequent, we define $FV(\Gamma) := \cup_{A \in \Gamma} FV(A)$. We use the following syntactic variables: A, B, C, F for formulas, Γ, Δ for sequents, and i, j, k, l, m, n for natural numbers. Next we define the notions of *weak* and *strong formulas* as follows. Every literal is a weak formula. If A and B are weak, then $A \wedge B$, and $A \vee B$ are weak. If $A(0)$ is weak, then $\forall x A(x)$ and $\exists x A(x)$ are weak. If $A(X)$ is weak and $X \notin PV(A(X))$, then $\forall X A(X)$ and $\exists X A(X)$ are weak formulas. If A is not weak, then A is strong. We define $PV(\Gamma) := \cup_{A \in \Gamma}(PV(A))$.

Example 1. $\exists X(X(t) \wedge \forall Y Y(t'))$ is weak, but $\exists X(X(t) \wedge \forall Y(X(t) \wedge Y(t')))$ is strong.

A quantifier in $A \wedge B, A \vee B, \forall x A(x), \exists x A(x)$ is *weak* (*strong*) if the corresponding quantifier in $A, B, A(0)$ is weak (strong). The indicated quantifier $\forall X$ or $\exists X$ is called *weak* (*strong*) if $\exists X A$ or $\forall X A$ is weak (strong). Any other quantifier in $\exists X A, \forall X A$ is *weak* (*strong*) if the corresponding quantifier in A is weak (strong).

If A is a formula which is not atomic, then its *negation* $\neg A$ is defined using de Morgan's laws. The set of true literals is denoted by TRUE. T denotes an expression of the form $\lambda x. A(x)$ called *abstraction* where $A(0)$ is a formula. If T is an abstraction of the form $\lambda x. A(x)$, then $A[X/T]$ denotes an expression obtained by replacing every $X(t)$ occurring in it by $A(t)$ (after renaming of bound variables if necessary). An abstraction $\lambda x. A(x)$ is called *arithmetical*, *weak*, or *strong* if the corresponding formula $A(0)$ is arithmetical, weak, or strong respectively. Note that if $\forall X A(X)$ is a weak formula and T is a weak abstraction, then $A[X/T]$ is also a weak formula.

Now, the notion of $rk(A)$ is defined as follows.
$rk(A) := 0$ if A is a weak formula.
$rk(A \wedge B) = rk(A \vee B) := max(rk(A), rk(B)) + 1$ if $A \wedge B$ is strong.
$rk(\forall x A(x)) = rk(\exists x A(x)) := rk(A(0)) + 1$ if $\forall x A(x)$ or $\exists x A(x)$ is strong.
$rk(\forall X A(X)) = rk(\exists X A(X)) := rk(A(X)) + 1$ if $\forall X A(X)$ or $\exists X A(X)$ is strong.

Next we define a formal language on which our infinitary systems are defined, but before introducing this language, we need an intermediate language called L^e.

Definition 1. The language L^e is obtained from L in the following way: (1) Terms and formulas of L are also terms and formulas of L^e. (2) If A is a formula of L, then A^e obtained by adding the superscript e is also a formula of L^e.

Remark 1. Informally, A^e is a formula in a derivation which is not traced into any cut-rule in the derivation. Such a formula is called "explicit" by Takeuti [21].

We adopt Buchholz and Schütte's ramified language [8] into the present setting by considering Takeuti's explicit / implicit distinction.

Definition 2. Language L^*

1. The *terms* of L^* are the same as terms of L^e.

2. The formulas of L^* are obtained by the following replacements from ones of L^e:

    (a) Any formula A^e is unchanged,
    (b) A formula A without e is replaced in the following way:
        i. Every free predicate variable X is replaced by X^n for some $n \in \omega$.

ii. Every strong predicate quantifier $\forall X, \exists X$ is replaced by $\forall X^\omega, \exists X^\omega$, respectively.

Let A be a formula in L^* and let A^- denote the result of deleting all superscripts e and n in A. We define the rank of A by $rk(A) := rk(A^-)$. We define $PV(A) := PV(A^-)$. A is called a *literal*, an *arithmetical formula*, a *weak formula*, or a *strong formula* if A^- is such a formula. T denotes an *abstraction* as in L. An abstraction $\lambda x.A(x)$ is called *arithmetical*, *weak*, or *strong* if $A(0)$ is arithmetical, weak, or strong respectively. If A is a formula without e which is not atomic, then its *negation* $\neg A^*$ is defined by de Morgan's laws as before.

Definition 3. Level for L^*

1. $lev(A^e) := 0$.
2. $lev(A) := 0$ if A is a literal, and $A \not\equiv X^n(t), \neg X^n(t)$ for some $n \in \omega$.
3. $lev(X^n(t)) = lev(\neg X^n(t)) := n$.
4. $lev(A \wedge B) = lev(A \vee B) := max(lev(A), lev(B))$.
5. $lev(\forall x A(x)) = lev(\exists x A(x)) := lev(A(0))$.
6. $lev(\forall X A(X)) := lev(A(X^0))$ if $\forall X A(X)$ is weak.
7. $lev(\exists X A(X)) := lev(A(X^0)) + 1$ if $\exists X A(X)$ is weak.
8. $lev(\forall X^\omega A(X)) := lev(\exists X^\omega A(X)) = \omega$ if $\forall X A(X)$ or $\exists X A(X)$ is strong.

Example 2. $lev(\exists X(X(t) \wedge Y^0(t'))) = 1$, $lev(\exists X^\omega(X(t) \wedge \forall Y(X(t) \wedge Y(t')))) = \omega$.

3 The Systems $\mathbf{BI}_0^\Omega$, $\mathbf{BI}_{n+1}^\Omega$, and $\mathbf{BI}^\Omega$

In this section, we introduce the infinitary systems with the Ω_{n+1}-rules based on L^*. Following Buchholz' notation from [7], only the *minor formulas*, and the *principal formulas* are shown explicitly in inference symbols. Any rule below is supposed to be closed under weakening and contains contraction.

If I be an inference symbol, then we write $\Delta(I)$ and $|I|$ to indicate the set of principal formulas of I and the index set of I, respectively. Moreover, $\bigcup_{i \in |I|}(\Delta_i(I))$ is the set of the minor formulas of I. If $d = I(d_i)_{i \in |I|}$, then d_i is the subderivation of d indexed by i. $\Gamma(d)$ denotes the end-sequent of d. If Γ is a sequent, then $lev(\Gamma) := max(lev(A)|A \in \Gamma)$. Eigenvariables of $\bigwedge_{\forall X A(X)}$ and $\widetilde{\Omega}_{n+1}$ may occur free only in the premises, but not in the conclusion. To denote that Y^n is the eigenvariable of an inference symbol I, we use the notation $!Y^n!$ and write

$$I \frac{\ldots \Delta_i(I) \ldots}{\Delta(I)} \; !Y^n!$$

where $i \in |I|$. We use the same notation for the eigenvariable Y (without superscript n) of an inference symbol I.

Since we are taking care of the explicit/implicit distinction, there are two rules deriving A or A^e in the cases of arithmetical rules. For a compact notation, we write $A^{[e]}$ for a formula A with a possible occurrence of e, that is, for both cases that A has e and not so. When we write $A^{[e]}, B^{[e]}$, there are four possibilities; A, B, A^e, B, A, B^e, or A^e, B^e. For the inference symbols, we use the following notation; if we write

$$I \frac{\ldots A_i^{[e]} \ldots}{A^{[e]}}$$

where $i \in |I|$, then we mean that $A_i^{[e]} \equiv A_i^e$ if and only if $A^{[e]} \equiv A^e$.

Definition 4. The systems $\mathrm{BI}_0^\Omega, \mathrm{BI}_{n+1}^\Omega (0 \leq n)$, and BI^Ω

1. BI_0^Ω consists of the following inference rules.

$$(\mathrm{Ax}_\Delta) \; \Delta$$

where either $\Delta = \{A^{[e]}\} \subseteq \mathrm{TRUE}$ or $\Delta^- = \{C, \neg C\}$ with the condition that

(a) C is an atomic formula in L, and
(b) Δ^- is the result of deleting all superscripts e and n of $D \in \Delta$.

$$(\bigwedge_{(A_0 \wedge A_1)^{[e]}}) \frac{A_0^{[e]} \quad A_1^{[e]}}{(A_0 \wedge A_1)^{[e]}} \qquad (\bigvee_{(A_0 \vee A_1)^{[e]}}^k) \frac{A_k^{[e]}}{(A_0 \vee A_1)^{[e]}} \text{ where } k \in \{0, 1\}$$

$$(\bigwedge_{(\forall x A(x))^{[e]}}) \frac{\ldots A(n)^{[e]} \ldots}{\forall x A(x)^{[e]}} \text{ for all } n \in \omega \qquad (\bigvee_{\exists x A(x)^{[e]}}^k) \frac{A(k)^{[e]}}{\exists x A(x)^{[e]}} \text{ where } k \in \omega$$

$$(\bigwedge_{\forall XA(X)}^{Y^n}) \frac{A(Y^n)}{\forall XA(X)} \;!Y^n! \text{ where } lev(\forall XA(X)) = n.$$

$$(\bigwedge_{\forall X^\omega A(X)}) \frac{\ldots A(Y^n) \ldots (n \in \omega)}{\forall X^\omega A(X)} \;!(Y^n)_{n\in\omega}!$$

$$(\bigwedge_{\forall XA(X)^e}^{Y}) \frac{A(Y)^e}{\forall XA(X)^e} \;!Y!$$

$$(\bigvee_{\neg \forall X^\omega A(X)}^{Y^n}) \frac{\neg A(Y^n)}{\neg \forall X^\omega A(X)} \qquad (\bigvee_{\neg \forall XA(X)^e}^{T}) \frac{\neg A(T)^e}{\neg \forall XA(X)^e} \text{ with arbitrary } T$$

$$(Cut_A) \frac{A \quad \neg A}{\emptyset} \; (A \in L^i)$$

where $L^i :=$ set of all L^*-formulas without e.

2. BI_{n+1}^Ω is obtained by adding the following rules to BI_n^Ω.

$$(\Omega_{\neg \forall XA}) \frac{\ldots \Delta_q \ldots (q \in |\forall XA(X)|)}{\neg \forall XA(X)} \text{ where } lev(\forall XA(X)) = n.$$

$$(\widetilde{\Omega}_{\neg \forall XA}^{Y^n}) \frac{A(Y^n) \quad \ldots \Delta_q \ldots (q \in |\forall XA(X)|)}{\emptyset} \;!Y^n! \text{ where } lev(\forall XA(X)) = n.$$

with

$|\forall XA(X)| := \{(d, Z^n) : d \text{ is a cut-free derivation in } \text{BI}_n^\Omega$
with $lev(\Gamma(d)) \leq n$, and Z is a predicate variable with $Z^n \notin FV(\Delta_{(d,Z^n)})\}$
with $\Delta_{(d,Z^n)} := \Gamma(d) \setminus \{A(Z^n)\}$.

3. $\text{BI}^\Omega := \bigcup_{n \in \omega} \text{BI}_n^\Omega$.

4 Complete cut-elimination theorem for BI^Ω

In this section, we prove the complete cut-elimination theorem for the infinitary system BI^Ω introduced in the previous section (Corollary 1). Moreover, we define the substitution operator $\mathcal{S}_T^{X^m}$ (Theorem 5) which will be needed in Section 6.

Definition 5. Cut-Degree

Let I be an inference symbol and d a derivation in BI^Ω.

1. $dg(I) := \begin{cases} rk(C) + 1 & \text{if } I = Cut_C; \\ 0 & \text{otherwise.} \end{cases}$

2. $dg(I(d_\tau)_{\tau \in |I|}) := sup(\{dg(I)\} \cup \{dg(d_\tau) | \tau \in |I|\})$.

We write
$$d \vdash_\alpha \Gamma$$
if $dg(d) \leq \alpha$ and $\Gamma(d) \subseteq \Gamma$. In what follows, we may assume that $\Gamma(d) = \Gamma$ without loss of generality unless otherwise noted.

Lemma 1. *If $d \vdash_\alpha \Gamma, C$ with literal $C \in L^i$, then there exists a derivation $d' \vdash_\alpha \Gamma, C^e$.*

Proof. Assume $d \vdash_\alpha \Gamma, C$. The proof is by induction on d. The crucial case is that d is Ax_Δ and $\Delta = \{C, \neg C^{[e]}\}$. Then set $d' := Ax_{\Delta'}$ with $\Delta' = \{C^e, \neg C^{[e]}\}$. Other cases are treated using the induction hypothesis (IH). □

We define an operator $\mathcal{R}_C$ which transforms an impredicative cut into $\tilde{\Omega}_{n+1}$, and does one-step reduction for other cuts in the standard way.

Theorem 1. *For $C \in L^i$ there is an operator $\mathcal{R}_C$ on derivations in BI^Ω such that if $d_0 \vdash_\alpha \Gamma, C$, $d_1 \vdash_\alpha \Gamma, \neg C$, and $rk(C) \leq \alpha$ with $\alpha \leq \omega$, then $\mathcal{R}_C(d_0, d_1) \vdash_\alpha \Gamma$.*

Proof. By double induction on $d_0(:= I_0(d_{0i})_{i \in |I_0|})$ and $d_1(:= I_1(d_{1j})_{j \in |I_1|})$. Note that the formulas C in $\Gamma(d_0)$ and $\neg C$ in $\Gamma(d_1)$ do not have the superscript e. If $C \notin \Delta(I_0)$ or $\neg C \notin \Delta(I_1)$, then the claim follows from IH. We consider only important cases which are different from [1, 7].

1. d_0 is an axiom $C, \neg C^{[e]}$.

    It follows that $\neg C^{[e]} \in \Gamma$, and $\Gamma, \neg C^{[e]} = \Gamma$. We define $\mathcal{R}_C(d_0, d_1) := d_1'$, which is obtained from d_1 by Lemma 1 if $\neg C^{[e]} \equiv \neg C^e$.

2. $C \in \Delta(I_0)$, and $\neg C \in \Delta(I_1)$.

It is impossible that both C and $\neg C$ are true literals.

(a) $C \equiv \forall X C_0(X)$. If $lev(C) = n$, then $lev(\neg C) = n+1$.
Now $d_0 = \bigwedge_C^{Y^n}(d_{00})$ and $d_1 = \Omega_{\neg C}(d_{1q})_{q \in |C|}$. Using IH and $\widetilde{\Omega}_{\neg C}^{Y^n}$, we define
$$\mathcal{R}_C(d_0, d_1) := \widetilde{\Omega}_{\neg C}^{Y^n}(\mathcal{R}_C(d_{00}, d_1), \mathcal{R}_C(d_0, d_{1q}))_{q \in |C|}.$$

(b) $C \equiv \forall X^\omega C_0(X)$. $d_0 = \bigwedge_C(d_{0i})_{i \in \omega}$ and $d_1 = \bigvee_{\neg C}(d_{10})$. In this case, we have $d_{10} \vdash_m \Gamma, \neg C_0(Y^k), \neg C$ for some $k \in \omega$. By IH, we have $\mathcal{R}_C(d_{0k}, d_1) \vdash_m \Gamma, C_0(Y^k)$ and $\mathcal{R}_C(d_0, d_{10}) \vdash_m \Gamma, \neg C_0(Y^k)$. We see that $rk(C) > rk(C_0(Y^k))$. Therefore we define
$$\mathcal{R}_C(d_0, d_1) := Cut_{C_0(Y^k)}(\mathcal{R}_C(d_{00}, d_1), \mathcal{R}_C(d_0, d_{10})).$$

This complete the proof. □

Iterating $\mathcal{R}_C$, we define an operator $\mathcal{E}$ which reduces cut-degree by 1.

Theorem 2. *There is an operator $\mathcal{E}$ on derivations in BI^Ω such that if $d \vdash_{m+1} \Gamma$, then $\mathcal{E}(d) \vdash_m \Gamma$.*

Proof. By induction on d. We consider only the crucial case $d = Cut_C(d_0, d_1)$ with $C \in L^i$. Other cases are treated using IH.
By IH, we have $\mathcal{E}(d_0) \vdash_m \Gamma, C$, and $\mathcal{E}(d_1) \vdash_m \Gamma, \neg C$. Define
$$\mathcal{E}(d) := \mathcal{R}_C(\mathcal{E}(d_0), \mathcal{E}(d_1)).$$

This complete the proof. □

Using $\mathcal{E}$, we can define an operator $\mathcal{E}_\omega$ which reduces cut-degree to 0.

Theorem 3. *There is an operator $\mathcal{E}_\omega$ on derivations in BI^Ω such that if $d \vdash_\omega \Gamma$, then $\mathcal{E}_\omega(d) \vdash_0 \Gamma$.*

Proof. By induction on d. We consider only the crucial case $d = Cut_C(d_0, d_1)$ with $C \in L^i$. In this case, $d_0 \vdash_\omega \Gamma, C$, and $d_1 \vdash_\omega \Gamma, \neg C$.
By IH, $\mathcal{E}_\omega(d_0) \vdash_0 \Gamma, C$, and $\mathcal{E}_\omega(d_1) \vdash_0 \Gamma, \neg C$. Let $rk(C) := m$, then we see $Cut_C(\mathcal{E}_\omega(d_0), \mathcal{E}_\omega(d_1)) \vdash_{m+1} \Gamma$. Let $\mathcal{E}^{m+1}$ be $m+1$ applications of the operator $\mathcal{E}$. We define
$$\mathcal{E}_\omega(d) := \mathcal{E}^{m+1}(Cut_C(\mathcal{E}_\omega(d_0), \mathcal{E}_\omega(d_1))) \vdash_0 \Gamma.$$

This complete the proof. □

Now we define the collapsing operator $\mathcal{D}_n$ eliminating $\tilde{\Omega}^{Y^m}_{\neg \forall X A(X)}$ with $m = lev(\forall X A(X))$ if $dg(d) = 0$, $lev(\Gamma(d)) \leq n$, and $n \leq m$.

Theorem 4. *There is an operator $\mathcal{D}_n$ such that if $d \vdash_0 \Gamma$ and $lev(\Gamma) \leq n$, then $\mathrm{BI}^\Omega_n \ni \mathcal{D}_n(d) \vdash_0 \Gamma$.*

Proof. By induction on d. We consider only the important cases. Other cases are treated using IH. Let I be the last inference symbol of d.

1. $I = \tilde{\Omega}^{Y^m}_{\neg \forall X A(X)}$ with $lev(\forall X A(X)) = m$ and $n \leq m$.

    In this case $d = \tilde{\Omega}^{Y^m}_{\neg \forall X A(X)}(d_\tau)_{\tau \in \{0\} \cup |\forall X A(X)|}$. Then $Y^m \notin FV(\Gamma)$, $d_0 \vdash_0 \Gamma, A(Y^m)$ and $d_q \vdash_0 \Gamma, \Delta_q$ for all $q \in |\forall X A(X)|$ with $lev(\Delta_q) \leq m$. By IH, $\mathrm{BI}^\Omega_m \ni \mathcal{D}_m(d_0) \vdash \Gamma, A(Y^m)$, and $Y^m \notin FV(\Gamma(\mathcal{D}_m(d_0)) \setminus \{A(Y^m)\})$. We define $q_0 := (\mathcal{D}_m(d_0), Y^m) \in |\forall X A(X)|$. Hence, using IH again, we can define
    $$\mathcal{D}_n(\tilde{\Omega}^{Y^m}_{\neg \forall X A(X)}(d_\tau)_{\tau \in \{0\} \cup |\forall X A(X)|}) := \mathcal{D}_n(d_{q_0}) \in \mathrm{BI}^\Omega_n.$$

2. $I = \tilde{\Omega}^{Y^m}_{\neg \forall X A(X)}$ with $lev(\forall X A(X)) = m$ and $m < n$.

    Using IH, we define the required derivation $\tilde{\Omega}^{Y^m}_{\neg \forall X A(X)}(\mathcal{D}_n(d_\tau))_{\tau \in \{0\} \cup |\forall X A(X)|}$, which is in BI^Ω_n since $m < n$.

3. Otherwise.

    By IH, $\mathcal{D}_n(d_i) \ni \mathrm{BI}^\Omega_n$ for $i \in |I|$. Then, we define $\mathcal{D}_n(d) := I(\mathcal{D}_n(d_i))_{i \in |I|} \in \mathrm{BI}^\Omega_n$. An important case is that $I = \bigvee^T_{\neg \forall X A(X)^e}$. In this case, $\Gamma(d)$ contains $\neg \forall X A(X)^e$, but $lev(\neg \forall X A(X)^e) = 0$ by Definition 3.

□

Remark 2. Note that $\mathcal{D}_n(d)$ is a cut-free derivation in BI^Ω_n.

Corollary 1. *If $d \in \mathrm{BI}^\Omega$ and $lev(\Gamma(d)) \leq n$, then there exists $d' \in \mathrm{BI}^\Omega_n$ such that $d' \vdash \Gamma(d)$.*

Proof. By Theorems 3 and 4, we have $\mathcal{D}_n(\mathcal{E}_\omega(d)) \in \mathrm{BI}^\Omega_n \vdash \Gamma$. □

An interpretation from L^e into L^* is a function which assigns a number $*(X) \in \omega$ to each predicate variable X. Given an interpretation $*$, for any L^e-formula A we define the L^*-formula A^* as follows: If $A \equiv B^e$, then $A^* :\equiv A$; otherwise A^* results from A by replacing every free predicate variable X by X^n with $n := *(X)$, and every strong predicate quantifier $\forall X, \exists X$ by $\forall X^\omega, \exists X^\omega$, respectively.

Lemma 2. *For any formula F of L and an interpretation $*$, the following sequents are cut-free provable in BI^Ω: $\{F^*, \neg F^*\}, \{F^*, \neg F^e\}, \{F^e, \neg F^e\}$.*

Proof. The proof is by induction on F. The interesting case would be $F \equiv \forall X F_0$. We have to consider three subcases. Now, we treat only the most interesting case, that is, aim to prove $\{F^e, (\neg F)^*\}$. Let $n + 1 := lev(\neg \forall X F_0^*)$ with $\neg \forall X F_0^* \equiv (\neg \forall X F_0)^*$. Moreover, we write $F_0^*(X^n)$ to denote a formula obtained from $F_0(X)$ by replacing X with X^n and each other free predicate variable Z with Z^m where $m := *(Z)$.

By IH, we have $BI^\Omega \vdash_0 F_0^e, \neg F_0^*(X^n)$. If we consider any $q \in |\forall X F_0|$, we obtain the following derivation:

$$\frac{\dfrac{\dfrac{q : \Delta_q, F_0^*(X^n) \quad F_0^e, \neg F_0^*(X^n)}{\ldots \Delta_q, F_0^e \ldots} Cut_{F_0}}{\dfrac{\neg \forall X F_0^*, F_0^e}{\neg \forall X F_0^*, \forall X F_0^e} \bigwedge_{\forall X F_0}} \Omega_{n+1}}$$

Then we obtain the required cut-free derivation by applying Theorem 3. □

If T^* is an abstraction in L^*, then we define the substitution operator $\mathcal{S}_{T^*}^{X^m}$ under some suitable conditions. Let T^{*-} be the result by eliminating all superscripts e and n occurring in T^*. Then, $(\Gamma, \Lambda^e)[X^m/T^*] := \Gamma[X^m/T^*], \Lambda[X/T^{*-}]^e$ where Λ^e is a set of explicit formulas.

Theorem 5. *There is an operator $\mathcal{S}_{T^*}^{X^m}$ such that if*

1. $BI_k^\Omega \ni d \vdash_0 \Gamma$,

2. $X^m \notin PV(\Gamma)$, *and*

3. $k \leq m$,

then $BI^\Omega \ni \mathcal{S}_{T^*}^{X^m}(d) \vdash_0 \Gamma[X^m/T^*]$.

Proof. By induction on d. Let d be $I(d_i)_{i \in |I|}$.

1. $k = 0$.

    If d is Ax_Δ with $\Delta^- = \{X(t), \neg X(t)\}$, then we apply Lemma 2. Otherwise, we define $\mathcal{S}_T^{X^m}(d) := I_{A[X^m/T]}(\mathcal{S}_T^X(d_i))_{i \in |I|}$. For example, assume that $d = \bigwedge_{\forall Y A(Y)}(d_0)$. Then, X^m does not occur free in $\forall Y A(Y)$ since $X^m \notin PV(\forall Y A(Y))$. Hence we can apply IH to d_0 and use $\bigwedge_{\forall Y A(Y)}$ to obtain the required derivation.

2. $k = n+1$. By induction on d.

   (a) $I = \Omega_{\neg\forall Y A(Y)}, \widetilde{\Omega}_{\neg\forall Y A(Y)}$ with $l = lev(\forall Y A(Y))$.
   We consider only $d = \Omega_{\neg\forall Y A(Y)}(d_q)_{q \in |\neg\forall Y A(Y)|}$ with $l \leq n$. We have $X^m \notin FV(\neg\forall Y A(Y))$ because $X^m \notin PV(\Gamma)$. If $X^m \in FV(\Delta_q)$, then $l \leq n < n+1 = k \leq m \leq lev(\Delta_q)$. This contradicts $lev(\Delta_q) \leq l$. Hence, $X^m \notin FV(\neg\forall Y A(Y), \Delta_q)$. Now we can apply IH to d_q and $\Omega_{\neg\forall Y A(Y)}$ to get the required derivation.

   (b) Otherwise.
   Use IH and apply the same inference rule. □

5 The systems BI_ω and BI_ω^e

In this section, we introduce a system BI_ω which corresponds to $\Pi_1^1\text{-}CA+BI$ with the ω-rule. To take care of the explicit/implicit distinction, we introduce an additional system BI_ω^e based on the language L^e.

Definition 6. The systems BI_ω and BI_ω^e

1. BI_ω consists of the following inference rules.

$$(\mathrm{Ax}_\Delta)\ \Delta$$

where $\Delta = \{A\} \subseteq \text{TRUE}$ or $\Delta = \{C, \neg C\}$ with atomic C

$$(\bigwedge_{A_0 \wedge A_1}) \frac{A_0 \quad A_1}{A_0 \wedge A_1} \qquad (\bigvee_{A_0 \vee A_1}^k) \frac{A_k}{A_0 \vee A_1} \text{ where } k \in \{0,1\}$$

$$(\bigwedge_{\forall x A(x)}) \frac{\ldots A(n) \ldots}{\forall x A(x)} \text{ for all } n \in \omega \qquad (\bigvee_{\exists x A(x)}^k) \frac{A(k)}{\exists x A} \text{ where } k \in \omega$$

$$(\bigwedge_{\forall X A(X)}^Y) \frac{A(Y)}{\forall X A(X)}\ !Y! \qquad (\mathrm{R}_A) \frac{A \quad \neg A}{\emptyset}$$

$$(\bigvee_{\neg \forall X A(X)}^T) \frac{\neg A(T)}{\neg \forall X A(X)}$$

with

(a) $\neg \forall X A(X)$ is a weak formula or

(b) $T = \lambda y.(Yy)$.

2. BI^e_ω is based on the language L^e and consists of the following inference rules:

$$(\mathrm{Ax}_\Delta)\ \Delta$$

where $\Delta = \{A^{[e]}\} \subseteq \mathrm{TRUE}$ or $\Delta = \{C^{[e]}, \neg C^{[e]}\}$ with atomic C

$$(\bigwedge_{(A_0 \wedge A_1)^{[e]}}) \frac{A_0^{[e]} \quad A_1^{[e]}}{(A_0 \wedge A_1)^{[e]}} \qquad (\bigvee^k_{(A_0 \vee A_1)^{[e]}}) \frac{A_k^{[e]}}{(A_0 \vee A_1)^{[e]}} \text{ where } k \in \{0,1\}$$

$$(\bigwedge_{\forall x A(x)^{[e]}}) \frac{\ldots A(n)^{[e]} \ldots \text{ for all } n \in \omega}{\forall x A(x)^{[e]}} \qquad (\bigvee^k_{\exists x A(x)^{[e]}}) \frac{A(k)^{[e]}}{\exists x A(x)^{[e]}} \text{ where } k \in \omega$$

$$(\bigwedge^Y_{\forall X A(X)^{[e]}}) \frac{A(Y)^{[e]}}{\forall X A(X)^{[e]}}\ !Y! \qquad (\mathrm{R}_A) \frac{A \quad \neg A}{\emptyset}\ (A \in L)$$

$$(\bigvee^T_{\neg \forall X A(X)^{[e]}}) \frac{\neg A(T)^{[e]}}{\neg \forall X A(X)^{[e]}}$$

with

(a) $\neg \forall X A(X)$ is a weak formula or

(b) $T = \lambda y.(Yy)$.

6 The complete cut-elimination theorem for BI_ω

Let $*$ denote an interpretation from L^e into L^* assigning a number $*(X) \in \omega$ to each X. We define the embedding function g^* from derivations in BI^e_ω into derivations BI^Ω depending on $*$ below (Theorem 6).

Recall that an "explicit" formula in L^e is obtained assigning the superscript e to the corresponding one in L, and $A^{[e]}$ means a "possible occurrence" of e. Similarly, set $\Gamma^{[e]} := \{A^{[e]} : A \in \Gamma\}$. For example, there are four cases if $\Gamma = \{B, C\}$, that is, $\{B, C\}, \{B^e, C\}, \{B, C^e\}$, and $\{B^e, C^e\}$.

The next lemma is easy to see:

Lemma 3. *If $BI_\omega \vdash \Gamma$, then $BI_\omega^e \vdash \Gamma^{[e]}$ for all cases of $\Gamma^{[e]}$.*

Proof. The proof is by induction on d in BI_ω. □

We define $\deg(d)$ where d is a derivation in BI_ω^e in such a way that $dg(g^*(d)) \leq \deg(d)$.

Definition 7. $\deg(d)$

Let d be a derivation in BI_ω^e.

1. $\deg(d) := max(rk(A(T)), \deg(d_0))$ if $I = \bigvee_{\neg \forall X A(X)}^T$ and $0 < lev(\neg \forall X A) < \omega$.
2. $\deg(d) := max(rk(C), \deg(d_0), \deg(d_1))$ If $I = R_C$.
3. $\deg(I(d_\tau)_{\tau \in |I|}) := \sup\{\deg(d_\tau) | \tau \in |I|\}$ otherwise.

To define the embedding function g^*, we need the following definition:

Definition 8. $*(X/n)$

Let be $*$ be an interpretation from L^e to L^*. A *variant interpretation of $*(X/n)$* of $*$ is defined by

$$*(X/n)(Y) := \begin{cases} n & \text{if } X \equiv Y, \\ *(Y) & \text{otherwise.} \end{cases}$$

If Γ is a set of formulas in L^e, then $\Gamma^* := \{A^* | A \in \Gamma\}$.

Theorem 6. *Let $*$ be an interpretation. Then there is an embedding function g^* such that if $BI_\omega^e \ni d \vdash \Gamma$, then $BI^\Omega \ni g^*(d) \vdash_{\deg(d)} \Gamma^*$.*

Proof. By induction on d. We consider only important cases.

1. $d = Ax_\Delta$.

    This case is obvious since Ax_{Δ^*} is again an axiom in BI^Ω.

2. $d = \bigwedge_{(A_0 \wedge A_1)^{[e]}}(d_0, d_1)$.

    Let $A \equiv (A_0 \wedge A_1)^{[e]}$. By IH, we have $g^*(d_i) \vdash_{\deg(d_i)} \Gamma^*, (A_i^{[e]})^*$. Hence, we define

    $g^*(d) := \bigwedge_{A^*}(g^*(d_0), g^*(d_1)) \vdash_{\deg(d)} \Gamma^*, A^*$.

3. $d = \bigwedge_{\forall X A(X)^e}^Y(d_0)$.

    Note that Y is the eigenvariable of the last inference rule. Using IH, we define

    $g^*(d) := \bigwedge_{\forall X A(X)^e}^Y(g^*(d_0)) \vdash_{\deg(d)} \Gamma^*, \forall X A(X)^e$.

4. $d = \bigwedge_{\forall XA(X)}^{Y}(d_0)$.

   In this case, $d_0 \vdash \Gamma, A(Y), \forall XA(X)$ where Y is an eigenvariable. We consider two subcases.

   (a) $\forall XA(X)$ is a weak formula.

   In what follows, we write $(\forall XA(X))^*$ as $\forall XA(X)^*$. Moreover, A^* denotes the formula obtained from A by replacing every free predicate variable Z except for X by Z^m with $m := *(Z)$.
   Let $n = lev(\forall XA(X)^*)$.
   By IH, we have
   $$g^{*(Y/n)}(d_0) \vdash_{\deg(d_0)} \Gamma, \forall XA(X)^*, A^*(Y^n).$$
   Therefore, we define
   $$g^*(d) := \bigwedge_{\forall XA(X)^*}^{Y^n}(g^{*(Y/n)}(d_0)) \vdash_{\deg(d)} \Gamma^*, \forall XA(X)^*.$$

   (b) Otherwise.
   Using IH, we define
   $$g^*(d) := \bigwedge_{\forall X^\omega A^*(X)}(g^{*(Y/n)}(d_0))_{n \in \omega} \vdash_{\deg(d)} \Gamma^*, \forall X^\omega A^*(X).$$

5. $d = \bigvee_{\neg \forall XA(X)}^{T}(d_0)$.
   $d_0 \vdash \Gamma, \neg \forall XA(X), \neg A(T)$.

   (a) $\neg \forall XA(X)$ is a weak formula.
   Let $lev(\neg \forall XA(X)^*) = n + 1$. We write $A^*[X/T]$ as $A^*(T)$.
   Using IH, we define
   $$g^*(d) := \Omega_{\neg \forall XA(X)^*}(\mathcal{R}_{A^*(T^*)}(\mathcal{S}_{T^*}^{X^n}(d_q)), g^*(d_0))_{q \in |\forall XA(X)^*|}.$$
   Then we see
   $$g^*(d) \vdash_{\deg(d)} \Gamma^*, \neg \forall XA(X).$$

   (b) Otherwise.
   Now $T = \lambda y.(Yy)$. By IH, $g^*(d_0) \vdash_{\deg(d_0)} \Gamma^*, \neg A^*(Y^n), \neg \forall X^\omega A^*(X)$ where $*$ assigns Y^n to Y. We define $g^*(d) := \bigvee_{\neg \forall X^\omega A(X)^*}(g^*(d_0)) \vdash_{\deg(d)} \Gamma^*, \neg \forall X^\omega A^*(X)$.

6. $d = \bigvee_{\neg \forall XA(X)^e}^{T}(d_0)$.

   Use IH and the same inference symbol. Note that $\bigvee_{\neg \forall XA(X)^e}^{T}$ is an inference symbol in BI^Ω.

7. $d = \mathrm{R}_A(d_0, d_1)$.

Using IH and Theorem 1, we define $g^*(d) := \mathcal{R}_{A^*}(g^*(d_0), g^*(d_1)) \vdash_{\deg(d)} \Gamma^*$.

$\square$

Now, $\mathbf{0}$ is the interpretation which assigns X^0 to each predicate variable X. Then, the embedding function based on $\mathbf{0}$ is denoted by $g^{\mathbf{0}}$ (cf. Theorem 6).

Now we are in position to prove the main theorem of this paper.

Theorem 7. *If $BI_\omega \ni d \vdash \Gamma$, then there exists d' such that $BI_\omega \ni d' \vdash_0 \Gamma$.*

Proof. Let d be a derivation in BI_ω. We define $\Gamma^e := \{A^e | A \in \Gamma\}$. Then, by Lemma 3, we obtain the derivation d^e such that $BI_\omega^e \ni d^e \vdash \Gamma^e$. By Theorem 6, we have $BI^\Omega \ni g^{\mathbf{0}}(d^e) \vdash_{\deg(d)} (\Gamma^e)^{\mathbf{0}}$. Note that $(\Gamma^e)^{\mathbf{0}} = \Gamma^e$ and $lev((\Gamma^e)^{\mathbf{0}}) = 0$. Using Theorems 3 and 4, we get $BI_0^\Omega \ni \mathcal{D}_0(\mathcal{E}_\omega(g^{\mathbf{0}}(d^e))) \vdash_0 \Gamma^e$. By deleting the superscript e, we obtain the required derivation d' such that $BI_\omega \ni d' \vdash_0 \Gamma$.

$\square$

References

[1] Ryota Akiyoshi and Grigori Mints. An extension of the Omega-rule. *Archive for Mathematical Logic*, 55:593–603, 2016.

[2] Toshiyasu Arai. Proof theory for theories of ordinals I: Recursively Mahlo ordinals. *Annals of Pure and Applied Logic*, 122:1–85, 2003.

[3] Toshiyasu Arai. Proof theory for theories of ordinals II: Π_3-reflection. *Annals of Pure and Applied Logic*, 129:39–92, 2004.

[4] Toshiyasu Arai. A sneak preview of proof theory of ordinals. *Annals of the Japan Association for Philosophy of Science*, 20:29–47, 2012.

[5] Wilfried Buchholz. Notation systems for infinitary derivations. *Archive for Mathematical Logic*, 30:277–296, 1991.

[6] Wilfried Buchholz. Explaining Gentzen's consistency proof within infinitary proof theory. In Georg Gottlob, Alexander Leitsch, and Daniele Mundici, editors, *Computational Logic and Proof Theory: 5th Kurt Gödel Colloquium, KGC'97*, volume 1289 of *Lecture Notes in Computer Science*, pages 4–17, Springer, Berlin, 1997.

[7] Wilfried Buchholz. Explaining the Gentzen-Takeuti reduction steps. *Archive for Mathematical Logic*, 40:255–272, 2001.

[8] Wilfried Buchholz and Kurt Schütte. *Proof Theory of Impredicative Subsystems of Analysis*, volume 2 of *Studies in Proof Theory, Monographs*. Bibliopolis, Naples, 1988.

[9] Jean-Yves Girard, Yves Lafont, and Paul Taylor. *Proofs and Types*, volume 7 of *Cambridge Tracts in Theoretical Computer Science*. Cambridge University Press, Cambridge, 1989.

[10] Grigori Mints. Finite investigations of transfinite derivations (in Russian). *Zapiski Nauchnyh Seminarov LOMI*, 49:67–122, 1975. English translation in [11].

[11] Grigori Mints. *Selected Papers in Proof Theory*, volume 3 of *Studies in Proof Theory, Monographs*. Bibliopolis, Naples, 1992.

[12] Grigori Mints. Reduction of finite and infinite derivations. *Annals of Pure and Applied Logic*, 104:167–188, 2000.

[13] Grigori Mints. Non-deterministic epsilon substitution method for PA and ID_1: effective proof. In Ulrich Berger, Hannes Diener, Peter Schuster and Monika Seisenberger, editors, *Logic, Construction, Computation*, volume 3 of *Ontos Mathematical Logic*, pages 325–342, De Gruyter, Berlin, 2013.

[14] Grigori Mints. Non-deterministic epsilon substitution method for PA and ID_1. In Reinhard Kahle and Michael Rathjen, editors, *Gentzen's Centenary*, pages 479–500, Springer, Berlin, 2015.

[15] Michael Rathjen. Proof theory of reflection. *Annals of Pure and Applied Logic*, 68:181–224, 1993.

[16] Michael Rathjen. An ordinal analysis of parameter free Π^1_2-comprehension. *Archive for Mathematical Logic*, 44:263–362, 2005.

[17] Michael Rathjen. An ordinal analysis of stability. *Archive for Mathematical Logic*, 44:1–62, 2005.

[18] William Tait. Applications of the cut-elimination theorem to some subsystems of classical analysis. In Akiko Kino, John Myhill, and Richard Vesley, editors, *Intuitionism and Proof Theory*, volume 60 of *Studies in Logic and the Foundations of Mathematics*, pages 475–488, North-Holland, Amsterdam, 1970.

[19] Gaisi Takeuti. On the fundamental conjecture of GLC V. *Journal of the Mathematical Society of Japan*, 10(2):121–134, 1958.

[20] Gaisi Takeuti. Consistency proofs of subsystems of classical analysis. *The Annals of Mathematics*, 86(2):299–348, 1967.

[21] Gaisi Takeuti. *Proof Theory*, volume 81 of *Studies in Logic and the Foundations of Mathematics*, 2nd edition, North-Holland, Amsterdam, 1987. Reissued by Dover Publications, 2013.

[22] Mariko Yasugi. Cut elimination theorem for second order arithmetic with the Π^1_1-comprehension axiom and the ω-rule. *Journal of the Mathematical Society of Japan*, 22(3):308–324, 1970.

Model Theory of Some Local Rings

Paola D'Aquino
*Department of Mathematics and Physics, Università della Campania
"L. Vanvitelli", Viale Lincoln 5, I-81100, Caserta, Italy*
paola.daquino@unicampania.it

Angus Macintyre[*]
*Queen Mary, University of London, School of Mathematical Sciences,
Mile End Road, London E1 4NS, UK*
a.macintyre@qmul.ac.uk

Abstract

Our understanding of the first-order theory of the class of all local rings $\mathbb{Z}/p^n\mathbb{Z}$ as p and n vary comes from the Ax-Kochen-Ershov analysis of the rings of p-adic integers. This analysis does not directly produce axioms. In this paper we give fairly explicit axioms for the class.

Keywords: Model Theory, Henselian Fields.

1 Introduction

1.1 Dedication

We dedicate this paper to Grisha's memory, with a feeling he might have found it congenial. Beneath its model-theoretic surface are some unresolved issues about extracting axioms for reducts from axioms for richer structures (in this case axioms for quotients from axioms for Henselian valuation rings) and we regret not having had a chance to discuss this with Grisha. We did not meet that often, but were inspired by his intellect and his charm. A. Macintyre is grateful for the chances he had to discuss with Grisha issues about proving hard number theory in Peano Arithmetic. Our work on Zilber's Problem [8] can be construed as showing that the natural quotient structures of models of Peano Arithmetic are completely axiomatized without any induction axioms, because of the work of Ax-Kochen and Ershov and Feferman-Vaught. Both of us fondly (but very sadly) remember our last meetings with him, at AIMS in Palo Alto, during a workshop on analogues of Hilbert's 10th Problem.

---

[*]Supported by a Leverhulme Emeritus Fellowship.

1.2 *p*-adics and the $\mathbb{Z}/p^n\mathbb{Z}$

In the mid 1960's Ax-Kochen [2, 3, 4] and Ershov [14, 15, 16] proved fundamental theorems about the logic of Henselian valued fields. These theorems seem assured of a permanent place among the most important in "applied model theory". Combined with Ax's work on the elementary theory of finite fields [1], they constitute an indispensable repertoire for those who try to connect logic to algebra and number theory.

The work of the above authors on *p*-adic fields and then on the theory of finite fields gives a rather indirect proof of the decidability of the class of all finite local rings of the form $\mathbb{Z}/p^n\mathbb{Z}$, as p and n vary. We discuss this below, but simply note for now that we have no recollection of ever having seen any other proof of the decidability. If there is no other proof, this is rather intriguing. In any case there cannot be a trivial proof, since one can interpret the theory of finite fields in the theory of the local rings mentioned above. We note in passing that one can show, by adding the Feferman-Vaught method [17] to the mix (or by a much later argument using model theory of adeles [13], which also depends on Feferman-Vaught) that the theory of all rings $\mathbb{Z}/n\mathbb{Z}$ is decidable. We note too that Rabin [25] showed that the theory of finite commutative rings is undecidable, from which it follows by Feferman-Vaught that the theory of finite local rings is undecidable. We believe that merely finding explicit axioms for the theory of the class of rings $\mathbb{Z}/p^n\mathbb{Z}$ is a problem with proof-theoretic resonance. We achieve a non-optimal solution to this in the present paper.

1.3 Formalism

The logic of valued fields has rather more formalism than one meets in most areas of applied logic. The basic structures involve a field K (the valued field), a field k (the residue field), an ordered abelian group Γ (the value group), the valuation ring V, the multiplicative group K^*, the valuation v from K^* to Γ, and the residue map res from V to k. All of these notions are interpretable in the one-sorted structure consisting of the field with the valuation ring V distinguished. In fact, in all the cases discussed below, V is definable [7]. However, V is not always definable, and there is sometimes a need to use the one-sorted $L_{rings,V}$ with a predicate, for the valuation ring, adjoined to the ring language. The basic language for the analysis below is L_{rings}, the usual first-order language for unital rings, with $+, \cdot, -, 0$ and 1. When we pass to sorted formalisms, L_{rings} is appropriate for both K and k. For Γ we have the standard language L_{OAG} (ordered abelian groups), with $+, -, 0$ and $<$. However, if Γ is discretely ordered, as it always will be in this paper, it is better to augment the preceding by a symbol "1" for the least positive element, giving L_{DOAG} (discretely ordered abelian groups).

Experience has shown that we get the clearest picture if we look at valued fields in a 3-sorted formalism, with (roughly) sorts corresponding to K, k and Γ, each with their natural formalism, and certain intra-sortal functions like v and res. The literature is a little casual on all this, because the natural maps are not total on the sorts. While there is little danger of error in current practice, we choose to make our intra-sortal maps total, at the cost of modifying slightly the K and Γ sorts and/or their formalisms. As is traditional we augment either of our languages for ordered abelian groups by adding a symbol ∞ for an element larger than all the others. There is no problem in defining an extension of $+$, preserving the universal axioms for $+$. But there is a problem for $-$, exactly as regards the interpretation of $\infty - \infty$. We choose to interpret this as having value 0, thereby losing some laws connecting $+$ and $\cdot$. This is however inconsequential, as the set Γ is definable, as is its ordered group structure. We call the resulting sort the enlarged Γ- sort. Now v is a regular intrasortal map from the K-sort to the enlarged Γ-sort, if we put $v(0) = \infty$ as usual.

What to do about the res map, since it is not totally defined on K? We could do something similar to the preceding by modifying the k-sort by adding another ∞ and using a place. But it is better to use an angular component map ac modulo k. There is no such map in general, but any valued field has an elementary extension with such a map, and for our analysis of definability this is enough. ac is total on K to k, and what is required of it is that it respects $\cdot$, maps 1 to 1, and 0 to 0, and agrees with res on the units of V, i.e the elements of value 0. We are going to be dealing only with cases where K has characteristic 0 and v is unramified, and then we take $ac(p) = 1$ if k has finite characteristic p.

The formalism just presented is called $L_{Denef-Pas}$ in honour of Denef and Pas who devised it [24]. It provides refined analysis of definitions, and has been used to great effect in work on motivic integration and uniformity of p-adic integrals. We already used this formalism in [11] on Henselizations of p-adic valuations (where p can be non-standard) on non-standard models of Peano Arithmetic. Another language with similar virtues was discovered earlier by Basarab [5], but we will not need it here. In this paper we use the above formalism to provide an analysis of definitions in certain uniserial local rings, making essential use both of Ax-Kochen-Ershov and Denef-Pas. It is curious that no such analysis has been undertaken long ago. Our interest in the topic came from a question of Boris Zilber about interpretability in quotient rings of nonstandard models of arithmetic. Prior to working on Zilber's Problem and writing the paper dedicated to A. Woods [11], we have looked at residue fields of models of Peano Arithmetic and considered p-adic issues in this setting for many years, see [23], [9], [10].

2 From Ax-Kochen, Ershov and Ax to the analysis of quotient rings

The most comprehensive result from the 1960's (holding in any of the formalisms mentioned above) is

Theorem 1. *Suppose K_1 and K_2 are Henselian and k_1 and k_2 are of characteristic 0. Then $K_1 \equiv K_2 \iff k_1 \equiv k_2$ and $\Gamma_1 \equiv \Gamma_2$.*

There is an important extension of this to give a result about elementary extensions [2, 3, 4, 16].

Of much more importance for number theory is the situation when k has characteristic $p \neq 0$. When also K has characteristic p, there is an important result due to Robinson when K is algebraically closed [26], beyond that there is only a series of elaborations of an idea going back to Kaplansky [20], (see [12], [21]). When K has characteristic 0, one understands very well some cases, those of finite ramification [16], but the general case remains mysterious. We have no need here for the general finitely ramified case, and so restrict ourselves to the basic case of $\mathbb{Q}_p$.

As valued field, $\mathbb{Q}_p$ is completely axiomatized by

1. Henselian.

2. Residue field of cardinality p.

3. Γ is a model of Presburger arithmetic with least positive element 1.

4. $v(p) = 1$.

See [4].

Moreover, the valuation ring V is existentially, and universally, definable in L_{rings} [7]. $\mathbb{Q}_p$ is model-complete in L_{rings}, and indeed has a useful quantifier-elimination in terms of power predicates [22].

Each $\mathbb{Q}_p$ is decidable [4]. When k has characteristic 0, K is decidable (in $L_{rings,V}$) if and only if k and Γ are decidable [2, 3, 4, 14, 15, 16].

Finally, we present the most important ingredient for our work, namely the work of Denef and Pas. The following (see [24]) is a variant of their main result, tailored to our needs.

Theorem 2. *There is a computable function DP from the set of $L_{Denef-Pas}$ formulas to itself, and a computable function β from the set of $L_{Denef-Pas}$ formulas to $\mathbb{N}$, such that for any $L_{Denef-Pas}$ formula Ψ, $DP(\Psi)$ has the same free variables as Ψ but has no bound variables of sort K, so that if K is a Henselian valued field in*

$L_{Denef-Pas}$, and k has characteristic not a prime less than $\beta(\Psi)$ then Ψ and $DP(\Psi)$ are equivalent in K.

In this paper we consider logical questions about the quotient rings $V/\{y : v(y) \geq \gamma\}$ (written henceforward as V_γ, where γ is a non-negative element of Γ).

Consider first the case $K = \mathbb{Q}_p$, so $V = \mathbb{Z}_p$ and $\Gamma = \mathbb{Z}$. A typical γ is a positive integer n and then V_γ is naturally the quotient ring by p^n, a finite, innocuous local ring. The interesting questions are about the class of such rings, for fixed p and varying n. Decidability and a uniform analysis of definitions are the key issues. That the class is decidable is certainly <u>not</u> obvious. It is not known to us who first proved decidability. The interesting thing is that decidability is an easy consequence of decidability of the valued field $\mathbb{Q}_p$, but seems not easy to prove directly. As far as the structure of definable sets is concerned, this is not at all obvious even given the Macintyre quantifier-elimination for the p-adics [22]. Equally, the axiomatization of the p-adics casts no immediate light on axiomatization of the quotient rings as n varies, and p is fixed.

Next one asks for a corresponding analysis as p and n vary. The $\mathbb{Z}/p^n\mathbb{Z}$ are finite local rings, the quotient rings of $\mathbb{Z}_p$ as p varies. Note that now one can interpret $\mathbb{F}_p$ uniformly, as residue fields, so any analysis of decidability and definability should be as difficult as Ax's great paper [1]. Note the curious fact that Ax asks at the end of his paper about the decidability of the class of all $\mathbb{Z}/m\mathbb{Z}$ as m varies, evidently unaware of the fact that decidability follows from his work and that of Feferman-Vaught.

For our purposes the key results of Ax are those on ultraproducts of finite fields and of p-adic fields [1]. The infinite fields which satisfy the theory of all finite fields are exactly the perfect fields F with absolute Galois group $\hat{\mathbb{Z}}$ and such that every absolutely irreducible curve over F has a point in F (the restriction to <u>curves</u> is a refinement by Geyer [18] of Ax's basic Lang-Weil analysis in [1]). Such fields are called <u>pseudofinite</u>. A startling result of Ax [1] is that characteristic 0 pseudofinite fields are exactly those elementarily equivalent to nonprincipal ultraproducts of <u>prime</u> fields. From this and Ax-Kochen-Ershov one readily proves that the class of <u>all</u> $\mathbb{Q}_p$ is decidable, thus giving, via interpretability, the decidability of the class of <u>all</u> rings $\mathbb{Z}/p^n\mathbb{Z}$, a result which could hardly have been known prior to [1].

There is a natural uniform quantifier-elimination for pseudofinite fields (originally by Kiefe, see [18]), and this can readily be made into a natural quantifier-elimination for the class of all $\mathbb{Q}_p$ [12]. But this alone does not give a nice quantifier-elimination for the class of all $\mathbb{Z}/p^n\mathbb{Z}$. The goal of this paper is to axiomatize the class of all $\mathbb{Z}/p^n\mathbb{Z}$ in L_{rings}.

Note at the outset that if K is Henselian and γ is a positive element of V then

V_γ is local, and the quotient map $V \to V_\gamma$ is local, so V_γ is Henselian. We have a more general project to study the model theory of finite local rings (which are automatically Henselian), and we have an ongoing project to consider nonstandard primes p and nonstandard integers n and analyze definitions in the corresponding quotient rings which turn out to be Henselian local [8].

3 Towards the Axioms

3.1 The basic rings

Henceforward K is an Henselian field, as in first section, with valuation ring V, maximal ideal μ of V, residue field k, value group Γ, valuation v and an angular component still to be specified. Let γ be a positive element of Γ and R the Henselian local ring V_γ. Clearly, the maximal ideal of R is $\mu + \{y : v(y) \geq \gamma\}$, and the residue field is naturally k.

3.2 Truncated valuations

The ring R carries a truncated valuation, in a sense which we now explain. For possible future reference we pass to a more general setting. Let $\Gamma^{\geq 0}$ be the nonnegative part of Γ. We say Δ is an initial segment of $\Gamma^{\geq 0}$ if Δ is a nonempty subset of $\Gamma^{\geq 0}$ and is closed downwards under the order on Γ. Let I be the set of elements of V whose value is either ∞ or bigger than every element of Δ. It is clearly an ideal in V, and $V \to V/I$ is a morphism of local rings. Moreover, since V is Henselian V/I is Henselian.

There are two cases. Firstly, suppose Δ is the set of all δ less than γ for some fixed γ. Then this is just the case mentioned in the previous subsection. This is the only case considered in this paper, but one should note that for Γ not discretely ordered the case where Δ consists of the $\delta \leq \gamma$ occurs too, and is rather different.

Suppose $\Delta \neq \Gamma$, and consider v_I from V/I to $\Delta \cup \{\infty\}$ given by

1. $v_I(x + I) = \infty$ if $x \in I$,

2. $v_I(x + I) = v(x)$ if $x \notin I$.

The function v_I is clearly well-defined. We refer to this map as the truncation of v to Δ.

Such maps occur naturally, for example in [19], where the author studies various important truncated discrete valuation rings. For Hiranouchi, however, these are Artinian local rings with principal maximal ideal, the former restriction being quite unnatural from a model-theoretic point of view (the second is, however, first-order). Certainly the rings $\mathbb{Z}/p^n\mathbb{Z}$ are examples of Hiranouchi's notion.

The most obvious laws satisfied by the above truncation (written now simply as "v", for convenience of notation) are:

1. v is a map from R onto $\Lambda \cup \{\infty\}$ where R is a local ring, and Λ is a set with an ordering $<$ with a least element 0, and $\infty > l$ for each $l \in \Lambda$;

2. $v(0) = \infty$ iff $x = 0$;

3. $v(x+y) \geq min(v(x), v(y))$;

4. $x|y \iff v(x) \leq v(y)$;

5. $v(x) = v(y) \iff x = yu$ for some unit u in R;

6. the maximal ideal μ is principal.

Property (5) is equivalent to saying that the ideal generated by x and the ideal generated by y coincide iff x and y have the same valuation. We now identify Γ inside R as, firstly, the set of principal nonzero ideals. (0) is identified with ∞, the maximal ideal μ corresponds to 1 in the Presburger model and if Λ has a top element τ then it corresponds to the (unique) principal ideal above (0). Since R comes from a valuation domain where *divisibility* is a linear order, the same is true for principal ideals and the linear order is given by reverse inclusion. The operation $+$ on Γ comes from $(x) + (y) = (xy)$, and $v(x) = (x)$ the ideal generated by x.

From (1)-(6) it follows that there is an operation $+$ on $\Lambda \cup \{\infty\}$ making it into the nonnegative part of a truncated ordered abelian group, by showing that $v(xy)$ depends only only on $v(x)$ and $v(y)$, and then defining $v(x) + v(y)$ as $v(xy)$. One gets the following laws:

1. $+$ is commutative and associative, with neutral element 0;

2. $x + \infty = \infty$;

3. If $x + y = x + z$ then either $y = z$ or $x + y = x + z = \infty$;

4. If $x < y$ then $y = x + z$ for some z.

One goes routinely to the notion of truncated ordered abelian group. The following is proved in [8], and was first done by Derakhshan and Macintyre in unpublished work on the adeles:

Theorem 3. *The nonnegative part of a truncated ordered abelian group is always given by taking an ordered abelian group Γ with ∞ adjoined, and an initial segment Λ of Γ, and taking $+$ to be defined by*

1. *If x and y and $x + y$ are in Λ, where $+$ is the group operation, then this is the $+$ in the sense of Γ with ∞ adjoined.*

2. *Otherwise $x + y$ is ∞.*

In this paper we are mainly interested in truncations of models of Presburger Arithmetic, and we refer to them as <u>Presburger truncations</u>. These can be of two kinds, depending on whether the initial segment has or does not have a last element (of course the truncation always has a last element $\{\infty\}$, but we are distinguishing the two cases where $\{\infty\}$ has or does not have an immediate predecessor). Our main theorems will be about the case when the initial segment has a last element, but we consider the other case as well.

We note that if a truncated ordered abelian group $\Lambda \cup \{\infty\}$ comes from a discretely ordered abelian group, and contains the least positive element, and R is a commutative ring with a truncated valuation onto $\Lambda \cup \{\infty\}$, then R is a local ring with principal maximal ideal (but R need not be Artinian). This is a simple exercise.

3.3 Presburger type of the penultimate element

In this subsection we assume that R has a truncated valuation onto a Presburger truncation $\Lambda \cup \{\infty\}$ and Λ has a last element τ. τ is the penultimate element of the Presburger truncation. In the case of $R = \mathbb{Z}_p^n$, $\tau = n - 1$. If we construe τ as living in a model of Presburger extending Λ then τ has model-theoretic type (relative to Presburger) given by the atomic formulas $x = n$ ($n \in \mathbb{N}$) and $x \equiv r$ modulo m ($m \in \mathbb{N}$, $m \geq 2$ and $0 \leq r < m$) satisfied by τ (this set is independent of the ambient Presburger model, and indeed depends only on the segment determined by τ, and thus depends only on Λ). Note that any formula of the form $x \geq n$ or $n \geq x$ is Boolean definable from the formulas we listed.

The essential point here is that for any non-negative elements γ and δ in a model of Presburger they satisfy the same formula iff they have the same Presburger type as given by the above simple conditions.

Now suppose that R is a local ring with truncated valuation to a Presburger truncation $\Lambda \cup \{\infty\}$ with penultimate element τ. In terms of principal ideals τ is the minimal ideal above (0), and this can easily be expressed in R. If δ is an element of Λ we define R_δ as the quotient ring of R by the ideal of elements of value at least δ (thereby generalizing a notion given earlier for a valuation ring V).

Lemma 1. *Uniformly in R one can express in R in L_{rings} the Presburger type of τ.*

Proof. If τ satisfies the formula $x = n$ one expresses in R that $\tau = n$ by saying that there is a chain of nonzero principal ideals of length n, and none of length $n + 1$.

If τ satisfies the formula $x \equiv r \pmod{m}$ one expresses in R that $\tau \equiv r \pmod{m}$ by saying that there is x with $v(x) = \tau$ and (x) is a minimal prime ideal and there is a descending chain of length r of principal ideals where the first one is μ, there is no ideal between two consecutive ones, and the last ideal is (w) with $v(w) = r$ and $x = wb^m$, for some b. □

3.4 Natural R and their ultraproducts

For us the natural local rings are $\mathbb{Z}/p^n\mathbb{Z}$ as p and $n \geq 1$ vary. Any K which is a model of the theory T of the class of all $\mathbb{Z}/p^n\mathbb{Z}$ is elementarily equivalent to an ultraproduct $\prod_D \mathbb{Z}/p^n\mathbb{Z}$ for some ultrafilter D on set of pairs (p, n) (hence it is a pseudofinite ring), and it is also elementarily equivalent to the ultraproduct $\prod_D \mathbb{Z}_p/p^n\mathbb{Z}_p$ since $\mathbb{Z}/p^n\mathbb{Z} \cong \mathbb{Z}_p/p^n\mathbb{Z}_p$. By Ax's result the residue fields of these are models of either prime finite fields or pseudofinite fields of characteristic 0. Let $FinPrim$ be the class of finite prime fields. Then Ax's result shows that $Th(FinPrim)$ is exactly the theory of the class of pseudofinite fields of characteristic 0, and also the theory of all characteristic zero ultraproducts of finite fields. So the residue fields of the preceding ultraproducts of finite local rings are models of $Th(FinPrim)$, and any model of $Th(FinPrim)$ is elementarily equivalent to the residue ring of one of the preceding ultraproducts.

We first want axioms for residue fields of models of T, and later axioms for the theory T along the line of Ax, Kochen, Ershov. We will use the result of Denef and Pas to get explicit axioms in our case with Presburger truncations as value sets, and not groups. The truncated value structure for $\mathbb{Z}/p^n\mathbb{Z}$ is $[0, n-1] \cup \{\infty\}$, and for $\prod_D \mathbb{Z}/p^n\mathbb{Z}$ is $\prod_D [0, n-1] \cup \{\infty\}$. The theory of the Presburger truncation is uniquely determined by the Presburger type of the last element different from ∞, and the same holds for the theory of Presburger truncation of $\prod_D \mathbb{Z}/p^n\mathbb{Z}$.

We consider R arising as follows. Let K be an Henselian valued field, of characteristic 0, with residue field k, valuation ring V, and value group Γ a model of Presburger Arithmetic. We require that k is a model of the theory of finite prime fields. We require too that if $k = \mathbb{F}_p$ then $v(p) = 1$. Then we select $\gamma > 0$ in V and we consider $R = V_\gamma$. This is of course a special case of the construction in 3.2, with $\Delta = \{\tau : \tau < \gamma\}$. Now R is local and has a truncated valuation to $\Delta \cup \{\infty\}$, a Presburger truncation. Our main objective is to find axioms for all such R. We already have some, but to get <u>all</u> we need the Denef-Pas Theorem for the field K.

So we pass to the many-sorted formalism, with sorts for K, k and Γ with ∞, adjoined as usual, and connecting maps v and ac. The latter is available in an elementary extension of K and we can just assume K has an ac. We do have to make some basic decisions about the choice of ac, as terms like $ac(1)$ have to get a definite meaning. Note that if one has a partial angular component on a subring it is possible to extend it to an angular component on an elementary extension.

If k has characteristic 0 we take $ac(n) = n$, for all $n \in \mathbb{Z}$. If k has characteristic p, and so is elementarily equivalent to $\mathbb{Q}_p$ by our choice of conditions on K, and $n = p^k n_1$, with n_1 prime to p, we can take $ac(n) = n_1$. Recall that the Denef-Pas analysis is uniform in choices of ac.

Now let Ψ be a sentence in the language of rings. Consider the subset of Γ consisting of the positive γ such that $V_\gamma \models \Psi$. This is definable in the Denef-Pas language by a formula Θ with a single free Γ-variable, and thus, by the Main Theorem, by the $DP(\Theta)$ formula with the same free variable, and no bound variables of sort K, except possibly when k is some $\mathbb{F}_p$ and $p < \beta(\Theta)$. For each of the remaining finitely many $p < \beta(\Theta)$ we can appeal to the nice theorem that in $\mathbb{Q}_p$ the value group is stably embedded to replace $DP(\Theta)$ by a Presburger formula (see Corollary 5.25 in [27]). The outcome is that we can replace, uniformly, $DP(\Theta)$ by a $DP^*(\Theta)$ with a single free Γ-variable, which defines, uniformly in all the K we have chosen, the subset of Γ consisting of the positive γ such that $V_\gamma \models \Psi$. Only finitely many $ac(n)$ occur, and these can be replaced by n except when k has characteristic p and p divides n. There are only finitely many such p, and thus we may computably avoid the use of ac by a slight modification of $DP^*(\Theta)$ with the same properties.

Let us now unpack the structure of the many-sorted $DP^*(\Theta)$. There are no bound K-variables, and no free ones, so no terms $ac(t)$ or $v(t)$ for nonclosed K-terms t. There are no intersortal functions between the k and Γ sorts, so we may readily show that $DP^*(\Theta)$ is (logically) equivalent to a Boolean combination of quantifier-free K-sentences, k-sentences and Γ formulas in a single variable γ. Since K is of characteristic 0, the truth value of open K-sentences is independent of K and so such sentences can be erased. We already erased the $ac(n)$, so the k-sentences are equivalent, for p computably sufficiently large (via the Lang-Weil estimates) to Boolean combinations of solvability statement saying that certain $f \in \mathbb{Z}[x]$ are solvable. For the remaining finitely many p we isolate each case by the clause $p = 0$ (remember k is a model of the theory of finite prime fields!), and have a disjunction of such equations over the finitely many p, outside our computable cofinite set, such that the given k sentence holds in $\mathbb{F}_p$.

As regards the Presburger formulas in the variable γ, these are Boolean combinations of various $\gamma = n$ and congruence conditions modulo various m, where $m, n \in \mathbb{N}$.

Now we have a serious result about the class of all R as above. We recall that we showed that the Presburger conditions on the penultimate element of $\Lambda \cup \{\infty\}$ are expressible in the ring R. Also, it is obvious that residue field conditions are expressible in R since the maximal ideal is definable. Thus we have:

Theorem 4.

1. *The elementary theory of a ring R as above is determined by the elementary theory of the residue field and the Presburger type of γ, and these can be given independently.*

2. *There are computable maps D_1, D_2, and D_3 defined on ring sentences Ψ so that $D_1(\Psi)$ is a finite set of polynomials in one variable over $\mathbb{Z}$, $D_2(\Psi)$ is a finite set of integers and $D_3(\Psi)$ is an integer ≥ 2, so that the truth value of Ψ in R is determined by which elements of $D_1(\Psi)$ are solvable in the residue field, which element of $D_2(\Psi)$, if any, is equal to γ, and what is the congruence class of γ modulo $D_3(\Psi)$.*

Proof. Both parts (except for the remark about independence in (1)) are immediate from the preceding discussion. For the independence, just take an allowed k (a model of the theory of finite prime fields), and a model Γ of Presburger with an element γ realizing a Presburger type s. We can assume Γ is an ultrapower of $\mathbb{Z}$, using an ultrafilter D on an index set I. If k is $\mathbb{F}_p$ let K be the ultra power of $\mathbb{Q}_p$ with respect to I and D. Then V_γ is the required local ring. $\square$

We note that the D_j's ($j = 1, 2, 3$) are, in our present knowledge, rapidly growing. For particular cases of known bounds for such problems, in connection with Artin's Conjecture, see [6].

4 Refining the Axioms

4.1 Which R come from a K?

The limitation in what we have done above is that the R we considered were assumed to be quotients of valuations rings of K subject to various conditions. Now we want to find axioms for general local rings R that will guarantee that they are at least elementarily equivalent to a local ring coming from a K. We have some quite explicit conditions on the R that interest us, namely:

1. R is a local ring, with a truncated valuation onto some Presburger truncation, with a last element below ∞.

2. R is Henselian.

3. The residue field is a model of the theory of finite prime fields.

We have stressed that these axioms are all first-order. It is fairly clear that the analysis above in terms of D_1, D_2, D_3 can be reformulated in terms of first-order axioms for the R coming from K, and we can then add these to the preceding (1), (2), (3). The problem is that we do not have a very explicit version of the D, and we believe that such an explicit version would require very explicit quantitative information about quantifier elimination in the theory of finite prime fields and in the theory of p-adic fields. Despite the obvious interest of having such information, none has been obtained.

Despite the lack of explicitness, it is useful to draw some consequences from what we done so far. Once more, start with an R, a local ring which is of the form V_γ, with V the valuation ring of a Henselian K as above. Let us interpret what we have got out of Denef-Pas. Let us look at R_1, a local ring got from R by dividing out a principal ideal generated by a single element x of value γ_1, with x in the maximal ideal of R. Now R and R_1 have the same residue field, namely k. The ring R_1 is of course a quotient of V, namely V_{γ_1} and the preceding analysis applies to it (and of course we use the uniformity of the preceding analysis). Thus if γ and γ_1 have the same Presburger type then R and R_1 are elementarily equivalent. Thus in any R as above, and for any ring sentence Ψ, the set of γ_1 such that Ψ holds in $R/\{y : v(y) \geq \gamma_1\}$ differs finitely (at certain standard elements) from a finite union of arithmetical progressions with modulus m. This "periodicity" seems not to have been observed before. It gives the following

Theorem 5. *Let Ψ be a ring sentence, as above. Then by the D_1, D_2, D_3 analysis there exist m and n in $\mathbb{N}$, a set E of residues modulo m and a subset B of $[1,n]$, each computable from Ψ, together with a finite set Pol of elements of $\mathbb{Z}[x]$ and a set Sol of subsets of Pol, also given computably from Ψ, so that (uniformly in the quotient rings R of Henselian K, as above) Ψ holds in R iff*

1. *For some $Y \in \mathrm{Sol}$ and all $f \in \mathrm{Pol}$ f is solvable in k iff $f \in Y$, and*

2. *If τ is the penultimate element of the Presburger truncation of R then either $\tau \in B$ or τ is congruent modulo m to an element of E.*

Suppose that we add, for each Ψ, an axiom giving the equivalence with the given (1) and (2) (note that this can be done computably). Let Σ be the resulting set of axioms, with no assumption being made that the models R of Σ come from Henselian fields.

Theorem 6. *Let R be a model of Σ. Then R is elementarily equivalent to a quotient of a Henselian field valued in a Presburger group and with residue field a model of $Th(FinPrim)$.*

Proof. If R is a model of Σ then it has the same residue field and value group as an unramified Henselian field K with the required properties and with an element having the same Presburger type as the penultimate element of the Presburger truncation of R. Now pick a sentence Ψ. The criterion for Ψ to hold in V_γ is exactly the same as that for Ψ to hold in R, and depends only on the residue field and the type of the penultimate element of the Presburger truncation of R. □

Of course we would like to replace elementary equivalence by isomorphism in the above Theorem. We have a construction that seems close to giving this refinement, but it is complicated, and we prefer to prepare a sequel in which this matter is resolved. For most purposes it is enough to know the quotients up to elementary equivalence.

5 Induction

Since we have shown that any R under consideration above is elementarily equivalent to one with value group $\mathbb{Z}$, it follows that the nonnegative part of the value group of R satisfies "definable induction" where we allow sets definable in the ring language using constants. This is known for Henselian fields with value group a $\mathbb{Z}$-group and residue field a model of the theory of prime fields, which are unramified (i.e., $v(p) = 1$) if the residue field has characteristic p (see [2, 3, 4]). This in turn relates to the stable embedding of the value group in such cases [27], and we easily see that this stable embedding is true in the truncated cases we have considered here.

The main point to emphasize is that though induction is not mentioned in the axiom, nor any pigeonhole principle, the induction scheme and an obvious pigeonhole scheme are derivable from the axioms Σ.

Acknowledgements. We are deeply grateful to the anonymous referee for pointing out many obscurities in earlier versions. We have tried to take account of all of the referee's suggestions.

References

[1] J. Ax. The elementary theory of finite fields. *Annals of Mathematics*, 88(2):239–271, 1968.

[2] J. Ax and S. Kochen. Diophantine problems over local fields: I. *American Journal of Mathematics*, 87:605–630, 1965.

[3] J. Ax and S. Kochen. Diophantine problems over local fields: II. A complete set of axioms for *p*-adic number theory. *American Journal of Mathematics*, 87:631–648, 1965.

[4] J. Ax and S. Kochen. Diophantine problems over local fields: III. Decidable fields. *American Journal of Mathematics*, 83:437–456, 1966.

[5] Ş. Basarab. Relative elimination of quantifiers for Henselian valued fields. *Annals of Pure and Applied Logic*, 53(1):51–74, 1991.

[6] S. S. Brown. Bounds on transfer principles for algebraically closed and complete discretely valued fields. *Memoirs of the American Mathematical Society*, 15(204):iv+92, 1978.

[7] R. Cluckers, J. Derakhshan, E. Leenknegt, and A. J. Macintyre. Uniformly defining valuation in henselian valued fields with finite or pseudofinite residue field. *Annals of Pure and Applied Logic*, 164(12):1236–1246, 2013.

[8] P. D'Aquino and A. Macintyre. *Quotient rings of models of arithmetic*, In preparation.

[9] P. D'Aquino and A. Macintyre. Non standard finite fields over $I\Delta_0 + \Omega_1$. *Israel Journal of Mathematics*, 17:311–333, 2000.

[10] P. D'Aquino and A. Macintyre. Quadratic forms in models of $I\Delta_0 + \Omega_1$, Part II: Local equivalence. *Annals of Pure and Applied Logic*, 162:447–456, 2011.

[11] P. D'Aquino and A. Macintyre. *Primes in models of $I\Delta_0 + \Omega_1$: Density in henselizations*. In *New Studies in Weak Arithmetics*, pages 85–92, CSLI Lecture Notes 211, CSLI Publications, Stanford, 2013.

[12] F. Delon. *Quelques propriétés des corps valués*. Université Paris VII, 1981. Thèse d'état.

[13] J. Derakhshan and A. J. Macintyre. Model theory of adeles 1. Submitted to *Annals of Pure and Applied Logic*.

[14] J. Ershov. On the elementary theory of maximal valued fields I. *Algebra i Logika*, 4(3):31–70, 1965. In Russian.

[15] J. Ershov. On the elementary theory of maximal valued fields II. *Algebra i Logika*, 5(1):5–40, 1966. In Russian.

[16] J. Ershov. On the elementary theory of maximal valued fields III. *Algebra i Logika*, 6(3):31–38, 1967. In Russian.

[17] S. Feferman and R. L. Vaught. The first order properties of products of algebraic systems. *Fundamenta Mathematicae*, 47:57–103, 1959.

[18] M. Fried and M. Jarden. *Field arithmetic*. Ergebnisse der Mathematik und ihrer Grenzgebiete (3), vol. 11, Springer-Verlag, Berlin, 1986.

[19] T. Hiranouchi. Ramification of truncated discrete valuation rings: a survey. *RIMS Kôkyûroku Bessatsu*, B19:35–43, 2010.

[20] I. Kaplansky. *Infinite abelian groups*. University of Michigan Press, Ann Arbor, 1954.

[21] F.-V. Kuhlmann. Quantifier elimination for Henselian fields relative to additive and multiplicative congruences. *Israel Journal of Mathematics*, 85(1-3):277–306, 1994.

[22] A. Macintyre. On definable subsets of p-adic fields. *Journal of Symbolic Logic*, 41:605–610, 1976.
[23] A. Macintyre. *Residue Fields of Models of P*. In L. E. Cohen et al., editors, *Logic, Methodology and Philosophy of Science VI*, pages 193–206, Studies in Logic and the Foundations of Mathematics, North-Holland, Amsterdam, 1982.
[24] J. Pas. Uniform p-adic cell decomposition and local zeta functions. *Journal für die reine und angewandte Mathematik*, 399:137–172, 1989.
[25] M. O. Rabin. *A simple method for undecidability proofs and some applications*. In Y. Bar-Hillel, editor, *Logic, Methodology and Philosophy of Science. Proceedings of the 1964 International Congress*, pages 58–68, Studies in Logic and the Foundations of Mathematics, North-Holland, Amsterdam, 1965.
[26] A. Robinson. *Complete theories*. North-Holland, 1956.
[27] L. van den Dries. *Lectures on the Model Theory of Valued Fields*. In D. Macpherson and C. Toffalori, editors, *Model Theory in Algebra, Analysis and Arithmetic*, pages 55–157, Lecture Notes in Mathematics 2111, Springer, 2014.

On the Complexity of Translations from Classical to Intuitionistic Proofs

Matthias Baaz*
Institute of Discrete Mathematics and Geometry,
Vienna University of Technology
Wiedner Hauptstraße 8/104, 1040 Vienna, Austria
baaz@logic.at

Alexander Leitsch
Institute of Computer Languages,
Vienna University of Technology
Favoritenstraße 9–11, 1040 Vienna, Austria
leitsch@logic.at

Abstract

We investigate various translations from classical to intuitionistic proofs in Gentzen calculus and analyze their computational complexity. In particular we investigate the Kolmogorov translation, (a new form of) the extended Glivenko translation and the optimized Kolmogorov translation. While the Kolmogorov translation maps cut-free proofs into cut-free proofs this does not hold for the extended Glivenko translation. For analyzing the complexity of translating cut-free proofs into cut-free proofs we use the method CERES (cut-elimination by resolution), a global proof analysis method. We establish an elementary bound on the complexity of translating cut-free classical proofs with only weak quantifiers into cut-free intuitionistic proofs via the Glivenko translation. We even prove a more general result for classical cut-free proofs of sequents of the form $A_1, \ldots, A_n \vdash$ with only weak quantifiers, from which the complexity of the extended Glivenko translation for the weak quantifier fragment easily follows.

Keywords: Classical Proofs, Intuitionistic Proofs, Translations, Complexity.

*Supported by the Austrian Science Fund proj. nr. I-2671-N35

1 Introduction

The most fundamental relation between classical and intuitionistic logic is given by the following facts:

- intuitionistic logic is contained in classical logic, in the sense that the set of provable intuitionistic sentences is a subset of the set of classical sentences,

- provable sentences A of classical logic can be translated into provable sentences B of intuitionistic logic such that A and B are classically equivalent.

The most important consequence of these facts is the equiconsistency of intuitionistic and classical theories; the classical theories have just to be replaced by their equivalent translations. This refutes the original claim that the consistency of intuitionistic (constructive) systems might be self-evident. Gödel even interpreted the existence of a translation of classical logic and Peano arithmetic into intuitionistic logic and Heyting arithmetic as a proof that intuitionistic logic and Heyting arithmetic are in fact richer than classical logic and Peano arithmetic, because in the intuitionistic framework one can distinguish formulas which are classically equivalent, while both frameworks have the same consistency strength [12].

Most historical embeddings are defined via stepwise translations: first the axioms are translated and the results shown to be provable in intuitionistic logic; second it is proven inductively that the intuitionistic provability of the translation is propagated through the rules [7, 11, 13, 15, 16, 18]. The original aim of these embeddings is to establish a relation between classical and intuitionistic *provability* but not between the proofs themselves. For a general framework to compare the provability of various translations see [10].

In this paper we investigate the impact of such embdeddings on the complexity of first-order proofs where we choose the sequent calculi **LK** and **LJ**. In particular we compare the computational complexity of transformations from classical cut-free proofs of sentences A to intuitionistic cut-free proofs of the translations B of A for various types of translations. We also define a new extension of the Glivenko translation to first-order logic resulting in an improvement of the Kuroda translation (for first-order logic). The complexity of the extended Glivenko translation of cut-free **LK**-proofs into cut-free **LJ**-proofs is shown to be elementary by using **CERES** (cut-elimination by resolution) [4].

The paper is organized as follows: In Section 2 we define the basic concepts for our analysis. In Section 3 we define various translations and investigate their computational complexity. While the Kolmogorov translation and the extended Glivenko translation can be computed in polynomial time, we show that an optimized form of the Kolmogorov translation is of nonelementary complexity. While the Kolmogorov translation maps cut-free proofs into cut-free proofs this does not hold for the extended Glivenko translation. So the complexity of translations into cut-free proofs would have to be established by the complexity of cut-elimination. Instead of stepwise cut-elimination a la Gentzen we use CERES (cut-elimination by resolution), a global proof analysis method, to establish an elementary bound on the complexity of translating cut-free classical proofs with only weak quantifiers into cut-free intuitionistic proofs. We prove, in fact, a more general result for classical cut-free proofs of sequents of the form $A_1, \ldots, A_n \vdash$ with only weak quantifiers, yielding the complexity of the extended Glivenko translation for the weak quantifier fragment as a corollary.

2 Preliminaries

Definition 1 (strong and weak quantifiers). If $(\forall x)$ occurs positively (negatively) in B then $(\forall x)$ is called a strong (weak) quantifier. If $(\exists x)$ occurs positively (negatively) in B then $(\exists x)$ is called a weak (strong) quantifier. If S is a sequent $A_1, \ldots, A_n \vdash B_1, \ldots, B_m$ then quantifiers occurring in S are strong (weak) according to their status in $(A_1 \wedge \cdots \wedge A_n) \rightarrow (B_1 \vee \cdots \vee B_m)$.

Definition 2 (complexity). If F is a formula in predicate logic then its complexity $\|F\|$ is the number of symbols occurring in F. Similarly the complexity of a proof φ, denoted by $\|\varphi\|$, is the number of symbol occurrences in φ.

Our complexity analysis aims at the distinction between proof transformations of *elementary* and those of *nonelementary* complexity. A specific role play the bound functions e and s:

Definition 3. Let $e: \mathbf{N} \times \mathbf{N} \to \mathbf{N}$ and $s: \mathbf{N} \to \mathbf{N}$ be defined as:

$$e(0, n) = n \text{ for } n \in \mathbf{N}, \ e(m+1, n) = 2^{e(m,n)} \text{ for } m, n \in \mathbf{N},$$
$$s(n) = e(n, 1) \text{ for } n \in \mathbf{N}.$$

A function $g: \mathbf{N} \to \mathbf{N}$ is called *elementary* if there exists a nondeterministic Turing machine M computing g and a number $k \in \mathbf{N}$ with $time_M(n) \leq e(k, n)$ for all $n \in \mathbf{N}$, where $time_M(n)$ is the computing time of M on input n.

Remark 1. It is not hard to show that the function s in Definition 3 is not elementary. The nonelementary complexity of cut-elimination is typically proven by using s as lower bound (see [21] and [19]).

The following definition applies to all proof transformations, though we will investigate mainly transformations from **LK** to **LJ** in this paper. The calculi **LK** and **LJ** are defined as in [9] with the only difference that we use multisets and do not need the exchange rules.

Definition 4. Let T be a mapping from proofs to proofs. T is called *elementary* if there exists a nondeterministic Turing machine M computing $T(\varphi)$ on input φ and an elementary function h s.t. for all proofs φ $time_M(\varphi) \leq h(\|\varphi\|)$.

If the computing time is bounded by an elementary function so is the size of the output: if T is an elementary proof transformation then there exists also an elementary function g such that for all proofs φ:

$$(\star) \qquad \|T(\varphi)\| \leq g(\|\varphi\|).$$

If $(\star)$ holds for T we speak about an elementary *output complexity*. However, for establishing upper bounds, output complexity (comparing just the lengths of φ and its transformation $T(\varphi)$) is not sufficiently informative. Consider, e.g., the following transformation to be defined below. Given a traditional embedding T from classical to intuitionistic logic (let us say the Kolmogorov translation) we define a new embedding T' in the following way: $T'(A) = A \to A$ if A is provable in intuitionistic logic and $T'(A) = T(A)$ otherwise. Note that, clearly, $T'(A)$ is logically equivalent to A in classical logic. The mapping can be extended to proofs by translating classical proofs of intuitionistically provable formulas A into (trivial intuitionistic) proofs of $A \to A$; if A is not intuitionistically provable we choose the original translation T. Then, obviously, there exists an elementary function g (the bound on the original transformation T) such that $(\star)$ holds also for T' and T' has elementary output complexity. On the other hand, T' is not even computable as it is undecidable whether a formula is provable in intuitionistic logic. Therefore there is no recursive bound on the computing time of T'. To exclude pathological transformations like this one, or at least to dismantle their complexities, we choose nondeterministic time complexity as the main complexity measure for our transformations. Note, however, that for establishing *lower bounds*, output complexity gives more information that just estimating the computing time of $T(\varphi)$. To make our results sharper we use time complexity for upper bounds and output complexity for lower bounds.

3 On elementary translations of classical to intuitionistic proofs

In this paper we basically investigate two translations from classical to intuitionistic logic: the Kolmogorov translation (double negation is used for any occurrence of a logical connective) and the Glivenko translation (the formula as a whole is doubly negated) and its extendion to first-order logic. We also investigate a new extended form of the Glivenko translation; this translation is largely defined as the Kuroda translation with the difference that only strong occurrences of $\forall x.A(x)$ is translated to $\forall x.\neg\neg A(x)$. Finally we introduce a so-called optimized translation where the translated formulas can be simplified via propositional tautologies. For a good survey of different negative translation see [8].

3.1 The Kolmogorov translation

We first define the translation of formulas and afterwards the translation of proofs. We denote that translation (on formulas and proofs) by Ψ_K.

Definition 5 (Kolmogorov translation for formulas [15]).

$$\psi_K(A) = \neg\neg A \text{ for atoms } A, \quad \psi_K(\neg B) = \neg\neg\neg\psi_K(B),$$
$$\psi_K(B \circ C) = \neg\neg(\psi_K(B) \circ \psi_K(C)) \text{ for } \circ \in \{\wedge, \vee, \rightarrow\},$$
$$\psi_K(Qx.B) = \neg\neg Qx.\psi_K(B) \text{ for } Q \in \{\forall, \exists\}.$$

Next we define a translation of sequents:

Definition 6 (Kolmogorov translation of sequents). By Definition 5, for every A there exists a B s.t. $\psi_K(A) = \neg\neg B$. So if $\psi_K(A) = \neg\neg B$ we define $\psi_K^*(A) = \neg B$ (i.e. for all A we get $\neg\psi_K^*(A) = \psi_K(A)$).

Let S be the sequent $A_1, \ldots, A_n \vdash B_1, \ldots, B_m$ and ψ_K for formulas as in Definition 5. Then we define

$$\psi_K(S) = \psi_K(A_1), \ldots, \psi_K(A_n), \psi_K^*(B_1), \ldots, \psi_K^*(B_m) \vdash,$$

For $\mathcal{A} = A_1, \ldots, A_n$ we write $\psi_K(\mathcal{A})$ for $\psi_K(A_1), \ldots, \psi_K(A_n)$ (the same for $\psi_K^*(\mathcal{A})$).

The translation of proofs via ψ_K, which we denote by Ψ_K, is defined stepwise via the last inference in the proof to be translated. A particular feature of this translation is the absence of the cut rule (except for the simulation of cut itself), hence cut-free proofs are directly translated to cut-free proofs.

Definition 7 (Kolmogorov translation of proofs). We first translate the (atomic) axioms:

K-ax: $\Psi_K(A \vdash A) = \varphi(A)$ where $\varphi(A)$ is an intuitionistic proof of the sequent $\neg\neg A, \neg A \vdash$. Note that, as A is an atom, $\neg\neg A, \neg A \vdash = \Psi_K(A), \Psi_K^*(A) \vdash$.

Now we assume inductively that we have already translated the proofs φ_1, φ_2 of S_1, S_2 to proofs $\Psi_K(\varphi_1)$ of $\Psi_K(S_1)$ and to $\Psi_K(\varphi_2)$ of $\Psi_K(S_2)$. Consider a binary rule joining φ_1, φ_2 to a proof φ of a sequent S. Then we construct a proof $\Psi_K(\varphi)$ of $\Psi_K(S)$. The case of unary rules is analogous. We illustrate the transformation on the rules $\wedge_r, \wedge_l - 1, \forall_r,$ cut which are the last inferences of φ. The other cases are analogous. Remember that $\neg \psi_K^*(A) = \psi_K(A)$ for all formulas A.

K-$\wedge_r$: Let φ be the proof

$$\frac{(\varphi_1) \quad\quad (\varphi_2)}{\mathcal{A} \vdash \mathcal{B}, C \wedge D} \wedge_r \quad \mathcal{A} \vdash \mathcal{B}, C \quad \mathcal{A} \vdash \mathcal{B}, D$$

Then we define $\Psi_K(\varphi)$ as

$$\frac{\dfrac{(\Psi_K(\varphi_1))}{\psi_K(\mathcal{A}), \psi_K^*(\mathcal{B}), \psi_K^*(C) \vdash} \neg_r \quad \dfrac{(\Psi_K(\varphi_2))}{\psi_K(\mathcal{A}), \psi_K^*(\mathcal{B}), \psi_K^*(D) \vdash} \neg_r}{\dfrac{\psi_K(\mathcal{A}), \psi_K^*(\mathcal{B}) \vdash \neg \psi_K^*(C) \wedge \neg \psi_K^*(D)}{\psi_K(\mathcal{A}), \psi_K^*(\mathcal{B}), \neg(\psi_K(C) \wedge \psi_K(D)) \vdash} \neg_l} \wedge_r$$

Note that $\neg \psi_K^*(C) \wedge \neg \psi_K^*(D) = \psi_K(C) \wedge \psi_K(D)$ and the end sequent of $\Psi_K(\varphi)$ is $\psi_K(\mathcal{A}), \psi_K^*(\mathcal{B}), \psi_K^*(C \wedge D) \vdash$ what is exactly what we need.

K-$\wedge_l - 1$: Let $\varphi =$

$$\frac{(\varphi_1)}{C \wedge D, \mathcal{A} \vdash \mathcal{B}} \wedge_l 1 \quad C, \mathcal{A} \vdash \mathcal{B}$$

Then $\Psi_K(\varphi) =$

$$\frac{\dfrac{(\Psi_K(\varphi_1)}{\psi_K(C), \psi_K(\mathcal{A}), \psi_K^*(\mathcal{B}) \vdash}}{\dfrac{\psi_K(C) \wedge \psi_K(D), \psi_K(\mathcal{A}), \psi_K^*(\mathcal{B}) \vdash}{\dfrac{\psi_K(\mathcal{A}), \psi_K^*(\mathcal{B}) \vdash \neg(\psi_K(C) \wedge \psi_K(D))}{\neg\neg(\psi_K(C) \wedge \psi_K(D)), \psi_K(\mathcal{A}), \psi_K^*(\mathcal{B}) \vdash} \neg_l} \neg_r} \wedge_l 1$$

By definition of ψ_K, the end-sequent of $\Psi_K(\varphi)$ is $\psi_K(C \wedge D), \psi_K(\mathcal{A}), \psi_K^*(\mathcal{B}) \vdash$.

K-$\forall_r$ Let $\varphi =$

$$\frac{(\varphi_1)}{\mathcal{A} \vdash \mathcal{B}, \forall x. A(x)} \forall_r \quad \mathcal{A} \vdash \mathcal{B}, A(u)$$

Then $\Psi_K(\varphi) =$

$$
\dfrac{\dfrac{\dfrac{\dfrac{(\Psi_K(\varphi_1))}{\psi_K(\mathcal{A}), \psi_K^*(\mathcal{B}), \psi_K^*(A(u)) \vdash}}{\psi_K(\mathcal{A}), \psi_K^*(\mathcal{B}) \vdash \psi_K(A(u))} \neg_r}{\psi_K(\mathcal{A}), \psi_K^*(\mathcal{B}) \vdash \forall x.\psi_K(A(x))} \forall_r}{\psi_K(\mathcal{A}), \psi_K^*(\mathcal{B}), \neg\forall x.\psi_K(A(x)) \vdash} \neg_l
$$

Note that the end-sequent of $\Psi_K(\varphi)$ is $\psi_K(\mathcal{A}), \psi_K^*(\mathcal{B}), \psi_K^*(\forall x.A(x)) \vdash$.

K-*cut*: Let $\varphi =$

$$
\dfrac{(\varphi_1) \quad (\varphi_2)}{\Gamma, \Pi \vdash \Delta, \Lambda} \; cut \qquad \Gamma \vdash \Delta, A \quad A, \Pi \vdash \Lambda
$$

Then we define $\Psi_K(\varphi)$ as

$$
\dfrac{\dfrac{(\Psi_K(\varphi_1))}{\psi_K(\Gamma), \psi_K^*(\Delta), \psi_K^*(A) \vdash}}{\psi_K(\Gamma), \psi_K^*(\Delta) \vdash \neg\psi_K^*(A)} \neg_r \qquad \dfrac{(\Psi_K(\varphi_2))}{\psi_K(A), \psi_K(\Pi), \psi_K^*(\Lambda) \vdash} \quad cut
$$
$$
\psi_K(\Gamma), \psi_K(\Pi), \psi_K^*(\Delta), \psi_K^*(\Lambda) \vdash
$$

Note that $\neg\psi_K^*(A) = \psi_K(A)$.

Proposition 1. *The transformation Ψ_K, the Kolmogorov transformation of **LK**-proofs into **LJ**-proofs is*

(a) *computable in polynomial time,*

(b) *transforms cut-free proofs into cut-free proofs.*

Proof. (a) Consider an **LK**-proof φ and its translation $\Psi_K(\varphi)$. Let K be a constant greater or equal to the maximal number of sequents added in a transformation step of Ψ_K and

$$\sigma = \max\{\|S\|\ S \text{ is a sequent in } \varphi\}, \; nodes(\varphi) = \text{ number of nodes in } \varphi.$$

For every sequent we have $\|\psi_K(S)\| \leq 4 * \|S\|$. We assume $K > 4$. Then it is easy to show by induction on the structure of φ that there exists a nondeterministic Turing machine M computing Ψ_K such that

$$time_M(\varphi) \leq K * \sigma * nodes(\varphi) \leq K * \|\varphi\|^2.$$

(b) Obvious as $\Psi_K(\varphi)$ contains cut only if there is a cut in φ; the transformation of all rules, except the cut rule, does not use cut. □

3.2 The extended Glivenko translation

The Glivenko translation [11, 18] is originally defined for propositional logic and is much simpler than the Kolmogorov translation: indeed if A is propositional and classically valid (provable in **LK**) then $\neg\neg A$ is intuitionistically valid (provable in **LJ**). The Glivenko translation cannot be simply applied to first-order formulas as the universal quantifier needs a special treatment (two additional "$\neg$"s have to be introduced). In this paper we improve the treatment of the universal quantifier by introducing the additional $\neg\neg$ only for positive polarity. However, for proof transformations, we have to distinguish whether a formula is an ancestor of a cut formula or not (in the cut rule one and the same formula occurs in two polarities).

Definition 8 (the extended Glivenko formula transformation). We define a mapping ψ_G from first-order formulas to first order formulas in the following way:

$\psi_G(A) = \neg\neg\psi_G^+(A)$ for all formulas A,
$\psi_G^+(A \wedge B) = \psi_G^+(A) \wedge \psi_G^+(B), \quad \psi_G^-(A \wedge B) = \psi_G^-(A) \wedge \psi_G^-(B),$
$\psi_G^+(A \vee B) = \psi_G^+(A) \vee \psi_G^+(B), \quad \psi_G^-(A \vee B) = \psi_G^-(A) \vee \psi_G^-(B),$
$\psi_G^+(A \to B) = \psi_G^-(A) \vee \psi_G^+(B), \quad \psi_G^-(A \to B) = \psi_G^+(A) \to \psi_G^-(B),$
$\psi_G^+(\neg A) = \neg\psi_G^-(A), \quad \psi_G^-(\neg A) = \neg\psi_G^+(A),$
$\psi_G^+(\exists x.A(x)) = \exists x.\psi_G^+(A(x)), \quad \psi_G^-(\exists x.A(x)) = \exists x.\psi_G^-(A(x)),$
$\psi_G^+(\forall x.A(x)) = \forall x.\neg\neg\psi_G^+(A(x)), \quad \psi_G^-(\forall x.A(x)) = \forall x.\psi_G^-(A(x)),$
$\psi_G^+(A) = \psi_G^-(A) = A$ for atoms A.

Remark 2. Note that our extension of the Glivenko translation [11] differs from the Kuroda translation [16] in the handling of the universal quantifier. While, in the Kuroda translation, $\psi_G^+(\forall x.A) = \psi_G^-(\forall x.A) = \forall x.\neg\neg A$ independent of the polarity, we have two different translations ψ_G^-, ψ_G^+ of $\forall x.A$ dependent on the polarity. For instance our translation of $\forall x.A \to \forall x.B$ (for A and B atoms) is

$\neg\neg(\forall x.A(x) \to \forall x.\neg\neg A(x))$, instead of $\neg\neg(\forall x.\neg\neg A(x) \to \forall x.\neg\neg A(x))$.

In order to extend the Glivenko transformation above to proofs we have to define translations for sequents first: if $S \colon \Gamma \vdash \Delta$ is a sequent we define

$$\Psi_G(S) = \vdash \psi_G(\bigwedge \Gamma \to \bigvee \Delta)$$

where $\bigwedge \Gamma$ denotes the conjunction of formulas in Γ, $\bigvee \Delta$ the disjunction of formulas in Δ. Note that

$\vdash \psi_G(\bigwedge \Gamma \to \bigvee \Delta) = \vdash \neg\neg(\bigwedge \psi_G^-(\Gamma) \to \bigvee \psi_G^+(\Delta))$, where

$\psi_G^-(A_1,\ldots,A_n) = \psi_G^-(A_1),\ldots,\psi_G^-(A_n), \psi_G^+(A_1,\ldots,A_n) = \psi_G^+(A_1),\ldots,\psi_G^+(A_n).$

We first define the proof transformation Ψ_G for cut-free proofs.

The Glivenko translation for cut-free proofs:

We define the proof transformation inductively via the last rule applied in a proof. We carry out the construction in detail for the **LK**-rules

$\vee_{r1}, \wedge_r, \to_r, \forall_r, \exists_l, \forall_l, \exists_r.$

- the case $\vee_{r1}$:

Let φ be the proof
$$\frac{(\varphi_1)}{\Gamma \vdash \Delta, A}{\Gamma \vdash \Delta, A \vee B} \vee_{r1}$$

We assume that we already have an **LJ**-proof

$$\Psi_G(\varphi_1) \text{ of } \vdash \neg\neg(\bigwedge \psi_G^-(\Gamma) \to (\bigvee \psi_G^+(\Delta) \vee \psi_G^+(A))).$$

We have to define an **LJ**-proof

$$\Psi_G(\varphi) \text{ of } \vdash \neg\neg(\bigwedge \psi_G^-(\Gamma) \to (\bigvee \psi_G^+(\Delta) \vee ((\psi_G^+(A)) \vee \psi_G^+(B)))).$$

To make things simpler we introduce the following abbreviations:

$$\bigwedge \psi_G^-(\Gamma) = G, \bigvee \psi_G^+(\Delta) = D, \psi_G^+(A) = F, \psi_G^+(B) = H.$$

Then the task is to use the **LJ**-proof $\Psi_G(\varphi_1)$ of $\vdash \neg\neg(G \to (D \vee F))$ for constructing an **LJ**-proof of

$$\vdash \neg\neg(G \to (D \vee (F \vee H))).$$

We define $\Psi_G(\varphi) =$

$$\frac{(\Psi_G(\varphi_1)) \qquad (\chi_{\vee_{r1}})}{\vdash \neg\neg(G \to (D \vee F)) \quad \neg\neg(G \to (D \vee F)) \vdash \neg\neg(G \to (D \vee (F \vee H)))}{\vdash \neg\neg(G \to (D \vee (F \vee H)))} \text{ cut}$$

where $\chi_{\vee_{r1}}$ is an obvious **LJ**-proof.

- The case $\wedge_r$:

Let $\varphi =$

$$\frac{(\varphi_1) \quad (\varphi_2)}{\Gamma \vdash \Delta, A \quad \Gamma \vdash \Delta, B} \wedge_r$$
$$\Gamma \vdash \Delta, A \wedge B$$

We assume inductively that we already have proofs

$$\Psi_G(\varphi_1) \text{ of } \vdash \neg\neg(\bigwedge \psi_G^-(\Gamma) \to (\bigvee \psi_G^+(\Delta) \vee \psi_G^+(A))) \text{ and}$$
$$\Psi_G(\varphi_2) \text{ of } \vdash \neg\neg(\bigwedge \psi_G^-(\Gamma) \to (\bigvee \psi_G^+(\Delta) \vee \psi_G^+(B))).$$

We have to construct a proof $\Psi_G(\varphi)$ of

$$\vdash \neg\neg(\bigwedge \psi_G^-(\Gamma) \to (\bigvee \psi_G^+(\Delta) \vee (\psi_G^+(A) \wedge \psi_G^+(B)))).$$

We define $G = \bigwedge \psi_G^-(\Gamma), D = \bigvee \psi_G^+(\Delta), F = \psi_G^+(A), H = \psi_G^+(B)$. Now

$\Psi_G(\varphi_1)$ proves $\vdash \neg\neg(G \to (D \vee F))$, $\Psi_G(\varphi_2)$ proves $\vdash \neg\neg(G \to (D \vee H))$.

To define $\Psi_G(\varphi)$ we construct a proof $\chi_{\wedge_r}$ of

$$\neg\neg(G \to (D \vee F)), \neg\neg(G \to (D \vee H)) \vdash \neg\neg(G \to (D \vee (F \wedge H)))$$

and use two cuts, one with $\Psi_G(\varphi_1)$, the other with $\Psi_G(\varphi_2)$.

It remains to define $\chi_{\wedge_r}$; $\chi_{\wedge_r} =$

$$\frac{\frac{\frac{\frac{\frac{(\chi')}{G, G, G \to (D \vee F), G \to (D \vee H) \vdash D \vee (F \wedge H)} c_l}{G, G \to (D \vee F), G \to (D \vee H) \vdash D \vee (F \wedge H)} \to_r}{G \to (D \vee F), G \to (D \vee H) \vdash G \to (D \vee (F \wedge H))}}{G \to (D \vee F), G \to (D \vee H), \neg(G \to (D \vee (F \wedge H))) \vdash} \neg_l}{\neg\neg(G \to (D \vee F)), \neg\neg(G \to (D \vee H)), \neg(G \to (D \vee (F \wedge H))) \vdash} (\neg_l + \neg_r)^*}{\neg\neg(G \to (D \vee F)), \neg\neg(G \to (D \vee H)) \vdash \neg\neg(G \to (D \vee (F \wedge H)))} \neg_l$$

where $\chi' =$

$$\frac{G \vdash G \quad \frac{G \vdash G \quad \frac{\frac{D \vdash D}{D \vdash D \vee (F \wedge H)} \vee_{r1} \quad \frac{\frac{\frac{D \vdash D}{D \vdash D \vee (F \wedge H)} \vee_{r1}}{H, D \vdash D \vee (F \wedge H)} w_l \quad \frac{(\text{obvious})}{H, F \vdash F \wedge H} \quad \frac{H, F \vdash D \vee (F \wedge H)}{H, F \vdash D \vee (F \wedge H)} \vee_{r2}}{H, D \vee F \vdash D \vee (F \wedge H)} \vee_l}{D \vee H, D \vee F \vdash D \vee (F \wedge H)} \vee_l}{G, G \to (D \vee H), D \vee F \vdash D \vee (F \wedge H)} \to_l}{G, G, G \to (D \vee F), G \to (D \vee H) \vdash D \vee (F \wedge H)} \to_l$$

- The case $\to_r$:

Let φ be the **LK**-proof
$$\frac{(\varphi_1)}{\Gamma \vdash \Delta, A \to B} \to_r$$

By induction hypothesis we may assume that there exists an **LJ**-proof
$$\Psi_G(\varphi_1) \text{ of } \vdash \neg\neg((\psi_G^-(A) \land \bigwedge \psi_G^-(\Gamma)) \to (\bigvee \psi_G^+(\Delta) \lor \psi_G^+(B))).$$

We have to construct an **LJ**-proof
$$\Psi_G(\varphi) \text{ of } \vdash \neg\neg(\bigwedge \psi_G^-(\Gamma)) \to (\bigvee \psi_G^+(\Delta) \lor (\psi_G^-(A) \to \psi_G^+(B))).$$

We abbreviate:
$$G = \bigwedge \psi_G^-(\Gamma), \ D = \bigvee \psi_G^+(\Delta), \ F = \psi_G^-(A), \ H = \psi_G^+(B).$$

Thus $\Psi_G(\varphi_1)$ is an **LJ**-proof of $\vdash \neg\neg((F \land G) \to (D \lor H))$, and we need an **LJ**-proof of
$$\vdash \neg\neg(G \to (D \lor (F \to H))).$$

We first prove the lemma
$$(S_1): \neg\neg((F \land G) \to (D \lor H)) \vdash \neg\neg(\neg\neg(F \land G) \to \neg\neg(D \lor H))$$

which is an instance of the (easily) LJ-provable sequent
$$\neg\neg(X \to Y) \vdash \neg\neg(\neg\neg X \to \neg\neg Y).$$

So let χ be the **LJ**-proof of S_1. A cut with $\Psi_G(\varphi_1)$ then gives the proof χ' of
$$\vdash \neg\neg(\neg\neg(F \land G) \to \neg\neg(D \lor H)).$$

Let χ'' be an **LJ**-proof of
$$\neg\neg(\neg\neg(F \land G) \to \neg\neg(D \lor H)) \vdash \neg\neg(G \to (D \lor (F \to H))).$$

Then a cut of χ' and χ'' eventually gives $\Psi_G(\varphi)$ with the end-sequent
$$\vdash \neg\neg(G \to (D \lor (F \to H))).$$

It remains to define χ''. $\chi'' =$

$$\cfrac{\cfrac{\cfrac{(\chi_2)}{\neg(G \to (D \lor (F \to H))) \vdash \neg\neg(F \land G)} \quad \cfrac{(\chi_3)}{\neg\neg(D \lor H), \neg(G \to (D \lor (F \to H))) \vdash}}{\cfrac{\neg(G \to (D \lor (F \to H))), \neg(G \to (D \lor (F \to H))), \neg\neg(F \land G) \to \neg\neg(D \lor H) \vdash}{\neg(G \to (D \lor (F \to H))), \neg\neg(F \land G) \to \neg\neg(D \lor H) \vdash} c_l}}{\neg\neg(\neg\neg(F \land G) \to \neg\neg(D \lor H)) \vdash \neg\neg(G \to (D \lor (F \to H)))} (\neg_l + \neg_r)^*$$

Where $\chi_2 =$

$$\cfrac{\cfrac{\cfrac{\cfrac{\cfrac{\cfrac{(\chi_4)}{\neg(F \wedge G), G, F \vdash H}}{\neg(F \wedge G), G \vdash F \to H} \to_r}{\neg(F \wedge G), G \vdash D \vee (F \to H)} \vee_{r2}}{\neg(F \wedge G) \vdash G \to (D \vee (F \to H))} \to_r}{\neg(F \wedge G), \neg(G \to (D \vee (F \to H))) \vdash} \neg_l}{\neg(G \to (D \vee (F \to H))) \vdash \neg\neg(F \wedge G)} \neg_r$$

and χ_4 is a short and obvious **LJ**-proof. We define $\chi_3 =$

$$\cfrac{\cfrac{\cfrac{\cfrac{(\chi_5)}{D, G \vdash D \vee (F \to H)}}{D \vdash G \to (D \vee (F \to H))} \to_r}{D, \neg(G \to (D \vee (F \to H))) \vdash} \neg_l \quad \cfrac{\cfrac{\cfrac{\cfrac{(\chi_6)}{H, G \vdash F \to H}}{H, G \vdash D \vee (F \to H)} \vee_{r2}}{H \vdash G \to (D \vee (F \to H))} \to_r}{H, \neg(G \to (D \vee (F \to H))) \vdash} \neg_l}{\cfrac{D \vee H, \neg(G \to (D \vee (F \to H))) \vdash}{\neg\neg(D \vee H), \neg(G \to (D \vee (F \to H))) \vdash} \neg_r + \neg_l} \vee_l$$

where χ_5, χ_6 are trivial **LJ**-proofs.

- The case $\forall_r$:

Let $\varphi =$

$$\cfrac{\cfrac{(\varphi_1)}{\Gamma \vdash \Delta, C(u)}}{\Gamma \vdash \Delta, \forall x.C(x)} \forall_r$$

By induction hypothesis we assume that we have an **LJ**-proof

$$\Psi_G(\varphi_1) \text{ of } \vdash \neg\neg(\psi_G^-(\Gamma) \to (\psi_G^+(\Delta) \vee \psi_G^+(C(u)))).$$

we define $C = \psi_G^-(\Gamma), D = \psi_G^+(\Delta), \hat{C} = \psi_G^+(C(u))$.

We first construct an **LJ**-proof

$$\chi_1 \text{ of } \neg\neg(C \to (D \vee \hat{C}(u))) \vdash \neg\neg((C \wedge \neg D) \to \hat{C}(u)).$$

by combining $\Psi_G(\varphi_1)$ and χ_1 by cut we obtain a proof

$$\chi_2 \text{ of } \vdash \neg\neg((C \wedge \neg D) \to \hat{C}(u)).$$

We use an **LJ**-proof χ_3 of $\neg\neg((C \wedge \neg D) \to \hat{C}(u)) \vdash \neg\neg(C \wedge \neg D) \to \neg\neg\hat{C}(u)$ and cut χ_2 with χ_3. The result is a proof

$$\chi_4 \text{ of } \vdash \neg\neg(C \wedge \neg D) \to \neg\neg\hat{C}(u).$$

We cut χ_4 with an obvious **LJ**-proof χ_5 of

$$\neg\neg(C \wedge \neg D) \to \neg\neg\hat{C}(u), \neg\neg(C \wedge \neg D) \vdash \neg\neg\hat{C}(u)$$

and obtain a proof χ_6 of $\neg\neg(C \wedge \neg D) \vdash \neg\neg\hat{C}(u)$.

Note that u is not free in $\neg\neg(C \wedge \neg D)$ (as u is not free in $\psi_G^-(\Gamma), \psi_G^+(\Delta)$). We define $\chi_7 =$

$$\cfrac{\cfrac{(\chi_6)}{\neg\neg(C \wedge \neg D) \vdash \neg\neg\hat{C}(u)}{\neg\neg(C \wedge \neg D) \vdash \forall x.\neg\neg\hat{C}(x)} \forall_r}{\vdash \neg\neg(C \wedge \neg D) \to \forall x.\neg\neg\hat{C}(x)} \to_r$$

Let χ_8 be an **LJ**-proof of

$$\neg\neg(X \wedge Y) \to Z \vdash \neg\neg((X \wedge Y) \to Z).$$

Tne using cut between χ_7 and $\chi_8(C, \neg D, \forall x.\neg\neg\hat{C}(x))$ gives a proof

$$\chi_9 \text{ of } \vdash \neg\neg((C \wedge \neg D) \to \forall x.\neg\neg\hat{C}(x)).$$

Let $\chi_{10}(X, Y, Z)$ be an **LJ**-proof of

$$\neg\neg((X \wedge \neg Y) \to Z) \vdash \neg\neg(X \to (Y \vee Z)).$$

Then a cut of χ_9 with $\chi_{10}(C, D, \forall x.\neg\neg\hat{C}(x))$ gives a proof

$$\chi_{11} \text{ of } \vdash \neg\neg(C \to (D \vee \forall x.\neg\neg\hat{C}(x))).$$

By definition the last sequent is

$$\neg\neg(\psi_G^-(\Gamma) \to (\psi_G^+(\Delta) \vee \forall x.\neg\neg\psi_G^+(C(x))))$$

which is $\psi_G(\Gamma \vdash \Delta, \forall x.A(x))$ (note that $\psi_G^+(\forall x.C(x)) = \forall x.\neg\neg\psi_G^+(C(x))$ by definition of ψ_G). Therefore χ_{11} is the desired proof $\Psi_G(\varphi)$.

- The case $\exists_l$:

Let $\varphi =$

$$\cfrac{\cfrac{(\varphi_1)}{C(u), \Gamma \vdash \Delta}}{\exists x.C(x), \Gamma \vdash \Delta} \exists_l$$

where u is not free in Γ, Δ. By induction hypothesis we assume that we have a proof $\Psi_G(\varphi_1)$ of
$$\neg\neg((\psi_G^-(C(u)) \wedge \psi_G^-(\Gamma)) \to \psi_G^+(\Delta)).$$
we define $\hat{C}(u) = \psi_G^-(C(u))$, $C = \psi_G^-(\Gamma)$, $D = \psi_G^+(\Delta)$.

Let $\chi_1(X, Y)$ be an **LJ**-proof of $\neg\neg(X \to Y) \vdash (\neg\neg X \to \neg\neg Y)$. We use $\chi_1(\hat{C}(u) \wedge C, D)$ and cut it with the proof $\Psi_G(\varphi_1)$. The result is a proof
$$\chi_2 \text{ of } \vdash \neg\neg(\hat{C}(u) \wedge C) \to \neg\neg D.$$

Then we cut χ_2 with an **LJ**-proof $\chi_3(\neg\neg(\hat{C}(u) \wedge C), \neg\neg D)$ for

$\chi_3(X, Y)$ being an LJ-proof of $X, X \to Y \vdash Y$ and obtain a proof
$$\chi_4 \text{ of } \neg\neg(\hat{C}(u) \wedge C) \vdash \neg\neg D.$$

Let $\chi_5(X, Y)$ be an **LJ**-proof of $X, Y \vdash \neg\neg(X \wedge Y)$. Then cutting $\chi_5(\hat{C}(u), C)$ with χ_4 gives a proof
$$\chi_6 \text{ of } \hat{C}(u), C \vdash \neg\neg D.$$

Let χ_7 be an **LJ**-proof of $\neg D \vdash \neg\neg\neg D$. Then we define $\chi_8 =$

$$\cfrac{\cfrac{(\chi_7)}{\neg D \vdash \neg\neg\neg D} \quad \cfrac{\cfrac{(\chi_6)}{\hat{C}(u), C \vdash \neg\neg D}}{\neg\neg\neg D, \hat{C}(u), C \vdash} \neg l}{\hat{C}(u), C, \neg D \vdash} \text{ cut}$$

and $\chi_9 =$

$$\cfrac{\cfrac{\cfrac{\cfrac{\cfrac{\cfrac{(\chi_8)}{\hat{C}(u), C, \neg D \vdash}}{\exists x.\hat{C}(x), C, \neg D \vdash} \exists_l}{\exists x.\hat{C}(x) \wedge C, \neg D \vdash} \wedge_l 1 + \wedge_l 2 + c_l}{\neg\neg(\exists x.\hat{C}(x) \wedge C), \neg D \vdash} \neg_r + \neg_l}{\neg\neg(\exists x.\hat{C}(x) \wedge C) \vdash \neg\neg D} \neg_r}{\vdash \neg\neg((\exists x.\hat{C}(x) \wedge C) \to \neg\neg D)} \to_r$$

Note that u does not occur free in C, D (which is $\psi_G^-(\Gamma), \psi_G^+(\Delta)$). Finally, let $\chi_{10}(X, Y)$ be an **LJ**-proof of $\vdash (\neg\neg X \to \neg\neg Y) \to \neg\neg(X \to Y)$. Then by cutting χ_9 and $\chi_{10}(\exists x.\hat{C}(x) \wedge C, D)$ we obtain a proof
$$\chi_{11} \text{ of } \neg\neg((\exists x.\hat{C}(x) \wedge C) \to D)$$

which (by $\psi_G^-(\exists x.C(x)) = \exists x.\psi_G^-(C(x))$) is an **LJ**-proof of $\psi_G(\exists x.C(x), \Gamma \vdash \Delta)$ and we define $\Psi_G(\varphi) = \chi_{11}$.

- The case $\forall_l$: analogous to $\exists_l$.
- The case $\exists_r$:

Let $\varphi =$
$$\dfrac{\begin{array}{c}(\varphi_1)\\ \Gamma \vdash \Delta, C(t)\end{array}}{\Gamma \vdash \Delta, \exists x.C(x)}\ \exists_r$$

we assume that we already have a proof $\Psi_G(\varphi_1)$ of $\vdash \neg\neg(\psi_G^-(\Gamma) \to (\psi_G^+(\Delta) \lor \psi_G^+(C(t))))$. Like in the case $\forall_r$ (proof χ_7) we obtain an **LJ**-proof

$$\sigma_1 \text{ of } \neg\neg(\psi_G^-(\Gamma) \land \neg\psi_G^+(\Delta)) \vdash \exists x.\neg\neg\psi_G^+(C(x))$$

Then we cut with of a proof $\chi_\exists$ (to be defined below) of $\exists x.\neg\neg\psi_G^+(C(x)) \vdash \neg\neg\exists.\psi_G^+(C(x))$ and obtain a proof

$$\sigma_2 \text{ of } \neg\neg(\psi_G^-(\Gamma) \land \neg\psi_G^+(\Delta)) \vdash \neg\neg\exists x.\psi_G^+(C(x)).$$

The remaining part of the derivation is the same as in case $\forall_r$ and we obtain a proof

$$\Psi_G(\varphi) \text{ of } \neg\neg(\psi_G^-(\Gamma) \to (\psi_G^+(\Delta) \lor \psi_G^+(\exists x.C(x)))).$$

We set $\hat{C}(x) = \psi_G^+(C(x))$ and define $\chi_\exists$:

$$\dfrac{\dfrac{\dfrac{\dfrac{\dfrac{\dfrac{\hat{C}(u) \vdash \hat{C}(u)}{\hat{C}(u) \vdash \exists x.\hat{C}(x)}\ \exists_r}{\hat{C}(u), \neg\exists x.\hat{C}(x) \vdash}\ \neg_l}{\neg\neg\hat{C}(u), \neg\exists x.\hat{C}(x) \vdash}\ \neg_r + \neg_l}{\neg\neg\hat{C}(u), \vdash \neg\neg\exists x.\hat{C}(x)}\ \neg_r}{\exists x.\neg\neg\hat{C}(x) \vdash \neg\neg\exists x.\hat{C}(x)}\ \exists_l$$

Note that a similar derivation for the universal quantifier is *impossible*: $\forall x.\neg\neg\hat{C}(x) \vdash \neg\neg\forall x.\hat{C}(x)$ is not derivable in **LJ**! This concludes the translations of cut-free proofs.

The extended Glivenko translation for proofs with cuts:

As the cut-formula appears in both polarities the translation of $\forall x.A(x)$ via ψ_G does not work:

we have $\psi_G^-(\forall x.A(x)) = \forall x.\psi_G^-(A(x))$ but $\psi_G^+(\forall x.A(x)) = \forall x.\neg\neg\psi_G^+(A(x))$.

So we define a translation ψ'_G (a variant of ψ_G) for formulas which are ancestors of a cut in a proof: $\psi'_G(A) = \neg\neg\psi''_G(A)$, where ψ''_G is not sensitive to polarity and, in particular, $\psi''_G(\forall x.A(x)) = \forall x.\neg\neg\psi''_G(A(x))$. We define the proof transformation Ψ^c_G corresponding to the extended Glivenko translation with cut.

Definition 9. Let $\Gamma \vdash \Delta$ be a sequent occurring in an **LK**-proof φ. Then Γ can be partitioned into formulas Γ_c which are ancestors of a cut-formula and formulas Γ_e which are ancestors of a formula in the end-sequent; in the same way we can partition Δ into Δ_c and Δ_e. Therefore any sequent $\Gamma \vdash \Delta$ occurring in φ can be translated as follows:
$$\psi_G(\Gamma_c, \Gamma_e \vdash \Delta_c, \Delta_e) =$$
$$\neg\neg((\bigwedge \psi'_G(\Gamma_c) \wedge \bigwedge \psi^-_G(\Gamma_e)) \to (\bigvee \psi'_G(\Delta_c) \vee \bigvee \psi^+_G(\Delta_e))).$$

It remains to simulate the cut rule. Let χ be a subproof of φ of the form

$$\frac{\begin{array}{cc}(\chi_1) & (\chi_2)\\ \Gamma_c, \Gamma_e \vdash \Delta_c, \Delta_e, A & A, \Pi_c, \Pi_e \vdash \Lambda_c, \Lambda_e\end{array}}{\Gamma_c, \Pi_c, \Gamma_e, \Pi_e \vdash \Delta_c, \Delta_e, \Lambda_c, \Lambda_e} \; cut$$

and assume we have already constructed the proofs $\Psi^c_G(\chi_1)$ and $\Psi^c_G(\chi_2)$. Then $\Psi^c_G(\chi_1)$ is a proof of
$$\neg\neg((\bigwedge \psi'_G(\Gamma_c) \wedge \bigwedge \psi^-_G(\Gamma_e)) \to (\bigvee \psi'_G(\Delta_c) \vee \bigvee \psi^+_G(\Delta_e) \vee \psi'_G(A)))$$
and $\Psi^c_G(\chi_2)$ is a proof of
$$\neg\neg((\psi'_G(A) \wedge \bigwedge \psi'_G(\Pi_c) \wedge \bigwedge \psi^-_G(\Pi_e)) \to (\bigvee \psi'_G(\Lambda_c) \vee \bigvee \psi^+_G(\Lambda_e))).$$

Let us define
$$\begin{aligned} B &\equiv \bigwedge \psi'_G(\Gamma_c) \wedge \bigwedge \psi^-_G(\Gamma_e)),\\ C &\equiv \bigvee \psi'_G(\Delta_c) \vee \bigvee \psi^+_G(\Delta_e),\\ D &\equiv \bigwedge \psi'_G(\Pi_c) \wedge \bigwedge \psi^-_G(\Pi_e)),\\ E &\equiv \bigvee \psi'_G(\Lambda_c) \vee \bigvee \psi^+_G(\Lambda_e),\\ A^* &\equiv \psi'_G(A).\end{aligned}$$

Then $\Psi^c_G(\chi_1)$ proves $\vdash \neg\neg(B \to (C \vee A^*))$ and $\Psi^c_G(\chi_2)$ proves $\vdash \neg\neg((A^* \wedge D) \to E)$. We define $\Psi^c_G(\chi)$ as

$$\frac{\dfrac{\begin{array}{cc}(\Psi^c_G(\chi_1)) & (\Psi^c_G(\chi_2))\\ \vdash \neg\neg(B \to (C \vee A^*)) & \vdash \neg\neg((A^* \wedge D) \to E)\end{array}}{\vdash \neg\neg(B \to (C \vee A^*)) \wedge \neg\neg((A^* \wedge D) \to E)} \wedge_r}{\vdash \neg\neg((B \wedge D) \to (C \vee E))} \; \varrho \quad cut$$

where ϱ is a **LJ**-proof of the sequent
$$\neg\neg(B \to (C \vee A^*)) \wedge \neg\neg((A^* \wedge D) \to E) \vdash \neg\neg((B \wedge D) \to (C \vee E)).$$
Note that $\neg\neg((B \wedge D) \to (C \vee E))$ is just $\psi'_G(\Gamma_c, \Pi_c, \Gamma_e, \Pi_e \vdash \Delta_c, \Delta_e, \Lambda_c, \Lambda_e)$.

Remark 3. If we had used the extended Glivenko translation without taking care of the cut status of the formula the end-sequent of ϱ could be of the form
$$\neg\neg(B \to (C \vee \neg\neg A^*)) \wedge \neg\neg((A^* \wedge D) \to E) \vdash \neg\neg((B \wedge D) \to (C \vee E)).$$
But this sequent is *not* provable in **LJ**.

Example 1. Let A, B, C be formulas and φ be a proof of the form
$$\frac{\overset{(\varphi_1)}{\forall x.A \vdash \forall x.B} \quad \overset{(\varphi_2)}{\forall x.B \vdash \forall x.C}}{\forall x.A \vdash \forall x.C} \ cut$$

Then $\Psi_G^c(\varphi) =$

$$\frac{\dfrac{\overset{(\Psi_G^c(\varphi_1))}{\vdash \neg\neg(\forall x.\psi_G^-(A) \to \forall x.\neg\neg\psi'_G(B))} \quad \dfrac{\overset{(\psi'_G(\varphi_2))}{\vdash \neg\neg(\forall x.\neg\neg\psi'_G(B) \to \forall x.\neg\neg\psi_G^+(C))}}{\vdash \neg\neg(\forall x.\psi_G^-(A) \to \forall x.\neg\neg\psi'_G(B)) \wedge \neg\neg(\forall x.\neg\neg\psi'_G(B) \to \forall x.\neg\neg\psi_G^+(C))} \wedge_r}{\vdash \neg\neg(\forall x.\psi_G^-(A) \to \forall x.\neg\neg\psi_G^+(C))} \ \varrho \ cut$$

where ϱ is an **LJ**-proof of
$$\neg\neg(\forall x.\psi_G^-(A) \to \forall x.\neg\neg\psi'_G(B)) \wedge \neg\neg(\forall x.\neg\neg\psi'_G(B) \to \forall x.\neg\neg\psi_G^+(C)) \vdash$$
$$\neg\neg(\forall x.\psi_G^-(A) \to \forall x.\neg\neg\psi_G^+(C)).$$

Theorem 1. *The transformation Ψ_G from proofs φ of S in **LK** to proofs $\Psi_G(\varphi)$ of $\psi_G(S)$ is computable in nondeterministic polynomial time.*

Proof. Let φ be a proof of S. The situation is similar to that of the Kolmogorov translation. Again we define a constant K which is bigger than the number of sequents added in a rule simulation and σ as the maximal complexity of a sequent in φ. Every simulation step of a rule in **LK** requires **LJ**-proofs with $\leq K$ rule applications. Moreover there exists a constant c s.t. $\|\Psi_G(S)\| \leq c * \|S\|$. Here we have to observe that the **LJ**-simulations have auxiliary proofs with axioms of the form $A \vdash A$, where A is nonatomic. But $A \vdash A$ has a proof from atomic axioms of a complexity linear in $\|A\|$ (see [3]). Therefore, like for Ψ_K, there exists a constant L and a nondeterministic Turing machine M computing Ψ_G such that
$$time_M(\varphi) \leq L * \sigma * nodes(\varphi) \leq L * \|\varphi\|^2.$$
□

Remark 4. In contrast to the Kolmogorov translation, cut-free proofs are translated to proofs with cut. We will deal with the complexity of translating cut-free **LK**-proofs of $\vdash A$ into cut-free **LJ**-proofs of $\psi_G(A)$ using another form of proof transformation in Section 5.

3.3 Optimized translations

Translations of formulas A (provable in **LK**) to formulas A^* (provable in **LJ**) can be simplified by first simplifying the formulas A under logical equivalence based on propositional tautologies; i.e. we first replace A by a tautologically equivalent A_1 and then define A_1^*. We consider the Kolmogorov translation ψ_K under such a form of optimization. We define a formula transformation s_f simplifying formulas:

- If A is an atom then $s_f(A) = A$.
- If A is a tautology we define $s_f(A) = \top$.
- If $\neg A$ is a tautology we set $s_f(A) = \bot$.
- If neither A nor $\neg A$ is a tautology and $A = B \circ C$ for $\circ \in \{\wedge, \vee, \to\}$ we define $s_f(A) = \sigma(s_f(B) \circ s_f(C))$ where σ is the following simplification operator:
  - $\sigma(F \wedge \bot) = \sigma(\bot \wedge F) = \bot, \sigma(F \wedge \top) = \sigma(\top \wedge F) = F$,
  - $\sigma(F \vee \bot) = \sigma(\bot \vee F) = F, \sigma(F \vee \top) = \sigma(\top \vee F) = \top$,
  - $\sigma(\bot \to F) = \top, \sigma(F \to \bot) = \neg F, \sigma(\top \to F) = F, \sigma(F \to \top) = \top$.
- If neither A nor $\neg A$ is a tautology and $A = \neg B$ then $s_f(A) = \sigma(\neg s_f(B))$ for $\sigma(\neg \top) = \bot, \sigma(\neg \bot) = \top$.

Definition 10 (optimized Kolmogorov translation). For all formulas A we define $\psi_{Ko}(A) = \psi_K(s_f(A))$.

Remark 5. Note that ψ_{Ko} fulfils all the required properties: $A \leftrightarrow \psi_K(s_f(A))$ is provable in **LK** and $\psi_K(s_f(A))$ is provable in **LJ**. Moreover, the formula transformation is computable in exponential time (which is the cost for computing s_f).

For a sequent of the form $S: \vdash A$ we define $\Psi_{Ko}(S) = \vdash \Psi_{Ko}(A)$. We have seen that Ψ_K is a polynomial nondeterminstic proof transformation mapping cut-free proofs in **LK** to cut-free proofs in **LJ**. While the formula transformation ψ_{Ko} is elementary (computable in exponential time) there exists no corresponding elementary transformation for cut-free proofs:

Lemma 1. *There exists a sequence of formulas F_n s.t.*

(1) There are cut-free **LK**-proofs of φ_n of $\vdash H_n$ which are computable in nondeterministic elementary time.

(2) For all cut-free **LK**-proofs ψ of $\vdash \psi_{Ko}(H_n)$ we have $\|\psi\| > 1/2 * s(n)$ (see Definition 3).

Proof. In [21] Statman proved the nonelementary complexity of cut-elimination. In particular, he defined a sequence of formulas G_n s.t. $\|G_n\| \leq k^n$ (for a constant k independent of n) and there are proofs φ_n of G_n (with cut formulas $A_1, \ldots, A_{2n+1}$) with $\|\varphi_n\| \leq h(\|G_n\|)$ for all n (where h is an elementary function independent of n), but $\|\psi\| > 1/2 * s(n)$ on *cut-free* proofs ψ of G_n. In [2] Statman's proof sequence is formalized in **LK**. As the cut formulas are closed, the proofs φ_n with cuts $A_1, \ldots, A_{2n+1}$ can be transformed into cut-free proofs of $A_1 \to A_1, \ldots, A_n \to A_n \vdash G_n$ and, finally, into cut-free proofs

$$\varphi_n^* \text{ of } \vdash H_n \text{ for } H_n = ((A_1 \to A_1) \land \cdots \land (A_n \to A_n)) \to G_n$$

The construction of φ_n^* can be done in nondeterministic polynomial time in terms of $\|\varphi_n\|$. Therefore there exists an elementary function g and a nondeterministic Turing machine M computing φ_n^* s.t.

$$time_M(\vdash H_n) \leq g(\|H_n\|).$$

Thus the sequents $\vdash H_n$ have "short" cut-free **LK**-proofs. Now let us consider $\vdash \psi_{Ko}(H_n)$. As ψ_{Ko} eliminates all tautologies we get

$$\vdash \psi_{Ko}(H_n) = \vdash G_n.$$

But, by definition of the G_n, there is no elementary bound on cut-free proofs of $\vdash G_n$ in terms of $\|G_n\|$. In particular there is no sequence of cut-free proofs ψ_n of $\vdash \psi_{Ko}(H_n)$ which can be computed in elementary time. $\square$

The following theorem states that the proof complexities of the translations ψ_K and ψ_{Ko} strongly differ.

Theorem 2. *There are sequences of formulas $\vdash H_n$ s.t. there exists cut-free **LJ**-proofs χ_n of $\vdash \psi_K(H_n)$ which can be computed in elementary time, but there exists no elementary bound on the lengths of cut-free **LJ**-proofs of $\vdash \psi_{Ko}(H_n)$.*

Proof. By (1) in Lemma 1 we obtain a sequence of cut-free **LK**-proofs of $\vdash H_n$ which can be computed in elementary time. By Proposition 1 we also obtain a sequence of cut-free **LJ**-proofs of $\vdash \psi_K(H_n)$ which can be computed in elementary time. By (2) in Lemma 1 there is no elementary bound on cut-free **LK**-proofs of $\vdash \psi_{Ko}(H_n)$, thus there is no such bound on the **LJ**-proofs of these sequents. $\square$

Remark 6. Theorem 2 does not only hold for the Kolmogorov translation but also for the extended Glivenko translation and, more generally, for any proof transformation obtained from a negative translation. If negative translations induce additional cuts (like in the extended Glivenko translation) we do not have a direct transformation of cut-free proofs into cut-free proofs; but in all cases where the complexity of the elimination of these additional cuts is elementary (and thus below the worst case complexity of cut-elimination) a similar result as this of Theorem 2 can be obtained. Negative translations are widely used in proof mining (see e.g. [14]). Theorem 2 indicates a potential impact of propositional optimizations (by increasing the complexity of proof normalization) on the results in the analysis of mathematical proofs.

4 The method CERES

In Section 3 we defined several proof transformations from **LK** to **LJ** based on *stepwise* translation of inferences. In this section we introduce a method of proof transformation (from **LK** to **LK** with only atomic cuts) which is radically different as it is *global* and takes into account the structure of the whole proof. This method, called CERES [4, 5], is a cut-elimination method that is based on resolution. We will apply CERES to investigate the complexity of translating cut-free proofs of $\vdash A$ (for A without strong quantifiers) into cut-free proofs of $\vdash \psi_G(A)$ (the Glivenko translation of A).

CERES roughly works as follows: The structure of a proof φ containing cuts is encoded in an unsatisfiable set of clauses $\mathrm{CL}(\varphi)$ (the *characteristic clause set* of φ). A resolution refutation of $\mathrm{CL}(\varphi)$ then serves as a skeleton for an *atomic cut normal form*, a new proof which contains at most atomic cuts. The corresponding proof theoretic transformation uses so-called proof projections $\varphi[C]$ for $C \in \mathrm{CL}(\varphi)$, which are simple cut-free proofs extracted from φ (proving end-sequent S extended by the atomic sequent C). In [5] it was shown that CERES outperforms reductive methods of cut-elimination (a la Gentzen or Tait) in computational complexity: there are infinite sequences of proofs where the computing time of CERES is nonelementarily faster than that of the reductive methods; on the other hand a nonelementary speed-up of CERES via reductive methods is shown impossible.

4.1 CERES in classical logic

In this section we describe the original CERES method which was designed for classical logic. Given an **LK**-proof φ of a skolemized sequent $\Gamma \vdash \Delta$, the main steps of (classical) CERES are:

1. Extraction of the characteristic clause set $\text{CL}(\varphi)$.

2. Construction of a resolution refutation of $\text{CL}(\varphi)$.

3. Extraction of a set of projections $\pi(C)$ for every $C \in \text{CL}(\varphi)$.

4. Merging of refutation and projections into a proof φ^* (a **CERES**-normal form) with only atomic cuts.

For extracting $\text{CL}(\varphi)$ we need the concept of formula ancestors, which is defined below.

Definition 11 (Formula ancestor). Let ν be a formula occurrence in a sequent calculus proof φ. If ν is a principal formula occurrence of an inference then the occurrences of the auxiliary formula (formulas) in the premises are ancestors of ν. If ν is principal formula of a weakening or occurs in an axiom then ν has no ancestor. If ν is not a principal occurrence then the corresponding occurrences in contexts of the (premise) premises are ancestors of ν. The ancestor relation is then defined as the reflexive transitive closure.

We will use the following proof as our running example to clarify the definitions below.

Example 2. Below we give a proof with two cuts where the cut ancestors are marked with $\star$.

$$\cfrac{\cfrac{\cfrac{\cfrac{\cfrac{P(a)^\star \vdash P(a)}{\neg P(a), P(a)^\star \vdash}\neg_l}{\neg P(a) \vdash (\neg P(a))^\star}\neg_r}{\neg P(a), (\neg\neg P(a))^\star \vdash}\neg_l}{\neg P(a), (\forall x.\neg\neg P(x))^\star \vdash}\forall_l \quad \cfrac{\cfrac{\cfrac{\cfrac{\cfrac{P(y) \vdash P(y)^\star}{P(y), (\neg P(y))^\star \vdash}\neg_l}{P(y) \vdash (\neg\neg P(y))^\star}\neg_r}{\forall x.P(x) \vdash (\neg\neg P(y))^\star}\forall_l}{\forall x.P(x) \vdash (\forall x.\neg\neg P(x))^\star}\forall_r}{\forall x.P(x) \vdash Q(b), (\forall x.\neg\neg P(x))^\star}w_r}{\cfrac{\cfrac{\neg P(a), \forall x.P(x) \vdash Q(b)}{\neg P(a) \vdash \forall x.P(x) \to Q(b)}\to_r \quad \cfrac{P(c) \vdash P(c)^\star \quad P(c)^\star \vdash P(c)}{P(c) \vdash P(c)}cut}{\neg P(a), P(c) \vdash (\forall x.P(x) \to Q(b)) \wedge P(c)}\wedge_r}cut$$

Intuitively, the clause set extraction consists in collecting all atomic ancestors of the cuts which occur in the axioms of the proof. The clauses are formed depending on how these atoms are related via binary inferences in the proof.

Definition 12 (Clause). A sequent $\Gamma \vdash \Delta$ is called a clause if Γ and Δ are multisets of atoms.

Definition 13 (Characteristic clause-set). Let φ be a proof of a skolemized sequent. The characteristic clause set is built recursively from the leaves of the proof until the end sequent. Let ν be a sequent in this proof. Then:

- If ν is an axiom, then $\mathrm{CL}(\nu)$ contains the sub-sequent of ν composed only of cut ancestors.

- If ν is the result of the application of a unary rule on a sequent μ, then $\mathrm{CL}(\nu) = \mathrm{CL}(\mu)$

- If ν is the result of the application of a binary rule on sequents μ_1 and μ_2, then we distinguish two cases:
  - If the rule is applied to ancestors of the cut formula, then $\mathrm{CL}(\nu) = \mathrm{CL}(\mu_1) \cup \mathrm{CL}(\mu_2)$
  - If the rule is applied to ancestors of the end sequent, then $\mathrm{CL}(\nu) = \mathrm{CL}(\mu_1) \times \mathrm{CL}(\mu_2)$

Where[1]:
$$\mathrm{CL}(\mu_1) \times \mathrm{CL}(\mu_2) = \{C \circ D \mid C \in \mathrm{CL}(\mu_1), D \in \mathrm{CL}(\mu_2)\}.$$

If ν_0 is the root node $\mathrm{CL}(\nu_0)$ is called the characteristic clause set of φ.

The clause set of our proof φ from Example 2 is

$$\mathrm{CL}(\varphi) = \{P(a) \vdash P(c);\ P(a), P(c) \vdash;\ \vdash P(y), P(c);\ P(c) \vdash P(y)\}$$

The next step is to obtain a resolution refutation of $CL(\varphi)$. It is thus important to show that this set is always refutable.

Theorem 3. *Let φ be a proof of a skolemized end-sequent. Then the characteristic clause set $\mathrm{CL}(\varphi)$ is refutable.*

Proof. In [1,4]; basically the proof consists in the construction of an **LK**-derivation of the empty sequent from $\mathrm{CL}(\varphi)$, thus obtaining a refutation of $\mathrm{CL}(\varphi)$. □

Definition 14 (Resolution calculus). The resolution calculus consists of the following rules:

$$\frac{\Gamma \vdash \Delta, A \quad \Gamma', A' \vdash \Delta'}{\Gamma\sigma, \Gamma'\sigma \vdash \Delta\sigma, \Delta'\sigma} R \quad \frac{\Gamma, A, A' \vdash \Delta}{\Gamma\sigma, A\sigma \vdash \Delta\sigma} C_l \quad \frac{\Gamma \vdash \Delta, A, A'}{\Gamma\sigma \vdash \Delta\sigma, A\sigma} C_r$$

Where σ is the most general unifier of A and A'. It is also required that $\Gamma \vdash \Delta, A$ and $\Gamma', A' \vdash \Delta'$ are variable disjoint. A resolution derivation from a set of clauses $\mathcal{C}$ is tree derivation based on the rules above where all clauses in the leaves are variants of clauses in $\mathcal{C}$. A resolution derivation of $\vdash$ from $\mathcal{C}$ is called a resolution refutation of $\mathcal{C}$.

---

[1]The operation $\circ$ represents the merging of sequents, i.e., $(\Gamma \vdash \Delta) \circ (\Gamma' \vdash \Delta') = \Gamma, \Gamma' \vdash \Delta, \Delta'$.

Example 3. We give a resolution refutation of $\text{CL}(\varphi)$ for φ in Example 2:

$$\dfrac{\dfrac{\dfrac{P(c) \vdash P(y) \quad P(a), P(c) \vdash}{P(c), P(c) \vdash} R}{P(c) \vdash} c_l \qquad \vdash P(z), P(c)}{\vdash P(z)} R \qquad \dfrac{\dfrac{P(a) \vdash P(c) \quad P(a), P(c) \vdash}{P(a), P(a) \vdash} R}{P(a) \vdash} c_l$$
$$\dfrac{\vdash P(z) \qquad \qquad P(a) \vdash}{\vdash} R$$

Each clause in the clause set will have a projection associated with it. A projection of a clause C is a derivation built from φ by taking the axioms in which the atoms of C occur and all the inferences that operate on end-sequent ancestors. As a result, the end-sequent of a projection will be the end-sequent of φ extended by the atoms of C.

Definition 15 (Projections). Let φ be a proof and ξ the last (lower most) inference with conclusion ν. We define $S(\nu)$ as the sequent occurring at node ν and $p(\nu)$ as the set of projections $\{\pi(C) | C \in \text{CL}(\nu)\}$. Each projection $\pi(C)$ is a cut-free proof of the sequent $S(\nu_0) \circ C$ where ν_0 is the root node and $S(\nu_0)$ the end-sequent.

- If ξ is an axiom, then $p(\nu) = \{\varphi\}$.

- If ξ is a unary rule with premise μ:

    - If ξ operates on a cut ancestor, then $p(\nu) = p(\mu)$.
    - If ξ operates on an end-sequent ancestor, then $p(\nu)$ is the set of:

    $$\dfrac{\pi(C_i)}{\zeta} \xi$$

    such that $\pi(C_i) \in p(\mu)$.

- If ξ is a binary rule with premises μ_1 and μ_2:

    - If ξ operates on a cut ancestor, then $p(\nu) = p(\mu_1) \cup p(\mu_2)$.
    - If ξ operates on an end-sequent ancestor, then $p(\nu)$ is the set of:

    $$\dfrac{\pi(C_i^1) \quad \pi(C_j^2)}{\zeta} \xi$$

    such that $\pi(C_i^1) \in p(\mu_1)$ and $\pi(C_j^2) \in p(\mu_2)$.

In each step, it might be necessary to weaken the auxiliary formulas of an inference. Moreover, if not all formulas of the end-sequent are present after constructing the projection, they are weakened as well.

Note that no rule operates on cut ancestors, therefore they occur as atoms in the end-sequent of the projections.

Example 4. The projection of φ (as defined in Example 2) to the clause $\vdash P(y), P(c)$ (where these atoms are marked with $\star$) is:

$$\dfrac{\dfrac{\dfrac{\dfrac{P(y) \vdash P(y)^\star}{\forall x.P(x) \vdash P(y)^\star} \forall_l}{\forall x.P(x) \vdash Q(b), P(y)^\star} w_r}{\dfrac{\vdash \forall x.P(x) \to Q(b), P(y)^\star}{\to_r} \quad \dfrac{\dfrac{P(c) \vdash P(c)^\star}{P(c) \vdash P(c)^\star, P(c)} w_r}{P(c) \vdash P(y)^\star, P(c)^\star, (\forall x.P(x) \to Q(b)) \wedge P(c)}} \wedge_r}{\neg P(a), P(c) \vdash P(y)^\star, P(c)^\star, (\forall x.P(x) \to Q(b)) \wedge P(c)} w_l$$

Given the projections and a grounded resolution refutation, it is possible to build a proof $\hat{\varphi}$ of $\Gamma \vdash \Delta$ with only atomic cuts.

Definition 16 (Context product). Let C be a sequent and φ be an **LK** derivation with end-sequent S such that no free variable in C occurs as eigenvariable in φ. We define the *context product* $C \star \varphi$ (which gives a derivation of $C \circ S$) inductively:

- If φ consists only of an axiom, then $C \star \varphi$ is composed by one sequent: $C \circ S$.

- If φ ends with a unary rule ξ:

$$\dfrac{\dfrac{\varphi'}{S'}}{S} \xi$$

then we assume that $C \star \varphi'$ is already defined and thus $C \star \varphi$ is:

$$\dfrac{\dfrac{C \star \varphi'}{C \circ S'}}{C \circ S} \xi$$

Since C does not contain free variables which are eigenvariables of φ, the context product is well defined also for $\xi = \forall_r, \exists_l$.

- If φ ends with a binary rule ξ:

$$\dfrac{\varphi_1 \quad \varphi_2}{\dfrac{S_1 \quad S_2}{S}} \xi$$

then assume that $C \star \varphi_1$ and $C \star \varphi_2$ are already defined. We define $C \star \varphi$:

$$\frac{\dfrac{C \star \varphi_1 \quad C \star \varphi_2}{\dfrac{C \circ S_1 \quad C \circ S_2}{C \circ C \circ S}\xi}}{C \circ S} c^*$$

if ξ is a multiplicative rule; in case ξ is additive no additional contractions are needed.

If we apply all most general unifiers in the resolution proof γ we obtain a proof in **LK** (in fact only contractions and cut remain). If $\gamma\sigma$ is such a proof and we apply a substitution replacing all variables by a constant symbol we obtain a ground resolution refutation. Note that after applying the most general unifiers to γ we obtain a derivation in **LK** where the resolution rule becomes a cut rule. For a formal definition see [1].

Example 5. Consider the resolution refutation γ in example 3 and apply the substitution $\sigma\colon \{x \leftarrow a, z \leftarrow a\}$; then we receive the ground resolution refutation $\gamma'\colon \gamma\sigma$ where $\gamma' =$

$$\frac{\dfrac{\dfrac{\dfrac{P(c) \vdash P(a) \quad P(a), P(c) \vdash}{P(c), P(c) \vdash} cut}{P(c) \vdash} c_l \quad \dfrac{\dfrac{P(a) \vdash P(c) \quad P(a), P(c) \vdash}{P(a), P(a) \vdash} cut}{P(a) \vdash} c_l}{\vdash P(a)} \quad \vdash P(a), P(c)}{\vdash} cut$$

Note that γ' is an **LK**-refutation of ground instances of clauses in $CL(\varphi)$.

Definition 17 (CERES normal form). Let φ be an **LK** proof of a skolemized sequent S, $CL(\varphi)$ its clause set and ϱ a grounded resolution refutation of $CL(\varphi)$. We first construct $\varrho' = S \star \varrho$. Note that this is a derivation of S from a set of axioms $C \circ S$, with $C \in CL(\varphi)$, which are exactly the end-sequents of the projections $\pi(C)$ of φ. Now we define $\varphi(\varrho)$ by replacing all axioms of ϱ' by the respective projections. By definition, $\varphi(\varrho)$ is an **LK** proof of S with only atomic cuts. We call it the *CERES normal form* of φ with respect to ϱ.

Example 6. Consider the characteristic clause set

$$CL(\varphi) = \{C_1\colon P(a) \vdash P(c); C_2\colon P(a), P(c) \vdash; C_3\colon \vdash P(y), P(c); C_4\colon P(c) \vdash P(y)\}$$

and the grounded resolution refutation γ' from Example 5. Then $\varphi(\gamma')$, the CERES normal form of φ, is defined as

$$\cfrac{\cfrac{\cfrac{\cfrac{\pi(C_4)\sigma \quad \pi(C_2)\sigma}{(P(c), P(c) \vdash) \circ \Gamma, \Gamma} \text{ cut}}{(P(c) \vdash) \circ \Gamma, \Gamma} c_l}{(\vdash P(a)) \circ \Gamma^3} \quad \pi(C_3)\sigma}{\cfrac{\Gamma^5}{\Gamma} c^*} \text{ cut} \quad \cfrac{\cfrac{\pi(C_1)\sigma \quad \pi(C_2)\sigma}{(P(a), P(a) \vdash) \circ \Gamma, \Gamma} \text{ cut}}{(P(a) \vdash) \circ \Gamma, \Gamma} c_l}{} \text{ cut}$$

where $\Gamma = \neg P(a), P(c) \vdash (\forall x.P(x) \rightarrow Q(b)) \wedge P(c)$ is the end-sequent of φ.

4.2 CERES in intuitionistic logic

For CERES in classical logic any resolution refutation of the characteristic clause set $\mathrm{CL}(\varphi)$ of a proof φ can serve as a skeleton for a CERES-normal form. In intuitionistic logic, however, this is impossible as the resulting CERES normal form may be classical. The example below [20] demonstrates that the result may be even *essentially* *classical*, in the sense that even by eliminating the cuts in the classical CERES-normal form we do not obtain an intuitionistic proof.

Example 7. Let φ be the following (propositional) **LJ**-proof:

[proof tree with sequents involving P, P^*, $\neg P$, $\neg\neg P \rightarrow P$, using rules $\neg l$, $\neg r$, $\vee_{r2}$, $\vee_l$, w_r, w_l, $\rightarrow_r$, and cut]

For φ in Example 7 we obtain

$$\mathrm{CL}(\varphi) = \{P \vdash P \ ; \ \vdash P \ ; \ P \vdash\}.$$

In most resolution refinements tautologies can be eliminated without loss of completeness. So we "ignore" the clause $P \vdash P$ and construct the following obvious resolution refutation $\gamma =$

$$\cfrac{\vdash P \quad P \vdash}{\vdash} R$$

The next step is to construct the projections $\varphi[\vdash P]$ and $\varphi[P \vdash]$. We will use so-called o-projections to facilitate the analysis. The only difference to the regular projections is the lack of a weakening of the left-side formula $P \vee \neg P$ in the end.

$$\cfrac{\cfrac{\cfrac{\cfrac{P \vdash P^*}{\vdash P^*, \neg P} \neg_r^\dagger}{\cfrac{\neg\neg P \vdash P^*}{\neg\neg P \vdash P^*, P} w_r} \neg_l}{\vdash P^*, \neg\neg P \to P} \to_r^\dagger \qquad \cfrac{\cfrac{P^* \vdash P}{P^*, \neg\neg P \vdash P} w_l}{P^* \vdash \neg\neg P \to P} \to_r$$

The inferences marked by † violate the restrictions of **LJ**. This makes it a classical non-intuitionistic derivation.

Putting the projections and the refutation γ together we obtain the following **CERES**-normal form:

$$\cfrac{\cfrac{\cfrac{\cfrac{\cfrac{P \vdash P^*}{\vdash P^*, \neg P} \neg_r^\dagger}{\cfrac{\neg\neg P \vdash P^*}{\neg\neg P \vdash P^*, P} w_r} \neg_l}{\vdash P^*, \neg\neg P \to P} \to_r^\dagger \quad \cfrac{\cfrac{P^* \vdash P}{P^*, \neg\neg P \vdash P} w_l}{P^* \vdash \neg\neg P \to P} \to_r}{\cfrac{\cfrac{\vdash \neg\neg P \to P, \neg\neg P \to P}{\vdash \neg\neg P \to P} c_r}{P \vee \neg P \vdash \neg\neg P \to P} w_l} \text{cut}$$

The proof is essentially classical as it contains a derivation of $\neg\neg P \to P$ which is not intuitionistically provable. If the full projections are used the situation is the same. The left formula $P \vee \neg P$ would be weakened in the projections and not used in the proof of $\neg\neg P \vdash P$ at all. We see that, after applying **CERES** based on the refutation γ, we got a proof with atomic cuts, but of a classical formula.

We have seen above that **CERES**, when applied to intuitionistic proofs, may yield **CERES**-normal forms which are classical. There are, however, subclasses of intuitionistic proofs where the usual **CERES** method works, provided we restrict the resolution calculus (i.e. we use resolution refinements) and apply a postprocessing procedure to the **CERES**-normal form. Indeed, if φ is a skolemized proof of a sequent $\Gamma \vdash$ (the succedent of the end-sequent is empty) and the resolution refutation belongs to the negative refinement of resolution then the **CERES** normal form (defined as in Definition 17 can be transformed into an intuitionistic proof.

Definition 18. An **LJ**-proof φ of S belongs to the class **LJ**$_-$ if S is skolemized and is of the form $\Gamma \vdash$.

Definition 19 (Negative resolution refinement). A resolution derivation is called *negative* if, in every application of the rule R, one of the clauses in the premise is negative and the only factoring rule is C_l applied to negative clauses, i.e. all rules are of the form:

$$\frac{\Gamma \vdash \Delta, A \quad B, \Gamma' \vdash}{\Gamma\sigma, \Gamma'\sigma \vdash \Delta\sigma} R \quad \frac{\Gamma, A, B \vdash}{\Gamma, A\sigma \vdash} C_l$$

where $\Gamma \vdash \Delta, A$ and $B, \Gamma' \vdash$ are variable disjoint and σ is a most general unifier of A and B. Negative resolution deductions are defined like in Definition 14.

Theorem 4. *The negative resolution refinement is complete.*

Proof. By Theorem 3.6.1. in [17] and by sign renaming. □

Example 8. The resolution refutation defined in Example 3 is a negative resolution refutation.

From Theorems 3 and 4 we conclude that there is always a negative resolution refutation of the clause set.

We have seen in Example 4 that, even for intuitionistic input proofs, the projections obtained from it might be classical. But, for proofs in **LJ**_, projections of negative clauses are always valid intuitionistic derivations.

Theorem 5 ([20]). *Let φ be an **LJ**-proof. Then the projections of negative clauses are valid **LJ** derivations.*

Proof. The projections are obtained by applying inferences from φ that operate on end-sequent ancestors. Since this is an **LJ**-proof, these are initially valid intuitionistic inferences. The only thing that changes on the projections' sequents (to which the inferences are applied) is the occurrence of extra atoms from the clause. Given the single conclusion restriction of **LJ**, the only time this is violated is when atoms occur on the right side of the sequent. As this is not the case for negative clauses, the rules in the projections of such clauses will be single conclusion and therefore the projection itself will be a valid **LJ** derivation. □

This procedure of obtaining a negative CERES normal form from an **LJ** proof φ is called *negative CERES* and we will denote the proof with atomic cuts obtained by $\hat{\varphi}$. The only modification of negative CERES over the CERES method is the enforcement of negative resolution.

Since we are using negative resolution and the end-sequent of φ is negative, every atomic cut in $\hat{\varphi}$ will have the shape:

$$\frac{\Gamma \vdash \Delta, A \quad A, \Gamma' \vdash}{\Gamma, \Gamma' \vdash \Delta} cut$$

Note that, since projections might be classical derivations, $\hat{\varphi}$ may also be a classical proof. Nevertheless it can be transformed again into an intuitionistic proof by removing the atomic cuts.

Since the original proof was in a single conclusion calculus, we know that every sequent with more than one formula on the right side must contain at most one end-sequent ancestor, the other formulas being atomic cut-ancestors. Therefore, if we can eliminate the atomic cuts maintaining always at most one end-sequent ancestor on the right side of every sequent, we will obtain an **LJ** proof. Now we show how to achieve this by insisting on a specific discipline for reductive cut-elimination.

Definition 20 (Left-shift cut-elimination [20]). Let $\hat{\varphi}$ be an **LJ**-proof with only atomic cuts. We call *left-shift cut-elimination* the process of removing the atomic cuts that, starting from the top most cuts down, (1) permutes the cut over all the rules of its left branch until reaching an axiom and (2) eliminates the cut by using the proof on its right branch. The permutation rules (1) are

$$\cfrac{\cfrac{(\varphi_1)}{\Gamma^* \vdash \Delta^*, A}\,\rho \quad (\varphi_2)}{\cfrac{\Gamma \vdash \Delta, A \quad \Gamma', A \vdash}{\Gamma, \Gamma' \vdash \Delta}\, cut} \Rightarrow \cfrac{\cfrac{(\varphi_1) \quad (\varphi_2)}{\Gamma^* \vdash \Delta^*, A \quad \Gamma', A \vdash}\, cut}{\cfrac{\Gamma^*, \Gamma' \vdash \Delta^*}{\Gamma, \Gamma' \vdash \Delta}\, \rho}$$

$$\cfrac{\cfrac{(\varphi_1) \quad (\varphi_2)}{\Gamma^* \vdash \Delta^*, A \quad \Gamma'^* \vdash \Delta'^*}\, \rho \quad (\varphi_3)}{\cfrac{\Gamma, \Gamma' \vdash \Delta, \Delta' A \quad A, \Gamma'' \vdash}{\Gamma, \Gamma', \Gamma'' \vdash \Delta, \Delta'}\, cut} \Rightarrow \cfrac{\cfrac{(\varphi_1) \quad (\varphi_3)}{\Gamma^* \vdash \Delta^*, A \quad A, \Gamma'' \vdash}\, cut \quad (\varphi_2)}{\cfrac{\Gamma^*, \Gamma'' \vdash \Delta^* \quad \Gamma'^* \vdash \Delta'^*}{\Gamma, \Gamma', \Gamma'' \vdash \Delta, \Delta'}\, \rho}$$

$$\cfrac{\cfrac{(\varphi_1) \quad (\varphi_2)}{\Gamma^* \vdash \Delta^* \quad \Gamma'^* \vdash \Delta'^*, A}\, \rho \quad (\varphi_3)}{\cfrac{\Gamma, \Gamma' \vdash \Delta, \Delta' A \quad A, \Gamma'' \vdash}{\Gamma, \Gamma', \Gamma'' \vdash \Delta, \Delta'}\, cut} \Rightarrow \cfrac{(\varphi_1) \quad \cfrac{(\varphi_2) \quad (\varphi_3)}{\Gamma'^* \vdash \Delta'^*, A \quad A, \Gamma'' \vdash}\, cut}{\cfrac{\Gamma^* \vdash \Delta^* \quad \Gamma'^*, \Gamma'' \vdash \Delta'^*}{\Gamma, \Gamma', \Gamma'' \vdash \Delta, \Delta'}\, \rho}$$

The elimination rule (2) is

$$\cfrac{A \vdash A \quad \Gamma', A \vdash}{\Gamma', A \vdash}\, cut \quad \cfrac{(\varphi)}{\Gamma', A \vdash}$$

Theorem 6 ([20]). *Let φ be a proof in **LJ**$_-$ and $\hat{\varphi}$ the negative CERES normal form obtained with negative CERES. Then eliminating the cuts from $\hat{\varphi}$ using left-shift cut-elimination yields an **LJ**-proof.*

Proof. Although φ is an **LJ** proof, each inference ρ in $\hat{\varphi}$ might be applied to a multiple conclusion sequent because of atomic cut-ancestors. By reductively eliminating the cuts, we make sure that the resulting sequents in the proof contain no atomic cut-ancestors on the right, but there is no guarantee that they will all be single conclusion. This can be ensured by two things: (1) φ is a proof of a negative sequent and (2) left-shift cut-elimination is used to eliminate the atomic cuts from $\hat{\varphi}$.

Let ρ be an inference in $\hat{\varphi}$ that was an instance of an inference in φ (which was originally applied to a single conclusion sequent). All the other inferences in $\hat{\varphi}$ will be eliminated after reductive cut-elimination. We have thus to show that after left-shift cut-elimination, every ρ will be applied to a single conclusion sequent.

First note that every inference ρ is applied to a sequent such that its right context contain at most one end-sequent ancestor, the other formulas being atomic cut ancestors. Now observe that, in the reduction rules of Definition 20, the ρ in the resulting derivation is always applied to a sequent whose right context contains *strictly less* formulas then in the original derivation. Moreover, these are all the rules necessary for eliminating the atomic cuts, as there is no right contraction of the cut-formulas because there is no right contraction in the negative resolution fragment. After eliminating all the cuts, every ρ will be applied to a sequent whose right context contains at most one end-sequent ancestor and no cut ancestors, exactly as it was in φ.

Second, upon actually eliminating the cut (see Definition 20), the derivation used is a negative projection which, by Theorem 5, is an **LJ** derivation.

The final proof is therefore a valid **LJ** proof. □

5 Transforming cut-free proofs into cut-free proofs

We have shown in Section 3 how to translate proofs in classical logic to proofs in intuitionistic logic. All these transformations were elementary. But if we have a classical cut-free proof φ of S its extended Glivenko translation $\Psi_G(\varphi)$ contains cuts. Clearly we can transform $\Psi_G(\varphi)$ into a cut-free intuitionistic proof ψ of $\Psi_G(S)$ using reductive cut-elimination. But the worst-case complexity of cut-elimination is nonelementary. A way to solve the problem could be to investigate the complexity of reductive cut-elimination in proofs of type $\Psi_G(\varphi)$. We choose a different approach by solving a more general proof transformation problem from **LK** to **LJ** by using the **CERES**-method and results from Section 4.2. The results in this section are based on [6], but the complexity analysis and the connection of the results with the Glivenko transformation are improved.

Below we show some complexity results about **CERES**.

Theorem 7. *There exists a nondeterministic Turing machine M and an elementary function h s.t., given a proof φ, a resolution refutation ϱ of $\mathrm{CL}(\varphi)$, M computes a*

CERES normal form $\hat{\varphi}$ of φ, s.t. $time_M(\varphi, \varrho) \leq h(\|\varphi\|, \|\varrho\|)$

Proof. We investigate the complexity of computing a CERES-normal form given the input proof φ and a resolution refutation ϱ of $\text{CL}(\varphi)$. Now let M be a nondeterministic Turing machine performing the following computation, given (φ, ϱ) as input:

(1) construct a resolution refutation ϱ of $\text{CL}(\varphi)$,

(2) compute a ground resolution refutation ϱ' of ϱ via substitution σ,

(3) instantiate the proof projections via σ,

(4) insert the instantiated proof projections into ϱ'.

Step (2) is computable in exponential time in $\|\varrho\|$ (computation of a global unifier). (3) can be performed in time $\leq \|\varphi\| r(\varrho')$ for any projection where $r(\varrho')$ is the size of a maximal term occurring in ϱ'; note that $r(\varrho') \leq \|\varrho'\|$. (4) can be computed in time $\leq \varrho' * p(\varphi, \varrho')$ where p is the maximal complexity of an instantiated projection. But $p(\varphi, \varrho') \leq \|\varphi\| r(\varrho')$. Putting things together we obtain an elementary function H s.t.
$$time_M(\varphi, \varrho) \leq H(\|\varphi\|, \|\varrho\|).$$
□

Theorem 8. *Let $\hat{\varphi}$ be a negative CERES normal form of an **LJ**- proof φ and let φ_0 be the cut-free **LJ**-proof obtained after applying left-shift cut-elimination to $\hat{\varphi}$. Then φ_0 can be computed in linear time.*

Proof. Given the transformations in Definition 20 (which are all the rules necessary for eliminating the atomic cuts), observe that the right-hand side uses only those derivations that were already present on the left, without duplicates. Thus left-shift cut-elimination does not increase the number of inference nodes in the proof. As the transformation rules (2) in Definition 20 even eliminates an inference, φ_0 contains less inferences than $\hat{\varphi}$, provided there is at least one cut in $\hat{\varphi}$. Still the rules in Definition 20 may mildly increase the symbolic size of a proof. Note that e.g. in the first rule we may have $\|\Gamma^*\| > \|\Gamma\|$ (ρ may be $\forall_l$ and a large term is eliminated (top-down) by the rule which now occurs twice in the result). But this increase happens for every rule ρ (coming from a left-hand-side of a cut) only once, and the material causing the increase is already present in the original proof. Therefore, there exists a constant c such that: $\|\varphi_0\| \leq c * \|\hat{\varphi}\|$ and φ_0 can be computed in linear time in $\|\varphi\|$.
□

Remark 7. The worst-case complexity for the elimination of atomic cuts is exponential in general. In fact, given a CERES normal form based on an arbitrary resolution refutation, cut-elimination may lead to an exponential increase in size. That cut-elimination is linear for negative CERES normal forms is due to the fact that there are no right contractions on the atoms of the cuts and proof duplication can be avoided. The price to pay is that negative CERES normal forms may be exponential in the minimal size of CERES normal forms.

Proposition 2. *Let φ be a proof of a skolemized sequent S. Then $\mathrm{CL}(\varphi)$ can be computed in exponential time.*

Proof. Consider the clause term $\Theta(\varphi)$ (for a definition see [5] and [1]). $\Theta(\varphi)$ is computable in linear time from φ. The evaluation of $\Theta(\varphi)$ to $\mathrm{CL}(\varphi)$ (which, basically is the computation of a conjunctive normal form from a negation normal form) can be done in exponential time. □

In [5] a comparison of reductive cut-elimination and CERES was given. It turned out that reductive cut-elimination is, in some sense, redundant w.r.t. CERES. The measure of redundancy is the well known subsumption principle from automated deduction.

Let Γ be a multiset of formulas; by $\mathrm{set}(\Gamma)$ we describe the set defined by the elements in Γ.

Definition 21 (subsumption). Let $C\colon \Gamma \vdash \Delta$ and $D\colon \Pi \vdash \Lambda$ be clauses. We define $C \subseteq D$ if $\mathrm{set}(\Gamma) \subseteq \mathrm{set}(\Pi)$ and $\mathrm{set}(\Delta) \subseteq \mathrm{set}(\Lambda)$. We define $C \leq_{ss} D$ if there exists a substitution ϑ s.t. $C\vartheta \subseteq D$. Let $\mathcal{C}, \mathcal{D}$ be sets of clauses; then $\mathcal{C} \leq_{ss} \mathcal{D}$ if for every clause $D \in \mathcal{D}$ there exists a $C \in \mathcal{C}$ s.t. $C \leq_{ss} D$.

Proposition 3. *Let φ be a proof of a skolemized sequent S and ψ be a proof obtained from φ via (one or more) cut-elimination steps of Gentzen's reductive method (without eliminating atomic cuts). Then $\mathrm{CL}(\varphi) \leq_{ss} \mathrm{CL}(\psi)$.*

Proof. In [1, 5]. □

The subsumption principle can be extended from sets of clauses to resolution deductions: let us assume that $\mathcal{C}$ and $\mathcal{D}$ are sets of clauses s.t. $\mathcal{C} \leq_{ss} \mathcal{D}$, $\mathcal{D}$ is unsatisfiable and δ a resolution refutation of $\mathcal{D}$. Then there exists a resolution refutation γ of $\mathcal{C}$ which "subsumes" δ. γ is in fact smaller than δ, i.e. $\|\gamma\| \leq \|\delta\|$. For a formal definition of subsumption among resolution derivations see Definition 6.6.4 in [1]. The subsumption property of resolution refutations will be used in the proof of Theorem 9.

We define the classical analogue to the class **LJ_**:

Definition 22. An **LK**-proof φ of S belongs to the class $\mathbf{LK_-}$ if S is skolemized and is of the form $\Gamma \vdash$.

Our aim is to construct an elementary translation from $\mathbf{LK_-}$ to $\mathbf{LJ_-}$ in preserving the end-sequent.

Definition 23. A proof transformation Φ is called *end-sequent preserving* if for all proofs φ of a sequent S $\Phi(\varphi)$ is a proof of S.

Theorem 9. *There exists an elementary end-sequent preserving proof transformation Φ from cut-free proofs in $\mathbf{LK_-}$ to cut-free proofs in $\mathbf{LJ_-}$. That means there exists an elementary function g and a nondeterministic Turing machine M computing Φ s.t. for all cut-free proofs φ in $\mathbf{LK_-}$ we have $\text{time}_M(\varphi) \leq g(\|\varphi\|)$.*

Proof. Let us consider a cut-free (classical) proof φ of the skolemized sequent $S\colon A_1, \ldots, A_n \vdash$ (i.e. there are no strong quantifiers in the A_i).

Now consider our proof transformation T for $T(\varphi) =$

$$\dfrac{\dfrac{\dfrac{\dfrac{(\varphi)}{A_1,\ldots,A_n \vdash} \wedge\colon l^*}{A \vdash}}{\vdash \neg A} \neg\colon r \qquad \dfrac{(\psi)}{\neg A, A_1, \ldots, A_n \vdash}}{A_1, \ldots, A_n \vdash} \text{cut}$$

where $A = A_1 \wedge \cdots \wedge A_n$ and ψ is a cut-free intuitionistic proof of length polynomial in $\|A\|$. $T(\varphi)$ proves the same end-sequent as φ and can be constructed in time polynomial in $\|\varphi\|$. Now observe that the cut-formula $\neg A$ on the left branch of the cut has only weak quantifiers, and only strong quantifiers on the right branch. Now we apply reductive cut-elimination to $T(\varphi)$, eliminate the quantifiers in the cuts (i.e. we break down the proof to propositional cuts) and obtain a proof χ; this transformation can be done in double exponential time - here we are doing cut-elimination in *classical logic*! Putting things together χ can be computed from φ in time $t(\|\varphi\|)$ for an elementary function t.

Now negative CERES comes into play: consider the characteristic clause set $\text{CL}(T(\varphi))$ and let $\mathcal{C}'$ be the characteristic clause set of χ. Note that $\mathcal{C}'$ is a set of ground clauses (indeed we may assume that in a proof containing no strong quantifiers only ground terms are introduced by the quantifier rules). As $\|\chi\| \leq t(\|\varphi\|)$ $\mathcal{C}'$ can be computed within $t'(\|\varphi\|)\colon c^{t(\|\varphi\|)}$ steps for a constant c by Proposition 2. As $\mathcal{C}'$ is ground, the computation of a *shortest* negative resolution refutation ϱ' can be done in nondeterministic exponential time in $\|\mathcal{C}'\|$ i.e. within $t''(\|\varphi\|)\colon d^{t'(\|\varphi\|)}$ for some constant d (note that the number of different negative clauses definable over the ground atoms is at most exponential). By Proposition 3 there exists a resolution refutation ϱ of $\mathrm{CL}(\psi)$ s.t. $\varrho \leq_{ss} \varrho'$ and, by definition of subsumption, $\|\varrho\| \leq \|\varrho'\|$. Moreover, ϱ is also a negative resolution refutation as negative clauses can only be subsumed by negative clauses or by the empty clause. Clearly also ϱ can be computed within nondeterministic exponential time $t''(\|\varphi\|)$.

So we refute $\mathrm{CL}(T(\varphi))$ with ϱ and get a CERES-normal form φ^*. By Theorem 7 we can compute φ^* in time $\leq h(\|T(\varphi)\|, \|\varrho\|)$ for an elementary function h and, as $\|\varrho\| \leq t''(\|\varphi\|)$ and $T(\varphi)$ is polynomial in $\|\varphi\|$ there exists an elementary function g' s.t. φ^* can be computed in time $\leq g'(\|\varphi\|)$.

As ϱ' is the shortest negative resolution refutation of $\mathcal{C}'$ there are no tautological clauses occurring in ϱ' (note that a shortest negative resolution refutation never contains tautologies!). As a consequence also ϱ does not contain tautological clauses. Now consider the proof $T(\varphi)$. As all inferences in φ (within $T(\varphi)$) go into the cut formula $\neg A$, the clauses of the characteristic clause sets coming from φ are all tautologies. But these tautologies are not used in ϱ! It follows that all projections used in the CERES normal form φ^* come from the intuitionistic part of the proof. But note that, in this case, φ^* can be transformed into an intuitionistic cut-free proof ψ via the method described in Theorem 6 in linear time. This is the last step of the transformation which gives a proof in $\mathbf{LJ}_-$ and, putting things together, the whole transformation can be done in elementary time g. $\square$

We illustrate the transformation of Theorem 9 with an example. Let φ be the **LK** proof:

$$
\cfrac{\cfrac{\cfrac{\cfrac{\cfrac{\cfrac{Pfa \vdash Pfa}{Pa, Pfa \vdash Pfa, Pffa} w}{Pa \vdash Pfa, Pfa \to Pffa} \to_r}{\vdash Pa \to Pfa, Pfa \to Pffa} \to_r}{\vdash \exists x.(Px \to Pfx), \exists x.(Px \to Pfx)} \exists_r}{\vdash \exists x.(Px \to Pfx)} c_r}{\neg \exists x.(Px \to Pfx) \vdash} \neg_l
$$

Then we can construct $\Xi = T(\varphi)$, which proves the same end-sequent but has a full intuitionistic proof on the right branch of the cut:

$$
\cfrac{
  \cfrac{\varphi}{\neg\exists x.(Px \to Pfx) \vdash}
  \quad
  \cfrac{
    \cfrac{
      \cfrac{
        \cfrac{
          \cfrac{
            \cfrac{
              \cfrac{Pa \vdash Pa \quad Pf\alpha \vdash Pf\alpha}{Pa \to Pf\alpha, Pa \vdash Pf\alpha} \to_l
            }{Pa \to Pf\alpha \vdash Pa \to Pf\alpha} \to_r
          }{Pa \to Pf\alpha \vdash \exists x.(Px \to Pfx)} \exists_r
        }{\exists x.(Px \to Pfx) \vdash \exists x.(Px \to Pfx)} \exists_l
      }{\neg\exists x.(Px \to Pfx), \exists x.(Px \to Pfx) \vdash} \neg_l
    }{\neg\exists x.(Px \to Pfx) \vdash \neg\exists x.(Px \to Pfx)} \neg_r
  }{\neg\exists x.(Px \to Pfx), \neg\neg\exists x.(Px \to Pfx) \vdash} \neg_l \quad
  \cfrac{}{\vdash \neg\neg\exists x.(Px \to Pfx)} \neg_r
}{\neg\exists x.(Px \to Pfx) \vdash} \text{cut}
$$

We apply the negative CERES method to this proof. The clause set extracted is the following:

$$CL(\Xi) = \{Pfa \vdash Pfa \;;\; \vdash P\alpha \;;\; Pf\alpha \vdash\}$$

Note that the tautological clause $Pfa \vdash Pfa$, which came from the classical part of Ξ can be eliminated. The only possible (negative) refutation is ϱ:

$$
\cfrac{
  \cfrac{\vdash P\alpha}{\vdash Pfa} \alpha \leftarrow fa \quad
  \cfrac{Pf\alpha \vdash}{Pfa \vdash} \alpha \leftarrow a
}{\vdash} R
$$

Since ϱ uses clauses that come from the intuitionistic side of Ξ, these are the only projections we need:

$$\pi(\vdash P\alpha):$$

$$
\cfrac{
  \cfrac{
    \cfrac{
      \cfrac{
        \cfrac{Pa \vdash Pa}{Pa \vdash Pa, Pf\alpha} w_r
      }{\vdash Pa, Pa \to Pf\alpha} \to_r
    }{\vdash Pa, \exists x.(Px \to Pfx)} \exists_r
  }{\neg\exists x.(Px \to Pfx) \vdash Pa} \neg_l
}{}
$$

$$\pi(Pf\alpha \vdash):$$

$$
\cfrac{
  \cfrac{
    \cfrac{
      \cfrac{
        \cfrac{Pf\alpha \vdash Pf\alpha}{Pf\alpha, Pa \vdash Pf\alpha} w_l
      }{Pf\alpha \vdash Pa \to Pf\alpha} \to_r
    }{Pf\alpha \vdash \exists x.(Px \to Pfx)} \exists_r
  }{Pf\alpha, \neg\exists x.(Px \to Pfx) \vdash} \neg_l
}{}
$$

Note that the projection of the negative clause is intuitionistic, but the other one is classical. Then we can compute the CERES normal form $\hat{\varphi}$:

$$
\cfrac{
  \cfrac{
    \cfrac{
      \cfrac{
        \cfrac{
          \cfrac{Pfa \vdash Pfa}{Pfa \vdash Pfa, Pffa} w_r
        }{\vdash Pfa, Pfa \to Pffa} \to_r
      }{\vdash Pfa, \exists x.(Px \to Pfx)} \exists_r
    }{\neg \exists x.(Px \to Pfx) \vdash Pfa} \neg_l
    \qquad
    \cfrac{
      \cfrac{
        \cfrac{
          \cfrac{
            \cfrac{Pfa \vdash Pfa}{Pfa, Pa \vdash Pfa} w_l
          }{Pfa \vdash Pa \to Pfa} \to_r
        }{Pfa \vdash \exists x.(Px \to Pfx)} \exists_r
      }{Pfa, \neg\exists x.(Px \to Pfx) \vdash} \neg_l
    }{}
  }{\neg\exists x.(Px \to Pfx), \neg\exists x.(Px \to Pfx) \vdash} cut
}{\neg\exists x.(Px \to Pfx) \vdash} c_l
$$

By performing left-shift cut-elimination, we obtain the **LJ** proof ψ:

$$
\cfrac{
  \cfrac{
    \cfrac{
      \cfrac{
        \cfrac{
          \cfrac{
            \cfrac{
              \cfrac{
                \cfrac{
                  \cfrac{Pfa \vdash Pfa}{Pfa, Pa \vdash Pfa} w_l
                }{Pfa \vdash Pa \to Pfa} \to_r
              }{Pfa \vdash \exists x.(Px \to Pfx)} \exists_r
            }{\neg\exists x.(Px \to Pfx), Pfa \vdash} \neg_l
          }{\neg\exists x.(Px \to Pfx), Pfa \vdash Pffa} w_r
        }{\neg\exists x.(Px \to Pfx) \vdash Pfa \to Pffa} \to_r
      }{\neg\exists x.(Px \to Pfx) \vdash \exists x.(Px \to Pfx)} \exists_r
    }{\neg\exists x.(Px \to Pfx), \neg\exists x.(Px \to Pfx) \vdash} \neg_l
  }{\neg\exists x.(Px \to Pfx) \vdash} c_l
}{}
$$

Now we can apply Theorem 9 to the extended Glivenko translation.

Corollary 1. *There exists a proof transformation T with the following properties:*

(1) *For any cut-free **LK**-proof φ of a sequent $\vdash A$ (where A is a formula without strong quantifiers) $T(\varphi)$ is a cut-free **LJ**- proof of $\vdash \psi_G(A)$,*

(2) *T can be computed in elementary time.*

Proof. We extend φ by a $\neg\colon l$ rule and obtain a cut-free proof φ' of $\neg A \vdash$. By Theorem 9 there exists a intuitionistic cut-free proof ψ' of $\neg A \vdash$ which can be computed in elementary time (in $\|\varphi'\|$). We obtain an intuitionistic cut-free proof χ of $\neg\neg A$ just by appending $\neg\colon r$ to ψ'. As A does not contain strong quantifiers we have $\psi_G(A) = \neg\neg A$ and so χ is a proof of $\vdash \psi_G(A)$. Obviously χ can be constructed in elementary time (in $\|\varphi\|$). □

6 Conclusion

We have analyzed the complexity of proof tranlations defined via the formula transformations of Kolmogorov and Glivenko. We proved using **CERES** that, for the Glivenko translation, cut-free **LK**-proofs of S without strong quantifiers can be translated into cut-free **LJ**-proofs of $\Psi_G(S)$ in elementary time. It remains an open question whether the same result could be obtained by reductive cut-elimination of the translations $\Psi_G(\varphi)$ within **LJ** and, even, whether reductive cut-elimination in this class is elementary at all. We did not investigate proof translations based on the Gödel-Gentzen translation and the question whether elementary translations of cut-free proofs into cut-free proofs via this translations exist. Also a methodological comparison of **CERES** and reductive cut-elimination methods on proof classes defined by negative translations is left to future research.

References

[1] M. Baaz and A. Leitsch. *Methods of Cut-Elimination*. Trends in Logic. Springer, 2011.

[2] M. Baaz and A. Leitsch. On skolemization and proof complexity. *Fundamenta Informaticae*, 20(4):353–379, December 1994.

[3] M. Baaz and A. Leitsch. Cut normal forms and proof complexity. *Annals of Pure and Applied Logic*, 97(1-3):127–177, 1999.

[4] M. Baaz and A. Leitsch. Cut-elimination and redundancy-elimination by resolution. *Journal of Symbolic Computation*, 29(2):149–176, 2000.

[5] M. Baaz and A. Leitsch. Towards a clausal analysis of cut-elimination. *J. Symb. Comput.*, 41(3-4):381–410, 2006.

[6] M. Baaz, A. Leitsch, and G. Reis. A note on the complexity of classical and intuitionistic proofs. In *30th Annual ACM/IEEE Symposium on Logic in Computer Science, LICS*, pages 657–666, 2015.

[7] M. Boudard and O. Hermant. Polarizing double-negation translations. In *LPAR-19*, volume 8312 of *LNCS*, pages 182–197, 2013.

[8] G. Ferreira and P. Oliva. On various negative translations. In *Proceedings Third International Workshop on Classical Logic and Computation, CL&C 2010, Brno, Czech Republic, 21-22 August 2010.*, pages 21–33, 2010.

[9] G. Gentzen. Untersuchungen über das logische Schließen I. *Mathematische Zeitschrift*, 39(1):176–210, dec 1935.

[10] F. Gilbert. A lightweight double-negation translation. In *LPAR-20*, volume XXX of *EPiC Series in Computer Science*, pages 1–13, 2015.

[11] V. Glivenko. Sur quelques points de la logique de M. Brouwer. *Bull. Acad. Royale de Belgique*, 15:183–188, 1929.

[12] K. Gödel. Zur intuitionistischen Arithmetik und Zahlentheorie. *Ergebnisse eines Mathematischen Kolloquiums*, 4:34–38, 1933.

[13] H. Ishihara. A note on the Gödel-Gentzen translation. *Mathematical Logic Quaterly*, 46(1):135–137, 2000.

[14] U. Kohlenbach. *Applied Proof Theory - Proof Interpretations and their Use in Mathematics*. Springer Monographs in Mathematics. Springer, 2008.

[15] A. N. Kolmogorov. On the principle of TERTIUM NON DATUR. *Mathematicheskii Shornik*, 32:646–667, 1924/1925.

[16] S. Kuroda. Intuitionistische Untersuchungen der formalistischen Logik. *Nagoya Mathematical Journal*, 3:35–47, 1951.

[17] A. Leitsch. *The Resolution Calculus*. Texts in Theoretical Computer Science. An EATCS Series. Springer, 1997.

[18] G. Mints. *A short introduction to intuitionistic logic*. Kluwer Academic Publishers, 2nd edition, 2002.

[19] V. P. Orevkov. Lower bounds for increasing complexity of derivations after cut elimination. *Zapiski Nauchnykh Seminarov Leningradskogo Otdeleniya Matematicheskogo Instituta*, 88:137–161, 1979.

[20] G. Reis. *Cut-elimination by Resolution in Intuitionistic Logic*. PhD thesis, Vienna University of Technology, 2014.

[21] R. Statman. Lower bounds on Herbrand's theorem. *Proceedings of the American Mathematical Society*, 75(1):104–107, 1979.

Unification for Multi-Agent Temporal Logics with Universal Modality

Stepan I. Bashmakov
Institute of mathematics and computer science, Siberian Federal University
79, pr. Svobodny, Krasnoyarsk 660041, Russia
krauder@mail.ru

Anna V. Kosheleva
Institute of space and informatics technologies, Siberian Federal University
26, ul. Kirenskogo, ULK building, Krasnoyarsk 660074, Russia
koshelevaa@mail.ru

Vladimir V. Rybakov
Department of Computing and Mathematics, Manchester Metropolitan University
John Dalton Building, Chester Street, Manchester M1 5GD, UK
vladimir_rybakov@mail.ru

Abstract

We investigate the unification problem for all logics with expressible universal modality. The main results are syntactic conditions for formulas to be not unifiable and theorems describing bases for inference rules passive in such logics. Then we apply these results to various logics, in particular to linear temporal logics with time states with agents logical operations, and even to some branching time logics with multi-agents logical operations.

Keywords: Unification, Multi-Agent Logics, Modal Logic, Temporal Logic, Passive Inference Rules.

We dedicate our paper to Grigori Mints, for his profound contributions to proof theory in general and also to the theory of non-classical logics, in particular. V. Rybakov knew Grigori from the beginning of 80s, when he met Grigori on several occasions, such as conferences in the Soviet Union (USSR), when they were both still working there. During the 1995/1996 academic year V. Rybakov visited Grigori as Fulbright Senior Scholar, (funded by Fulbright Foundation, from Washington DC. USA) at Stanford University and they worked together on dynamic logical systems. He remembers his collaboration with Grigori with the greatest pleasure.

1 Introduction

Logical systems modeling reasoning and multi-agent environments, computation truth values and vote taking decisions are popular areas nowadays in Information Sciences and Knowledge Representation (cf. e.g. M. Kracht [28], Francesco Belardinelli, Alessio Lomuscio [9], W. van der Hoek and M. Wooldridge [44]). In this paper we would like to contribute to these areas with our recent results concerning the syntactic description of unification.

Unification is one of the important tools in automated deduction; as a concept it was originated in Computer Science (cf. F. Baader and W. Snyder [5]). Later the concept of unifiability was applied and studied for various non-classical logics. For example, the problem of unification in intuitionistic logic and in propositional modal logics over $K4$ was investigated by S. Ghilardi using the technique of projective formulas, – [17, 19, 16, 18, 20] (this is an application of ideas from the field of projective algebras). In these papers the problem of constructing finite complete sets of unifiers was solved for the logics considered and efficient algorithms were found. Unification in the field of Computer Science appeared initially in the form of the possibility of transforming two different terms into syntactically equivalent ones (by the replacing its variables, cf. [31, 27]), that eventually changed course to the study of semantic equivalence (cf. Baader et al. [5, 1]). For the majority of non-classical logics (modal, intuitionistic, temporal, etc.) there are special dual equational theories of algebraic systems, so their unification problems are interpreted into the corresponding logic-unificational counterparts ([3, 4, 2]). The unification problem can be generalized to a more difficult question: whether the formula can be converted into a theorem after replacing only some of the variables (keeping the rest, as a set of parameters, intact). This problem has been studied and solved for some modal and intuitionistic logics (cf. e.g. V. Rybakov [32, 33, 34] for the case of intuitionistic logic itself and modal logics S4 and Grz).

The approach based on the ideas of projective formulas proved to be useful and effective in dealing with admissibility and the basis of admissible rules (cf. Jerabek [24, 25, 26], Iemhoff, Metcalfe [22, 23]). If algorithms for the construction of computable finite sets of unifiers are found, it directly gives a solution of the admissibility problem.

Temporal logic is also a very dynamic area of mathematical logic and computer science (cf. Gabbay and Hodkinson [15, 13, 14]). In particular, LTL (linear temporal logic) has significant applications in the field of Computer Science (cf. Manna, Pnueli [29, 30], Vardi [46, 45]). The solution to the problem of admissibility for rules in LTL was found by Rybakov [36] (cf. also [35]), the basis of admissible rules in LTL was constructed by Babenyshev and Rybakov in [6] (and for the case without the operator Until – in [35]).

The solution of the unification problem for formulas with coefficients in LTL has been found by Rybakov [37, 40] and its analogs were also re-settled for basic modal and intuitionistic logic in [38, 39]. In particular, in [37], it was proved that not all formulas unifiable in LTL are projective, and [40] proved the projectivity of any unified formulas in LTL_u (to recall, LTL_u is a fragment of LTL, with only the operator Until, no NEXT). In the paper of Dzik and Wojtylak [11] the same result was obtained for the modal linear logic $S4.3$.

In [41], V. Rybakov found a description of all non-unifiable formulas in a broad class of modal logics: in the extensions of S4 and $[K4 + \Box\bot \equiv \bot \in \mathcal{L}]$ and also constructed finite bases for rules which are passive in these logics. Using results from [7], following closely this technique, in our paper [8] we find a criterion for non-unifiability of formulas in the linear temporal logic of knowledge with multi-agent relations – $LFPK$, and construct a basis for inference rules which are passive in this logic.

So, we obtained theorems syntactically describing non-unifiable formulas and basis for passive inference rules in linear temporal logic with multi-agent logical operations in time-point states (cf. [8]). Verifying and analyzing our proofs we recently observed that the results might be transferred to a wide class of logics - all logics where the universal modality might be modeled by any possible terms composed from native logical operations (recall that the universal modality, first investigated in Goranko and Passy [21], is regarded nowadays as a standard constructor in modal logic; see, e.g., Blackburn et al. [10]). As a result, actually, all schemes of proofs from [8] may be transferred to this more general case.

The main results of this our paper are Theorems 2 and 3 describing syntactic conditions for formulas to be not unifiable and bases for passive inference rules for all logics with an expressible universal modality. In the final sections we apply these theorems to various logics, in particular to linear temporal logics with time states with agents logical operations, and even to some branching time logics.

2 Definitions, notation, logics with universal modality

We first recall relational semantics for modal and temporal logics. A relational frame (n-frame) F is a tuple $\langle W, R_1, \ldots, R_n \rangle$, where W is a non-empty set (the base), and for all i, $R_i \subseteq W \times W$. We use the notation $|F|$ for the set W; and $a \in F$ will abbreviate $a \in |F|$. A frame F is said to be *rooted* if there exists $a \in |F|$ such that, for any $b \neq a$ in $|F|$, there are $a_1, \ldots, a_{m-1}, a_m \in |F|$ such that $a R_{i_1} a_1 R_{i_2} a_2 \ldots R_{i_{m-1}} a_{m-1} R_{i_m} a_m = b$, where R_{i_j} are some accessibility relations from F. A *valuation* V of a set P of proposition letters in F is a mapping $V : P \to 2^{|F|}$, i.e. $V(q) \subseteq |F|$ for any $q \in P$. An n-frame F together with a valuation V of some set of letters P is called a *(Kripke – Hintikka) relational model* (based on F). The notation $(F, a) \Vdash_V q$ means $a \in V(q)$; if $(F, a) \Vdash_V q$ we say that the letter q is *true at* a with respect to V.

The language of multi-modal propositional logics consists of a countable set of proposition letters (denoted by Latin letters, possibly with subscripts), Boolean logical operations, and a finite set of unary modal operations $\Box_i$, $1 \leq i \leq n$. The formation rules for formulas are standard. The formula $\Diamond_i \alpha$ is the abbreviation for the formula $\neg \Box_i \neg \alpha$. A *multi-modal logic* (or, to be more precise, an n-*modal logic*) is a set $\mathcal{L}$ of formulas containing all classical tautologies, the axioms $\Box_i(p \to q) \to (\Box_i p \to \Box_i q)$ for all i, and closed under substitutions, Modus Ponens, and the rules of necessitation: for all formulas A if $A \in \mathcal{L}$ then $\Box_i A \in \mathcal{L}$ as well (for every i). In the sequel multi-modal logics are called just "logics".

For any model with a valuation V, the *truth relation* with respect to V is extended to all Boolean formulas built from the set of letters P in a standard way. Computation of the truth values for modal operations are as follows:

$$(F, a) \Vdash_V \Box_j A \leftrightarrow \forall b \in |F|(a R_j b \Rightarrow (F, b) \Vdash_V A)).$$

For a frame F and a formula A we write $F \Vdash A$ if for any valuation V on F and any $a \in |F|$ $(F, a) \Vdash_V A$ holds.

For any class of frames K, $L(K) := \{A \mid \forall F \in K, F \Vdash V\}$ is the multi-modal logic generated by the class K. The majority of popular modal logics coincide with $L(K)$ for some K. Such logics are said to be Kripke complete; there are some modal logics which are not Kripke complete, though they are sophisticate examples constructed in order to disprove the conjecture of Kripke completeness for all logics (cf. K. Fine, S. K. Thomason, J. van Benthem [12, 42, 43].

Temporal logics are similar to modal logics, but with the assumption that one of the accessibility relations from the frames F generating these logics, e.g. R_1, is responsible for modeling the passing of time. In this case two modalities for R_1 are reserved, one – is the $\Box_1$ itself, and another one $\Box_1^{-1}$ where the second one is based on the relation R_1^{-1}, the converse to R_1. The first relation is referred as always in the future, and the second one - as always in the past (and they are usually denoted as $\Box_F$ and $\Box_P$).

We may extend multi-modal logics and temporal (single modal or multi-modal) ones to logics possessing a universal modality as follows. Assume that the language of a logic is extended by a new modal operation $\Box_U$, and that the rule for computation for truth values of formulas with applications of $\Box_U$ is as follows:

$$\forall a, (F,a) \Vdash_V \Box_U A \leftrightarrow [\forall b \in |F|(F,b) \Vdash_U A].$$

In other words, $\Box_U A$ says that the formula A is true always and everywhere, so it acts as a universal quantifier and is therefore called the universal modality.

Definition 1. A logic $\mathcal{L}$ is said to be one with universal modality if its language contains the modality $\Box_U$, and there is a class of frames K, such that $\mathcal{L} = L(K)$.

Now we recall some definitions and already known results related to the notion of unification and to "passive inference rules", as they first appeared in [41] and then in [7].

Definition 2. A formula $A(p_1, \ldots, p_n)$ is *unifiable* in a logic $\mathcal{L}$ iff there is a tuple of formulas $B_1, \ldots, B_n$ such that $A(B_1, \ldots, B_n) \in \mathcal{L}$, $(B_1, \ldots, B_n)$ is said to be its unifier).

Definition 3. Some formulas $A(p_1, \ldots, p_n)$ and $B(p_1, \ldots, p_n)$ are said to be *unifiable* in a logic $\mathcal{L}$ iff the formula

$$[A(p_1,\ldots,p_n) \to B(p_1,\ldots,p_n)] \wedge [(B(p_1,\ldots,p_n) \to A(p_1,\ldots,p_n))]$$

(the latter formula is usually abbreviated by $A(p_1,\ldots,p_n) \equiv B(p_1,\ldots,p_n)$) is unifiable in $\mathcal{L}$ and the corresponding unifier $B_1,\ldots,B_n$ is said to be the unifier for formulas A and B.

We consider below only Kripke complete logics $\mathcal{L}$ (that is $\mathcal{L} = L(K)$ for some class of frames K) with the property that $\neg\Box_j\bot \in \mathcal{L}$, for all j ($\bot = p \wedge \neg p$, that is, any of its frames do not have maximal R_j irreflexive worlds (and minimal irreflexive worlds, – for temporal logics)). The restriction to only Kripke-complete logics follows from the proof technique below, where it is indeed necessary. The proof technique only works for such logics. The property $\neg\Box_j\bot \in \mathcal{L}$ is necessary for our proofs as well (for the inductive steps in the proofs would work).

In this case, if a logic is decidable, generally speaking, it is an easy task to verify whether a formula or two formulas are unifiable in this logic. It is immediate to see that it is sufficient to look for unifiers among formulas $\top$ and $\bot$, cf. e.g.

Corollary 1 (Corollary 2.7 from [41]). *For all superintuitionistic logics and modal logics extending logics $S4$ or $K4 + \Box\bot \equiv \bot$, unifiers for unifiable formulas can be effectively computed; if they (unifiers) exist then some substitution replacing letters by formulas $\top$ or $\bot$ will be a unifier.*

The presence of the equivalence $\Box\bot \equiv \bot$ above is essential too, since otherwise a possibly increasing sequence of operations $\Box$ applied to $\bot$ will not have a clear visible computable bound.

This corollary is rather evident, however if we wish to characterize all formulas which are not unifiable, to obtain a general mathematical theorem describing all such formulas, it is not so immediate. The following result was known:

Theorem 1 (2.10 from [41]). *For any modal logic $\mathcal{L}$ extending $S4$ and any modal formula α, α is not unifiable in $\mathcal{L}$ iff the formula*

$$\Box\alpha \to \left[\bigvee_{p \in Var(\alpha)} \Diamond p \wedge \Diamond\neg p \right]$$

is provable in $\mathcal{L}$ (that is this formula belongs to $\mathcal{L}$, as to the set of its theorems).

Though such logics do not have universal modality. Recall that an inference rule is an expression $r := A_1, \ldots, A_n/B$ where B and all A_i are some formulas in the language of a certain logic. The letters of these formulas are called its variables.

We aim to characterize the inference rules which are passive, those whose premises are not unifiable. Recall that:

Definition 4. Let $r := A_1, \ldots, A_n/B$ be an inference rule, r is said to be passive for a logic L if for any substitution g of formulas instead of variables in r, we never have $g(A_1) \in L \& \ldots \& g(A_n) \in L$. In other words, r is a passive rule if the formulas from its premise do not have common unifiers.

We would like to characterize such rules in a syntactic way, to find some bases for them. That is we actually wish to describe formulas which are not unifiable pure syntactically, in a sense – to axiomatize them.

Definition 5. For any given rule $r := A_1, \ldots, A_n/B$, r is a consequence of a sequence of rules $r_1 := A_1/B_1, \ldots, r_n := A_n/B_n$ in a logic $\mathcal{L}$ if there is a derivation in $\mathcal{L}$ of the conclusion B from the premises of the rule r, as a hypothesis, by means of rules from $r_1, \ldots, r_n$, theorems of $\mathcal{L}$ and postulated rules of $\mathcal{L}$ (e.g. modus ponens (for classical propositional logic or the intuitionistic logics) and Goedel necessitation rule $A/\Box_i A$ for modal logics).

Definition 6. A set of rules B_r is a basis for a set of rules S_r in a logic $\mathcal{L}$ if any rule $r \in S_r$ is a consequence of some rules from B_r in $\mathcal{L}$.

3 A criterion of non-unifiability

Let $\mathcal{L}$ be a Kripke complete logic (that is $\mathcal{L} = L(K)$ for some class of frames K) with the property $\neg \Box_j \bot \in \mathcal{L}$, for all j (that is all frames $F \in K$ do not have maximal irreflexive w.r.t. any R_j worlds (and minimal irreflexive – for the case of temporal logics)). Let $\mathcal{L}$ has the universal modality $\Box_U$.

Theorem 2. *A formula A is non-unifiable in $\mathcal{L}$ iff the formula*

$$\Box_U A \to \left[\bigvee_{p \in Var(A)} \Diamond_U p \wedge \Diamond_U \neg p \right]$$

is a theorem in $\mathcal{L}$.

Proof. The proof will go by reduction to absurdum. Assume that

$$\Box_U A \to \left[\bigvee_{p \in Var(A)} \Diamond_U p \wedge \Diamond_U \neg p \right] \in \mathcal{L}$$

but, at the same time, the formula A is unifiable in $\mathcal{L}$.

Then by definition of the unifier, there is a substitution g such that $g(A) \in \mathcal{L}$. Because $\mathcal{L}$ is closed under substitutions, we obtain

$$g\left(\Box_U A \to \left[\bigvee_{p \in Var(A)} \Diamond_U p \wedge \Diamond_U \neg p \right]\right) \in \mathcal{L}.$$

We know that $\mathcal{L} = L(K)$ for some class of frames K. Take an arbitrary frame $F \in K$ for $\mathcal{L}$. Consider the valuation V for all letters q of formulas $g(p)$, where $p \in Var(A)$, on the F, where $V(q) = \emptyset$. It is easy to show by induction on the length of formulas B constructed out of letters q that:

$$\forall b \in F, \forall c \in F : b \Vdash_V B \Leftrightarrow c \Vdash_V B.$$

The inductive step for operations $\Box_j$ follows from our assumption that $\neg \Box_j \bot \in \mathcal{L}$. Consequently,

$$\forall b \in F : b \not\Vdash_V \bigvee_{p \in Var(A)} \Diamond_U g(p) \wedge \Diamond_U g(\neg p).$$

At the same time,
$$\forall b \in F : b \Vdash g(\Box_U A)$$

since $g(A) \in \mathcal{L}$. Thereby,

$$\forall b \in F : b \not\Vdash_V g(\Box_U A \to \left[\bigvee_{p \in Var(A)} \Diamond_U p \wedge \Diamond_U \neg p\right])$$

which contradicts the hypothesis:

$$g(\Box_U A \to \left[\bigvee_{p \in Var(A)} \Diamond_U p \wedge \Diamond_U \neg p\right]) \in \mathcal{L}.$$

In the opposite direction, assume that the formula A is non-unifiable in $\mathcal{L}$, but at the same time

$$\Box_U A \to \left[\bigvee_{p \in Var(A)} \Diamond_U p \wedge \Diamond_U \neg p\right] \notin \mathcal{L}$$

Then there is a certain $\mathcal{L}$-frame $F \in K$, that disproves this formula:

$$\exists a \in F : \langle F, a \rangle \not\Vdash_V \Box_U A \to \left[\bigvee_{p \in Var(A)} \Diamond_U p \wedge \Diamond_U \neg p\right].$$

That is $\langle F, a \rangle \Vdash_V \Box_U A$ and $\langle F, a \rangle \not\Vdash_V \left[\bigvee_{p \in Var(A)} \Diamond_U p \wedge \Diamond_U \neg p\right]$. Because we have that $\langle F, a \rangle \not\Vdash_V \left[\bigvee_{p \in Var(A)} \Diamond_U p \wedge \Diamond_U \neg p\right]$ and $\Box_U$ is the universal modality, it immediately follows that for all $p \in Var(A)$ either (1) $\forall b \in F(b \Vdash_V p)$ or (2) $\forall b \in F(b \Vdash_V \neg p)$.

Choose the substitution g for all of variables p from the formula A as follows: $\forall p \in Var(A) : g(p) = \top$ if (1) holds and $g(p) = \bot$ in the case if (2) is the case. Using that $\langle F, a \rangle \Vdash_V \Box_U A$ we immediately obtain that g is a unifier for the formula A (using again that $\mathcal{L} = L(K)$ and $\neg \Box_j \bot \in \mathcal{L}$, for all j). Therefore, the formula A is unifiable in $\mathcal{L}$. □

4 Passive inference rules

Below we always assume that $\mathcal{L}$ is a logic with the properties which were required in the previous section (that is: $\mathcal{L}$ is a Kripke complete logic: $\mathcal{L} = L(K)$ for some class of frames K and $\neg \Box_j \bot \in \mathcal{L}$, for all j) and let $\mathcal{L}$ to have the universal modality.

Theorem 3. *The rules*

$$r_n := \frac{\bigvee_{1 \leq i \leq n} \Diamond_U p_i \wedge \Diamond_U \neg p_i}{\bot}$$

form a basis for all passive inference rules in any such logic $\mathcal{L}$.

Proof. Evidently we have

$$\Box_U \left[\bigvee_{p \in Var(A)} \Diamond_U p \wedge \Diamond_U \neg p \right] \to \left[\bigvee_{p \in Var(A)} \Diamond_U p \wedge \Diamond_U \neg p \right]$$

is a theorem of $\mathcal{L}$. Hence by Theorem 2 the formula

$$A := \left[\bigvee_{p \in Var(A)} \Diamond_U p \wedge \Diamond_U \neg p \right]$$

is not unifiable in $\mathcal{L}$, thus, any rule r_n is passive.

Let us assume that a rule $R_1 := A_1, \ldots, A_n / B$ is passive for $\mathcal{L}$. Then the rule $R_2 := A_1 \wedge \cdots \wedge A_n / B$ is also passive and the formula $A_1 \wedge \cdots \wedge A_n$ is not unifiable in $\mathcal{L}$. Applying Theorem 2, we conclude:

$$(a) \quad \Box_U(A_1 \wedge \cdots \wedge A_n) \to \left[\bigvee_{p \in Var(A_1 \wedge \cdots \wedge A_n)} \Diamond_U p \wedge \Diamond_U \neg p \right] \in \mathcal{L}.$$

Applying Gödel's rule w.r.t. $\Box_U$ to the premise of R_2 we may derive the formula $\Box_U(A_1 \wedge \cdots \wedge A_n)$. Using this, (a) and the modus ponens rule we derive the formula $[\bigvee_{p \in Var(A_1 \wedge \cdots \wedge A_n)} \Diamond_U p \wedge \Diamond_U \neg p]$.

From this formula, applying the rule r_n, where n is the number of variables in the conjunction of $A_1 \wedge \cdots \wedge A_n$, we can derive the formula $\bot$. Using that $\bot \to B \in \mathcal{L}$ and modus ponens, we derive B. Thus, all rules r_n form a basis for all rules passive in $\mathcal{L}$. $\square$

5 Applications: Temporal logics of agents knowledge with universal modality

Now we approach the central part of the paper – the application to Multi-Agent Temporal logics as well as other ones where the universal modality is not present in the language, but can be modeled by compound formulas in the native language of the given logics.

Consider first the standard linear temporal logic **LTL**, with operations *Until* and *Since*. The language of **LTL** extends the language of Boolean logic by operations **N** (next), **U** (until) and **S** (since). The formulas of **LTL** are built up from a set *Prop* of atomic propositions (synonymously – propositional letters). The set of all formulas is closed w.r.t. applications of Boolean operations, the unary operation **N** (next) and the binary operations **U** (until) and **S** (since).

The semantics for **LTL** uses *infinite transition systems (runs, computations)*, which we describe in terms of linear Kripke structures based on natural numbers. These structures can be represented as quadruples

$$M := \langle N, \leq, \text{Next}, V \rangle,$$

where N is the set of all natural numbers, $\leq$ is the standard order on N, Next is the binary relation, where a Next b means b is the number next to a. We can also consider here the operation *Previous* which is the opposite to *Next*. All the following results will be valid for this case as well). A valuation V of any set of letters S assigns truth values to elements of S. So, for any $p \in S$, $V(p) \subseteq N$, $V(p)$ is the set of all n from N where p is true (w.r.t. V).

The triple $\langle N, \leq, \text{Next} \rangle$ is a Kripke frame which we will denote by $\mathbf{F}$. For any Kripke frame the truth values can be extended from propositions of S to arbitrary formulas constructed from these propositions as follows:

$$\forall p \in Prop \ (\mathbf{F}, a) \Vdash_V p \Leftrightarrow [a \in N \wedge a \in V(p)];$$

$$(\mathbf{F}, a) \Vdash_V A \wedge B \Leftrightarrow [[(\mathbf{F}, a) \Vdash_V A] \wedge [(\mathbf{F}, a) \Vdash_V B]];$$

$$(\mathbf{F}, a) \Vdash_V \neg A \Leftrightarrow not[(\mathbf{F}, a) \Vdash_V A];$$

$$(\mathbf{F}, a) \Vdash_V \mathbf{N} A \Leftrightarrow \forall b[(a \text{ Next } b) \Rightarrow (\mathbf{F}, b) \Vdash_V A];$$

$$(\mathbf{F}, a) \Vdash_V A \ \mathbf{U} B \Leftrightarrow \exists b[(a \leq b) \wedge ((\mathbf{F}, b) \Vdash_V B) \wedge$$

$$\forall c[(a \leq c < b) \Rightarrow (\mathbf{F}, c) \Vdash_V A]];$$

$$(\mathbf{F}, a) \Vdash_V A \ \mathbf{S} \ B \Leftrightarrow \exists b[(b \leq a) \wedge ((\mathbf{F}, b) \Vdash_V B) \wedge$$

$$\forall c[(b \leq c < a) \Rightarrow (\mathbf{F}, c) \Vdash_V A]].$$

Using operations **U**, **S** and **N** we can define all standard temporal and modal operations. For instance, $\mathbf{F}A$ (A holds eventually, which, in terms of modal logic, means A is possible (denotation $\Diamond^+ A$)), can be described as $true \mathbf{U} A$. Therefore, in this language, we can also define the modal operation $\Box$ (as $\Box^+ A := \neg \Diamond^+ \neg A$). Modal operation $\Diamond^-$ directed to past may be defined as $\Diamond^- A := true \mathbf{S} A$, respectively $\Box^- A := \neg \Diamond^- \neg A$.

The logic **LTL** is the set of all formulas which are true at all such models. It is clear that the universal modality may be expressed in this logic as $\Box_U p := \Box^+ p \wedge \Box^- p$. Therefore we may directly transfer the results from the previous section to this logic.

Theorem 4. *Theorems 2 and 3 hold for* **LTL**.

Now we wish to obtain our earlier results from [8] using the theorems of the previous section; that is, we want to describe non-unifiability and passive rules for the linear temporal logic with Multi-Agents modalities for Multi-Agent Knowledge.

First we recall the definitions and notation from that earlier paper. The alphabet of the language for the logic $\mathcal{L}^{LFPK}$ includes a countable set of propositional variables $P := \{p_1, \ldots, p_n, \ldots\}$, brackets (,) default Boolean logical operations and a variety of unary modal operators $\{\Box_F, \Box_P, \Box_1, \ldots, \Box_n\}$. The name $LFPK$ is supposed to abbreviate the sequence of words *logic, future, past, knowledge*.

The formation rules for formulas are: every propositional variable $p \in P$ is a well-formed formulae (wff), and if A is a wff, then so are $\Box_F A, \Box_P A, \Box_i A$, for $i \in I$. Logical operations $\Diamond_F, \Diamond_P, \Diamond_i$ are defined using the logical operations $\Box_F, \Box_P, \Box_i$ as usual $\Diamond_F = \neg \Box_F \neg, \Diamond_P = \neg \Box_P \neg, \Diamond_i = \neg \Box_i \neg$.

The meanings of the modal operations described are as follows. For $\Box_P A$: A is true at all previous and at the current point in time; for $\Box_F A$: A is true at the given time point and will be true at all future points. The formula $\Box_i A$ means that A is true at all informational states which are available for the agent i in a current time state.

Semantics for the language of $\mathcal{L}^{LFPK}$ models linear and discrete streams of the computational process, at which each point in time is associated with an integer number $n \in \mathbb{Z}$.

Definition 7. Temporal k-modal Kripke-frame is a tuple

$$\mathcal{T} = \langle W_T, R_1, R_2, \ldots, R_k \rangle,$$

where W_T is a non-empty set of worlds, $R_1, \ldots, R_k$ are some binary relations on W_T, where $R_2 = R_1^{-1} := \{(a,b) | (b,a) \in R_1\}$ is the converse relation to R_1.

Definition 8. Let $F = \langle W_F, R_1, \ldots, R_k \rangle$ is a Kripke-frame, then $\forall R_i$ R_i-cluster (if exists) is the subset $C^{R_i} \in W_F$ such that $\forall v, z \in C^{R_i} : (vR_iz)\&(zR_iv)$ and $\forall z \in W_F$, $\forall v \in C^{R_i} : ((vR_iz\&zR_iv) \Rightarrow z \in C^{R_i})$. For any relation R_i, $C^{R_i}(v)$ is the R_i-cluster s.t. $v \in C^{R_i}$ or the cluster, generated by the element v.

Definition 9. $LFPK$-frame is a temporal $(n+2)$-modal Kripke-frame

$$T = \langle Z_T, R_F, R_P, R_1, \ldots, R_n \rangle,$$

where $R_P = R_F^{-1}$ and:
 a. Z_T is the disjoint union of clusters of states
  C^t, $t \in \mathbb{Z}$ ($\mathbb{Z}$ is the set of all integer numbers), and $C^{t_1} \bigcap C^{t_2} = \emptyset$ if $t_1 \neq t_2$;
 b. $\forall t_1, t_2 \in \mathbb{Z}$, if $t_1 \leq t_2$ then $\forall a \in C^{t_1}, \forall b \in C^{t_2}(aR_Fb)$ and (bR_Pa).
  None other relations via R_P and R_F are allowed.
 c. $R_1, \ldots, R_n$ are some equivalence relations in each separate cluster C^t.

Definition 10. A model M_T on a $LFPK$-frame T is a tuple $M_T = \langle T, V \rangle$, where V is a valuation of a set of propositional letters $p \in P$ on T, i.e $\forall p \in P$ $[V(p) \subseteq Z_T]$.

Given a model $M_T = \langle T, V \rangle$, where T is a $LFPK$-frame Z_T we compute truth values of formulas at states $w \in Z_T$ as follows:

 a. $\langle T, w \rangle \Vdash_V p \Leftrightarrow w \in V(p)$;
 b. $\langle T, w \rangle \Vdash_V \Box_F A \Leftrightarrow \forall z \in Z_T(wR_Fz \Rightarrow \langle T, z \rangle \Vdash_V A)$;
 c. $\langle T, w \rangle \Vdash_V \Box_P A \Leftrightarrow \forall z \in Z_T(wR_Pz \Rightarrow \langle T, z \rangle \Vdash_V A)$;
 d. $\forall i \in I, \langle T, w \rangle \Vdash_V \Box_i A \Leftrightarrow \forall z \in Z_T(wR_iz \Rightarrow \langle T, z \rangle \Vdash_V A)$.
 e. $\langle T, w \rangle \Vdash_V A \vee B \Leftrightarrow [(\langle T, w \rangle \Vdash_V A) \text{ or } (\langle T, w \rangle \Vdash_V B)]$;
 f. $\langle T, w \rangle \Vdash_V A \wedge B \Leftrightarrow [(\langle T, w \rangle \Vdash_V A) \text{ and } (\langle T, w \rangle \Vdash_V B)]$;
 l. $\langle T, w \rangle \Vdash_V A \to B \Leftrightarrow [(\langle T, w \rangle \Vdash_V B) \text{ or } \text{not}(\langle T, w \rangle \Vdash_V A)]$;
 i. $\langle T, w \rangle \Vdash_V \neg A \Leftrightarrow [\text{not}(\langle T, w \rangle \Vdash_V A)]$;

If a formula A is true at any element of a frame T w.r.t. any valuation V, we say A is true at the frame T and write $T \Vdash A$.

Definition 11. The temporal Linear Future/Past logic $LFPK$ (of agents knowledge) is the set of all $LFPK$-formulas valid (true) on all frames: $LFPK := \{A \mid A \in Fml(\mathcal{L}^{LFPK}), \forall T(T \Vdash A)\}$. If a formula A is a member of $LFPK$, then we say that A is a *theorem* of $LFPK$.

Is is immediate to see that the formula $\Box_F p \wedge \Box_P p$ models the universal modality in $LFPK$. Therefore again we may directly transfer the results from the previous section to this logic.

Theorem 5 ([8]). *Theorems 2 and 3 hold for $LFPK$.*

And now we would like to obtain an yet more general result. We will consider some semantic models for not just linear but branching time. Such models look as follows. Let n be a given fixed natural number. Any such model M is compound from some arbitrary set S of models M_i based on some $LFPK$ frames which are glued in the following way.

The model M is based on all models from S, all these models are sub-models of M, and M has no states which do not belong to any model from S. For any two different models M_{i_1} and M_{i_2} from S, there are some two clusters C_a and C_b from M_{i_1} and M_{i_2} respectively, such that there is a zig-zag passageway of length at most n in the model M by time to future and to past from C_a into C_b.

Note that such models might be very complicated and differently compound, even with possible common whole intervals of states. The truth values of formulas with $\Box_F$ and $\Box_P$ may be calculated in such models M as usual in temporal/modal models. the only distinction with our previous case is that the time sometimes may be branching, though not compulsory branching in each cluster. Since we have bounded by n time-zigzag, the formula

$$(\Box_F \Box_P)^{n+1} p \wedge (\Box_P \Box_F)^{n+1} p$$

represents the universal modality in all such models (for fixed n).

Let $L(n)$ be the logic generated by a (any given) class of arbitrary models constructed as described above. We call this logic branching time multi-agent logic with bounded time zigzag. Then $L(n)$ has expressible universal modality, and therefore

Theorem 6. *Theorems 2 and 3 hold for $L(n)$.*

6 Conclusion

Our paper describes (algorithmically and syntactically) formulas which are not unifiable in a wide class modal, temporal and multi-agent logics from areas close to Information Sciences and general Computer Science. The important case for future investigation is the case of similar logics but without the request for logics to be Kripke complete. We may see that our Theorem 1 does not requires Kripke completeness, but later the methods which we use here need logics to be Kripke complete. Second open problem is to extend such results to the branching time temporal logics without restriction to bounded time-zig-zag. The next interesting question is to attempt to model in our framework the agents' accessibility relations, which are not equivalence relations but some other, as linear, some branching structure with hierarchy, etc, that is to consider the case when agents accessibility relations are more complicated, and to extend our results to such logics.

References

[1] F. Baader and S. Ghilardi. Unification in modal and description logics. *Logic Journal of IGPL*, 19:705–730, 2011.

[2] F. Baader and B. Morawska. Unification of concept terms in description logics. *Journal of Symbolic Computation*, 31:277–305, 2001.

[3] F. Baader and B. Morawska. Unification in the description logic EL. *Logical Methods in Computer Science*, 6:1–31, 2010.

[4] F. Baader and P. Narendran. Unification in a description logic with transitive closure of roles. *Logic for Programming, Artificial Intelligence, and Reasoning, LPAR 2001*, 2250:217–232, 2001.

[5] F. Baader and W. Snyder. Unification Theory. In *Handbook of Automated Reasoning. I*, pages 447–533. Elsevier Science Publishers, 2001.

[6] S. Babenyshev and V. Rybakov. Linear temporal logic LTL: basis for admissible rules. *Journal of Logic and Computation*, 21:157–177, 2011.

[7] S. I. Bashmakov. Unification and inference rules in the multi-modal logic of knowledge and linear time LTK. *J. Siberian Federal University. Mathematics and Physics*, 9:148–156, 2016.

[8] S. I. Bashmakov, A. V. Kosheleva, and V. V. Rybakov. Non-unifiability in linear temporal logic of knowledge with multi-agent relations. *Siberian Electronic Mathematical Reports*, 13:656–663, 2016.

[9] F. Belardinelli and A. Lomuscio. Interactions between knowledge and time in a first-order logic for multi-agent systems: completeness results. *Journal of Artificial Intelligence Research*, 45:1–45, 2012.

[10] P. Blackburn, J. van Benthem, and F. Wolter, editors. *Handbook of Modal Logic*. Elsevier, 2007.

[11] W. Dzik and P. Wojtylak. Projective unification in modal logic. *Logic Journal of IGPL*, 20:121–153, 2012.

[12] K. Fine. An incomplete logic containing S4. *Theoria*, 40:23–29, 1974.

[13] D. M. Gabbay and I. M. Hodkinson. An axiomatization of the temporal logic with Until and Since over the real numbers. *Journal of Logic and Computation*, 1:229–260, 1990.

[14] D. M. Gabbay and I. M. Hodkinson. Temporal Logic in Context of Databases. In *Logic and Reality, Essays on the Legacy of Arthur Prior*. Oxford University Press, 1995.

[15] D. M. Gabbay, I. M. Hodkinson, and M. A. Reynolds. *Temporal Logic – Mathematical Foundations and Computational Aspects*. Clarendon Press, Oxford, 1994.

[16] G. Ghilardi. Unification in Intuitionistic Logic. *Journal of Symbolic Logic*, 64:859–880, 1999.

[17] S. Ghilardi. Unification through Projectivity. *J. Logic and Computation*, 7:733–752, 1997.

[18] S. Ghilardi. Best solving modal equations. *Annals of Pure and Applied Logic*, 102:183–198, 2000.

[19] S. Ghilardi. Unification, finite duality and projectivity in varieties of Heyting algebras. *Annals of Pure and Applied Logic*, 127:99–115, 2004.

[20] S. Ghilardi and L. Sacchetti. Filtering Unification and Most General Unifiers in Modal Logic. *Journal of Symbolic Logic*, 69:879–906, 2004.

[21] V. Goranko and S. Passy. Using the universal modality: Gains and questions. *Journal of Logic and Computation*, 2:5–30, 1992.

[22] R. Iemhoff. On the admissible rules of intuitionistic propositional logic. *Journal of Symbolic Logic*, 66:281–294, 2001.

[23] R. Iemhoff and G. Metcalfe. Proof theory for admissible rules. *Annals of Pure and Applied Logic*, 159:171–186, 2009.

[24] E. Jerabek. Admissible rules of modal logics. *Journal of Logic and Computation*, 15:411–431, 2005.

[25] E. Jerabek. Independent bases of admissible rules. *Logic Journal of the IGPL*, 16:249–267, 2008.

[26] E. Jerabek. Rules with parameters in modal logic I. *Logic Journal of the IGPL*, CoRR abs/1305.4912, 2013.

[27] D. Knuth and P. Bendix. *Simple Word Problems in Universal Algebras*. Pergamon Press, 1970.

[28] M. Kracht. Judgment and Consequence Relations. *Journal of Applied Non-classical Logic*, 20:223–235, 2010.

[29] Z. Manna and A. Pnueli. *The Temporal Logic of Reactive and Concurrent Systems: Specification*, volume 1. Springer, 1992.

[30] Z. Manna and A. Pnueli. *Temporal Verification of Reactive Systems: Safety*, volume 1. Springer, 1995.

[31] A. Robinson. A machine oriented logic based on the resolution principle. *J. of the ACM*, 12:23–41, 1965.

[32] V. V. Rybakov. Problems of Substitution and Admissibility in the Modal System Grz and in Intuitionistic Propositional Calculus. *Annals of Pure and Applied Logic*, 50:71–106, 1990.

[33] V. V. Rybakov. Rules of Inference with Parameters for Intuitionistic Logic. *J. of Symbolic Logic*, 57:912–923, 1992.

[34] V. V. Rybakov. *Admissible Logical Inference Rules*, volume 136 of *Studies in Logic and the Foundations of Mathematics*. Elsevier Sci. Publ., North-Holland, 1997.

[35] V. V. Rybakov. Logical Consecutions in Discrete Linear Temporal Logic. *J. of Symbolic Logic*, 70:1137–1149, 2005.

[36] V. V. Rybakov. Linear temporal logic with until and next, logical consecutions. *Annals of Pure and Applied Logic*, 155:32–45, 2008.

[37] V. V. Rybakov. Writing out Unifiers in Linear Temporal Logic. *J. Logic Computation*, 22:1199–1206, 2012.

[38] V. V. Rybakov. Unifiers in transitive modal logics for formulas with coefficients (meta-variables). *Logic Journal of the IGPL*, 21:205–215, 2013.

[39] V. V. Rybakov. Writing out unifiers for formulas with coefficients in intuitionistic logic. *Logic Journal of the IGPL*, 21:187–198, 2013.

[40] V. V. Rybakov. Projective formulas and unification in linear temporal logic LTLU. *Logic Journal of the IGPL*, 22:665–672, 2014.

[41] V. V. Rybakov, M. Terziler, and C. Gencer. An essay on unification and inference rules for modal logics. *Bulletin of the Section of Logic*, 28:145–157, 1999.

[42] S. K. Thomason. An incomplete theorem in modal logic. *Theoria*, 40:30–34, 1974.

[43] J. van Benthem. Two simple incomplete modal logics. *Theoria*, 44:25–37, 1978.

[44] W. van der Hoek and M. Wooldridge. Logics for Multi-Agent Systems. In G. Weiss, editor, *Multi-Agent Systems*, pages 671–810. MIT Press, 2nd edition, 2013.

[45] M. Vardi. Reasoning about the past with two-way automata. *Automata, Languages and Programming*, pages 628–641, 1998.

[46] M. Y. Vardi. An automata-theoretic approach to linear temporal logic. In *Logics for concurrency*, pages 238–266. Springer, 1996.

Implicit Dynamic Function Introduction and Ackermann-like Function Theory

MARCOS CRAMER
University of Luxembourg,
6, rue Richard Coudenhove-Kalergi, 1359 Luxembourg
marcos.cramer@uni.lu

Abstract

We discuss a feature of the natural language of mathematics – the implicit dynamic introduction of functions – that has, to our knowledge, not been captured in any formal system so far. If this feature is used without limitations, it yields a paradox analogous to Russell's paradox. Hence any formalism capturing it has to impose some limitations on it. We sketch two formalisms, both extensions of Dynamic Predicate Logic, that innovatively do capture this feature, and that differ only in the limitations they impose onto it. One of these systems is based on *Ackermann-like Function Theory*, a novel foundational theory of functions that is inspired by Ackermann Set Theory and that interprets ZFC.

Keywords: Dynamic Predicate Logic, Function Introduction, Ackermann Set Theory, Function Theory.

1 Dynamic predicate logic

Dynamic predicate logic (DPL) [7] is a formalism whose syntax is identical to that of standard first-order predicate logic (PL), but whose semantics is defined in such a way that the dynamic nature of natural language quantification is captured in the formalism:

1. If a farmer owns a donkey, he beats it.
2. PL: $\forall x \, \forall y \, (farmer(x) \wedge donkey(y) \wedge owns(x,y) \rightarrow beats(x,y))$

3. DPL: $\exists x\,(farmer(x) \wedge \exists y\,(donkey(y) \wedge owns(x,y))) \rightarrow beats(x,y)$

In PL, 3 is not a sentence, since the rightmost occurrences of x and y are free. In DPL, a variable may be bound by a quantifier even if it is outside its scope. The semantics is defined in such a way that 3 is equivalent to 2. So in DPL, 3 captures the meaning of 1 while being more faithful to its syntax than 2.

1.1 DPL semantics

We present DPL semantics in a way slightly different but logically equivalent to its definition by Groenendijk and Stokhof in [7]. Structures and assignments are defined as for PL: A structure S specifies a domain $|S|$ and an interpretation a^S for every constant, function or relation symbol a in the language. An S-assignment is a function from variables to $|S|$. Let G_S denote the set of S-assignments. Given two assignments g, h, we define $g[x]h$ to mean that g differs from h at most in what it assigns to the variable x. Given a DPL term t, we recursively define

$$[t]_S^g = \begin{cases} g(t) & \text{if } t \text{ is a variable,} \\ t^S & \text{if } t \text{ is a constant symbol,} \\ f^S([t_1]_S^g, \ldots, [t_n]_S^g) & \text{if } t \text{ is of the form } f(t_1, \ldots, t_n). \end{cases}$$

Groenendijk and Stokhof [7] define an interpretation function $[\![\cdot]\!]_S$ from DPL formulae to subsets of $G_S \times G_S$. We instead recursively define for every $g \in G_S$ an interpretation function $[\![\cdot]\!]_S^g$ from DPL formulae to subsets of G_S:[1]

1. $[\![\top]\!]_S^g := \{g\}$
2. $[\![t_1 = t_2]\!]_S^g := \{h \mid h = g \text{ and } [t_1]_S^g = [t_2]_S^g\}$[2]
3. $[\![R(t_1, \ldots, t_2)]\!]_S^g := \{h \mid h = g \text{ and } ([t_1]_S^g, \ldots, [t_2]_S^g) \in R^S\}$
4. $[\![\neg\varphi]\!]_S^g := \{h \mid h = g \text{ and there is no } k \in [\![\varphi]\!]_S^h\}$
5. $[\![\varphi \wedge \psi]\!]_S^g := \{h \mid \text{there is a } k \text{ s.t. } k \in [\![\varphi]\!]_S^g \text{ and } h \in [\![\psi]\!]_S^k\}$
6. $[\![\varphi \rightarrow \psi]\!]_S^g := \{h \mid h = g \text{ and for all } k \text{ s.t. } k \in [\![\varphi]\!]_S^h, \text{ there is a } j \text{ s.t. } j \in [\![\psi]\!]_S^k\}$
7. $[\![\exists x\,\varphi]\!]_S^g := \{h \mid \text{there is a } k \text{ s.t. } k[x]g \text{ and } h \in [\![\varphi]\!]_S^k\}$

$\varphi \vee \psi$ and $\forall x\,\varphi$ are defined to be a shorthand for $\neg(\neg\varphi \wedge \neg\psi)$ and $\exists x\,\top \rightarrow \varphi$ respectively.

---

[1] This can be viewed as a different currying of the uncurried version of Groenendijk and Stokhof's interpretation function.

[2] The condition $h = g$ in cases 2, 3, 4 and 6 implies that the defined set is either $\emptyset$ or $\{g\}$.

2 Implicit dynamic introduction of function symbols

Functions are often dynamically introduced in an implicit way in mathematical texts. For example, [10] introduces the additive inverse function on the reals as follows:

(a) For each a there is a real number $-a$ such that $a + (-a) = 0$. [10, p. 1]

Here the natural language quantification "there is a real number $-a$" *locally* (i.e. inside the scope of "For each a") introduces a new real number to the discourse. But since the choice of this real number depends on a and we are universally quantifying over a, it *globally* (i.e. outside the scope of "For each a") introduces a function "$-$" to the discourse.

The most common form of implicitly introduced functions are functions whose argument is written as a subscript, as in the following example:

(b) Since f is continuous at t, there is an open interval I_t containing t such that $|f(x) - f(t)| < 1$ if $x \in I_t \cap [a, b]$. [10, p. 62]

If one wants to later explicitly call the implicitly introduced function a function, the standard notation with a bracketed argument is preferred:

(c) Suppose that, for each vertex v of K, there is a vertex $g(v)$ of L such that $f(\operatorname{st}_K(v)) \subset \operatorname{st}_L(g(v))$. Then g is a simplicial map $V(K) \to V(L)$, and $|g| \simeq f$. [8, p. 19]

When no uniqueness claims are made about the object locally introduced to the discourse, implicit function introduction presupposes the existence of a choice function, i.e. presupposes the Axiom of Choice. We hypothesise that the naturalness of such implicit function introduction in mathematical texts contributes to the widespread feeling that the Axiom of Choice must be true.

Implicitly introduced functions generally have a restricted domain and are not defined on the whole universe of the discourse. In the example (c), g is only defined on vertices of K and not on vertices of L. Implicit function introduction can also be used to introduce multi-argument functions, but for the sake of simplicity and brevity, we restrict ourselves to unary functions in this paper.

If the implicit introduction of functions is allowed without limitations, one can derive a contradiction:

(d) For every function f, there is a natural number $g(f)$ such that

$$g(f) = \begin{cases} 0 & \text{if } f \in \operatorname{dom}(f) \text{ and } f(f) \neq 0, \\ 1 & \text{if } f \notin \operatorname{dom}(f) \text{ or } f(f) = 0. \end{cases}$$

Then g is defined on every function, i.e. $g(g)$ is defined. But from the definition of g, $g(g) = 0$ iff $g(g) \neq 0$.

This contradiction is due to the *unrestricted function comprehension* that is implicitly assumed when allowing implicit introductions of functions without limitations. Unrestricted function comprehension could be formalised as an axiom schema as follows:

Axiom Schema 1 (Unrestricted function comprehension). For every formula $\varphi(x, y)$, the following is an axiom: $\forall x \, \exists y \, \varphi(x, y) \to \exists f \, \forall x \, \varphi(x, f(x))$

The inconsistency of unrestricted function comprehension is analogous to the inconsistency of unrestricted set comprehension, i.e. Russell's paradox.

Russell's paradox led to the abandonment of unrestricted comprehension in set theory. Two radically different approaches have been undertaken for restricting set comprehension: Russell himself restricted it through his Ramified Theory of Types, which was later simplified to Simple Type Theory (STT), mainly known via Church's formalisation in his simply typed lambda calculus [2]. On the other hand, the risk of paradoxes like Russell's paradox also contributed to the development of ZFC (Zermelo-Fraenkel set theory with the Axiom of Choice), which allows for a much richer set theoretic universe than the universe of simply typed sets. Since all the axioms of ZFC apart from the Axiom of Extensionality, the Axiom of Foundation and the Axiom of Choice are special cases of comprehension, one can view ZFC as an alternative way to restrict set comprehension.

Similarly, the above paradox must lead to the abandonment of unrestricted function comprehension. The type-theoretic approach is easily adapted to functions, so we will first sketch the system that formalises this approach, *Typed Higher-Order Dynamic Predicate Logic*. For an untyped approach, there is no clear way to transfer the limitations that ZFC puts onto set comprehension to the case of function comprehension. However, there is an axiomatization of set theory (with classes) called *Ackermann set theory* that is a conservative extension of ZFC. It turns out that the limitations that Ackermann set theory poses on set comprehension can be transferred to the case of function comprehension, and hence to the case of implicit dynamic function introduction.

The need to deal with implicit function introduction arose for us in the context of the *Naproche project*, a project aiming at automatic formalisation of natural language mathematics [3, 5, 6]. It has been implemented in the Naproche system using type restrictions as in Typed Higher-Order Dynamic Predicate Logic, and we plan to implement it using the less strict restrictions of the untyped Higher-Order Dynamic Predicate Logic in a future version of the system.

3 Typed higher-order dynamic predicate logic

In this section, we extend DPL to a system called *Typed Higher-Order Dynamic Predicate Logic* (THODPL), which formalises implicit dynamic function introduction, and also allows for explicit quantification over functions. THODPL has variables typed by the types of STT. In the below examples we use x and y as variables of the basic type i, and f as a variable of the function type $i \to i$. A complex term is built by well-typed application of a function-type variable to an already built term, e.g. $f(x)$ or $f(f(x))$.

The distinctive feature of THODPL syntax is that it allows not only variables but any well-formed terms to come after quantifiers. So (1) is a well-formed formula:

$$\forall x \, \exists f(x) \, R(x, f(x)) \tag{1}$$

$$\forall x \, \exists y \, R(x, y) \tag{2}$$

$$\exists f \, (\forall x \, R(x, f(x))) \tag{3}$$

The semantics of THODPL is to be defined in such a way that (1) has the same truth conditions as (2). But unlike (2), (1) dynamically introduces the function symbol f to the context, and hence turns out to be equivalent to (3).

We now sketch how these desired properties of the semantics can be achieved. In THODPL semantics, an assignment assigns elements of $|S|$ to variables of type i, functions from $|S|$ to $|S|$ to variables of type $i \to i$ etc. Additionally, an assignment can also assign an object (or function) to a complex term. For example, any assignment in the interpretation of $\exists f(x) \, R(x, f(x))$ has to assign some object to $f(x)$. The definition of $g[x]h$ can now naturally be extended to a definition of $g[t]h$ for terms t. The definition of $[t]_S^g$ has to be adapted in the natural way to account for function variables.

Just as in the case of DPL semantics, we recursively define an interpretation $[\![\cdot]\!]_S^g$ from DPL formulae to subsets of G_S (the cases 1-5 of the recursive definition are as in Section 1.1):

6. $[\![\varphi \to \psi]\!]_S^g := \{h | h$ differs from g in at most some function variables $f_1, \ldots, f_n$ (where this choice of function variables is maximal), and there is a variable x such that for all $k \in [\![\varphi]\!]_S^g$, there is an assignment $j \in [\![\psi]\!]_S^k$ such that $j(f_i(x)) = h(f_i)(k(x))$ for $1 \leq i \leq n$, and if $n > 0$ then $k[x]g \}$

7. $[\![\exists t \, \varphi]\!]_S^g := \{h | $ there is a k s.t. $k[t]g$ and $h \in [\![\varphi]\!]_S^k\}$

In order to make case 6 of the definition more comprehensible, let us consider its role in determining the semantics of (1), i.e. of $\exists x\ \top \to \exists f(x)\ R(x, f(x))$: First note that $[\![\exists f(x)\ R(x, f(x))]\!]_S^k$ is the set of assignments j satisfying $R(x, f(x))$ (i.e. for which $[\![R(x, f(x))]\!]_S^j$ is non-empty) such that $j[f(x)]k$. Furthermore note that $[\![\exists x\ \top]\!]_S^g$ is the set of assignments k such that $k[x]g$. So by case 6 with $n = 1$,

$[\![\exists x\ \top \to \exists f(x)\ R(x, f(x))]\!]_S^g = \{h \mid h[f]g$ and there is a variable x such that for all k such that $k[x]g$, there is an assignment j satisfying $R(x, f(x))$ such that $j[f(x)]k$ and $j(f(x)) = h(f)(k(x))$, and $k[x]g\}$

$= \{h \mid h[f]g$ and for all k such that $k[x]g$, there is an assignment j satisfying $R(x, f(x))$ such that $j[f(x)]k$ and $j(f(x)) = h(f)(k(x))\}$

$= \{h \mid h[f]g$ and for all k such that $k[x]h$, k satisfies $R(x, f(x))\}$

$= [\![\exists f\ (\forall x\ R(x, f(x)))]\!]_S^g$

The type restrictions THODPL imposes may be too strict for some applications: Mathematicians sometimes do make use of functions that do not fit into the corset of strict typing, e.g. a function defined on both real numbers and real functions. To overcome this restriction, we will introduce an untyped variant HODPL in Section 6. But for this, we require some foundational preliminaries.

4 Ackermann set theory

Ackermann set theory [1] postulates not only sets, but also proper classes which are not sets.[3] The sets are distinguished from the proper classes by a unary predicate M (from the German word "Menge" for "set").

Ackermann presented a pure version of his theory without urelements, and a separate version with urelements, which we will present here. The language of Ackermann set theory contains three predicates: A binary predicate $\in$, a unary predicate M and a unary predicate U for urelements. We introduce L(x) ("x is limited") as an abbreviation for M(x) $\vee$ U(x). The idea is that sets and urelements are objects of limited size, and are distinguished from the more problematic classes of unlimited size.

The axioms of Ackermann set theory with urelements are as follows:

---

[3]Note, however, that unlike the more well-known class theory NBG, Ackermann set theory also allows for proper classes that contain proper classes.

- *Extensionality axiom:* $\forall x \, \forall y \, (\forall z \, (z \in x \leftrightarrow z \in y) \to x = y)$

- *Class comprehension axiom schema:* Given a formula $F(y)$ (possibly with parameters[4]) that does not have x among its free variables, the following is an axiom:
  $\forall y \, (F(y) \to \mathrm{L}(y)) \to \exists x \, \forall y \, (y \in x \leftrightarrow F(y))$

- *Set comprehension axiom schema:* Given a formula $F(y)$ (possibly with parameters that are limited[5]) that does not have x among its free variables and does not contain the symbol M, the following is an axiom:
  $\forall y \, (F(y) \to \mathrm{L}(y)) \to \exists x \, (\mathrm{M}(x) \wedge \forall y \, (y \in x \leftrightarrow F(y)))$

- *Elements and subsets of sets are limited:*
  $\forall x \, \forall y \, (\mathrm{M}(y) \wedge (x \in y \vee \forall z \, (z \in x \to z \in y)) \to \mathrm{L}(y))$

So unlimited set comprehension is replaced by two separate comprehension schemata, one for class comprehension and one for set comprehension. In both cases, the comprehension is restricted by the constraint that only limited objects satisfy the property that we are applying comprehension to. But for set comprehension, we have the additional constraint that the property may not be defined using the setness predicate or using a proper class as parameter. Ackermann justified this approach by appeal to a definition of "set" from Cantor's work [1].

If an Axiom of Foundation for sets is added, Ackermann set theory turns out to be – in what it says about sets – precisely equivalent to ZF [9]. But this equivalence is not a triviality: It is especially hard to establish Replacement for the sets of Ackermann set theory.

5 Ackermann-like function theory

Now we transfer the ideas of a comprehension limited in this way from set comprehension to function comprehension. For this a dichotomy similar to that between sets and classes has to be imposed on functions. We propose the terms *function* and *map* respectively for this dichotomy, and call the theory resulting from these limitations on function comprehension *Ackermann-like Function Theory* (AFT). AFT

---

[4]This means that F may actually be of the form $F(\bar{z}, y)$, and that these parameters are universally quantified in the axiom:
$\forall \bar{z} \, (\forall y \, (F(\bar{z}, y) \to \mathrm{M}(y)) \to \exists x \, \forall y \, (y \in x \leftrightarrow F(\bar{z}, y)))$

[5]Formally, with the parameters made explicit, the set comprehension axiom schema reads as follows:
$\forall z_1, \ldots, z_n \, (\mathrm{L}(z_1) \wedge \cdots \wedge \mathrm{L}(z_n) \to (\forall y \, (F(z_1, \ldots, z_n, y) \to \mathrm{L}(y)) \to \exists x \, (\mathrm{M}(x) \wedge \forall y \, (y \in x \leftrightarrow F(z_1, \ldots, z_n, y)))))$

can be shown to be equiconsistent with Ackermann set theory and hence with ZFC (see Theorem 4 below).

The language of Ackermann-like function theory (L_{AFT}) contains

- a unary predicate F for functions,
- a unary predicate U for urelements,
- a constant symbol u for undefinedness, and
- a binary function symbol a for function application.

Instead of $a(f,t)$ we usually simply write $f(t)$. We write L(x) instead of U(x)$\vee$F(x). The undefinedness constant u is needed for formalising the idea that a function is only defined for certain values and undefined for others. In this language, the unrestricted function comprehension schema would be as follows:

Axiom Schema 2 (Unrestricted function comprehension in L_{AFT}). Given a variable z and formulae $P(z)$ and $R(z,x)$ (possibly with parameters), the following is an axiom: $\forall z \, (P(z) \to \exists x \, R(z,x)) \to \exists f \, (\neg \text{U}(f) \wedge \forall z \, ((P(z) \to R(z, f(z))) \wedge (\neg P(z) \to f(z) = u)))$.

Analogously to the case of Ackermann set theory, AFT has separate comprehension schemata for maps and functions. The restriction that is imposed on both schemata now is $\forall z \, \forall x \, (R(z,x) \to \text{L}(z) \wedge \text{L}(x))$. In the function comprehension schema, in which F(f) appears among the conclusions we may draw about f, the additional restriction is that the formula $R(z,x)$ may not contain the symbol F and may not have unlimited objects as parameters.

Additionally to these comprehension schemata, AFT has

- a function extensionality axiom,
- an axiom stating that any value a function takes and any value a function is defined at is limited, and
- an axiom stating that submaps of functions are functions.

In AFT one can interpret Ackermann set theory with Foundation, and hence ZFC (see Theorems 1 and 3 below). Since the map and function comprehension schemata presuppose the existence of choice maps and choice functions, the Axiom of Choice naturally comes out true in these interpretations.

We now state the main theorems about AFT. Their proofs can be found in the author's PhD thesis [5, pp. 58–62].

Theorem 1 (Theorem 4.2.7 in [5, p. 58]). *AFT interprets Ackerman set theory with urelements and the Axiom of Choice.*

Theorem 2 (Theorem 4.2.20 in [5, p. 61]). *Ackermann set theory with the Axiom of Foundation and the Axiom of Global Choice interprets AFT.*

Theorem 3 (Theorem 4.2.8 in [5, p. 59]). *AFT interprets ZFC.*

Theorem 4 (Corollary in [5, p. 62]). *AFT is equiconsistent with ZFC.*

6 Higher-order dynamic predicate logic

Now we are ready to sketch the untyped *Higher-Order Dynamic Predicate Logic* (HODPL). The restriction we impose on implicit function introduction are those imposed by AFT. AFT gives us untyped maps, which always have a restricted domain. So instead of using types to syntactically restrict the possible arguments for a given function term, we implement a semantic restriction on function application by integrating a formal account of presuppositions into the HODPL.[6] HODPL syntax thus allows for any term to be applied to any number of arguments to form a new term.

Besides the binary "=", HODPL has two unary logical relation symbols, U for urelements and F for functions. HODPL syntax does not depend on a signature, as we do not allow for constant, function and relation symbols other than "=", U and F. These can be mimicked by variables that respectively denote a non-function, denote a normal function or denote a function that only takes two predesignated urelements ("booleans") as values.

The domain of a structure always has to be a model of AFT. The possibility of presupposition failure is implemented in HODPL semantics by making the interpretation function partial rather than total. For conveniently talking about partial functions, we use the notation $\text{def}(f(x))$ to abbreviate that f is defined on x.

We define the partial interpretation function $[\![\cdot]\!]_S^g \subseteq G_S \times G_S$ by specifying its domain and its values trough a simultaneous recursion (the cases 3-8 of the second part are as in THODPL):

- Domain of $[\![\cdot]\!]_S^g$:
    1. $\text{def}([\![U(t)]\!]_S^g)$ iff $[\![t]\!]_S^g \neq u^S$.
    2. $\text{def}([\![F(t)]\!]_S^g)$ iff $[\![t]\!]_S^g \neq u^S$.

---

[6]See [4] for an introduction to presuppositions in mathematical texts.

3. def($[\![\top]\!]_S^g$).
4. def($[\![t_1 = t_2]\!]_S^g$) iff $[\![t_1]\!]_S^g \neq u^S$ and $[\![t_2]\!]_S^g \neq u^S$.
5. def($[\![\neg\varphi]\!]_S^g$) iff def($[\![\varphi]\!]_S^g$).
6. def($[\![\varphi \wedge \psi]\!]_S^g$) iff def($[\![\varphi]\!]_S^g$) and for all $h \in [\![\varphi]\!]_S^g$, def($[\![\psi]\!]_S^h$).
7. def($[\![\varphi \to \psi]\!]_S^g$) iff def($[\![\varphi]\!]_S^g$) and for all $h \in [\![\varphi]\!]_S^g$, def($[\![\psi]\!]_S^h$).
8. def($[\![\exists t\, \varphi]\!]_S^g$) iff for all h s.t. $h[t]g$, def($[\![\varphi]\!]_S^h$).

- Values of $[\![\cdot]\!]_S^g$:

  1. $[\![\mathrm{U}(t)]\!]_S^g := \{h \mid g = h \text{ and } [\![t]\!]_S^g \in \mathrm{U}^S\}$
  2. $[\![\mathrm{F}(t)]\!]_S^g := \{h \mid g = h \text{ and } [\![t]\!]_S^g \in \mathrm{F}^S\}$

One can define a sound proof system for HODPL that can prove everything provable in AFT: In the author's PhD thesis, a proof system for an extension of HODPL is defined [5, pp. 108–113] and proven to be sound [5, pp. 147, 148] and complete [5, pp. 156–176]. The details of this proof system are beyond the scope of this paper.

7 Conclusion

We have studied a feature of the natural language of mathematics that has previously not been studied by other logicians or linguists, the *implicit dynamic function introduction*, exemplified by constructs of the form "for every x there is an $f(x)$ such that ...". If this feature is used without limitations, it yields a paradox analogous to Russell's paradox. Hence any formalism capturing it has to impose some limitations on it. We have sketched two higher-order extensions of Dynamic Predicate Logic, *Typed Higher-Order Dynamic Predicate Logic* (THODPL) and *Higher-Order Dynamic Predicate Logic* (HODPL), which capture this feature, and which differ only in the limitations they impose onto it. HODPL is based on *Ackermann-like Function Theory*, a novel foundational theory of functions that is inspired by Ackermann Set Theory and that interprets ZFC.

References

[1] W. Ackermann. Zur Axiomatik der Mengenlehre. *Mathematische Annalen*, 131:336–345, 1956.

[2] A. Church. A formulation of the simple theory of types. *Journal of Symbolic Logic*, 5:56–68, 1940.

[3] M. Cramer, B. Fisseni, P. Koepke, D. Kühlwein, B. Schröder, and J. Veldman. The Naproche Project – Controlled Natural Language Proof Checking of Mathematical Texts. In *CNL 2009 Workshop, LNAI 5972*, pages 170–186, 2010.

[4] M. Cramer, D. Kühlwein, and B. Schröder. Presupposition Projection and Accommodation in Mathematical Texts. In *Semantic Approaches in Natural Language Processing: Proceedings of the Conference on Natural Language Processing 2010 (KONVENS)*, pages 29–36. Universaar, 2010.

[5] Marcos Cramer. *Proof-checking mathematical texts in controlled natural language*. PhD thesis, Universität Bonn, 2013.

[6] Marcos Cramer. The Naproche system: Proof-checking mathematical texts in controlled natural language. *Sprache und Datenverarbeitung – International Journal for Language Data Processing*, 38(1-2):9–33, 2014.

[7] J. Groenendijk and M. Stokhof. Dynamic Predicate Logic. *Linguistics and Philosophy*, 14(1):39–100, 1991.

[8] M. Lackenby. Topology and Groups. Retrieved December 1, 2016, from `http://people.maths.ox.ac.uk/lackenby/tg050908.pdf`, 2008. Lecture Notes.

[9] W. Reinhardt. Ackermann's set theory equals ZF. *Annals of Mathematical Logic*, 2:189–249, 1970.

[10] W. Trench. *Introduction to Real Analysis*. Prentice Hall, 2003.

Locology and Localistic Logic: Mathematical and Epistemological Aspects

Michel De Glas
SPHERE, CNRS-Université Paris Diderot
5, rue Thomas Mann 75205 Paris Cedex 13 France
`michel.deglas@univ-paris-diderot.fr`

Abstract

The object of this paper is to present and thoroughly study a new logic, called localistic logic, the essential features of which are as follows. First, it relies upon a rejection of the positive paradox axiom and a weakening of the deduction theorem. Second, the localistic logic provides locology with a logical framework. Third, the concepts of prelocus and locus provide logic (and locology) with a categorical substratum.

Keywords: Locology, Localistic Logic, Prelocus, Locus, Constructivism.

Introduction

One may point out, in modern mathematics, many mathematical, logical and philosophical oppositions to Cantor's transfinite "paradise". As is well known, Kronecker, Poincaré, Brouwer, Weyl, Feferman, and some others are particularly reluctant to accept Cantor's conception of the continuum ("The actual infinite is not required for the mathematics of the physical world", Feferman says).

Surprisingly enough, topology has never really been touched by the criticisms on set theory and actual infinity, although it incorporates many problematic notions of set theory. For instance, unless one chooses to consider non-T_1 topological spaces (i.e. spaces of little mathematical significance and of practically no use in applied domains), boundaries are lines of Lebesgue-measure zero. Next, contrary to what intuition suggests, the operators of interior and closure are idempotent. Moreover, the concept of neighbourhood, which is supposed to model the notion of proximity

---

Talk at the conference "Philosophy, Mathematics, Linguistics: Aspects of Interaction 2012" (PhML-2012), held on May 22–25, 2012 at the Euler International Mathematical Institute.

or nearness, is somehow transitive. It is not difficult to prove that all these counter-intuitive and mathematically hard to accept situations are immediate consequences of the actual infinity, particularly of the atomic nature of the continuum.

The various attempts to generalize point-set topology take place in the course of "point elimination". It is on this road that one can meet abstract spaces (first studied by Hausdorff who took the notion of open set as a primitive in the study of continuity in such spaces), Heyting algebras (which arose from the epistemological deliberations of Brouwer), pointless topology (where open-set lattices are taken as primitive notions, irrespective of whether they are composed of points), point-wise, or formal, topology (an intuitionistic approach to topology, based upon Martin-Löf's type theory, which proves to be slightly more restrictive than pointless topology). A further step in the process of (pointless) abstraction may be taken by considering the category of locales (whose objects are complete lattices equipped with the infinite distributive law, and whose morphisms are maps preserving finite meets and arbitrary joins), which, according to many category-theorists is the structure within which pointless topology must be developed. Whatever one may think of the latter assertion, an essential feature of the results available is that they all invoke non constructive principles: the localistic framework allows to give classical theorems of topology constructive proofs. What one can gain by doing constructive topology is that there are contexts in which one may like to do "topology" but one does not wish to assume the law of excluded middle or the axiom of choice. Such contexts are called topoï. Apart from this alleged constructive aspect, there are nevertheless several results which say that, from one point of view (i.e., when one works with spatial locales), working with locales is doing nothing more than a disguised version of classical point-set topology. One may, of course, consider non spatial locales but very little work has been done on specific applications of such tools. Furthermore, large parts of the theory of locales can be internalized in any topos and a topos is nothing but a category which is sufficiently "like" the category of sets for one to carry out set-theoretical constructions inside it. Point-wise (or formal) topology is related to pointless topology by the adjunction (in the category-theoretic sense of the word) between the category of locales and the category of topological spaces. In the case of spatial locales, the adjunction reduces to an adjoint equivalence between the category of spatial locales and the category of sober topological spaces. Thus the point-wise and the pointless approaches are essentially equivalent as soon as one wishes to deal with spatiality.

The overall gain possibly provided by pointless or point-wise topology is thus quite limited. The basic reason is that, despite the generalization provided by the "elimination" of points and whatever the level of abstraction is, the algebraic structures implied by these approaches are essentially the same as those defined in a point-set topological framework.

Locology [1,2,5–7] has been elaborated by the author as an alternative to topology in order to provide new mathematically and philosophically acceptable and fruitful solutions to the above mentioned problems. It allows in particular, from the giving of a reflexive (and possibly symmetric) relation, which may be seen as a relation of resemblance or as the measure of a granularity over some carrier set, to redefine most concepts of topology in a more satisfactory way: the concepts of *core* and *shadow*, which are substituted for that of interior and closure, are not idempotent; to any subset in a locological space may be associated its *frontier* and its *boundary* (the former being divided into its inner and outer parts), the distinction between the two entities being of prime importance both from a mathematical and an epistemological viewpoint (mathematically speaking, a frontier has a certain "thickness"; epistemologically speaking, it allows to distinguish between punctuality and indivisibility); the relevant algebraic structure is that of a complete and complemented, but not distributive, lattice with a semi-implication. The distinction, in locology, between boundaries (which have no analogue in the world of "real" entities) and frontiers allows, in particular, to revisit some fundamental problems left open by topology (and mereotopology): that of contiguity and contact [2]. These problems originate from the set-theoretical and topological definition of the continuum and the consecutive failure in the treatment of boundaries. It also leads to formalize, in an essentially new way, the key concepts of categorization [7]. Locological spaces encompass Poincaré-Zeeman tolerances spaces [9, 10, 13], Choquet's pretopological spaces [3, 12], and mathematical morphology. The study of locological concepts and the structure thus implied allow to understand why these three (independent) streams of research have not been followed up.

The anti-realism at the root of the rejection of actual infinity and the Cantorian conception of the continuum is, as is well known, intimately related with the anti-realism (or anti-platonism) in logic which leads to substructural logics, in particular intuitionistic logic. However, the topology/locology alternative, sketched above, suggests, first, that the criticisms addressed to topology translate to intuitionistic logic, and, second that a new logic of which locology would be the "geometric" counterpart is needed.

Localistic logic, the definition and the study of which are the object of Section 2, meets this requirement. We first prove that, contrary to the intuitionists' claim, the law of excluded middle is in no way a principle of omniscience and is perfectly compatible with a constructive view of logic and mathematics, and that the excluded middle and the reductio ad absurdum are not in general mutually dependent. Next, localistic logic allows to revisit the question of the admissibility, from a constructivist viewpoint, of the thinning on the left (also called the positive paradox axiom), i.e.

$A \to (B \to A)$ and the thinning on the right, i.e. $A \to (\sim A \to B)$, which are admitted by both classical and intuitionistic logics. It is worth noticing that this question was raised by the first intuitionists: Kolmogorov rejected $A \to (\sim A \to B)$ but accepted $A \to (B \to A)$; Glivenko, whose axiomatization was the one adopted by Gentzen, raised the same question but eventually followed Heyting in keeping both thinning on the right and thinning on the left. This question has also been tackled by relevant logics, the first axiomatization of which was actually proposed by another Russian intuitionist, Orlov. The rejection of both thinning on the right and thinning on the left by relevant logics is the main departure from intuitionistic logic. However, the various versions of relevant logics fall short of an interesting semantics (i.e. a semantics where the truth-values may be expressed in terms of classes of objects).

Localistic logic leads to rejection of $A \to (B \to A)$ on the basis that, if A may be derived from a set Γ of hypotheses ($\Gamma \vdash A$) then there is no reason that, for any B, $B \to A$ may be derived *gratis prodeo* ($\Gamma \vdash B \to A$). However, this may hold for some Γ's, in particular for $\Gamma = \emptyset$: if $\vdash A$ then $\vdash B \to A$ (a theorem may be derived from anything). What relevantists did not actually realize is that a theorem is more than a formula deduced from an empty set of hypotheses. As far as thinning on the right is concerned, the localistic argument is as follows: $A \to (\sim A \to B)$ being related to the reductio ad absurdum, its (in)admissibility depends on some further assumptions. A study of propositional and predicate logics is performed in Section 2. An essential feature is, as alluded to above, *the weakening of the deduction theorem*.

Section 3 is devoted to the study of the categorical substratum of localistic logic. It is shown that the *theory of (pre)loci* provides localistic logic with a category-theoretic basis. However, the role played by localistic logic for loci theory is not quite analogous to that played by topoï theory for intuitionistic logic. Indeed localistic logic cannot be seen as an internal logic of a locus. On the contrary, it may be seen as *emerging* from a (pre)locus. We may prove however the equivalence between locus-validity and localistic provability.

1 Locology

Let X be a set and let λ be a reflexive relation on X: $x\lambda x$, for any x in X. The relation λ is to be thought of as a resemblance or an indistinguishability relation on X. The set $\lambda[x] = \{y : x\lambda y\}$ of λ-relatives of x, the elements of which may be seen as being close to (or resembling, or being indistinguishable from) x, is called the halo of x. As λ may be defined as a map $X \longrightarrow \wp(X)$, $x \longmapsto \lambda[x]$, we denote by $\lambda(A)$ the set

$$\lambda(A) = \bigcup_{x \in A} \lambda[x]$$

so that $\lambda[x] = \lambda(\{x\})$. Next, one defines the following two operators h and s which associate to any A in $\wp(X)$ its *core* and its *shadow* respectively. More precisely, let $h : \wp(X) \longrightarrow \wp(X)$ be the operator which associates to any A its core

$$h(A) = \{x \in X : \lambda[x] \subseteq A\}.$$

Immediate properties follow:

(1) $h(A) \subseteq A, h(X) = X$,
(2) If $A \subseteq B$ then $h(A) \subseteq h(B)$,
(3) $h \circ h(A) \subseteq h(A)$,
(4) $h(A \cup B) \supseteq h(A) \cup h(B)$,
(5) $h(\bigcap_i A_i) = \bigcap_i h(A_i)$.

It is worth emphasizing that, contrary to the properties of an interior operator in topology, h is not idempotent (unless λ is assumed to be transitive). On the other hand, property (5) holds for infinite intersections.

The *shadow* operator is defined in a dual way. It associates to any $A \in \wp(X)$ its shadow $s(A)$ defined by

$$s(A) = \{x \in X : \lambda[x] \cap A \neq \emptyset\}.$$

The operators h and s are interdefinable since clearly

$$s(A) = \overline{h(\overline{A})},$$

where $\overline{A}$ denotes the complement of A. Immediate properties of s then come out:

(1) $s(A) \supseteq A$, $s(\emptyset) = \emptyset$,
(2) If $A \subseteq B$, then $s(A) \subseteq s(B)$,
(3) $s \circ s(A) \supseteq s(A)$,
(4) $s(A \cap B) \subseteq s(A) \cap s(B)$,
(5) $s(\bigcup_i A_i) = \bigcup_i s(A_i)$.

Like h, the operator s is not idempotent and, unlike topology, equality (5) holds for infinite unions.

The idea of using a reflexive (and symmetric) relation to recapture the intuitive notion of indistinguishability is not new. That of having recourse to non idempotent "interior" or "closure" operators is not without predecessors either. The former idea can be traced back to Poincaré's works on the physical continuum. As claimed by Poincaré [9], "the raw result of experience may be expressed by the relation $A = B, B = C, A < C$, which may be taken as a formula of the physical continuum". Here, $A = B$ is to be understood as "A and B are indistinguishable", and $A = B$ is then a reflexive and symmetric relation over the collection of entities under study. This approach was exploited by Zeeman [13] in his works on tolerance spaces.

The idea of a non idempotent closure operator can be traced back to Choquet's paper on pretopology [3]. Such an operator is nowadays referred to as a Čech closure operator [12]. Depending upon the properties it is equipped with (isotony, accretivity, sub-linearity, ...) the resulting spaces are called extended topologies [8], neighbourhood spaces [4], Smyth spaces [11], or pretopologies.

However these two streams of research have not been followed up. The basic reasons seems to be the following. The Poincaré-Zeeman approach is essentially geometric and is then deprived of an algebraic (and a logical) content: there is nothing, in tolerance spaces, which can play the role of the lattice of open sets in a topological space. In a symmetric way, the approaches pertaining to the stream initiated by Choquet have a poor geometric content. Furthermore, they lead to very poor algebraic structures too: as a generalization of the Kuratowski closure algebra, the algebra $\{cl(A) : A \subset X\}$, where cl denotes the generalized closure operator, fails, for instance, to be sup-complete, so that the disjunction of objects of the algebra cannot be defined. If A and B are "closed" sets, nothing can be said of the entity "A and B" (apart from the fact that, in general, $A \cap B$ is not "closed"). These limitations are insurmountable.

Owing to the property (5) of h and s, the corresponding algebras, as will be seen below, have much stronger properties. Many results may then be derived, most of which are not derivable in a generalized closure space or in a tolerance space.

We consider the two families:

$$\mathcal{L} = \{h(A) : A \subseteq X\},$$

$$\mathcal{K} = \{s(A) : A \subseteq X\}.$$

It is clear that $\mathcal{L} = \{\bar{A} : A \in \mathcal{K}\}$ and $\mathcal{K} = \{\bar{A} : A \in \mathcal{L}\}$. In view of the properties of h, $(\mathcal{L}, \cap)$ is a complete and bounded inf-semi-lattice. However, for A and B in $\mathcal{L}$, $A \cup B$ may not be an element of $\mathcal{L}$. Indeed, given A and B there may not exist $C \in \wp(X)$ such that $h(C) = A \cup B$. Hence $(\mathcal{L}, \cup)$ is not a sup-semi-lattice. We may however define, for A and B in $\mathcal{L}$:

$$A \sqcup B = \bigcap \{C \in \mathcal{L} : C \supseteq A, C \supseteq B\},$$

the existence of which is guaranteed by the inf-completeness and the boundedness of $\mathcal{L}$. $A \sqcup B$ is thus the least upper bound of the set of objects of $\mathcal{L}$ which contain

$A \cup B$. Thus $(\mathcal{L}, \cap, \sqcup)$ is a complete lattice. But it fails to be distributive. Indeed we may have $A \cap B = A \cap C$ and $A \sqcup B = A \sqcup C$ and $B \neq C$, the equality $B = C$ being a necessary and sufficient condition for a lattice to be distributive. Furthermore

$$h \circ \lambda(A) = \bigcap \{B \in \mathcal{L} : B \supseteq A\},$$

for any $A \subseteq X$. Hence

$$\lambda \circ h(A) \subseteq A \subseteq h \circ \lambda(A),$$

where the equality holds on the right-hand side iff $A \in \mathcal{L}$, and then

$$A \sqcup B = h \circ \lambda(A \cup B),$$

for any $A, B \in \mathcal{L}$.

The lattice $(\mathcal{L}, \cap, \sqcup)$ is called a *locology* and X a *locological space*. Despite the differences between locology and topology (non-distributivity of $\mathcal{L}$, completeness of $\mathcal{L}$, non-idempotency of h and s), the consequences of which are of prime importance, it is quite clear that the objects of $\mathcal{L}$ bear some resemblance with open sets in a topological space. However, $\mathcal{L}$ may be defined as

$$\mathcal{L} = \{A \subseteq X : A = h \circ \lambda(A)\}.$$

The operator $h \circ \lambda$ is accretive, order-preserving, and idempotent. Hence $h \circ \lambda$ is an algebraic closure operator, i.e. a closure operator in $\wp(X)$ viewed as an ordered set. Of course, it is not a topological closure operator since e.g. $h \circ \lambda(A \cup B) \neq h \circ \lambda(A) \cup h \circ \lambda(B)$. This means that the objects of $\mathcal{L}$ have something in common with *closed sets* in topology.

A dual analysis may be carried out for the algebra

$$\mathcal{K} = \{s(A) : A \subseteq X\}.$$

Indeed, if one defines, for $A, B \in \mathcal{K}$,

$$A \sqcap B = \bigcup \{C \in \mathcal{K} : C \subseteq A, C \subseteq B\},$$

then $(\mathcal{K}, \sqcap, \cup)$ is a complete, but not distributive lattice, where, furthermore:

$$\overline{h \circ \lambda(\overline{A})} = \bigcup \{C \in \mathcal{K} : B \subseteq A\},$$

$$A \sqcap B = \overline{h \circ \lambda(\overline{A \cap B})},$$

and, if λ is symmetric,

$$\overline{h \circ \lambda(\overline{A})} = \lambda \circ h(A)$$

$$A \cap B = \lambda \circ h(A \cap B)$$

The fact that $\mathcal{K}$ may be rewritten as

$$\mathcal{K} = \{A \subseteq X : A = \overline{h \circ \lambda(\bar{A})}\} = \{A \subseteq X : \bar{A} = h \circ \lambda(\bar{A})\}$$

and, if λ is symmetric, as

$$\mathcal{K} = \{A \subseteq X : A = \lambda \circ h(A)\}$$

shows that the objects of $\mathcal{K}$ have a superficial similarity with closed sets in a topological space and a deeper resemblance with *open* sets in a topological space.

This may seem paradoxical at first glance. It is, however, a key point of locology. To more accurately specify this, one first has to revisit the concept of a boundary. The critical analysis of the concept of a boundary in topology leads us to actually define two different concepts: to any region A, one associates its *frontier* and its *boundary*, the former being divided into its inner and outer parts.

To any $A \subseteq X$, one associates its *inner frontier* $\partial_{in}(A) = \lambda(\bar{A}) \cap A$ and its *outer frontier* $\partial_{out}(A) = \lambda(A) \cap \bar{A}$. The *frontier* of A is then $\partial(A) = \partial_{in}(A) \cup \partial_{out}(A) = \lambda(A) \cap \lambda(\bar{A})$. Among many properties, which follow from their definition, a remarkable property is the idempotency of ∂_{in} and ∂_{out} which follows from

$$h[\partial_{in}(A)] = h[\partial_{out}(A)] = \emptyset.$$

The epistemological significance of these equalities is that locology allows us to distinguish between punctuality and indivisibility (these two notions being unduly identified to each other in topology). Indeed, $\partial_{in}(A)$ and $\partial_{out}(A)$ may be considered, on the one hand, as indivisible since they have empty cores and, on the other hand, as having a certain "thickness" (unless λ coincides with the diagonal of the carrier set X).

One may now define the concept of a boundary. The *boundary* of $A \subseteq X$ is defined to be the core of its frontier

$$\beta(A) = h(\partial(A)).$$

Since $\partial(A) = \lambda(A) \cap \lambda(\bar{A})$, one has

$$\beta(A) = h \circ \lambda(A) \cap h \circ \lambda(\bar{A}).$$

Hence

(1) $\beta(A) = \beta(\bar{A}) \in \mathcal{L}$,

(2) $A \smallsetminus \beta(A) = \lambda \circ h(A)$,
(3) $A \cup \beta(A) = h \circ \lambda(A)$,
(4) $A \in \mathcal{L}$ iff $\beta(A) \subseteq A$,
(5) $A \in \mathcal{K}$ iff $A \cap \beta(A) = \varnothing$.

Thus, as already alluded to above, objects of $\mathcal{L}$ and $\mathcal{K}$ have properties in common with closed sets and open sets in topology, respectively: any A in $\mathcal{L}$ contains its boundary; any A in $\mathcal{K}$ is disjoint from its boundary.

This shows that locology allows us, not only, to define purely locological concepts (the core h, the shadow s, the frontiers ∂_{in} and ∂_{out}) which have no counterpart in topology, but also concepts that may be seen as "quasi topological" (the operators $h \circ \lambda$ and $\lambda \circ h$, the boundary β) with common features and essential differences with their topological pendants. It may also be shown that topology is the limit case of locology corresponding to an infinitely small granularity.

Given a locological space X, one may define in $\mathcal{L}$ the unary operator $\neg$ by setting $\neg A = h(\bar{A})$, the core of the complement of A. The operator $\neg$ clearly satisfies

(1) $A \cap \neg A = \varnothing$,
(2) $A \cup \neg A \neq X$,
(3) $A \sqcup \neg A = X$,
(4) $\neg\neg A = A$ iff λ is symmetric.

Hence, $\neg$ is a complementation in $\mathcal{L}$ and an orthocomplementation iff λ is symmetric (these properties being not equivalent in a non distributive lattice). Anticipating on the next section, this translates into logical terms as follows. First, (1)-(3) show that, contrary to the intuitionists' claim, *the law of excluded middle is in no way a principle of omniscience*: for any $A \in \mathcal{L}$, A and $\neg A$ are disjoint and their disjunction $A \sqcup \neg A$ covers the universe, but there may exist objects of X that do not belong to either of A and $\neg A$. Second, (3) and (4) show that the law of excluded middle and the reductio ad absurdum may not be interdependent.

Next, for any A and B in $\mathcal{L}$, let $\Rightarrow$ be the binary operator defined by

$$A \Rightarrow B = h(\bar{A} \cup B),$$

the essential properties of which are

(1) $A \Rightarrow B = X$ iff $A \subseteq B$,
(2) $A \Rightarrow \varnothing = \neg A$,
(3) $A \cap (A \Rightarrow B) \subseteq B$,

(4) $(A \Rightarrow B) \cap (A \Rightarrow C) \subseteq A \Rightarrow (B \cap C)$,
(5) $(A \Rightarrow B) \cap (B \Rightarrow C) \subseteq A \Rightarrow C$,
(6) $\dfrac{C \subseteq A \Rightarrow B}{A \cap C \subseteq B}$.

However, the reciprocal to (6), i.e.
$$\dfrac{A \cap C \subseteq B}{C \subseteq A \Rightarrow B}$$
holds only if $\mathcal{L}$ is distributive (in which case $\mathcal{L}$ is a Boolean algebra) i.e. only if λ is transitive. Therefore, $\Rightarrow$ is not, strictly speaking, an implication. It implies, in particular, that
$$\dfrac{A = X}{B \Rightarrow A = X}$$
holds but
$$A \Rightarrow (B \Rightarrow A) \neq X.$$

Anticipating, once again, on the following section, this inequality translates into the non-validity of the *positive paradox axiom* (thinning on the left).

Similarly, although
$$\dfrac{A \Rightarrow C = X \quad B \Rightarrow C = X}{(A \sqcup B) \Rightarrow C = X}$$
holds, one generally has
$$(A \Rightarrow C) \cap (B \Rightarrow C) \nsubseteq (A \sqcup B) \Rightarrow C.$$

The properties of $\Rightarrow$ (called, from now on, a *semi-implication*) and of the disjunction $\sqcup$, as compared to those enjoyed by the corresponding operators in a Boolean or a Heyting algebra (hence in classical and intuitionistic logics), are essential features of the locological framework from a logico-algebraic viewpoint.

One may then define, as a natural abstraction of the above algebraic structure, the concept of Λ-algebra. A Λ-*algebra* is a 5-tuple $(\mathcal{L}, \wedge, \vee, \neg, \Rightarrow)$ such that

(1) $(L, \wedge, \vee)$ is a lattice with a least element 0 and a greatest element 1.

(2) $(L, \Rightarrow)$ satisfies
   $a \Rightarrow b = 1$ iff $a \leq b$,
   $a \wedge (a \Rightarrow b) \leq b$,
   $(a \Rightarrow b) \wedge (b \Rightarrow c) \leq (a \Rightarrow c)$,
   $(a \Rightarrow b) \wedge (a \Rightarrow c) \leq a \Rightarrow (b \wedge c)$.

(3) $(L, \neg)$ satisfies
$\neg a = a \Rightarrow 0$,
$\neg\neg a = a$.

where $a \leq b$ iff $a \wedge b = a$.

Theorem 1.1. *In a Λ-algebra L, the following properties hold:*

(a) $\neg$ *is an order-reversing involution.*

(b) $a \wedge \neg a = 0$.

(c) $\neg(a \vee b) = \neg a \wedge \neg b$; $\neg(a \wedge b) = \neg a \vee \neg b$.

(d) $a \Rightarrow b$ *is increasing wrt a and decreasing wrt b.*

(e) $a \vee \neg a = 1$.

(f) $(a \Rightarrow b) \wedge (a \Rightarrow c) = a \Rightarrow (b \wedge c)$.

(g) $(a \vee b) \Rightarrow c \leq (a \Rightarrow c) \wedge (b \Rightarrow c)$.

(h) *if $c \leq a \Rightarrow b$ then $a \wedge c \leq b$.*

(i) *if $a \wedge c \leq b$ then $1 \Rightarrow c \leq a \Rightarrow b$.* □

From now on, we will consider locologies whose underlying relation λ is symmetric, i.e., locologies that are orthocomplemented, as lattices.

Theorem 1.2. (a) *Any locology is a Λ-algebra.* (b) *Any Λ-algebra $(L, \wedge, \vee, \neg, \Rightarrow)$ may be embedded into a locology on some set.*

Proof. (a) is obvious. To prove (b), consider the MacNeille completion L^* of L, i.e.

$$L^* = \{A^* : A \in L\} \subset \wp(L)$$

where A^* is defined, for any $A \in L$, by

$$M_A = \{m : a \leq m, \text{for any } a \in A\},$$

$$A^* = \{x : x \leq m, \text{for any } m \in A\}.$$

If $A = \emptyset$ then $M_A = L$ and $A^* = 0$ where 0 is the least element of A. Hence $(L^*, \subseteq, \cap)$ is a complete inf-semi-lattice with a least element 0. It may then be equipped with the structure of a complete lattice $(L^*, \subseteq, \cap, \vee^*)$ by setting

$$A^* \vee^* B^* = \bigcap \{C \in L^* : C \supseteq A^*, C \supseteq B^*\}.$$

The image under the transformation $A \longmapsto A^*$ of a singleton a of L is the subset $\{b \in L : b \le a\}$ which will be denoted $(a \downarrow)$. Let $\mathcal{T}$ be the set $\{(a \downarrow) : a \in L\}$. The mapping $a \longmapsto (a \downarrow) \in \mathcal{T}$ is obviously one-to-one and onto. Since $a \le b$ in L iff $(a \downarrow) \subseteq (b \downarrow)$ in $\mathcal{T}$, identifying $\{a\}$ with a leads to considering a mapping $f : L \longrightarrow \mathcal{T} \subseteq L^*$, $a \longmapsto (a \downarrow)$. One may then easily prove that f is a monomorphism. □

2 Localistic logic

The language of propositional localistic logic (LL for short) has an alphabet consisting of proposition symbols: $p_0, p_1, \ldots$, connectors: $\wedge, \vee, \rightarrow, \leftrightarrow, \bot$, and auxiliary symbols: (,). The set Φ of formulas is the smallest set X such that
    (1) $p_i \in X$, $i \in \mathbb{N}$, $\bot \in X$,
    (2) If $\phi, \psi \in X$, then $\phi \wedge \psi$, $\phi \vee \psi$, $\phi \rightarrow \psi \in X$.

The axioms and the inference rules for propositional LL are instances of one of the following forms, where $\sim \phi$ stands for $\phi \rightarrow \bot$ and $\phi \leftrightarrow \psi$ stands for $(\phi \rightarrow \psi) \wedge (\psi \rightarrow \phi)$:

Axioms

    A1 $\bot \rightarrow \phi$

    A2a $\phi \wedge \psi \rightarrow \phi$

    A2b $\phi \wedge \psi \rightarrow \psi$

    A3 $((\phi \rightarrow \psi) \wedge (\phi \rightarrow \chi)) \rightarrow (\phi \rightarrow (\psi \wedge \chi))$

    A4 $(\phi \rightarrow \psi) \wedge (\psi \rightarrow \chi) \rightarrow (\phi \rightarrow \chi)$

    A5a $\phi \rightarrow \phi \vee \psi$

    A5b $\psi \rightarrow \phi \vee \psi$

    A6 $(\phi \wedge (\phi \rightarrow \psi)) \rightarrow \psi$

    A7 $\sim\sim \phi \leftrightarrow \phi$

Inference rules

R1 $\quad \dfrac{\phi \quad \phi \to \psi}{\psi}$

R2 $\quad \dfrac{\phi}{\psi \to \phi}$

R3 $\quad \dfrac{\phi \to \chi \quad \psi \to \chi}{(\phi \vee \psi) \to \chi}$

A formula is said to be *provable*, denoted $\vdash_{LL} \phi$ or simply $\vdash \phi$, iff there exists a sequence $\phi_1, \phi_2, \ldots, \phi_n$ of formulas such that $\phi_n = \phi$ and, for any $i \leq n$, ϕ_i is either an axiom or follows form earlier formulas in the sequence by a rule of inference from {R1, R2, R3}. A *valuation* v is a mapping $v : \phi \longrightarrow (L, \wedge, \vee, \neg, \Rightarrow)$, where L is a Λ-algebra, such that

$$v(\bot) = 0,$$
$$v(\phi \wedge \psi) = v(\phi) \wedge v(\psi),$$
$$v(\phi \vee \psi) = v(\phi) \vee v(\psi),$$
$$v(\phi \to \psi) = (\phi) \Rightarrow v(\psi),$$
$$v(\sim \phi) = \neg v(\phi).$$

Let L be a Λ-algebra. A formula $\phi \in \Phi$ is *L-valid* iff, for any algebra $v, v(\phi) = 1$, the greatest element of L.

Theorem 2.1. $\vdash \phi$ *iff* ϕ *is L-valid for any Λ-algebra L.*

The above completeness theorem deals only with the equivalence between *provability* and validity in a Λ-algebra. One now has to consider deducibility from a set Γ of formulas (which, as usual, may be thought of as hypotheses). Unlike the classical and the intuitionistic cases, the extension from provability ($\vdash \phi$) to deducibility from Γ ($\Gamma \vdash \phi$) is far from obvious. It leads, in particular, to the non-validity of the deduction theorem.

We say that ϕ is *deducible* from a set Γ of formulas, denoted $\Gamma \vdash \phi$, iff either ϕ is provable ($\vdash \phi$) or there exists a sequence $\phi_1, \ldots, \phi_n = \phi$ of formulas such that each ϕ_i is either an axiom or a formula of Γ or follows from earlier formulas in the sequence by the inference rule R1.

We say that ϕ is Γ-valid, denoted $\Gamma \models \phi$ iff, for any L and any valuation v, there exists $\Gamma_0 \subseteq \Gamma, \Gamma_0$ finite, such that

$$\bigwedge_{\gamma \in \Gamma_0} v(\gamma) \leq v(\phi).$$

Theorem 2.2. *If $\phi \vdash \psi$, then $\vdash \phi \to \psi$.*

Proof. Two cases have to be considered. If $\vdash \psi$ then $\vdash \phi \to \psi$ by R2. If $\not\vdash \psi$ then $\phi \vdash \psi$ means that ψ may be deduced from ϕ, A1-A7 plus R1. As R1 is the only inference rule, unless ψ and ϕ are the same formula (in which case the statement of the theorem is trivially true), $\phi \vdash \phi \to \psi$. But this holds iff $\vdash \phi \to \psi$. □

We seem to be well on the way to a proof of the deduction theorem. Indeed, from theorem 2.2 which asserts

$$\frac{\phi \vdash \psi}{\vdash \phi \to \psi}$$

and consequently

$$\frac{\phi_1, \phi_2, \ldots, \phi_n \vdash \psi}{\vdash (\phi_1 \wedge \phi_2 \wedge \cdots \wedge \phi_n) \to \psi}$$

we might expect that, for any Γ,

$$\frac{\Gamma \vdash \phi}{\Gamma \vdash \psi \to \phi}.$$

However, we have the

Theorem 2.3. *$\Gamma \vdash \phi$ does not entail $\Gamma \vdash \psi \to \phi$.*

Proof. Suppose that $\Gamma \vdash \phi$ implies $\Gamma \vdash \psi \to \phi$ for any ψ. Then, applying theorem 2.2 above and theorem 2.7 below, $\Gamma \vDash \phi$ implies $\Gamma \vDash \psi \to \phi$ for any ψ. A necessary and sufficient condition is that $v(\psi) \Rightarrow v(\phi) \geq v(\phi)$ for any valuation v. Such an inequality generally does not hold. □

We must emphasize the contrast between the validity of

$$\frac{\psi \vdash \phi}{\vdash \psi \to \phi}$$

and the non-validity of

$$\frac{\Gamma, \psi \vdash \phi}{\Gamma \vdash \psi \to \phi}$$

i.e. of the deduction theorem. The validity of the former means that either $\vdash \phi$ and then $\vdash \psi \to \phi$ (a theorem may be deduced from anything) or $\not\vdash \phi$ in which case $\psi \vdash \phi$ iff $\vdash \psi \to \phi$, i.e. iff $\psi \to \phi$ has already been proved. The validity of the latter would mean that if ϕ may be deduced from $\Gamma \cup \{\psi\}$ then, irrespective of whether ψ (or $\psi \to \phi$) appears or not in the deduction of ϕ, $\psi \to \phi$ may be deduced, *gratis prodeo*, from Γ. Such a scheme is not allowed in a localistic framework. Although highly non constructive, this derivation is possible in intuitionistic logic.

It must be clear that the weakening of the deduction theorem (or, equivalently, the non-validity of the positive paradox axiom $\phi \to (\psi \to \phi)$) is the logical counterpart of the non-distributivity of a Λ-algebra, which itself is the translation in algebraic terms of the non-idempotency of the shadow and the core operators in a locological space.

A set Γ of formulas is said to be consistent if $\Gamma \nvdash \bot$. Otherwise, it is said to be inconsistent. Γ is said to be complete iff, for any formula ϕ, $\Gamma \vdash \phi$ or $\Gamma \vdash \sim \phi$.

Theorem 2.4. (a) *If $\Gamma \cup \{\phi\}$ is consistent then $\Gamma \nvdash \sim \phi$.* (b) *If Γ is complete, then the reciprocal to (a) holds.*

Proof. (a) Let $\Gamma \cup \{\phi\}$ be consistent and suppose that $\Gamma \vdash \sim \phi$. Then $\Gamma, \phi \vdash \sim \phi$. But, since $\Gamma, \phi \vdash \phi$, then $\Gamma, \phi \vdash \bot$, a contradiction.

(b) If $\Gamma \nvdash \sim \phi$, since Γ is complete, $\Gamma \vdash \phi$. Suppose that $\Gamma, \phi \vdash \bot$. Then $\Gamma \vdash \bot$, and $\Gamma \vdash \sim \phi$, a contradiction. □

Theorem 2.5. *Γ is consistent iff Γ is satisfiable.*

Proof. (a) If Γ is satisfiable, then there exists a valuation v such that $v(\gamma) = 1$ for any $\gamma \in \Gamma$. Suppose Γ is not consistent. Then $\Gamma \vdash \bot$ and therefore $\Gamma_0 \vdash \bot$ for some $\Gamma_0 \subseteq \Gamma$, Γ_0 finite. Setting $\Gamma_0 = \{\gamma_1, \gamma_2, \ldots, \gamma_n\}$ leads to $\gamma_1, \gamma_2, \ldots, \gamma_n \vdash \bot$, i.e. $\vdash \sim(\gamma_1 \wedge \gamma_2 \wedge \cdots \wedge \gamma_n)$ and $\vDash \sim (\gamma_1 \wedge \gamma_2 \wedge \cdots \wedge \gamma_n)$. Thus, $v(\gamma_1 \wedge \gamma_2 \wedge \cdots \wedge \gamma_n) = 0$, a contradiction.

(b) Let Γ be consistent. From part (a) of theorem 2.4, if γ is consistent then $\nvdash \sim \gamma$ and, consequently, $\nvDash \sim \gamma$. Suppose Γ is not satisfiable, in which case there exists $\gamma \in \Gamma$ such that, for any $v, v(\gamma) < 1$. By induction on the length of γ, $v(\gamma) = 0$ for any v, i.e. $v(\sim \gamma) = 1$, for any v. Thus $\vDash \sim \gamma$, a contradiction. □

Theorem 2.6. *If $\Gamma \vdash \phi$ then $\Gamma \vDash \phi$.*

Proof. If $\vdash \phi$ then $\vDash \phi$, whence $\Gamma \vDash \phi$. If $\Gamma \vdash \phi$ (and $\nvdash \phi$), then there exists $\Gamma_0 = \{\gamma_1, \gamma_2, \ldots, \gamma_n\} \subseteq \Gamma$ such that $\Gamma_0 \vdash \phi$. Then $\vdash (\gamma_1 \wedge \gamma_2 \wedge \cdots \wedge \gamma_n) \to \phi$ and $\vDash (\gamma_1 \wedge \gamma_2 \wedge \cdots \wedge \gamma_n) \to \phi$. Thus $\Gamma_0 \vDash \phi$ and $\Gamma \vDash \phi$. □

Theorem 2.7. *If $\Gamma \vDash \phi$, then $\Gamma \vdash \phi$.*

Proof. (a) If Γ is finite, the reciprocal to theorem 2.6 clearly holds. Indeed, if $\Gamma = \{\gamma_1, \gamma_2, \ldots, \gamma_n\}$ then $\gamma_1, \gamma_2, \ldots, \gamma_n \vDash \phi$ entails $\vDash (\gamma_1 \wedge \gamma_2 \wedge \cdots \wedge \gamma_n) \to \phi$ hence $\vdash (\gamma_1 \wedge \gamma_2 \wedge \cdots \wedge \gamma_n) \to \phi$ i.e. $\gamma_1, \gamma_2, \ldots, \gamma_n \vdash \phi$.

(b) If Γ is infinite, there exists $\Gamma_0 \subseteq \Gamma$, Γ_0 finite, such that
$$\bigwedge_{\gamma \in \Gamma_0} v(\gamma) \leq v(\phi),$$
i.e. $\Gamma_0 \vDash \phi$. Applying (a) leads to $\Gamma_0 \vdash \phi$. Thus $\Gamma \vdash \phi$. □

3 Categorical substratum

The aim of this section is to provide localistic logic (and locology) with a categorical substratum which would play to some extent the role played by topoï theory and set theory for intuitionistic and classical logic respectively. As a Λ-algebra is a non distributive lattice the disjunction of which is weaker than the intuitionistic and the classical ones, it is a priori quite clear that the required categorical framework must exhibit a weakened form of exponentiation.

3.1 Preloci

A category $\mathcal{C}$ is said to have semi-exponentiation if

(1) any pair $\langle A, B \rangle$ of objects of $\mathcal{C}$ has product $A \times B$,

(2) for any pair $\langle A, B \rangle$ of objects, there is an object B^A and an arrow $e: B^A \times A \longrightarrow B$ such that, for any $g: C \times A \longrightarrow B$, there exists at most one arrow $\hat{g}: C \longrightarrow B^A$ such that the diagram

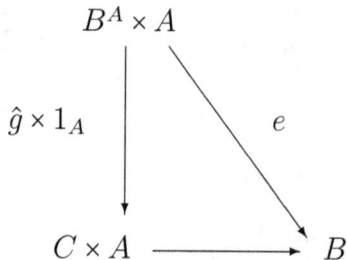

commutes. If, for a given g, $\hat{g}$ exists, we will write
$$\frac{g: C \times A \longrightarrow B}{\hat{g}: C \longrightarrow B^A}$$
where g and $\hat{g}$ may be omitted.

(3) The following rules hold

(a) $\dfrac{A \longrightarrow B}{1 \longrightarrow B^A}$,

(b) $\dfrac{C^B \times B^A \times A \longrightarrow C}{C^B \times B^A \longrightarrow C^A}$,

(c) $\dfrac{B^A \times C^A \times A \longrightarrow B \times C}{B^A \times C^A \longrightarrow (B \times C)^A}$.

The arrow e is called the *evaluation arrow*. The arrow $\hat{g}$, if it exists, is called the *exponential adjoint* of g.

Theorem 3.1. *The correspondence $g \longmapsto \hat{g}$ is bijective.* □

Although the correspondence $g \longmapsto \hat{g}$ is bijective, $\mathrm{Hom}(C \times A, B)$ and $\mathrm{Hom}(C, B^A)$ are not isomorphic. As $\hat{g}$ may not exist, the correspondence is not necessarily a total function.

A category $\mathcal{A}$ which has

(1) a terminal object 1,

(2) a pullback and a pushout for each pair of arrows,

(3) semi-exponentiation

will be called a *prelocus* (plural: *preloci*). Clearly, a prelocus has initial object 0, for any pair $\langle A, B \rangle$ of objects, a product defined by an object $A \times B$ together with projections $\pi_{A,B} : A \times B \longrightarrow A$ and $\pi'_{A,B} : A \times B \longrightarrow B$ and a coproduct given by an object $A + B$ and injection arrows $\jmath_{A,B} : A \longrightarrow A + B$ and $\jmath'_{A+B} : B \longrightarrow A + B$.

Theorem 3.2. *In a prelocus $\mathcal{A}$, the following properties hold*
(1) $0 \cong 0 \times A$, *for any object A.*
(2) *If there exists an arrow $A \longrightarrow 0$, then $A \cong 0$ and the arrow is a mono.* □

3.2 The algebra of subobjects in a prelocus

Let $\mathcal{A}$ be a prelocus and let X be an object of $\mathcal{A}$. First recall that a subobject of X is defined as follows. Given two monos $f : A \twoheadrightarrow X$ and $g : B \twoheadrightarrow X$, one sets $f \subseteq g$ iff there exists $h : A \twoheadrightarrow B$ such that $f = g \circ h$. Then, the relation $\simeq$ defined by

$f \simeq g$ iff $f \subseteq g$ and $g \subseteq f$ is an equivalence on the set of monos with codomain X. Furthermore, if $f \simeq g$, there exists an iso $k : B \longrightarrow X$ with inverse $h : A \longrightarrow X$ such that $f = g \circ h$ and $g = f \circ k$. The equivalence class of f modulo $\simeq$ is denoted $[f]$ and is said to be a subobject of X. The set $\mathrm{Sub}(X)$ of subobjects of X is thus

$$\mathrm{Sub}(X) = \{[f] : f : A \twoheadrightarrow X, \text{some } A\}.$$

We will usually write "the subobject f" when we mean "the subobject $[f]$".

3.2.1 Conjunction

Let $f : A \twoheadrightarrow X$ and $g : B \twoheadrightarrow X$ be two subobjects of X. The *conjunction* of f and g is defined to be the pullback of f and g, i.e. the subobject $f \cap g : A \times_X B \twoheadrightarrow X$ such that

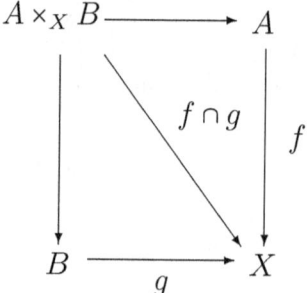

is a pullback square. The subobject $f \cap g$ is thus defined up to isomorphism.

3.2.2 Disjunction

The *disjunction* of two subobjects $f : A \twoheadrightarrow X$ and $g : B \twoheadrightarrow X$ of X is the subobject $f \cup g$ of X such that the diagram

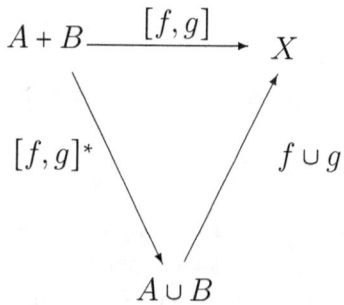

is an epi-mono factorization. In other words, $f \cup g$ is the image of the coproduct arrow $[f,g]$ (the least subobject of X through which $[f,g]$ factors) and $A \cup B \cong [f,g]^*(A+B), [f,g]^*$ being an epi.

Theorem 3.3. $(\mathrm{Sub}(X), \subseteq, \cap, \cup)$ *is a lattice with a least element* 0_X *and a greatest element* 1_X. □

However, $\mathrm{Sub}(X)$ is not, in general, distributive. Indeed, let $f : A \twoheadrightarrow X$ and $g : B \twoheadrightarrow X$ be subobjects of X such that

$$f \cap g \simeq f \cap h \simeq 0_X,$$

$$f \cup g \simeq f \cup h.$$

We then have the following commutative diagrams

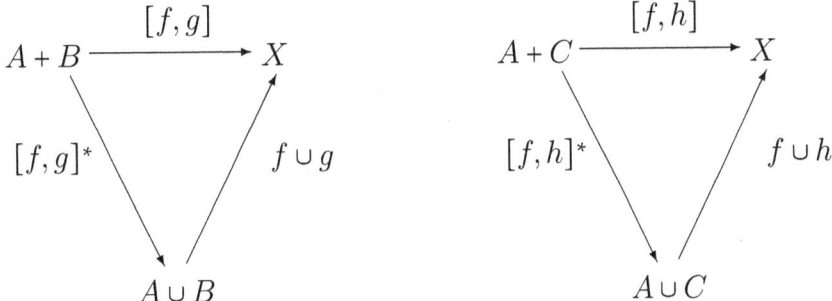

i.e. $f \cup g \simeq [f,g]$ and $f \cup h \simeq [f,h]$. But, $g \not\simeq h$ since $B \not\simeq C$. Thus, in general, $\mathrm{Sub}(X)$ is not distributive.

3.2.3 Semi-implication

Let $f : A \twoheadrightarrow X$ and $g : B \twoheadrightarrow X$ be two subobjects of X. We define $f \Rightarrow g : B^A \twoheadrightarrow X$ as a subobject of X such that the diagram

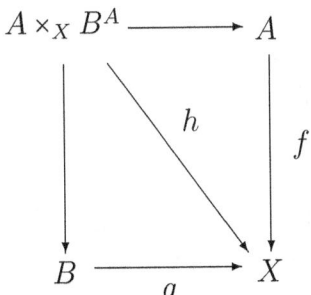

where $h = f \cap (f \Rightarrow g)$, commutes. From the definition of $f \cap g$, we have the following commutative diagram

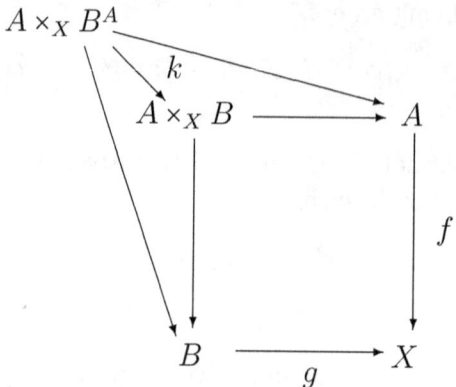

Hence $(f \cap g) \circ k \approx f \cap (f \Rightarrow g)$. Thus

$$f \cap (f \Rightarrow g) \subseteq f \cap g.$$

The existence of such a subobject is guaranteed by the fact that, in any lattice L, for any x, y such that $y \leq x$, there exists $z \in L$ such that $x \wedge z \leq y$. Clearly there are several possible choices.

Theorem 3.4. *The following holds*

(a) *If $h \subseteq f \Rightarrow g$ then $h \cap f \subseteq g$,*
(b) *If $f \Rightarrow g \simeq 1_X$ then $f \subseteq g$.* □

The converse to (a) does not hold. Thus $f \Rightarrow g$ is not an implication. That is why it is called a *semi-implication*.

Theorem 3.5. *For any subobjects $f : A \twoheadrightarrow X, g : B \twoheadrightarrow X$ and $h : C \twoheadrightarrow X$ of X, the following holds:*

(a) $(f \Rightarrow g) \cap (g \Rightarrow h) \subseteq f \Rightarrow h$,
(b) $(f \Rightarrow g) \cap (f \Rightarrow h) \subseteq f \Rightarrow (g \cap h)$,
(c) *If $f \subseteq g$ then $f \Rightarrow g \simeq 1_X$.* □

Let $f : A \twoheadrightarrow X$ be a subobject of X in a prelocus $\mathcal{A}$ and let $x : 1 \longrightarrow X$. If there exists $k : 1 \longrightarrow A$ such that the diagram

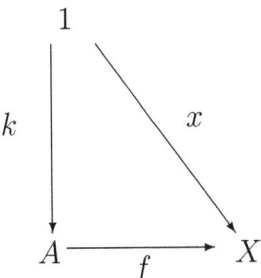

commutes, we say that x is an element of f, denoted $x \in f$.

Theorem 3.6. *In any prelocus, for any object X, we have in $\mathrm{Sub}(X)$, $x \in f \cap g$ iff $x \in f$ and $x \in g$.*

Proof. (a) If $x \in f \cap g$, then there exists k such that $x = (f \cap g) \circ k$. Since $f \cap g \subseteq f$, there exists j such that $f \cap g = f \circ j$. Thus $x = f \circ j \circ k$, i.e. $x \in f$. Similarly, $x \in g$.
(b) Suppose that $x \in f$ and $x \in g$ and consider the diagram

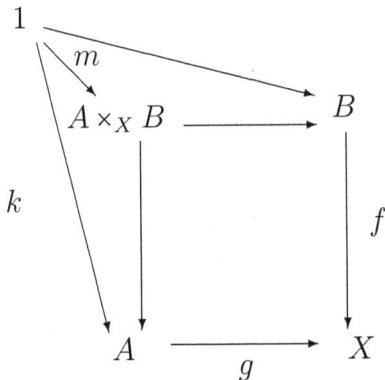

By definition of $f \cap g$, the inner square is a pullback, so the arrow m does exist making the whole diagram commute. Hence $(f \cap g) \circ m = f \circ k = x$. Thus $x \in f \cap g$. □

3.2.4 Complementation

In a prelocus, one can associate to any $f : A \twoheadrightarrow X$ in $\mathrm{Sub}(X)$ the subobject $\neg f : A \twoheadrightarrow X$, defined by $\neg f \simeq f \Rightarrow 0_X$ and called the *complement* of f.

Theorem 3.7. *For any X and any $f : A \twoheadrightarrow X$ and $g : B \twoheadrightarrow X$ in $\mathrm{Sub}(X)$, we have*
  (a) $f \cap \neg f \simeq 0_X$,
  (b) *If $f \subseteq g$ then $\neg g \subseteq \neg f$.*

Proof. (a) follows from the definition of the semi-implication. (b) Given $f \subseteq g$, then, for any $h : C \twoheadrightarrow X$, $g \Rightarrow h \subseteq f \Rightarrow h$. In particular $g \Rightarrow 0_X \subseteq f \Rightarrow 0_X$ i.e. $\neg g \subseteq \neg f$. □

3.3 Loci

A prelocus is called a *locus* if the commutativity of one of the two diagrams

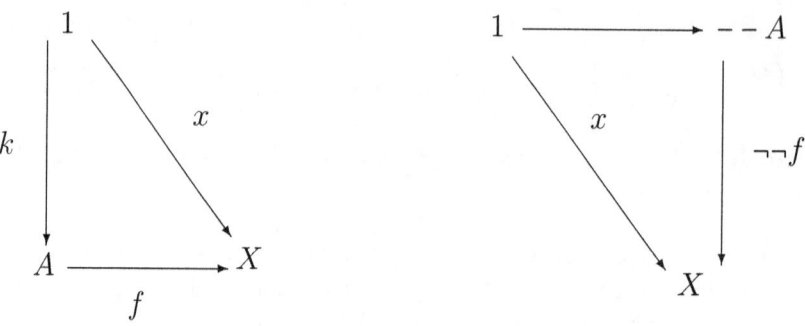

implies that of the other and then of the square

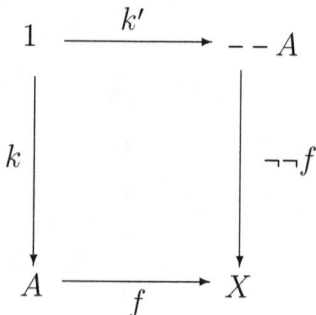

for any object X. In other words, a prelocus $\mathcal{A}$ is a locus iff, for any object X of $\mathcal{A}$, and any $x : 1 \twoheadrightarrow X$, the following equivalence holds true: $(x \in f)$ iff $(x \in \neg\neg f)$.

Theorem 3.8. *For any object X in a locus $\mathcal{A}$ and any $f : A \twoheadrightarrow X$ and $g : B \twoheadrightarrow X$ in $\mathrm{Sub}(X)$, we have*
  (a) $f \simeq \neg\neg f$.
  (b) *If* $\neg f \subseteq \neg g$ *then* $g \subseteq f$.
  (c) $f \cup \neg f \simeq 1_X$.
  (d) $\neg(f \cup g) \simeq \neg f \cap \neg g$.
  (e) $\neg(f \cap g) \simeq \neg f \cup \neg g$. □

Theorem 3.9. *For any object X in a locus $\mathcal{A}$, $(\mathrm{Sub}(X), \subseteq, \cap, \cup, \neg, \Rightarrow)$ is a Λ-algebra.*
$\square$

Remark. In a locus, the following do *not* hold true:
(a) If $f \cap g \simeq 0_X$ then $g \subseteq \neg f$,
(b) If $x \in \neg f$ then not $x \in f$,
(c) If $x \in f \cup g$ then $x \in f$ or $x \in g$,
although the converse implications hold true.

The relationships between loci and Λ-algebras may be made more precise. In a lattice L, when considered as a poset category, there exists an arrow $a \longrightarrow b$ between two elements of L iff $a \leq b$. Since, furthermore, in a Λ-algebra, $x \leq a \Rightarrow b$ entails $x \wedge a \leq b$ (the converse being generally false), the existence of an arrow $x \longrightarrow (a \Rightarrow b)$ implies that of an arrow $x \wedge a \longrightarrow b$. This is reminiscent of the situation in a locus where there is a bijection between a subset of $\mathrm{Hom}(x, b^a)$ and $\mathrm{Hom}(x \times a, b)$. Now, in a Λ-algebra $a \wedge x = x \wedge a$ is the product $x \times a$ and $a \Rightarrow b$ provides us with the exponential b^a. The evaluation arrow $b^a \times a \longrightarrow b$ is the unique arrow $(a \Rightarrow b) \wedge a \longrightarrow b$ which appears in the definition of the semi-implication. Conversely, semi-exponentiation provides semi-implication. Thus, categorically, a Λ-algebra is nothing but a category with a terminal object, with pullbacks and pushouts for any pair of arrows, with products and coproducts for any pair of objects and with a semi-exponentiation. Thus any Λ-algebra is a locus.

3.4 Locus-validity

The above remark on the links between loci and Λ-algebras leads us to consider the concept of locus-validity and its relation with localistic provability. Given a set Φ of formulas, defined via the formation rules given in section 2.1, and a locus $\mathcal{A}$, a formula $\phi \in \Phi$ is said to be $\mathcal{A}$-valid, denoted $\mathcal{A} \vDash \phi$, iff, for any object X of $\mathcal{A}$ and for any valuation $v: \phi \longrightarrow \mathrm{Sub}(X), v(\phi) = 1_X$.

Theorem 3.10. $\vdash_{LL} \phi$ *iff ϕ is $\mathcal{A}$-valid for any locus $\mathcal{A}$.*

Proof. (a) If $\vdash_{LL} \phi$ then ϕ is L-valid for any Λ-algebra L. Then, in particular, ϕ is $\mathrm{Sub}(X)$-valid, for any X in $\mathcal{A}$ and any $\mathcal{A}$, i.e. $\mathcal{A} \vDash \phi$ for any $\mathcal{A}$.
(b) If $\mathcal{A} \vDash \phi$ for any locus $\mathcal{A}$, then ϕ is $\mathrm{Sub}(X)$-valid for any X in $\mathcal{A}$ and for any $\mathcal{A}$. Suppose that $\nvdash_{LL} \phi$. Then $\nvDash_{LL} \phi$ i.e. there exists a Λ-algebra L such that ϕ is not L-valid. But any Λ-algebra being a locus, this leads to a contradiction.
$\square$

Clearly, in a locus with exponentiation we have $f \Rightarrow (g \Rightarrow f) \simeq 1_X$ and $f \subseteq g \Rightarrow h$ iff $f \cap g \subseteq h$, for any object X and any $f: A \twoheadrightarrow X$, $g: B \twoheadrightarrow X$ and $h: C \twoheadrightarrow X$

in Sub(X). This, in turn, leads us to compare the locus-theoretic and the topos-theoretic frameworks, in particular from a logico-algebraic viewpoint. The question of whether the algebraic operators $\cap, \cup, \neg$ and $\Rightarrow$ in the algebra Sub(X) of subobjects of some X in a locus (or, equivalently, in view of theorem 3.10, the logical connectors in localistic logic) may be - or should be - internalized is of prime interest. Indeed, a remarkable feature of the above analysis and results is that loci theory makes no use of the concept of subobject classifier and that the logical connectors or the corresponding algebraic operators have no internal counterpart.

Far from being a drawback, the impossibility to internalize the logical connectors and then to consider localistic logic as an internal logic of some (hence any) locus, is a highly desirable result. First, it means that LL must be seen as emerging from a locus-theoretic structure and it asserts the pre-eminence of the (categorical) structure over the logic which emerges from it. The links between topoï theory and intuitionistic logic (IL) convey the opposite - and highly controversial - view. Second, the definition of a subobject classifier in **Set**, which allows to recapture the definition of the Boolean topos **Set** as a special case of the general definition of a topos, is rather artificial. Finally, the equivalence between IL-provability and topos validity hides a situation which seems somehow anomalous. On the one hand, topos validity and IL-provability only depend on the algebraic - hence external to the topos - structure of the algebra Sub(1) of subobjects of the terminal object 1, Sub(1) being not an actual object in a topos. On the other hand, from an internal viewpoint - which should prevail since IL is defined, via truth-arrows, as an internal logic -, what actually matters is Ω^X, for any X, i.e. the internal version of the notion of power set, of which Sub(X) is the external version. But Ω^X plays no role in the definition of the validity/provability. The divorce between the internal and the external versions culminates in a rather counter-intuitive result: there are non-Boolean topoï, i.e. topoï where Sub(Ω) is not a Boolean algebra, which do validate classical logic. The usual claim that topoï theory is to IL what set theory is to classical logic (CL) and, therefore, that topoï theory is the *right* generalization of set theory in some sense, is, to say the least, questionable. Such a situation is just impossible in a locus-theoretic framework. Indeed, if we define a Boolean locus as a locus such that, for any object X, Sub(X) is a Boolean algebra (i.e. such that, for any $f : A \twoheadrightarrow X, g : B \twoheadrightarrow X$ and $h : C \twoheadrightarrow X, f \Rightarrow (g \Rightarrow f) \simeq 1_X$ and $f \cap g \subseteq h$ iff $f \subseteq g \Rightarrow h$), a formula ϕ is CL-valid iff ϕ is valid in any Boolean locus.

References

[1] J.-P. Barthélemy, M. De Glas, J. Desclés, and J. Petitot. Logique et dynamique de la cognition. *Intellectica*, 23:219–301, 1996.

[2] O. Breysse and M. De Glas. A New Approach to the Concept of Boundary and Contact: Toward an Alternative to Mereotopology. *Fondamenta Informaticae*, 78:217–238, 2007.

[3] G. Choquet. Convergences. *Annales de l'Université de Grenoble*, 23:55–112, 1947.

[4] M. M. Day. Convergence, Closure, and Neighborhoods. *Duke Math. J.*, 11:181–199, 1947.

[5] M. De Glas. Locological Spaces. *Cognitiva*, 90:437–445, 1990.

[6] M. De Glas. Subintuitionnistic Logics and Locology. *Logic Colloquium*, 1997.

[7] M. De Glas and J.-L. Plane. *Une approche formelle de la typicité*, volume 20. École Polytechnique, Paris, 2005. ISSN 0984-5100.

[8] P. C. Hammer. Extended Topology: Continuity. *Portugaliae Mathematica*, 25:77–93, 1964.

[9] H. Poincaré. *La science et l'hypothèse*. Flammarion, 1902.

[10] H. Poincaré. *La valeur de la science*. Flammarion, 1905.

[11] M. B. Smyth. Semi-Metric, Closure Spaces, and Digital Topology. *Th. Comp. Sci.*, pages 257–276, 1995.

[12] E. Čech. *Topological Spaces*. Wiley, 1966.

[13] E. C. Zeeman and O. P. Buneman. Tolerance Spaces and the Brain. In C. H. Waddington, editor, *Towards a Theoretical Biology*, pages 140–151. Oxford University Press, 1968.

Received 14 April 2016

Secure Multiparty Computation without One-way Functions

Dima Grigoriev[*]
CNRS, Mathématiques, Université de Lille
59655, Villeneuve d'Ascq, France
dmitry.grigoryev@math.univ-lille1.fr

Vladimir Shpilrain[†]
Department of Mathematics, The City College of New York
New York, NY 10031, USA
shpil@groups.sci.ccny.cuny.edu

Abstract

We describe protocols for secure computation of the sum, product, and some other functions of two or more elements of an arbitrary constructible ring, without using any one-way functions. One of the new inputs that we offer here is that, in contrast with other proposals, we conceal intermediate results of a computation. For example, when we compute the sum of k numbers, only the final result is known to the parties; partial sums are not known to anybody. Other applications of our method include voting/rating over insecure channels and a rather elegant and efficient solution of the "two millionaires problem".

We also give a protocol, without using a one-way function, for the so-called "mental poker", i.e., a fair card dealing (and playing) over distance.

Finally, we describe a secret sharing scheme where an advantage over Shamir's and other known secret sharing schemes is that nobody, including the dealer, ends up knowing the shares (of the secret) owned by any particular player.

It should be mentioned that computational cost of our protocols is negligible to the point that all of them can be executed without a computer.

In memory of Grigori Mints

Part of this research was presented by the first author at the conference "Philosophy, Mathematics, Linguistics: Aspects of Interaction 2012" (PhML-2012), held on May 22–25, 2012 at the Euler International Mathematical Institute.

[*]Research was supported by the RSF grant 16-11-10075.

[†]Research was partially supported by the NSF grant CNS-1117675 and by the ONR (Office of Naval Research) grant N000141512164.

1 Introduction

As a society, we have become dependent on information technology for many aspects of our daily life, and as a consequence, dependent upon cryptography. The need for developing various cryptographic tools to address new challenges in storing and processing information is therefore clear. One of these challenges, namely how to securely and efficiently process information owned by several different parties, is addressed in this paper.

The problem of secure multi-party computation was originally suggested by Yao [19] in 1982. The concept usually refers to computational systems in which several parties wish to jointly compute some value based on individually held secret bits of information, but do not wish to reveal their secrets to anybody in the process. For example, two individuals, each possessing some secret numbers, x and y, respectively, may wish to jointly compute some function $f(x, y)$ without revealing any information about x or y other than what can be reasonably deduced by knowing the actual value of $f(x, y)$.

Secure computation was formally introduced by Yao as secure two-party computation. His "two millionaires problem" (cf. our Section 3) and its solution gave way to a generalization to multi-party protocols, see e.g. [4], [7]. Secure multi-party computation provides solutions to various real-life problems such as distributed voting, private bidding and auctions, sharing of signature or decryption functions, private information retrieval, etc.

In this paper, we showcase several protocols, originally offered in [13], for secure computation of various functions (including the sum and product) of three or more elements of an arbitrary constructible ring, without using encryption or any one-way functions whatsoever. We require in our scheme that there are k secure channels for communication between the $k \geq 3$ parties, arranged in a cycle. We also show that less than k secure channels is not enough.

Unconditionally secure multiparty computation was previously considered in [4] and elsewhere. A new input that we offer here is that, in contrast with [4] and other proposals, we conceal "intermediate results" of a computation. For example, when we compute a sum of k numbers n_i, only the final result $\sum_{i=1}^{k} n_i$ is known to the parties; partial sums are not known to anybody. This is not the case in [4] where each partial sum $\sum_{i=1}^{s} n_i$ is known to at least some of the parties. This difference is important because, by the "pigeonhole principle", at least one of the parties may accumulate sufficiently many expressions in n_i to be able to recover at least some of the n_i other than his own.

Here we show how our method works for computing the sum (Section 2) and the product (Section 2.2) of private numbers. We ask what other functions can be securely computed without revealing intermediate results.

Other applications of our method include voting/rating over insecure channels (Section 2.4) and a rather elegant solution of the "two millionaires problem" (Section 3).

In Section 5, we consider a cryptographic primitive known as "mental poker", i.e., fair card dealing (and playing) over distance. Several protocols for doing this, most of them using encryption, have been suggested, the first by Shamir, Rivest, and Adleman [18], and subsequent proposals include [5] and [9]. As with bit commitment, fair card dealing between just two players over distance is impossible without a one-way function since commitment is part of any meaningful card dealing scenario. However, this turns out to be possible if the number of players is $k \geq 3$. What we require though is that there are k secure channels for communication between players, arranged in a cycle. We also show that our protocol can, in fact, be adapted to deal cards to just 2 players. Namely, if we have 2 players, they can use a "dummy" player (e.g. a computer), deal cards to 3 players, and then just ignore the "dummy"'s cards, i.e., "put his cards back in the deck". An assumption on the "dummy" player is that he cannot generate any randomness, so randomness has to be supplied to him by the two "real" players. Another assumption is that there are secure channels for communication between either "real" player and the "dummy". We believe that this model is adequate for 2 players who want to play online but do not trust the server. "Not trusting" the server exactly means not trusting with generating randomness. Other, deterministic, operations can be verified at the end of the game; we give more details in Section 5.2.

We note that the only known (to us) proposal for dealing cards to $k \geq 3$ players over distance without using one-way functions was published in [1], but their protocol lacks the simplicity, efficiency, and some of the functionalities of our proposal; this is discussed in more detail in our Section 6. Here we just mention that computational cost of our protocols is negligible to the point that they can be easily executed without a computer.

Finally, in Section 7, we propose a secret sharing scheme where an advantage over Shamir's [17] and other known secret sharing schemes is that nobody, including the dealer, ends up knowing the shares (of the secret) owned by any particular players. The disadvantage though is that our scheme is a (k,k)-threshold scheme only.

2 Secure computation of a sum

In this section, our scenario is as follows. There are k parties $P_1, \ldots, P_k$; each P_i has a private element n_i of a fixed constructible ring R. The goal is to compute the sum of all n_i without revealing any of the n_i to any party $P_j, j \ne i$.

One obvious way to achieve this is well studied in the literature (see e.g. [8, 9, 12]): encrypt each n_i as $E(n_i)$, send all $E(n_i)$ to some designated P_i (who does not have a decryption key), have P_i compute $S = \sum_i E(n_i)$ and send the result to the participants for decryption. Assuming that the encryption function E is *homomorphic*, i.e., that $\sum_i E(n_i) = E(\sum_i n_i)$, each party P_i can recover $\sum_i n_i$ upon decrypting S.

This scheme requires not just a one-way function, but a one-way function with a trapdoor since both encryption and decryption are necessary to obtain the result.

What we suggest in this section is a protocol that does not require any one-way function, but involves secure communication between some of the P_i. So, *our assumption* here is that there are k secure channels of communication between the k parties P_i, arranged in a cycle. *Our result* is computing the sum of private elements n_i without revealing any individual n_i to any $P_j, j \ne i$. Clearly, this is only possible if the number of participants P_i is greater than 2. As for the number of secure channels between P_i, we will show that it cannot be less than k, by the number of parties.

2.1 The protocol (computing the sum)

1. P_1 initiates the process by sending $n_1 + n_{01}$ to P_2, where n_{01} is a random element ("noise").

2. Each P_i, $2 \le i \le k-1$, does the following. Upon receiving an element m from P_{i-1}, he adds his $n_i + n_{0i}$ to m (where n_{0i} is a random element) and sends the result to P_{i+1}.

3. P_k adds $n_k + n_{0k}$ to whatever he has received from P_{k-1} and sends the result to P_1.

4. P_1 subtracts n_{01} from what he got from P_k; the result now is the sum $S = \sum_{1 \le i \le k} n_i + \sum_{2 \le i \le k} n_{0i}$. Then P_1 publishes S.

5. Now all participants P_i, except P_1, broadcast their n_{0i}, possibly over insecure channels, and compute $\sum_{2 \le i \le k} n_{0i}$. Then they subtract the result from S to finally get $\sum_{1 \le i \le k} n_i$.

Thus, in this protocol we have used k (by the number of the parties P_i) secure channels of communication between the parties. If we visualize the arrangement as a graph with k vertices corresponding to the parties P_i and k edges corresponding to secure channels, then this graph will be a k-cycle. Other arrangements are possible, too; in particular, a union of disjoint cycles of length ≥ 3 would do. (In that case, the graph will still have k edges.) Two natural questions that one might now ask are: (1) is any arrangement with less than k secure channels possible? (2) with k secure channels, would this scheme work with any arrangement other than a union of disjoint cycles of length ≥ 3? The answer to both questions is "no". Indeed, if there is a vertex (corresponding to P_1, say) of degree 0, then any information sent out by P_1 will be available to everybody, so other participants will know n_1 unless P_1 uses a one-way function to conceal it. If there is a vertex (again, corresponding to P_1) of degree 1, this would mean that P_1 has a secure channel of communication with just one other participant, say P_2. Then any information sent out by P_1 will be available at least to P_2, so P_2 will know n_1 unless P_1 uses a one-way function to conceal it. Thus, every vertex in the graph should have degree at least 2, which implies that every vertex is included in a cycle. This immediately implies that the total number of edges is at least k. If now a graph Γ has k vertices and k edges, and every vertex of Γ is included in a cycle, then every vertex has degree exactly 2 since by the "handshaking lemma" the sum of the degrees of all vertices in any graph equals twice the number of edges. It follows that our graph is a union of disjoint cycles.

2.2 Secure computation of a product

Now we show how to use the general ideas of the protocol for computing the sum (see Section 2.1) to securely compute a product. Again, there are k parties $P_1, \ldots, P_k$; each P_i has a private (nonzero) element n_i of a fixed constructible ring R. The goal is to compute the product of all n_i without revealing any of the n_i to any party $P_j, j \neq i$. Requirements on the ring R are going to be somewhat more stringent here than they were in Section 2. Namely, we require that R does not have zero divisors and, if an element r of R is a product $a \cdot x$ with a known a and an unknown x, then x can be efficiently recovered from a and r. Examples of rings with these properties include the ring of integers and any constructible field.

The protocol (computing the product)

1. P_1 initiates the process by sending $n_1 \cdot n_{01}$ to P_2, where n_{01} is a random nonzero element ("noise").

2. Each P_i, $2 \leq i \leq k-1$, does the following. Upon receiving an element m from P_{i-1}, he multiplies m by $n_i \cdot n_{0i}$ (where n_{0i} is a random element) and sends the result to P_{i+1}.

3. P_k multiplies by $n_k \cdot n_{0k}$ whatever he has received from P_{k-1} and sends the result to P_1. This result is the product $P = \Pi_{1 \le i \le k} \, n_i \cdot \Pi_{2 \le i \le k} \, n_{0i}$.

4. P_1 divides what he got from P_k by his n_{01}; the result now is the product $P = \Pi_{1 \le i \le k} \, n_i \cdot \Pi_{2 \le i \le k} \, n_{0i}$. Then P_1 publishes P.

5. Now all participants P_i, except P_1, broadcast their n_{0i}, possibly over insecure channels, and compute $\Pi_{2 \le i \le k} \, n_{0i}$. Then they divide P by the result to finally get $\Pi_{1 \le i \le k} \, n_i$.

2.3 Effect of coalitions

Suppose now we have $k \ge 3$ parties with k secure channels of communication arranged in a cycle, and suppose 2 of the parties secretly form a coalition. Our assumption here is that, because of the circular arrangement of secure channels, a secret coalition is only possible between parties P_i and P_{i+1} for some i, where the indices are considered modulo k; otherwise, attempts to form a coalition (over insecure channels) will be detected. If two parties P_i and P_{i+1} exchanged information, they would, of course, know each other's elements n_i, but other than that, they would not get any advantage if $k \ge 4$. Indeed, we can just "glue these two parties together", i.e., consider them as one party, and then the protocol is essentially reduced to that with $k - 1 \ge 3$ parties. On the other hand, if $k = 3$, then, of course, two parties together have all the information about the third party's element.

For an arbitrary $k \ge 4$, if $n < k$ parties want to form a (secret) coalition to get information about some other party's element, all these n parties have to be connected by secure channels, which means there is a j such that these n parties are $P_j, P_{j+1}, \ldots, P_{j+n-1}$, where indices are considered modulo k. It is not hard to see then that only a coalition of $k - 1$ parties $P_1, \ldots, P_{i-1}, P_{i+1}, \ldots, P_k$ can suffice to get information about the P_i's element.

2.4 Ramification: voting/rating over insecure channels

In this section, our scenario is as follows. There are k parties $P_1, \ldots, P_k$; each P_i has a private integer n_i. There is also a computing entity B (for Boss) who shall compute the sum of all n_i. The goal is to let B compute the sum of all n_i without revealing any of the n_i to him or to any party $P_j, j \ne i$.

The following example from real life is a motivation for this scenario.

Example 1. Suppose members of the board in a company have to vote for a project by submitting their numeric scores (say, from 1 to 10) to the president of the company. The project gets a green light if the total score is above some threshold value T. Members of the board can discuss the project between themselves and exchange information privately, but none of them wants his/her score to be known to either the president or any other member of the board.

In the protocol below, we are again assuming that there are k channels of communication between the parties, arranged in a cycle: $P_1 \to P_2 \to \ldots \to P_k \to P_1$. On the other hand, communication channels between B and any of the parties are not assumed to be secure.

2.5 The protocol (rating over insecure channels)

1. P_1 initiates the process by sending $n_1 + n_{01}$ to P_2, where n_{01} is a random number.

2. Each P_i, $2 \leq i \leq k-1$, does the following. Upon receiving a number m from P_{i-1}, he adds his $n_i + n_{0i}$ to m (where n_{0i} is a random number) and sends the result to P_{i+1}.

3. P_k adds $n_k + n_{0k}$ to whatever he has received from P_{k-1} and sends the result to B.

4. P_k now starts the process of collecting the "adjustment" in the opposite direction. To that effect, he sends his n_{0k} to P_{k-1}.

5. P_{k-1} adds $n_{0(k-1)}$ and sends the result to P_{k-2}.

6. The process ends when P_1 gets a number from P_2, adds his n_{01}, and sends the result to B. This result is the sum of all n_{0i}.

7. B subtracts what he got from P_1 from what he got from P_k; the result now is the sum of all n_i, $1 \leq i \leq k$.

3 Application: the "two millionaires problem"

The protocol from Section 2, with some adjustments, can be used to provide an elegant and efficient solution to the "two millionaires problem" introduced in [19]: there are two numbers, n_1 and n_2, and the goal is to solve the inequality $n_1 \geq ? n_2$ without revealing the actual values of n_1 or n_2.

To that effect, we use a "dummy" as the third party. Our concept of a "dummy" is quite different from a well-known concept of a "trusted third party"; importantly, our "dummy" is not supposed to generate any randomness; it just does what it is told to. Basically, the only difference between our "dummy" and a usual calculator is that there are secure channels of communication between the "dummy" and either "real" party. One possible real-life interpretation of such a "dummy" would be an online calculator that can combine inputs from different users. Also note that in our scheme below *the "dummy" is unaware of the committed values* of n_1 or n_2, which is useful in case the two "real" parties do not want their private numbers to ever be revealed. This suggests yet another real-life interpretation of a "dummy", where he is a mediator between two parties negotiating a settlement.

Thus, let A (Alice) and B (Bob) be two "real" parties, and D (Dummy) the "dummy". Suppose A's number is n_1, and B's number is n_2.

3.1 The protocol (comparing two numbers)

1. A splits her number n_1 as a difference $n_1 = n_1^+ - n_1^-$. She then sends n_1^- to B.

2. B splits his number n_2 as a difference $n_2 = n_2^+ - n_2^-$. He then sends n_2^- to A.

3. A sends $n_1^+ + n_2^-$ to D.

4. B sends $n_2^+ + n_1^-$ to D.

5. D subtracts $(n_2^+ + n_1^-)$ from $(n_1^+ + n_2^-)$ to get $n_1 - n_2$, and announces whether this result is positive or negative.

Remark 1. Perhaps a point of some dissatisfaction in this protocol could be the fact that the "dummy" ends up knowing the actual difference $n_1 - n_2$, so if there is a leak of this information to either party, this party would recover the other's private number n_i. This can be avoided if n_1 and n_2 are represented in the binary form and compared one bit at a time, going left to right, until the difference between bits becomes nonzero. However, this method, too, has a disadvantage: the very moment the "dummy" pronounces the difference between bits nonzero would give an *estimate* of the difference $n_1 - n_2$ *to the real parties*, not just to the "dummy".

We note that the original solution of the "two millionaires problem" given in [19], although lacks the elegance of our scheme, does not involve a third party, whereas our solution does. On the other hand, the solution in [19] uses encryption, whereas our solution does not, which makes it by far more efficient. Finally, we mention that since our paper [13] was published, we have come up with several other solutions of the "two millionaires problem" without using either one-way functions or a dummy [14], [11]. Some of those solutions use simple laws of (classical) physics instead.

4 Secure computation of symmetric functions

In this section, we show how our method can be easily generalized to allow secure computation of any expression of the form $\sum_{i=1}^{k} n_i^r$, where n_i are parties' private numbers, k is the number of parties, and $r \geq 1$ an arbitrary integer. We simplify our method here by removing the "noise", to make the exposition more transparent. Otherwise, the protocol is the same as the protocol for secure computation of a sum in Section 2.

4.1 The protocol (computing the sum of powers)

1. P_1 initiates the process by sending a random element n_0 to P_2.

2. Each P_i, $2 \leq i \leq k-1$, does the following. Upon receiving an element m from P_{i-1}, he adds his n_i^r to m and sends the result to P_{i+1}.

3. P_k adds his n_k^r to whatever he has received from P_{k-1} and sends the result to P_1.

4. P_1 subtracts $(n_0 - n_1^r)$ from what he got from P_k; the result now is the sum of all n_i^r, $1 \leq i \leq k$.

Now that the parties can securely compute the sum of any powers of their n_i, they can also compute any symmetric function of n_i. However, in the course of computing a symmetric function from sums of different powers of n_i, at least some of the parties will possess several different polynomials in n_i, so chances are that at least some of the parties will be able to recover at least some of the n_i. On the other hand, because of the symmetry of all expressions involved, there is no way to tell which n_i belongs to which party.

4.2 Open problem

Now it is natural to ask:

Problem 1. What other functions (other than the sum and the product) can be securely computed without revealing intermediate results to any party?

To be more precise, we note that one intermediate result is inevitably revealed to the party who finishes computation, but this cannot be avoided in any scenario. For example, after the parties have computed the sum of their private numbers, each party also knows the sum of all numbers except his own. What we want is that no other intermediate results are ever revealed.

To give some insight into this problem, we consider a couple of examples of computing simple functions different from the sum and the product of the parties' private numbers.

Example 2. We show how to compute the function $f(n_1, n_2, n_3) = n_1 n_2 + n_2 n_3$ in the spirit of the present paper, without revealing (or even computing) any intermediate results, i.e., without computing $n_1 n_2$ or $n_2 n_3$.

1. P_2 initiates the process by sending a random element n_0 to P_3.
2. P_3 adds his n_3 to n_0 and sends $n_3 + n_0$ to P_1.
3. P_1 adds his n_1 to $n_0 + n_3$ and sends the result to P_2.
4. P_2 subtracts n_0 from $n_0 + n_3 + n_1$ and multiplies the result by n_2. This is now $n_1 n_2 + n_2 n_3$.

Example 3. The point of this example is to show that functions that can be computed by our method do not have to be homogeneous (in case the reader got this impression based on the previous examples).

The function that we compute here is $f(n_1, n_2, n_3) = n_1 n_2 + g(n_3)$, where g is any computable function.

1. P_1 initiates the process by sending a random element a_0 to P_2.
2. P_2 multiplies a_0 by his n_2 and sends the result to P_3.
3. P_3 multiplies $a_0 n_2$ by a random element c_0 and sends the result to P_1.
4. P_1 multiplies $a_0 n_2 c_0$ by his n_1, divides by a_0, and sends the result, which is $n_1 n_2 c_0$, back to P_3.
5. P_3 divides $n_1 n_2 c_0$ by c_0 and adds $g(n_3)$, to end up with $n_1 n_2 + g(n_3)$.

Note that in this example, the parties used more than just one loop of transmissions in the course of computation. Also, information here was sent "in both directions" in the circuit.

Remark 2. Another collection of examples of multiparty computation without revealing intermediate results can be obtained as follows. Suppose, without loss of generality, that some function $f(n_1, \ldots, n_k)$ can be computed by our method in such a way that the last step in the computation is performed by the party P_1, i.e., P_1 is the one who ends up with $f(n_1, \ldots, n_k)$ while no party knows any intermediate result $g(n_1, \ldots, n_k)$ of this computation. Then, obviously, P_1 can produce any function of the form $F(n_1, f(n_1, \ldots, n_k))$ (for a computable function F) as well. Examples include $n_1^r + n_1 n_2 \cdots n_k$ for any $r \geq 0$; $n_1^r + (n_1 n_2 + n_3)^s$ for any $r, s \geq 0$, etc.

5 Mental poker

"Mental poker" is the common name for a set of cryptographic problems that concerns playing a fair game over distance without the need for a trusted third party. One of the ways to describe the problem is: how can 2 players deal cards fairly over the phone? Several protocols for doing this have been suggested, including [5, 9, 18] and [1]. As with bit commitment, it is rather obvious that fair card dealing to two players over distance is impossible without a one-way function, or even a one-way function with trapdoor. However, it turns out to be possible if the number of players is at least 3, assuming, of course, that there are secure channels for communication between at least some of the players. In our proposal, we will be using k secure channels for $k \geq 3$ players $P_1, \ldots, P_k$, and these k channels will be arranged in a cycle: $P_1 \to P_2 \to \ldots \to P_k \to P_1$.

To begin with, suppose there are 3 players: P_1, P_2, and P_3 and 3 secure channels: $P_1 \to P_2 \to P_3 \to P_1$.

The first protocol, Protocol 1 below, is for distributing *all* integers from 1 to m to the players in such a way that each player gets about the same number of integers. (For example, if the deck that we want to deal has 52 cards, then two players should get 17 integers each, and one player should get 18 integers.) In other words, Protocol 1 allows one to randomly split a set of m integers into 3 disjoint sets.

The second protocol, Protocol 2, is for collectively generating random integers modulo a given integer M. This very simple but useful primitive can be used: **(i)** for collectively generating, uniformly at random, a permutation from the group S_m. This will allow us to assign cards from a deck of m cards to the m integers distributed by Protocol 1; **(ii)** introducing "dummy" players as well as for "playing" after dealing cards.

5.1 Protocol 1

For notational convenience, we are assuming below that we have to distribute integers from 1 to $r = 3s$ to 3 players.

To begin with, all players agree on a parameter N, which is a positive integer of a reasonable magnitude, say, 10.

1. each player P_i picks, uniformly at random, an integer (a "counter") c_i between 1 and N, and keeps it private.

2. P_1 starts with the "extra" integer 0 and sends it to P_2.

3. P_2 sends to P_3 either the integer m he got from P_1, or $m+1$. More specifically, if P_2 gets from P_1 the same integer m less than or equal to c_2 times, then he sends m to P_3; otherwise, he sends $m+1$ and keeps m (i.e., in the latter case m becomes one of "his" integers). Having sent out $m+1$, he "resets his counter", i.e., selects, uniformly at random between 1 and N, a new c_2. He also resets his counter if he gets the number m for the first time, even if he does not keep it.

4. P_3 sends to P_1 either the integer m he got from P_2, or $m+1$. More specifically, if P_3 gets from P_2 the same integer m less than or equal to c_3 times, then he sends m to P_1; otherwise, he sends $m+1$ and keeps m. Having sent out $m+1$, he selects a new counter c_3. He also resets his counter if he gets the number m for the first time, even if he does not keep it.

5. P_1 sends to P_2 either the integer m he got from P_3, or $m+1$. More specifically, if P_1 gets from P_3 the same integer m less than or equal to c_1 times, then he sends m to P_2; otherwise, he sends $m+1$ and keeps m. Having sent out $m+1$, he selects a new counter c_1. He also resets his counter if he gets the number m for the first time, even if he does not keep it.

6. This procedure continues until one of the players gets s integers (not counting the "extra" integer 0). After that, a player who already has s integers just "passes along" any integer that comes his way, while other players keep following the above procedure until they, too, get s integers.

7. The protocol ends as follows. When all $3s$ integers, between 1 and $3s$, are distributed, the player who got the last integer, $3s$, keeps this fact to himself and passes this integer along as if he did not "take" it.

8. The process ends when the integer $3s$ makes $N+1$ "full circles".

We note that the role of the "extra" integer 0 is to prevent P_3 from knowing that P_2 has got the integer 1 if it happens that $c_2 = 1$ in the beginning.

We also note that this protocol can be generalized to arbitrarily many players in the obvious way, if there are k secure channels for communication between k players, arranged in a cycle.

5.2 Protocol 2

Now we describe a protocol for generating random integers modulo some integer M collectively by 3 players. As in Protocol 1, we are assuming that there are secure channels for communication between the players, arranged in a cycle.

1. P_2 and P_3 uniformly at random and independently select private integers n_2 and n_3 (respectively) modulo M.

2. P_2 sends n_2 to P_1, and P_3 sends n_3 to P_1.

3. P_1 computes the sum $m = n_2 + n_3$ modulo M.

Note that neither P_2 nor P_3 can cheat by trying to make a "clever" selection of their n_i because the sum, modulo M, of any integer with an integer uniformly distributed between 0 and $M-1$, is an integer uniformly distributed between 0 and $M-1$.

Finally, P_1 cannot cheat simply because he does not really get a chance: if he miscalculates $n_2 + n_3$ modulo M, this will be revealed at the end of the game. (All players keep contemporaneous records of all transactions, so that at the end of the game, correctness could be verified.)

To generalize Protocol 2 to arbitrarily many players $P_1, \ldots, P_k$, $k \geq 3$, we can just engage 3 players at a time in running the above protocol. If, at the same time, we want to keep the same circular arrangement of secure channels between the players that we had in Protocol 1, i.e., $P_1 \to P_2 \to \ldots P_k \to P_1$, then 3 players would have to be P_{i+1}, P_i, P_{i+2}, where i would run from 1 to k, and the indices are considered modulo k.

Protocol 2 can now be used to collectively generate, uniformly at random, a permutation from the group S_m. This will allow us to assign cards from a deck of m cards to the m integers distributed by Protocol 1. Generating a random permutation from S_m can be done by taking a random integer between 1 and m (using Protocol 2) sequentially, ensuring that there is no repetition. This "brute-force" method will require occasional retries whenever the random integer picked is a repeat of an integer already selected. A simple algorithm to generate a permutation from S_m uniformly randomly without retries, known as the *Knuth shuffle*, is to start with the identity permutation or any other permutation, and then go through the positions 1 through

($m-1$), and for each position i swap the element currently there with an arbitrarily chosen element from positions i through m, inclusive (again, Protocol 2 can be used here to produce a random integer between i and m). It is easy to verify that any permutation of m elements will be produced by this algorithm with probability exactly $\frac{1}{m!}$, thus yielding a uniform distribution over all such permutations.

After this is done, we have m cards distributed uniformly randomly to the players, i.e., we have:

Proposition 1. *If m cards are distributed to k players using Protocols 1 and 2, then the probability for any particular card to be distributed to any particular player is $\frac{1}{k}$.*

5.3 Using "dummy" players while dealing cards

We now show how a combination of Protocol 1 and Protocol 2 can be used to deal cards to just 2 players. If we have 2 players, they can use a "dummy" player (e.g. a computer), deal cards to 3 players as in Protocol 1, and then just ignore the "dummy"'s cards, i.e., "put his cards back in the deck". We note that the "dummy" in this scenario would not generate randomness; it will be generated for him by the other two players using Protocol 2. Namely, if we call the "dummy" P_3, then the player P_1 would randomly generate c_{31} between 1 and N and send it to P_3, and P_2 would randomly generate c_{32} between 1 and N and send it to P_3. Then P_3 would compute his random number as $c_3 = c_{31} + c_{32}$ modulo N.

Similarly, "dummy" players can help k "real" players each get a fixed number s of cards, because Protocol 1 alone is only good for distributing *all* cards in the deck to the players, dealing each player about the same number of cards. We can introduce m "dummy" players so that $(m+k) \cdot s$ is approximately equal to the number of cards in the deck, and position all the "dummy" players one after another as part of a circuit $P_1 \to P_2 \to \ldots P_{m+k} \to P_1$. Then we use Protocol 1 to distribute all cards in the deck to $(m+k)$ players taking care that each "real" player gets exactly s cards. As in the previous paragraph, "dummy" players have "real" ones generate randomness for them using Protocol 2.

After all cards in the deck are distributed to $(m+k)$ players, "dummy" players send all their cards to one of them; this "dummy" player now becomes a "dummy dealer", i.e., he will give out random cards from the deck to "real" players as needed in the course of a subsequent game, while randomness itself will be supplied to him by "real" players using Protocol 2.

6 Summary of the properties of our card dealing (Protocols 1 and 2)

Here we summarize the properties of our Protocols 1 and 2 and compare, where appropriate, our protocols to the card dealing protocol of [1].

1. **Uniqueness of cards.** Yes, by the very design of Protocol 1.

2. **Uniform random distribution of cards.** Yes, because of Protocol 2; see our Proposition 1 in Section 5.2.

3. **Complete confidentiality of cards.** Yes, by the design of Protocol 1.

4. **Number of secure channels for communication between $k \geq 3$ players:** k, arranged in a cycle.
   By comparison, the card dealing protocol of [1] requires $3k$ secure channels.

5. **Average number of transmissions between $k \geq 3$ players:** $O(\frac{N}{2}mk)$, where m is the number of cards in the deck, and $N \approx 10$. This is because in Protocol 1, the number of circles (complete or incomplete) each integer makes is either 1 or the minimum of all the counters c_i at the moment when this integer completes the first circle. Since the average of c_i is at most $\frac{N}{2}$, we get the result because within one circle (complete or incomplete) there are at most k transmissions. We note that in fact, there is a precise formula for the average of the minimum of c_i in this situation: $\frac{\sum_{j=1}^{N} j^k}{N^k}$, which is less than $\frac{N}{2}$ if $k \geq 2$.
   By comparison, in the protocol of [1] there are $O(mk^2)$ transmissions.

6. **Total length of transmissions between $k \geq 3$ players:** $\frac{N}{2}mk \cdot \log_2 m$ bits. This is just the average number of transmissions times the length of a single transmission, which is a positive integer between 1 and m.
   By comparison, total length of transmissions in [1] is $O(mk^2 \log k)$.

7. **Computational cost of Protocol 1:** negligible (because computation amounts to selecting random integers from a small interval).
   By comparison, the protocol of [1] requires computing products of up to k permutations from the group S_k to deal just one card; the total computational cost therefore is $O(mk^2 \log k)$.

7 Secret sharing

Secret sharing refers to method for distributing a secret amongst a group of participants, each of whom is allocated a share of the secret. The secret can be reconstructed only when a sufficient number of shares are combined together; individual shares are of no use on their own.

More formally, in a secret sharing scheme there is one dealer and k players. The dealer gives a secret to the players, but only when specific conditions are fulfilled. The dealer accomplishes this by giving each player a share in such a way that any group of t (for threshold) or more players can together reconstruct the secret but no group of fewer than t players can. Such a system is called a (t, k)-threshold scheme (sometimes written as a (k, t)-threshold scheme).

Secret sharing was invented by Shamir [17] and Blakley [2], independent of each other, in 1979. Both proposals assumed secure channels for communication between the dealer and each player. In our proposal here, the number of secure channels is equal to $2k$, where k is the number of players, because in addition to the secure channels between the dealer and each player, we have k secure channels for communication between the players, arranged in a cycle: $P_1 \to P_2 \to \ldots \to P_k \to P_1$.

The advantage of our scheme over Shamir's and other known secret sharing schemes is that nobody, including the dealer, ends up knowing the shares (of the secret) owned by any particular players. The disadvantage is that our scheme is a (k, k)-threshold scheme only.

We start by describing a subroutine for distributing shares by the players among themselves. More precisely, k players want to split a given number in a sum of k numbers, so that each summand is known to one player only, and each player knows one summand only.

7.1 The Subroutine (distributing shares by the players among themselves)

Suppose a player P_i receives a number M that has to be split in a sum of k private numbers. In what follows, all indices are considered modulo k.

1. P_i initiates the process by sending $M - m_i$ to P_{i+1}, where m_i is a random number (could be positive or negative).

2. Each subsequent P_j does the following. Upon receiving a number m from P_{j-1}, he subtracts a random number m_j from m and sends the result to P_{j+1}. The number m_j is now P_j's secret summand.

3. When this process gets back to P_i, he adds m_i to whatever he got from P_{i-1}; the result is his secret summand.

Now we get to the actual secret sharing protocol.

7.2 The protocol (secret sharing (k,k)-threshold scheme)

The dealer D wants to distribute shares of a secret number N to k players P_i so that, if P_i gets a number s_i, then $\sum_{i=1}^{k} s_i = N$.

1. D arbitrarily splits N in a sum of k integers: $N = \sum_{i=1}^{k} n_i$.

2. The loop: at Step i of the loop, D sends n_i to P_i, and P_i initiates the above Subroutine to distribute shares n_{ij} of n_i among the players, so that $\sum_{j=1}^{k} n_{ij} = n_i$.

3. After all k steps of the loop are completed, each player P_i ends up with k numbers n_{ji} that sum up to $s_i = \sum_{j=1}^{k} n_{ji}$. It is obvious that $\sum_{i=1}^{k} s_i = N$.

Acknowledgement

Both authors are grateful to Max Planck Institut für Mathematik (MPI), Bonn for its hospitality during the work on this paper. The first author is also grateful to MCCME, Moscow for excellent working conditions and inspiring atmosphere.

References

[1] I. Bárány and Z. Füredi. Mental poker with three or more players. *Inform. and Control*, 59:84–93, 1983.

[2] G. R. Blakley. Safeguarding cryptographic keys. *Proceedings of the National Computer Conference*, 48:313–317, 1979.

[3] G. Brassard, C. Crépeau, and J.-M. Robert. *All-or-nothing disclosure of secrets.* Advances in Cryptology – CRYPTO '86, pages 234–238, Lecture Notes Comp. Sci. 263. Springer, 1986.

[4] D. Chaum, C. Crépeau, and I. Damgård. *Multiparty unconditionally secure protocols (extended abstract)*, In Proceedings of the Twentieth ACM Symposium on the Theory of Computing, pages 11–19, ACM, 1988.

[5] C. Crépeau. *A zero-knowledge poker protocol that achieves confidentiality of the players' strategy or how to achieve an electronic poker face*, Advances in Cryptology – CRYPTO '86, pages 239–247, Lecture Notes Comp. Sci. 263, Springer, 1986.

[6] I. Damgård, M. Geisler, and M. Kroigard. Homomorphic encryption and secure comparison. *Int. J. Appl. Cryptogr.*, 1:22–31, 2008.

[7] I. Damgård, Y. Ishai. *Scalable secure multiparty computation*, Advances in Cryptology – CRYPTO 2006, pages 501–520, Lecture Notes Comp. Sci. 4117, Springer, Berlin, 2006.

[8] O. Goldreich. *Foundations of Cryptography: Volume 1, Basic Tools*. Cambridge University Press, 2007.

[9] S. Goldwasser and S. Micali. *Probabilistic Encryption and How to Play Mental Poker Keeping Secret All Partial Information*, In Proceedings of the 14th Annual ACM symp. on Theory of computing, ACM-SIGACT, pages 365–377, May 1982.

[10] S. Goldwasser and S. Micali. Probabilistic encryption. *J. Comput. System Sci.*, 28:270–299 1984.

[11] D. Grigoriev, L. Kish, and V. Shpilrain. *Yao's millionaires' problem with computationally unbounded parties and public-key encryption without computational assumptions*, preprint.

[12] D. Grigoriev and I. Ponomarenko. Constructions in public-key cryptography over matrix groups. *Contemp. Math., Amer. Math. Soc.*, 418:103–119, 2006.

[13] D. Grigoriev and V. Shpilrain. Secrecy without one-way functions. *Groups, Complexity, and Cryptology*, 5:31–52, 2013.

[14] D. Grigoriev and V. Shpilrain. Yao's millionaires' problem and decoy-based public key encryption by classical physics. *J. Foundations Comp. Sci.*, 25:409–417, 2014.

[15] R. Impagliazzo and M. Luby. *One-way functions are essential for complexity based cryptography*. In FOCS'89, pages 230–235, IEEE Computer Society, 1989.

[16] A. Menezes, P. van Oorschot, and S. Vanstone. *Handbook of Applied Cryptography*. CRC-Press, 1996.

[17] A. Shamir. How to share a secret. *Comm. ACM* 22:612–613, 1979.

[18] A. Shamir, R. Rivest, and L. Adleman. *Mental poker*, Technical Report LCS/TR-125, Massachusetts Institute of Technology, April 1979.

[19] A. C. Yao. *Protocols for secure computations* (Extended Abstract). 23rd annual symposium on foundations of computer science (Chicago, Ill., 1982), pages 160–164, IEEE, New York, 1982.

IF Logic and Linguistic Theory

Jaakko Hintikka
*Department of Philosophy, Boston University
745 Commonwealth Avenue, Boston MA 02215, USA
and
Collegium for Advanced Studies, University of Helsinki
Fabianinkatu 24, 00014 University of Helsinki, Finland*

Abstract

Originally, modern symbolic logic was supposed to be a disambiguated and streamlined version of the logic of natural language. It has nevertheless failed to provide a full account of several telltale semantical phenomena of ordinary language, including Peirce's paradox, "donkey sentences" and more generally conditionals and different kinds of anaphora. It is shown here by reference to examples how these phenomena can be treated by means of IF logic and its semantical basis, game-theoretical semantics. Furthermore, methodological questions like compositionality and logical form will be discussed.

1 Frege-gate

The relations between symbolic logic and linguistic theorizing have been (and still are) complicated, close and confused. Symbolic logic was first thought, typically if not universally, as a minor regimentation and smoothlining of ordinary language. In another direction, mathematicians were formulating much of their reasoning in terms of ordinary prose, not in terms of manipulation of equations or other complexes of symbols. In fact mathematicians like Cauchy or Weierstrass were using – as they had to do – an explicit but unformalized logic of quantifiers in the guise of the so-called epsilon-delta technique, expressed in such ordinary language terms as "given such-and-such a number", "one can find" etc. (See here and in the following Hintikka [3, 5]).

But then a huge scientific scandal, a veritable Frege-gate, took place without anyone's noticing. Frege undertook to formalize our entire logic, to present a notation (a

---

Talk at the conference "Philosophy, Mathematics, Linguistics: Aspects of Interaction 2012" (PhML-2012), Euler International Mathematical Institute, May 22–25, 2012.

Schrift) for all our concepts. Yet he failed to understand his fellow mathematicians' quantifier logic, and instead gave his followers a flawed logic that is only a part of the full story. Subsequent logicians unfortunately followed Frege and used this defective logic as their basic working logic. This alone would not been serious, for Frege's logic of quantifiers (by which I mean what is nowadays called first-order logic) is correct as far as its expressive powers go. The catastrophic mistake the logicians made was to think in effect that it is the full logic of quantifiers. The first specific disaster this caused was the bunch of paradoxes of set theory, which prompted the entire crisis of the foundations of set theory. This in turn led to further catastrophes, such as Zermelo-Fraenkel first-order set theory and the wishful belief that such results as Gödel's, Tarski's or Paul Cohen's tell us something about the limitations of logic and axiomatization or about the continuum hypothesis (Hintikka [4]).

This "Frege-gate" scandal came to light only recently when it was pointed out that the logic that mathematicians were using already hundred years ago was not the received first-order logic, but the richer logic that had been meanwhile rediscovered and systematized under the title "independence-friendly logic" (IF logic). (see e.g. Mann et al. [7]). However, the Frege-gate scandal has not hit headlines yet even in logic journals.

2 IF logic and linguistics

In this paper, I will discuss one aspect of the new problem situation, viz., its impact on linguistic theorizing. That there must be such an impact is obvious. To mention only one indication, at one time Chomsky thought that his syntactical counterparts to logical forms, the LF's, were essentially like formulas of (the received) first-order logic (see e.g. Chomsky, [1, p. 197]; [2, p. 67]). If they are not adequate representations of logical and, *a fortiori*, semantical forms of ordinary language sentences, we do not only need a better logic, but also a better syntactical theory.

Now IF logic, at least in its simplest version, has been around for a while and has even become an established research area in logic. Hence there has in fact been some discussion of its role in natural language. Much ingenuity has been expended on the first examples of purportedly IF sentences in ordinary language. They have been mostly so-called branching quantifier sentences like

$$(2.1) \quad \begin{matrix} (\forall x)(\exists y) \\ \\ (\forall z)(\exists u) \end{matrix} \Big\rangle F(x,y,z,u).$$

Its meaning can be expressed by the IF sentence

(2.2) $(\forall x)(\forall z)(\exists y/\forall z)(\exists u/\forall x)F(x,y,z,u)$.

This meaning cannot be expressed by a first-order quantifier sentence without the independence indication slash.

Examples from ordinary language were presented and discussed. An example was

(2.3) Every villager has a friend and every townsman has a relative who know each other.

Here choice of a friend is independent of the choice of a relative and *vice versa*.

Such examples are sufficiently complicated for confusing some philosophers. However, it has turned out that the examples are only the tip of an iceberg. Other examples look syntactically simple but still turn out to be semantically rather complex, e.g.

(2.4) Everybody has a different friend.

Its logical form can be seen to be

(2.5) $(\forall x_1)(\forall x_2)(\exists y_1/\forall x_2)(\exists y_2/\forall x_1)(((x_1=x_2) \leftrightarrow (y_1=y_2)) \& F(y_1,x_1) \& F(y_2,x_2))$.

What was explained in these early linguistic applications of IF logic are particular examples, rather than general semantical or syntactical phenomena. In this paper, we concentrate on one particular relatively unexplored semantical phenomenon, viz., informational independence involving propositional connectives instead of (or in addition to) quantifiers.

3 Peirce's paradox

Ironically, the shortcomings of the usual ("Fregean") first-order logic were known already at the time of its formulation to Frege's co-inventor Charles S. Peirce (see Peirce [8, 4.546 and 4.580]). He pointed out a problem about the following pair of English sentences:

(3.1) Someone is such that, if he fails in business, he commits suicide.

(3.2) Someone is such that if everybody fails in business, he commits suicide.

Their respective logical forms seem to be

(3.3) $(\exists x)(F(x) \supset S(x))$,

(3.4) $(\exists x)((\forall y)F(y) \supset S(x))$.

Here (3.4) is equivalent to

(3.5) $(\exists x)((\exists y) \sim F(y) \vee S(x))$.

But something is paradoxical here. Formulas (3.1) and (3.2) obviously mean something different whereas, as Peirce pointed out, in the usual first-order logic (3.3) and (3.4) are logically equivalent.

Various *ad hoc* explications have been proposed, but they remain just that: adhockey. Yet game-theoretical semantics yields a diagnosis of the problem without any further assumptions or considerations. The problem is how the conditional (3.4) can be as strong as (3.3).

An answer is found by examining the meaning of (3.1) or (3.2) in game-theoretical terms. What (3.5) says is that it is true. That truth means in the existence of a winning strategy for the verifier ("myself") in the semantical game associated with (3.2). The first part of this strategy is a specification of the value c of x in $(\exists x)$. In order for it to be part of a winning strategy, there has to be a similar winning strategy in the game with

(3.6) $(\exists y) \sim F(y) \vee S(c)$.

The next step in a play of the game is the verifier's choice of one of the disjuncts. Whether or not this makes (3.6) true does depend on what the world is like.

If the world is such that everybody fails in business, the right choice of c is one of the people who commit suicide. But the world might be such that there are no such persons, so that the choice of $x = c$ must make the other disjunct true, in other words must satisfy $\sim F(x)$. This is guaranteed only if x satisfies $\sim F(x)$, in other words if it is a case that

(3.7) $\sim S(x) \supset \sim F(x)$.

In other words only

(3.8) $F(x) \supset S(x)$.

In that case, (3.2) can be true only if its antecedent is false, in other words only if not everybody fails in business. Hence the choice of x must provide a counter-example to everybody's failing in business. And the choice $x = c$ provides such an counter-example only if

(3.9) $\sim S(c) \supset \sim F(c)$.

The existence of such a counter-example means the truth of (3.3). Hence (the truth of) (3.5) implies the (the truth of) (3.3), which is Peirce's paradox.

In still other words, (3.3) is true only if there is an x such that if he fails in business, he commits suicide. Depending on what the world is like in (3.1) the verifier might have to choose $\sim F(c)$ or $S(c)$. In other words, c depends on the world. This means that the x in (3.5) or (3.4) is not the same individual independently of what the world is like. It is not really a choice of an "individual individual" as is required in (3.1) and (3.2).

4 Peirce's paradox and independence

This is clear interpretationally. But what does it mean in terms of the semantical games that convey our sentences their meaning? What is the right logic translation of (3.2)?

The analysis carried out above shows that the choice of a disjunct ("of a world") must be neutral with respect to the choice of objects. Hence the solution is to make $\vee$ independent of $(\exists x)$. Instead of (3.4) one should have

(4.1) $(\exists x)((\forall y) F(y) (\vee/\exists x) S(x))$.

Thus the true representation of (3.2) is not (3.4) but (4.1). It cannot be formulated in IF logic in the usual narrow sense, but it can be formulated if this logic is amplified by allowing extra independencies between quantifiers and connectives. This opens up a new dimension of the entire hierarchy of different logics, besides further illustrating the inadequacy of Frege's logic.

5 Hierarchies of IF logics

In IF logic in the narrowest sense – which is the one in which it currently being used in the literature – the only extra kind of independence allowed is an independence of existential-force quantifiers of universal-force quantifiers within the formal scope of which they occur. (Quantifiers, whose scopes are not nested are automatically independent.) Only strong negations, $\sim$, are admitted. If we admit sentence-initial contradictory negations, $\neg$, we obtain richer and more satisfactory logic which is usually called extended independence-friendly logic (EIF) logic. It should perhaps be considered as the "real" basic IF logic. If we allow arbitrary extra independencies (existential quantifiers on existentials, universal quantifiers on universals, and universal quantifiers on existentials) we obtain a still much stronger logic that might be called generalized IF logic.

Here we are dealing with yet another way of enriching the basic or extended IF logic. This way is to allow extra independencies between quantifiers and propositional connectives. From the Peircean example and from others it is seen that this dimension of expressive enrichment is independent of quantifier independencies.

6 Simple donkey sentences

This new dimension also facilitates analysis of many interesting linguistic phenomena. One instructive example is constituted by the so-called donkey sentences. The interpretation of these sentences is a routine question discussed in the linguistic literature on definite and indefinite pronouns. The simplest example has the same form as the following sentence:

(6.1)   If Peter owns a donkey, he beats it.

This is *prima facie* of the following form

(6.2) $(\exists x)(D(x) \& O(p, x)) \supset B(p, x)$.

This would have to be equivalent with

(6.3) $(\forall x)(\sim D(x) \vee \sim O(p, x)) \vee B(p, x)$.

But (6.2) is ill-formed in that the last x is not bound to (is outside the scope of) $(\exists x)$. But the alternative

(6.4) $(\exists x)((D(x) \& O(p,x)) \supset B(p,x))$

says only that there is at least one animal such that if it is a donkey and is owned by Peter, he beats it. The true semantical form of (6.1) seems to be intuitively

(6.5) $(\forall x)((D(x) \& O(p,x)) \supset B(p,x))$.

But why? How come (6.1) should be translated as (6.5)? An indefinite article has the force of an existential quantifier. So why does it seem to have here the force of an universal one?

The answer can be obtained by analyzing the meaning of (6.1) the same way as the meaning of Peirce's paradoxical sentence (3.1) was analyzed earlier. The crucial point is that the choice of x must be independent of the choice between different relevant semantics codified in the second $\vee$ in (6.3). The solution is now to make the quantifier and the connective $\vee$ independent of each other. Here the covert logic translation of (6.1) will be

(6.6) $(\exists x)(D(x) \& O(p,x))(\supset /\exists x)B(p,x)$

which is equivalent with

(6.7) $(\forall x)(\sim D(x) \vee \sim O(p,x))(\vee /\forall x)B(p,x)$.

When is there a winning strategy for the verifier in the game with (6.7), as (6.7) says? In that strategy, since $\vee$ is independent of $(\forall x)$, the falsifier chooses a value d for x. The resulting sentence

(6.8) $(\sim D(d) \vee \sim O(p,d)) \vee B(p,d)$

must be true, i.e. the verifier must be able to choose a true disjunct. Such a choice is possible for any d if it is the case that for any donkey d owned by Peter it is true that he beats it i.e. that $B(p,d)$ is true. But this is obviously just what (6.1) says.

7 Complex donkey sentences

This shows that the extensive literature designed to account for donkey sentences is, if not wrong, then at least redundant. Many purported explanations do not work for more complex donkey sentences like

(7.1) If you give each child a gift for Christmas, some child will open it to-day.

Here even a merely linguistic account of the role of the anaphoric phenomenon "it" is very tricky. No usual IF logic expression captures the meaning of (7.1) either. Yet its logic translation in terms of connective independence is possible.

The right translation is perhaps best seen if we first eliminate the existential quantifier in terms of its Skolem function and express (7.1) as

(7.2) $(\exists g)(\forall x)(G(x, g(x)) \supset (\exists z)O(z, g(z)))$.

This is a second-order sigma one-one sentence. It is possible to translate such sentences to the corresponding IF first-order language, but not without independent connectives. Here is a translation:

(7.3) $(\forall x_1)(\forall x_2)(\exists y_1 /\forall x_2)(\exists y_2 /\forall x_1)((x_1 = x_2) \supset (y_1 = y_2))$ &

$G(x_1, y_1) \& G(x_2, y_2)(\supset /\forall x_1, \forall x_2)(\exists z)((z = x_1) \vee (z = x_2) \supset (O(z, y_1) \& O(z, y_2)))$.

This explains the meaning of (7.1).

8 Conditional reasoning

This is in explicit terms what the idea of "remembering" a strategy used in earlier subgame amounts to.

In general we have found an important distinction. It may be called a distinction between deductive reasoning and conditional reasoning. A deductive conclusion B from a premise A is a proposition that is true as soon as A is true. In the language of possible world semantics, B is true in each world in which A is true.

But the premise A does more than put forward a truth condition. It presents a situation, a fragment of one particular possible world, maybe a world in which Peter owns a donkey. We can then ask what else must be true in that particular world. This is a different question from asking what is true of all the worlds in which the premise A is true, for instance all worlds in which Peter owns some donkey or other. We are asking about the fate of that particular donkey postulated by the premise. Does Peter beat it?

What has been shown in this paper is how this question can be spelled out in sample cases by means of quantifiers independent of propositional connectives. These independencies are the gist of conditionality. It cannot be captured by ordinary "conditional" sentences of the form $(A \supset B)$ or by ordinary logical consequence relations. It is also the gist of the linguistic phenomenon of conditionality.

What is especially striking in all these examples is that the extra-connective-independence is not just one formally possible explanation of certain semantical phenomena, but the overwhelmingly natural one. This naturalness is easily converted into generality. When (in game-theoretical terms) a quantifier invites a player to choose an individual, the choice must not depend on what there may turn up later in the game. Thus the normal logic translation of disjunctive "or" appears to be, not $\vee$, but $(\vee / Q_1 x_1, Q_2 x_2, \ldots)$, where $(Q_i x_i)$ are the quantifiers within whose scope $\vee$ occurs in the translation.

9 Conditionality explained

These case studies illustrate *ipso facto* some of the explanatory possibilities in linguistic theorizing that are opened here. Consider, for example, the equivalence of (7.1) and (7.2). I have much earlier presented a semi-formal analysis of conditionals in a game-theoretical framework (Hintikka & Kulas [6]). It worked, but it was not purely logical. I had to resort to pre-formal ideas, e.g. the idea that a player in a semantical game could "remember" a strategy from another subgame. Such semiformal ideas can now be replaced by purely logical ones. For instance, look at (7.2). The Skolem function g there codifies (a part of a) strategy. This is used in a subgame with the antecedent of (7.1). From (7.2) one can see how it figures also as a strategy function (partial) in a game with the consequent.

In (7.2), this transfer of a strategy becomes the possibility of making use of the connection between x and y (subscripts do not matter) that was introduced in the antecedent also in the consequent. This is precisely what is made possible by the independence of $\vee$ of the quantifiers $(\forall x_1), (\forall x_2)$.

This shows how by means of independences involving connectives we can capture the very conditionality of conditionals. This means that by means of such independences we can develop a viable general theory of conditionals.

10 Explaining anaphora

Even more generally much of any first-order logic can be thought as framework for a semantical representation of such phenomena as co-reference and anaphora. Not all such logics can be applied directly to the analysis of these phenomena in natural languages, mainly owing to the syntactical differences between them and natural language.

Certain general advantages of the kind of treatment of anaphora based on IF logic over some typical linguistic theories can presently be pointed out. Linguistic approaches to anaphora and co-reference often rely on the head-anaphora relation as one of their explanatory concepts. Of course linguists are aware that there are examples where there is no head to be found for a given anaphora or where the head and the anaphora cannot be said to be literally co-referential, that is, refer to one and the same entity. But such cases are typically considered somehow exceptional, not automatically explainable by the normal operation of anaphora.

We have already analyzed such an apparently anomalous case. In the complex donkey sentence (7.1), the obviously anaphoric pronoun "it" is not literally co-referential with any other phrase in the sentence. (It is not a "pronoun of laziness" either.) Yet (7.1) has an explicit logical form (7.2).

An explanation is implicit in what has been said earlier. We can interpret "it" because it is co-referential with an object that is functionally determined by other referring phrases in the same sentence or the same discourse. The functions that effect this determination are sometimes expressed in the sentence in question by a separate phrase. But they need not be. As we saw in our analysis of complex donkey sentences, existential quantifiers can introduce such dependencies through their Skolem functions. Sometimes the dependence is mediated by background information that the actual or hypothetical speaker if assumed to possess.

Hence a purely syntactical approach to the phenomena of anaphora and co-reference, such as Chomsky's government and binding theory, is bound to be incomplete account these phenomena.

11 Limits of compositionality

There is another general methodological moral in the story of this paper. The mode of operation of independent connectives illustrates a phenomenon that is as prevalent as it is important both in natural and formal languages. It is non-compositionality. (For a collection of articles on different aspects of compositionality, see Werning et al. [9]).[1]

Compositionality is rightly understood tantamount to semantical context-independence. Now we have seen in this paper how the logical force of a connective is different according to what quantifiers in its context it depends on. Of course

---

[1] I take this opportunity to correct a group of mistakes. On page 10 the authors say that Hodges has refuted "Hintikka's claim that Independence-Friendly logic is non-compositional". I have never made such a claim *simpliciter*, and on the contrary suggested a way in which any logic can in principle be given a compositional "semantics". What is the case (also according to Hodges) is that IF first-order logic cannot have a compositional semantics on the first order level.

a similar non-compositionality is obvious (though it was not to Frege) already in the dependence of quantifiers on other quantifiers. The main reason why this context dependence has not been emphasized more is that in the received first-order logic quantificational dependencies are expressed by the syntactical device of nesting scopes. But the only thing the necessity of so doing shows is the inadequacy of traditional first-order logic in semantic theorizing.

References

[1] N. Chomsky. *Essays on Form and Interpretation*. North-Holland, Amsterdam, 1977.

[2] N. Chomsky. *Knowledge of Language. Its Nature, Origin and Use*. Praeger, New York, 1986.

[3] J. Hintikka. What is the Significance of Incompleteness Results? Retrieved February 27, 2016 from http://people.bu.edu/hintikka/Papers_files/What_is_the_significance_of_the_incompleteness_%20results_JHintikka_0211_032211.pdf.

[4] J. Hintikka. IF Logic, Definitions and the Vicious Circle Principle. *Journal of Philosophical Logic*, Volume 41(Issue 2):505–517, April 2012.

[5] J. Hintikka. Which Mathematical Logic is the Logic of Mathematics? *Logica Universalis*, Volume 6(Special Issue 3 commemorating Jean van Heijenoort):459–475, December 2012.

[6] J. Hintikka and J. Kulas. *Anaphora and Definite Descriptions: Two Applications of Game-Theoretical Semantics*. Synthese language library; v. 17. D. Reidel, Dordrecht, Holland, 1985.

[7] A. Mann, G. Sandu, and M. Sevenster. *Independence-Friendly Logic: A Game-Theoretic Approach*. Cambridge University Press, Cambridge, 2011.

[8] C. S. Peirce. Exact Logic. The Simplest Mathematics. In Ch. Hartshorne and P. Weiss, editors, *Collected Papers of Charles Sanders Peirce*, volume 3 and 4. Harvard University Press, Cambridge MA, 1933.

[9] M. Werning, E. Machery, and G. Schurz, editors. *The compositionality of meaning and content. Volumes I & II*. Ontos verlag, Heusenstamm, 2005.

Commentary on Jaakko Hintikka's "IF Logic and Linguistic Theory"

Gabriel Sandu
*Department of Philosophy, History, Culture and Arts, University of Helsinki
Unioninkatu 40 A, 00014 Helsinki, Finland*
sandu@mappi.helsinki.fi

In the paper Hintikka gives various arguments for the need of an extension of Independence-Friendly logic (IF-logic) with informationally independent disjunctions, i.e. connectives of the form

$$(\vee/\forall x)$$

that I will render more simply as $(\vee/x)$. Actually such an extension has been studied in Sandu and Väänänen [4], Hella and Sandu [2] and Mann, Sandu and Sevenster [3] but no application to natural language has been given. Thus I welcome Hintikka's endeavour. He introduces the case for informationally independent connectives by first offering a solution to what he calls Peirce's paradox which consists in the equivalence of

Hintikka compares

(3.1) Someone is such that if he fails in business, he commits suicide.

with

(3.2) Someone is such that if everybody fails in business, he commits suicide.

when they are represented in ordinary first-order logic as

$$\exists x(F(x) \to S(x))$$

and

$$\exists x(\forall y F(y) \to S(x))$$

respectively. (I will use '$\to$' instead of Hintikka's '$\supset$'). Hintikka analyzes the equivalence between these two sentences in game-theoretical semantics. This is a good idea, although I prefer a more straightforward game-theoretical argument than the one he offers. We establish the logical equivalence between

$$\exists x(\neg F(x) \vee S(x))$$

and
$$\exists x(\exists y \neg F(y) \vee S(x))$$

by showing that the Verifier has a winning strategy in one game if and only if she has a winning strategy in the other game (on any underlying model). As usual, these claims are established by a copy cat strategy argument. (Again a notational point: Hintikka makes a distinction between game-theoretical negation that he symbolizes by '$\sim$' and contradictory negation that he symbolizes by '$\neg$'. I will simply use the latter given that for ordinary first-order formulas the two are equivalent.)

Suppose there is a winning strategy for the Verifier in the first game. It consists of the choice of an individual, $x = a$ and the choice of a disjunct, left or right. Given that the strategy is winning, then, if left is chosen, a must satisfy $\neg F(x)$ and if right is chosen, then x must satisfy $S(x)$. Here is a winning strategy for Verifier in the second game. If in the first game Verifier chooses left, then in the second game she chooses $x = a$, then left, and then $y = a$. Given that a satisfies $\neg F(x)$ then this is a winning strategy. If in the first game Verifier chooses right, then in the second game Verifier chooses $x = a$ then right. Given that a satisfies $S(x)$, then this is a winning strategy in the second game.

For the converse, suppose the Verifier has a winning strategy in the second game. It is: choose $x = a$; then choose left or right. If left, choose $y = b$; if right, do nothing. Given that this is a winning strategy, then if right is chosen, x must satisfy $S(x)$. If left is chosen, then b must satisfy $\neg F(y)$. Here is a winning strategy for Verifier in the first game. If Verifier chooses right in the second game, then choose $x = a$ and then right in the first game. Then a satisfies $S(x)$ and thus this is a winning strategy. If Verifier chooses left and then $y = b$ in the second game, then in the first game she chooses $x = b$ and then left. Clearly given that b must satisfy $\neg F(y)$ this is a winning strategy.

Actually $\exists x(\exists y \neg F(y) \vee S(x))$ is logically equivalent with

$$(\exists y)\neg F(y) \vee \exists x S(x)$$

and thus "Peirce's paradox" is seen to be an instance of the more general law

$$\exists x(A(x) \vee B(x)) \equiv \exists x A(x) \vee \exists y B(y) \equiv \exists x A(x) \vee \exists x B(x).$$

Hintikka's suggestion in the paper is to block the paradox by blocking the above equivalence in this particular case, that is, by taking the logical form of (3.2) to be (there is a misprint in the text):

$$\exists x(\neg \forall y F(y) \, (\vee/x) \, S(x))$$

that is
$$\exists x(\exists y \neg F(y) \,(\vee/x)\, S(x))$$
where $(\vee/x)$ means that when Verifier chooses a disjunct, she does not know the value chosen earlier for x. Now apart from creating interpretational problems of its own, the proposal will not help him. Informally the proposal says that the choice of a disjunct should take place before the choice of a value of x takes place. But this renders the last sentence logically equivalent with
$$\exists y \neg F(y) \vee \exists x S(x)$$
which is, as pointed out above, logically equivalent with $\exists x(\neg F(x) \vee S(x))$. We are back to square one! I guess Hintikka has in mind another way to analyze the informational independence of Verifier of its own move than the one I proposed (games of imperfect information), that is, a proposal that does not render $\exists x(\exists y \neg F(y) \,(\vee/x)\, S(x))$ equivalent with $\exists x(\neg F(x) \vee S(x))$. I remember he once in conversation objected to the equivalence between $\exists x(\exists y/x)x = y$ with $\exists x \exists y x = y$ which holds in IF-logic. Fausto Barbero [1, forthcoming] has a notion of independence which does not render the two equivalent. It might be that Hintikka is relying in his proposal on a notion of independence on the basis of which $\exists x(A \,(\vee/x)\, B$ is not equivalent with $\exists x A \vee \exists x B$ but this is something for future work.

Based on his attempted solution to Peirce's paradox, Hintikka suggests also a new way to analyze simple donkey sentences like

(6.1)   If Peter owns a donkey, he beats it.

He takes the force of this sentence to be that of

(6.5) $\forall x(D(x) \wedge O(p, x)) \to B(p, x))$.

He asks: How do we get from (6.1) to (6.5)? One way to proceed is to take literally the surface structure of (6.1) where the indefinite is in the "scope" of the implication, and translate the indefinite "a donkey" by an existential quantifier, as standardly done. The result is, as Hintikka correctly points out:

(6.2) $\exists x(D(x) \wedge O(p, x)) \to B(p, x)$

which is equivalent, as he points out with

(6.3) $\forall x(\neg D(x) \vee \neg O(p, x)) \vee B(p, x)$.

But Hintikka is right to point out that (6.2) (and consequently (6.3)) is ill formed given that the last occurrence of the variable x is not bound. On the other side, if we try to bind the variable x by the existential quantifier, we get

(6.4) $\quad \exists x (D(x) \wedge O(p,x)) \to B(p,x))$

which, as Hintikka correctly points out, says only that "there is at least one animal such that if it is a donkey and is owned by Peter, he beats it." So it seems we cannot obtained the true logical form of (6.1) which is (6.5).

Hintikka proposes an answer which is to go back to (6.2) and to take the implication to be independent of the existential quantifier

(6.6) $\quad \exists x (D(x) \wedge O(p,x)) (\to /\exists x) B(p,x)$

or, if we operate instead on (6.3) which he takes to be equivalent to (6.2), he takes disjunction to be independent of the universal quantifier:

(6.7) $\quad \forall x (\neg D(x) \vee \neg O(p,x)) (\vee /\forall x) B(p,x).$

We are then told that the existence of a winning strategy for the Verifier in (6.7) means that for any choice d by the Falsifier, the sentence

(6.8) $\quad (\neg D(d) \vee \neg O(p,d)) \vee B(p,d)$

must be true. And this yields (6.5).

Hintikka's analysis is ingenious but it does not get through, as it stands. I claim that the independence $(\to /\exists x)$ of implication from the existential quantifier in (6.6), or, equivalently the independence $(\vee /\forall x)$ of disjunction from the universal quantifier (6.7), does not make sense. The reason for this, focusing on the latter, is simply that in IF-logic as it currently stands, for a move to be informationally independent from another, the first must be in the syntactical scope of the second. Or this is not the case in (6.7).

Finally Hitnikka motivates the use of informationally independent disjunctions by its role in the logical representation of complex donkey sentences like

(7.1)   If you give each child a gift for Christmas, some child will open it to-day.

that Hintikka represents in second-order logic by

(7.2) $\quad \exists g \forall x (G(x, g(x)) \to \exists z O(z, g(z))).$

He then tells us that (7.2) can be represented on the first-order level by the IF sentence (7.3) which involves informationally independent disjunctions.

Hintikka's claim is not true. (7.3) is a second-order existential formula and as such known to be equivalent, by standard results of Walkoe [5], to an ordinary IF-formula which does not involve informationally independent disjunctions. Let me reproduce the procedure by which the IF-formula is obtained (I am grateful to Fausto Barbero here).

1. First in (7.2) we push the existential quantifier in front of the conditional and then Skolemize it:

$$\exists f \exists g \forall x (G(x, g(x)) \to O(f(x), g(f(x)))).$$

2. Next we eliminate the nesting of functions to obtain

$$\exists f \exists g \forall x \forall y (y = f(x) \to (G(x, g(x)) \to O(y, g(y)))).$$

3. Third we want each function to have a unique set of arguments (so we replace the second g with a new h):

$$\exists f \exists g \exists h \forall x \forall y (x = y \to g(x) = h(y) \wedge \\ \wedge [(y = f(x) \to (G(x, g(x)) \to O(y, h(y))))]).$$

4. Finally we replace each function by its appropriate pair of quantifiers and obtain the IF-formula which is the logical form of (7.1):

$$\forall x \forall y (\exists u/y)(\exists v/y, u)(\exists w/x, u, v)(x = y \to v = w \wedge \\ \wedge [(y = u \to (G(x, v) \to O(y, w)))]).$$

References

[1] F. Barbero. Cooperation in games and epistemic readings of Independence-Friendly sentences. Forthcoming in *Journal of Logic, Language, and Information*.

[2] L. Hella and G. Sandu. Partially Ordered Connectives and Finite Graphs. In M. Krynicki, M. Mostowski, and L. W. Szczerba, editors, *Quantifiers: Logics, Models and Computation. Volume Two: Contributions*, pages 79–88. Springer Netherlands, 1995.

[3] A. Mann, G. Sandu, and M. Sevenster. *Independence-Friendly Logic: A Game-theoretic Approach*. Cambridge University Press, Cambridge, 2011.

[4] G. Sandu and J. Väänänen. Partially ordered connectives. *Mathematical Logic Quarterly*, 38(1):361–372, 1992.

[5] J. Walkoe. Finite partially-ordered quantification. *J. Symbolic Logic*, 35(4):535–555, 1970.

Medieval Modalities and Modern Methods: Avicenna and Buridan

Wilfrid Hodges
Herons Brook, Sticklepath, Okehampton, Devon EX20 2PY, United Kingdom

Spencer Johnston
Department of Philosophy, University of York, Heslington, York, United Kingdom

In respectful memory of Grisha Mints

The two authors of this paper independently found themselves applying methods of modern logic to medieval modal systems—Johnston to the divided modal syllogisms of the 14th century French Scholastic Jean Buridan, and Hodges to the modal work of the 11th century scholar Ibn Sīnā, known in the West as Avicenna, who worked in Persia and wrote mostly in Arabic. Quite late in the day we realised that there was a mathematical equivalence between things in our work, and we put our heads together on this. There is a curious twist: what Johnston did is mathematically equivalent to work of Avicenna, not of Hodges. That adds a piquant question to the issue: What on earth are Kripke structures doing in an 11th century text? The question naturally leads on to another one: What are Kripke structures doing in *any* modal enquiry? It's unlikely we could answer the question about Avicenna in any depth without having some view on the role of Kripke structures in general.

We would dearly have liked to discuss all this with Grisha Mints. Not least, this is because Johnston's work included a detailed comparison of Buridan's proof theory with natural deduction methods. Avicenna had a proof theory too—it went in a completely different direction both from Buridan's and from anything we know of in today's proof theory.

We thank Stephen Read, both for his work on Buridan which we use below, and for supervising the PhD of the second author; and Saloua Chatti for useful consultations. We also thank the members of a Workshop of the Medieval Philosophy Network, organised by Anna Marmodoro and John Marenbon at the Warburg Institute in April 2016, who contributed to a helpful discussion of some of the issues in this paper. Finally we thank the two journal referees. One of them urged us to clarify and expand our account of Avicenna's sentence forms; we are particularly grateful to this referee for a close analysis of what was unclear. We added an Appendix in response to this referee's request for supporting texts.

That creates a further issue where Grisha's advice would have been invaluable. Avicenna raised a range of new questions which seem to need a proof-theoretic answer, but as far as we know, neither his own proof theory nor anything in the modern literature will provide the required answers. So Hodges [9] concocted a proof theory that does the job, though inelegantly. Grisha would surely have seen how to improve it. But the present paper is not about these proof-theoretic questions.

The collection of facts that we bring together here is complex in several ways. First there is the difference between Avicenna in Persia and Buridan three hundred years later in Western Europe. Second there is the difference between medieval methods of logic and the modern ones that we apply to the medieval authors. And third there are three kinds of fact in play: (1) textual facts about what each of the medieval authors wrote and what their words meant; (2) mathematical facts about the formalisms involved; and (3) interpretative facts about the reasons why the medievals did certain things, or about what our reasons are for doing certain things. In §1 below we survey the pieces of the jigsaw; we also examine one of the mathematical equivalences involved, using Grisha's textbook [17]. In §2 we examine the textual facts about Buridan and the modern formalism that the second author brought to bear on Buridan's discussions. In §3 we do the same for Avicenna and the first author. Then in §4 we ask the interpretative questions and suggest some answers. Finally §5 is an Appendix with supporting texts.

1 Overview

Here is a picture of this paper:

(1)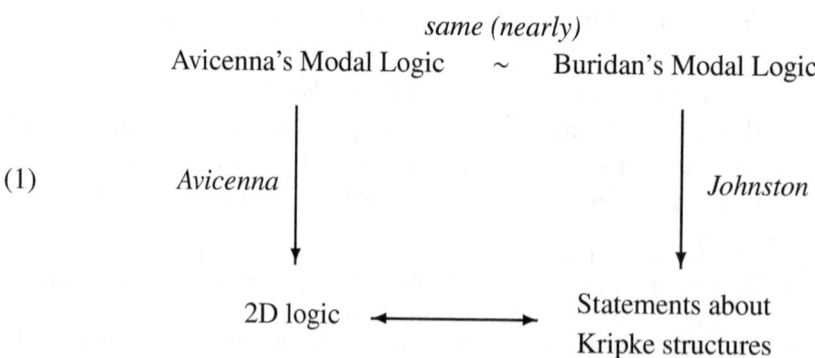

In the top row of the picture, Avicenna's Modal Logic and Buridan's Modal Logic are both of the kind called divided alethic modal logic. This means that they are about logical inferences between sentences of the following forms:

(2) $$\left\{\begin{array}{l}\text{Every } A \\ \text{Some } A\end{array}\right\} \text{ is } \left\{\begin{array}{l}\text{necessarily} \\ \text{possibly} \\ \text{contingently}\end{array}\right\} \left\{\begin{array}{l}\text{a } B. \\ \text{not a } B.\end{array}\right\}$$

where A and B are distinct. Both Avicenna and Buridan inherited these sentence forms from Aristotle's *Prior Analytics* [2]. Some shorter names for these sentences will be helpful. Traditionally

(3)
'a' stands for 'Every ... is ...',
'e' stands for 'Every ... is not ...',
'i' stands for 'Some ... is ...',
'o' stands for 'Some ... is not ...'.

If we use '*nec*', '*pos*' and '*con*' for necessary, possible and contingent, then we can write for example

(4)
$(a\text{-}nec)(A, B)$ for 'Every A is necessarily a B',
$(o\text{-}pos)(C, D)$ for 'Some C is possibly not a D'

and so on. In practice we will often ignore the 'contingently' sentences, since their theory is largely parasitic on that of the 'possibly' sentences.

Buridan adds a further kind of form by allowing 'now' as an alternative to 'necessarily/possibly/contingently'; he calls the resulting sentences *de inesse ut nunc* sentences. Strictly these further forms are not modal at all, but since Buridan integrates them with the modal forms we will count them in as part of 'Buridan's (divided, alethic) modal logic'. The 'same (nearly)' at the top of picture (1) represents two facts: Buridan considered a larger collection of sentences than Avicenna did, and within the class of sentences common to Avicenna and Buridan there is evidence that they had different views on the proper logical representation of necessity (on which see (23) and (24) below).

The expressions A, B in the sentences above are called 'terms'; A is the 'subject' and B the 'predicate' of the sentence. Both Avicenna and Buridan concentrate on logical relationships of the form 'ϕ_1 and ϕ_2 entail ϕ_3' with three sentences; these are known as 'syllogistic moods'. In fact Buridan considers no logical relationships more complicated than this. Avicenna spreads his net wider, but we ignore his extensions in the present paper.

Avicenna is on the left of picture (1) because he was earlier than Buridan. But no line of influence from him to Buridan is known, and we won't assume that there was any such influence. In §§2 and 3 we will take Buridan before Avicenna because the facts to be reported there for Buridan are less controversial than those for Avicenna.

On the righthand side of picture (1), the vertical arrow represents an action of Johnston and a set of facts. Johnston's action was to describe a family of Kripke structures and a translation from sentences of Buridan's modal logic to set-theoretic statements about Kripke structures. The set of facts are to the effect that this family of Kripke structures and translations validates all Buridan's claims for the validity or invalidity of the syllogistic moods of his modal logic. Details are in §2 below.

On lefthand side of picture (1), the 2D logic, short for 'Avicenna's two-dimensional logic', is a logic invented and studied by Avicenna. We formalise it as a fragment of a two-sorted first-order logic with two sorts *object* and *time*, where every relation symbol is binary and has first argument of sort *object* and second argument of sort *time*. There is no consensus of scholars about how Avicenna intended this logic to relate to his alethic modal logic. The arrow in the picture represents a set of facts relating the two logics, together with our conjecture that Avicenna knew this set of facts and intended it to be used to relate the two logics in a certain way. The facts we will set out in §3, but we leave the question of his intentions to §4, where we will argue that the comparison with Buridan strengthens our interpretation.

The picture is tied together by the horizontal arrow at the bottom. This arrow represents a mathematical fact about Kripke structures and modal sentences. The fact can be derived as a generalisation of a result in Grisha's introductory textbook of modal logic [17], and the rest of this section will be devoted to deriving it.

In Chapter Two of [17] Grisha introduces 'classical monadic predicate logic'—a logic which contains Aristotle's categorical syllogisms, as Hilbert and Ackermann [7] ii.3 point out. On page 25 he identifies within monadic predicate logic a subclass of formulas which he calls 'modal-like'; we can suppose that they contain at most one free variable. His first example is

(5) $\quad (P(y) \vee Q(y)) \wedge \forall x(\neg P(x) \wedge \neg Q(x) \wedge \forall x \exists x(P(x) \vee Q(x)))$.

He describes how to translate modal-like formulas into modal propositional logic so that 'all information ... is completely preserved' (p. 26). The modal translation of the formula above is

(6) $\quad (p \vee q) \wedge \Box(\neg p \wedge \neg q \wedge \Box \Diamond (p \vee q))$.

The modal sentences that contain or are within the scope of a modal operator in (6) are exactly those that come from a subformula of (5) containing no variable that is free in (5). On page 40 he shows that the modal sentences derived in this way from modal-like predicate formulas are exactly the sentences of modal propositional logic. On page 42 he shows that for every first-order structure M in the language of a modal-like formula such as (5) there is a corresponding S5 Kripke structure $\tilde{M}$ with the property that (5) is satisfied by an element of M if and only if (6) is true at some world in $\tilde{M}$. In fact $\tilde{M}$ can be taken to have a universal accessibility relation in the sense that every world is accessible from every world. As he says (p. 41), there is 'a very close semantic connection' between the logic of modal-like predicate sentences and the S5 logic. Essentially, the quantifiers of the modal-like predicate sentences are taken to range over 'worlds'.

The horizontal arrow in our picture above is Grisha's 'very close semantic connection', but lifted to monadic modal predicate logic, i.e. modal predicate logic where the relation symbols are all monadic. An example of a sentence of this form is

(7) $$\forall z(Pz \to \Box Qz).$$

If we run Grisha's translation backwards, starting with monadic modal predicate logic instead of modal propositional logic, then we will reach a form of first-order logic where, besides the variables ranging over worlds, there will be variables doing the job of z in the sentence (7). It will be convenient to use different sorts of variables to do the different jobs, so from now on we will use Greek variables to range over worlds. Thus the predicate sentence corresponding to (7) will be

(8) $$\forall z(Pz\alpha \to \forall \beta Qz\beta).$$

In short, the equivalence is now between monadic modal predicate sentences and a certain class of two-sorted predicate sentences where each relation symbol is binary, with one variable of the first sort and one of the second. We will call the two sorts *object* and *world*. The variable z in (8), which is inherited from the quantified variable in (7), has sort *object*, and the world variables α and β have sort *world*. Just as before, every sortal structure M for the language of (8) translates into an S5 Kripke structure $\tilde{M}$ for the modal language of (7) (again with universal accessibility), and a modal sentence is true at some world in $\tilde{M}$ if and only if its sortal predicate translation is satisfied by some element of the sort *world* in M.

Since the sortal structure M will have just one domain D of objects, the universe of each world in $\tilde{M}$ will be the same; we can take it to be D again. But we can loosen things up by allowing that in each world, some objects may be actual and other objects may be non-actual. Formally we do this by allowing the predicate formulas to include a distinguished predicate symbol O read as 'actual'; so $Ox\alpha$ means that x is actual in world α. In the modal language this binary predicate symbol goes over to a monadic predicate symbol, which we can write again as O, so that 'Ox' is read as 'x is actual'.

We will call this equivalence between modal-like binary two-sorted predicate logic and monadic modal predicate logic the 'basic modal equivalence'. Of course we claim no originality for the observation that this equivalence exists; presumably it's common knowledge among modal logicians.

2 Buridan

This section is closely based on part of the second author's PhD thesis [16], though we have changed some notation to make the comparison with the Avicennan material easier.

Buridan's divided modal logic is a fragment of his treatment of the alethic modals necessity, possibility, contingency, and non-contingency, treating only those propositions where the modal occurs as either an adverb or a verb that has modal force (e.g. verbs like 'can'), together with assertoric propositions ([4] pp. 95–96). The core idea that underpins Buridan's treatment of divided modal propositions is that the subject in these propositions is ampliated (i.e. the class of objects that the subject term is taken for is expanded) to include those things which either do or could fall under the subject. For example, according to Buridan, 'the proposition "B can be A" is equivalent to "That which is or can be B can be A." ' ([4] p. 97) In the presence of the assumption that whatever is the case can be the case, this can be simplified to 'Something can be B and can be A'. When Buridan says that something 'can be B', what exactly does he have in mind? The following passage from George Hughes explains the intuition behind the use of the actual and the possible:

(9) > A short digression seems in order here. For a long time I was puzzled about what Buridan could mean by talking about possible but non-actual things of a certain kind. Did he mean by a 'possibly A', I wondered, an actual object which is not in fact A, but might have been or might become, A?...But this interpretation will not do; for Buridan wants to talk, e.g., about possible horses; and it seems quite clear that he does not believe that there are, or even could be, things which are not in fact horses but which might become horses. What I want to suggest here, very briefly, is that we might understand what he says in terms of modern 'possible world semantics'. Possible world theorists are quite accustomed to talking about possible worlds in which there are more horses than there are in the actual world. And then, if Buridan assures us that by 'Every horse can sleep' he means 'Everything that is or can be a horse can sleep' we could understand this to mean that for everything that is a horse in any possible world, there is a (perhaps other) possible world in which it is asleep. It seems to me, in fact, that in his modal logic he is implicitly working with a kind of possible worlds semantics throughout. ([15] p. 9)

We relativise Buridan's statements to worlds α. In line with the quotation above, we consider that an object x in the world α can in some sense be a man even if x is not actual in α; to avoid confusion with anything that Buridan might understand to be implied by 'is a man', we express our notion in symbols as $v(\alpha, man, x)$. We write $O(\alpha, x)$ for 'x is actual in world α', and $D(x)$ for 'x is an object'. We translate:

(10)
$$\begin{aligned} x \text{ is a } P \text{ in world } \alpha &\mapsto O(\alpha, x) \wedge v(\alpha, P, x) \\ x \text{ is a } non\text{-}P \text{ in world } \alpha &\mapsto D(x) \wedge \neg(O(\alpha, x) \wedge v(\alpha, P, x)). \end{aligned}$$

Then 'x can be a man' translates to 'There is a world α such that x is a man in α', and likewise 'x is necessarily a man' translates to 'For every world α, x is a man in α'. These translations justify the following definitions of V, M and L, with M and L representing possibility and necessity:

(11)
$$\begin{aligned} V(\alpha, P, x) &\equiv O(\alpha, x) \wedge v(\alpha, P, x); \\ M(\alpha, P, x) &\equiv \exists \beta V(\beta, P, x); \\ L(\alpha, P, x) &\equiv \forall \beta V(\beta, P, x). \end{aligned}$$

It will be convenient to write

(12)
$$\begin{aligned} V(\alpha, P) &\equiv \{x : V(\alpha, P, x)\}, \\ M(\alpha, P) &\equiv \{x : M(\alpha, P, x)\}, \\ L(\alpha, P) &\equiv \{x : L(\alpha, P, x)\}. \end{aligned}$$

These formulas can be interpreted in a suitable kind of Kripke structure, which we call a 'Buridan modal model', as follows. A Buridan modal model is a tuple $\mathfrak{M} = \langle D, W, R, O, v \rangle$ such that:

D and W are non-empty sets. D is the domain of objects and W is a set of worlds.

$R = W^2$.

$O : W \to \mathcal{P}(D)$.

$v : W \times PRED \to \mathcal{P}(D)$

where $PRED$ is the set of monadic predicate symbols P. Here R is the accessibility relation. Since R is universal, by the standard Kripke semantics the interpretations of $M(\alpha, P)$ and $L(\alpha, P)$ in $\mathfrak{M}$ don't depend on α, and we can simplify these terms to $M(P)$, $L(P)$. We write $V(\alpha, P)^{\mathfrak{M}}$ for the interpretation of $V(\alpha, P)$ in the Buridan modal model $\mathfrak{M}$; and likewise $M(P)^{\mathfrak{M}}$ and $L(P)^{\mathfrak{M}}$. Informally, we can think of $V(\alpha, P)^{\mathfrak{M}}$, $M(P)^{\mathfrak{M}}$, and $L(P)^{\mathfrak{M}}$ as respectively giving the class of objects that are P at α in $\mathfrak{M}$, can be P in $\mathfrak{M}$, and are necessarily P in $\mathfrak{M}$.

With these definitions in place, we can translate Buridan's sentences into conditions on a Buridan modal model $\mathfrak{M}$ as follows. For the non-modal sentences the conditions are on a

Buridan modal model $\mathfrak{M}$ at a world α:

(13)
$$\begin{aligned}
(a\text{-}nec)(A,B) &\mapsto M(A)^{\mathfrak{M}} \subseteq L(B)^{\mathfrak{M}} \text{ and } M(A)^{\mathfrak{M}} \neq \emptyset; \\
(e\text{-}nec)(A,B) &\mapsto M(A)^{\mathfrak{M}} \cap M(B)^{\mathfrak{M}} = \emptyset; \\
(i\text{-}nec)(A,B) &\mapsto M(A)^{\mathfrak{M}} \cap L(B)^{\mathfrak{M}} \neq \emptyset; \\
(o\text{-}nec)(A,B) &\mapsto M(A)^{\mathfrak{M}} \not\subseteq M(B)^{\mathfrak{M}} \text{ or } M(A)^{\mathfrak{M}} = \emptyset; \\
(a\text{-}pos)(A,B) &\mapsto M(A)^{\mathfrak{M}} \subseteq M(B)^{\mathfrak{M}} \text{ and } M(A)^{\mathfrak{M}} \neq \emptyset; \\
(e\text{-}pos)(A,B) &\mapsto M(A)^{\mathfrak{M}} \cap L(B)^{\mathfrak{M}} = \emptyset; \\
(i\text{-}pos)(A,B) &\mapsto M(A)^{\mathfrak{M}} \cap M(B)^{\mathfrak{M}} \neq \emptyset; \\
(o\text{-}pos)(A,B) &\mapsto M(A)^{\mathfrak{M}} \not\subseteq L(B)^{\mathfrak{M}} \text{ or } M(A)^{\mathfrak{M}} = \emptyset; \\
(a\text{-}now)(A,B) &\mapsto V(\alpha,A)^{\mathfrak{M}} \subseteq V(\alpha,B)^{\mathfrak{M}} \text{ and } V(\alpha,A)^{\mathfrak{M}} \neq \emptyset; \\
(e\text{-}now)(A,B) &\mapsto V(\alpha,A)^{\mathfrak{M}} \cap V(\alpha,B)^{\mathfrak{M}} = \emptyset; \\
(i\text{-}now)(A,B) &\mapsto V(\alpha,A)^{\mathfrak{M}} \cap V(\alpha,B)^{\mathfrak{M}} \neq \emptyset; \\
(o\text{-}now)(A,B) &\mapsto V(\alpha,A)^{\mathfrak{M}} \not\subseteq V(\alpha,B)^{\mathfrak{M}} \text{ or } V(\alpha,A)^{\mathfrak{M}} = \emptyset;
\end{aligned}$$

If ϕ is a sentence on the left, we write $\phi^{\mathfrak{M}}$ (or $\phi^{\mathfrak{M},\alpha}$ for the last four sentences) for the condition on the right that translates ϕ.

Formally, we define a syllogism, S, to be a triple $\langle \Phi, \phi, \psi \rangle$ such that:[1]

1. Φ, ϕ, and ψ are sentences of modal logic;

2. there are exactly three terms that occur in at least one of Φ, ϕ, and ψ;

3. The predicate of ψ occurs in Φ;

4. The subject of ψ occurs in ϕ;

5. Φ and ϕ share a common term that does not occur in ψ.

We say that a syllogism $\langle \Phi, \phi, \psi \rangle$ is 'semantically valid' if for every Buridan modal model $\mathfrak{M}$ and every world α of $\mathfrak{M}$,

(14)   If $\Phi^{(\mathfrak{M},\alpha)}$ and $\phi^{(\mathfrak{M},\alpha)}$ hold, then $\psi^{(\mathfrak{M},\alpha)}$ holds.

Now Buridan himself makes detailed claims about which syllogisms are 'valid' (*ualent* in his Latin). The following tables, due to Stephen Read ([4] pp. 41–44), summarise the syllogisms that Buridan lists as valid. Read uses L for 'necessary', M for 'possible', Q

---

[1] This definition is standard; see for example [26]. The sentences Φ, ϕ, ψ are known respectively as the major premise, the minor premise and the conclusion of the syllogism.

	L	M	X	Q
L	L	L, M Celarent X	*Darii, Ferio* L, *Barbara, Celarent* X	L, M *Celarent* X
M	M	M	*Darii, Ferio* M	M
X	M *Celarent* X	∅	*Darii, Ferio* M,	∅
Q	M, Q	M, Q	*Darii, Ferio* Q	Q

Table 1: Valid First Figure Syllogisms (by Buridan)

	L	M	X	Q
L	L	L,M *Cesare* X *Camestres* X	*Festino* L *Camestres* X *Baroco* X	L,M *Cesare* X *Camestres* X
M	L,M, *Cesare* X *Camestres* X	∅	∅	∅
X	M, *Cesare* X *Camestres* X	∅	*Festino* M	∅
Q	M, *Cesare* X *Camestres* X	∅	∅	∅

Table 2: Valid Second Figure Syllogisms (by Buridan)

for 'contingent', and he lists under X the *de inesse ut nunc* sentences with 'now'. The lefthand column lists the major premise and the top row lists the minor premise. We assume known the classification of syllogisms by mood and figure, together with their Latin names (*Celarent* etc.).

Besides the validity claims in these tables, Buridan also says explicitly that a number of other syllogisms are not valid.

Fact 2.1. *All the syllogisms that Buridan claims to be valid are semantically valid, and none of the syllogisms that Buridan claims to be invalid are semantically valid.*

	L	M	X	Q
L	L,X	L,M	Darapti, Felapton X Datisi, Ferison X Darapti, Felapton L Datisi, Ferison L	L,M
M	M	M	Darapti, Felapton M Datisi, Ferison M	M
X	Darapti X Disamis X	Darapti M Disamis M	Datisi M	Darapti M Disamis M
Q	M, Q	Q	Disamis, Bocardo M Datisi, Ferison Q	Q

Note: The entry for major premise L and minor premise X corrects the table in [4]; see the errata at
http://www.st-andrews.ac.uk/~slr/Buridan_errata.html.

Table 3: Valid Third Figure Syllogisms (by Buridan)

This fact is proved in [16]. The valid cases are proved by direct argument or reduction to other valid cases, and the invalid cases are proved by building explicit countermodels.

Given the complexity of the tables above, we can see that this agreement between Buridan's claims and the facts of semantic validity is highly significant in the statistical sense. But we leave for the moment the question what it signifies, and turn our attention to Avicenna.

3 Avicenna

The four main surviving logical works of Avicenna's mature period (say 1024 to 1034) are *Qiyās*, *Mašriqiyyūn*, *Dānešnāmeh* and *Pointers*. *Dānešnāmeh* doesn't deal with modal logic, so we ignore it here. As for *Pointers*, Avicenna warns his readers that the book will give no benefit to 'those not endowed with blazing sagacity, training and practice' ([6] p. 48). He deliberately makes it difficult for people to use it as an introduction to his views, though people still attempt this.

There remain *Qiyās* and *Mašriqiyyūn*, the latter written some four or five years later than the former. *Qiyās* is much longer, but this is partly because the manuscript of *Mašriqiyyūn* was stolen and destroyed soon after it was written. It may be that what we have is an author's draft of the first eighty-odd pages. Ibn Sīnā himself advises that *Qiyās* contains more details but was also written with some ulterior motives; *Mašriqiyyūn* tells it like it is (cf. Text M in the Appendix and [6] pp. 44f). Both *Qiyās* and *Mašriqiyyūn* contain what Avicenna considers to be the first steps in his logic. They set up a new logic of his own invention, consisting of sentence-forms that are like those of Aristotle but with added temporal operators. *Qiyās* i.3 states the forms by giving examples in Arabic, and *Mašriqiyyūn* goes over the same ground but with abstract descriptions of the sentence forms. (See §5.2 below for the relevant texts.) The resulting logic is what we called two-dimensional or 2D logic in §1 above. (The name comes from Oscar Mitchell, a student of C. S. Peirce who proposed a similar extension of Aristotle's logic in the 1880s; see [13].)

2D logic is not defined as precisely as we would require of a logic today. There is some ambiguity about exactly what sentences should be included; also some of the distinctions that Avicenna makes seem to be linguistic rather than logical. We will concentrate on some central forms, where there is little doubt about the truth-conditions that Avicenna has in mind, and hence little doubt about the appropriate formalisations. Chatti [5] is a good recent survey of some of the same material.

Three groups of sentences of 2D logic will interest us. These are the groups that Avicenna himself calls respectively 'necessary' (*ḍarūrī*), 'general absolute' (*muṭlaq ^{c}āmm*) and 'at a time' (*zamānī*); we abbreviate these to d, t and z respectively. In each group there are an a sentence, an e sentence, an i sentence and an o sentence. So for example we have a sentence $(a\text{-}d)(A, B)$; its subject and predicate are A and B, its 'assertoric form' is (a) and its 'avicennan form' is (d), and its 'form' is $(a\text{-}d)$. The logic of the sentence forms (d) and (t) forms the '*dt* fragment'.

The forms are as follows, written in a two-sorted first-order language with object variable x, time variable τ and time constant δ, and a distinguished relation symbol E; $Ex\tau$ is read as 'x exists at time τ'.

name	sentence
$(a\text{-}d)(A, B)$	$(\forall x(\exists \tau Ax\tau \to \forall \tau(Ex\tau \to Bx\tau)) \land \exists x \exists \tau Ax\tau)$
$(e\text{-}d)(A, B)$	$\forall x(\exists \tau Ax\tau \to \forall \tau(Ex\tau \to \neg Bx\tau))$
$(i\text{-}d)(A, B)$	$\exists x(\exists \tau Ax\tau \land \forall \tau(Ex\tau \to Bx\tau))$
$(o\text{-}d)(A, B)$	$(\exists x(\exists \tau Ax\tau \land \forall \tau(Ex\tau \to \neg Bx\tau)) \lor \forall x \forall \tau \neg Ax\tau)$
$(a\text{-}t)(A, B)$	$(\forall x(\exists \tau Ax\tau \to \exists \tau(Ex\tau \land Bx\tau)) \land \exists x \exists \tau Ax\tau)$
$(e\text{-}t)(A, B)$	$\forall x(\exists \tau Ax\tau \to \exists \tau(Ex\tau \land \neg Bx\tau))$
$(i\text{-}t)(A, B)$	$\exists x(\exists \tau Ax\tau \land \exists \tau(Ex\tau \land Bx\tau))$
$(o\text{-}t)(A, B)$	$(\exists x(\exists \tau Ax\tau \land \exists \tau(Ex\tau \land \neg Bx\tau)) \lor \forall x \forall \tau \neg Ax\tau)$
$(a\text{-}z)(A, B)$	$(\forall x(Ax\delta \to Bx\delta) \land \exists x Ax\delta)$
$(e\text{-}z)(A, B)$	$\forall x(Ax\delta \to \neg Bx\delta)$
$(i\text{-}z)(A, B)$	$\exists x(Ax\delta \land Bx\delta)$
$(o\text{-}z)(A, B)$	$(\exists x(Ax\delta \land \neg Bx\delta) \lor \forall x \neg Ax\delta)$

	d	t
d	d	d
t	t	t

Table 4: Valid First Figure Syllogisms (by Avicenna)

Examples that Avicenna himself offers in Text D below include:

(*a-d*): Every human is an animal. (Literally: Everything that is sometimes a human is an animal for as long as it exists.)

(*a-t*): Everything that breathes in breathes out. (Literally: Everything that sometimes breathes in breathes out sometime during its existence.)

Avicenna has a tendency to treat the sentences with avicennan form (*z*) as if they had a wide-scope time quantifier; for example he regards (*a-z*)(*A, B*) as close to

$$\exists \tau (\forall x (Ax\tau \to Bx\tau) \wedge \exists x Ax\tau)$$

with examples like 'There is a time when all humans are Muslims'. He says in Text G that 'many precautions' need to be taken when we handle sentences like these, and he promises a full treatment of them in his Appendices—which as far as we know were never written.

There is certainly more to be said about the passages in which Avicenna does discuss (*z*) sentences. The resemblance between (*z*) sentences and Buridan's *de inesse ut nunc* sentences is clear to see.

We can repeat the definition of syllogism from the previous section, but we should add some riders to it. Avicenna doesn't talk of validity. He asks first whether a given pair Φ, ϕ of 2D sentences entails some 2D sentence with the right terms as in the previous section; if it does, he says that the pair is 'productive'. For each productive pair, the 'conclusion' is the *strongest* 2D sentence, again with the right terms, that can be deduced. This is an unambiguous notion; in the *dt* fragment of 2D logic there always is a strongest such sentence. Counting conclusions this way, we can draw up tables like Read's tables for Buridan. Again the lefthand column is for the major premise, although when he writes out syllogisms, Avicenna follows the Arabic custom of putting the minor premise first.

More precisely there are two ways of drawing up these tables. The first way is as with Buridan in §2, letting the tables show the moods that Avicenna himself declares valid. Street lists these at the end of [23], with L for d and X for t. We checked and confirmed Street's list,

	d	t
d	d	d
t	d	

Table 5: Valid Second Figure Syllogisms (by Avicenna)

	d	t
d	d	d
t	t	t

Table 6: Valid Third Figure Syllogisms (by Avicenna)

widening the scope to include *Al-muktaṣar al-awsaṭ*, a relatively early work from around 1014. Apart from the telegraphic *Pointers*, Avicenna always gives the same list. The second way is to calculate what syllogisms Avicenna ought to have declared valid, given his own explanations of the forms of the sentences involved. The situation is different from that with Buridan, because Buridan explained his sentences using modal notions and nothing like a Kripke structure, so that the Johnston semantics involves a new set of concepts introduced by Johnston. With Avicenna the sortal first-order formulas merely report what Avicenna himself said the sentences mean, or more strictly the truth-conditions that he intended. So we could draw up the tables to show which syllogisms are valid in first-order logic.

In parallel with §2 above, we have drawn up the tables in the first way, i.e. to report what moods Avicenna himself describes as valid.

Fact 3.1. *In the dt fragment, the 2D syllogisms that Avicenna lists as productive are exactly those that are productive, and in every case he gives correctly the strongest conclusion.*

This is proved in [9] as a corollary of a characterisation of all the minimal inconsistent sets of 2D sentences; see §10.3 'Productive two-premise moods' in [9].

The tables for Avicenna are very much simpler than those for Buridan. The main simplification is that the conditions for validity in each figure are independent of the mood (provided the mood is valid in assertoric logic) and depend only on the choices of d and t. (This fact about the tables is a consequence of the more general Orthogonality Principle for (d) and (t) sentences, [9] §10.2.) After seeing Johnston's results, Hodges checked what would happen to the tables if we added the (z) sentences. The result is that everything becomes much more complicated and requires 'many precautions', as Avicenna foresaw. We can be thankful that at least Buridan wasn't discouraged from diving in. But maybe life's problems weighed less heavily on Buridan than they did on Avicenna.

Another reason why Buridan's tables are more complicated than Avicenna's is that Buridan doesn't limit himself to strongest conclusions. For example in first figure with necessary major premise and possible minor premise he can deduce a necessary conclusion; but he lists also a possible conclusion, though this follows from the necessary one. However, there are a few cases where Buridan lists two conclusions, and we can show that in the Johnston semantics there are two distinct strongest conclusions. One example is

(15)   Some *B* is now an *A*.
        Every *B* is necessarily a *C*.

In the Johnston semantics the second sentence quantifies over possible objects that are *B*s, whereas the first quantifies only over things that are *B*s actual in the present world. So we can deduce that some actual *A* is necessarily a *C*. But none of Buridan's sentence forms express this; he can say only that some possible *A* is necessarily a *C*, or that some actual *A* is actually a *C*. These two conclusions account for the L and the X in third figure *Datisi*.

Our result for Avicenna is in one way stronger than our result for Buridan: we show not only that Avicenna's claims about productivity and conclusions are correct, but also that these claims are complete. Avicenna detected all the cases that arise. This is actually not true for Buridan. For example in first figure *Darii* and *Ferio* he omits that we can get an X conclusion from an L major premise and an X minor premise; the case is like (15), where we can deduce that some actual *A* is necessarily a *C*, but this time Buridan catches the L and misses the X. But overall he makes very few omissions.

So far we have been talking only about Avicenna's 2D logic, not his alethic modal logic with 'necessarily', 'possibly' and 'contingently'. The facts about his alethic modal logic are rather remarkable. As above, we are leaving aside the contingent propositions.

Fact 3.2. *(a) Avicenna's claims for validity of alethic modal syllogisms are exactly the same as his claims of validity for 2D syllogisms, except that they replace d by 'necessarily' and t by 'possibly'.*

(b) Avicenna's Arabic name for (d) sentences is the same word as his name for the 'necessarily' sentences, namely ḍarūrī (which just means 'necessary').

What on earth is going on here? Is Avicenna really using the same word 'necessary' both for alethic modal necessary sentences and for 2D (*d*) sentences expressing that something is true of *x* 'so long as *x* exists'? We know that Avicenna was well aware of the difference between 'necessary' and 'permanent'. See for example Text I, where he says:

(16)   Being permanent is not the same as being necessary. ... But it is not for the logician as logician to know the truth about this.

What can he mean?

We can say straight away that the problem is how to make sense of the alethic modal logic, not how to make sense of the 2D logic. In *Qiyās* Avicenna presents us with a textbook of his new 2D logic: he defines the sentence forms, and he shows how to prove the valid syllogisms. His proofs, though sometimes missing some details, are accurate up to the most rigorous modern standards, and in several cases they are completely new. Where possible he copies Aristotle's proofs from *Prior Analytics* i.3,4, but where these methods won't deliver the results he finds other methods that will. One example is a case of second figure *Baroco*, where he introduces a method that involves defining a new term (an ecthesis). The received wisdom of his time was that no such proof is possible for *Baroco*—his predecessor al-Fārābī thought he had found such a proof, but his explanation suggests he had missed the point. Avicenna gave an ecthetic proof of *Baroco* that avoided al-Fārābī's infelicities, and showed how to use it to plug the gap in the Aristotelian methods. For one case of second figure *Camestres* he could find no proof along these lines, so he invented a new method that he called 'incorporating in the predicate'. Not only did this method work, but it could be made the basis of an entirely new approach to proving syllogisms in logics more complicated than Aristotle's categorical syllogisms. In the century after Avicenna, the Persian genius Suhrawardī made it the main method of his logic. After trying other methods, the present first author came to the conclusion that it was the best method to use in [9] for proving metatheorems about the *dt* fragment. This is not even a full list of the accomplishments of Avicenna's 2D logic. But it's enough to make the point that Avicenna's results on validity of 2D syllogisms need no support from modal arguments.

By contrast the arguments that Avicenna deploys to justify the alethic modal first figure syllogisms are frankly appalling, if we are to take them at face value as logical inferences. Some are just word-play. Others use methods that he rightly condemns elsewhere in *Qiyās*. They betray no overall vision or plan. (See §5.5 and Texts J, K, L for documentation. In §4 and §5.5 below we indicate how these arguments might be justified, but not as logical inferences.)

The agreement stated in Fact 3.2 (a) between the 2D results and the alethic modal results is far too non-trivial to be an accident, and it can hardly be the result of the nonsensical proofs that Avicenna offers in the modal case. The only reasonable explanation is that Avicenna uses the translation

(17) $\quad$ necessary $\mapsto$ (d),
$\qquad$ possible $\mapsto$ (t)

to *read off* from the 2D logic what syllogisms he should count as valid in the modal case. One thing that he certainly does do in several places is to take a modal syllogism and *claim to justify* it by translating it to a valid 2D syllogism by the translation (17). Text J below is a clear example of this, and Text L is at least a prima facie example.

The translation (17) is the vertical arrow on the lefthand side of picture (1). Its existence and properties are mathematical facts. Avicenna's reasons for using it are one of the main interpretative questions to be discussed in §4 below.

4 Bringing the pieces together

We have explained the lefthand and righthand sides of picture (1), so we are now in a position to explain what the Basic Modal Equivalence of §1 tells us about them.

Johnston found for each of Buridan's sentences ϕ a translation $\phi^{(\mathfrak{M},\alpha)}$ which is a statement about any given Buridan modal model $\mathfrak{M}$ and world α of $\mathfrak{M}$. We didn't say it earlier, but each of these statements $\phi^{(\mathfrak{M},\alpha)}$ can be written as a modal sentence ϕ^{mod}, so that for any $\mathfrak{M}$ and α, ϕ^{mod} is true in $\mathfrak{M}$ at α if and only if $\phi^{(\mathfrak{M},\alpha)}$ is true. For example if ϕ is $(a\text{-}nec)(A, B)$ then ϕ^{mod} is

(18) $\qquad (\forall x(\Diamond(Ox \wedge Ax) \to \Box(Ox \wedge Bx)) \wedge \exists x \Diamond(Ox \wedge Ax)),$

and if ϕ is $(i\text{-}now)(A, B)$ then ϕ^{mod} is

(19) $\qquad \exists x(Ox \wedge Ax \wedge Bx)$

The other sentences are equally straightforward to find. So the Basic Modal Equivalence converts each of these into a two-sorted first-order sentence. In fact Johnston's formulations already go halfway to this two-sorted sentence, so let's now go the whole way. We get

(20)
$$
\begin{aligned}
(a\text{-}nec)(A,B) &\mapsto (\forall x(\exists \tau(Ox\tau \wedge Ax\tau) \to \forall \tau(Ox\tau \wedge Bx\tau)) \\
&\qquad \wedge \exists x \exists \tau(Ox\tau \wedge Ax\tau)) \\
(e\text{-}nec)(A,B) &\mapsto \forall x(\exists \tau(Ox\tau \wedge Ax\tau) \to \forall \tau \neg(Ox\tau \wedge Bx\tau)) \\
(i\text{-}nec)(A,B) &\mapsto \exists x(\exists \tau(Ox\tau \wedge Ax\tau) \wedge \forall \tau(Ox\tau \wedge Bx\tau)) \\
(o\text{-}nec)(A,B) &\mapsto (\exists x(\exists \tau(Ox\tau \wedge Ax\tau) \wedge \forall \tau \neg(Ox\tau \wedge Bx\tau)) \\
&\qquad \vee \forall x \forall \tau \neg(Ox\tau \wedge Ax\tau)) \\
(a\text{-}pos)(A,B) &\mapsto (\forall x(\exists \tau(Ox\tau \wedge Ax\tau) \to \exists \tau(Ox\tau \wedge Bx\tau)) \\
&\qquad \wedge \exists x \exists \tau(Ox\tau \wedge Ax\tau)) \\
(e\text{-}pos)(A,B) &\mapsto \forall x(\exists \tau(Ox\tau \wedge Ax\tau) \to \exists \tau \neg(Ox\tau \wedge Bx\tau)) \\
(i\text{-}pos)(A,B) &\mapsto \exists x(\exists \tau(Ox\tau \wedge Ax\tau) \wedge \exists \tau(Ox\tau \wedge Bx\tau)) \\
(o\text{-}pos)(A,B) &\mapsto (\exists x(\exists \tau(Ox\tau \wedge Ax\tau) \wedge \exists \tau \neg(Ox\tau \wedge Bx\tau)) \\
&\qquad \vee \forall x \forall \tau \neg(Ox\tau \wedge Ax\tau)) \\
(a\text{-}now)(A,B) &\mapsto (\forall x(Ox\delta \wedge Ax\delta) \to Bx\delta) \wedge \exists x(Ox\delta \wedge Ax\delta)) \\
(e\text{-}now)(A,B) &\mapsto \forall x(Ox\delta \wedge Ax\delta) \to \neg Bx\delta) \\
(i\text{-}now)(A,B) &\mapsto \exists x(Ox\delta \wedge Ax\delta) \wedge Bx\delta) \\
(o\text{-}now)(A,B) &\mapsto (\exists x(Ox\delta \wedge Ax\delta) \wedge \neg Bx\delta) \vee \forall x \neg(Ox\delta \wedge Ax\delta))
\end{aligned}
$$

Apart from using O instead of E, the sentences on the right here are remarkably like the corresponding sentences of Avicenna's 2D logic, where the correspondence is[2]

(21)
$$
\begin{aligned}
(nec) &\sim (d) \\
(pos) &\sim (t) \\
(now) &\sim (z).
\end{aligned}
$$

Of course the differences are interesting too; we will see that in most but not all cases these differences can be ironed out. We will refer to this translation from Buridan's sentences to two-sorted sentences, and the slight variants of it that we will consider below, as the 'Avicenna-Johnston semantics for Buridan's modal logic'. The sentences on the right in (20), and their variants below, are the 'Avicenna-Johnston sentences'.

Fact 4.1. *The logical relationships between the Avicenna-Johnston sentences are not affected if we remove all parts of the form '$Ox\tau\wedge$' or '$Ox\delta\wedge$'.*

Sketch proof: Suppose first that T is a set of Avicenna-Johnston sentences as above, and $\mathfrak{M}$ is a model of T. Let T' be the result of removing all the O's from T as described.

---
[2]Thom in [25] uses X to stand for *(now)* in the case of Buridan but for *(t)* in the case of Avicenna. Sometimes, as at the bottom of his page 174, he uses this correlation as a basis for comparing Avicenna and Buridan. But comparing *de inesse ut nunc* sentences with *(t)* sentences is rather meaningless.

Then we get a model $\mathfrak{N}'$ of T by taking $\mathfrak{N}$ and re-interpreting each relation symbol A so that $\mathfrak{N}' \models Aa\beta$ if and only if $\mathfrak{N} \models (Oa\beta \wedge Aa\beta)$ (for any object a and time/world β).

Suppose next that T and T' are as above, and $\mathfrak{K}$ is a model of T'. Then we get a model $\mathfrak{K}'$ of T by adding a relation O so that $Oa\beta$ holds everywhere. □

So without affecting what moods come out as valid, we can replace the first version of Avicenna-Johnston semantics by the following simpler form:

(22)
$$
\begin{aligned}
(a\text{-}nec)(A, B) &\mapsto (\forall x(\exists \tau Ax\tau \to \forall \tau Bx\tau) \wedge \exists x \exists \tau Ax\tau) \\
(e\text{-}nec)(A, B) &\mapsto \forall x(\exists \tau Ax\tau \to \forall \tau \neg Bx\tau) \\
(i\text{-}nec)(A, B) &\mapsto \exists x(\exists \tau Ax\tau \wedge \forall \tau Bx\tau) \\
(o\text{-}nec)(A, B) &\mapsto (\exists x(\exists \tau Ax\tau \wedge \forall \tau \neg Bx\tau) \vee \forall x \forall \tau \neg Ax\tau) \\
(a\text{-}pos)(A, B) &\mapsto (\forall x(\exists \tau Ax\tau \to \exists \tau Bx\tau) \wedge \exists x \exists \tau Ax\tau) \\
(e\text{-}pos)(A, B) &\mapsto \forall x(\exists \tau Ax\tau \to \exists \tau \neg Bx\tau) \\
(i\text{-}pos)(A, B) &\mapsto \exists x(\exists \tau Ax\tau \wedge \exists \tau Bx\tau) \\
(o\text{-}pos)(A, B) &\mapsto (\exists x(\exists \tau Ax\tau \wedge \exists \tau \neg Bx\tau) \vee \forall x \forall \tau \neg Ax\tau) \\
(a\text{-}now)(A, B) &\mapsto (\forall x(Ax\delta \to Bx\delta) \wedge \exists x Ax\delta) \\
(e\text{-}now)(A, B) &\mapsto \forall x(Ax\delta \to \neg Bx\delta) \\
(i\text{-}now)(A, B) &\mapsto \exists x(Ax\delta \wedge Bx\delta) \\
(o\text{-}now)(A, B) &\mapsto (\exists x(Ax\delta \wedge \neg Bx\delta) \vee \forall x \neg Ax\delta)
\end{aligned}
$$

Can we perform a similar reduction on Avicenna's 2D sentences? It turns out that we can, with an important restriction.

Fact 4.2. *The logical relationships between 2D sentences in the dt fragment are not affected if in all the sentences we remove all parts of the form '$\exists x \tau \wedge$' or '$\exists x \tau \to$'.*

Supersketch proof: This is harder than the previous result. It rests on the fact that the sentences never correlate what holds at one object at a time τ with what holds at another object at that same time τ; so we can manipulate the time frames of the elements separately to ensure that everything always exists; in which case the statement of existence becomes redundant. The proof is in §12.2 of [9]. This argument doesn't work for the (z) sentences since the time δ is fixed across all objects. □

It follows that the criterion for the semantic validity of a Buridan syllogism involving only 'necessarily' and 'possibly' is equivalent to the criterion for the validity of the corresponding 2D syllogism with 'necessarily' replaced by (d) and 'possibly' replaced by (t). Given that Avicenna and Buridan did calculations that agree with these criteria for their respective logics, we should be able to check that the L and M parts of Read's tables above agree exactly with the corresponding tables for Avicenna, with (d) for L and (t) for M, except where Buridan includes non-optimal conclusions. And indeed this is the case.

We can show that Fact 4.2 can't be extended to include the sentences with avicennan form (z). If it could, then the Buridan 'now' sentences would go over into (z) sentences. But there are examples to show that this fails, for example the following from §12.2 of [9]. The Buridan syllogism

(23)
Some B is necessarily an A.
Every B is necessarily a C.
Therefore some A is now a C.

is valid, and Read's table for Third Figure witnesses this by showing X in the top left square corresponding to L in both premises. But the corresponding 2D syllogism

(24)
Some B is an A throughout its existence.
Every B is a C throughout its existence.
Therefore at time δ some A is a C.

is invalid, since it could happen that none of the things that are sometimes a B exist at time δ.

Now we come back to the second author's discussion of Buridan's modal logic. What does his semantics for Buridan's logic show? We have to tread carefully. Suppose someone claims:

(25) Claim One. Buridan is a reliable logician, because he gets correct answers about which syllogisms are valid.

As it stands this is a non-sequitur, because it assumes that we know what answers are correct. We could claim to know it if we already knew that our semantics for Buridan's logic correctly reflects his intentions. Suppose someone claims:

(26) Claim Two. Our semantics correctly reflects Buridan's intentions, because it gives the same answers as he does about which syllogisms are valid.

As it stands this is a non-sequitur too, for more than one reason. First, it assumes that Buridan calculated correctly which syllogisms are valid. So we have a circular argument. But second—and our calculations above bring this to the fore—there can be two different semantics that express different intentions but happen to give the same verdicts on which syllogisms are valid.

To expand this second point: Fact 4.1 above implies that the Johnston semantics for Buridan's modal logic would give exactly the same verdicts on which syllogisms are valid if we added the restriction that all objects are actual in all worlds. (Any countermodel using the notion of actuality can be replaced by a countermodel where the same work is done by the other relation symbols.) Does it follow, as the argument of Claim Two would imply, that our evidence shows that Buridan assumes every object is necessarily actual? Clearly not. We can't even say we have shown that Buridan assumes something like this; other things that we know about modal logic make it clear that the assumption is quite mad. (The point is not always appreciated; see the last paragraph of §5.4 for an example in the literature.)

Another example that emerges from the comparison of Avicenna and Buridan is the fact that Avicenna speaks of times where the corresponding items in Johnston's semantics for Buridan are possible worlds. Switching between times and worlds makes no difference to the formalism at all. So could we argue, noting that Johnston's semantics works equally well with times instead of worlds, that Buridan really had in mind a temporal logic rather than an alethic modal one? Or vice versa, could we argue that Ibn Sīnā really meant worlds when he said times? Any answer to these questions must consult what the authors themselves said. In the case of Buridan there is almost nothing in his text to suggest that he means a temporal logic. (Granted, he does say *ut nunc* 'as now'.) In the case of Avicenna the question has some bite, because in his propositional logic where he uses the notion that something is the case 'always' or 'sometimes', he says explicitly that 'always' is not meant to cover just times; see Movahed [18] pp. 7–23 on this.

A further point is that two different semantics may give the same verdict over most syllogisms, but differ on some small group. It may not be obvious where to look for this group. We saw this with Fact 4.2; if we took the 2D sentences as giving the semantics of the corresponding Buridan sentences, we would get a different verdict from the Avicenna-Johnston semantics on certain syllogisms involving 'now' sentences, but no difference would show up using the 'necessary' and 'possible' sentences alone.

We have no snap answers to these problems. Claim Two is definitely dangerous and one should be aware of that. But in many cases where we give a semantics to an author's logic, we can read the author's statement of intentions, and any other evidence from the author, as a guide to the appropriate semantics. We used the quotation (9) from George Hughes in this spirit. And of course then we can argue as in Claim One to show that our author is a reliable logician, if the author's statements of validity agree with what we calculate to be valid using the semantics that we derived from the author's indications.

There are two things about set-theoretic semantics that make this a feasible enterprise. The first thing is that a set-theoretic semantics is normally objective, in the sense that its definitions are based on elementary set theory, and in consequence there is no room for dispute about the properties of the semantics. One of the earliest logicians to recognise this objectivity of elementary set theory was George Bentham, who recommended converting arguments into set theoretic form as a way of detecting fallacies. He may have taken this from his uncle Jeremy Bentham, whose unpublished notes on logic he used. In any case the fact is now well recognised. The second thing is that we have ways of calculating in set theory, which are not simply paraphrases of informal arguments. In consequence the causes of error are likely to be different; so if our author informally calculates that a syllogism is valid, and we show it by a formal calculation, there is less chance that we and the author have made the same mistake. All this is common sense.

We can add that a semantics for Buridan's modal logic is also a prima facie semantics for other people's accounts of Buridan's modal logic. So it can be used to correct misunderstandings. For example it has been claimed that Buridan forgot to include among the valid syllogisms *Barbara* with both premises X and conclusion M. But we can show that this mood is not semantically valid on the Johnston semantics, which makes it much less likely that Buridan omitted it by mistake.

Turning back to Avicenna, does our experience with Buridan throw any light on what Avicenna might be doing with his modal and 2D logics?

Yes, it does. By starting his logic in *Qiyās* and *Mašriqiyyūn* with the 2D sentences, he has adopted a set of notions that can be represented in elementary set theory. So the logic is objective, and it allows exact calculations. Here we are using modern language, but there are things in Avicenna's text that point in the same direction. One is his repeated insistence that he quantifies only over 'actuals' (*bil-fi^{c}l*), for example in Text A. Another is the interest that he expresses, in *Qiyās* i.2, 16.8–10, in those sciences that are 'integrated and orderly', so that they are 'unlikely to lead to error' and the experts have few differences of opinion about them. There is an obvious contrast with modal logic, where we expect many philosophical disagreements.

Note one significant difference between what Johnston does to Buridan and what Avicenna does with his 2D logic and his alethic modal logic. Buridan has a body of claims about modal validity, and Johnston can use the semantics to check them. But Avicenna has no body of modal theorems to start with. Rather the opposite: he proceeds as if he is casting around for some way of finding theorems. Not having any direct access to alethic modal theorems, he proves some theorems in 2D logic and then borrows them into modal logic.

Does it make sense for him to do this? Yes, we can defend this procedure in either of two distinct ways, as follows.

First, suppose we are trying to find the laws of necessity. There are various kinds of necessity, and maybe they obey different laws. But if the bare notion of necessity—let us call it abstract necessity—obeys some laws, then all the more specific kinds of necessity should obey these laws too. One particular kind of necessity is permanence, and happily we can handle permanence in 2D logic where its laws can be found in an objective way. These laws contain all the laws of abstract necessity. Do they contain anything more?

If we wanted to show that every law of permanence is also a law of necessity, then thanks to the axiomatic form of Avicenna's 2D logic, we have a procedure that we can follow. It will suffice to show that abstract necessity obeys the axioms of permanence. These axioms consist mainly of conversions and first figure syllogisms, and it so happens that these are exactly where Avicenna concentrates his arguments for validity of modal syllogisms. But these axioms are precisely that, axioms, so Avicenna is not in a position to *derive* them logically. In practice he is forced to fall back on a kind of conceptual analysis, which consists of inviting the reader to play with the notions until the axioms feel natural. This fact could go a long way towards accounting for the less than cogent treatment of the first figure alethic syllogisms; it may be not so much bad logic as a device for creating intuitions.

He never says that this is what he is doing. But if we look at what he says about the procedures of discovery in science in general, and in logic in particular, then much of it makes good sense. (This point is expanded in [8].) So this first defence of Ibn Sīnā's procedure fits well with Ibn Sīnā's known general view of logic.

The second defence is a frank anachronism. The various formal equivalences given in this paper show that 2D logic is formally correct as a description of a Kripke semantics for Ibn Sīnā's alethic modal logic. (Ibn Sīnā's own interpretation of 2D logic in terms of time is irrelevant to this fact.) So when Ibn Sīnā claims to justify a modal syllogism by translating it by (17) into a 2D syllogism, what he is doing is formally the same as giving a semantic proof of the modal syllogism in terms of a Kripke semantics. His attempted proofs of the modal axioms would, if they worked, show that the alethic modal logic is sound for the Kripke semantics. Again this is not what Ibn Sīnā says he is doing—but then in the 11th century, how could he?

So in both the Buridan case and the Avicenna case, a set-theoretic semantics is being used to support a non-set-theoretic logic, and the properties of the set-theoretic semantics that make this possible are similar in the two cases. But the further details of the two cases are quite different.

References

[1] Asad Q. Ahmed. *Avicenna's Deliverance: Logic*. Oxford University Press, Oxford, 2011.

[2] Aristotle. *Prior Analytics, trans. and ed. R. Smith*. Hackett, Indianapolis, 1989.

[3] Allan Bäck. Avicenna's conception of the modalities. *Vivarium*, 30:217–225, 1992.

[4] John Buridan. *Treatise on Consequences, trans. and introduced by Stephen Read*. Fordham University Press, New York, 2015. Errata at http://www.st-andrews.ac.uk/~slr/Buridan_errata.html.

[5] Saloua Chatti. Existential import in Avicenna's modal logic. *Arabic Sciences and Philosophy*, 26:45–71, 2016.

[6] Dimitri Gutas. *Avicenna and the Aristotelian Tradition: Introduction to Reading Avicenna's Philosophical Works, Second Edition*. Brill, Leiden, 2014.

[7] D. Hilbert and W. Ackermann. *Grundzüge der Theoretischen Logik*. Springer, Berlin, 1928. Trans. of second edition (1937) by L. M. Hammond et al. as *Principles of Mathematical Logic*, Chelsea, New York 1950.

[8] Wilfrid Hodges. A Manual of Ibn Sīnā's Formal Logic. In preparation.

[9] Wilfrid Hodges. Mathematical Background to the Logic of Ibn Sīnā. Submitted; a draft is online at wilfridhodges.co.uk/arabic44.pdf.

[10] Wilfrid Hodges. Ibn Sina on analysis: 1. Proof search. or: Abstract State Machines as a tool for history of logic. In N. Dershowitz A. Blass and W. Reisig, editors, *Fields of Logic and Computation: Essays Dedicated to Yuri Gurevich on the Occasion of his 70th Birthday*, Lecture Notes in Computer Science 6300, pages 354–404. Springer, Heidelberg, 2010.

[11] Wilfrid Hodges. Affirmative and negative in Ibn Sina. In Catarina Dutilh Novaes and Ole Thomassen Hjortland, editors, *Insolubles and Consequences: Essays in honour of Stephen Read*, pages 119–134. College Publications, London, 2012.

[12] Wilfrid Hodges. Ibn Sina, Frege and the grammar of meanings. *Al-Mukhatabat*, 5:29–60, 2013.

[13] Wilfrid Hodges. The move from one to two quantifiers. In Arnold Koslow and Arthur Buchsbaum, editors, *The Road to Universal Logic, Festschrift for the 50th Birthday of Jean Yves Béziau, Vol. I*, pages 221–240. Birkhäuser, 2015.

[14] Wilfrid Hodges. Ibn Sina on reductio ad absurdum. *Review of Symbolic Logic*, to appear, 2016.

[15] George Hughes. The modal logic of John Buridan. In *Atti del Convegno Internationale di Storia della Logica: Le teorie delle modalita*, pages 93–111, 1989.

[16] Spencer C. Johnston. *Essentialism, Nominalism, and Modality: the Modal Theories of Robert Kilwardby and John Buridan*. PhD thesis, University of St Andrews, 2015.

[17] Grigori Mints. *A Short Introduction to Modal Logic*. CSLI, Stanford, 1992.

[18] Zia Movahed. *Reflections on the Logic of Ibn-Sīnā and Suhrawardi*. Hermes, Tehran, 2016.

[19] Nicholas Rescher and Arnold van der Nat. The theory of modal syllogistic in medieval Arabic philosophy. In *Nicholas Rescher, Studies in Modality*, American Philosophical Quarterly Monograph 8, pages 17–56. Blackwell, Oxford, 1974.

[20] Ibn Sina. *Manṭiq al-mašriqiyyīn (Easterners)*. Al-Maktaba al-Salafiyya, Cairo, 1910.

[21] Ibn Sina. *Kitāb al-najāt (Deliverance)*, ed. M. Danishpazhuh. Tehran University Press, Tehran, 1945.

[22] Ibn Sina. *Al-qiyās (Syllogism)*, ed. S. Zayed. Cairo, 1964.

[23] Tony Street. An outline of Avicenna's syllogistic. *Archiv für Geschichte der Philosophie*, 84:129–160, 2002.

[24] Tony Street. Suhrawardī on modal syllogisms. In Anna Akasoy and Wim Raven, editors, *Islamic Thought in the Middle Ages: Studies in Text, Transmission and Translation, in Honour of Hans Daiber*, pages 163–178. Brill, Leiden, 2008.

[25] Paul Thom. *Medieval Modal Systems: Problems and Concepts*. Ashgate, Aldershot, 2003.

[26] Sara L. Uckelman and Spencer Johnston. A simple semantics for Aristotelian apodeictic syllogistics. *Advances in Modal Logic*, 8:454–469, 2010.

A Appendix

One of the referees asked for textual evidence to support our attribution of various views to Avicenna. The main purpose of this Appendix is to give texts and put them in context. Except where stated, the translations are by the first author, from the texts of *Qiyās* [22], *Mašriqiyyūn* [20] and *Najāt* [21]. Normally he wouldn't publish such a substantial amount of translation without checking it in detail with a logically informed native Arabic speaker, but under the present publication schedule there was no time for this. He will try to get better-authenticated translations onto his website as soon as possible.

Since the paper was first submitted, an excellent scholarly treatment of Avicenna's definitions of his sentence forms has appeared, namely Chatti [5]; see also Chatti's article *Avicenna (Ibn Sina): Logic* in the *Internet Encyclopedia of Philosophy*.

By giving so many texts, we are in danger of bringing up issues that will distract the reader from the main business of the present paper. Unfortunately this is life with Avicenna. There is material for many more papers.

A.1 Avicenna's assertoric logic

Aristotle's formal logic falls into two parts, which today one describes as 'assertoric' (or 'categorical') and 'modal'. The modal sentences and inference rules are distinguished by the fact that they use 'alethic' modalities *necessary*, *possible* and *contingent*, while the assertoric logic has no modalities. Among Avicenna's various logics, he has one which corresponds to Aristotle's assertoric logic and one which corresponds to Aristotle's modal logic; we carry over the names 'assertoric' and 'modal' to these two logics. (The name 'assertoric' is not Avicenna's; when he needs a name for assertoric logic he tends to call it 'standard', *mašhūr*. Some of the recent literature on Avicenna uses 'assertoric' without defining it, but apparently as a synonym for Avicenna's term 'absolute', *muṭlaq*; we avoid this usage.)

Assertoric logic has four sentence forms, which Avicenna expresses with Arabic or Persian sentences that translate as

(27)
- Every B is an A.   (We write $(a)(B, A)$.)
- No B is an A.   (We write $(e)(B, A)$.)
- Some B is an A.   (We write $(i)(B, A)$.)
- Not every B is an A.   (We write $(o)(B, A)$.)

The notations $(a)(B, A)$ etc., which can be shortened to (a), (e), (i), (o), are derived from a Latin convention and are not found in Avicenna. He himself describes (a) as 'universal affirmative', (e) as 'universal negative', (i) as 'existential affirmative' and (o) as 'existential negative'. The letters B, A can be replaced by any other letters, normally subject to the convention that the two letters in an assertoric sentence form are distinct. (A rare counterexample is implied near the end of Text G below.) The first letter (B above) is called the 'subject' and the second letter (A above) is called the 'predicate'.

An assertoric sentence (or assertoric proposition) is got by replacing the letters in an assertoric sentence form by appropriate natural language text. The text put for B above, or more accurately the meaning of this text, is called the 'subject' of the sentence; likewise 'predicate' with A. In practice Avicenna uses the letters B, A to mark holes where text can be put. But from various passages, including his descriptions of logic as a theoretical science, it seems that his theoretical account is different: apparently he regards the letters in a logical inference rule as universally quantified by quantifiers (perhaps implicit) that range over 'well-defined meanings' (*maʿānī maʿqūla*). So for example the inference rule that the Latins knew as *Barbara*, and Avicenna knew as the first mood of the first figure, should be read as

(28) For all well-defined meanings C, B and A, if it is assumed that 'Every C is a B' and that 'Every B is an A', then these assumptions yield the conclusion that 'Every C is an A'.

The two assumptions are called 'premises' of the inference rule. 'Conclusion' here means the strongest sentence with subject C and predicate A that can be inferred from the two premises. (So for example Avicenna says that these premises have no conclusion with subject A and predicate C—he rejects the 'indirect moods'.)

Avicenna discusses the truth conditions of the assertoric sentence forms. One important point is that when B is empty (i.e. there are no Bs), he takes the affirmative forms $(a)(B, A)$ and $(i)(B, A)$ to be false and the negative forms $(e)(B, A)$ and $(o)(B, A)$ to be true (cf. [11]). This allows the following first-order translations of his sentence forms.

(29)
$(a)(B, A):$ $(\forall x(Bx \to Ax) \land \exists xBx).$
$(e)(B, A):$ $\forall x(Bx \to \neg Ax).$
$(i)(B, A):$ $\exists x(Bx \land Ax).$
$(o)(B, A):$ $(\exists x(Bx \land \neg Ax) \lor \forall x \neg Bx).$

It seems beyond reasonable doubt that these first-order formulations have the same truth conditions as Avicenna's assertoric forms in (27) above. But they differ from Avicenna's forms in two other ways. First, they use modern symbolism. Second, the formulas for (a) and (e) invove an analysis in terms of the truth-table or Philonic conditional $\to$, and we don't know that Avicenna was aware of this conditional.

Following Aristotle, Avicenna describes a sentence ψ as a 'contradictory negation' (*naqīd*) of a sentence ϕ if ψ is logically equivalent to $\neg \phi$. So $(a)(B, A)$ and $(o)(B, A)$ are contradictory negations of each other, and $(e)(B, A)$ and $(i)(B, A)$ are contradictory negations of each other.

Aristotle in *Prior Analytics* i.4–6 listed all the valid moods (i.e. two-premise inference rules) of assertoric logic, classified them into three 'figures', and derived the second- and third-figure moods from the first-figure moods and some further elementary logical procedures. This forms Aristotle's proof theory of assertoric logic. Avicenna accepts all of Aristotle's assertoric proof theory, and repeats it in detail in all his major surviving accounts of logic (except the late and telegraphic *Pointers*). He also makes some advances in this logic. One is that in place of Aristotle's one-by-one refutations of invalid moods, Avicenna adopts syntactic rules for recognising valid moods; this continues a trend begun by the Roman Empire logicians and continued further by the later Latin logicians who introduced the laws of distribution. Another is that he develops the theory of compound assertoric syllogisms, i.e. inferences that involve a series of applications two-premise rules. His major achievement here is a recursive proof search algorithm for these syllogisms, which he sets out in *Qiyās* ix; cf. [10].

Besides expounding and developing assertoric logic in its own right, Avicenna also uses it as a template for developing other logics, notably his two-dimensional logic.

A.2 The two-dimensional sentences

At the beginnings of his treatments of deductive logic in the encyclopedic *Qiyās* ('Syllogism') and the slightly later *Mašriqiyyūn* ('Easterners'), Avicenna sets out methodically a collection of sentence forms that broadly resemble the assertoric forms—for example each has a subject and a predicate—but allowing time arguments to occur in the terms. The account in *Qiyās* i.3 mainly consists of a list of sample sentences with some comments on them. The account in *Mašriqiyyūn* consists of formal descriptions of the sentence forms. The two accounts correlate neatly. Avicenna discusses separately what happens to the time argument in the subject and what happens to it in the predicate.

For the subject, Avicenna's clearest account is in the following text.

Text A: *Qiyās* i.3, 20.14–21.12.

> Also we must understand that when we say 'every white thing', it doesn't mean 'everything that fits the description "white" permanently'. In fact the phrase 'everything white' is broader than the phrase 'everything that is permanently white'. "White" includes both "white at a certain time" and "permanently white". The phrase 'every white thing' means 'each single thing /21/ that fits the description "white" permanently or not permanently, and regardless of whether it is a subject for "white" and it fits the description "white", or it is "white" itself'.
>
> This description is not the same as describing [the subject] as 'possibly such-

and-such', or 'what could legitimately be such-and-such'. When we say 'Every white-coloured thing', its sense is definitely not 'everything that could legitimately be coloured white'. Rather it means 'everything that in actuality fits the description "white", where besides being actual, it can be so for some time which is indeterminate or determinate or permanent'.

This actuality is not just the kind of actual existence that material things have. In some cases the reference to the subject doesn't place it as something satisfied in material things. An example is 'Every spherical object whose surface consists of twenty triangular faces'. This description is not one that a thing satisfies on the basis of existing [in the material world]. Rather, [a thing satisfies it] by being thought of as actually fitting the description, on the basis that the intellect describes it as actually satisfying [the defining condition], regardless of whether the thing exists [in the material world] or not. And the phrase "Every white thing" means every single thing that is described in the intellect as actually satisfying the condition that it is white, either permanently or at some time, regardless of which time that is. This takes care of the subject side [of the proposition].

In short, for the sentence forms under discussion here, Avicenna treats the subject term B as standing for 'thing which is an actual B at some time'. In symbols the Bx of the formulations in (29) becomes $\exists t Bxt$, where t is understood to range over times. We note also that with his emphasis on actuality, Avicenna excludes merely possible Bs; he doesn't ampliate to the possible, at least in the sentence forms under discussion. But he makes the point that mental constructs can count as actual if the intellect has actually made the construction, even though these constructs are never objects in the world.

For the predicate side it will be best to switch to *Mašriqiyyūn*. The translations below should be treated with caution, because the text of *Mašriqiyyūn* is not in a reliable state.

Text B: *Mašriqiyyūn* 65.1–11.

As a matter of usage, languages pretty much determine that the sentence 'B is a C' expresses that a thing is a C sometime while it fits the description B. What the unqualified meaning determines is called an 'absolute' (*muṭlaq*) proposition. If a condition is made in it mentally which excludes the strict necessity that we are about to mention, but does include those cases where the content holds, not so long as the essence continues to be satisfied, but rather at some time or under some condition and some case, [it is called] 'temporary' (*wujūdī*).

> Today people don't distinguish between the absolute proposition and the temporary. When the sentence is understood to mean that [every] B is a C while its essence continues to be satisfied, [the proposition is said to be] 'necessary' (*ḍarūrī*). When the meaning is [that it is a C] so long as it fits the description B, [the proposition is said to be] 'adherent' (*lāzim*). ... The two [kinds of proposition] are different. Thus there is a difference between the sentence
>
>> A thing that moves changes so long as its essence continues to be satisfied.
>
> (which means that the thing that fits the description 'moving' is changing so long as its essence is satisfied), and the sentence
>
>> A thing which fits the description 'moving' changes as long as it continues to move.
>
> Of course there is a difference—the first [sentence] is false and the second is true.

The word *lāzim* has a range of meanings, among them 'necessary'. In Text B we probably have to treat it as a technical term. We note that the definitions that Avicenna gives for *ḍarūrī* and *lāzim* sentences make no explicit mention of anything being 'necessary', so presumably Avicenna is here regarding permanence (in either of these forms) as a kind of necessity. The world *wujūdī* is also a technical term, and our translation 'temporary' fits its application rather than its etymology.

Before discussing further details of this passage, we move to *Mašriqiyyūn* 68.3–13. Here Avicenna discusses the same predicate forms as above, with one new form added, but in each case specifically for universal affirmative sentences.

Text C: *Mašriqiyyūn* 68.3–13.

> We consider the most general (a^camm) universal affirmative absolute (*muṭlaq*) proposition, for example when we say
>
>> Every B is a C.
>
> This proposition means that for everything that is taken to fit the description B in actuality, without there being any condition about the 'fit in actuality' being permanent or not permanent, each such thing fits the description C in actuality, without any further elaboration.

Next we consider the universally quantified necessary (*ḍarūrī*) proposition. This is like when you say

> Necessarily every *B* is a *C*.

meaning that for everything that fits the description *B* in actuality, regardless of whether it fits the description *B* permanently or not permanently, each such thing fits the description that for as long as its essence is satisfied, it is a *C*. An example is when you say

> Necessarily everything that moves is a body.

Next we consider the adherent (*lāzim*) proposition. This is like when you say

> Every *B* is a *C*.

—whether or not you say 'necessarily'—meaning that for everything that either permanently or not permanently fits the description *B*, each such thing also fits the description of being a *C* for as long as it continues to fit the description *B*. It is not [implied] that it also fits the description of being a *C* for as long as its essence continues to be satisfied.

Next we consider the congruent (*muwāfiq*) proposition. This is like when you say

> (30)   Every *B* is a *C*.

meaning that it is a *C* sometime when it is a *B*, but without adding that it is a *C* permanently for as long as it is a *B*, or that it is so but not permanently [for as long as it is a *B*].

The four universal affirmative forms that Avicenna here describes as *ḍarūrī*, *lāzim*, *muwāfiq* and *muṭlaq aᶜamm* (or elsewhere *muṭlaq ᶜāmm*) are the forms that in [9] are labelled (a-d), (a-ℓ), (a-m) and (a-t) respectively. Note also that Avicenna makes clear that a sentence can have the form *ḍarūrī* without using the word 'necessary' or the curious piece of terminology about 'while its essence is satisfied'. The logician has to look at the sentence in context to see what the user intended, and then find the appropriate logical form.

In fact 'while its essence is satisfied' is not a phrase used in any normal medieval Arabic discourse, and clearly Avicenna intends it as a term of art. The sense is not controversial:

if E is your individual essence, then to say that E is satisfied at time t is just to say that you exist at time t. This is clear from many examples that Avicenna gives. (Avicenna may have used this circumlocution to give emphasis to a notion that he wanted to introduce into logic. Also he may have wanted to underline the point that logic is about propositions that we can believe or assume, and you as a physical person are not part of anything that can be believed or assumed; if a proposition involves you, it can do so only by including an object in the mental world that corresponds to you, and this is your individual essence. This explanation is speculative—Avicenna doesn't explain himself.)

A few pages later in *Mašriqiyyūn*, Avicenna adjusts these sentence forms to the cases (*e*), (*i*) and (*o*). Rather than follow these details (which are relatively routine), we move back now to the examples of sentence forms given in *Qiyās* i.

Text D: *Qiyās* i.3, 21.14–23.7.

> So we should say something about the affirmative universally quantified absolute proposition, and pin down the difference between the absolute and the necessary. We say: There are sentences that are all affirmative but behave in different ways. We say:
>
> > God is alive.
>
> and mean that he is permanently [alive]; he never stopped being alive and he never will. But we say:
>
> > Every whiteness is a colour.
>
> and
>
> > Every human is an animal.
>
> meaning not that every single thing which is white is a colour which always was and will be [a colour], or that every human is an animal and always was and always will be [an animal]. Rather, we are just saying /22/ that everything that fits the description "whiteness", and that is [properly] said to be a whiteness, is a colour so long as its essence continues to be satisfied. And likewise everything [properly] said to be human [is not an animal in the sense] that it always was and always will be an animal; but rather so long as its essence and substance continue to be satisfied. And when we say:
>
> > Everything that moves is a body.

we don't mean that everything that moves is a body just so long as it continues to move, but rather we just mean that even if it hadn't been moving, it would be a body for so long as its essence continued to be satisfied. There is a difference between this and the previous case: in the previous case the phrase 'so long as its essence is satisfied' and the phrase 'so long as it remains white' don't describe different situations, whereas in the present case the situations described by the phrase 'everything that fits the description "moves" so long as its essence continues to be satisfied' and by the phrase '[everything etc.] so long as it is moving' are different. And when we say

> Every white thing has a colour which opens out to the eye.

and we don't mean that everything [properly] called white has a colour that opens out to the eye as long as its essence is satisfied, but rather, as long as it fits the description 'white'. When a thing fits the description 'white' and then ceases to be white, its essence doesn't lapse, even though this description no longer fits it.

When we say:

> Everyone who travels from Rayy to Baghdad reaches Kermanshah.

(for example), we don't mean that [he reaches Kermanshah] while [his essence] continues to be satisfied or throughout the time while he is moving to Baghdad. Rather [we mean] that there is some specific time at which he is described as reaching Kermanshah. ... Also we say

> Everything that watches sleeps.

with the meaning that everything that fits the description of watching is asleep at some specific time. [When we say]

> Everything that breathes in breathes out.

we mean that everything that fits the description 'breathing in' breathes out, not so long as its essence continues to be satisfied, or so long as it is breathing out; rather [we mean that] there is a time at which it fits the description 'breathing out'.

Two terminological remarks are in order here. First, the word 'watches' is to 'is awake' as 'sleeps' is to 'is asleep'. In English this usage is now obsolete, but it is needed to represent the Arabic. Second, the mention of 'substance' (*jawhar*)—the only place where Avicenna uses this word in this context—is an acknowledgment of al-Fārābī's usage of *jawhar* to mean essence. It is not a departure from Avicenna's consistent position that logic doesn't make category distinctions, for example between substances and accidents. In fact animals are substances, but whitenesses are qualities and hence accidents.

The examples in Text D correlate well with the sentence forms in Texts B and C. Every human is a living being all the time he or she exists. Every white thing has physical properties of whiteness all the time that it's white. Every traveller from Rayy to Baghdad is in Kermanshah at some time while he or she is a traveller from Rayy to Baghdad. Everything that breathes in does sometimes breathe out. Also Avicenna says here that he is explaining 'necessary' (*ḍarūrī*), and by this he certainly means the form (*a-d*). But In *Qiyās* we have to wait another twelve pages for him to state this as a definition of *ḍarūrī*, as follows.

Text E: *Qiyās* i.4, 33.8–10.

> Among the propositions in this group, the purely necessary (*ḍarūrī*) propositions are those in which the predicate is asserted or denied for so long as the essence of whatever fits the subject description continues to be satisfied.

There are reminders of this definition at *Qiyās* ii.3, 99.14; iii.1, 126,15f; iii.3, 156.12; iv.3, 202.11f, 203.6–8.

The two-dimensional sentences are certainly more complicated than the assertoric ones, and using them to validate natural language arguments is correspondingly more hazardous. But there seems to be no ambiguity at all in their truth conditions, except perhaps about what happens when the subject term is empty. Chatti's paper [5] is largely about this last question. Here we note briefly that there are three prima facie sources of relevant evidence. One is Avicenna's own explicit statements about existential assumptions, if he makes any. The second is his argument for these assumptions in the assertoric case; we can consider whether it should carry over to the two-dimensional case. The third is his proof theory for two-dimensional logic; does it contain moves that depend on the existential assumptions? (This third form of evidence depends on our correctly identifying the two-dimensional sentences within Avicenna's proof theory.)

Taking all these sources into account, we feel confident that Avicenna meant his statements about assertoric sentences with empty subject terms to carry over to two-dimensional logic too. But there is room for further discussion. Chatti [5] rightly argues that the second kind of evidence can be used to make a case that Avicenna should have required the subject to be nonempty also in (*e-t*) and (*i-t*); but it seems that the rest of the evidence overrules this reading.

The texts above, and others along similar lines, provide the basis for the symbolisations given for the (d) and (t) sentences in §3 above. We can add corresponding symbolisations for the *lāzim* and *muwāfiq* sentences:

name	sentence
$(a\text{-}\ell)(A, B)$	$(\forall x(\exists \tau Ax\tau \to \forall \tau(Ax\tau \to Bx\tau)) \land \exists x \exists \tau Ax\tau)$
$(e\text{-}\ell)(A, B)$	$\forall x(\exists \tau Ax\tau \to \forall \tau(Ax\tau \to \neg Bx\tau))$
$(i\text{-}\ell)(A, B)$	$\exists x(\exists \tau Ax\tau \land \forall \tau(Ax\tau \to Bx\tau))$
$(o\text{-}\ell)(A, B)$	$(\exists x(\exists \tau Ax\tau \land \forall \tau(Ax\tau \to \neg Bx\tau)) \lor \forall x \forall \tau \neg Ax\tau)$
$(a\text{-}m)(A, B)$	$(\forall x(\exists \tau Ax\tau \to \exists \tau(Ax\tau \land Bx\tau)) \land \exists x \exists \tau Ax\tau)$
$(e\text{-}m)(A, B)$	$\forall x(\exists \tau Ax\tau \to \exists \tau(Ax\tau \land \neg Bx\tau))$
$(i\text{-}m)(A, B)$	$\exists x(\exists \tau Ax\tau \land \exists \tau(Ax\tau \land Bx\tau))$
$(o\text{-}m)(A, B)$	$(\exists x(\exists \tau Ax\tau \land \exists \tau(Ax\tau \land \neg Bx\tau)) \lor \forall x \forall \tau \neg Ax\tau)$

The reader may notice that some of these formulas, for example $(a\text{-}\ell)$, can be simplified so as to remove a repetition of A.

The authors claim no originality for these formalisations. Almost equivalent formalisations, for the sentence forms of types (d), (ℓ), (m) and (t), were given by Rescher and van der Nat [19] in 1974. They used a notation of their own devising, and didn't include the condition on empty subject terms.

A.3 (z) sentences and wide time scope

Avicenna defines several other temporal sentence forms. One that we discuss in the paper above is where the time is taken to be a certain fixed time, for example the present. Avicenna describes it in *Mašriqiyyūn* as follows.

Text F: *Mašriqiyyūn* 68.16–19.

> Next we consider the 'as-of-now' (*ḥāḍir*) proposition. This is like when you say
>
> Every human is a Muslim.
>
> at a time when it happens to be the case that there is no human unbeliever. It's plausible to say that such sentences, for example
>
> Every animal is human.
>
> would be true if they were [uttered] at such a time. The [existence] condition for this affirmative proposition is that the subject term is satisfied [at the relevant time].

Avicenna has two other names for these sentences: *waqtī* (*Mašriqiyyūn* 65.14) and *zamānī* (*Mašriqiyyūn* 72.7), both of which carry the meaning 'at a given time'. In [9] the sentences are called (z) after *zamānī*.

Avicenna tends to conflate three kinds of sentence: (1) sentences $\phi(now)$ which express that something is the case now, (2) sentences $\phi(\delta)$ which express that something is the case at a given time δ, and (3) sentences $\exists\delta\phi(\delta)$ which express that for some time δ, something is the case at δ. For him the main difference between (2) and (3) is how determinate the time δ is in the speaker's mind; this is not the only place where he allows linguistic considerations to mix with logical ones. As noted in the paper, he does discuss the logical roles of sentences of these three kinds, but this part of his work has yet to be analysed with modern tools. Text G below is a sample.

Text G: *Qiyās* 134.11–136.6.

> An example is that when we say, at some time when there is no white colour and no red colour or colour intermediate [between white and black] (assuming this is possible):
>
>> Every colour is black.
>
> [In this meaning] this proposition would be true at that time, but not a necessary (*ḍarūrī*) truth. Neither would it be meant that every individual that fits the description 'colour' has 'black colour' true but not necessarily true of it, so that that individual can continue to have its essence satisfied, and be a colour, but cease to be [the colour] black. That would be as if we had judged that
>
>> Each thing fitting the description 'colour' at that time is not black permanently and for as long as its essence continues to be satisfied—far from it!
>
> So in fact the non-necessary truth of this sentence of ours just has to do with truth [on] the quantifier, and not with whether the non-necessary predication applies to a single individual or to all of them.
>
> Likewise in the negative proposition the assertion is not about whether the subject term is satisfied; rather it is about the satisfaction of the truth of the universally quantified denial. Even if the subject term in an affirmative proposition has to be satisfied if the quantifier is to be true, the position with the negative proposition has to be as we said. In fact if at some particular time no colour is

white or intermediate [between black and white], and all colours are black and there is no [non-black] colour at all, it is true that

> No [non-black] colour is the colour white, at a certain time.

namely at that time. This is because an unsatisfied [subject] doesn't satisfy the description 'white colour' or have any affirmative property. When the affirmation is not true the [corresponding] denial must be true. If we take care about what we say, and pay regard to the satisfaction of truth on the quantifier, we can apply [the rule of] conversion to this proposition.

If [the Peripatetic logicians] were to follow the path I have presented, they would discover for themselves the great number of different kinds of proposition ... Thus when we have

> Every eclipse of the moon is a black colour.

and

> No eclipse of the moon at time t is a black colour.

because there isn't an eclipse of the moon [at time t], then

> No eclipse of the moon at time t is an eclipse [of the moon].

One gets in the same way that no person is a person, and likewise with all sorts of things. ... One doesn't consider whether the subject term of the negative proposition is satisfied. In future we will take this view for granted.

We have been lengthy and repetitious about this topic, so as to give the student a feeling for what the topic is about, and for the many precautions that need to be taken into account when this approach is adopted ...

References to 'the quantifier' in this and related passages are to the quantifier over objects, not the quantification over times. Avicenna's notion is that when the time quantification is taken with wide scope, it is semantically attached to the object quantifier; see for example [12]. Movahed [18] Ch. 3 on wide scope modalities in Avicenna is also relevant.

Although Avicenna consistently takes the position that his object quantifiers should be read as quantifying over actuals (as for example in Text A above), Text G seems to show him quantifying over possible *times*. This fits with Movahed's observation ([18] p. 14) that at least some of Avicenna's quantifications over 'times' should be read as quantifications over situations or circumstances.

The next text illustrates Avicenna's view that the temporal qualifications in the two-dimensional sentence forms should be seen as kinds of modality, regardless of how they may be expressed in natural Arabic.

Text H: *Mašriqiyyūn* 71.3f.

> Being *ḍarūrī*, being *lāzim* and being *waqtī* are each a modality, but /123/ sometimes in some such sentences there is no [explicit] modal [expression] to signify the modality.

A.4 The alethic modal sentences

Avicenna's chief alethic modal sentences are, on his own account in *ᶜIbāra*, the same as the assertoric sentences but with an alethic modality attached either to the copula or to the quantifier. There is evidence that the sentences where he counts the modality as attached to the copula are those where the scope of the modality includes just the predicate, and the negation if there is one; attachment to the quantifier means that the modality has wide scope taking in the whole sentence. (See Movahed [18] Chapter 2.) This distinction matches the distinction between (*d*) and (*t*) on the one hand, and wide-scope time quantification as in the previous subsection on the other hand.

In *Qiyās* Avicenna tends to give modal sentences with the modality expressed by a phrase 'with necessity' or 'with possibility' or 'with contingency', and this phrase can appear at the beginning or the end or in the middle of the sentence. There seems to be no correlation between the place of the phrase and the distinction between modality on the copula and modality on the quantifier.

The first author's present impression is that for any modal proposition there are three forms to be considered in Avicenna's view:

- The vernacular form is how an Arabic speaker would naturally express the proposition.

- The semantic form is a structure describing how the meanings of the parts of the proposition relate to each other.

- The logical form is the form that Avicenna uses to represent the proposition in discussions of formal logic.

Probably the copula/quantifier distinction refers to the semantic form, and the phrases 'with necessity' etc. are part of the logical form. Avicenna seems happy to accept a wide variety of vernacular forms.

In any event the alethic modal logic operates with sentence forms that come from assertoric forms (a), (e) and (i) and (o) by adding necessity, possibility or contingency, and we can label the resulting forms

(31) $(a\text{-}nec)$, $(e\text{-}nec)$, $(a\text{-}pos)$ etc., etc.

Avicenna defines 'possibly ϕ' as 'not necessarily not ϕ'; he is emphatic that 'possibly' should be defined in terms of 'necessarily' and not the other way round (cf. *Qiyās* 170.7 and the surrounding discussion). This definition allows us to calculate that the contradictory negation of $(a\text{-}nec)(A, B)$ is $(o\text{-}pos)(A, B)$, and so on. In this paper we will ignore contingency, which is the modal counterpart of the 2D *wujūdī*.

But none of this tells us what Avicenna takes 'necessary' and 'possible' to mean, let alone what the modal sentence forms containing them mean. For these meanings we have to look at his various comments on the notions of necessity and possibility. (Bäck [3] is an introduction.) Be aware that not everything that Avicenna says about these notions is relevant to his alethic modal logic. For example in several places he discusses a notion of possibility that refers to the future; he takes this notion from the Peripatetic tradition, but the notion doesn't appear in any of his formal logic.

There is no simple array of texts that will show what Avicenna intended by his alethic modalities. But one text worth noting is the following, in which Avicenna distinguishes between permanence and necessity.

Text I: *Qiyās* 48.9–18.

> We come to the existentially quantified affirmative case, as in the sentence
>
> (32)   Some B is an A.
>
> taken as general absolute. In this case the facts are obscure. Are
>
> (33)   [No B is an A,] necessarily.
>
> and
>
> (34)   [No B is an A,] contingently.

both opposed to (32)? It's plausible that it is not correct to say that

(35) Something which is contingent for each individual could fail to be true of any of them ever.

If (35) is not correct, then a thing that is contingent will become true of some individuals and not of others; and then the truth of (34) is a special case of the truth of (32), so the two are not contradictory negations of each other. It remains the case that (33) is opposed to (32). And even if (35) is correct, it is still the case that

(36)   [No B is an A,] permanently.

is [contradictory] opposite of (32).

But being permanent is not the same as being necessary. [A thing is] necessarily what it is by its nature, and this requires that if it is false of an individual then it is permanently false of that individual; while [a thing is] permanent either by its nature or because it just happens to be. It is not for the logician as logician to know the truth about this.

Then let us take it that the [contradictory] opposite of the negative [general absolute, i.e.

(37)   Some B is not an A.

taken as general absolute], is the permanent, [i.e.

(38)   Every B is permanently an A.]

This has the effect that if the only things that are permanent are those that are necessary, then that's how it is; but if there are things that are permanent but not necessary, then the [permanent which is] contingent would come with the contradictory negation [of the general absolute].

No doubt readers will want to make up their own minds about this passage; but let us throw in comments on two points.

The first point is Avicenna's statement about what logicians need to know. Avicenna says that being permanent is not the same as being (alethic) necessary, but he adds that it is not for logicians to know the truth about this. We can straight away point out two situations in which it certainly would be for logicians to know about the difference between permanence and alethic necessity; and so prima facie Avicenna is ruling out these two situations.

The first situation is that the logical laws of permanence are different from those of alethic necessity. In this case certainly the logician would need to know the difference. Also because of the relations between general absolute and permanent on the one hand, and those between possible and necessary on the other, the logician would need to know any place where the laws of the *dt* fragment of two-dimensional logic fail to correspond to the laws of the (*nec*)-(*pos*) part of alethic modal logic. It should follow that Avicenna is telling us there is no such failure of correspondence: the laws map across to each other exactly.

This is too much to extract safely from just a pair of lines of Avicenna's text. But it leads immediately to the question whether Avicenna does ever point to any failure of correspondence between these two sets of laws. And the answer seems to be that he never does. This is a strong fact. We can say more: sometimes he *justifies* an alethic syllogism simply by showing that the corresponding 2D syllogism under translation (17) holds.

Text J: *Najāt* 73.9–12.

> As for mixing contingency and Necessity [premises] in the first figure... Let us explain this in another way that is easier to understand. We say that if every *B* is *A* perpetually by Necessity, then that thing of which *B* is said is perpetually *A*. So if *B* is said of *J*, it will always be *A*, not [just] for as long as it is described as *B*. For the Necessity that we intend in these figures is other than of this [type]; and we have already explained this. Rather, [the qualifying condition is] 'as long as the essence of *J*, described as *B*, is satisfied'. So if a certain *J* comes to be *B*, it was already *A*, even before its coming to be *B*. And so [it will continue to be] after its coming to be [*B*] and after [the latter's] passing away from it. (Trans. Ahmed [1] p. 57f, with slight adjustments.)

Here Avicenna justifies the inference

(39) Every *C* is contingently a *B*; every *B* is necessarily an *A*; therefore every *C* is necessarily an *A*.

by inferring from 'contingently' to 'possibly', and then translating across into 2D logic by (17). This move makes sense only if we know, or postulate, that laws of 2D logic carry over to laws of alethic modal logic under this translation.

The second situation is that the logician needs to handle an argument which involves both an alethic proposition and a 2D proposition, and the validity of the argument rests on some non-obvious relationship between the two kinds of proposition. To see whether

Avicenna does avoid situations of this kind, we first need to be able to allocate his *ḍarūrī* sentences between (*d*) and (*nec*). This leads directly to the problem of his *ḵabṭ*, which we address in the next section below.

The second point for comment is Avicenna's statement that the necessary properties of a thing are those which it has 'by its nature'—which in Avicenna's terminology is equivalent to 'by its essence'. Does this mean that Avicenna's alethic modal logic should be taken to be essentialist in the sense that its laws are intended to express properties of essences?

There are plenty of places where Avicenna claims that some statement is a necessary truth because of some feature of the essence of the subject term. But he distinguishes between the properties that a thing has 'by its nature' and those properties that are parts of its nature. For example fire burns by its nature or essence, but burning is not part of the essence of fire; it is deducible from the essence by a logical deduction using suitable middle terms (*Burhān* 83.5–8). Likewise 'having its internal angles add up to two right angles' is not a part of the essence of triangle, but it is deducible from the essence of triangle and hence is an 'essential accident' of triangle.

So it may very well be the case that for the reading of 'necessary' that he has in mind in Text I, Avicenna believes that every necessary property of a thing is 'by its nature' in the sense that it is logically deducible from the essence of the thing. But this belief puts no constraints on the laws of logic. In fact we know of no place, in all his treatments of modal logic, where Avicenna justifies a rule of inference by reference to properties of essences. (This puts him in a totally different logical world from the viewpoint of al-Fārābī's *Burhān*.)

In [24] Street offers a different kind of reason for thinking that Avicenna's modal logic is essentialist. He argues that since Avicenna's divided modal logic can be translated into sentences of monadic predicate S5 in such a way that the inferences claimed valid by Avicenna are those valid in S5, it follows that the divided modal logic is essentialist. This is an incomprehensible argument, since there are plenty of interpretations of □ and ◇ that exactly validate monadic predicate S5 but have nothing to do with properties of essences. In fact permanence for □, and its dual for ◇, are one such interpretation. We mention this as an example of the second difficulty that we raise with Claim Two of §4 of our paper, namely that 'there can be two different semantics that express different intentions but happen to give the same verdicts on which syllogisms are valid'.

A.5 The *ḵabṭ*

A reader who starts at the beginning of *Qiyās* and proceeds section by section will first meet the two-dimensional sentences, then a discussion of their contradictory negations, then a discussion of their conversions, and finally a discussion of the syllogistic moods that consist of them. This is the 'two-dimensional textbook' referred to in §3 above. But already in the discussion of contradictory negations the reader will face a disturbing fact: there are two quite different kinds of sentence form that Avicenna describes as *ḍarūrī*. The first is the (*d*) two-dimensional sentences, and the second is the alethic modal sentences of the form (*nec*).

A century or so after Avicenna, Abū al-Barakāt in his treatment of logic summed up the situation neatly with two short phrases: there is necessity in the sense of permanence (*dawām*) and there is necessity in the sense of inevitability (*lā budda*). Thus (*d*) is permanence and (*nec*) is inevitability. Fakr al-Dīn Rāzī picked up Barakāt's distinction and coined a slogan. Logicians in the past, he said (and he clearly had Avicenna in mind), stumbled around in the dark because of their failure to distinguish between permanence and inevitability. Following Rāzī, we can speak of Avicenna's 'stumbling' (*kabṭ*).

Rāzī's comments contain a strong suggestion that Avicenna himself didn't always know whether he was talking about permanence or inevitability. Is this fair?

Go back to our imagined reader of *Qiyās*. She will learn in *Qiyās* i that *ḍarūrī* means (*d*). Then (*nec*) will come briefly into the picture, but only in connection with possibility. For almost all of *Qiyās* i–ii she will have no reason not to take Ibn Sīnā at his word and read *ḍarūrī* as (*d*).

In *Qiyās*, and in fact in all his major treatments of modal logic except that in the late *Pointers*, Avicenna follows Aristotle's arrangement of the material, but with absolute in place of assertoric. This produces the following pattern:

1. Conversion of absolutes.
2. Conversion of modals.
3. Moods with both premises absolute.
4. Moods with both premises necessary.
5. First figure moods, one premise absolute and one necessary.
6. Second figure moods, one premise absolute and one necessary.
7. Third figure moods, one premise absolute and one necessary.
8. First figure moods, both premises possible.
9. First figure moods with one premise possible and one absolute.
10. First figure moods with one premise possible and one necessary.
11. Second figure moods with both premises possible.
12. Second figure moods with one premise possible and one absolute.
13. Second figure moods with one premise possible and one necessary.
14. Third figure moods with both premises possible.
15. Third figure moods with one premise possible and one absolute.
16. Third figure moods with one premise possible and one necessary.

Apart from a bumpy patch at 2, our reader will take all of 1–7 to be about 2D logic. This is the two-dimensional textbook. In *Qiyās* it consists of books i, ii and the first three sections of iii, a clean initial segment of the whole book.

The problems of *kabṭ* are concentrated in the first sections of *Qiyās* iv. As we note in the paper, Avicenna presents the two-dimensional logic as an axiomatic system. For those inferences that are derived from the axioms, similar derivations are available for the corresponding alethic inferences; Avicenna moves through most of these quickly. The problems arise when Avicenna has to justify the alethic inferences that correspond to the two-dimensional *axioms*—in particular the conversions and the first-figure syllogisms. And these are exactly the places where something seems to go badly wrong, as if someone had switched off the light.

The referee encouraged us to quote texts to illustrate this darkness. Two features of the darkness are hard to illustrate with single texts. One is the general sense of aimlessness. In his calculations in the two-dimensional textbook, Avicenna was usually precise, efficient and methodical. But as far as we know, nobody has detected a cogent overall logical strategy in *Qiyās* iv. The second feature is that in *Qiyās* iv Avicenna seems to have given no genuine solutions of genuine logical problems, and made no original suggestions that he or anybody else could pursue. Seven and a half for effort but zero for results. If anybody can prove us wrong about this, we would be hugely pleased to hear it.

Turning to more specific faults that we can illustrate, we begin with an example of a piece of reasoning that is hardly more than word-play. Avicenna considers the mood

(40)   Every C is possibly a B.
Every B is possibly an A.
Therefore every C is possibly an A.

His claim about this mood is not that it is valid, but that it is 'perfect', i.e. so obviously and immediately valid that no proof is needed for it. He argues for this claim as follows.

Text K: *Qiyās* iv.1, 183.1–4.

> One should learn that it often happens that something is clear for people to see, but people want to force the explanation in a particular direction and this compels them to deviate from what is clear. Just as it is clear that things that are true of what is true of something are true of that thing, so likewise it is clear that a thing that is possibly possible is possible. There is no clear way of making this obvious fact more obvious than it already is.

The analogy with 'true' doesn't work—'truly ϕ' implies ϕ but 'possibly ϕ' doesn't imply ϕ.

One might guess that there is some kind of S5 argument here involving the inference $\Diamond\Diamond\phi \vdash \Diamond\phi$. Movahed [18] p. 49 follows up this lead and shows that the most obvious way of applying the S5 rule needs a further rule, namely

$$(\phi \to \psi) \vdash (\Diamond \phi \to \Diamond \psi),$$

and even with this rule his proof runs to nine lines. This is no help at all for showing that the mood is perfect.

Next we give an example of an argument that Avicenna himself has rightly condemned elsewhere. In *Prior Analytics* i.15 Aristotle gives an argument by reductio ad absurdum, where he can be read as claiming that the contradiction is caused by just one of the premises. Avicenna in his treatment of reductio ad absurdum in *Qiyās* mentions an argument that looks very much like Philoponus's version of this argument, and he expresses the view that looking for one premise that causes the contradiction 'gives us no new information' (cf. [14]). It's a little surprising to find Avicenna reporting this same argument at *Qiyās* 192.9–11 without any comment on this deficiency; but at least he does make it clear that he is reporting Aristotle's argument, not giving one of his own. There is less excuse for the following text, where he claims to be offering his own argument.

Text L: *Qiyās* 202.3–8.

> I say: Both the affirmative and the negative moods whose major premises are necessity propositions entail a necessity conclusion. An example of the affirmative case is:
>
> > Every C is a B with possibility;
> > and (★) every B is an A with necessity;
> > so every C is an A with necessity.
>
> Otherwise it is possible for some C not to be an A. And so let us posit that
>
> > (★★) Some C is not an A.
>
> is true. Then this [and the major premise (★)] form a productive syllogism in the second figure, yielding the conclusion that possibly some C is not a B; or rather,
>
> > It is not possible that every C is a [B].
>
> This is an absurdity. It follows not from the premise [(★)] that was counted as true, but from the one [(★★)] that was considered dubious.

The red herring about a single premise causing the absurdity is in the final sentence.

It is of course very bad practice in the history of ideas to dismiss a text as erroneous when there is any chance that we might have misunderstood it. But part of our claim in this paper is that there are better explanations of what Avicenna is doing in *Qiyās* iv, that absolve him from the charge of accepting fallacious or irrelevant logical arguments. We can point to three other things that he is doing instead.

The first is that he justifies alethic arguments by translating them through (17) into sound arguments of 2D logic. Text J was an instance of this in *Najāt*. It seems very likely that Text L contains another instance, at the point where Avicenna cites *Baroco* (his 'productive syllogism in the second figure') with necessary major premise. If 'necessary' is read here as (*nec*), then this is a mood that he hasn't yet justified; moreover he would need to check that the justification of it doesn't involve the very mood that he is proving, or another first-figure mood proved by the same means. So probably he intends 'necessary' as (*d*) here, and he is relying on the fact that he has already proved *Baroco* with (*t*) minor premise and (*d*) major premise.

It transpires that Street in [23] noticed this translation strategy but misinterpreted what he saw. On his page 141 he remarks that some of Avicenna's proofs 'contain a move in which a possible proposition is supposed to be an absolute proposition'. In other words, a (*pos*) proposition is translated into the corresponding (*t*) proposition. But first, Street manoeuvres from 'absolute' to 'assertoric' and hence to 'actual'; this is an irrelevance, based on a conjectured analogy with a move made by Alexander of Aphrodisias. And second, Street fails to notice that when the (*pos*) premise is translated to a (*t*) premise, a (*nec*) premise is simultaneously translated to a (*d*) premise. As a result, where he should be noticing a translation from an argument in alethic modal logic to an argument in 2D logic, he finds instead a dubious logical move which he describes as 'supposing a possible actual'.

The translation is not a move in a logical argument. We discussed at the end of §4 two ways in which it can be justified, noting that the first of these justifications fits Avicenna's own general scheme, and the second is one that is an accepted practice of metalogic today.

The second thing that Avicenna is doing in *Qiyās* iv is to discuss axioms in a way that he hopes will generate an intuition of the truth of the axioms. This is different from deriving the axioms by logical procedures. It will be discussed more fully in [8].

There is a residue of unhelpful quotations of arguments from Aristotle. This is the third thing that Avicenna does in *Qiyās* iv, and in this case we can quote his own justification for doing it. Text M is taken from the Prologue of *Mašriqiyyūn*; Avicenna is describing some defects of his writings before *Mašriqiyyūn*, and these writings include *Qiyās*.

Text M: *Mašriqiyyūn* 3.14–4.1.

We perfected what [the Peripatetics] meant to say but fell short of doing, never reaching their aim in it; and we pretended not to see what they were mistaken about, devising reasons for it and pretexts, while we were conscious of its real nature and aware of its defect. If ever we spoke out openly our disagreement with them, then it concerned matters which it was impossible to tolerate; the greater part [of these matters], however, we concealed with the veils of feigned neglect: ... in many matters with whose difficulty we were fully acquainted, we followed a course of accommodation [with the Peripatetics] rather than one of disputation ... (trans. Gutas [6] p. 38f)

This goes a long way towards accounting for features of Text L and similar passages.

On the Existence of Alternative Skolemization Methods

Rosalie Iemhoff*
Utrecht University
Janskerkhof 13, 3512 BL Utrecht, the Netherlands
r.iemhoff@uu.nl

In memory of Grigori Mints

Abstract

It is shown that no intermediate predicate logic that is sound and complete with respect to a class of frames, admits a strict alternative Skolemization method. In particular, this holds for intuitionistic predicate logic and several other well-known intermediate predicate logics. The result is proved by showing that the class of formulas without strong quantifiers as well as the class of formulas without weak quantifiers is sound and complete with respect to the class of constant domain Kripke models.

Keywords: Skolemization, Herbrand's Theorem, Intermediate Logics, Kripke Models.
MSC: 03B10, 03B55, 03F03.

1 Introduction

The insight that certain quantifier combinations can be reduced in complexity by introducing fresh function symbols, goes back to Thoralf Skolem's work at the beginning of the twentieth century [18]. This insight has been used in the metamathematical study of logics, but it also has practical applications, since it provides, in combination with Herbrand's Theorem, a connection between propositional and predicate logic that is one of the key ingredients in automated theorem proving and logic programming. Because of the elegance and usefulness of the Skolemization method, one might hope to be able to use it also in nonclassical settings, such as intermediate predicate logics. Grigori Mints was one of the first to study Skolemization

---

*I thank Matthias Baaz for the many discussions on Skolemization that we have had over the years, and two referees for their useful comments on an earlier draft of this paper.

and Herbrand's Theorem in nonclassical logic, and from the references it can be seen that it remained a point of interest for him throughout his life [13, 14, 15, 16, 17].

As it turns out, to many nonclassical logics, including intuitionistic predicate logic, the Skolemization method does not apply. This gave rise to the search for alternative methods, that, in combination with Herbrand's Theorem, result in a connection between propositional and predicate intermediate logics, similar to the one for classical logic. In the case of intuitionistic logic, a partial solution that only applies to the fragment without universal quantifiers has been obtained, by extending the logic with an existence predicate [2, 4], and in [12] it has been shown that for intermediate logics with the finite model property this *existence Skolemization method* applies to the full logic. In [3] an alternative method for full intuitionistic predicate logic IQC has been developed, but at the cost of extending the language considerably. There have appeared various results on the Skolemization method and Herbrand's Theorem in substructural logics, and in some cases, when the latter does not hold, an alternative *approximate Herbrand Theorem* has been obtained [8, 1, 6, 7, 9, 10]. For intermediate logics with the finite model property, an alternative Skolemization method called *parallel Skolemization* has been developed [5], and in [9] a similar method has been developed for substructural logics.

In this paper we take the opposite approach and try to establish, given an intermediate logic, what alternative Skolemization methods cannot exist for it. For this, we first need to define what an alternative Skolemization method is, as will be done in Section 5, where the notion of a strict method will be defined as well. In Section 6 it will be shown that no intermediate logic that is sound and complete with respect to a class of frames, admits a strict alternative Skolemization method. In particular, this holds for IQC, QD_n, QKC, QLC, and all tabular logics.

As the reader will see, none of the theorems in this short paper are complex. In fact, the proof of the main result is surprisingly simple. Nevertheless, what is obtained improves our understanding of Skolemization in nonclassical logics to such an extend that I think it worthwhile to publish it separately in this note.

2 Preliminaries

We consider intermediate predicate logics, which are predicate logics between intuitionistic predicate logic IQC and classical predicate logic CQC. The language $\mathcal{L}$ consists of infinitely many variables, which are denoted by $x, y, z, x_i, y_i, \dots$, infinitely many predicate symbols, function symbols (of every arity infinitely many), and the connectives $\wedge, \vee, \to$, the truth constants $\top, \bot$, the quantifiers $\forall, \exists$, and $\neg \varphi$ is defined as $\varphi \to \bot$. Constants are included in the language and treated as nullary function symbols. Terms and formulas are defined as usual. We use $\bar{x}$ as an abbreviation of $x_1, \ldots, x_n$, where the n will always be clear from the context. For example, $\forall x_1 \exists y_1 \forall x_2 \exists y_2 \varphi(\bar{x}, \bar{y})$ is short for $\forall x_1 \exists y_1 \forall x_2 \exists y_2 \varphi(x_1, x_2, y_1, y_2)$. Given a logic L, $\vdash_L$ denotes valitidy in L.

Important in this paper is the distinction between strong and weak quantifiers, where the former are exactly those quantifier occurrences that become universal under classical prenexification: A quantifier occurrence in φ is *strong* if it is a positive occurrence of a universal quantifier or a negative occurrence of an existential quantifier, and it is called *weak* otherwise. Let $\mathcal{F}_{ns}$ and $\mathcal{F}_{nw}$ denote the set of formulas without strong and weak quantifiers, respectively. Identifying a logic with its set of theorems, the *strong quantifier free fragment* of a logic consists of those theorems of the logic that do not contain strong quantifiers, and likewise for weak quantifiers.

3 Kripke models

Kripke models are defined as in Section 5.11 of [19], although we use slightly different notation. First, we define, given a set D, the notion of an *interpretation I in D*, which is such that for every n-ary relation symbol R and every n-ary function symbol f in the language, $I(R) \subseteq D^n$ and $I(f)$ is a function from D^n to D. Interpretation I in D is extended to all terms by letting it be the identity on variables, and by inductively defining for an n-ary function symbol f and terms $t_1, \ldots, t_n$: $I(f(t_1, \ldots, t_n)) = I(f)(I(t_1), \ldots, I(t_n))$. Given a term $t(x_1, \ldots, x_m)$ and a sequence $d_1, \ldots, d_m$ of elements in D, we denote by $I(t)(d_1, \ldots, d_m)$ the result of replacing x_i in $I(t)$ by d_i. Note that $I(t)(d_1, \ldots, d_m) \in D$. Given a set D, let $\mathcal{L}(D)$ be the language to which the elements of D are added as constants.

A *Kripke model* is defined to be a tuple $(K, \preccurlyeq, \mathcal{D}, \mathcal{I}, \Vdash)$, where

- K is a set and $\preccurlyeq$ a partial order on it with a least element, the *root*;

- $\mathcal{D} = \{D_k \mid k \in K\}$ is a collection of sets;

- $\mathcal{I} = \{I_k \mid k \in K\}$, where I_k is an interpretation in D_k;

- $\Vdash$ is a relation between elements of K and atomic formulas in $\mathcal{L}(D_k)$.

Moreover, such a Kripke model must satisfy the following persistency requirements for any relation symbol R and any function symbol f in the language, where the graph of an n-ary function $f : D^n \to D$ is defined as $\{(\bar{e}, d) \in D^{n+1} \mid f(\bar{e}) = d\}$ and denoted by $\operatorname{graph}(f)$:

- $k \preccurlyeq l$ implies $D_k \subseteq D_l$;

- $k \preccurlyeq l$ implies $I_k(R) \subseteq I_l(R)$;

- $k \preccurlyeq l$ implies $\mathrm{graph}(I_k(f)) \subseteq \mathrm{graph}(I_l(f))$;

- for any n-ary predicate φ, any $\bar{d} = d_1, \ldots, d_m$ in D, and terms $t_1(\bar{x}), \ldots, t_n(\bar{x})$ which free variables are among $\bar{x} = x_1, \ldots, x_m$: if $k \Vdash \varphi(I(t_1)(\bar{d}), \ldots, I(t_n)(\bar{d}))$ and $k \preccurlyeq l$, then $l \Vdash \varphi(I(t_1)(\bar{d}), \ldots, I(t_n)(\bar{d}))$.

The forcing relation $\Vdash$ is extended to all formulas in the usual way.

If no confusion is possible, the model $(K, \preccurlyeq, \mathcal{D}, \mathcal{I}, \Vdash)$ is denoted by K. The model has *constant domains* if all elements of $\mathcal{D}$ are equal. Note that the Kripke models are in general *not* required to have constant domains. Given a class of Kripke models $\mathcal{K}$, let $\mathcal{K}_{cd}$ denote the set of those models in $\mathcal{K}$ that have constant domains.

4 Skolemization

The most popular consequence of the Skolemization method is the statement that in classical predicate logic CQC, a prenex formula

$$\forall x_1 \exists y_1 \ldots \forall x_n \exists y_n \varphi(\bar{x}, y_1, \ldots, y_n)$$

is satisfiable if and only if its *Skolemization*

$$\forall x_1 \ldots x_n \varphi(\bar{x}, f_1(x_1), f_2(x_1, x_2), \ldots, f_n(x_1, \ldots, x_n))$$

is satisfiable, where f_i is a function symbol of arity i that does not occur in φ. This is equivalent to the statement that for such function symbols f_i:

$$\vdash_{\mathsf{CQC}} \exists x_1 \forall y_1 \ldots \exists x_n \forall y_n \varphi(\bar{x}, y_1, \ldots, y_n)$$
$$\Leftrightarrow$$
$$\vdash_{\mathsf{CQC}} \exists x_1 \ldots x_n \varphi(\bar{x}, f_1(x_1), f_2(x_1, x_2), \ldots, f_n(x_1, \ldots, x_n)).$$

This formulation in terms of derivability rather than satisfiability is the one used in this paper.

Less well-known is the fact that Skolemization also applies to infix formulas, formulas that are not necessarily in prenex normal form. To state this result one needs to distinguish strong from weak quantifiers, defined in Section 2.

The *Skolemization*, φ^s, of a formula φ is the result of replacing every strong quantifier occurrence $Qx\psi(x, \bar{y})$ by $\psi(f(\bar{y}), \bar{y})$, where f is a fresh function symbol and the variables $\bar{y}$ are the variables of the weak quantifiers in the scope of which $Qx\psi(x, \bar{y})$ occurs. In formal terms: The *Skolemization*, φ^s, of a formula φ is such that φ^s does not contain strong quantifiers and there exist formulas $\varphi = \varphi_1, \ldots, \varphi_n = \varphi^s$ such that every φ_{i+1} is the result of replacing the leftmost strong quantifier occurrence $Qx\psi(x,\bar{y})$ in φ_i by $\psi(f_i(\bar{y}),\bar{y})$, where the $f_1, \ldots, f_{n-1}$ are distinct fresh function symbols that do not occur in φ and $\bar{y}$ are the variables of the weak quantifiers in the scope of which $Qx\psi(x,\bar{y})$ occurs.

The following is an example of Skolemization.

$$(\forall u \exists v \varphi(u,v) \to \forall x \exists y \forall z \psi(x,y))^s = \forall u \varphi(u, f_1(u)) \to \exists y \psi(f_2, y, f_3(y)).$$

Note that f_2 is a constant, as the corresponding quantifier $\forall x$ is not in the scope of any weak quantifiers.

Classical logic admits Skolemization:

$$\vdash_{\mathsf{CQC}} \varphi \iff \vdash_{\mathsf{CQC}} \varphi^s.$$

Note that the result for prenex formulas given above is a special case of this theorem.

Interestingly, many of the standard nonclassical logics do not admit Skolemization. For example, in IQC and the predicate versions of LC and KC[1] there are various counterexamples, such as the following formulas, in which φ ranges over predicates, and which are not derivable in the logics, though their Skolemization (at the right) is.[2]

DNS $\forall x \neg\neg \varphi(x) \to \neg\neg \forall x \varphi(x)$ $\forall x \neg\neg \varphi(x) \to \neg\neg \varphi(c)$

SMP $\neg\neg \exists x \varphi(x) \to \exists x \neg\neg \varphi(x)$ $\neg\neg \varphi(c) \to \exists x \neg\neg \varphi(x)$

CD $\forall x (\varphi(x) \vee \psi) \to \forall x \varphi(x) \vee \psi$ $\forall x (\varphi(x) \vee \psi) \to \varphi(c) \vee \psi$

As mentioned above, in this paper we are not concerned with developing alternative methods but rather with proving that certain alternatives cannot obtain for certain logics. The question then is what one requires of such an alternative method, and the answer to that question clearly depends on the application one has in mind. Starting point in this paper is the idea that an alternative method $(\cdot)^a$ should produce a formula without strong quantifiers and that a logic L *admits* this method if

$$\vdash_{\mathsf{L}} \varphi \iff \vdash_{\mathsf{L}} \varphi^a. \tag{1}$$

In this way, the alternative method provides a connection between the propositional fragment of L and L itself, at least in case the logic admits some form of a Herbrand

---

[1]By this we mean the predicate logics QLC and QAJ from [11], axiomatized by $\forall \bar{x}\big((\varphi(x) \to \psi(\bar{x})) \vee (\psi(\bar{x}) \to \varphi(\bar{x}))\big)$ and $\forall \bar{x}\big(\neg\varphi(\bar{x}) \vee \neg\neg\varphi(\bar{x})\big)$, respectively.

[2]These principles can be found in [11]: DNS is shown to be equivalent, over IQC, to the principle KF, which is $\neg\neg\forall x(\varphi(x) \vee \neg\varphi(x))$; the strong Markov principle SMP appears under the name Ma.

Theorem, by which we mean a translation $(\cdot)^h$ such that $\varphi^h \in \mathcal{F}_{nw}$ and for all $\varphi \in \mathcal{F}_{ns}$:
$$\vdash_L \varphi \Leftrightarrow \vdash_L \varphi^h.$$

Therefore, requirement (1) seems a reasonable one. However, if no further requirements are made, then the notion trivializes in the sense that every logic with $\top$ and $\bot$ admits at least one alternative Skolemization method:

$$\varphi^a \equiv_{def} \begin{cases} \top & \text{if } \vdash_L \varphi \\ \bot & \text{if } \nvdash_L \varphi. \end{cases}$$

This is the reason that alternative methods are required to be *computable* as well.

We show in this paper that under the mild condition of strictness, to be defined in Section 5, there is no intermediate logic except CQC that is sound and complete with respect to a class of frames and that admits a strict, alternative Skolemization method. Thus implying that the logic IQC, the predicate versions of the Gabbay-deJongh logics, the predicate version of DeMorgan logic and Gödel–Dummett logic, as given in [11], and all tabular logics, which are the logics of a single frame, do not admit any strict, alternative Skolemization method.

5 Alternative Skolemization methods

An *alternative Skolemization method* is a computable total translation $(\cdot)^a$ from formulas to formulas such that for all formulas φ, φ^a does not contain strong quantifiers. A logic L *admits* the alternative Skolemization method if

$$\vdash \varphi \Leftrightarrow \vdash \varphi^a. \tag{2}$$

The method is *strict* if for every Kripke model K of L and all formulas φ:

$$K \nVdash \varphi^a \Rightarrow K \nVdash \varphi. \tag{3}$$

Clearly, standard Skolemization is an alternative Skolemization method, and CQC admits that method since $\varphi \to \varphi^s$ holds in CQC. An example of a different alternative Skolemization method is the one where occurrences of strong quantifiers $Qx\psi(x,\bar{y})$ are replaced by $\psi(f(\bar{y})) \vee \psi(g(\bar{y}))$ for fresh distinct f and g. Note that this method, a special case of the *parallel Skolemization method* introduced in [5], is strict, as is parallel Skolemization. On the other hand, the *existence Skolemization method* from [2, 4] is not strict.

Note that the form of Skolemization that we consider here does not take into account the identity axioms for Skolem functions as is usually done in the setting of model theory. This strengthens our results in the sense that if the problematic direction from right to left in (2) fails to hold, it does so too if we allow the identity axioms for Skolem functions on the right.

The requirement of computability alone does not suffice to prove that intuitionistic logic does not admit alternative Skolemization methods, as the following translation satisfies (2): $\varphi^a = (\psi_1 \to \psi_2)$, where ψ_1 consists of a conjunction of defining axioms for suitable primitive recursive functions that imply ψ_2, which is a coded statement that φ is provable in IQC, exactly whenever φ is provable in IQC. Since ψ_1 and ψ_2 can be defined in such a way that the first is a universal and the second an existential formula, the translation thus defined is an alternative Skolemization method. It is, however, not strict.

6 The strong and the weak quantifier fragments

Given a Kripke model K (recalling that they are assumed to be rooted), $K^{\downarrow}$ denotes the Kripke model that is the result of replacing every domain in K by the domain at the root of K and $K^{\uparrow}$ denotes the Kripke model that is the result of replacing every domain in K by the union of all domains in K. For predicates $P(\bar{x})$ and nodes k, we put $K^{\downarrow}, k \Vdash P(\bar{d})$ precisely if $\bar{d}$ consists of elements in D and $K, k \Vdash P(\bar{d})$, and we put $K^{\uparrow}, k \Vdash P(\bar{d})$ precisely if $\bar{d}$ consists of elements in D_k and $K, k \Vdash P(\bar{d})$.

Lemma 6.1. Let K be a rooted Kripke model, which root has domain D. Then the following holds for all k in K. Recall that $\bar{d}$ is short for $d_1, \ldots, d_n$, and $\bar{d} \in D$ means that $d_i \in D$ for all $i \leq n$.

1. For all formulas $\varphi(\bar{x}) \in \mathcal{F}_{nw}$, for all $\bar{d} \in D$: $K, k \Vdash \varphi(\bar{d}) \Rightarrow K^{\downarrow}, k \Vdash \varphi(\bar{d})$.

2. For all formulas $\varphi(\bar{x}) \in \mathcal{F}_{nw}$, for all $\bar{d} \in D_k$: $K, k \nVdash \varphi(\bar{d}) \Rightarrow K^{\uparrow}, k \nVdash \varphi(\bar{d})$.

3. For all formulas $\varphi(\bar{x}) \in \mathcal{F}_{ns}$, for all $\bar{d} \in D_k$: $K, k \Vdash \varphi(\bar{d}) \Rightarrow K^{\uparrow}, k \Vdash \varphi(\bar{d})$.

4. For all formulas $\varphi(\bar{x}) \in \mathcal{F}_{ns}$, for all $\bar{d} \in D$: $K, k \nVdash \varphi(\bar{d}) \Rightarrow K^{\downarrow}, k \nVdash \varphi(\bar{d})$.

Proof. The four properties are proved simultaneously, by formula induction. For atomic formulas $\varphi(\bar{x})$ the lemma follows by definition. The case where φ is a conjunction or disjunction follows immediately from the induction hypothesis.

Suppose $\varphi(\bar{x}) = \varphi_1(\bar{x}) \to \varphi_2(\bar{x})$. For 1., assume $\varphi \in \mathcal{F}_{nw}$ and $K, k \Vdash \varphi(\bar{d})$ for some $\bar{d} \in D$, and consider $l \succcurlyeq k$ such that $K^{\downarrow}, l \Vdash \varphi_1(\bar{d})$. Because φ_1 does not contain strong quantifiers, it follows from 4. that $K, l \Vdash \varphi_1(\bar{d})$. Hence $K, l \Vdash \varphi_2(\bar{d})$, and thus $K^{\downarrow}, l \Vdash \varphi_2(\bar{d})$ by 1. and the fact that φ_2 does not contain weak quantifiers.

For 2., assume $\varphi \in \mathcal{F}_{nw}$ and $K, k \not\Vdash \varphi(\bar{d})$ for some $\bar{d} \in D_k$ and consider $l \succcurlyeq k$ such that $K, l \Vdash \varphi_1(\bar{d})$ and $K, l \not\Vdash \varphi_2(\bar{d})$. Because φ_1 does not contain strong quantifiers, it follows from 3. that $K^{\uparrow}, l \Vdash \varphi_1(\bar{d})$. Because φ_2 does not contain weak quantifiers, $K^{\uparrow}, l \not\Vdash \varphi_2(\bar{d})$ follows from 2. Thus $K^{\uparrow}, k \not\Vdash \varphi_1(\bar{d}) \to \varphi_2(\bar{d})$. The proofs of 3. and 4. are analogous.

Suppose $\varphi(\bar{y}) = \forall x \psi(x, \bar{y})$. For 1., assume $\varphi \in \mathcal{F}_{nw}$ and $K, k \Vdash \forall x \psi(x, \bar{e})$ for some $\bar{e} \in D$ and consider $l \succcurlyeq k$ and $d \in D$. By the induction hypothesis and the fact that ψ does not contain weak quantifiers and D is the domain at the root of K, it follows that $K^{\downarrow}, l \Vdash \psi(d, \bar{e})$. Hence $K^{\downarrow}, k \Vdash \forall x \psi(x, \bar{e})$. For 2., assume $\varphi \in \mathcal{F}_{nw}$ and $K, k \not\Vdash \forall x \psi(x, \bar{e})$ for some $\bar{e} \in D_k$ and consider $l \succcurlyeq k$ and $d \in D_l$ such that $K, l \not\Vdash \psi(d, \bar{e})$. By the induction hypothesis and the fact that ψ does not contain weak quantifiers, it follows that $K^{\uparrow}, l \not\Vdash \psi(d, \bar{e})$. Hence $K^{\uparrow}, k \not\Vdash \forall x \psi(x, \bar{e})$. Cases 3. and 4. do not apply, as φ contains a strong quantifier.

Suppose $\varphi = \exists x \psi(x)$. Cases 1. and 2. do not apply, as φ contains a weak quantifier. For 3., assume $\varphi \in \mathcal{F}_{ns}$ and $K, k \Vdash \exists x \psi(x, \bar{e})$ for some $\bar{e} \in D_k$ and consider $d \in D_k$ such that $K, k \Vdash \psi(d)$. By the induction hypothesis and the fact that ψ does not contain strong quantifiers, it follows that $K^{\uparrow}, k \Vdash \psi(d)$. Hence $K^{\uparrow}, k \Vdash \exists x \psi(x, \bar{e})$. For 4., assume $\varphi \in \mathcal{F}_{ns}$ and $K, k \not\Vdash \exists x \psi(x, \bar{e})$ for some $\bar{e} \in D$. Thus for all $d \in D$, $K, k \not\Vdash \psi(d, \bar{e})$. Since ψ does not contain strong quantifiers the induction hypothesis gives $K^{\downarrow}, k \not\Vdash \psi(d, \bar{e})$ for all $d \in D$. Hence $K^{\downarrow}, k \not\Vdash \exists x \psi(x, \bar{e})$. □

Theorem 6.2. Let L be a logic that is sound and complete with respect to a class of Kripke models $\mathcal{K}$ which is closed under $\downarrow$ and $\uparrow$, then the strong quantifier free fragment of L is sound and complete with respect to $\mathcal{K}_{cd}$. And so is the weak quantifier free fragment of L.

Proof. For the first case, suppose that φ is a formula without strong quantifiers that is not derivable. Thus there is a model K in $\mathcal{K}$ that refutes φ. Let $\text{Sub}_{\text{neg}}(\varphi)$ and $\text{Sub}_{\text{pos}}(\varphi)$ denote the formulas that occur in φ negatively and positively, respectively. It suffices to show that

1. For all $\psi(\bar{x}) \in \text{Sub}_{\text{pos}}(\varphi)$, for all $\bar{d}$ in D: $K, k \not\Vdash \psi(\bar{d}) \Rightarrow K^{\downarrow}, k \not\Vdash \psi(\bar{d})$.

2. For all $\psi(\bar{x}) \in \text{Sub}_{\text{neg}}(\varphi)$, for all $\bar{d}$ in D: $K, k \Vdash \psi(\bar{d}) \Rightarrow K^{\downarrow}, k \Vdash \psi(\bar{d})$.

This follows from the previous lemma, using the fact that $\text{Sub}_{\text{pos}}(\varphi) \subseteq \mathcal{F}_{ns}$ and $\text{Sub}_{\text{neg}}(\varphi) \subseteq \mathcal{F}_{nw}$.

The second case is similar, using $K^{\uparrow}$ instead of $K^{\downarrow}$. □

$$
\begin{array}{lll}
1 & \Vdash \varphi(d) \quad \not\Vdash \varphi(e) \quad \Vdash \psi & D_1 = \{d, e\} \\
| & & \\
0 & \Vdash \varphi(d) \quad \not\Vdash \psi & D_0 = \{d\}
\end{array}
$$

Figure 1: Model that refutes CD

Corollary 6.3. Except for CQC, there is no intermediate logic that is sound and complete with respect to a class of frames and that admits a strict, alternative Skolemization method.

Proof. Consider an intermediate logic L that is sound and complete with respect to a class of frames, that is not equal to CQC, and that admits an alternative Skolemization method $(\cdot)^a$ that is strict. We show how this leads to a contradiction. Let $\mathcal{K}$ be the class of Kripke models based on the frames in the given class.

First, we show that L is sound and complete with respect to the class $\mathcal{K}_{cd}$ of models in $\mathcal{K}$ that have constant domains:

$$\vdash_L \varphi \Leftrightarrow \forall K \in \mathcal{K}_{cd}(K \Vdash \varphi).$$

The direction from left to right is trivial. The other direction is easy too: If $\not\vdash \varphi$, then $\not\vdash \varphi^a$, and so $K \not\Vdash \varphi^a$ for some $K \in \mathcal{K}$. Therefore $K^\downarrow \not\Vdash \varphi^a$ by Lemma 6.1. Thus $K^\downarrow \not\Vdash \varphi$ by strictness, and since $K^\downarrow \in \mathcal{K}_{cd}$, this completes the argument.

Having proven that L is sound and complete with respect to $\mathcal{K}_{cd}$, it follows that the constant domain formula CD (Section 4) holds in L, as it holds in all models with constant domains. However, if L ≠ CQC, then its class of frames contains at least one frame in which at least one node has a successor. Since on such a frame there exists a model that refutes CD, as in Figure 1, CD does not hold in L. The desired contradiction has been obtained. □

Let QD_n be the intermediate predicate logic of the frames of branching at most n, let QKC be the logic of the frames with one maximal node, and QLC be the logic of linear frames.

Corollary 6.4. The logics IQC, QD_n, QKC, QLC, and all tabular logics, do not admit any strict, alternative Skolemization method.

Note that the constant domain logics, such as the Gödel logics, are not covered by Corollary 6.4, as they are not complete with respect to a class of frames, but with respect to the constant domain models on a certain class of frames.

We close with a short observation about logics that do not admit any strict alternative Skolemization method. Suppose that for such a logic there is an alternative method $(\cdot)^a$ that it admits, and that the proof of this fact is semantical, showing that for every countermodel K to φ there is a countermodel K' to φ^a and vice versa. Then from the fact that the method cannot be strict, and thus cannot satisfy (3), it follows that not in all cases one can take K for K', as one could do in CQC.

References

[1] M. Baaz, A. Ciabattoni, and C. G. Fermüller. *Herbrand's Theorem for Prenex Gödel Logic and its Consequences for Theorem Proving*, pages 201–216. Springer, Berlin, Heidelberg, 2001.

[2] M. Baaz and R. Iemhoff. The Skolemization of existential quantifiers in intuitionistic logic. *Annals of Pure and Applied Logic*, 142(1):269–295, 2006.

[3] M. Baaz and R. Iemhoff. On Skolemization in constructive theories. *Journal of Symbolic Logic*, 73(3):969–998, 2008.

[4] M. Baaz and R. Iemhoff. Eskolemization in Intuitionistic Logic. *Journal of Logic and Computation*, 21(4):625–638, 2011.

[5] M. Baaz and R. Iemhoff. Skolemization in intermediate logics with the finite model property. *Logic Journal of the IGPL*, 24(3):224–237, 2016.

[6] M. Baaz and G. Metcalfe. *Herbrand Theorems and Skolemization for Prenex Fuzzy Logics*, pages 22–31. LNCS 5028. Springer, Berlin, Heidelberg, 2008.

[7] M. Baaz and G. Metcalfe. Herbrand's Theorem, Skolemization and Proof Systems for First-Order Łukasiewicz Logic. *Journal of Logic and Computation*, 20(1):35–54, 2010.

[8] M. Baaz and R. Zach. *Hypersequents and the Proof Theory of Intuitionistic Fuzzy Logic*, pages 187–201. LNCS 1862. Springer, Berlin, 2000.

[9] P. Cintula, D. Diaconescu, and G. Metcalfe. *Skolemization for Substructural Logics*, pages 1–15. LNCS 9450. Springer, Berlin, Heidelberg, 2015.

[10] P. Cintula and G. Metcalfe. *Herbrand Theorems for Substructural Logics*, pages 584–600. LNCS 8312. Springer, Berlin, Heidelberg, 2013.

[11] D. M. Gabbay, V. Shehtman, and D. Skvortsov. *Quantification in Nonclassical Logic*, volume 153 of *Studies in Logic and the Foundations of Mathematics*. Elsevier Science, UK, 2009.

[12] R. Iemhoff. The eskolemization of universal quantifiers. *Annals of Pure and Applied Logic*, 162(3):201–212, 2010.

[13] G. E. Mints. An analogue of Hebrand's theorem for the constructive predicate calculus. *Sov. Math. Dokl.*, 3:1712–1715, 1962.

[14] G. E. Mints. Skolem's method of elimination of positive quantifiers in sequential calculi. *Sov. Math. Dokl.*, 7(4):861–864, 1966.

[15] G. E. Mints. The Skolem method in intuitionistic calculi. *Proc. Steklov Inst. Math.*, 121:73–109, 1972.

[16] G. E. Mints. Resolution strategies for the intuitionistic predicate logic. In *Constraint Programming. Proceedings of the NATO Advanced Study Institute*, Comput. Syst. Sci. 131, pages 289–311. Springer, 1994.

[17] G. E. Mints. *Axiomatization of a Skolem function in intuitionistic logic*, pages 105–114. CSLI Lect. Notes 91. 2000.

[18] T. Skolem. Logisch-kombinatorische Untersuchungen über die Erfüllbarkeit oder Beweisbarkeit mathematischer Sätze nebst einem Theoreme über dichte Mengen. In *Skrifter utgit av Videnskapsselskapet i Kristiania. I, Matematisk-naturvidenskabelig klasse*, volume 1920 bd.1, pages 1–36. Kristiania : I Kommission hos J. Dybwad, 1920. http://www.biodiversitylibrary.org/item/52015#page/111/mode/1up.

[19] A. S. Troelstra and D. van Dalen. *Constructivism in mathematics. An introduction*, volume 1 of *Studies in Logic and the Foundations of Mathematics 121*. Elsevier Science, 1988.

The Logical Cone

Reinhard Kahle
*CMA and DM, FCT, Universidade Nova de Lisboa
P-2829-516 Caparica, Portugal*
kahle@mat.uc.pt

In Memory of Grisha Mints[1]

Abstract

This paper presents an outline of a new account of counterfactuals. It is based on a proof-theoretic perspective that allows a controlled replacement of axioms questioned in the antecedence of a counterfactual.

Keywords: Counterfactuals, Soccer.

1 Introduction

The dominant formal treatment of counterfactuals, due to Stalnaker [11] and Lewis [6], is given in the setting of possible world semantics. As appealing as it might be, possible world semantics not only raises ontological worries, it also makes use of an entirely intuitive neighborhood relation that—in our judgment—makes it impossible to evaluate specific counterfactual statements.[2] Here we propose an alternative account for counterfactuals, which could be dubbed *proof-theoretical*.[3] The idea is to take a rather limited set of examples of counterfactuals placed in a "real world"

---

[1] When this paper was presented at the conference *Philosophy, Mathematics, Linguistics: Aspects of Interaction 2014 (PhML-2014)* in St. Petersburg in April 2014, Grisha Mints showed great interest in the approach; of course, I do not claim that he agreed, but he acknowledged that the current state of the logical analysis of counterfactuals is unsatisfactory and that new approaches are desirable. Here we would like to show such a possible new approach.

[2] See [10] to see how messy an analysis of counterfactuals in terms of a neighborhood relation can become, This paper give also a neat overview of a classical example (Fine's bomb) discussed by Fine, Lewis, and others.

[3] This work can be considered as a contribution to the programme of *proof-theoretic semantics* [5], understood in a rather broad sense. An account related to *necessity* was proposed in [3]; see also [4].

environment, i.e., that may easily be encountered in "everyday" conversations. This limitation will allow us to evaluate the counterfactuals in a context of rather precisely given rules; the "real world" environment will give us some feedback about the common sense understanding of these counterfactuals. As a matter of fact, our analysis aims more for a descriptive or empirical analysis of counterfactuals, rather than a normative one.

As examples we choose counterfactual statements about the results of soccer matches.[4] There are two reasons for this choice. Firstly, results of soccer matches (and their consequences) follow quite well-defined rules, which can be formalized easily and allow, therefore, for an uncontroversial formal treatment. Secondly, results of soccer matches (and their consequences) are subject to quite profound discussions involving counterfactual conditionals—as you can easily experience if you have a discussion with a soccer fan.

In the next section we give a list of examples, taken from the European Championship in 2012, together with an informal evaluation; in section 3 we prepare the formal environment in which to analyze the examples and argue, in the following section, for a special treatment of axioms in this context. Sections 5 and 6 serve to introduce a narrow and a wide notion of the *logical cone*. In the following section we propose a specific understanding of the communicative function of counterfactuals in view of our previous analysis. In the final section we give a short conclusion with directions for further research.

2   "If the team had won this match, ..."

Spain won the European Championship in 2012 with a 4–0 win over Italy in the final.[5] Now, consider the following statement:

(1) If Italy had won the final, it would have been European Champion.

There should be no discussion that it is simply *true*. In view of the following examples, also its variation

(1') If Italy had won the final, it would have become European Champion.

has to be considered *true*.

We see the difference in (1) and (1') that, in the former case, the hypothetical fact in the consequent is an immediate consequence of the hypothetical fact in the

---

[4]The examples can, of course, be replaced by some from other sports, in particular for our North American fellows.

[5]All results of the tournament can be found on Wikipedia (http://en.wikipedia.org/wiki/UEFA_Euro_2012) or on the official UEFA pages (http://www.uefa.com/).

antecedent.⁶ while the latter case "leaves space" for further hypothetical facts taking place "between" the antecedent and the consequent (see the next example).

Based on the fact that Germany lost its semifinal against Italy, we may consider the statement:

(2) If Germany had won the semifinal, it would have become European Champion.

Let us first note that its variation corresponding to (1):

(2') If Germany had won the semifinal, it would have been European Champion.

is clearly *false*. Germany would not have been *immediately* European Champion after winning the semifinal, as the final in which Germany would now be one of the teams would still have to be played.

But also (2) would—taken literally—widely be considered as false, simply because there is no reason why Germany should have (also) won the final. But, of course, to consider it as false would not mean that one would consider its "consequent-negation" as true

(2*) If Germany had won the semifinal, it would not have become European Champion.

In fact, a natural reply to (2) is neither "that's true" nor "that's false" but the affirmation "that we don't know"; one could even remove the apparent epistemic aspect by replying "that would not be decided". The fact that (2) leaves the outcome of the consequent open can be made explicit by stating:

(3) If Germany had won the semifinal, it could have become European Champion.

In contrast to (2), this counterfactual should be considered as *true*.⁷

Now let us consider the following statements (bearing in mind that Ireland was already eliminated at the group stage of the European Championship):

(4) If Germany had won the semifinal, Ireland could have become European Champion.

The statement (4) is surely to be considered as *false*, simply because of the fact that Ireland, at the time Germany played its semifinal, was already eliminated and therefore *could not become European Champion*, regardless of what had happen in a semifinal match.

Interesting are the following two examples (taking into account that Portugal lost its semifinal with Spain a day before Germany played Italy):

---

⁶In the following we will use "antecedent" and "consequent" directly for the hypothetical facts expressed in the antecedent and in the consequent of a counterfactual, respectively.

⁷But we will come back to (2) in the last section.

(5) If Germany had won the semifinal, Portugal could have become European Champion.

(6) If Portugal had won the semifinal, Germany could have become European Champion.

(5) could be considered as false, in the way we rejected (4), as Portugal was already eliminated a day before the match of Germany took place. However, in the case of (6), it is arguable that the Germany–Italy match might have taken another course, and therefore (6) could be true.

3 Formal treatment

For a formal treatment of counterfactuals as given in the previous section we assume that one can formalize the usual rules for soccer results and the European Championship, together with the results of the particular matches in a sufficiently rich logical framework.

Among the rules we should find formulas expressing, for instance:

(L1) If team A scores more goals than team B in the match of A against B, team A wins this match.

(L2) The team that wins the final of the European Championship is the European Champion.

(L3) The teams that play in the final are the winners of the two semifinals.

The results should be expressed by formulas corresponding to the outcome of the matches, for instance:

(C1) Spain won the final against Italy.

(C2) Italy won the semifinal against Germany.

(C3) Spain won the semifinal against Portugal.

Both the rules and results can be considered as *axioms* which allow us to reason about the European Championship. With them, it should be more or less straightforward to prove facts like:

- Spain became European Champion.
- Ireland was eliminated at the group stage.

For our analysis it will be important that we split the rules and results into two different groups of axioms. This separation recalls some features of the separation of general laws and specific conditions in the well-known Hempel–Oppenheim scheme of scientific explanations [2, 15], and we will borrow from it the designations L and C for the axioms of rules and axioms of results, respectively. As a matter of fact, what is usually considered in a soccer discussion in the antecedent of a counterfactual is only the *result* of a particular match, but not a rule. The rules should provide a frame for the discussion, which is *generally* not put into question by the counterfactual situation proposed in the antecedent (see the end of the next section).

If we consider again example (1), it is easy to observe that we only need L-axioms, in fact just rule (L2), together with the hypothetical fact that Italy won the final, to prove that Italy would be European Champion. In this sense it would be an "immediate" consequence, and we can obtain a proof of the consequent if we just replace the C-axiom (C1) by one which expresses that Italy won the final.

But in the other examples, (2)–(6), the result of a semifinal is put into question and this raises the question, of which other results are also put into question. It should be clear that axiom (C1) does not make any sense any longer, as the teams in the final are now different. Thus, questioning (C2) or (C3) *implies* that (C1) cannot be an axiom any longer.[8] Replacing, in (2)–(5), the axiom (C2) by an axiom (C2') stating that Germany won the semifinal, does not give us any hint how to replace (C1). We would have to retract this axiom without replacing it. As a consequence, the resulting axiom system would be incomplete with respect to the winner of the final. This explains why we should consider neither (2) nor (2*) as true, but should consider (2') as false, and our analysis is in line with the reply that the "European Champion would not be decided".

In (3) we weakened the consequent to be (only) *possible*. Possibility is here simply understood as underivability of the negation.[9] With this understanding of possibility, (3) turns out to be true: the antecedent of (3), together with (L3) allows one to derive that Germany would play in the final. This implies the removal of (C1) without any replacement. The usual soccer rules will also not allow one to *derive* the result of the final (that is, a proof, in one way or another, of the result of the final), thus also not the negation of the consequent.

The statemente (4), however, is *false*, as changing the result of the semifinal does

---

[8]This implication can be made explicit by deriving formally the teams which play the final from the results of the semifinals by invoking the rule (L3).

[9]This is in line with (the first part of) Frege's observation: "If a proposition is advanced as possible, either the speaker is suspending judgment by suggestion that he knows no laws from which the negation of the proposition would follow or he says that the generalization of this negation is false." [1, p. 13].

not imply any change of the results at the group stage. A proof of the elimination of Ireland is not affected by a change of (C2), as this axiom does not imply any change of the matches played on the group stage. The change of (C2) to (C2') will not affect the proof of the fact that Ireland was already eliminated.

As in the informal analysis, we could consider (5) to be false on the same basis as (4): the semifinal between Portugal and Spain was played the day before the semifinal between Germany and Italy; thus changing the result of the latter match could not change the result of the former one.

But what about (6)? If (C3) is replaced by an axiom expressing that Portugal won against Spain, we already know that (C1) has to be retracted as axiom. (C2), however, is not related by any axiom to the result of other semifinal and there is *no need* to retract (C2) as axiom. But as the Germany–Italy match was played *after* the Portugal–Spain match, we already noted, that it is *arguable* that the former match had taken another course. Soccer fans might invoke (strange) arguments like: knowing they would face Portugal, the German players would have been so excited that their performance would been sufficient to beat Italy, or conversely, Italy would have feared facing Cristiano Ronaldo so much that they would have lost their semifinal. Although such arguments are far-fetched, it is reasonable to consider at least the possibility that when questioning facts at time t_0, one may also question facts at time t for $t > t_0$ (or $t \geq t_0$). We will see below how this possibility can be included in our analysis, even reopening the question whether (5) should, indeed, be considered as false.

4 On L- and C-axioms

We start with an *affirmative counterfactual* of the form:

($\star$) If ϕ were the case, then ψ would be the case.

Its analysis should take place in a formal system describing the actual situation. The non-logical axioms of the formal system are collected in a set $\mathcal{T}$, and its deductive closure will be designated by $\overline{\mathcal{T}}$. To be a *counterfactual*, ($\star$) presupposes that the negation[10] of ϕ, here designated by $\phi^\neg$, is actually the case, i.e., $\phi^\neg \in \overline{\mathcal{T}}$.

To argue in the counterfactual situation—to be given by a set of axioms $\mathcal{T}'$ with its deductive closure $\overline{\mathcal{T}'}$—we need to have $\phi^\neg \notin \overline{\mathcal{T}'}$, but $\phi \in \overline{\mathcal{T}'}$. A consistent replacement of $\phi^\neg$ by ϕ would be rather complicated, if $\phi^\neg$ could be an arbitrary formula in $\overline{\mathcal{T}}$. It would require tracing all axioms of $\mathcal{T}$ which are involved in all

---

[10] We do not want to be too formal here, but the negation of ϕ should something like $\neg\phi$ modulo *double negation elimination*.

possible derivations of $\phi^\neg$, and to replace some of them (which?) by some other axioms (which?) to obtain an axiom system $\mathcal{T}'$ with $\phi \in \overline{\mathcal{T}'}$. However, if we assume that $\phi^\neg$ is an *axiom* in $\mathcal{T}$, its replacement by ϕ is a quite reasonable operation.

Although it seems to be a technical condition here, we think that it is natural to demand that the antecedent (and its negation) has to be considered as an axiom in *some* representation of the hypothetical (actual, respectively) situation. It would be going too far here to give a full justification of this claim, but one may note that—with the formulation of a counterfactual like $(\star)$—ϕ is definitely not questionable in (a description of) the hypothetical situation; it is, of course, the role of the antecedent to fix ϕ as starting point of any further argument. In this perspective $\phi^\neg$ just "inherits" this status—or, to put it the other way around: to evaluate a counterfactual, the actual situation has to be described in a way that the fact put into question, i.e., $\phi^\neg$, is considered as starting point of any further discussion, i.e., as an axiom.[11]

As counterfactuals suggest the existence of arguments—or, on the formal side, proofs—of the consequent from the antecedent, we take a closer look to the setting in which such arguments or proofs can be performed. The first aspect, that $\phi^\neg$ should be settled on the level of axioms, was just discussed. If we replace $\phi^\neg$ by ϕ in the hypothetical situation, we would like to have, of course, that $\phi^\neg$ does not enter again in the discourse, making it inconsistent. Thus, the representation of the actual situation should be "sufficiently independent". As a matter of fact, this is more complicated as it sounds: if we consider the antecedent of (2), for instance, we would like to replace (C2) by its negation. However (L3), together with (C1)—expressing that Italy was playing the final—implies that Italy won its semifinal.

To overcome this problem we introduce the separation of L- and C-axioms, as indicated above: L-axioms should collect general rules, or "laws", which construct the frame in which a counterfactual should be evaluated; this frame is fixed and is not up for debate. C-axioms, however, are the "facts" that might be subject to changes—not only the one explicitly questioned in the antecedent.

In general, there are no apriori criteria that say what is an L-axiom and what is a C-axiom. In the soccer example, it seems to be quite clear that L-axioms should be the rules, while the C-axioms are the results. In other contexts, however, this separation is probably simply stipulated.

The specific feature of the L-axioms is that they should not be questioned in the discussion of the counterfactual. Soccer fans do not like replies to (1)

"If Italy had won the final, it would have been European Champion."

of the form:

---
[11] A similar argument is put forward in our analysis of necessity in [3].

"No, in this case—if Italy had won the final—UEFA would have changed the rules so that the loser of the final would be European Champion."

In contrast, the C-axioms are those which are subject to revisions. And we will discuss in the next section how those that should be retracted in a hypothetical situation are chosen.

5 The logical cone

As a *specification* we may say that the *logical cone of* $\phi^\neg$ should consist of exactly those formulae that should be removed from $\overline{\mathcal{T}}$ in the hypothetical situation proposed by the antecedent of the counterfactual $(\star)$.

Intuitively, it should be sufficient to retract those (C-)axioms which "depend" on $\phi^\neg$. In our examples, for instance, an axiom about the result of the final depends on the axioms about the results of the semifinals (as they determine the teams playing in the final). Note, however, that the results of the two semifinals do not depend on each other (we will come back to this issue in the next section).

For the time being, we will dispense with a formal definition of the dependency of one C-axiom on another,[12] but use it as an undefined, intuitive notion (and note that an axiom depends on itself). Given χ,[13] we say that the set of C-axioms that depend on χ, designated by $\mathcal{D}_\chi$, are in the *logical cone* of χ. More generally, the (narrow) *logical cone* of a C-axiom χ in $\mathcal{T}$ consists of those formulae of $\overline{\mathcal{T}}$ which need for their proofs axioms from $\mathcal{D}_\chi$; in formal terms, the (narrow) logical cone of χ is the set $\{\theta \mid \theta \in \overline{\mathcal{T}} \text{ and } \theta \notin \overline{\mathcal{T} \setminus \mathcal{D}_\chi}\}$.[14] It should be clear, that χ should be in its own logical cone, i.e., that $\chi \notin \overline{\mathcal{T} \setminus \mathcal{D}_\chi}$ (otherwise, χ would have been redundant in $\mathcal{T}$). $\overline{\mathcal{T} \setminus \mathcal{D}_\chi}$ (and in abuse of language also $\mathcal{T} \setminus \mathcal{D}_\chi$) may well be called a *background theory*, as it consists of those formulas (axioms) which should not be influenced by a change of χ.

Our (narrow) reading of the counterfactual:

$(\star)$ If ϕ were the case, then ψ would be the case.

is now that

$(\star)$ is true, if and only if, ψ follows from $(\mathcal{T} \setminus \mathcal{D}_{\phi^\neg}) \cup \{\phi\}$.

---

[12]This is, admittedly, in part because we encounter some technical problems in making it precise; it can, however, also be justified by the fact that the determination of a logical cone is subject to stipulation; this will be discussed in connection with the *umbra* in the next section.

[13]In the following, the role of ϕ and $\phi^\neg$ would have to be reversed; therefore, we use χ here.

[14]The same idea of dependence of θ on χ to formalize "χ is necessary for θ" is discussed in [3].

The intuitive idea is that the removal of the logical cone "gives space" to prove ψ, even if it contradicts a fact in $\overline{\mathcal{T}}$. As $\phi^{\neg}$ will not be in its own logical cone, we get that $\overline{(\mathcal{T} \setminus \mathcal{D}_{\phi^{\neg}}) \cup \{\phi\}}$ is consistent whenever $\overline{\mathcal{T}}$ is consistent.

With this analysis,

(1) If Italy had won the final, it would have been European Champion.

turns out to be true: if ϕ expresses *Italy wins the final*, then $\mathcal{D}_{\phi^{\neg}}$ is essentially (C1) (i.e., *Spain won the final against Italy*) and (L2) allows one to derive in $(\mathcal{T} \setminus \mathcal{D}_{\phi^{\neg}}) \cup \{\phi\}$ that *Italy is European Champion*.

But as noted above, for the other examples (2)–(6), we cannot expect to have an affirmative counterfactual of the type $(\star)$ turn out to be true. In all these cases, ϕ is the result of a semifinal; following the intuitive description, the (actual) result of the final (C1) will be in the logical cone of $\phi^{\neg}$, i.e., $\mathcal{D}_{\phi^{\neg}}$ contains (C1). In this case, however, $(\mathcal{T} \setminus \mathcal{D}_{\phi^{\neg}}) \cup \{\phi\}$ will no longer contain any axiom that could give us the result of the final, necessary to verify the consequent. In this view, (2), (2'), and (2*) are indeed *false* in our reading.

Thus, we were considering instead what we like to call a *possibility counterfactual* where the consequent is qualified as (only) possible.[15] If the possibility of χ in a theory $\mathcal{T}$ is formally rendered as $\mathcal{T}$ does not prove $\neg\chi$, we propose as analysis of

$(\star\star)$ If ϕ were the case, then ψ could be the case.

the following:

$(\star\star)$ is true, if and only if, $\neg\psi$ does not follow from $(\mathcal{T} \setminus \mathcal{D}_{\phi^{\neg}}) \cup \{\phi\}$.

With this analysis,

(3) If Germany had won the semifinal, it could have become European Champion.

turns out to be true: If ϕ expresses *Germany wins its semifinal*, then $\mathcal{D}_{\phi^{\neg}}$ should contain (C2) and (C1), and $(\mathcal{T} \setminus \mathcal{D}_{\phi^{\neg}}) \cup \{\phi\}$ should clearly not prove that Germany became *not* European Champion—it should be incomplete with respect to this question.

(4), however, is false, as $(\mathcal{T} \setminus \mathcal{D}_{\phi^{\neg}}) \cup \{\phi\}$—in fact, already the background theory $(\mathcal{T} \setminus \mathcal{D}_{\phi^{\neg}})$ alone—proves that Ireland will not be European Champion, using only

---

[15] Possibility counterfactuals are discussed by Lewis under the name of 'might' counterfactual, [6, §1.5]. But as he *defines* them in terms of his 'would' counterfactuals (the counterpart of our affirmative counterfactuals), we prefer another name here. The choice between 'might' and 'could' is, of course, rather a question of taste than of significance.

the results of the group stage, which are not affected by the change of a semifinal result.

Also (5) and (6) turn out to be false in this reading, as—according to our informal account of dependency—the result of one semifinal does not depend on the result of the other, and, thus, the teams in the consequents of (5) and (6), respectively, would not be in the final. But, as noted, (6) is, at least, arguable, and we will discuss in the next section how this can be incorporated in an enlargement of our analysis.

6 The umbra and the wide logical cone

Our intuitive characterization of $\mathcal{D}_\chi$ presupposes some kind of dependency between χ and the elements of $\mathcal{D}_\chi$ in the way that these elements should be incompatible with the negation of χ. As argued in the discussion of (6), it is, however, reasonable to permit also changes of facts which are not directly incompatible with χ, but for which one can "construct a scenario" that would change them after a change of χ—just as described in the far-fetched arguments given above for (6).

C-axioms of such facts are not elements of the (narrow) logical cone of χ. We can say, however, that they are in the *umbra* of χ.

But how to determine the element of the umbra of a formula? From the analysis (6) we could consider the temporal aspect, i.e., any fact taking place later than—or, not earlier than—χ could be considered as an element of the umbra. But we are not inclined to give a fixed characterization of the umbra. Instead, we suggest that it is actually *stipulated* by the context (or the utterer) of the counterfactual.

Such a stipulation can be checked explicitly by questioning the umbra of a given counterfactual. The utterer of (6) might be asked a question like: "If you change the result of Portugal's semifinal, you also allow changing the result of Germany's semifinal, don't you?" Or even better, one can demand that elements of the umbra are explicitly mentioned. For instance, one could ask the utterer of (6), whether (s)he means that:[16]

> (6') If Portugal had won the semifinal *and the result of Germany's semifinal could have been different*, Germany could have become European Champion.

If the umbra is, indeed, stipulated, it is also conceivable, that, when one questions the result of *one* semifinal, the result of the other semifinal is added to the umbra,

---

[16]If you consider this counterfactual as odd—as it seems to depend just on the change in the umbra—you can consider also the following one:

> If Portugal had won the semifinal and the result of Germany's semifinal could *therefore* have been different, Germany could have become European Champion.

independently of the historical timing, but just because the matches are played on the "same level". In this sense, it would not be surprising that soccer fans would come to the conclusion that (5) could be considered as true, *if* the change of the result of Germany's semifinal also allows changing the result of Portugal's semifinal, despite the fact that the match was played before. Again, this could be made explicit by questioning whether Portugal's semifinal should be considered to be in the umbra of Germany's semifinal.

Letting $\mathcal{U}_\chi$ be $\mathcal{D}_\chi$ augmented by the elements of the umbra of χ, the *wide logical cone* of a C-axiom χ in $\mathcal{T}$ consists of those formulae of $\overline{\mathcal{T}}$ that need for their proofs axioms from $\mathcal{U}_\chi$. With this notion, we can propose a *wide* reading of the counterfactual:

($\star$) If ϕ were the case, then ψ would be the case.

as

The counterfactual ($\star$) is true, if and only if, ψ follows from $(\mathcal{T} \setminus \mathcal{U}_{\phi^\neg}) \cup \{\phi\}$.

and analogously for ($\star\star$).

On the basis of example (5) and (6) we prefer the wide reading of counterfactuals over the narrow one. As this includes the stipulation of the umbra, we can overcome the problem of the definition of $\mathcal{D}_\chi$ by simply treating it in a similar way. Instead of demanding a formal definition, we let the elements of $\mathcal{D}_\chi$ be subjects to confirmations, i.e., one may *ask* whether a certain formula is considered to be dependent on the antecedent of a counterfactual.

7 "..., the team would have become European Champion" revisited

We will finish this paper by reconsidering affirmative counterfactuals, as given in (2). On the base of our analysis, given so far, only rather trivial, somehow non-informative affirmative counterfactuals will turn out to be true; probably no soccer fan would find the example (1) of particular interest. In contrast, statements like (2) may occur naturally in a soccer discussion, despite being controversial.

We believe that affirmative counterfactuals—which are not trivially true because the consequent follows immediately from the antecedent by L-axioms—have actually a specific status in a discussion. They are not meant as propositions with a fixed truth value, but rather as assertions which commit oneself *implicitly* to hidden assertions which would make the affirmative counterfactual true. This requires, in fact, that the corresponding possibility counterfactual needs to be true (in the sense of our analysis).

Let illustrate this using the example

(2) If Germany had won the semifinal, it would have become European Champion.

The corresponding possibility counterfactual is true according to our analysis: if we replace (C2) by (C2'), expressing that Germany won the semifinal, we have to retract (C1), and the negation of the consequent is not provable. In this theory, however, it is also not provable that Germany is European Champion. But it is possible to augment the theory by a new axiom (C1'), expressing that Germany wins the final. In fact, such an axiom would be necessary to obtain the consequent.[17] Thus, our claim is that the utterer of (2) *implicitly* commits h(im/er)self to this additional axiom (C1'), which, if added, makes the counterfactual true in terms of our analysis.

This analysis can be cross-checked: it is easy to question the utterer of (2): "Oh, you mean, if Germany had won its semifinal, it would also have won the final, don't you?" and one would probably get an affirmative answer.[18]

Thus, our analysis leads to the consequence that, in general, affirmative counterfactuals are not bivalent propositions, but rather assertions which hide implicit presuppositions in a subjective line of argument. This explains, at least, why counterfactuals are often so controversial.

8 Conclusion

We propose a new account of counterfactuals which is based on the notion of a logical cone.

First, we have to separate the description $\mathcal{T}$ of the "actual world" into two classes of axioms, L-axioms for fixed rules and facts and C-axioms for those which might be altered.

Secondly, we define the logical cone of a (C-)axiom ϕ, which consists intuitively of all formulas in the deductive closure $\overline{\mathcal{T}}$ of $\mathcal{T}$ which depend on ϕ.

Now we may add the negation of ϕ to the background theory obtained by removing the logical cone of ϕ from $\overline{\mathcal{T}}$ and investigate whether the consequence of an

---

[17] Again, the notion of necessity expressed in this condition can be made formal along the proposal given in [3]. There, we invoke a stipulated *variety of alternatives*, and it will be essential, for instance, that this variety consists of potential C-axioms, but does not involve L-axioms.

[18] Soccer discussions continue, in fact, along these lines, and may continue with a discussion like: "Thus, you mean Germany would have been better than Spain, don't you?" etc., and sooner or later people start to disagree...

affirmative counterfactual follows—or, for possibility counterfactuals, whether it is consistent with the background theory.

Further, we notice that, in many instances, affirmative counterfactuals are only defensible under hidden assumptions which are left out. These assumptions correspond to traces in the logical cone of the antecedence leading to the consequence. Such hidden assumptions, however, can be uncovered by asking for them.

In the present paper we provide a qualitative outline of our approach. A further elaboration has to specify better the status and dependency of the C-axioms. Also, we have to address how certain properties of combinations of counterfactuals behave in the proof-theoretic setting.[19] But one may already observe that the background theory $\mathcal{T} \setminus \mathcal{D}_\chi$ gives rise to a variety of possible worlds which may allow one to compare our approach with the usual semantic ones. Here, we see several advantages for the proof-theoretic account. First of all, it comes without any ontological burden. Secondly, the logical cone allows one to identify the sentences which are, indeed, affected by the antecedent, leaving "unreachable" sentences out of consideration.[20] In this way, there is also no need for "avoidance of big miracles" (see [10]), as they should be ruled out by the L-axioms. Furthermore, along the lines of §7, arguments for counterfactuals are easily "checkable" by inquiry; they correspond to traces in the logical cone, which allows us to dispense with any kind of "neighborhood relation", needed in possible worlds semantics. Finally, the axiomatic setting gives us "full control" over the background theory and the way it is modified by the alternatives of an antecedent of a counterfactual. In general, we see it as one of the defects of approaches using possible worlds semantics that they usually do not provide criteria for determining the possible worlds, but just argue on the base of a given variety of them.[21]

Let us finish by recalling Wehmeier [12, 13, 14], who pointed out the fact that the grammatical mood plays an essential role in counterfactuals; it is important, for instance, to distinguish the evaluation of definite descriptions in the actual world versus the (or any) counterfactual world(s). In our account such a distinction can be mirrored by evaluating such descriptions in the theory $\mathcal{T}$ of the actual situation or in the one for the counterfactual situation, $(\mathcal{T} \setminus \mathcal{D}_{\phi^\neg}) \cup \{\phi\}$ (if not already in the background theory $\mathcal{T} \setminus \mathcal{D}_{\phi^\neg}$). Also Wehmeier's *subjunctive-indicative conditionals*

---

[19]One may consult [9] for a discussion of such combinations in the semantic approaches growing out from Lewis's account.

[20]The term "logical cone" was, of course, chosen in analogy to Minkowski's *light cone* in relativity theory, [7, § III].

[21]It is worth mentioning here that, in particular, *modal logic* (see, for instance, [8]) as syntactic counterpart of possible worlds is just investigating the relation between *given* possible worlds, but not contributing to the question how to determine a variety of possible worlds, see also [4].

[14, § 2] fit in our context. The given example is: "If everybody who voted for Christa had voted for Barbara, Anna wouldn't have been elected chair." In our context, a corresponding example would be: "If every team that lost its semifinal had won it, Spain wouldn't have become Champion." The indicative part ("every team that lost its semifinal") just specifies the (teams and matches for which the) axioms of the actual situation that would have to be retracted ((C2) and (C3) in our example); the subjunctive part ("[each of these teams] had won [its semifinal]") tells us how to replace them. In this analysis, the antecedent is used as some kind of "instruction" how to modify an existing axiom system, rather than a formula on the object level which has to be integrated in one or the other way in the existing theory. But it is the proof-theoretic setting that allows to perform such an "instruction" in a fully controlled way.

In general, we see good potential in our proof-theoretic approach for all kind of intensional phenomena, as axiomatic frameworks allow for a more fine-grained analysis of logical relationships than approaches based on "structureless" worlds.

Acknowledgments

Research supported by the Portuguese Science Foundation, FCT, through projects: *Hilbert's Legacy in the Philosophy of Mathematics*, PTDC/FIL-FCI/109991/2009, *The Notion of Mathematical Proof*, PTDC/MHC-FIL/5363/2012, *Hilbert's 24th problem*, PTDC/MHC-FIL/2583/2014, and the Centro de Matemática e Aplicações, UID/MAT/00297/2013. I am grateful to Jesse Alama, Bartosz Więckowski, Alan J. Cain, and Kai Wehmeier for valuable comments on earlier drafts of this paper.

References

[1] G. Frege. *Begriffsschrift*, a formula language, modeled upon that of arithmetic, for pure thought. In J. van Heijenoort, editor, *From Frege to Gödel*, pages 1–82. Harvard University Press, Cambridge, Massachusetts, 1967.

[2] C. G. Hempel and P. Oppenheim. Studies in the logic of explanation. *Philosophy of Science*, 15:135–175, 1948.

[3] R. Kahle. A proof-theoretic view of necessity. *Synthese*, 138(3):659–673, 2006. Special issue on *Proof-theoretic semantics*.

[4] R. Kahle. Modalities without worlds. In S. Rahman, G. Primiero, and M. Marion, editors, *The Realism-Antirealism Debate in the Age of Alternative Logics*, Logic, Epistemology and the Unity of Science, pages 101–118. Springer, 2012.

[5] R. Kahle and P. Schroeder-Heister. Proof-theoretic semantics — Introduction. *Synthese*, 148(3):503–506, 2006.

[6] D. Lewis. *Counterfactuals*. Oxford: Blackwell Publishers and Cambridge: Harvard University Press, 1973. Reprinted with revisions, 1986. Wiley.

[7] H. Minkowski. Raum und zeit. *Jahresbericht der Deutschen Mathematiker-Vereinigung*, 18:75–88, 1909.

[8] G. Mints. *A Short Introduction to Modal Logic*, volume 30 of *CSLI Lecture Notes*. CSLI Publications, 1992.

[9] D. Nute. Conditional logic. In D. Gabbay and F. Guenthner, editors, *Handbook of Philosophical Logic*, volume II: Extensions of Classical Logic, pages 387–439. Kluwer, 1984.

[10] J. Schaffer. Counterfactuals, causal independence and conceptual circularity. *Analysis*, 64(4):299–309, 2004.

[11] R. C. Stalnaker. A theory of conditionals. *Americal Philosophical Quarterly*, pages 98–112, 1968.

[12] K. Wehmeier. In the mood. *Journal of Philosophical Logic*, 33(6):607–630, 2004.

[13] K. Wehmeier. Modality, mood, and descriptions. In R. Kahle, editor, *Intensionality*, volume 22 of *Lecture Notes in Logic*, pages 187–216. ASL and AK Peters, 2005.

[14] K. Wehmeier. Subjunctivity and conditionals. *Journal of Philosophy*, CX(3):117–142, 2013.

[15] J. Woodward. Scientific explanation. In E. N. Zalta, editor, *The Stanford Encyclopedia of Philosophy*. Winter 2011 edition, 2011.

Parsing and Generation as Datalog Query Evaluation

Makoto Kanazawa
National Institute of Informatics, 2-1-2 Hitotsubashi, Chiyoda-ku, Tokyo, 101-8430, Japan

Abstract

Parsing and generation (or surface realization) are two of the most important tasks in the processing of natural language by humans and by computers. This paper studies both tasks in the style of formal language theory, using typed λ-terms to represent meanings. It is shown that the problems of parsing and surface realization for grammar formalisms with "context-free" derivations, coupled with a kind of Montague semantics (satisfying a certain restriction) can be reduced in a uniform way to Datalog query evaluation. This makes it possible to apply to parsing and surface realization known efficient evaluation methods for Datalog. Moreover, the reduction has the following complexity-theoretic consequences for all such formalisms: (i) the decision problem of recognizing grammaticality (surface realizability) of an input string (logical form) is in LOGCFL; and (ii) the search problem of computing all derivation trees (in the form of shared forest) from an input string or input logical form is in functional LOGCFL. These bounds are tight. The reduction is carried out by way of "context-free" grammars on typed λ-terms, a relaxation of the second-order fragment of de Groote's abstract categorial grammar. The method works whenever a grammar uses only "almost linear" λ-terms.

Keywords: Generation, Surface Realization, Parsing, Datalog, LOGCFL, Montague Semantics, Abstract Categorial Grammar, Typed Lambda Calculus, Almost Linear Lambda Term.

1 Introduction

The representation of context-free grammars (augmented with features) in terms of definite clause programs is well-known. In the case of a bare-bone CFG, the

---

This paper originally evolved from an inspiration I got from a discussion with Sylvain Salvati on his work [63]. I was supported by the Japan Society for the Promotion of Science under the Grant-in-Aid for Scientific Research (KAKENHI), Grant Numbers 19500019 and 21500025.

corresponding program is in the function-free subset of logic programming, known as *Datalog*. For example, determining whether a string John found a unicorn belongs to the language of the CFG in (1) is equivalent to deciding whether the Datalog program in (2) together with the *database* in (3) can derive the goal or *query* (4):[1]

$$
\begin{array}{lll}
\text{S} \to \text{NP VP} & \text{NP} \to \text{John} & \text{Det} \to \text{a} \\
\text{VP} \to \text{V NP} & \text{V} \to \text{found} & \text{N} \to \text{man} \\
\text{V} \to \text{V and V} & \text{V} \to \text{caught} & \text{N} \to \text{unicorn} \\
\text{NP} \to \text{Det N} & \text{V} \to \text{is} &
\end{array}
\qquad(1)
$$

$$
\begin{array}{ll}
\text{S}(i,j) :\!-\ \text{NP}(i,k), \text{VP}(k,j). & \text{V}(i,j) :\!-\ \text{caught}(i,j). \\
\text{VP}(i,j) :\!-\ \text{V}(i,k), \text{NP}(k,j). & \text{V}(i,j) :\!-\ \text{is}(i,j). \\
\text{V}(i,j) :\!-\ \text{V}(i,k), \text{and}(k,l), \text{V}(l,j). & \text{Det}(i,j) :\!-\ \text{a}(i,j). \\
\text{NP}(i,j) :\!-\ \text{Det}(i,k), \text{N}(k,j). & \text{N}(i,j) :\!-\ \text{man}(i,j). \\
\text{NP}(i,j) :\!-\ \text{John}(i,j). & \text{N}(i,j) :\!-\ \text{unicorn}(i,j). \\
\text{V}(i,j) :\!-\ \text{found}(i,j). &
\end{array}
\qquad(2)
$$

$$\text{John}(0,1).\ \text{found}(1,2).\ \text{a}(2,3).\ \text{unicorn}(3,4). \qquad(3)$$

$$?\!-\ \text{S}(0,4). \qquad(4)$$

In the Datalog representation, terminals and nonterminals of the CFG are interpreted as binary predicates on positions within the input string. The database representing a string can be viewed as a certain type of directed graph (called a *string graph*). We depict a string graph by a diagram like (5), where circles represent nodes (string positions) and boxes are labels of directed edges, which, by convention, point from left to right.

$$\text{⓪}\!-\!\boxed{\text{John}}\!-\!\text{①}\!-\!\boxed{\text{found}}\!-\!\text{②}\!-\!\boxed{\text{a}}\!-\!\text{③}\!-\!\boxed{\text{unicorn}}\!-\!\text{④} \qquad(5)$$

By *naive* (or *seminaive*) bottom-up evaluation (see, e.g., [76] or [1]), the answer to a query like (4) can be computed in polynomial time in the size of the database, for any fixed Datalog program. This method of evaluation generates all facts derivable from the program together with the input database in the order of the height of the Datalog derivation tree, until no new fact is derivable. By recording ground instances of rules used to derive facts, a packed representation of the complete set of Datalog derivation trees for a given query can also be obtained in polynomial time using this technique. Since a Datalog derivation tree uniquely determines a grammar derivation tree and vice versa (Figure 1), the translation gives a reduction

---

[1]The term *query* means different things in logic programming/Prolog and relational database theory/finite model theory. The use of the term in this paper follows the logic programming/Prolog tradition.

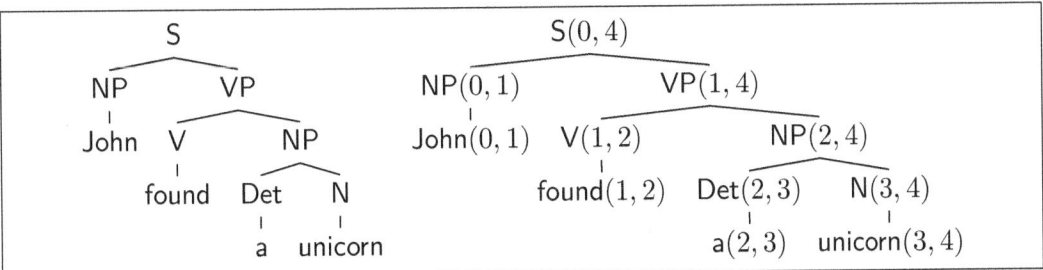

Figure 1: A CFG derivation tree (left) and a Datalog derivation tree (right).

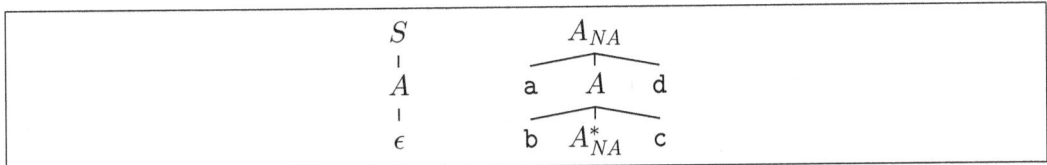

Figure 2: A TAG with one initial tree (left) and one auxiliary tree (right)

of context-free recognition and parsing to query evaluation in Datalog. This is of course all well known and well understood, even though the Datalog parlance is not universally adopted.

In this paper, I extend this reduction in two directions. First, I show that a similar reduction to Datalog is possible for more powerful grammar formalisms that have "context-free" derivations, such as *(multi-component) tree-adjoining grammars* [37, 80], *IO macro grammars* [24], and *(parallel) multiple context-free grammars* [66]. For instance, the tree-adjoining grammar in Figure 2 is represented by the Datalog program in (6).

$$S(i_1, i_3) :\!- A(i_1, i_3, i_2, i_2). \qquad (6)$$
$$A(i_1, i_8, i_4, i_5) :\!- \mathsf{a}(i_1, i_2), \mathsf{b}(i_3, i_4), \mathsf{c}(i_5, i_6), \mathsf{d}(i_7, i_8), A(i_2, i_7, i_3, i_6).$$
$$A(i_1, i_2, i_1, i_2).$$

Second, I extend the technique to the problem of *tactical generation (surface realization)* for such "context-free" grammar formalisms supplemented with a kind of *Montague semantics* [57], under a certain restriction to be made precise below. The method of reduction is uniform in both cases, and essentially relies on the encoding of different formalisms in terms of *abstract categorial grammars* [17].

The reduction to Datalog makes it possible to apply to parsing and generation sophisticated evaluation techniques for Datalog queries; in particular, an application of *generalized supplementary magic-sets* rewriting [8] automatically yields Earley-style algorithms for both parsing and generation. The reduction can also be used to obtain a tight upper bound, namely *LOGCFL*, on the computational complexity of the problem of recognition of input strings as well as of the problem of checking surface realizability of input logical forms.[2] This means that, in rough complexity-theoretic terms, these problems are no more difficult than the recognition problem for context-free languages.

With regard to parsing and recognition of input strings, polynomial-time algorithms and the LOGCFL upper bound on the computational complexity are already known for the grammar formalisms covered by our results [22]. Also, efficient tabular algorithms have already been obtained for many of these formalisms, and a general perspective on tabular parsing, in the names of *deductive parsing* [69] and *parsing schemata* [70], which can be equivalently expressed in terms of Datalog, is already available. Nevertheless, I believe that my method of reduction to Datalog is of independent interest, as it shows that efficient tabular parsing (recognition) algorithms are automatically obtained from various types of grammars in a uniform way. Concerning generation, where the input is a structured expression involving *binding*, the present results seem to be entirely new.[3]

Since the precise statement of my method of reduction and the proof of its correctness are quite technical, I first give an informal exposition of the method in Section 2. I develop the theory formally, complete in all details, in Section 3. I then discuss some consequences and extensions of the main results in Section 4, before giving a brief conclusion in Section 5.

The main results of the present paper were first announced in [42]; Sections 1 and 2.1, part of Section 2.2, and Section 4.3 are based on that paper.

---

[2]LOGCFL is the class of decision problems that can be reduced to some context-free language by a deterministic Turing machine operating in logarithmic space, and lies between the complexity classes NL and AC^1 (see [35]). Since LOGCFL is a subclass of NC, problems in LOGCFL are efficiently parallelizable. There are context-free languages that are complete for LOGCFL under log-space reduction (see [27]).

[3]As I explain below, the present method primarily applies to *exact generation* only, where the input logical form is supposed to exactly match the logical form produced by the grammar.

2 An informal exposition

2.1 Context-free grammars on λ-terms

Let us consider an augmentation of the CFG (1) with Montague semantics, which uses λ-terms as representations of meanings:[4]

$$
\begin{aligned}
&\mathsf{S}(X_1 X_2) \rightarrow \mathsf{NP}(X_1) \ \mathsf{VP}(X_2) &&(7)\\
&\mathsf{VP}(\lambda x. X_2(\lambda y. X_1 y x)) \rightarrow \mathsf{V}(X_1) \ \mathsf{NP}(X_2)\\
&\mathsf{V}(\lambda y x. \wedge^{t \rightarrow t \rightarrow t}(X_1 y x)(X_2 y x)) \rightarrow \mathsf{V}(X_1) \ \text{and} \ \mathsf{V}(X_2)\\
&\mathsf{NP}(X_1 X_2) \rightarrow \mathsf{Det}(X_1) \ \mathsf{N}(X_2)\\
&\mathsf{NP}(\lambda u. u \ \mathbf{John}^e) \rightarrow \text{John}\\
&\mathsf{V}(\mathbf{find}^{e \rightarrow e \rightarrow t}) \rightarrow \text{found}\\
&\mathsf{V}(\mathbf{catch}^{e \rightarrow e \rightarrow t}) \rightarrow \text{caught}\\
&\mathsf{V}(=^{e \rightarrow e \rightarrow t}) \rightarrow \text{is}\\
&\mathsf{Det}(\lambda uv. \exists^{(e \rightarrow t) \rightarrow t}(\lambda y. \wedge^{t \rightarrow t \rightarrow t}(uy)(vy))) \rightarrow \text{a}\\
&\mathsf{N}(\mathbf{man}^{e \rightarrow t}) \rightarrow \text{man}\\
&\mathsf{N}(\mathbf{unicorn}^{e \rightarrow t}) \rightarrow \text{unicorn}
\end{aligned}
$$

Here, the left-hand side of each rule is annotated with a λ-term that tells how the meaning of the left-hand side is composed from the meanings of the right-hand side nonterminals, represented by upper-case variables $X_1, X_2, \ldots$. Note that λ-terms may contain any number of constants, whose types are indicated by superscripts.[5] In such a grammar, the meaning of a sentence is computed from its derivation tree. For example, given the derivation tree of John found a unicorn (the left tree in Figure 1), we can decorate each nonterminal node with a λ-term in accordance with the grammar rule being applied at that node, obtaining the decorated tree in Figure 3. The λ-term decorating the root node,

$$(\lambda u. u \ \mathbf{John})(\lambda x.(\lambda uv. \exists(\lambda y. \wedge (uy)(vy))) \ \mathbf{unicorn} \ (\lambda y. \mathbf{find} \ y \ x)),$$

β-reduces to the λ-term

$$\exists(\lambda y. \wedge(\mathbf{unicorn} \ y)(\mathbf{find} \ y \ \mathbf{John})) \qquad (8)$$

encoding the first-order logic formula representing the meaning of the sentence (i.e., its logical form):

$$\exists y(\mathbf{unicorn}(y) \wedge \mathbf{find}(\mathbf{John}, y)).$$

---

[4]Grammars like this one are basically *generalized phrase structure grammars* [25] without features or metarules.

[5]We follow standard notational conventions in typed λ-calculus, rather than Montague's [57]. Thus, an application $M_1 M_2 M_3$ (written without parentheses) associates to the left, $\lambda x. \lambda y. M$ is abbreviated to $\lambda xy.M$, and $\alpha \rightarrow \beta \rightarrow \gamma$ stands for $\alpha \rightarrow (\beta \rightarrow \gamma)$.

$$\text{S}\Big((\lambda u.u\ \textbf{John})(\lambda x.(\lambda uv.\exists(\lambda y.\wedge(uy)(vy)))\ \textbf{unicorn}\ (\lambda y.\textbf{find}\ y\ x))\Big)$$

A tree:
- S has children NP($\lambda u.u$ **John**) and VP($\lambda x.(\lambda uv.\exists(\lambda y.\wedge(uy)(vy)))$ **unicorn** $(\lambda y.\textbf{find}\ y\ x)$)
- NP → John
- VP has children V(**find**) and NP$\big((\lambda uv.\exists(\lambda y.\wedge(uy)(vy)))\ \textbf{unicorn}\big)$
- V → found
- NP has children Det$(\lambda uv.\exists(\lambda y.\wedge(uy)(vy)))$ and N(**unicorn**)
- Det → a
- N → unicorn

Figure 3: A decorated derivation tree of a CFG with Montague semantics.

Thus, computing the logical form(s) of a sentence—the task of *semantic interpretation*[6]—involves parsing and λ-term normalization. Conversely, to find a sentence expressing a given logical form—the task of *surface realization*—it suffices to find a derivation tree whose root node is decorated with a λ-term that β-reduces to the given logical form; the desired sentence can simply be read off from the derivation tree. At the heart of both tasks is the computation of the derivation tree(s) that yield the input. In the case of surface realization, this may be viewed as parsing the input λ-term with a "context-free" grammar that generates a set of λ-terms (in β-normal form), which is obtained from the given CFG with Montague semantics by stripping off terminal symbols:

$$\begin{aligned}
&\mathsf{S}(X_1 X_2) :\!- \mathsf{NP}(X_1), \mathsf{VP}(X_2). \\
&\mathsf{VP}(\lambda x.X_2(\lambda y.X_1 yx)) :\!- \mathsf{V}(X_1), \mathsf{NP}(X_2). \\
&\mathsf{V}(\lambda yx.\wedge^{t\to t\to t}(X_1 yx)(X_2 yx)) :\!- \mathsf{V}(X_1), \mathsf{V}(X_2). \\
&\mathsf{NP}(X_1 X_2) :\!- \mathsf{Det}(X_1), \mathsf{N}(X_2). \\
&\mathsf{NP}(\lambda u.u\ \textbf{John}^e). \\
&\mathsf{V}(\textbf{find}^{e\to e\to t}). \\
&\mathsf{V}(\textbf{catch}^{e\to e\to t}). \\
&\mathsf{V}(=^{e\to e\to t}). \\
&\mathsf{Det}(\lambda uv.\exists^{(e\to t)\to t}(\lambda y.\wedge^{t\to t\to t}(uy)(vy))). \\
&\mathsf{N}(\textbf{man}^{e\to t}). \\
&\mathsf{N}(\textbf{unicorn}^{e\to t}).
\end{aligned} \qquad (9)$$

Determining whether a given logical form is surface realizable with the original grammar (7) is equivalent to recognition with the resulting *context-free λ-term grammar* (CFLG) (9). As with CFG recognition/parsing, solving the problem of recognition for CFLGs almost amounts to solving the problem of parsing; so algorithms and

---

[6]This is sometimes called "semantic parsing" or "parsing to logical form".

complexity results for the former translate into algorithms and complexity results for the problem of surface realization.

In a CFLG such as (9), there is a mapping f from nonterminals to their semantic types:

$$f = \begin{cases} \mathsf{S} \mapsto t, \\ \mathsf{NP} \mapsto (e \to t) \to t, \\ \mathsf{VP} \mapsto e \to t, \\ \mathsf{V} \mapsto e \to e \to t, \\ \mathsf{Det} \mapsto (e \to t) \to (e \to t) \to t, \\ \mathsf{N} \mapsto e \to t \end{cases}.$$

A rule that has B on the left-hand side and $B_1, \ldots, B_n$ as right-hand side nonterminals has its left-hand side annotated with a well-formed λ-term M that has type $f(B)$ under the type environment $X_1 : f(B_1), \ldots, X_n : f(B_n)$, or in symbols:

$$\vdash X_1 : f(B_1), \ldots, X_n : f(B_n) \Rightarrow M : f(B).$$

For example, in the case of the third rule of (9), we have

$$\vdash X_1 : e \to e \to t, X_2 : e \to e \to t \Rightarrow \lambda yx.\wedge^{t \to t \to t}(X_1 yx)(X_2 yx) : e \to e \to t. \quad (10)$$

What we are calling a context-free λ-term grammar is nothing but an alternative notation for an *abstract categorial grammar* [17] whose abstract vocabulary is second-order, with the restriction to *linear* λ-terms removed.[7] In the linear case, Salvati [62] showed the recognition/parsing complexity to be in P, and exhibited an algorithm similar to Earley parsing for TAGs. Second-order linear ACGs are known to be expressive enough to encode well-known mildly context-sensitive grammar formalisms in a straightforward way, including TAGs and (non-deleting) multiple context-free grammars (also known as *linear context-free rewriting systems*) [18, 19].

For example, the following linear CFLG is an encoding of the TAG in Figure 2, where $f(S) = o \to o$ and $f(A) = (o \to o) \to o \to o$ (see [18] for details of this encoding):

$$S(\lambda y.X_1(\lambda z.z)y) :\!\!-\, A(X_1). \quad (11)$$
$$A(\lambda xy.\mathsf{a}^{o \to o}(X_1(\lambda z.\mathsf{b}^{o \to o}(x(\mathsf{c}^{o \to o} z))) (\mathsf{d}^{o \to o} y))) :\!\!-\, A(X_1).$$
$$A(\lambda xy.xy).$$

In encoding a string-generating grammar, a CFLG uses o as the type of string position and $o \to o$ as the type of string. Each terminal symbol is represented by a

---

[7]A λ-term is a λI-*term* if each occurrence of λ binds at least one occurrence of a variable. A λI-term is *linear* if no subterm contains more than one free occurrence of the same variable.

constant of type $o \to o$, and a string $a_1 \ldots a_n$ is encoded by the λ-term

$$/a_1 \ldots a_n/ = \lambda z.a_1^{o\to o}(\ldots(a_n^{o\to o}z)\ldots),$$

which has type $o \to o$.[8]

A string-generating grammar coupled with Montague semantics may be represented by a *synchronous CFLG*, a pair of CFLGs with matching rule sets, as in Figure 4.[9] The transduction between strings and logical forms in either direction consists of parsing the input λ-term with the source-side grammar and normalizing the λ-term(s) constructed in accordance with the target-side grammar from the derivation tree(s) output by parsing.

2.2 Reduction to Datalog

We can show that under a weaker condition than linearity, a CFLG can be represented by a Datalog program. The presentation in this section is informal and not fully precise; formal definitions and rigorous proof of correctness are deferred to Section 3.

We use the grammar (9) as an example, which is repeated below:

$$\begin{aligned}
&\mathsf{S}(X_1X_2) :\!\!- \mathsf{NP}(X_1), \mathsf{VP}(X_2). \\
&\mathsf{VP}(\lambda x.X_2(\lambda y.X_1yx)) :\!\!- \mathsf{V}(X_1), \mathsf{NP}(X_2). \\
&\mathsf{V}(\lambda yx.\wedge^{t\to t\to t}(X_1yx)(X_2yx)) :\!\!- \mathsf{V}(X_1), \mathsf{V}(X_2). \\
&\mathsf{NP}(X_1X_2) :\!\!- \mathsf{Det}(X_1), \mathsf{N}(X_2). \\
&\mathsf{NP}(\lambda u.u\, \mathbf{John}^e). \\
&\mathsf{V}(\mathbf{find}^{e\to e\to t}). \\
&\mathsf{V}(\mathbf{catch}^{e\to e\to t}). \\
&\mathsf{V}(=^{e\to e\to t}). \\
&\mathsf{Det}(\lambda uv.\exists^{(e\to t)\to t}(\lambda y.\wedge^{t\to t\to t}(uy)(vy))). \\
&\mathsf{N}(\mathbf{man}^{e\to t}). \\
&\mathsf{N}(\mathbf{unicorn}^{e\to t}).
\end{aligned} \qquad (9)$$

Note that all λ-terms in this grammar are *almost linear* in the sense of satisfying the following conditions:

- every occurrence of λ binds at least one occurrence of a variable (i.e., they are λI terms), and

---

[8] It is known that the class of string languages generated by linear CFLGs under this encoding coincides with the class of multiple context-free languages [63]. The class of tree languages generated by linear CFLGs has been characterized by Kanazawa [45].

[9] The use of a pair of ACGs with a common abstract vocabulary as a synchronous grammar has already been advocated by de Groote [17].

```
S(λz.Y₁(Y₂z), X₁X₂) :- NP(Y₁, X₁), VP(Y₂, X₂).
VP(λz.Y₁(Y₂z), λx.X₂(λy.X₁yx)) :- V(Y₁, X₁), NP(Y₂, X₂).
V(λz.Y₁(/and/(Y₂z)), λyx.∧ᵗ→ᵗ→ᵗ(X₁yx)(X₂yx)) :- V(Y₁, X₁), V(Y₂, X₂).
NP(λz.Y₁(Y₂z), X₁X₂) :- Det(Y₁, X₁), N(Y₂, X₂).
NP(/John/, λu.u Johnᵉ).
V(/found/, find^(e→e→t)).
V(/caught/, catch^(e→e→t)).
V(/is/, =^(e→e→t)).
Det(/a/, λuv.∃^((e→t)→t)(λy.∧ᵗ→ᵗ→ᵗ(uy)(vy))).
N(/man/, man^(e→t)).
N(/unicorn/, unicorn^(e→t)).
```

```
S(λz.Y₁(Y₂z)) :- NP(Y₁), VP(Y₂).         S(X₁X₂) :- NP(X₁), VP(X₂).
VP(λz.Y₁(Y₂z)) :- V(Y₁), NP(Y₂).          VP(λx.X₂(λy.X₁yx)) :- V(X₁), NP(X₂).
V(λz.Y₁(/and/(Y₂z))) :- V(Y₁), V(Y₂).     V(λyx.∧ᵗ→ᵗ→ᵗ(X₁yx)(X₂yx)) :- V(X₁), V(X₂).
NP(λz.Y₁(Y₂z)) :- Det(Y₁), N(Y₂).         NP(X₁X₂) :- Det(X₁), N(X₂).
NP(/John/).                                NP(λu.u Johnᵉ).
V(/found/).                                V(find^(e→e→t)).
V(/caught/).                               V(catch^(e→e→t)).
V(/is/).                                   V(=^(e→e→t)).
Det(/a/).                                  Det(λuv.∃^((e→t)→t)(λy.∧ᵗ→ᵗ→ᵗ(uy)(vy))).
N(/man/).                                  N(man^(e→t)).
N(/unicorn/).                              N(unicorn^(e→t)).
```

Figure 4: The grammar in (7) expressed as a synchronous CFLG (top), with its two components separated out. The first component is a linear CFLG encoding the CFG (1), and the second component is the CFLG (9).

- for every subterm N, if a variable x occurs free more than once in N, x has an atomic type,

where the type of an occurrence of a variable is determined by the typing assigned to the λ-term by the grammar. The reduction to Datalog is guaranteed to be correct only when the grammar is *almost linear* in this sense.

The key to our construction is the *principal typing* of an almost linear λ-term. In this informal exposition, we represent principal typings graphically by means of *hypergraphs* of a certain kind. A hypergraph is a generalization of a directed graph where an edge (called a *hyperedge*) may be incident on any number of nodes, depending on its label.[10]

---

[10]The connection between CFLGs and hypergraphs goes beyond the present informal exposition. See [45] for the relation between linear CFLGs and *hyperedge replacement grammars*, a context-free grammar formalism generating sets of hypergraphs.

For example, take the λ-term

$$\lambda yx.\wedge^{t\to t\to t}(X_1 yx)(X_2 yx) \qquad (12)$$

annotating the left-hand side of the third rule of the grammar (9). Recall that the function f mapping nonterminals to their types gives a typing of the λ-term annotating the left-hand side of each rule. The typing assigned to the λ-term (12) is expressed by the typing judgment (10):

$$\vdash X_1 : e \to e \to t, X_2 : e \to e \to t \Rightarrow \lambda yx.\wedge^{t\to t\to t}(X_1 yx)(X_2 yx) : e \to e \to t. \qquad (10)$$

(Note that the bound variables x and y both have type e in this typing.) Given the typing judgment (10), we can build the hypergraph for the λ-term (12):

$$(13)$$

In a diagram like this, circles represent nodes, and circles with numbers attached to them are *external nodes* of the hypergraph. Each hyperedge is represented by a box with a label inside and *tentacles* connecting it to the nodes that it is incident on. The tentacles of a hyperedge are ordered; in this paper, we adopt the convention that they are ordered clockwise starting from the 12 o'clock position. Thus, the hyperedge with label X_2 in (13) has three tentacles, with the first tentacle leading to the node right above it, the second to the node right below it, and the third to the node right below the hyperedge with label X_1. We call the first node in the sequence of nodes that a hyperedge is incident on the *result node* of the hyperedge.

In general, the hypergraph graph(M) for a typed almost linear λ-term M is constructed by induction on the structure of M, as follows. If α is a type, let $|\alpha|$ be the number of occurrences of atomic types in α.[11]

---

[11] In this paper, we greatly overload the notation $|\cdot|$. In addition to the use just defined, we use it to mean the number of nodes of a tree, the length of a string, and the number of components of a tuple. It should be clear from the context which meaning is intended.

For a variable or a constant a of type α, graph(a) consists of $|\alpha|$ nodes $v_1, \ldots, v_{|\alpha|}$, all of which are external nodes, and a single hyperedge labeled by a, which is incident on $v_1, \ldots, v_{|\alpha|}$, in this order. Given the typing in (10), we have:

$$\text{graph}(\wedge) = \begin{array}{c}\wedge\end{array} \qquad \text{graph}(X_1) = \begin{array}{c}X_1\end{array} \qquad \text{graph}(X_2) = \begin{array}{c}X_2\end{array}$$

$$\text{graph}(y) = \begin{array}{c}y\end{array} \qquad \text{graph}(x) = \begin{array}{c}x\end{array}$$

If M is an application $M_1 M_2$, where M_1 and M_2 are of type $\alpha \to \beta$ and α, respectively, graph(M) is constructed from the union of graph(M_1) and graph(M_2) by identifying the last $|\alpha|$ external nodes of graph(M_1) with the external nodes of graph(M_2); the remaining external nodes of graph(M_1) become the external nodes of M. If M_1 and M_2 share a free variable x (which must be of atomic type since M is almost linear), then the x-labeled hyperedge in graph(M_1) and the x-labeled hyperedge in graph(M_2), as well as the nodes that they are incident on, are also identified.

$$\text{graph}(X_1 y) = \qquad \text{graph}(X_1 y x) =$$

$$\text{graph}(X_2 y) = \qquad \text{graph}(X_2 y x) =$$

$$\mathrm{graph}(\wedge(X_1 yx)) = \quad [\text{diagram}] \qquad \mathrm{graph}(\wedge(X_1 yx)(X_2 yx)) = \quad [\text{diagram}]$$

Finally, if M is a λ-abstraction $\lambda x.M_1$, then $\mathrm{graph}(M)$ is obtained from $\mathrm{graph}(M_1)$ by appending the sequence of nodes that the x-labeled hyperedge is incident on to the sequence of external nodes.

$$\mathrm{graph}(\lambda x.\wedge(X_1 yx)(X_2 yx)) = \quad [\text{diagram}]$$

$$\mathrm{graph}(\lambda yx.\wedge(X_1 yx)(X_2 yx)) = \quad [\text{diagram}] \qquad = (13)$$

There are several important points to note about this construction:

- If M has type α, $\mathrm{graph}(M)$ has $|\alpha|$ external nodes.

- For each free variable x in M, there is exactly one hyperedge labeled by x in $\mathrm{graph}(M)$.

- When M is in η-long β-normal form, graph(M) is what is called a *term graph* (see [59]) with external nodes; in particular, for each node v in graph(M), there is exactly one hyperedge whose result node is v.

To convert an almost linear CFLG rule

$$B(M) :\!- B_1(X_1), \ldots, B_n(X_n)$$

into a Datalog rule, we take graph(M) and name its nodes with Datalog variables (for which we use $i_1, i_2, i_3, \ldots$). In the case of the third rule of the grammar (9),

$$\mathsf{V}(\lambda yx.\wedge^{t\to t\to t}(X_1 yx)(X_2 yx)) :\!- \mathsf{V}(X_1), \mathsf{V}(X_2), \tag{14}$$

we get:

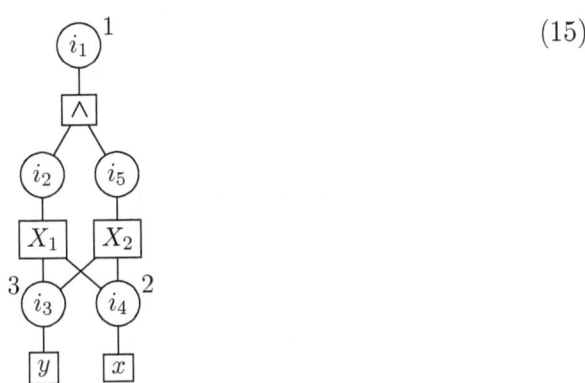
(15)

Then we do three things to the CFLG rule:

(i) replace the left-hand side λ-term M by the sequence of external nodes of graph(M),

(ii) replace each right-hand side variable X_i by the sequence of nodes that the X_i-labeled hyperedge is incident on in graph(M), and

(iii) for each hyperedge in graph(M) labeled by a constant b, add to the right-hand side of the rule an atom $b(\vec{v})$, where $\vec{v}$ is the sequence of nodes that the hyperedge is incident on.

Applying this procedure to (14) produces the following result:

$$\mathsf{V}(i_1, i_4, i_3) :\!- \wedge(i_1, i_5, i_2), \mathsf{V}(i_2, i_4, i_3), \mathsf{V}(i_5, i_4, i_3).$$

For another example, consider the ninth rule of the CFLG in Figure 9:

$$\mathsf{Det}(\lambda uv.\exists^{(e\to t)\to t}(\lambda y.\wedge^{t\to t\to t}(uy)(vy))).$$

The hypergraph for this λ-term is

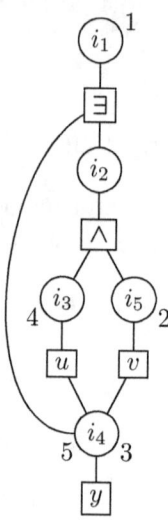

and the corresponding Datalog rule is

$$\mathsf{Det}(i_1, i_5, i_4, i_3, i_4) :\!- \exists(i_1, i_2, i_4), \wedge(i_2, i_5, i_3).$$

Applying the same procedure to all the rules in (9), we get the following Datalog program:

$$\mathsf{S}(i_1) :\!- \mathsf{NP}(i_1, i_2, i_3), \mathsf{VP}(i_2, i_3). \tag{16}$$
$$\mathsf{VP}(i_1, i_4) :\!- \mathsf{V}(i_2, i_4, i_3), \mathsf{NP}(i_1, i_2, i_3).$$
$$\mathsf{V}(i_1, i_4, i_3) :\!- \wedge(i_1, i_5, i_2), \mathsf{V}(i_2, i_4, i_3), \mathsf{V}(i_5, i_4, i_3).$$
$$\mathsf{NP}(i_1, i_4, i_5) :\!- \mathsf{Det}(i_1, i_4, i_5, i_2, i_3), \mathsf{N}(i_2, i_3).$$
$$\mathsf{NP}(i_1, i_1, i_2) :\!- \mathbf{John}(i_2).$$
$$\mathsf{V}(i_1, i_3, i_2) :\!- \mathbf{find}(i_1, i_3, i_2).$$
$$\mathsf{V}(i_1, i_3, i_2) :\!- \mathbf{catch}(i_1, i_3, i_2).$$
$$\mathsf{Det}(i_1, i_5, i_4, i_3, i_4) :\!- \exists(i_1, i_2, i_4), \wedge(i_2, i_5, i_3).$$
$$\mathsf{N}(i_1, i_2) :\!- \mathbf{man}(i_1, i_2).$$
$$\mathsf{N}(i_1, i_2) :\!- \mathbf{unicorn}(i_1, i_2).$$

The construction of the database representing the input λ-term is similar, but slightly more complex. A simple case is the λ-term (8), where each constant occurs just once:

$$\exists(\lambda y.\wedge(\mathbf{unicorn}\ y)(\mathbf{find}\ y\ \mathbf{John})) \tag{8}$$

This is an almost linear λ-term in η-long β-normal form, from which we obtain the

S$((\lambda u.u\ \textbf{John})(\lambda x.(\lambda uv.\exists(\lambda y.\wedge(uy)(vy)))\ \textbf{unicorn}\ (\lambda y.\textbf{find}\ y\ x)))$
NP$(\lambda u.u\ \textbf{John})$ VP$(\lambda x.(\lambda uv.\exists(\lambda y.\wedge(uy)(vy)))\ \textbf{unicorn}\ (\lambda y.\textbf{find}\ y\ x)))$
V(**find**) NP$((\lambda uv.\exists(\lambda y.\wedge(uy)(vy)))\ \textbf{unicorn})$
Det$(\lambda uv.\exists(\lambda y.\wedge(uy)(vy)))$ N(**unicorn**)

Figure 5: The CFLG derivation tree for (8)

following hypergraph:

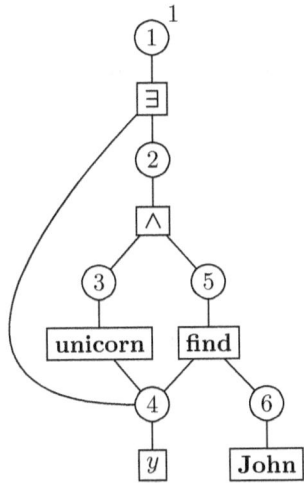

The hyperedges of this hypergraph that are labeled by constants in the λ-term constitute the facts in the database representing the λ-term:

$$\exists(1,2,4). \quad \wedge(2,5,3). \quad \textbf{unicorn}(3,4). \quad \textbf{find}(5,6,4). \quad \textbf{John}(6). \tag{17}$$

(Note that here, we are using database constants $1, 2, 3, \ldots$, rather than Datalog variables, to name nodes.) The external nodes of the hypergraph (of which there is only one in this example) determine the query:

$$?-\mathsf{S}(1). \tag{18}$$

The λ-term (8) is in the language of the CFLG (9). Correspondingly, the answer to the query (18) against the program in (16) and the database in (17) is "yes". Figures 5 and 6 show the associated CFLG and Datalog derivation trees.

The situation becomes more complex when the input λ-term contains more than one occurrence of the same constant. Such is the case with the λ-term (19) (this is

$$\begin{array}{c}
\text{S}(1)\\
\overbrace{\rule{4cm}{0pt}}\\
\text{NP}(1,1,6) \quad\quad \text{VP}(1,6)
\end{array}$$

Figure 6: The Datalog derivation tree for the query (18) against the database in (17) and the program in (16).

the λ-term associated with **John found and caught a unicorn** by the grammar (7)):

$$\exists(\lambda y.\wedge(\mathbf{unicorn}\ y)(\wedge(\mathbf{find}\ y\ \mathbf{John})(\mathbf{catch}\ y\ \mathbf{John}))). \tag{19}$$

Let us apply the same procedure to (19) as we did to (8). The hypergraph for (19) is the following:

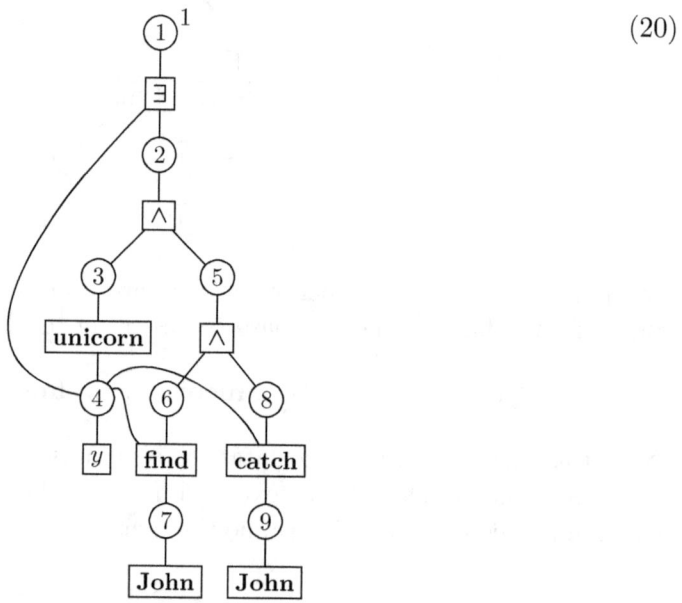

(20)

From this hypergraph, we would get the database (21) and the query (22):

$\exists(1,2,4).\ \wedge(2,5,3).\ \mathbf{unicorn}(3,4).\ \wedge(5,8,6).\ \mathbf{find}(6,7,4).\ \mathbf{John}(7).$ (21)
$\mathbf{catch}(8,9,4).\ \mathbf{John}(9).$

$$?\text{−}\mathsf{S}(1). \tag{22}$$

PARSING AND GENERATION AS DATALOG QUERY EVALUATION

$$\text{S}\big((\lambda u.u\ \text{John})(\lambda x.(\lambda uv.\exists(\lambda y.\wedge(uy)(vy)))\ \text{unicorn}\ (\lambda y.(\lambda yx.\wedge(\text{find}\ y\ x)(\text{catch}\ y\ x))\ y\ x))\big)$$

$$\text{NP}(\lambda u.u\ \text{John}) \quad \text{VP}\big(\lambda x.(\lambda uv.\exists(\lambda y.\wedge(uy)(vy)))\ \text{unicorn}\ (\lambda y.(\lambda yx.\wedge(\text{find}\ y\ x)(\text{catch}\ y\ x))\ y\ x)\big)$$

$$\text{V}\big(\lambda yx.\wedge(\text{find}\ y\ x)(\text{catch}\ y\ x)\big) \quad \text{NP}\big((\lambda uv.\exists(\lambda y.\wedge(uy)(vy)))\ \text{unicorn}\big)$$

$$\text{V(find)} \quad \text{V(catch)} \quad \text{Det}\big(\lambda uv.\exists(\lambda y.\wedge(uy)(vy))\big) \quad \text{N(unicorn)}$$

Figure 7: The CFLG derivation tree for (19).

It turns out, however, that (21) is not the correct database corresponding to the input λ-term (19). Even though (19) is generated by the CFLG in (9) with the derivation tree in Figure 7, the answer to the query (22) against the database (21) and the program (16) is "no", as the reader can easily verify.

To obtain the desired database, we need to modify (20) by identifying the two hyperedges labeled by **John** and the nodes they are incident on, as follows:

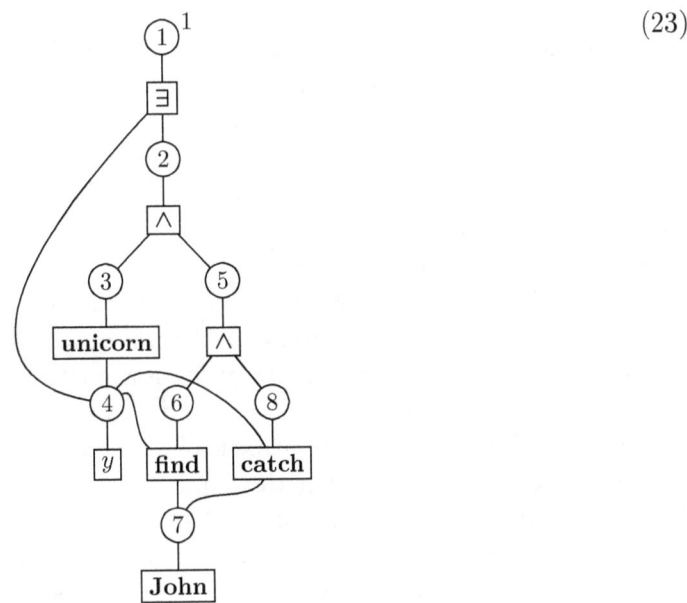

(23)

This gives the database (24).

$\exists(1,2,4).\ \wedge(2,5,3).\ \wedge(5,8,6).\ \textbf{unicron}(3,4).\ \textbf{find}(6,7,4).\ \textbf{John}(7).$ (24)
$\textbf{catch}(8,7,4).$

Against this database and the program in Figure 16, the query (22) is correctly answered "yes". Figure 8 shows the associated Datalog derivation tree for this query.

```
                               S(1)
           ┌─────────────────────┴─────────────────────┐
       NP(1,1,7)                                    VP(1,7)
           │                      ┌──────────────────┴──────────────────┐
        John(7)              V(5,7,4)                                NP(1,5,4)
                    ┌───────────┴───────────┐                ┌───────────┴───────────┐
                 ∧(5,8,6)   V(6,7,4)     V(8,7,4)       Det(1,5,4,3,4)            N(3,4)
                              │             │         ┌───────┴───────┐              │
                           find(6,7,4)  catch(8,7,4) ∃(1,2,4)      ∧(2,5,3)      unicorn(3,4)
```

Figure 8: The Datalog derivation tree for the query (22) against the database (24) and the program in (16).

Note that the database (24) can also be obtained from the following non-β-normal λ-term, which β-reduces to (19):

$$\exists(\lambda y.\wedge(\mathbf{unicorn}\ y)((\lambda x.\wedge(\mathbf{find}\ y\ x)(\mathbf{catch}\ y\ x))\ \mathbf{John})). \tag{25}$$

The hypergraph for (25) is identical to (23), except for the presence of an additional hyperedge labeled by x (incident on the node named "7").

The general rule is that the input λ-term should first be β-expanded to an almost linear λ-term that is the most "compact" in the sense of containing the fewest occurrences of constants, before the hypergraph and the associated database and query are extracted out of it. This explains why the two hyperedges labeled by $\wedge$ in (23) cannot be identified, because there is no almost linear λ-term with just one occurrence of $\wedge$ that β-reduces to (19). On the level of hypergraphs, the necessary operation is similar to the conversion of term graphs to their *fully collapsed* form (see [59]). This is by no means an accurate formulation, however, because the "fully collapsed form" does not always correspond to an almost linear λ-term, and there is some subtlety involved in the treatment of hyperedges labeled by bound variables.[12] A precise method of converting the input λ-term N to the desired almost linear λ-term $N°$ will be given by Algorithm 1 in Section 3.7.[13]

Note that the way we obtain a database from an input λ-term generalizes the standard database representation of a string: from the λ-term encoding

---

[12]For example, the algorithm β-expands $d(b(\lambda u.a(uc)))(b(\lambda v.a(vc)))$ to $(\lambda x.dxx)(b(\lambda u.a(uc)))$, but does not β-expand $d(\lambda u.a(uc))(\lambda u.a(uc))$ to $(\lambda x.dxx)(\lambda u.a(uc))$, which would correspond to the fully collapsed form.

[13]The input λ-terms we have used as examples are both almost linear. Since the class of almost linear λ-terms is not closed under β-reduction, a β-normal λ-term generated by an almost linear CFLG is not necessarily almost linear. Thus, in general, the input λ-term has to be β-expanded to an almost linear λ-term before any hypergraph can be obtained by the method outlined above, even when no constant occurs more than once in the input λ-term.

$/a_1 \ldots a_n/ = \lambda z.a_1^{o\to o}(\ldots(a_n^{o\to o}z)\ldots)$ of a string $a_1 \ldots a_n$, we obtain the database $\{a_1(0,1), \ldots, a_n(n-1,n)\}$ and the query ?− $\mathsf{S}(0,n)$, as the reader may verify.

2.3 An outline of the proof of correctness

Let us give a rough idea of the proof of correctness of our reduction, presented informally in Section 2.2.

For the reader familiar with the notion of a *principal typing*, it should be clear how the hypergraph graph(M) for an almost linear λ-term M corresponds to a principal (i.e., most general) typing of M, where occurrences of constants are treated like mutually distinct free variables. For instance, corresponding to the hypergraph (20) for the almost linear λ-term (19), we have the principal typing

$$\exists : (4 \to 2) \to 1, \wedge_1 : 3 \to 5 \to 2, \mathbf{unicorn} : 4 \to 3, \wedge_2 : 6 \to 8 \to 5,$$
$$\mathbf{find} : 4 \to 7 \to 6, \mathbf{John}_1 : 7, \mathbf{catch} : 4 \to 9 \to 8, \mathbf{John}_2 : 9 \Rightarrow 1. \quad (26)$$

Note that distinct occurrences of $\wedge$ and of **John** in (19) are regarded as distinct free variables. In the case of the λ-term (25), which has just one occurrence of **John**, we have

$$\exists : (4 \to 2) \to 1, \wedge_1 : 3 \to 5 \to 2, \mathbf{unicorn} : 4 \to 3, \wedge_2 : 6 \to 8 \to 5,$$
$$\mathbf{find} : 4 \to 7 \to 6, \mathbf{John} : 7, \mathbf{catch} : 4 \to 7 \to 8 \Rightarrow 1 \quad (27)$$

as its principal typing, corresponding to (23).[14]

What is special about almost linear λ-terms is that when an almost linear λ-term with constants (in η-long form) is "maximally compact" in the sense that it has no β-equal almost linear λ-term with fewer occurrences of constants, its principal typing exactly characterizes the set of almost linear λ-terms (in η-long form) that are β-equal to it. More precisely, let M be such a maximally compact almost linear λ-term in η-long form and let $\Gamma \Rightarrow \alpha$ be its principal typing. Then we have the following equivalence for every almost linear λ-term M' in η-long form:

M' has a typing $\Gamma' \Rightarrow \alpha$ for some subset Γ' of Γ

if and only if M' is β-equal to M. (28)

---

[14]The exact correspondence between graph(M) and a principal typing of M requires M to be in η-long form. Note that this notion of typing of a λ-term with constants is different from the notion of typing expressed by judgments like (10), where constants have fixed, pre-assigned types. In the rigorous presentation of Section 3, typings like (26) and (27) will be replaced by typings of pure λ-terms that result by replacing distinct occurrences of constants by distinct free variables.

The main ingredients of the proof of this property of almost linear λ-terms are the following:

- A principal typing of an almost linear λ-term is *negatively non-duplicated* in the sense that each atomic type has at most one negative occurrence in it (cf. [2]).

- All λ-terms that share a negatively non-duplicated typing are $\beta\eta$-equal [3]. This is a generalization of the *Coherence Theorem* (see [56]).

- The leftmost β-reduction from an almost linear λ-term is non-erasing and *almost non-duplicating* in the sense that for each β-redex $(\lambda x.P)Q$ that is contracted, x can occur free more than once in P only when the type of x is atomic.

- If there is a non-erasing, almost non-duplicating β-reduction from a pure (i.e., constant-free) λ-term M to N, every typing of N is a typing of M. This is a generalization of the *Subject Exapnsion Theorem* (see [31]).

Now let **P** be the Datalog program constructed from the given almost linear CFLG $\mathscr{G}$, and let N be the input λ-term (in η-long β-normal form). Suppose that our algorithm first β-expands N to an almost linear λ-term N°. Let $\Gamma \Rightarrow \alpha$ be a principal typing of N°, and let D and $?-S(\overline{\alpha})$ be the database and query constructed from this typing.

Suppose that there is a Datalog derivation tree T for the query $?-S(\overline{\alpha})$ against the program **P** and the database D. Given the one-one correspondence between the rules of $\mathscr{G}$ and the rules of **P**, the Datalog derivation tree T determines a CFLG derivation tree T'. (See Figures 5, 6, 7, 8 for examples.) The former, however, contains more information than the latter. Each ground instance ρ of a Datalog rule used in T corresponds to a typing of the λ-term in the corresponding CFLG rule. For instance, the ground instance

$$\mathsf{V}(5,7,4) :- \wedge(5,8,6), \mathsf{V}(6,7,4), \mathsf{V}(8,7,4)$$

of the third rule of (16) that is used in the Datalog derivation tree in Figure 8 gives the following typing judgment:

$$\vdash \wedge : 6 \to 8 \to 5,\ X_1 : 4 \to 7 \to 6,\ X_2 : 4 \to 7 \to 8 \Rightarrow \lambda yx.\wedge(X_1 yx)(X_2 yx) : 4 \to 7 \to 5.$$

Piecing together all these typing judgments corresponding to ground instances of rules used in T gives a typing judgment

$$\vdash \Gamma' \Rightarrow P : \alpha',$$

where P is the (non-β-normal) almost linear λ-term at the root node of T'. Since α' and Γ' correspond to the root node and the leaf nodes of T, respectively, we must have $\alpha' = \alpha$ and $\Gamma' \subseteq \Gamma$. By the special property (28) of almost linear λ-terms, it follows that P is $\beta\eta$-equal to N° and hence to N, which implies that T' is a derivation tree for N.

Let us now consider the converse direction and suppose that a derivation tree T' of $\mathscr{G}$ has its root node labeled by $S(P)$ and P β-reduces to N. By the one-one correspondence between the rules of $\mathscr{G}$ and the rules of $\mathbf{P}$, T' determines a "skeletal" Datalog derivation tree made up of non-ground instances of rules of $\mathbf{P}$, where predicates have Datalog variables as arguments, instead of database constants. The question is whether one can replace these Datalog variables with database constants from D in such a way that leaf nodes will correspond to facts in D, so that the derivation tree will become a derivation tree for $S(\overline{\alpha})$ against $\mathbf{P}$ and D. This is possible precisely when P has a typing $\Gamma' \Rightarrow \alpha$ with $\Gamma' \subseteq \Gamma$. By the special property (28) again, this must be so since P is almost linear and is $\beta\eta$-equal to N and hence to N°.

2.4 The scope of the present method

The present method of reduction to Datalog is directly applicable only to formalisms expressible in almost linear CFLGs. Almost linear λ-terms suffice to represent formulas in a logical language with quantification over individual variables only, so when the meaning representation language used in a surface realization problem is such a language, the input to the corresponding CFLG recognition problem will always be an almost linear λ-term. For instance, in the extensional subfragment of Montague's [57] fragment of English, the translations of English sentences will fall within such a language. Consequently, it is possible to extend the grammar (7) to one that covers a large portion of Montague's [57] fragment while keeping the semantic half of the grammar almost linear. However, even when almost linear λ-terms suffice to encode the target logical forms, we sometimes need grammar rules that are not almost linear.[15]

For example, suppose we add to the synchronous grammar in Figure 4 the following rules:

$\mathsf{NP}(\lambda z.Y_1(/\mathsf{and}/(Y_2 z)), \lambda u.\wedge^{t \to t \to t}(X_1(\lambda x.ux))(X_2(\lambda x.ux)))$:− $\mathsf{NP}(Y_1, X_1), \mathsf{NP}(Y_2, X_2)$.
$\mathsf{VP}(/\mathsf{sang}/, \mathbf{sing}^{e \to t})$.
$\mathsf{NP}(/\mathsf{Bill}/, \lambda u.u\, \mathbf{Bill}^e)$.

---

[15]This is already evidenced in the grammar of Montague [57], which has a rule similar to the first of the three rules below.

With these rules, the grammar can now generate John and Bill sang, with the logical form
$$\wedge(\text{sing John})(\text{sing Bill}). \tag{29}$$

Let us see how we might convert to Datalog the "semantic half" of the three synchronous rules above:

$$\mathsf{NP}(\lambda u.\wedge^{t\to t\to t}(X_1(\lambda x.ux))(X_2(\lambda x.ux))) :\!- \mathsf{NP}(X_1), \mathsf{NP}(X_2). \tag{30}$$
$$\mathsf{VP}(\mathbf{sing}^{e\to t}).$$
$$\mathsf{NP}(\lambda u.u\,\mathbf{Bill}^e).$$

Recall that $f(\mathsf{NP}) = (e \to t) \to t$, so the type of the variables X_1 and X_2 in the first rule of (30) are $(e \to t) \to t$ and the type of u is $e \to t$. This means that the λ-term M on the left-hand side of this rule is not almost linear. The method we described was not meant to apply to a case like this, but suppose we extend it to cover this case. We would get the following hypergraph.[16]

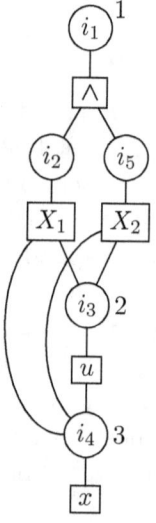

Thus, from the three CFLG rules in (30), we get the following Datalog rules:

$$\mathsf{NP}(i_1, i_3, i_4) :\!- \wedge(i_1, i_5, i_2), \mathsf{NP}(i_2, i_3, i_4), \mathsf{NP}(i_5, i_3, i_4). \tag{31}$$
$$\mathsf{VP}(i_1, i_2) :\!- \mathbf{sing}(i_1, i_2).$$
$$\mathsf{NP}(i_1, i_1, i_2) :\!- \mathbf{Bill}(i_2).$$

---

[16] This graph corresponds to the principal typing of the λ-term M.

As for the λ-term (29), there are two conceivable hypergraphs that can be associated with it:

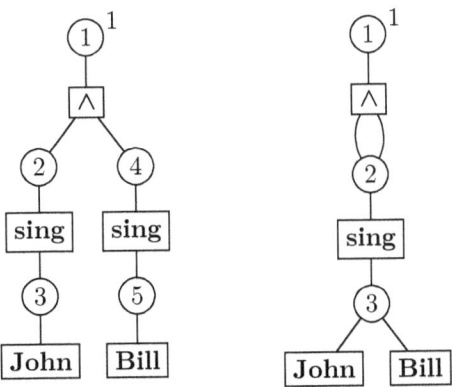

The first graph is what we obtain with the method described above. The second graph is the result of identifying the two edges labeled by **sing** and the nodes they are incident on. The corresponding databases are:

$$\wedge(1,4,2). \quad \text{sing}(2,3). \quad \text{sing}(4,5). \quad \text{John}(3). \quad \text{Bill}(5). \tag{32}$$

$$\wedge(1,2,2). \quad \text{sing}(2,3). \quad \text{John}(3). \quad \text{Bill}(3). \tag{33}$$

Against the database (32) and the program consisting of the rules in (16) and (31), the query

$$?- S(1).$$

is answered "no". Against the database (33) and the same program, the same query is answered "yes", but there are too many Datalog derivation trees for this query. In addition to the correct one corresponding to the CFLG derivation tree for (29), there are three others, corresponding to the CFLG derivation trees for the following λ-terms:

$$\wedge(\text{sing John})(\text{sing John})$$
$$\wedge(\text{sing Bill})(\text{sing John})$$
$$\wedge(\text{sing Bill})(\text{sing Bill})$$

This means that if (33) is used to solve the task of finding sentences expressing meaning (29), the output obtained contains not just John and Bill sang, but also John and John sang, Bill and John sang, and Bill and Bill sang. Thus, neither (32) nor (33) gives a correct reduction of surface realization to Datalog query evaluation.

As for applications to parsing and recognition, the present method directly applies to string-generating grammars with no copying operation, like multiple context-free grammars, but not to formalisms like macro grammars [24] and parallel multiple context-free grammars [66], where derivations involve copying of strings. To represent grammar rules that duplicate strings, a CFLG must use multiple occurrences of the same variable of type $o \to o$, and so cannot be almost linear. An almost linear CFLG can represent tree grammars with copying operations, however, because trees are represented by λ-terms of atomic type o. It turns out that this provides an indirect way of applying the present method to grammars with string copying, using as input a representation of a finite set of trees that yield a given input string. This point will be elaborated in Section 4.2.

2.5 The present approach to generation

In this section, I clarify some basic assumptions I make in this work about the meaning representation language and the task of surface realization. These assumptions do not concern the formal result about the reduction of almost linear CFLGs to Datalog, but rather the kind of application of the formal result to grammars for natural language I have in mind.

In Montague's [57] work, the meaning representation language, which incorporates a form of λ-calculus, is just a convenient tool used to give a model-theoretic semantics to the object language, and can in principle be dispensed with. In contrast, this work assumes that the level of semantic representation is crucial and that grammar rules specifically refer to λ-terms as structured, "syntactic" objects. Any computation on meanings must be performed on some form of representation or other; using λ-terms as semantic representations seems to be a convenient choice.

The formalism of λ-calculus can be used in different ways for different purposes. The example grammar I have given uses λ-terms to more or less directly represent formulas of the language of some logic (subsuming at least first-order logic), using appropriately typed constants for logical and non-logical symbols of the language. Binding of a variable by a quantifier is represented by an application of the constant representing the quantifier to a λ-abstract.[17] A pleasant consequence of this is that two formulas that are related by renaming of bound variables translate into α-equivalent λ-terms and are treated as the same. However, since constants are just uninterpreted symbols, all other cases of logically equivalent pairs of formulas come out as distinct λ-terms.

This use of λ-calculus, as an alternative syntax for the language of some logic, is of course not the only way to use λ-calculus as a meaning representation language. For example, logical connectives and quantifiers may be defined in terms of equality (at different types), à la Henkin [29].[18] It is also common to represent truth values,

---

[17]Following Church [14], Barwise and Cooper [7], and Lloyd [53], among many others.

[18]For example, the universal quantifier (over individuals) may be defined as $\forall^{(e \to t) \to t} =$

Boolean functions, etc., with pure (i.e., constant-free) λ-terms, using $\tau \to \tau \to \tau$ as the type of truth values.[19] One can even represent finite models as λ-terms and cast sentence meanings as functions from finite models to truth values [30]. These more sophisticated uses of λ-calculus, however, almost always take us outside of the realm of almost linear λ-terms, so the main result of this paper will not be applicable.[20]

The main result of this paper applies to surface realization as understood to be the problem of finding a sentence such that the logical form associated with it by the grammar exactly matches the input logical form. This means that the question of whether or not the input logical form is surface realizable depends on the exact shape of the input. If we take our example grammar (7), the answer is different for each of the following pairs:

(34) a. $\exists(\lambda y.\wedge(\text{\textbf{unicorn }} y)(\text{\textbf{find }} y \text{ \textbf{John}}))$
b. $\exists(\lambda y.\wedge(\text{\textbf{find }} y \text{ \textbf{John}})(\text{\textbf{unicorn }} y))$

(35) a. $\exists(\lambda y.\wedge(\text{\textbf{unicorn }} y)(\wedge(\text{\textbf{find }} y \text{ \textbf{John}})(\text{\textbf{catch }} y \text{ \textbf{John}})))$
b. $\exists(\lambda y.\wedge(\wedge(\text{\textbf{unicorn }} y)(\text{\textbf{find }} y \text{ \textbf{John}}))(\text{\textbf{catch }} y \text{ \textbf{John}}))$

(36) a. $\exists(\lambda x.\wedge(\text{\textbf{man }} x)(\exists(\lambda y.\wedge(\text{\textbf{unicorn }} y)(\text{\textbf{find }} y \text{ } x))))$
b. $\exists(\lambda y.\wedge(\text{\textbf{unicorn }} y)(\exists(\lambda x.\wedge(\text{\textbf{man }} x)(\text{\textbf{find }} y \text{ } x))))$

(37) a. $\exists(\lambda y.\wedge(\text{\textbf{man }} y)(= y \text{ \textbf{John}}))$
b. **man John**

It is generally agreed in computational linguistics that the input to surface realization should not be informed by the particularities of the grammar and that ideally, both members of these pairs should lead to the same result, since they are obviously logically equivalent [68]. While accounting for the full range of logical equivalence is clearly intractable, capturing commutativity and associativity of conjunction is considered particularly important in machine translation applications, and partly for this reason it is popular in computational linguistics to use a "flat" and "unordered" meaning representation language where equivalences like (34) and (35) are built in (see, e.g., [15] or [51]). Another motivation for flat semantics is the need for compact "underspecified" representation of a range of different scope readings of sentences with multiple scope-taking operators. Generation from flat semantics

---

$\lambda u^{e \to t}. =^{(e \to t) \to (e \to t) \to t} u \, (\lambda x^e. =^{e \to e \to t} x \, x)$

[19]The truth values "true" and "false" are encoded by the λ-terms $\lambda xy.x$ and $\lambda xy.y$, respectively. These are known as *Church Booleans*.

[20]Surface realization in such a setting is still decidable since Salvati [65] proves that recognition is decidable for general CFLGs. It is an open question how far the class of almost linear λ-terms can be extended without making the resulting CFLG recognition problem intractable.

has been shown to be NP-hard [50], however, so adopting a flat representation language is (for all we know) incompatible with polynomial-time algorithms for surface realization.

Typed λ-terms, with "hierarchical" and "ordered" structures, do not seem to be particularly well suited to encoding of flat semantics, but it is possible to adapt to λ-calculus the idea of Koller et al. [49], who have proposed to use regular tree grammars that generate finite sets of trees as a formalism for underspecification. Trees cannot properly represent variable binding, so a reasonably compact description of a "regular" set of λ-terms will improve upon Koller et al.'s [49] proposal.[21] It turns out that in certain cases, a *database* serves as such a compact representation. In Section 4.2, I present a result extending the main result to handle certain regular sets of λ-terms as input to the recognition problem for almost linear context-free λ-term grammars. Notwithstanding this possibility of accommodating underspecification, I believe that thorough understanding of the simpler problem of "exact" surface realization should take precedence.

The underlying theme of this work is that the problem of surface realization can and should be studied in the style of formal language theory, just like parsing. For this purpose, the problem of surface realization should be formulated in abstract, general terms. The primary goals of any such study would be to identify the computational complexity class for which the problem is complete, and to provide natural, efficient algorithms (insofar as is allowed by the complexity lower bound) to solve the problem. The formalism in which the input to surface realization is encoded should be sufficiently rich to support constructs (e.g., variable binding) that are necessary to express natural language meaning, but should not be tied to one particular logical language. Typed λ-calculus seems to fit this role very well; it has a wide variety of uses, its formal properties have been extensively studied, and its use is also fairly common in computational linguistics. All other things being equal, a general, mathematically elegant, and well-understood formalism should be preferred over ad hoc, application-specific, ill-understood alternatives.

3 Formal development

3.1 Preliminaries

3.1.1 Datalog

A *database schema* is a pair $\mathcal{D} = (R, U)$, where R is a finite set of predicates, each of fixed arity, and U is a (possibly infinite) set of database constants. A *ground fact*

---

[21] See [64] for a definition of a regular or *recognizable* set of typed λ-terms.

over $\mathcal{D}$ is
$$p(\vec{s}),$$
where p is a predicate in R of arity k, and $\vec{s}$ is a k-tuple of constants in U, for some k. A *database* over $\mathcal{D}$ is a finite set of ground facts over $\mathcal{D}$. If D is a database, the *universe* of D, written U_D, is the finite set of constants appearing in D.

We assume that we are given a countably infinite supply of variables. A *Datalog program* over R is a finite set of *rules*, which are function-free definite clauses of the form
$$p_0(\vec{x}_0) :- p_1(\vec{x}_1), \ldots, p_n(\vec{x}_n),$$
where $n \geq 0$, p_i are predicates, each of fixed arity, and $\vec{x}_i$ are tuples of variables (not necessarily distinct) of appropriate length, matching the predicate's arity. A predicate together with its arguments constitutes an *atom*. The left-hand side of a rule (the part to the left of :−) is called the *head*, and the right-hand side the *body*. The atoms that constitute the body are the *subgoals* of the rule.[22] The predicates in a program **P** are divided into the *intensional* predicates and the *extensional* predicates. A predicate is an intensional predicate if it appears in the head of some rule, and an extensional predicate otherwise. An *extensional database* for **P** is a database D for a schema $\mathcal{D} = (R, U)$ for some U, where R consists of the extensional predicates of **P**. We call ground facts in an extensional database *extensional facts*. We follow the logic programming parlance and call a negative Horn clause a *query*.[23] In this paper, we are mainly interested in simple (i.e., non-conjunctive) ground queries of the form
$$?- p(\vec{s}),$$
where $\vec{s}$ is a tuple of constants from U_D (of appropriate length).

Given a Datalog program **P** and an extensional database D, a ground fact $p(\vec{s})$ is *derivable* from **P** and D, written
$$\mathbf{P} \cup D \vdash p(\vec{s}),$$
if and only if either $p(\vec{s}) \in D$ or there is a ground instance
$$p(\vec{s}) :- p_1(\vec{s}_1), \ldots, p_n(\vec{s}_n)$$

---

[22]In Datalog, it is often required that the variables in the head of a rule all appear in the body, but we do not assume this restriction. In particular, we allow rules with empty body (i.e., facts) in Datalog programs.

[23]In relational database theory and finite model theory, the term *query* sometimes means a function that maps a finite relational structure to a finite relational structure. A query in this sense may be expressed by a pair $(\mathbf{P}, R')$ consisting of a Datalog program **P** and a subset R' of its intensional predicates [16]. See [1] for a similar use of the term "datalog query". The logic programming parlance was used by Ullman [77] in the context of Datalog query evaluation.

of a rule in **P** such that
$$\mathbf{P} \cup D \vdash p_i(\vec{s}_i)$$
for each $i = 1, \ldots, n$. A *derivation tree* is a tree whose nodes are labeled by ground facts in accordance with the above inductive definition. That is to say, a derivation tree for $p(\vec{s})$ from **P** and D is either a tree with a single node labeled by an extensional fact $p(\vec{s}) \in D$, or a tree of the form

$$\begin{array}{c} p(\vec{s}) \\ \overbrace{T_1 \cdots T_n} \end{array}$$

where there exists some ground instance $p(\vec{s}) :\!- p_1(\vec{s}_1), \ldots, p_n(\vec{s}_n)$ of a rule in **P** and T_i is a derivation tree for $p_i(\vec{s}_i)$ for $i = 1, \ldots, n$.

It is easy to see that for a fixed Datalog program **P**, the problem of determining, given a database D and a fact q, whether $\mathbf{P} \cup D \vdash q$ holds can be solved in polynomial time in the size of (D, q). For some Datalog program, this problem is known to be P-complete (see [48] for an overview of complexity issues). Among the most basic polynomial-time algorithms for this problem are *naive* and *seminaive* bottom-up evaluation (see [1] or [76]). In these methods, derived facts that share the same predicate are grouped together into a relation, and relational algebra operations are used to expedite the iterative, bottom-up computation of the relations. In the application of Datalog to recognition and parsing, however, the number of derivable facts is usually not large, so it is not so unreasonable to process one fact at a time. Under this simplification, seminaive bottom-up evaluation can be expressed by the following pseudocode. If π is a rule, we write ground(π, U) to denote the set of ground instances of π using only constants from U.

1: **procedure** SEMINAIVE(**P**, D)
2:     $D^0 \leftarrow \varnothing$
3:     $D^1 \leftarrow D \cup \{\, p(\vec{s}) \mid p(\vec{s}) \in \text{ground}(\pi, U_D) \text{ for some } \pi \in \mathbf{P} \,\}$
4:     $\Delta^1 \leftarrow D^1$
5:     $i \leftarrow 1$
6:     **while** $\Delta^i \neq \varnothing$ **do**
7:     $\displaystyle \Delta^{i+1} \leftarrow \left\{ p(\vec{s}) \,\middle|\, \begin{array}{l} \pi = p(\vec{x}) :\!- p_1(\vec{x}_1), \ldots, p_n(\vec{x}_n) \in \mathbf{P}, \\ p_1(\vec{s}_1), \ldots, p_{j-1}(\vec{s}_{j-1}) \in D^i,\ p_j(\vec{s}_j) \in \Delta^i, \\ p_{j+1}(\vec{s}_{j+1}), \ldots, p_n(\vec{s}_n) \in D^{i-1} \text{ for some } j \in [1, n], \\ \text{and } p(\vec{s}) :\!- p_1(\vec{s}_1), \ldots, p_n(\vec{s}_n) \in \text{ground}(\pi, U_D) \end{array} \right\} - D^i$
8:     $D^{i+1} \leftarrow D^i \cup \Delta^{i+1}$
9:     $i \leftarrow i + 1$
10:    **end while**
11:    **return** D^i

12: **end procedure**

In this algorithm, D^i is the set of ground facts whose derivation trees have minimal height $i - 1$.

Derivation trees are assembled from ground instances of rules. If, in addition to derived ground facts, we record ground instances of rules used to derive facts, we can obtain a packed representation of all derivation trees for ground facts derivable from the given program and the input database:[24]

1: **procedure** SEMINAIVE-PARSE($\mathbf{P}, D$)
2: $\quad D^0 \leftarrow \varnothing$
3: $\quad D^1 \leftarrow D \cup \{ p(\vec{s}) \mid p(\vec{s}) \in \text{ground}(\pi, U_D) \text{ for some } \pi \in \mathbf{P} \}$
4: $\quad G^1 \leftarrow D^1$
5: $\quad \Delta^1 \leftarrow D^1$
6: $\quad i \leftarrow 1$
7: $\quad$ **while** $\Delta^i \neq \varnothing$ **do**
8: $\quad\quad \Delta^{i+1} \leftarrow \left\{ p(\vec{s}) \;\middle|\; \begin{array}{l} \pi = p(\vec{x}) :\!- p_1(\vec{x}_1), \ldots, p_n(\vec{x}_n) \in \mathbf{P}, \\ p_1(\vec{s}_1), \ldots, p_{j-1}(\vec{s}_{j-1}) \in D^i,\; p_j(\vec{s}_j) \in \Delta^i, \\ p_{j+1}(\vec{s}_{j+1}), \ldots, p_n(\vec{s}_n) \in D^{i-1} \text{ for some } j \in [1, n], \\ \text{and } p(\vec{s}) :\!- p_1(\vec{s}_1), \ldots, p_n(\vec{s}_n) \in \text{ground}(\pi, U_D) \end{array} \right\} - D^i$
9: $\quad\quad G^{i+1} \leftarrow \left\{ \pi' \;\middle|\; \begin{array}{l} \pi = p(\vec{x}) :\!- p_1(\vec{x}_1), \ldots, p_n(\vec{x}_n) \in \mathbf{P}, \\ p_1(\vec{s}_1), \ldots, p_{j-1}(\vec{s}_{j-1}) \in D^i,\; p_j(\vec{s}_j) \in \Delta^i, \\ p_{j+1}(\vec{s}_{j+1}), \ldots, p_n(\vec{s}_n) \in D^{i-1} \text{ for some } j \in [1, n], \\ \text{and } \pi' = p(\vec{s}) :\!- p_1(\vec{s}_1), \ldots, p_n(\vec{s}_n) \in \text{ground}(\pi, U_D) \end{array} \right\} \cup G^i$
10: $\quad\quad D^{i+1} \leftarrow D^i \cup \Delta^{i+1}$
11: $\quad\quad i \leftarrow i + 1$
12: $\quad$ **end while**
13: $\quad$ **return** G^i
14: **end procedure**

In the implementation of SEMINAIVE-PARSE, the operations in lines 8 and 9 should be performed simultaneously. In this algorithm, the final value of G^i records all rule instances whose subgoals are derivable facts, and constitutes a *propositional Horn clause program*.[25]

There is a natural way to associate an *alternating Turing machine* operating in logarithmic space with each Datalog program [67, 48], and this is useful for the complexity analysis of Datalog programs. Alternating Turing machines (ATMs) [13] are a generalization of non-deterministic Turing machines. The set of states of

---

[24]The algorithms SEMINAIVE and SEMINAIVE-PARSE can also be written in the style of *chart parsing* [69, 71]. The set Δ^i will correspond to the agenda. See Section 4.3 below.

[25]At the end of the execution of SEMINAIVE-PARSE, we have $D^i = D^{i-1}$, but not necessarily $G^i = G^{i-1}$; it would require one more iteration for G^i to stabilize.

an ATM is partitioned into *existential* and *universal* sates. If a configuration is in an existential state, at least one of the successor configurations must lead to acceptance, whereas if a configuration is in a universal sate, all of its successor configurations must lead to acceptance. A *computation tree* of an ATM $\mathscr{M}$ is a finite rooted directed tree whose nodes are configurations of $\mathscr{M}$ such that the root node is an initial configuration, each existential configuration has just one of its successor configurations as its child, and each universal configuration has all of its successor configurations as its children. An *accepting* computation tree is a computation tree whose leaves are all accepting configurations. An ATM $\mathscr{M}$ operates (simultaneously) in space $S(n)$ and tree size $Z(n)$ if on each input x of length n accepted by $\mathscr{M}$, there is an accepting computation tree of size at most $Z(n)$ in which each configuration uses at most space $S(n)$. Ruzzo [61] characterizes the complexity class LOGCFL as the class of problems for which there is an ATM operating in logarithmic space and in polynomial tree size.

A log-space-bounded ATM $\mathscr{M}_\mathbf{P}$ simulating a Datalog program $\mathbf{P}$ may behave as follows. The input to $\mathscr{M}_\mathbf{P}$ is a pair (D, q) of an extensional database D for $\mathbf{P}$ and a ground fact q; $\mathscr{M}_\mathbf{P}$ accepts (D, q) if and only if $\mathbf{P} \cup D \vdash q$. This ATM uses $k+1$ work tapes, where k is at least as large as the maximal arity of the predicates in $\mathbf{P}$ and the maximal number of variables in rules of $\mathbf{P}$. Each of the first k work tapes serves as a pointer to a position on the input tape where an occurrence of a constant starts. The last work tape is used to check identity of two occurrences of constants (which we assume to be coded as binary strings). Part of $\mathscr{M}_\mathbf{P}$'s finite control is used to store a predicate or a rule in $\mathbf{P}$. We call the combination of this part of the finite control and the first k work tapes the "storage area". The storage area of $\mathscr{M}_\mathbf{P}$ either stores a ground fact $p(\vec{s})$, using the work tapes to store the sequence $\vec{s}$ of constants, or a ground instance of a rule $\pi = p(\vec{x}) \coloneq p_1(\vec{x}_1), \ldots, p_n(\vec{x}_n)$, using the work tapes to store a ground substitution for the variables in π. The machine starts by copying the ground fact q on the input tape onto its storage area. Whenever $\mathscr{M}_\mathbf{P}$ has a ground fact q' in the storage area, it tries to verify $\mathbf{P} \cup D \vdash q'$. If q' is an extensional fact, it verifies that q' appears in the database on the input tape and accepts. If q' is an intensional fact, the machine uses existential branching and guesses a ground instance $\pi\theta$ of a rule $\pi = p(\vec{x}) \coloneq p_1(\vec{x}_1), \ldots, p_n(\vec{x}_n)$ in $\mathbf{P}$ whose head matches q', and places $\pi\theta$ in the storage area. The machine then uses universal branching and for all $i = 1, \ldots, n$, places $p_i(\vec{x}_i)\theta$ in the storage area, and repeats the procedure. It should be clear that if there is a derivation tree T for $\mathbf{P} \cup D \vdash q$, then the ATM $\mathscr{M}_\mathbf{P}$ on input (D, q) has an accepting computation tree of size $|T| \cdot O(f(n))$, where $|T|$ is the size of T, f is a polynomial, and n is the size of the input (D, q).

Lemma 3.1. *Let $\mathbf{P}$ be a Datalog program and $g(n)$ be a polynomial. The following* *problem is in LOGCFL:*

$$\{\,(D, q, \mathbf{1}^m) \mid \text{there is a derivation tree for } \mathbf{P} \cup D \vdash q \text{ of size} \leq g(m)\,\}$$

Proof. The idea is from [26]. We modify $\mathcal{M}_\mathbf{P}$ by including bounds on the size of Datalog derivation trees in each configuration. The modified ATM starts by computing $g(m)$. This computation and the storage of the resulting value (in binary) can both be done within logarithmic space. When the machine is in a configuration storing an extensional fact q' and a bound b (a natural number in binary), it checks that q' appears in D and $b \geq 1$, and accepts. When the machine is in a configuration storing an intensional fact $p(\vec{s})$ and a bound b, it checks that $b > 1$ and guesses a ground instance $p(\vec{s}) := p_1(\vec{s}_1), \ldots, p_n(\vec{s}_n)$ of some rule, together with bounds $b_1, \ldots, b_n$ on the size of the derivations trees for $p_1(\vec{s}), \ldots, p_n(\vec{s}_n)$, such that $b_1 + \cdots + b_n = b - 1$. It then uses universal branching to write $p_i(\vec{s}_i)$ and b_i in the storage area and try to find a derivation tree for $p_i(\vec{s}_i)$ of size $\leq b_i$. It is clear that the size of any accepting computation tree of this ATM on input of size n is bounded by some polynomial in n. □

We call a node in a derivation tree an *extensional node* if it is labeled by an extensional fact (i.e., facts from the database), and an *intensional node* otherwise. A derivation tree is called *tight* [79] if no fact occurs more than once on any of its paths. Note that whenever T is a derivation tree for $\mathbf{P} \cup D \vdash p(\vec{s})$ that is not tight, one can turn T into a tight derivation tree for $\mathbf{P} \cup D \vdash p(\vec{s})$ by deleting some nodes from T.

The following elementary lemma will be useful later.

Lemma 3.2. *Let $\mathbf{P}$ be a Datalog program. Then there is a polynomial $g(n)$ such that whenever there is a derivation tree for $\mathbf{P} \cup D \vdash p(\vec{s})$ with l extensional nodes, there is a derivation tree $\mathbf{P} \cup D \vdash p(\vec{s})$ with $n \leq l$ extensional nodes whose size does not exceed $g(n)$.*

Proof. Let k be the number of intensional predicates in $\mathbf{P}$, r be the maximal arity of intensional predicates in $\mathbf{P}$, and m be the maximal number of subgoals of rules in $\mathbf{P}$.

If p is an extensional predicate, $p(\vec{s})$ must be in D and there is a one-node derivation tree for $\mathbf{P} \cup D \vdash p(\vec{s})$. In the following, we assume that p is an intensional predicate.

We first show that there is a constant c (depending on $\mathbf{P}$) such that if $\mathbf{P} \vdash p(\vec{s})$, then there is a derivation tree for $p(\vec{s})$ with at most c nodes. (Note that $\mathbf{P} \vdash p(\vec{s})$ means that $p(\vec{s})$ is derivable without using any extensional facts.) Let T be a smallest derivation tree for $\mathbf{P} \vdash p(\vec{s})$. Without loss of generality, we can assume that all constants that appear in T appear in $p(\vec{s})$, so that there are at most r of them. This is because if T contains other constants, they can be safely replaced by constants in $\vec{s}$. Since T must be a tight derivation tree, the height of T is bounded by $kr^r - 1$. Therefore, the size of T is bounded by m^{kr^r} (if $m \geq 2$) or kr^r (if $m \leq 1$).

Now suppose $\mathbf{P} \cup D \vdash p(\vec{s})$ and let T be a smallest derivation tree for $\mathbf{P} \cup D \vdash p(\vec{s})$ with $n \leq l$ extensional nodes. As before, we can assume without loss of generality that all constants in T occur in $p(\vec{s})$ or in facts labeling extensional nodes, so that there are at most $(n+1)r$ of them. The intensional nodes of T may be divided into the following three types:

Type 0 Intensional nodes that are not ancestors of any extensional nodes.

Type 1 Intensional nodes that have just one child that is an ancestor of some extensional node.

Type 2 Intensional nodes that have two or more children that are ancestors of extensional nodes.

Since the case of $n = 0$ has already been taken care of, assume $n \geq 1$. It is easy to see that the number of intensional nodes of type 2 is at most $n - 1$.

To find a bound on the number of type 1 nodes, note first that all children of type 1 nodes are type 0 nodes, except one, which is either an extensional node, a type 1 node, or a type 2 node. We call two type 1 nodes *equivalent* if they are related by the smallest equivalence relation extending the child-of relation restricted to type 1 nodes. Each equivalence class of type 1 nodes is linearly ordered by the child-of relation, and its minimal element is the parent of an extensional node or of a type 2 node. Since T must be tight by the minimality of T, the size of each equivalence class of type 1 nodes cannot exceed $k((n+1)r)^r$. Since there are at most $2n - 1$ equivalence classes of type 1 nodes, the number of type 1 nodes is bounded by $(2n - 1)k((n+1)r)^r$.

We finally turn to type 0 nodes. Note that all children of type 0 nodes are type 0 nodes. We call a type 0 node *maximal* if it is not a child of a type 0 node. Since we are assuming $n \geq 1$, any maximal type 0 node has a parent, which is either a type 1 node or a type 2 node. This implies that either there is no type 0 node or $m \geq 2$. Note that there may be up to $m - 1$ or $m - 2$ maximal type 0 nodes that share the same parent ($m - 1$ if the parent is type 1, $m - 2$ if the parent is type 2). Type 0 nodes that are not maximal are in a unique subtree rooted at a maximal type 0 node. Since we have seen that such a subtree has at most m^{kr^r} nodes, there are at most $((n-1)(m-2) + (2n-1)k((n+1)r)^r(m-1))m^{kr^r}$ type 0 nodes in total.

Therefore, the number of nodes of T is bounded by

$$2n - 1 + (2n-1)k((n+1)r)^r + ((n-1)(m-2) + (2n-1)k((n+1)r)^r(m-1))m^{kr^r}$$

when $n \geq 1$, which is $O(n^{r+1})$. □

3.1.2 Untyped λ-calculus with constants

In this and the next two sections, we review some basic concepts in λ-calculus we will need in what follows, introducing some nonstandard notions and notations along the way. For a more thorough introduction to the subject, see [6], [31], [73], or [32]. Like Sorensen and Urzyczyn [73], we make an explicit distinction between λ-terms and notations that represent them. It is important for our purposes to be completely precise about basic notions such as "subterm occurrence", "substitution", "β-reduction", "descendant", etc.

Following Statman [74], we consider a λ-term as an abstract object—namely, a binary tree equipped with some additional structure. We use a fixed scheme of naming nodes in a tree with strings of 0s and 1s. A *binary tree domain* is a finite, prefix-closed subset $\mathcal{T}$ of $\{0,1\}^*$ such that $w1 \in \mathcal{T}$ implies $w0 \in \mathcal{T}$. A node of the form wi with $i \in \{0,1\}$ is a child of the node w. A node is a leaf, a unary node, or a binary node according to whether it has 0, 1, or 2 children. We write $\mathcal{T}^{(0)}, \mathcal{T}^{(1)}, \mathcal{T}^{(2)}$, for the sets of leaves, unary nodes, and binary nodes, respectively, of $\mathcal{T}$. We write $v \leq w$ to mean v is a prefix of w, and $v < w$ to mean $v \leq w$ and $v \neq w$. The *lexicographic order* on $\{0,1\}^*$ is the strict total order $\prec$ extending $<$ such that $u0t \prec u1t'$ for every $u, t, t' \in \{0,1\}^*$. We say that v is *to the left of* w if $v \prec w$. We let $|w|$ denote the length of the string w. If $w \in \mathcal{T}$, then the *height* of w in $\mathcal{T}$ is $\max\{|v| \mid wv \in \mathcal{T}\}$. Note that $v < w$ implies that the height of v is greater than the height of w.

We assume that we are given a fixed countably infinite set $\mathcal{V} = \{\boldsymbol{v}_0, \boldsymbol{v}_1, \boldsymbol{v}_2, \dots\}$ of *variables*. Let C be a finite set of *constants*. A λ-*term* over C is a structure $(\mathcal{T}, f, b)$, where

- $\mathcal{T}$ is a binary tree domain,
- f is a function from a subset of $\mathcal{T}^{(0)}$ to $C \cup \mathcal{V}$,
- b is a function from $\mathcal{T}^{(0)} - \text{dom}(f)$ to $\mathcal{T}^{(1)}$ such that for all $w \in \text{dom}(b)$, $b(w) < w$.

Let $M = (\mathcal{T}, f, b)$ be a λ-term over C. If $c \in C$ and $f(w) = c$, we say that c *occurs at* w in M, and call the node w an *occurrence* of c in M. If $x \in \mathcal{V}$ and $f(w) = x$, then we say that x *occurs free* at w in M, and call w a *free occurrence* of x in M. For $w \in \text{dom}(b)$, we call $b(w)$ the *binder* of w. The set of variables that occur free in M is written $\text{FV}(M)$; its elements are the *free variables* of M. When $\text{FV}(M) = \varnothing$, M is a *closed* λ-term (over C). When no constant occurs in M, M is called a *pure* λ-term.

Let $M = (\mathcal{T}_M, f_M, b_M)$ and $N = (\mathcal{T}_N, f_N, b_N)$ be λ-terms (over C). Then the *application* of M to N is the λ-term $MN = (\mathcal{T}, f, b)$ defined as follows:

$$\mathcal{T} = \{\epsilon\} \cup 0\mathcal{T}_M \cup 1\mathcal{T}_N,$$
$$f = \{(0w, f_M(w)) \mid w \in \text{dom}(f_M)\} \cup \{(1w, f_N(w)) \mid w \in \text{dom}(f_N)\},$$
$$b = \{(0w, 0b_M(w)) \mid w \in \text{dom}(b_M)\} \cup \{(1w, 1b_N(w)) \mid w \in \text{dom}(b_N)\}.$$

It is easy to see that the map $(M, N) \mapsto MN$ is one-to-one and every λ-term whose root is a binary node is an application.

Let M be as above. For each variable $x \in \mathcal{V}$, we define the λ-term $\lambda x.M = (\mathcal{T}, f, b)$ by:

$$\mathcal{T} = 0\mathcal{T}_M,$$
$$f = \{(0w, f_M(w)) \mid w \in \text{dom}(f_M) \text{ and } f_M(w) \neq x\},$$
$$b = \{(0w, 0b_M(w)) \mid w \in \text{dom}(b_M)\} \cup \{(0w, \epsilon) \mid w \in \text{dom}(f_M) \text{ and } f_M(w) = x\}.$$

A λ-term of the form $\lambda x.M$ is called a λ-*abstract*. Clearly, any λ-term P whose root is a unary node is a λ-abstract; indeed, given any variable $x \notin \text{FV}(P)$, P can be written uniquely as $\lambda x.M$.

A λ-*expression* over C is an expression built up from variables, constants, parentheses, the dot ".", and the symbol λ by the following rules:[26]

- If $c \in C$, then c is a λ-expression over C.

- If $x \in \mathcal{V}$, then x is a λ-expression over C.

- If M, N are λ-expressions over C, then (MN) is a λ-expression over C.

- If M is a λ-expression over C and $x \in \mathcal{V}$, then $(\lambda x.M)$ is a λ-expression over C.

Then each λ-expression represents a λ-term, under the convention that a constant or variable $a \in \mathcal{C} \cup \mathcal{V}$ represents the λ-term

$$(\{\epsilon\}, \{(\epsilon, a)\}, \varnothing).$$

---

[26] A λ-expression is called a *pre-term* by Sorensen and Urzyczyn [73].

It is clear that a λ-expression has the same tree structure as the λ-term it represents.

If $M = (\mathcal{T}, f, b)$ is a λ-term, a *writing* of M [74] is a function $\ell \colon \mathcal{T}^{(1)} \to \mathcal{V}$ satisfying the following conditions:

- If $u, v \in \mathcal{T}^{(1)}, w \in \mathcal{T}^{(0)}, u < v < w$, and $b(w) = u$, then $\ell(u) \neq \ell(v)$.
- If $u \in \mathcal{T}^{(1)}, v \in \mathcal{T}^{(0)}, u < v$, and $v \in \mathrm{dom}(f)$, then $\ell(u) \neq f(v)$.

It is clear that every λ-term has a writing; in particular, there is always a writing ℓ of M such that ℓ is one-to-one and $\mathrm{ran}(\ell) \cap \mathrm{FV}(M) = \varnothing$.[27]

Given a λ-term $M = (\mathcal{T}, f, b)$ together with a writing ℓ, we can define a function $\mathrm{sub}_{M,\ell}$ from $\mathcal{T}$ to λ-expressions as follows:

$$\mathrm{sub}_{M,\ell}(w) = \begin{cases} f(w) & \text{if } w \in \mathrm{dom}(f), \\ \ell(b(w)) & \text{if } w \in \mathrm{dom}(b), \\ \lambda x.\, \mathrm{sub}_{M,\ell}(w0) & \text{if } w \in \mathcal{T}^{(1)} \text{ and } \ell(w) = x, \\ (\mathrm{sub}_{M,\ell}(w0)\ \mathrm{sub}_{M,\ell}(w1)) & \text{if } w \in \mathcal{T}^{(2)}. \end{cases}$$

Then it is easy to see that $\mathrm{sub}_{M,\ell}(\epsilon)$ is a λ-expression representing M. The λ-term represented by $\mathrm{sub}_{M,\ell}(w)$ is usually called the *subterm* of M occurring at w; but "subterm" is only defined relative to a writing ℓ of M.

We use usual abbreviations in writing λ-expressions. We omit the outermost parentheses from λ-expressions and write MNP for $(MN)P$, $\lambda x.MN$ for $\lambda x.(MN)$, and $\lambda x_1 x_2 \ldots x_n.M$ for $\lambda x_1.(\lambda x_2.\ldots(\lambda x_n.M)\ldots)$.

We define the operation of *substitution* of a λ-term for a free variable in another λ-term. Let $M = (\mathcal{T}_M, f_M, b_M)$ and $N = (\mathcal{T}_N, f_N, b_N)$ be λ-terms and x be a variable in $\mathcal{V}$. Let $X = \{\, v \in \mathcal{T}_M^{(0)} \mid f_M(v) = x \,\}$. The result of *substituting* N for x in M is the λ-term $M[x := N] = (\mathcal{T}, f, b)$ defined by

$\mathcal{T} = \mathcal{T}_M \cup X\mathcal{T}_N$,
$f = \{\, (w, f_M(w)) \mid w \in \mathrm{dom}(f_M) - X \,\} \cup \{\, (vw, f_N(w)) \mid v \in X, w \in \mathrm{dom}(f_N) \,\}$,
$b = b_M \cup \{\, (vw, vb_N(w)) \mid v \in X, w \in \mathrm{dom}(b_N) \,\}$.

It follows from this definition that for all λ-terms P, Q, N, all $y \in \mathcal{V} - \{x\}$, and all $z \in \mathcal{V} - (\{x\} \cup \mathrm{FV}(N))$, we have

$$x[x := N] = N,$$
$$y[x := N] = y,$$

---

[27]Such a writing corresponds to what Loader [54] calls a *regular* λ-term.

$$(PQ)[x := N] = P[x := N]\,Q[x := N],$$
$$(\lambda x.P)[x := N] = \lambda x.P,$$
$$(\lambda z.P)[x := N] = \lambda z.(P[x := N]).$$

The *simultaneous substitution* of λ-terms $N_1, \ldots, N_k$ for pairwise distinct variables $x_1, \ldots, x_k$ in a λ-term M is defined similarly, and is written $M[x_1{:=}N_1, \ldots, x_k{:=}N_k]$. We write $M[x_1, \ldots, x_k]$ to indicate that $\{x_1, \ldots, x_k\} \subseteq \mathrm{FV}(M[x_1, \ldots, x_k])$, and write $M[N_1, \ldots, N_k]$ for $(M[x_1, \ldots, x_k])[x_1 := N_1, \ldots, x_k := N_k]$.

Let $M = (\mathcal{T}, f, b)$ be a λ-term. Suppose that $w \in \mathcal{T}^{(2)}$ is a binary node of M such that $w0 \in \mathcal{T}^{(1)}$. Such a node w is called a β-*redex*. Note that for every writing ℓ of M, the λ-term represented by $\mathrm{sub}_{M,\ell}(w)$ is of the form $(\lambda x.P)N$. Let $X = \{\, v \mid w0v \in \mathcal{T}^{(0)}, b(w0v) = w0 \,\}$. (The set of leaves of M whose binder is $w0$ is $w0X$.) We write
$$M \xrightarrow{w}_\beta M'$$
if $M' = (\mathcal{T}', f', b')$, where

$$\mathcal{T}' = \{\, u \in \mathcal{T} \mid w \not\leq u \,\} \cup \{\, wv \mid w00v \in \mathcal{T} \,\} \cup \{\, wvu \mid v \in X, w1u \in \mathcal{T} \,\},$$
$$f = \{\, (u, f(u)) \mid u \in \mathrm{dom}(f), w \not\leq u \,\} \cup \{\, (wv, f(w00v)) \mid w00v \in \mathrm{dom}(f) \,\} \cup$$
$$\qquad \{\, (wvu, f(w1u)) \mid v \in X, w1u \in \mathrm{dom}(f) \,\},$$
$$b' = \{\, (u, b(u)) \mid u \in \mathrm{dom}(b), w \not\leq u \,\} \cup$$
$$\qquad \{\, (wv, b(w00v)) \mid w00v \in \mathrm{dom}(b), w \not\leq b(w00v) \,\} \cup$$
$$\qquad \{\, (wv, wv') \mid w00v \in \mathrm{dom}(b), b(w00v) = w00v' \,\} \cup$$
$$\qquad \{\, (wvu, b(w1u)) \mid v \in X, w1u \in \mathrm{dom}(b), w \not\leq b(w1u) \,\} \cup$$
$$\qquad \{\, (wvu, wvu') \mid v \in X, w1u \in \mathrm{dom}(b), b(w1u) = w1u' \,\}.$$

See Figure 9. If ℓ is a writing of M and $(\lambda x.P)N$ is the λ-term represented by $\mathrm{sub}_{M,\ell}(w)$, then for every writing ℓ' of M' such that ℓ' agrees with ℓ on $\{\, u \in \mathcal{T}^{(1)} \mid u < w \,\}$, $\mathrm{sub}_{M',\ell'}(w)$ represents $P[x := N]$.

From here on, we will let λ-expressions denote λ-terms, rather than themselves, unless we explicitly indicate otherwise, keeping in mind that distinct λ-expressions may represent the same λ-term. For example, if $M = c(\lambda y.d((\lambda x.yxx)(yzz)))$, then the node 101 of M is a β-redex, and $M \xrightarrow{101}_\beta c(\lambda y.d(y(yzz)(yzz))) = c(\lambda y.d((yxx)[x := yzz]))$.

We write $M \to_\beta M'$ if $M \xrightarrow{w}_\beta M'$ for some β-redex w in M. We say that M β-*reduces to* M' (or M' β-*expands to* M) and write $M \twoheadrightarrow_\beta M'$ if there is a finite sequence of λ-terms $M_0, M_1, \ldots, M_n$ ($n \geq 0$) such that
$$M = M_0 \to_\beta M_1 \to_\beta \cdots \to_\beta M_n = M'.$$

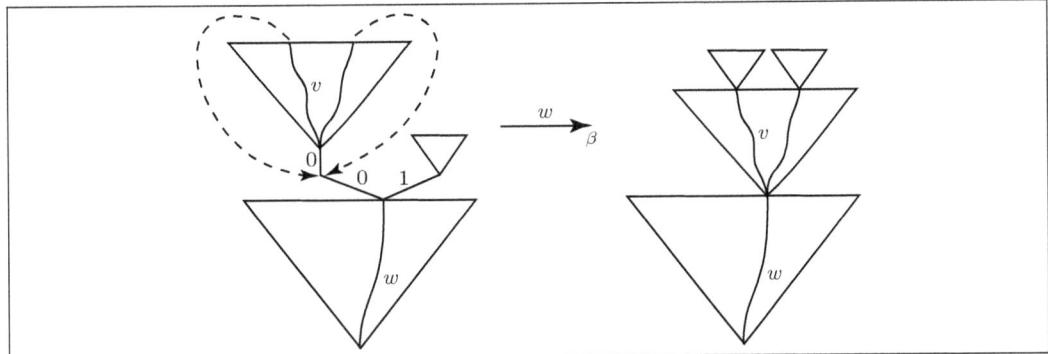

Figure 9: A one-step β-reduction. The dotted arrows represent the binding map.

If M and M' are related by the symmetric transitive closure of the relation $\twoheadrightarrow_\beta$, we say M is β-equal to M' and write $M =_\beta M'$.

Theorem 3.3 (Church-Rosser Theorem). *If $M \twoheadrightarrow_\beta N$ and $M \twoheadrightarrow_\beta P$, then there exists a Q such that $N \twoheadrightarrow_\beta Q$ and $P \twoheadrightarrow_\beta Q$.*

See [6] for a proof.

A λ-term is called β-*normal* if it does not contain a β-redex. If a λ-term β-reduces to a β-normal λ-term, the latter is called the β-*normal form* of the former. By the Church-Rosser Theorem for β-reduction, any λ-term M has at most one β-normal form. If a λ-term M has a β-normal form, we denote it by $|M|_\beta$.

If $M \twoheadrightarrow_\beta M'$, each node of M' is a *descendant* of a unique node (its *ancestor*) of M. For example, in $M = (\lambda x.yxx)(zw) \to_\beta y(zw)(zw) = M'$, both occurrences of z in M' are descendants of the unique occurrence of z in M. We give the definition of the ancestor-descendant relation for one-step β-reduction as follows.[28]

Let $M = (\mathcal{T}, f, b), M' = (\mathcal{T}', f', b')$, and suppose w is a β-redex in M and $M \xrightarrow{w}_\beta M'$. We write $(M, u) \overset{w}{\blacktriangleright} (M', u')$ to mean that the node u' of M' is a descendant of the node u of M. Let $u \in \mathcal{T}$. There are four cases to consider:

Case 1. $w \not\leq u$. Then $(M, u) \overset{w}{\blacktriangleright} (M', u')$ if and only if $u' = u$.

Case 2. $u = w$ or $u = w0$. Then there is no u' such that $(M, u) \overset{w}{\blacktriangleright} (M', u')$.

Case 3. $u = w00s$. Case 3a. If $u \in \text{dom}(b)$ and $b(u) = w0$, then there is no u' such that $(M, u) \overset{w}{\blacktriangleright} (M', u')$. Case 3b. Otherwise, $(M, u) \overset{w}{\blacktriangleright} (M', u')$ if and only if $u' = ws$.

Case 4. $u = w1s$. Then $(M, u) \overset{w}{\blacktriangleright} (M', u')$ if and only if $u' = wvs$ for some v such that $w00v \in \text{dom}(b)$ and $b(w00v) = w0$.

---

[28]See [10] for a formal definition of the ancestor-descendant relation using the technique of labeling bracket pairs, originally due to Newman [58].

It is clear that each node of M' is a descendant of a unique node of M. In Cases 1 and 3b, the node u of M has just one descendant in M'. In Case 4, it has as many descendants in M' as there are leaves in M whose binder is $w0$. We write $(M, u) \overset{w}{\blacktriangleright}_k (M', u')$ to mean that the node u' of M' is the k-th among the descendants of the node u of M under the lexicographic ordering of the nodes of M'.

Here are some important properties of the ancestor-descendant relation. The proof is by straightforward inspection.

Lemma 3.4. *Let $M = (\mathcal{T}, f, b)$ and $M' = (\mathcal{T}', f', b')$, and suppose $(M, u) \overset{w}{\blacktriangleright} (M', u')$.*

(i) $u \in \mathcal{T}^{(i)}$ *if and only if* $u' \in \mathcal{T}'^{(i)}$ *for* $i = 0, 1, 2$.

(ii) $u \in \text{dom}(f)$ *if and only if* $u' \in \text{dom}(f')$.

(iii) $u \in \text{dom}(b)$ *if and only if* $u' \in \text{dom}(b')$.

(iv) *If* $u \in \text{dom}(b)$, *then* $(M, b(u)) \overset{w}{\blacktriangleright} (M', b'(u'))$.

We write $(M, v) \overset{w_1,\ldots,w_n}{\blacktriangleright} (M', v')$ if there are sequences $M_0, M_1, \ldots, M_n$ and $v_0, v_1, \ldots, v_n$ such that $(M, v) = (M_0, v_0), (M', v') = (M_n, v_n)$, and for $1 \le i \le n$, $(M_{i-1}, v_{i-1}) \overset{w_i}{\blacktriangleright} (M_i, v_i)$. The following theorem says that if $M \twoheadrightarrow_\beta M'$ and M' is in β-normal form, the ancestor-descendant relation between the nodes of M and the nodes of M' does not depend on the β-reduction sequence from M to M'.

Theorem 3.5. *If* $(M, u) \overset{w_1,\ldots,w_n}{\blacktriangleright} (|M|_\beta, v)$ *and* $(M, u') \overset{v_1,\ldots,v_m}{\blacktriangleright} (|M|_\beta, v)$, *then* $u = u'$.

Proof. The proof is via an equivalent definition of the ancestor-descendant relation in terms of *simply labeled λ-calculus* $\lambda_{\mathscr{A}}$ [10]. This calculus defines β-reduction on *labeled λ-terms*, where each node carries a label, and the label of a node is passed to the node's descendants. If u is the only node labeled by a in a labeled λ-term M, the set of descendants of u in $|M|_\beta$ consists of those nodes labeled by a, which is independent of the β-reduction path from M to $|M|_\beta$ because $\lambda_{\mathscr{A}}$, being an orthogonal combinatory reduction system, enjoys the Church-Rosser Property (see [10] for details). □

A unary node w of $M = (\mathcal{T}, f, b)$ is an η-*redex* if $w0$ is a binary node and $w01$ is the only node whose binder is w. If ℓ is a writing of M, then the λ-term represented by $\text{sub}_{M,\ell}(w)$ is of the form $\lambda x.Px$, where $x \notin \text{FV}(P)$. We write

$$M \overset{w}{\to}_\eta M'$$

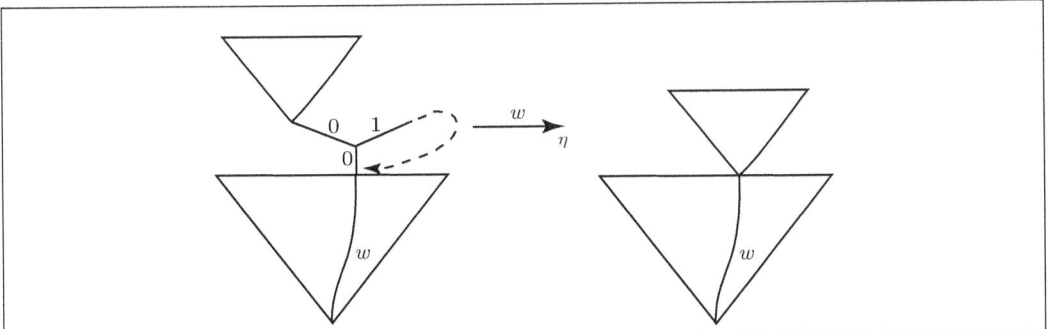

Figure 10: A one-step η-reduction. The node $w01$ is the unique node whose binder is w.

if $M' = (\mathcal{T}', f', b')$, where

$\mathcal{T}' = \{\, u \in \mathcal{T} \mid w \not\leq u \,\} \cup \{\, wv \mid w00v \in \mathcal{T} \,\},$
$f' = \{\, (u, f(u)) \mid u \in \mathrm{dom}(f), w \not\leq u \,\} \cup \{\, (wv, f(w00v)) \mid w00v \in \mathrm{dom}(f) \,\},$
$b' = \{\, (u, b(u)) \mid u \in \mathrm{dom}(b), w \not\leq u \,\} \cup$
$\qquad \{\, (wv, b(w00v)) \mid w00v \in \mathrm{dom}(b), b(w00v) < w \,\} \cup$
$\qquad \{\, (wv, wv') \mid w00v \in \mathrm{dom}(b), b(w00v) = w00v' \,\}.$

See Figure 10. If ℓ is a writing of M and $\lambda x.Px$ is the λ-term represented by $\mathrm{sub}_{M,\ell}(w)$, then for every writing ℓ' of M' such that ℓ' agrees with ℓ on $\{\, u \in \mathcal{T}^{(1)} \mid u < w \,\}$, the λ-term represented by $\mathrm{sub}_{M',\ell'}(w)$ is P. The notions of η-reduction, η-expansion, and η-equality are defined analogously to β-reduction, β-expansion, and β-equality. We write $M \twoheadrightarrow_\eta M'$ to mean M η-reduces to M' and $M =_\eta M'$ to mean M is η-equal to M'. The transitive closure of the union of $\twoheadrightarrow_\beta$ and $\twoheadrightarrow_\eta$ is written $\twoheadrightarrow_{\beta\eta}$, and similarly for $=_{\beta\eta}$.

The following are lemmas needed to prove the Church-Rosser Theorem for $\beta\eta$-reduction (see [6] for a proof):

Lemma 3.6 (η-Postponmenet Theorem). *If $M \twoheadrightarrow_\eta Q \twoheadrightarrow_\beta T$, then there exists a λ-term P such that $M \twoheadrightarrow_\beta P \twoheadrightarrow_\eta T$.*

Lemma 3.7 (Commuting Lemma). *If $M \twoheadrightarrow_\beta P$ and $M \twoheadrightarrow_\eta Q$, then there exists a λ-term T such that $P \twoheadrightarrow_\eta T$ and $Q \twoheadrightarrow_\beta T$.*

The following lemma is straightforward (see [31]):

Lemma 3.8. *If M is in β-normal form and $M \twoheadrightarrow_\eta M'$, then M' is in β-normal form.*

A λ-term M is a λI-*term* if every unary node of M binds at least one leaf. A λ-term M is *affine* if every variable occurs free in M at most once, and every unary node of M binds at most one leaf. A λ-term is *linear* if it is an affine λI-term. The class of λI-terms and the class of affine λ-terms are both closed under β-reduction and η-equality.

We introduce some nonstandard notations. The *sequence of constants in* $M = (\mathcal{T}, f, b)$, denoted $\overrightarrow{\mathrm{Con}}(M)$, is $\overrightarrow{\mathrm{Con}}(M, \epsilon)$, where $\overrightarrow{\mathrm{Con}}(M, w)$ is defined as follows:

$$\overrightarrow{\mathrm{Con}}(M, w) = \begin{cases} () & \text{if } w \in \mathrm{dom}(b), \\ () & \text{if } w \in \mathrm{dom}(f) \text{ and } f(w) \in \mathcal{V}, \\ (f(w)) & \text{if } w \in \mathrm{dom}(f) \text{ and } f(w) \in \mathcal{C}, \\ \overrightarrow{\mathrm{Con}}(M, w0) & \text{if } w \in \mathcal{T}^{(1)}, \\ \overrightarrow{\mathrm{Con}}(M, w0){\frown}\overrightarrow{\mathrm{Con}}(M, w1) & \text{if } w \in \mathcal{T}^{(2)}, \end{cases}$$

where $\frown$ denotes juxtaposition. The *sequence of free variables of* M, denoted $\overrightarrow{\mathrm{FV}}(M)$, is $\overrightarrow{\mathrm{FV}}(M, \epsilon)$, where $\overrightarrow{\mathrm{FV}}(M, w)$ is defined as follows:

$$\overrightarrow{\mathrm{FV}}(M, w) = \begin{cases} () & \text{if } w \in \mathrm{dom}(b), \\ (f(w)) & \text{if } w \in \mathrm{dom}(f) \text{ and } f(w) \in \mathcal{V}, \\ () & \text{if } w \in \mathrm{dom}(t) \text{ and } f(w) \in \mathcal{C}, \\ \overrightarrow{\mathrm{FV}}(M, w0) & \text{if } w \in \mathcal{T}^{(1)}, \\ \overrightarrow{\mathrm{FV}}(M, w0){\frown}\overrightarrow{\mathrm{FV}}(M, w1) & \text{if } w \in \mathcal{T}^{(2)}, \end{cases}$$

If $\overrightarrow{\mathrm{Con}}(M) = (c_1, \ldots, c_n)$ and $\{x_1, \ldots, x_n\} \cap \mathrm{FV}(M) = \varnothing$ (with $x_1, \ldots, x_n$ pairwise distinct), we let $\widehat{M}[x_1, \ldots, x_n]$ denote the pure λ-term such that $(x_1, \ldots, x_n)$ is a subsequence of $\overrightarrow{\mathrm{FV}}(\widehat{M}[x_1, \ldots, x_n])$ and $M = \widehat{M}[c_1, \ldots, c_n]$. For example, if $M = \lambda y.c(y(c(zd)))$, then $\overrightarrow{\mathrm{Con}}(M) = (c, c, d)$, $\overrightarrow{\mathrm{FV}}(M) = (z)$, $\widehat{M}[x_1, x_2, x_3] = \lambda y.x_1(y(x_2(zx_3)))$, and $\overrightarrow{\mathrm{FV}}(\widehat{M}[x_1, x_2, x_3]) = (x_1, x_2, z, x_3)$.

3.1.3 Simply typed λ-calculus with constants

Given a set A of *atomic types*, we let $\mathscr{T}(A)$ denote the set of *types* built up from atomic types using $\to$ as the sole type constructor. In other words, $\mathscr{T}(A)$ is the smallest set extending A such that

$$\alpha, \beta \in \mathscr{T}(A) \quad \text{implies} \quad (\alpha \to \beta) \in \mathscr{T}(A).$$

We omit the outermost parentheses in writing types, and write $\alpha \to \beta \to \gamma$ to mean $\alpha \to (\beta \to \gamma)$.

For $\alpha \in \mathscr{T}(A)$, we write $|\alpha|$ to denote the number of occurrences of atomic types in α. The notation $\overline{\alpha}$ denotes the sequence of atomic types (with repetitions) that appear in α *from right to left*, defined as follows:

$$\overline{p} = (p) \quad \text{if } p \in A,$$
$$\overline{\alpha \to \beta} = \overline{\beta}\,\widehat{}\,\overline{\alpha}.$$

As before, $\widehat{}$ denotes juxtaposition of sequences. For example, $\overline{p \to p \to q} = (q, p, p)$. Note that the length of $\overline{\alpha}$ is $|\alpha|$.

The set of *positions* within α, denoted $\langle \alpha \rangle$, is defined as follows:

$$\langle p \rangle = \{\epsilon\},$$
$$\langle \alpha \to \beta \rangle = \{\epsilon\} \cup 1\langle \alpha \rangle \cup 0\langle \beta \rangle.$$

Then for every type α, $\langle \alpha \rangle$ is a binary tree domain that has no unary nodes. The *subtype* of α that occurs at position $w \in \langle \alpha \rangle$, subtype$(\alpha, w)$ in symbols, is defined as follows:

$$\text{subtype}(\alpha, \epsilon) = \alpha,$$
$$\text{subtype}(\alpha \to \beta, 1w) = \text{subtype}(\alpha, w),$$
$$\text{subtype}(\alpha \to \beta, 0w) = \text{subtype}(\beta, w).$$

The *polarity* of position w, pol(w), is 1 if the number of occurrences of 1 in w is even, -1 otherwise. We say that β occurs positively (negatively) at position w in α if subtype$(\alpha, w) = \beta$ and pol$(w) = 1$ (pol$(w) = -1$).

A *type substitution* is a mapping σ from $\mathscr{T}(A)$ to $\mathscr{T}(A')$, written in postfix notation, satisfying the condition $(\alpha \to \beta)\sigma = \alpha\sigma \to \beta\sigma$. A *type relabeling* is a type substitution that sends atomic types to atomic types. Note that $\langle \alpha \rangle = \langle \beta \rangle$ if and only if there exist a type γ and type relabelings σ_1 and σ_2 such that $\alpha = \gamma\sigma_1$ and $\beta = \gamma\sigma_2$. If $|\alpha| = n$ and $q_1, \ldots, q_n \in A$, then we let $\langle \alpha \rangle(q_1, \ldots, q_n)$ denote the unique type β in $\mathscr{T}(A)$ such that $\overline{\beta} = (q_1, \ldots, q_n)$ and $\langle \alpha \rangle = \langle \beta \rangle$. For any type β, we have $\langle \beta \rangle(\overline{\beta}) = \beta$.

A *higher-order signature* is a triple (A, C, τ), where A is a finite set of atomic types, C is a finite set of constants, and τ is a mapping from C to $\mathscr{T}(A)$. We write $\Lambda(\Sigma)$ for the set of λ-terms over C.

A *type environment* is a finite partial function from $\mathcal{V}$ to $\mathscr{T}(A)$. A type environment $\Gamma = \{(x_1, \alpha_1), \ldots, (x_n, \alpha_n)\}$ is usually written as a list $x_1 : \alpha_1, \ldots, x_n : \alpha_n$.

Let Γ be a type environment and $M = (\mathcal{T}, f, b) \in \Lambda(\Sigma)$. A function $t\colon \mathcal{T} \to \mathscr{T}(A)$ is a *type decoration* of M *under* Γ if $\operatorname{dom}(\Gamma) = \operatorname{FV}(M)$ and

$$t(w) = \begin{cases} \Gamma(f(w)) & \text{if } w \in \operatorname{dom}(f) \text{ and } f(w) \in \mathcal{V}, \\ \tau(f(w)) & \text{if } w \in \operatorname{dom}(f) \text{ and } f(w) \in C, \\ \gamma & \text{if } w \in \operatorname{dom}(b) \text{ and } t(b(w)) = \gamma \to \delta, \\ \gamma \to \delta & \text{for some } \gamma \text{ if } w \in \mathcal{T}^{(1)} \text{ and } t(w0) = \delta, \\ \gamma \to \delta & \text{if for some } v \in \mathcal{T}^{(2)},\ w = v0,\ t(v) = \delta,\ \text{and } t(v1) = \gamma. \end{cases}$$

If t is a type decoration of M (under Γ), we call (M, t) a *typed λ-term* over Σ (under Γ).

A typed λ-term (M, t) can be visualized in the form of a *natural deduction*: each unary and binary node w is labeled with its type $t(w)$, each node $w \in \operatorname{dom}(f)$ is labeled with $a{:}\gamma$, where $f(w) = a$ and $t(w) = \gamma$, and each node $w \in \operatorname{dom}(b)$ is labeled with $[\gamma]^v$, where $b(w) = v$ and $t(w) = \gamma$. For example, the following figure depicts a typed λ-term (M, t) under the type environment $z\colon p$, where $M = (\lambda y.y(yz))(\lambda x.x)$:

$$\cfrac{\cfrac{[p \to p]^0 \quad \cfrac{[p \to p]^0 \quad z:p}{p}}{(p \to p) \to p}\, 0 \qquad \cfrac{[p]^1}{p \to p}\, 1}{p}$$

To aid legibility, we have also placed the label v next to the horizontal line right above each unary node v.[29]

Another familiar representation of a typed λ-term is by means of a λ-expression together with a type superscript on each of its subexpression. For instance, one way of representing the above example of a typed λ-term is

$$((\lambda y^{p \to p}.(y^{p \to p}(y^{p \to p} z^p)^p)^p)^{(p \to p) \to p}(\lambda x^p.x^p)^{p \to p})^p.$$

We call an expression of the form $\Gamma \Rightarrow \alpha$, where Γ is a type environment and α is a type, a *sequent*. A sequent $\Gamma \Rightarrow \alpha$ is a *typing* of M if there is a type decoration t of M under Γ such that $t(\epsilon) = \alpha$. In this case, we write

$$\vdash_\Sigma \Gamma \Rightarrow M : \alpha.$$

---

[29]The resulting figure is identical to the natural deduction as defined in, e.g., [75], except that we use strings in $\{0,1\}^*$, rather than variables, as markers for closed assumptions, and we label open assumptions with variables or constants. Hindley [31] also uses node addresses as assumption markers in natural deductions, albeit in a different way.

and say that t is a type decoration for the *typing judgment* $\Gamma \Rightarrow M : \alpha$. When Γ is empty, we omit the symbol $\Rightarrow$ and write $\vdash_\Sigma M : \alpha$. Reference to Σ is dropped when M is pure.

We say that an (untyped) λ-term M is *typable* if it has a typing. It is known that every typable λ-term has a β-normal form. A sequent is said to be *inhabited* if there is a pure λ-term M (an *inhabitant*) such that $\vdash \Gamma \Rightarrow M : \alpha$. A sequent is inhabited if and only if it is a theorem of intuitionistic logic.[30]

Let $M = (\mathcal{T}, f, b) \in \Lambda(\Sigma)$ and t be a type decoration of M. If ℓ is a writing of M and $w \in \mathcal{T}$, then it is clear that

$$t_w(v) = t(wv) \quad \text{for } wv \in \mathcal{T}$$

determines a type decoration t_w for $\text{sub}_{M,\ell}(w)$, and we have

$$\vdash_\Sigma \{(x, t(wv)) \mid f(wv) = x\} \cup \{(\ell(b(wv)), t(wv)) \mid b(wv) < w\} \Rightarrow \text{sub}_{M,\ell}(w) : t(w).$$

An important property of a typed λ-term in β-normal form is the so-called *subformula property*:

Theorem 3.9. *Let $M = (\mathcal{T}, f, b)$ be a pure untyped λ-term in β-normal form. If t is a type decoration for $x_1 : \alpha_1, \ldots, x_n : \alpha_n \Rightarrow M : \alpha_0$, then for every $w \in \mathcal{T}$, $t(w)$ is a subtype of α_i for some $i \in \{0, \ldots, n\}$.*

Proof. The theorem is a consequence of the following statement, which is easy to see: for every $w \in \mathcal{T}$, if $w \neq \epsilon$ and $w \notin \text{dom}(f)$, then there exists a $v \in \mathcal{T}$ such that $t(v) = t(w) \rightarrow \alpha$ or $t(v) = \alpha \rightarrow t(w)$ for some α. $\square$

In general, the same typing of a λ-term may have more than one type decoration. See [31] for the proof of the following theorem:

Theorem 3.10. *If $M \in \Lambda(\Sigma)$ is a λI-term, any typing of M has a unique type decoration.*

Thus, a λI-term M together with a typing of M can be treated in the same way as a typed λ-term.

A typing $\Gamma \Rightarrow \alpha$ of M is a *principal typing* of M if for every typing $\Gamma' \Rightarrow \alpha'$ of M, there is a type substitution σ such that $\Gamma' \Rightarrow \alpha' = (\Gamma \Rightarrow \alpha)\sigma$. We call a type decoration t of M (under some type environment) a *principal type decoration of M* if for every type decoration t' of M (under some type environment), there is a type substitution σ such that $t' = \sigma \circ t$. Clearly, the typing determined by a principal type decoration is a principal typing.

---

[30]We use the symbol $\Rightarrow$ in the same way as Mints [56] does. This is the way Hindley [31] uses the symbol $\mapsto$. Although $\vdash_\Sigma \Gamma \Rightarrow M : \alpha$ implies $\text{dom}(\Gamma) = \text{FV}(M)$, it is always possible to weaken the antecedent in the sense that $\vdash_\Sigma \Gamma \Rightarrow M : \alpha$ implies $\vdash_\Sigma \Gamma, x : \beta \Rightarrow (\lambda y.M)x : \alpha$, where $x, y \notin \text{FV}(M)$.

Theorem 3.11 (Principal Type Theorem). *If M is typable, then M has a principal typing and a principal type decoration.*

See [31] for a proof.

Let $M = (\mathcal{T}_M, f_M, b_M)$ and $N = (\mathcal{T}_N, f_N, b_N)$ be λ-terms and x be a variable in $\mathrm{FV}(M)$. Let $X = \{\, v \in \mathcal{T}_M^{(0)} \mid f_M(v) = x \,\}$. Let $M[x := N] = (\mathcal{T}, f, b)$ be the result of substituting N for x in M. The following lemmas are straightforward:

Lemma 3.12. *Suppose that t_M and t_N are type decorations for $\Gamma_1, x : \beta \Rightarrow M : \alpha$ and $\Gamma_2 \Rightarrow N : \beta$, respectively, and that Γ_1 and Γ_2 agree on $(\mathrm{FV}(M) - \{x\}) \cap \mathrm{FV}(N)$. Then we can define a type decoration t for $\Gamma_1 \cup \Gamma_2 \Rightarrow M[x := N] : \alpha$ by*

$$t(w) = \begin{cases} t_M(w) & \text{if } w \in \mathcal{T}_M, \\ t_N(v') & \text{if } w = vv' \text{ for some } v \in X \text{ and } v' \in \mathcal{T}_N. \end{cases}$$

Lemma 3.13. *Suppose that t is a type decoration for $\Gamma \Rightarrow M[x := N] : \alpha$ such that for some type β, $t(v) = \beta$ for every $v \in X$. Pick a $v \in X$. Then we can define type decorations t_M and t_N for $\Gamma_1, x : \beta \Rightarrow M : \alpha$ and $\Gamma_2 \Rightarrow N : \beta$, respectively, by*

$$t_M(w) = t(w) \quad \text{for all } w \in \mathcal{T}_M,$$
$$t_N(w) = t(vw) \quad \text{for all } w \in \mathcal{T}_N,$$

where Γ_1 and Γ_2 are the restrictions of Γ to $\mathrm{FV}(M)$ and to $\mathrm{FV}(N)$, respectively.

Let $M[x_1, \ldots, x_n]$ be a pure λ-term such that $\mathrm{FV}(M[x_1, \ldots, x_n]) = \{x_1, \ldots, x_n\}$. For any $c_1, \ldots, c_n \in C$, we have

$\vdash_\Sigma M[c_1, \ldots, c_n] : \alpha \quad$ if and only if $\quad \vdash x_1 : \tau(c_1), \ldots, x_n : \tau(c_n) \Rightarrow M[x_1, \ldots, x_n] : \alpha$.

Let (M, t) be a typed λ-term. If $M \overset{w}{\twoheadrightarrow}_\beta M'$, then t, in conjunction with the ancestor-descendant relation, induces a type decoration t' of M', defined by

$$t'(v') = t(v) \quad \text{if } (M, v) \overset{w}{\blacktriangleright} (M', v').$$

This is denoted by $(M, t) \overset{w}{\twoheadrightarrow}_\beta (M', t')$. Note that even though we do not have $(M, w) \overset{w}{\blacktriangleright} (M', w)$, it is always the case that $t'(w) = t(w)$, since $t(w) = t(w00)$ and $(M, w00) \overset{w}{\blacktriangleright} (M', w)$.

Theorem 3.14 (Subject Reduction Theorem). *If $\vdash_\Sigma \Gamma \Rightarrow M : \alpha$ and $M \twoheadrightarrow_\beta M'$, then $\vdash_\Sigma \Gamma' \Rightarrow M' : \alpha$, where Γ' is the restriction of Γ to $\mathrm{FV}(M')$.*

See, e.g., [31] for a proof.

Let $M = (\mathcal{T}, f, b)$ and suppose $M \xrightarrow{w}_\beta M'$. This β-reduction step is called *erasing* if there is no $v \in \mathcal{T}^{(0)}$ such that $b(v) = w0$, and *duplicating* if for some $v, v' \in \mathcal{T}^{(0)}$, $v \neq v'$ and $b(v) = b(v') = w0$. (The right child $w1$ of the β-redex w has no descendant in an erasing β-reduction step, and has more than one in a duplicating β-reduction step.) A β-reduction from M to M' is *non-erasing* (*non-duplicating*) if it consists entirely of non-erasing (non-duplicating) β-reduction steps.

Theorem 3.15 (Subject Expansion Theorem). *If $\vdash_\Sigma \Gamma \Rightarrow M' : \alpha$ and $M \twoheadrightarrow_\beta M'$ by non-erasing, non-duplicating β-reduction, then $\vdash_\Sigma \Gamma \Rightarrow M : \alpha$.*

See [31]. As a special case, if M is linear and $M \twoheadrightarrow_\beta M'$, then $\vdash_\Sigma \Gamma \Rightarrow M' : \alpha$ implies $\vdash_\Sigma \Gamma \Rightarrow M : \alpha$.

As with β-reduction, the η-reduction relation between untyped λ-terms induces the η-reduction relation between typed λ-terms. A typed λ-term (M, t), where $M = (\mathcal{T}, f, b)$, is in η-*long form* if every node $w \in \mathcal{T}$ satisfies the following condition:

- $t(w) = \beta \to \gamma$ for some β, γ implies that either $w \in \mathcal{T}^{(1)}$ or $w = v0$ for some $v \in \mathcal{T}^{(2)}$.

If (M, t) has a node w that does not satisfy this condition, there is a unique typed λ-term (M', t') such that $(M', t') \xrightarrow{w}_\eta (M, t)$. Both nodes w and $w00$ of (M', t') satisfy the condition, and $t'(w0) = \gamma$, $t'(w01) = \beta$, both of which are shorter than $\beta \to \gamma$. Thus, every typed λ-term can be converted to one in η-long form by a sequence of η-expansion steps applied to nodes that do not satisfy this condition. It is easy to see that the resulting λ-term is unique; we call it the η-*long form* of the original λ-term.

We say that an untyped λ-term $M \in \Lambda(\Sigma)$ is in η-*long form relative to* $\Gamma \Rightarrow \alpha$ if there is a type decoration t of M under Γ such that $t(\epsilon) = \alpha$ and (M, t) is in η-long form. We say that M is in η-long form if M is η-long relative to some typing (or, equivalently, relative to its principal typing).

The following lemmas are from [34]:

Lemma 3.16. *Let M and N be λ-terms and x be a variable in $\mathrm{FV}(M)$. Suppose that t_M and t_N are type decorations for $\Gamma_1, x{:}\beta \Rightarrow M{:}\alpha$ and $\Gamma_2 \Rightarrow N{:}\beta$, respectively, and that Γ_1 and Γ_2 agree on $(\mathrm{FV}(M) - \{x\}) \cap \mathrm{FV}(N)$. Let t be the type decoration for $\Gamma_1 \cup \Gamma_2 \Rightarrow M[x := N] : \alpha$ defined according to Lemma 3.12. If (M, t_M) and (N, t_N) are in η-long form, then $(M[x := N], t)$ is in η-long form.*

Lemma 3.17. *If M is in η-long form relative to $\Gamma \Rightarrow \alpha$ and $M \twoheadrightarrow_\beta M'$, then M' is in η-long form relative to $\Gamma' \Rightarrow \alpha$, where Γ' is the restriction of Γ to $\mathrm{FV}(M')$.*

Thus, the β-normal form of an η-long λ-term is η-long.

We refer to an occurrence of a type β in a sequent $x_1\colon \alpha_1,\ldots,x_n\colon \alpha_n \Rightarrow \alpha_0$ or a typing judgment $x_1\colon \alpha_1,\ldots,x_n\colon \alpha_n \Rightarrow M\colon \alpha_0$ by a pair (i,v), with $0 \leq i \leq n$ and $v \in \langle \alpha_i \rangle$, such that $\mathrm{subtype}(\alpha_i, v) = \beta$. We say that an occurrence (i,v) is *positive* (resp., *negative*) and write $\mathrm{pol}(i,v) = +1$ ($\mathrm{pol}(i,v) = -1$) if either $i = 0$ and $\mathrm{pol}(v) = 1$ ($\mathrm{pol}(v) = -1$) or $i \geq 1$ and $\mathrm{pol}(v) = -1$ ($\mathrm{pol}(v) = 1$). For example, in $x\colon p, y\colon p \to q \Rightarrow q$, the pairs $(1, \epsilon)$ and $(2, 0)$ refer to the first occurrences of p and q, respectively, which are both negative, and the pairs $(2, 1)$ and $(0, \epsilon)$ refer to the second occurrences of p and q, respectively, which are both positive. A sequent or typing judgment is *balanced* if every atomic type has at most one positive and at most one negative occurrence in it.

Theorem 3.18 (Coherence Theorem). *All inhabitants of a balanced sequent are $\beta\eta$-equal. In particular, if $\Gamma \cup \Gamma' \Rightarrow \alpha$ is a balanced sequent and both $\vdash \Gamma \Rightarrow M\colon \alpha$ and $\vdash \Gamma' \Rightarrow M'\colon \alpha$ hold, then $M =_{\beta\eta} M'$.*

See [56] for a proof.

According to Hirokawa [33], the first of the following theorems is due to Belnap [9]. See [33] for the proof of the second.

Theorem 3.19 ([9]). *If M is a pure affine λ-term, then the principal typing of M is balanced.*

Theorem 3.20 ([33]). *If a pure λ-term M in β-normal form has a balanced typing, then M is affine.*

Theorem 3.19 together with the Coherence Theorem (Theorem 3.18) implies that a pure affine λ-term is uniquely determined by its principal typing up to $\beta\eta$-equality.

3.1.4 Links in typed λ-terms

It will be convenient for our purposes to introduce a strengthening of the notion of η-long form. We say that a typed λ-term (M, t) with $M = (\mathcal{T}, f, b)$ is in *strict η-long form* if every node $w \in \mathcal{T}$ satisfies the following condition:

- if $t(w) = \beta \to \gamma$, then either (i) $w \in \mathcal{T}^{(1)}$ and $b(v) = w$ for some $v \in \mathcal{T}^{(0)}$, (ii) $w \in \mathcal{T}^{(1)}$ and β is an atomic type, or (iii) $w = v0$ for some $v \in \mathcal{T}^{(2)}$.

Note that if M is a λI-term and (M,t) is in η-long form, then (M,t) is in strict η-long form. For every typed λ-term (M,t) in η-long form, there is a typed λ-term (M',t') in strict η-long form such that both $(M',t') \twoheadrightarrow_\beta (M,t)$ and $(M',t') \twoheadrightarrow_\eta (M,t)$. Unlike η-long form, strict η-long form is not preserved under β-reduction, but we have the following:

Lemma 3.21. *Lemma 3.16 holds with "strict η-long form" in place of "η-long form".*

As with η-long form, we speak of an untyped λ-term being in strict η-long form relative to a typing.

Clearly, if $M \in \Lambda(\Sigma)$ is a closed λ-term and $\overrightarrow{\mathrm{Con}}(M) = (c_1, \ldots, c_n)$, then M is in (strict) η-long form relative to α if and only if $\widehat{M}[x_1, \ldots, x_n]$ is in (strict) η-long form relative to $x_1 : \tau(c_1), \ldots, x_n : \tau(c_n) \Rightarrow \alpha$.

Lemma 3.22. *Let M be a pure λ-term, and suppose that t is a type decoration of M such that (M, t) is in strict η-long form. Let $\tilde{t}$ be a principal type decoration of M. Then there is a type relabeling σ such that $t = \sigma \circ \tilde{t}$.*

Proof. It is easy to see that if an atomic type p occurs anywhere in (M, t), then it must be that there is a node of M that is assigned type p by t, or else there is a unary node of M that is assigned a type of the form $p \to \gamma$. In both cases, the relevant node must be assigned a type of the same shape by $\tilde{t}$. □

Lemma 3.22 implies the following:

Remark 3.23. Suppose that $M \in \Lambda(\Sigma)$ is a λ-term in strict η-long form relative to $x_1 : \gamma_1, \ldots, x_n : \gamma_n \Rightarrow \gamma_0$, $\overrightarrow{\mathrm{Con}}(M) = (d_1, \ldots, d_m)$, and $y_1 : \beta_1, \ldots, y_m : \beta_m, x_1 : \alpha_1, \ldots, x_n : \alpha_n \Rightarrow \alpha_0$ is a principal typing of $\widehat{M}[y_1, \ldots, y_m]$. Then $\widehat{M}[y_1, \ldots, y_m]$ is in strict η-long form relative to $y_1 : \beta_1, \ldots, y_m : \beta_m, x_1 : \alpha_1, \ldots, x_n : \alpha_n \Rightarrow \alpha_0$, and moreover, we have

$$\langle \beta_i \rangle = \langle \tau(d_i) \rangle \quad \text{for } i = 1, \ldots, m,$$
$$\langle \alpha_i \rangle = \langle \gamma_i \rangle \quad \text{for } i = 0, \ldots, n.$$

Let (M, t) be a pure typed λ-term, where $M = (\mathcal{T}, f, b)$. We associate with (M, t) a certain directed graph $G_{(M,t)} = (V_{(M,t)}, E_{(M,t)})$.[31] The set $V_{(M,t)}$ of vertices of $G_{(M,t)}$ consists of all triples of one of the forms

$$(w, v, \uparrow) \quad \text{and} \quad (w, v, \downarrow),$$

where $w \in \mathcal{T}$ and $v \in \langle t(w) \rangle^{(0)}$. (Recall that $\langle t(w) \rangle^{(0)}$ is the set of leaves of $\langle t(w) \rangle$, that is, the set of positions where atomic types occur in $t(w)$.) Triples $(w, v, \uparrow)$ and $(w, v, \downarrow)$ correspond to the same position in $t(w)$. The existence of an edge from $(w, v, -)$ to $(w', v', -)$ (where "$-$" is to be filled by $\uparrow$ or $\downarrow$) implies that the same

---

[31] Our graph is essentially the natural deduction counterpart of the *logical flow graph* of Buss [12]. See [41] for an equivalent definition.

atomic type must occur at v in $t(w)$ and at v' in $t(w')$ (i.e., subtype$(t(w), v) =$ subtype$(t(w'), v')$). The last component of the triples indicates the "direction of travel", which is explained below. The set $E_{(M,t)}$ of edges of $G_{(M,t)}$ is defined as follows:

$((w, v, \uparrow), (w', v', \uparrow)) \in E_{(M,t)}$ iff either $w \in \mathcal{T}^{(1)}$, $w0 = w'$, and $v = 0v'$, or
$$w \in \mathcal{T}^{(2)}, w0 = w', \text{ and } 0v = v'.$$

$((w, v, \downarrow), (w', v', \downarrow)) \in E_{(M,t)}$ iff either $w' \in \mathcal{T}^{(1)}$, $w = w'0$, and $0v = v'$, or
$$w' \in \mathcal{T}^{(2)}, w = w'0, \text{ and } v = 0v'.$$

$((w, v, \uparrow), (w', v', \downarrow)) \in E_{(M,t)}$ iff either $w \in \mathcal{T}^{(1)}$, $w = b(w')$ and $v = 1v'$, or
$$w' \in \mathcal{T}^{(1)}, b(w) = w' \text{ and } 1v = v'.$$

$((w, v, \downarrow), (w', v', \uparrow)) \in E_{(M,t)}$ iff for some $u \in \mathcal{T}^{(2)}$,
either $w = u0$, $w' = u1$, and $v = 1v'$, or
$$w = u1, w' = u0, \text{ and } 1v = v'.$$

Note that the edges in $E_{(M,t)}$ come in pairs: given an edge in $E_{(M,t)}$, one can interchange source and destination, then reverse the direction of the arrows in the third component of both vertices, and obtain another edge in $E_{(M,t)}$.

The meaning of the graph $G_{(M,t)}$ becomes easy to grasp when it is superimposed on the natural deduction representing (M, t). Each pair of edges is represented by a single curve connecting two occurrences of an atomic type; the two edges in the pair correspond to the two ways of traversing the curve, with the direction of traversal at each end point of the curve matching the direction of the arrow in the third component of the tuple $(w, v, -)$ corresponding to that point. Thus, $((w, v, \downarrow), (w', v', \downarrow))$ is an edge of the graph $G_{(M,t)}$ if there is a curve that departs downward from the atomic type occurrence at position v in the type labeling the node w of the natural deduction tree for (M, t) and reaches from above the atomic type occurrence at position v' in the type labeling the node w'; similarly for other combinations of $\uparrow$ and $\downarrow$. See Figure 11 for an example.

It is easy to see that for any pure typed λ-term (M, t), if there is a directed path from (w, v, d) to (w', v', d'), where $d, d' \in \{\uparrow, \downarrow\}$, then pol$(v) =$ pol(v') if and only if $d = d'$.

Note that the graph depicted in Figure 11 contains a directed cycle:

$(0, 10, \downarrow) - (1, 0, \uparrow) - (10, \epsilon, \uparrow) - (1, 1, \downarrow) - (0, 11, \uparrow) - (000, 1, \downarrow) - (001, \epsilon, \uparrow) -$
$$(0010, 0, \uparrow) - (0, 10, \downarrow)$$

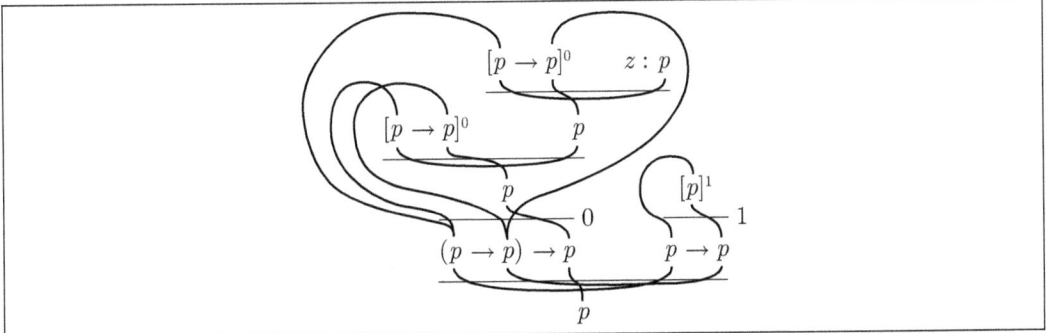

Figure 11: A natural deduction with links.

It is not hard to see that any cycle must involve the two children $w0, w1$ of a binary node w and positions u, v of $t(w1)$ such that

- $\mathrm{pol}(u) = -\mathrm{pol}(v)$,

- there is a directed path from $(w0, 1u, \uparrow)$ to $(w0, 1v, \downarrow)$ inside the subtree rooted at $w0$, and

- there is a directed path from $(w1, v, \uparrow)$ to $(w1, u, \downarrow)$ inside the subtree rooted at $w1$.

This implies that there exists an $n \geq 0$ such that $w0^n$ is a β-redex.

Lemma 3.24. *If (M, t) is a pure typed λ-form in β-normal form, then $G_{(M,t)}$ contains no directed cycle.*

Let $M = (\mathcal{T}, f, b)$ be a pure untyped λ-term with $\mathrm{FV}(M) = \{x_1, \ldots, x_n\}$, and let t be a type decoration for $x_1 : \alpha_1, \ldots, x_n : \alpha_n \Rightarrow M : \alpha_0$. We augment the graph $G_{(M,t)}$ with the nodes of the form

$$(i, v, d)$$

where $0 \leq i \leq n$, $v \in \langle \alpha_i \rangle^{(0)}$, $d \in \{\uparrow, \downarrow\}$, and the edges

$$((i, v, \uparrow), (w, v, \downarrow)) \quad \text{and} \quad ((w, v, \uparrow), (i, v, \downarrow))$$

with $1 \leq i \leq n$ and $f(w) = x_i$, and

$$((0, v, \uparrow), (\epsilon, v, \uparrow)) \quad \text{and} \quad ((\epsilon, v, \downarrow), (0, v, \downarrow)).$$

We refer to the resulting extended graph as $\overline{G}_{(M,t)}$. Note that when $\overline{G}_{(M,t)}$ has a directed path from $(i, v, \uparrow)$ to (w, u, d) with $w \in \mathcal{T}$, we have $\mathrm{pol}(i, v) = \mathrm{pol}(u)$ if and only if $d = \uparrow$, and likewise when $\overline{G}_{(M,t)}$ has a directed path from (w, u, d) to $(i, v, \downarrow)$.

In terms of $\overline{G}_{(M,t)}$, we define two binary relations on the set

$$\{ (i, v) \mid 0 \leq i \leq n, v \in \langle \alpha_i \rangle^{(0)} \}$$

of occurrences of atomic types in $x_1 : \alpha_1, \ldots, x_n : \alpha_n \Rightarrow \alpha_0$. We say that (i, v) is *linked* to (i', v') in (M, t) if $\overline{G}_{(M,t)}$ contains a directed path from $(i, v, \uparrow)$ to $(i', v', \downarrow)$. We say that (i, v) is *connected* to (i', v') in (M, t) if $\overline{G}_{(M,t)}$ contains an *undirected* path from (i, v, d) to (i', v', d') for some $d, d' \in \{\uparrow, \downarrow\}$. Note that the relation of being linked is symmetric, but not necessarily transitive; the relation of being connected is symmetric and transitive. Clearly, if (i, v) is connected to (i', v'), then $\mathrm{subtype}(\alpha_i, v) = \mathrm{subtype}(\alpha_{i'}, v')$.

The following is clear from the definitions of $\overline{G}_{(M,t)}$ and of principal typing:

Lemma 3.25. *Let M be a pure λ-term and t be a principal type decoration of M, with the associated principal typing $x_1 : \alpha_1, \ldots, x_n : \alpha_n \Rightarrow \alpha_0$. Then (i, v) and (i', v') are connected in (M, t) if and only if $\mathrm{subtype}(\alpha_i, v) = \mathrm{subtype}(\alpha_{i'}, v')$.*

It is clear that the graph $\overline{G}_{(M,t)}$, where $M = (\mathcal{T}, f, b)$, is completely determined by M and $\{ (w, \langle t(w) \rangle) \mid w \in \mathcal{T} \}$. This means that if σ is a type relabeling, $\overline{G}_{(M,t)} = \overline{G}_{(M, \sigma \circ t)}$. Thus, Lemmas 3.22 and 3.25 give

Lemma 3.26. *Let M be a pure λ-term and t be a type decoration of M under the type environment $x : \gamma_1, \ldots, x_n : \gamma_n$ such that (M, t) is in strict η-long form. Let $x_1 : \alpha_1, \ldots, x_n : \alpha_n \Rightarrow \alpha_0$ be a principal typing of M. Then (i, v) and (i', v') are connected in (M, t) if and only if $\mathrm{subtype}(\alpha_i, v) = \mathrm{subtype}(\alpha_{i'}, v')$.*

Moreover, we have

Lemma 3.27. *Let M be a pure λI-term in β-normal form and t be a type decoration of M under the type environment $x : \gamma_1, \ldots, x_n : \gamma_n$ such that (M, t) is in η-long form. Let $x_1 : \alpha_1, \ldots, x_n : \alpha_n \Rightarrow \alpha_0$ be a principal typing of M. Then (i, v) and (i', v') are related by the transitive closure of the relation of being linked in (M, t) if and only if $\mathrm{subtype}(\alpha_i, v) = \mathrm{subtype}(\alpha_{i'}, v')$.*

Proof. Since $\overline{G}_{(M,t)}$ does not contain any directed cycles, the fact that M is a λI-term implies that every directed path can be extended to one that starts in a node of the form $(i, v, \uparrow)$ and ends with one that ends in a node of the form $(i', v', \downarrow)$. □

The usefulness of the notion of being linked will become clear later.

3.2 Context-free λ-term grammars

A *context-free λ-term grammar* (CFLG) is a quintuple $\mathcal{G} = (\mathcal{N}, \Sigma, f, \mathcal{P}, S)$, where $\mathcal{N}$ is a finite alphabet of *nonterminals*, $\Sigma = (A, C, \tau)$ is a higher-order signature, f is a function from $\mathcal{N}$ to $\mathcal{T}(A)$, S is a distinguished member of $\mathcal{N}$, and $\mathcal{P}$ is a finite set of *rules* of the form:

$$B(M) :\!\!- B_1(X_1), \ldots, B_n(X_n),$$

where $X_1, \ldots, X_n$ are pairwise distinct variables and M is a λ-term in $\Lambda(\Sigma)$ that is in η-long form relative to

$$X_1 : f(B_1), \ldots, X_n : f(B_n) \Rightarrow f(B).$$

It is not required that M be in β-normal form. The language of a CFLG $\mathcal{G} = (\mathcal{N}, \Sigma, f, \mathcal{P}, S)$ is defined in terms of the predicate $\vdash_\mathcal{G}$. For a nonterminal $B \in \mathcal{N}$ and a closed λ-term $P \in \Lambda(\Sigma)$,

$$\vdash_\mathcal{G} B(P)$$

holds if and only if there exist a rule

$$B(M) :\!\!- B_1(X_1), \ldots, B_n(X_n)$$

in $\mathcal{P}$ and closed λ-terms Q_i ($i = 1, \ldots, n$) such that

$$P = M[X_1 := Q_1, \ldots, X_n := Q_n],$$
$$\vdash_\mathcal{G} B_i(Q_i).$$

When this holds, we have a *derivation tree* for $B(P)$ of the form

$$\begin{array}{c} B(P) \\ \overbrace{T_1 \ \ldots \ T_n} \end{array}$$

where T_i is a derivation tree for $B_i(Q_i)$ ($i = 1, \ldots, n$). Note that $\vdash_\mathcal{G} B(P)$ implies $\vdash_\Sigma P : f(B)$.

The language of $\mathcal{G}$ is

$$L(\mathcal{G}) = \{\, |N|_\beta \mid \vdash_\mathcal{G} S(N) \,\}.$$

Thus, the language of a CFLG is a set of closed β-normal λ-terms that are in η-long form relative to a certain type (namely $f(S)$) (cf. Lemmas 3.16 and 3.17).

In example grammars we have given, we have not always adhered to the condition that λ-terms in rules be in η-long form. Any rule with a non-η-long λ-term M should be understood as an abbreviation for the "official" rule that has the η-long form of M instead. The reason that we only allow λ-terms in η-long form in the language of a CFLG is that we do not wish to distinguish between λ-terms that are η-equal.

Example 3.28. The earlier example CFLG (9) in official notation is $\mathscr{G} = (\mathscr{N}, \Sigma, f, \mathscr{P}, \mathsf{S})$, where

$$\mathscr{N} = \{\mathsf{S}, \mathsf{NP}, \mathsf{VP}, \mathsf{V}, \mathsf{Det}, \mathsf{N}\},$$
$$\Sigma = (A, C, \tau),$$
$$A = \{e, t\},$$
$$C = \{\wedge, \mathbf{John}, \mathbf{find}, \mathbf{catch}, =, \exists, \mathbf{man}, \mathbf{unicorn}\},$$

$$\tau = \begin{cases} \wedge \mapsto t \to t \to t, \\ \mathbf{John} \mapsto e, \\ \mathbf{find} \mapsto e \to e \to t, \\ \mathbf{catch} \mapsto e \to e \to t, \\ = \mapsto e \to e \to t, \\ \exists \mapsto (e \to t) \to t, \\ \mathbf{man} \mapsto e \to t, \\ \mathbf{uncorn} \mapsto e \to t \end{cases},$$

$$f = \begin{cases} \mathsf{S} \mapsto t, \\ \mathsf{NP} \mapsto (e \to t) \to t, \\ \mathsf{VP} \mapsto e \to t, \\ \mathsf{V} \mapsto e \to e \to t, \\ \mathsf{Det} \mapsto (e \to t) \to (e \to t) \to t, \\ \mathsf{N} \mapsto e \to t \end{cases},$$

and $\mathscr{P}$ consists of the following rules:

$\mathsf{S}(X_1(\lambda x.X_2 x)) :\!\!- \mathsf{NP}(X_1), \mathsf{VP}(X_2).$
$\mathsf{VP}(\lambda x.X_2(\lambda y.X_1 y x)) :\!\!- \mathsf{V}(X_1), \mathsf{NP}(X_2).$
$\mathsf{V}(\lambda y x.\wedge(X_1 y x)(X_2 y x)) :\!\!- \mathsf{V}(X_1), \mathsf{V}(X_2).$
$\mathsf{NP}(\lambda u.X_1(\lambda x.X_2 x)(\lambda x.u x)) :\!\!- \mathsf{Det}(X_1), \mathsf{N}(X_2).$
$\mathsf{NP}(\lambda u.u\,\mathbf{John}).$

$\mathsf{V}(\lambda yx.\mathbf{find}\ y\ x)$.
$\mathsf{V}(\lambda yx.\mathbf{catch}\ y\ x)$.
$\mathsf{V}(\lambda yx.\mathbf{=}\ y\ x)$.
$\mathsf{Det}(\lambda uv.\exists(\lambda y.\wedge(uy)(vy)))$.
$\mathsf{N}(\lambda x.\mathbf{man}\ x)$.
$\mathsf{N}(\lambda x.\mathbf{unicorn}\ x)$.

3.3 Datalog programs associated with CFLGs

We associate with a CFLG $\mathscr{G} = (\mathscr{N}, \Sigma, f, \mathscr{P}, S)$, where $\Sigma = (A, C, \tau)$, a Datalog program program($\mathscr{G}$), whose set of intensional predicates is $\mathscr{N}$ and whose set of extensional predicates is C. The arity of $B \in \mathscr{N}$ is $|f(B)|$, and the arity of $d \in C$ is $|\tau(d)|$.

In order to facilitate the definition of program($\mathscr{G}$) and the statement of the next lemma, we adopt the following conventions:

Convention 1. If
$$B(M) :\!- B_1(X_1), \ldots, B_n(X_n)$$
is a rule in $\mathscr{P}$, then M is in strict η-long form relative to
$$X_1 : f(B_1), \ldots, X_n : f(B_n) \Rightarrow f(B).$$

Convention 2. If
$$B(M) :\!- B_1(X_1), \ldots, B_n(X_n)$$
is a rule in $\mathscr{P}$, then $\overrightarrow{\mathrm{FV}}(M) = (X_1, \ldots, X_n)$ and all occurrences of constants in M precede the occurrences of $X_1, \ldots, X_n$.

It is easy to transform a rule that does not obey these conventions into an equivalent one that does by changing M to the strict η-long form of $(\lambda X_1 \ldots X_n.M) X_1 \ldots X_n$, so adopting this convention does not lead to any loss of generality. It is also possible to complicate the definition of program($\mathscr{G}$) and the statement and proof of the lemma to make the following results not depend on the conventions.

We now give the definition of program($\mathscr{G}$), assuming Conventions 1 and 2. Consider a rule
$$\pi = B_0(M) :\!- B_1(X_1), \ldots, B_n(X_n),$$
in $\mathscr{P}$. Let
$$\overrightarrow{\mathrm{Con}}(M) = (d_1, \ldots, d_m),$$

and let

$$y_1 : \beta_1, \ldots, y_m : \beta_m, X_1 : \alpha_1, \ldots, X_n : \alpha_n \Rightarrow \widehat{M}[y_1, \ldots, y_m] : \alpha_0$$

be a principal typing of $\widehat{M}[y_1, \ldots, y_m]$. (Recall that $\widehat{M}[y_1, \ldots, y_m]$ is a pure λ-term such that $\widehat{M}[d_1, \ldots, d_m] = M$.) Note that by Convention 2, $\overrightarrow{\mathrm{FV}}(\widehat{M}[y_1, \ldots, y_m]) = (y_1, \ldots, y_m, X_1, \ldots, X_n)$. By Convention 1 and Remark 3.23,

$$\begin{aligned} \langle \beta_i \rangle &= \langle \tau(d_i) \rangle \quad \text{for } i = 1, \ldots, m, \\ \langle \alpha_i \rangle &= \langle f(B_i) \rangle \quad \text{for } i = 0, \ldots, n. \end{aligned} \tag{38}$$

The Datalog rule ρ_π corresponding to π is defined as

$$B_0(\overline{\alpha_0}) :\!- d_1(\overline{\beta_1}), \ldots, d_m(\overline{\beta_m}), B_1(\overline{\alpha_1}), \ldots, B_n(\overline{\alpha_n}),$$

where atomic types in $\overline{\alpha_i}, \overline{\beta_i}$ are considered Datalog variables. Clearly, ρ_π does not depend on the choice of variables $y_1, \ldots, y_m$. Also, the choice of atomic types in $\overline{\alpha_i}, \overline{\beta_i}$ is immaterial. So it does not matter which principal typing of $\widehat{M}[y_1, \ldots, y_m]$ we use.[32]

The Datalog program associated with $\mathscr{G}$ is defined as

$$\mathrm{program}(\mathscr{G}) = \{\, \rho_\pi \mid \pi \in \mathscr{P} \,\}.$$

Remark 3.29.

$$B_0(\vec{s}_0) :\!- d_1(\vec{t}_1), \ldots, d_m(\vec{t}_m), B_1(\vec{s}_1), \ldots, B_n(\vec{s}_n)$$

is an instance of ρ_π if and only if

$$\vdash y_1 : \langle \tau(d_1) \rangle(\vec{t}_1), \ldots, y_m : \langle \tau(d_m) \rangle(\vec{t}_m), X_1 : \langle f(B_1) \rangle(\vec{s}_1), \ldots, X_n : \langle f(B_n) \rangle(\vec{s}_n)$$
$$\Rightarrow \widehat{M}[y_1, \ldots, y_m] : \langle f(B_0) \rangle(\vec{s}_0).$$

Example 3.30. For the CFLG $\mathscr{G}$ of Example 3.28, $\mathrm{program}(\mathscr{G})$ consists of the rules in (16) in Section 2.2. For example, let M_3 be the λ-term in the third rule π_3 of $\mathscr{G}$. We have $\overrightarrow{\mathrm{Con}}(M_3) = (\wedge)$, and the following is a principal typing of $\widehat{M_3}[z_1]$:

$$z_1 : i_2 \to i_5 \to i_1, X_1 : i_3 \to i_4 \to i_2, X_2 : i_3 \to i_4 \to i_5 \Rightarrow \lambda y x. z_1(X_1 y x)(X_2 y x) : i_3 \to i_4 \to i_1.$$

---

[32]This definition of ρ_π is applicable to arbitrary CFLG rules satisfying Conventions 1 and 2. When M is almost linear, the definition of ρ_π given here is equivalent to the definition given in Section 2.2 in terms of the hypergraph representation $\mathrm{graph}(M)$ of a principal typing of M.

We thus obtain
$$\rho_{\pi_3} = \mathsf{V}(i_1, i_4, i_3) := \wedge(i_3, i_5, i_2), \mathsf{V}(i_2, i_4, i_3), \mathsf{V}(i_5, i_4, i_3).$$

Note that the hypergraph representation (15) of M_3 encodes the same information as its principal typing.

The following is a key fact about $\text{program}(\mathscr{G})$ that holds of any CFLG $\mathscr{G}$ satisfying Conventions 1 and 2. It basically says that under the correspondence between π and ρ_π defined above, a CFLG derivation tree plus a typing (of a certain kind) for the associated λ-term corresponds to a Datalog derivation tree, and vice versa. Its proof is quite straightforward, if rather tedious. If $\vec{u}$ is a tuple (sequence) of constants, we let $|\vec{u}|$ denote its length, i.e., the number of its components.

Lemma 3.31. *Let $\mathscr{G} = (\mathscr{N}, \Sigma, f, \mathscr{P}, S)$ with $\Sigma = (A, C, \tau)$ be a CFLG, and let U be some set of constants. Let $e_1, \ldots, e_l \in C$, $B \in \mathscr{N}$, and $\vec{u}_1, \ldots, \vec{u}_l, \vec{s}$ be sequences of constants from U such that $|\vec{u}_i| = |\tau(e_i)|$ and $|\vec{s}| = |f(B)|$. The following are equivalent:*

(i) *There exists $P \in \Lambda(\Sigma)$ such that*
$$\vdash_{\mathscr{G}} B(P),$$
$$\overrightarrow{\mathrm{Con}}(P) = (e_1, \ldots, e_l),$$
$$\vdash z_1 : \langle \tau(e_1) \rangle(\vec{u}_1), \ldots, z_l : \langle \tau(e_l) \rangle(\vec{u}_l) \Rightarrow \widehat{P}[z_1, \ldots, z_l] : \langle f(B) \rangle(\vec{s}).$$

(ii) *There exists a derivation tree T for*
$$\text{program}(\mathscr{G}) \cup \{ e_i(\vec{u}_i) \mid 1 \leq i \leq l \} \vdash B(\vec{s})$$
such that $(e_1(\vec{u}_1), \ldots, e_l(\vec{u}_l))$ lists the labels of the extensional nodes of T in the order from left to right.

Proof. (i) $\Rightarrow$ (ii). Induction on the derivation of $\vdash_{\mathscr{G}} B(P)$. Assume that $\vdash_{\mathscr{G}} B(P)$ is inferred from
$$\vdash_{\mathscr{G}} B_i(P_i) \quad (i = 1, \ldots, n)$$
using a rule
$$\pi = B(M) := B_1(X_1), \ldots, B_n(X_n)$$
such that
$$P = M[X_1 := P_1, \ldots, X_n := P_n].$$

Let $m = |\overrightarrow{\mathrm{Con}}(M)|$ and $l_i = |\overrightarrow{\mathrm{Con}}(P_i)|$ for $i = 1, \ldots, n$. Then $l = m + l_1 + \cdots + l_n$ and

$$\widehat{P}[z_1, \ldots, z_l] = \widehat{M}[z_1, \ldots, z_m]$$
$$[X_1 := \widehat{P_1}[z_{h(1,1)}, \ldots, z_{h(1,l_1)}], \ldots, X_n := \widehat{P_n}[z_{h(n,1)}, \ldots, z_{h(n,l_n)}]],$$
$$\overrightarrow{\mathrm{Con}}(M) = (e_1, \ldots, e_m),$$
$$\overrightarrow{\mathrm{Con}}(P_i) = (e_{h(i,1)}, \ldots, e_{h(i,l_i)}),$$

where
$$h(i, j) = m + l_1 + \cdots + l_{i-1} + j.$$

By assumption,
$$\vdash z_1 : \langle \tau(e_1) \rangle(\vec{u}_1), \ldots, z_l : \langle \tau(e_l) \rangle(\vec{u}_l) \Rightarrow \widehat{P}[z_1, \ldots, z_l] : \langle f(B) \rangle(\vec{s}). \qquad (39)$$

By Lemma 3.13, any type decoration for (39) splits into type decorations for

$$z_1 : \langle \tau(e_1) \rangle(\vec{u}_1), \ldots, z_m : \langle \tau(e_m) \rangle(\vec{u}_m), X_1 : \alpha_1, \ldots, X_n : \alpha_n$$
$$\Rightarrow \widehat{M}[z_1, \ldots, z_m] : \langle f(B) \rangle(\vec{s})$$

and

$$z_{h(i,1)} : \langle \tau(e_{h(i,1)}) \rangle(\vec{u}_{h(i,1)}), \ldots, z_{h(i,l_i)} : \langle \tau(e_{h(i,l_i)}) \rangle(\vec{u}_{h(i,l_i)})$$
$$\Rightarrow \widehat{P_i}[z_{h(i,1)}, \ldots, z_{h(i,l_i)}] : \alpha_i \qquad (40)$$

($i = 1, \ldots, n$). In order to apply the induction hypothesis to (40), we need
$$\langle \alpha_i \rangle = \langle f(B_i) \rangle \quad \text{for each } i = 1, \ldots, n. \qquad (41)$$

If $\widehat{P}[z_1, \ldots, z_l]$ is not λI, type decorations for (39) need not be unique, and indeed there may be one for which (41) fails. We show that a desirable type decoration for (39) can be obtained from a principal typing of $\widehat{P}[z_1, \ldots, z_l]$ by type relabeling.

Since Convention 1 ensures that P_i is in strict η-long form relative to $f(B_i)$ and M is in strict η-long form relative to $X_1 : f(B_1), \ldots, X_n : f(B_n) \Rightarrow f(B)$, it follows that $\widehat{P_i}[z_{h(i,1)}, \ldots, z_{h(i,l_i)}]$ is in strict η-long form relative to $z_{h(i,1)} : \tau(e_{h(i,1)}), \ldots, z_{h(i,l_i)} : \tau(e_{h(i,l_i)}) \Rightarrow f(B_i)$ and $\widehat{M}[z_1, \ldots, z_m]$ is in strict η-long form relative to $z_1 : \tau(e_1), \ldots, z_m : \tau(e_m), X_1 : f(B_1), \ldots, X_n : f(B_n) \Rightarrow f(B)$. By Lemma 3.21, there is a type decoration $t_{\widehat{P}[z_1, \ldots, z_l]}$ for

$$z_1 : \tau(e_1), \ldots, z_l : \tau(e_l) \Rightarrow \widehat{P}[z_1, \ldots, z_l] : f(B)$$

that is obtained by combining the type decoration for

$$z_1 : \tau(e_1), \ldots, z_m : \tau(e_m), X_1 : f(B_1), \ldots, X_n : f(B_n) \Rightarrow \widehat{M}[z_1, \ldots, z_m] : f(B)$$

and the type decoration for

$$z_{h(i,1)} : \tau(e_{h(i,1)}), \ldots, z_{h(i,l_i)} : \tau(e_{h(i,l_i)}) \Rightarrow \widehat{P_i}[z_{h(i,1)}, \ldots, z_{h(i,l_i)}] : f(B_i)$$

for $i = 1, \ldots, n$, such that $(\widehat{P}[z_1, \ldots, z_l], t_{\widehat{P}[z_1,\ldots,z_l]})$ is in strict η-long form.

Let $\widetilde{t}_{\widehat{P}[z_1,\ldots,z_l]}$ be a principal type decoration for $\widehat{P}[z_1, \ldots, z_l]$ with the associated principal typing

$$z_1 : \delta_1, \ldots, z_l : \delta_l \Rightarrow \gamma.$$

By Lemma 3.13, $\widetilde{t}_{\widehat{P}[z_1,\ldots,z_l]}$ splits into type decorations for

$$z_1 : \delta_1, \ldots, z_m : \delta_m, X_1 : \gamma_1, \ldots, X_n : \gamma_n \Rightarrow \widehat{M}[z_1, \ldots, z_m] : \gamma \tag{42}$$

and

$$z_{h(i,1)} : \delta_{h(i,1)}, \ldots, z_{h(i,l_i)} : \delta_{h(i,l_i)} \Rightarrow \widehat{P_i}[z_{h(i,1)}, \ldots, z_{h(i,l_i)}] : \gamma_i \quad (i = 1, \ldots, n). \tag{43}$$

By Lemma 3.22, we must have

$$\langle \delta_i \rangle = \langle \tau(e_i) \rangle \quad \text{for } i = 1, \ldots, l,$$
$$\langle \gamma \rangle = \langle f(B) \rangle,$$
$$\langle \gamma_i \rangle = \langle f(B_i) \rangle \quad \text{for } i = 1, \ldots, n.$$

By (39), there is a type substitution σ such that

$$\delta_i \sigma = \langle \tau(e_i) \rangle (\vec{u}_i),$$
$$\gamma \sigma = \langle f(B) \rangle (\vec{s})$$

that leaves atomic types that do not appear in $\delta_1, \ldots, \delta_l, \gamma$ unchanged. Then σ is a type relabeling, and there are sequences $\vec{s}_1, \ldots, \vec{s}_n$ of atomic types such that

$$\gamma_i \sigma = \langle f(B_i) \rangle (\vec{s}_i) \quad \text{for } i = 1, \ldots, n.$$

Without loss of generality, we may assume that $\vec{s}_1, \ldots, \vec{s}_n$ are sequences of constants from U. (Otherwise we may replace any constants not in U by constants in U.)

Applying σ to (42) and (43), we get

$$\vdash z_1 : \langle \tau(e_1)\rangle(\vec{u}_1), \ldots, z_m : \langle \tau(e_m)\rangle(\vec{u}_m), X_1 : \langle f(B_1)\rangle(\vec{s}_1), \ldots, X_n : \langle f(B_n)\rangle(\vec{s}_n)$$
$$\Rightarrow \widehat{M}[z_1, \ldots, z_l] : \langle f(B)\rangle(\vec{s}), \quad (44)$$

and

$$\vdash z_{h(i,1)} : \langle \tau(e_{h(i,1)})\rangle(\vec{u}_{h(i,1)}), \ldots, z_{h(i,l_i)} : \langle \tau(e_{h(i,l_i)})\rangle(\vec{u}_{h(i,l_i)})$$
$$\Rightarrow \widehat{P_i}[z_{h(i,1)}, \ldots, z_{h(i,l_i)}]\langle f(B_i)\rangle(\vec{s}_i) \quad (45)$$

for $i = 1, \ldots, n$.

By (45), the induction hypothesis applies to P_i, giving a Datalog derivation tree T_i for

$$\operatorname{program}(\mathcal{G}) \cup \{e_{h(i,j)}(\vec{u}_{h(i,j)}) \mid 1 \leq j \leq l_i\} \vdash B_i(\vec{s}_i) \quad (46)$$

such that $(e_{h(i,1)}(\vec{u}_{h(i,1)}), \ldots, e_{h(i,l_i)}(\vec{u}_{h(i,l_i)}))$ lists the labels of the extensional nodes of T_i from left to right.

By (44) and Remark 3.29,

$$B(\vec{s}) := e_1(\vec{u}_1), \ldots, e_m(\vec{u}_m), B_1(\vec{s}_1), \ldots, B_n(\vec{s}_n) \quad (47)$$

is an instance of ρ_π. Combining (46) and (47), we obtain a Datalog derivation tree T for

$$\operatorname{program}(\mathcal{G}) \cup \{e_i(\vec{u}_i) \mid 1 \leq i \leq l\} \vdash B(\vec{s}),$$

such that $(e_1(\vec{u}_1), \ldots, e_l(\vec{u}_l))$ lists the labels of the extensional nodes of T from left to right.

(ii) $\Rightarrow$ (i). Induction on T. Assume that T is of the form

$$\overbrace{e_1(\vec{u}_1) \quad \cdots \quad e_m(\vec{u}_m) \quad T_1 \quad \cdots \quad T_n}^{p(\vec{s})}$$

and the root node of T is obtained by an application of an instance

$$B(\vec{s}) := e_1(\vec{u}_1), \ldots, e_m(\vec{u}_m), B_1(\vec{s}_1), \ldots, B_n(\vec{s}_n)$$

of some ρ_π, where $m \leq l$ and

$$\pi = B(M) := B_1(X_1), \ldots, B_n(X_n), \quad (48)$$
$$\overrightarrow{\operatorname{Con}}(M) = (e_1, \ldots, e_m). \quad (49)$$

Let l_i be the number of extensional nodes of T_i. Then $l = m + l_1 + \cdots + l_n$, and for $i = 1, \ldots, n$, T_i is a derivation tree for

$$\text{program}(\mathscr{G}) \cup \{e_{h(i,1)}(\vec{u}_{h(i,1)}), \ldots, e_{h(i,l_i)}(\vec{u}_{h(i,l_i)})\} \vdash B_i(\vec{s}_i),$$

where $h(i,j) = m + l_1 + \cdots + l_{i-1} + j$, and $(e_{h(i,1)}(\vec{u}_{h(i,1)}), \ldots, e_{h(i,l_i)}(\vec{u}_{h(i,l_i)}))$ lists the labels of the extensional nodes of T_i from left to right.

By Remark 3.29, we have

$$\vdash z_1 : \langle \tau(e_1) \rangle(\vec{u}_1), \ldots, z_m : \langle \tau(e_m) \rangle(\vec{u}_m), X_1 : \langle f(B_1) \rangle(\vec{s}_1), \ldots, X_n : \langle f(B_n) \rangle(\vec{s}_n)$$
$$\Rightarrow \widehat{M}[z_1, \ldots, z_m] : \langle f(B) \rangle(\vec{s}). \quad (50)$$

By induction hypothesis, for $i = 1, \ldots, n$, there exists $P_i \in \Lambda(\Sigma)$ such that

$$\vdash_{\mathscr{G}} B_i(P_i), \quad (51)$$
$$\overrightarrow{\text{Con}}(P_i) = (e_{h(i,1)}, \ldots, e_{h(i,l_i)}), \quad (52)$$

and

$$\vdash z_{h(i,1)} : \langle \tau(e_{h(i,1)}) \rangle(\vec{u}_{h(i,1)}), \ldots, z_{h(i,l_i)} : \langle \tau(e_{h(i,l_i)}) \rangle(\vec{u}_{h(i,l_i)})$$
$$\Rightarrow \widehat{P}_i[z_{h(i,1)}, \ldots, z_{h(i,l_i)}] : \langle f(B_i) \rangle(\vec{s}_i). \quad (53)$$

Let
$$P = M[X_1 := P_1, \ldots, X_n := P_n].$$

Then by (48), (51), (49), and (52),

$$\vdash_{\mathscr{G}} B(P),$$
$$\overrightarrow{\text{Con}}(P) = (e_1, \ldots, e_m, e_{h(1,1)}, \ldots, e_{h(1,l_1)}, \ldots, e_{h(n,1)}, \ldots, e_{h(n,l_n)})$$
$$= (e_1, \ldots, e_l).$$

We have

$$\widehat{P}[z_1, \ldots, z_l] =$$
$$\widehat{M}[z_1, \ldots, z_m][X_1 := \widehat{P}_1[z_{h(1,1)}, \ldots, z_{h(1,l_1)}], \ldots, X_n := \widehat{P}_n[z_{h(n,1)}, \ldots, z_{h(n,l_n)}]].$$

By Lemma 3.12 applied to (50) and (53), we get

$$\vdash z_1 : \langle \tau(e_1) \rangle(\vec{u}_1), \ldots, z_m : \langle \tau(e_m) \rangle(\vec{u}_m),$$

$$z_{h(1,1)} : \langle \tau(e_{h(1,1)}) \rangle (\vec{u}_{h(1,1)}), \ldots, z_{h(1,l_1)} : \langle \tau(e_{h(1,l_1)}) \rangle (\vec{u}_{h(1,l_1)}),$$
$$\vdots$$
$$z_{h(n,1)} : \langle \tau(e_{h(n,1)}) \rangle (\vec{u}_{h(n,1)}), \ldots, z_{h(n,l_n)} : \langle \tau(e_{h(n,l_n)}) \rangle (\vec{u}_{h(n,l_n)})$$
$$\Rightarrow \widehat{P}[z_1, \ldots, z_l] : \langle f(B) \rangle (\vec{s}).$$

Hence P satisfies the required properties. □

Example 3.32. Let $\mathscr{G}$ be the CFLG of Example 3.28. Let

$$P = (\lambda u.u\ \mathbf{John})(\lambda x.$$
$$(\lambda x.$$
$$(\lambda u.$$
$$(\lambda uv.\exists(\lambda y.\wedge(uy)(vy)))$$
$$(\lambda x.$$
$$(\lambda x.\mathbf{unicorn}\ x)$$
$$x)(\lambda x.ux))$$
$$(\lambda y.$$
$$(\lambda yx.\mathbf{find}\ y\ x)$$
$$y\ x))$$
$$x).$$

Then $\vdash_{\mathscr{G}} \mathsf{S}(P)$. The derivation tree for $\mathsf{S}(P)$ (in abbreviated notation) was shown in Figure 5 in Section 2.2. We have $\overrightarrow{\mathsf{Con}}(P) = (\mathbf{John}, \exists, \wedge, \mathbf{unicorn}, \mathbf{find})$ and

$$\widehat{P}[z_1, z_2, z_3, z_4, z_5] = (\lambda u.uz_1)(\lambda x.$$
$$(\lambda x.$$
$$(\lambda u.$$
$$(\lambda uv.z_2(\lambda y.z_3(uy)(vy)))$$
$$(\lambda x.$$
$$(\lambda x.z_4 x)$$
$$x)(\lambda x.ux))$$
$$(\lambda y.$$
$$(\lambda yx.z_5 yx)$$
$$yx))$$
$$x).$$

By one direction of Lemma 3.31, whenever we have

$$\vdash z_1 : u_{1,1}, z_2 : (u_{2,3} \to u_{2,2}) \to u_{2,1},$$
$$z_3 : u_{3,3} \to u_{3,2} \to u_{3,1}, z_4 : u_{4,2} \to u_{4,1}, z_5 : u_{5,3} \to u_{5,2} \to u_{5,1}$$
$$\Rightarrow \widehat{P}[z_1, z_2, z_3, z_4, z_5] : s, \quad (54)$$

we must have

$$\text{program}(\mathscr{G}) \cup \{\mathbf{John}(u_{1,1}), \exists(u_{2,1}, u_{2,2}, u_{2,3}), \wedge(u_{3,1}, u_{3,2}, u_{3,3}),$$
$$\mathbf{unicorn}(u_{4,1}, u_{4,2}), \mathbf{find}(u_{5,1}, u_{5,2}, u_{5,3})\} \vdash \mathsf{S}(s). \quad (55)$$

The Datalog derivation tree for (55) will have the same shape as the one in Figure 6 in Section 2.2. Conversely, whenever (55) has a derivation tree of this shape, we must have (54), by (the proof of) the other direction of Lemma 3.31

Let $\Sigma = (A, C, \tau)$ be a higher-order signature and U be some set of database constants. We write $\mathcal{D}_{\Sigma,U}$ for the database schema (C, U), where each $d \in C$ has arity $|\tau(d)|$. Let D be a database over $\mathcal{D}_{\Sigma,U}$ and $\alpha \in \mathscr{T}(A)$. We define a set $\Lambda(D, \alpha)$ of closed λ-terms over Σ as follows:

$$\Lambda(D, \alpha) = \left\{ M \in \Lambda(\Sigma) \;\middle|\; \begin{array}{l} \mathrm{FV}(M) = \varnothing, \overrightarrow{\mathrm{Con}}(M) = (d_1, \ldots, d_n), \{d_1(\vec{s}_1), \ldots, d_n(\vec{s}_n)\} \subseteq D, \\ \vdash z_1 : \langle \tau(d_1) \rangle(\vec{s}_1), \ldots, z_n : \langle \tau(d_n) \rangle(\vec{s}_n) \Rightarrow \widehat{M}[z_1, \ldots, z_n] : \alpha \end{array} \right\}.$$

Example 3.33. Let Σ' be the extension of the higher-order signature Σ in Example 3.28 with an additional constant $\neg$ of type $t \to t$. Let $U = \{a, b, 0, 1\}$, and consider the following database D over $\mathcal{D}_{\Sigma',U}$:

$$\mathbf{man}(1, a), \;\; \mathbf{man}(0, b), \;\; \mathbf{unicorn}(0, a), \;\; \mathbf{unicorn}(1, b),$$
$$\wedge(1, 1, 1), \;\; \wedge(0, 1, 0), \;\; \wedge(0, 0, 1), \;\; \wedge(0, 0, 0), \;\; \neg(0, 1), \;\; \neg(1, 0),$$
$$\exists(1, 1, a), \;\; \exists(1, 1, b).$$

The set $\Lambda(D, 1)$ contains, e.g.,

$$\wedge(\exists(\lambda x.\mathbf{man}\; x))(\exists(\lambda y.\mathbf{unicorn}\; y)),$$
$$\exists(\lambda x.\wedge(\mathbf{man}\; x)(\neg(\mathbf{unicorn}\; x))),$$

but not, e.g.,

$$\exists(\lambda x.\wedge(\mathbf{man}\; x)(\mathbf{unicorn}\; x)).$$

Lemma 3.34. *Let* $M, M' \in \Lambda(\Sigma)$.

(i) *If* $M \twoheadrightarrow_\beta M'$, *then* $M \in \Lambda(D, \alpha)$ *implies* $M' \in \Lambda(D, \alpha)$.

(ii) *If* $M' \twoheadrightarrow_\beta M$ *by non-erasing non-duplicating β-reduction, then* $M \in \Lambda(D, \alpha)$ *implies* $M' \in \Lambda(D, \alpha)$.

Proof. Let $M \in \Lambda(D, \alpha)$ and $\overrightarrow{\mathrm{Con}}(M) = (d_1, \ldots, d_n)$. By the definition of $\Lambda(D, \alpha)$, for some $\vec{s}_1, \ldots, \vec{s}_n$ such that $\{d_1(\vec{s}_1), \ldots, d_n(\vec{s}_n)\} \subseteq D$,

$$\vdash z_1 : \langle \tau(d_1) \rangle(\vec{s}_1), \ldots, z_1 : \langle \tau(d_n) \rangle(\vec{s}_n) \Rightarrow \widehat{M}[z_1, \ldots, z_n] : \alpha.$$

(i). Suppose $M \twoheadrightarrow_\beta M'$. Let $m = |\overrightarrow{\mathrm{Con}}(M')|$ and $g \colon \{1, \ldots, m\} \to \{1, \ldots, n\}$ be the function such that the ith occurrence of a constant in M' is a descendant of the $g(i)$th occurrence of a constant in M. Then $\overrightarrow{\mathrm{Con}}(M') = (d_{g(1)}, \ldots, d_{g(m)})$ and

$$\widehat{M}[z_1, \ldots, z_n] \twoheadrightarrow_\beta \widehat{M'}[z_{g(1)}, \ldots, z_{g(m)}].$$

By the Subject Reduction Theorem (Theorem 3.14),

$$\vdash \{ z_{g(i)} : \langle \tau(d_{g(i)}) \rangle(\vec{s}_{g(i)}) \mid 1 \le i \le m \} \Rightarrow \widehat{M'}[z_{g(1)}, \ldots, z_{g(m)}] : \alpha,$$

and thus

$$\vdash y_1 : \langle \tau(d_{g(1)}) \rangle(\vec{s}_{g(1)}), \ldots, y_m : \langle \tau(d_{g(m)}) \rangle(\vec{s}_{g(m)}) \Rightarrow \widehat{M'}[y_1, \ldots, y_m] : \alpha.$$

This shows $M' \in \Lambda(D, \alpha)$.

(ii). Suppose $M' \twoheadrightarrow_\beta M$ by non-erasing non-duplicating β-reduction. Then $|\overrightarrow{\mathrm{Con}}(M')| = n$ and there is a permutation g of $\{1, \ldots, n\}$ such that the ith occurrence of a constant in M' is the ancestor of the $g(i)$th occurrence of a constant in M. We have $\overrightarrow{\mathrm{Con}}(M') = (d_{g(1)}, \ldots, d_{g(n)})$ and $\widehat{M'}[z_{g(1)}, \ldots, z_{g(n)}] \twoheadrightarrow_\beta \widehat{M}[z_1, \ldots, z_n]$ by non-erasing non-duplicating β-reduction. By the Subject Expansion Theorem (Theorem 3.15),

$$\vdash z_1 : \langle \tau(d_1) \rangle(\vec{s}_1), \ldots, z_n : \langle \tau(d_n) \rangle(\vec{s}_n) \Rightarrow \widehat{M'}[z_{g(1)}, \ldots, z_{g(n)}] : \alpha.$$

Therefore, $M' \in \Lambda(D, \alpha)$. □

The next lemma is an immediate consequence of Lemma 3.31:

Lemma 3.35. *Let* $\mathscr{G} = (\mathcal{N}, \Sigma, f, \mathscr{P}, S)$ *be a CFLG. Let U be some set of database constants, D be a database over $\mathcal{D}_{\Sigma, U}$, and $\vec{s}$ be a sequence of constants from U such that $|\vec{s}| = |f(S)|$. The following are equivalent:*

$$\boxed{\begin{array}{c} \{\, P \in \Lambda(\Sigma) \mid \vdash_{\mathscr{G}} S(P) \,\} \cap \Lambda(D, \langle f(S)\rangle(\vec{s})) \neq \varnothing \iff \mathrm{program}(\mathscr{G}) \cup D \vdash S(\vec{s}) \\ \Downarrow \\ L(\mathscr{G}) \cap \Lambda(D, \langle f(S)\rangle(\vec{s})) \neq \varnothing \end{array}}$$

Figure 12: A general property of $\mathrm{program}(\mathscr{G})$.

(i) *There exists some* $P \in \Lambda(D, \langle f(S)\rangle(\vec{s}))$ *such that* $\vdash_{\mathscr{G}} S(P)$.

(ii) $\mathrm{program}(\mathscr{G}) \cup D \vdash S(\vec{s})$.

Lemma 3.36. *Let* $\mathscr{G}, U, D, \vec{s}$ *be as in Lemma 3.35. If* $\mathrm{program}(\mathscr{G}) \cup D \vdash S(\vec{s})$, *then* $L(\mathscr{G}) \cap \Lambda(D, \langle f(S)\rangle(\vec{s})) \neq \varnothing$.

Proof. By Lemma 3.34, part (i) and Lemma 3.35. □

See Figure 12. The converse of Lemma 3.36 does not hold in general, but we shall see below some special cases where it does hold (Theorems 3.40, 3.65, and 4.3).

3.4 Databases determined by λ-terms

Let $M \in \Lambda(\Sigma)$ be a closed λ-term in strict η-long form relative to γ such that $\overrightarrow{\mathrm{Con}}(M) = (d_1, \ldots, d_m)$. Define

$$\mathrm{database}(M) = \{\, d_i(\overline{\beta_i}) \mid 1 \leq i \leq m \,\},$$
$$\mathrm{tuple}(M) = \overline{\alpha}$$

where

$$y_1 : \beta_1, \ldots, y_m : \beta_m \Rightarrow \widehat{M}[y_1, \ldots, y_m] : \alpha$$

is a principal typing of $\widehat{M}[y_1, \ldots, y_m]$.[33] Here, atomic types that occur in $\beta_1, \ldots, \beta_m, \alpha$ are regarded as database constants. Note that by Remark 3.23, $\langle \gamma \rangle(\mathrm{tuple}(M)) = \alpha$ and $\Lambda(\mathrm{database}(M), \langle \gamma \rangle(\mathrm{tuple}(M)))$ is well-defined. The following is obvious from the above definition:

Lemma 3.37. *If* $M \in \Lambda(\Sigma)$ *is a closed λ-term in strict η-long form relative to γ, then* $M \in \Lambda(\mathrm{database}(M), \langle \gamma \rangle(\mathrm{tuple}(M)))$.

---

[33] When M is almost linear, the definition of $(\mathrm{database}(M), \mathrm{tuple}(M))$ here is equivalent to the definition in terms of $\mathrm{graph}(M)$ given in Section 2.2.

Note that if M is in strict η-long form relative to γ, then $|M|_\beta$ is in η-long form relative to γ and belongs to $\Lambda(\mathrm{database}(M), \langle\gamma\rangle(\mathrm{tuple}(M)))$ (Lemma 3.34). We shall see below that in some special cases $|M|_\beta$ is the only η-long β-normal λ-term in $\Lambda(\mathrm{database}(M), \langle\gamma\rangle(\mathrm{tuple}(M)))$ (Lemmas 3.41 and 3.54).

Example 3.38. Consider the λ-term (8) from Sections 2.1–2.2:

$$M = \exists(\lambda y. \wedge (\mathbf{unicorn}\ y)(\mathbf{find}\ y\ \mathbf{John})).$$

Using the principle typing

$$z_1:(4\to 2)\to 1, z_2:3\to 5\to 2, z_3:4\to 3, z_4:4\to 6\to 5, z_5:6 \Rightarrow z_1(\lambda y.z_2(z_3 y)(z_4 y z_5)):1$$

of $\widehat{M}[z_1, z_2, z_3, z_4, z_5]$, we obtain

$$\mathrm{database}(M) = \{\exists(1,2,4), \wedge(2,5,3), \mathbf{unicorn}(3,4), \mathbf{find}(5,6,4), \mathbf{John}(6)\},$$
$$\mathrm{tuple}(M) = (1).$$

Lemma 3.54 below implies that M is the only λ-term in $\Lambda(\mathrm{database}(M), 1)$ that is in η-long β-normal form relative to the type t.

3.5 From CFLGs to Datalog: The case of linear CFLGs

We first treat the special case of linear CFLGs because the reduction to Datalog as well as the proof of its correctness can be made much simpler in this case than in the more general case of almost linear CFLGs.

The crucial property is the following:

Lemma 3.39. *Let $\Sigma = (A, C, \tau)$ be a higher-order signature, U be a set of database constants, D be a database over $\mathcal{D}_{\Sigma,U}$, and $\alpha \in \mathcal{T}(A)$. For every linear closed λ-term $M \in \Lambda(\Sigma)$, $M \in \Lambda(D, \alpha)$ if and only if $|M|_\beta \in \Lambda(D, \alpha)$.*

Proof. The "only if" direction is by Lemma 3.34, part (i). Since the β-reduction $M \twoheadrightarrow_\beta |M|_\beta$ must be non-erasing and non-duplicating, the "if" direction follows by part (ii) of the same lemma. □

Theorem 3.40. *Let $\mathscr{G} = (\mathscr{N}, \Sigma, f, \mathscr{P}, S)$ be a linear CFLG. Let U be some set of database constants, D be a database over $\mathcal{D}_{\Sigma,U}$, and $\vec{s}$ be a sequence of constants from U such that $|\vec{s}| = |f(S)|$. The following are equivalent:*

(i) $L(\mathscr{G}) \cap \Lambda(D, \langle f(S)\rangle(\vec{s})) \neq \varnothing.$

(ii) $\mathrm{program}(\mathscr{G}) \cup D \vdash S(\vec{s}).$

Proof. In view of Lemmas 3.35 and 3.36, it suffices to show that $\vdash_{\mathscr{G}} S(P)$ and $|P|_\beta \in \Lambda(D, \langle f(S)\rangle(\vec{s}))$ imply $P \in \Lambda(D, \langle f(S)\rangle(\vec{s}))$. But this is immediate from Lemma 3.39, since $\vdash_{\mathscr{G}} S(P)$ implies that P is linear. □

Lemma 3.41. *If M is an affine λ-term in strict η-long form relative to γ, then $|M|_\beta$ is the only λ-term in $\Lambda(\mathrm{database}(M), \langle \gamma \rangle(\mathrm{tuple}(M)))$ that is in η-long β-normal form relative to γ.*

Proof. Let $\overrightarrow{\mathrm{Con}}(M) = (d_1, \ldots, d_m)$, and let

$$y_1 : \beta_1, \ldots, y_m : \beta_m \Rightarrow \alpha \tag{56}$$

be a principal typing of $\widehat{M}[y_1, \ldots, y_m]$. Then $\mathrm{database}(M) = \{\, d_i(\overline{\beta_i}) \mid 1 \leq i \leq m \,\}$. Note that $\widehat{M}[y_1, \ldots, y_m]$ is a pure affine λ-term. By Theorem 3.19, (56) is a balanced typing.

We know from Lemmas 3.17, 3.34 and 3.37 that $|M|_\beta$ is in η-long form relative to γ and that $|M|_\beta \in \Lambda(\mathrm{database}(M), \langle \gamma \rangle(\mathrm{tuple}(M)))$. Suppose $M' \in \Lambda(\mathrm{database}(M), \langle \gamma \rangle(\mathrm{tuple}(M)))$ and $|\overrightarrow{\mathrm{Con}}(M')| = n$. Then there is a function $g : \{1, \ldots, n\} \to \{1, \ldots, m\}$ such that $\overrightarrow{\mathrm{Con}}(M') = (d_{g(1)}, \ldots, d_{g(n)})$ and

$$\vdash z_1 : \beta_{g(1)}, \ldots, z_n : \beta_{g(n)} \Rightarrow \widehat{M'}[z_1, \ldots, z_n] : \alpha.$$

Substituting $y_{g(i)}$ for z_i, we get

$$\vdash \{\, y_{g(i)} : \beta_{g(i)} \mid 1 \leq i \leq n \,\} \Rightarrow \widehat{M'}[y_{g(1)}, \ldots, y_{g(n)}] : \alpha.$$

By the Coherence Theorem (Theorem 3.18), $\widehat{M'}[y_{g(1)}, \ldots, y_{g(n)}] =_{\beta\eta} \widehat{M}[y_1, \ldots, y_n]$, and so $M' =_{\beta\eta} M$. It follows that if M' is in η-long β-normal form relative to γ, then $M' = |M|_\beta$. □

Theorem 3.42. *Let $\mathscr{G} = (\mathcal{N}, \Sigma, f, \mathscr{P}, S)$ be a linear CFLG. Suppose that $N \in \Lambda(\Sigma)$ is a linear λ-term in η-long β-normal form relative to $f(S)$. Then the following are equivalent:*

(i) $N \in L(\mathscr{G})$.

(ii) $\mathrm{program}(\mathscr{G}) \cup \mathrm{database}(N) \vdash S(\mathrm{tuple}(N))$.

Proof. Immediate from Lemma 3.41 and Theorem 3.40. □

Let us analyze the computational complexity of this reduction.

Lemma 3.43. *Given a linear λ-term $N \in \Lambda(\Sigma)$ in η-long β-normal form relative to $f(S)$, the pair $(\text{database}(N), S(\text{tuple}(N)))$ can be computed by a deterministic log-space-bounded Turing machine.*

Proof (sketch). We sketch how a deterministic log-space-bounded Turing machine $\mathscr{M}$ that has multiple heads on the input tape can compute $(\text{database}(N), S(\text{tuple}(N)))$, relying on the fact that an extra head can be simulated by a log-space-bounded work tape. We assume that the input λ-term N is given in the form of a λ-expression; the λ symbol, parentheses, and constants in N are represented by individual symbols of the input alphabet of $\mathscr{M}$, and variables are represented by strings of the form vl, where v is a special symbol and l is a natural number written in binary (i.e., a string over $\{0,1\}$).

Let $\overrightarrow{\text{Con}}(N) = (c_1, \ldots, c_n)$. The output of $\mathscr{M}$ will be of the form

$$c_1(k_{1,1}, \ldots, k_{1,r_1}), \ldots, c_n(k_{n,1}, \ldots, k_{n,r_n}), S(k_{0,1}, \ldots, k_{0,r_0}),$$

where $r_i = |\tau(c_i)|$ for $i = 1, \ldots, n$ and $r_0 = |f(S)|$, and each $k_{i,j}$ is a natural number in binary. Let $v_{i,j}$ be the jth leaf (counting from the right) of $\langle \tau(c_i) \rangle$ if $1 \leq i \leq n$ and $1 \leq j \leq r_i$, and let $v_{0,j}$ be the jth leaf (from right) of $\langle f(S) \rangle$ for $1 \leq j \leq r_0$. If either $1 \leq i \leq n$ and $\text{pol}(v_{i,j}) = 1$ or $i = 0$ and $\text{pol}(v_{i,j}) = -1$, then for some $p \leq \sum_{i=0}^{r_i} r_i$, the pair $(i, v_{i,j})$ represents the pth negative atomic type occurrence in

$$z_1 : \tau(c_1), \ldots, z_n : \tau(c_n) \Rightarrow \widehat{N}[z_1, \ldots, z_n] : f(S). \tag{57}$$

In this case, $k_{i,j}$ will be the binary representation of p (which can be computed in logarithmic space). If either $1 \leq i \leq n$ and $\text{pol}(v_{i,j}) = -1$ or $i = 0$ and $\text{pol}(v_{i,j}) = 1$, then the pair $(i, v_{i,j})$ represents a positive atomic type occurrence in (57). In this case, $k_{i,j}$ will be $k_{i',j'}$, where (i', j') is the unique pair such that $(i, v_{i,j})$ is linked to $(i', v_{i',j'})$ in $(\widehat{N}[z_1, \ldots, z_n], t)$, where t is the type decoration that is determined by the typing (57). (Uniqueness is guaranteed by the linearity of N.)

For each pair (i, j) for which $(i, v_{i,j})$ is positive, the machine $\mathscr{M}$ computes the corresponding pair (i', j') by starting from $(i, v_{i,j}, \uparrow)$ and following edges of $\overline{G}_{(\widehat{N}[z_1, \ldots, z_n], t)}$. The machine does this without explicitly computing the type decoration t. In order to represent a vertex (w, v, d) of $G_{(\widehat{N}[z_1, \ldots, z_n], t)}$, the machine $\mathscr{M}$ can place one of its heads at the beginning of the subexpression of N occurring at node w, and store (v, d) in its finite control. This is possible because the fact that $\widehat{N}[z_1, \ldots, z_n]$ is β-normal implies that for all nodes w of $\widehat{N}$, $t(w)$ is a subtype of some type in $\{\tau(c_1), \ldots, \tau(c_n), f(S)\} \subseteq \{\tau(c) \mid c \in C\} \cup \{f(S)\}$ by the subformula property, and there are only finitely many possible values of v. Traversal of edges in $G_{(\widehat{N}[z_1, \ldots, z_n], t)}$, which is deterministic because $\widehat{N}[z_1, \ldots, z_n]$ is linear, can easily be done using an extra head. For example, suppose $\mathcal{M}$ is in a configuration representing $(w, 1v, \uparrow)$, where w is a unary node (λ-abstract) and $t(w) = \gamma \to \delta$. The machine's head is at the first symbol of a string of the form $(\lambda vl.P)$. In order to switch to a configuration representing $(w', v, \downarrow)$, where $b(w') = w$, $\mathcal{M}$ can use an extra head to locate the occurrence of vl bound by the lambda. For another example, suppose $\mathcal{M}$ is in a configuration representing $(w1, v, \downarrow)$, where w is a binary node (application) and $t(w0) = \gamma \to \delta$. The machine's head is at the first symbol of Q in a string of the form (PQ). In order to switch to a configuration representing $(w0, 1v, \uparrow)$, the machine can move the head to the first symbol of P by counting in binary unmatched closing parentheses encountered along the way, which requires no more than logarithmic space. □

Thus, for every linear CFLG $\mathcal{G}$, the set $L(\mathcal{G})$ is log-space-reducible to $\{(D, q) \mid \text{program}(\mathcal{G}) \cup D \vdash q\}$. Since for every Datalog program $\mathbf{P}$, the language $\{(D, q) \mid \mathbf{P} \cup D \vdash q\}$ is in P, it immediately follows that $L(\mathcal{G})$ is in P for every linear CFLG $\mathcal{G}$, a fact first proved by Salvati [62]. A more careful analysis gives a tight complexity upper bound:

Theorem 3.44. *For every linear CFLG $\mathcal{G}$, $L(\mathcal{G})$ belongs to LOGCFL.*

Proof. Let $\mathcal{G} = (\mathcal{N}, \Sigma, f, \mathcal{P}, S)$, and let $g(n)$ be the polynomial associated with program($\mathcal{G}$) by Lemma 3.2. We show that whenever $N \in L(\mathcal{G})$, there is a derivation tree for program($\mathcal{G}$) $\cup$ database(N) $\vdash S(\text{tuple}(N))$ of size $\leq g(|\overrightarrow{\text{Con}}(N)|)$. The proof of this claim is by a more careful use of Lemma 3.31 than in the proof of Theorem 3.42.

Let $N \in \Lambda(\Sigma)$ be a linear λ-term in η-long β-normal form relative to $f(S)$. Assume $N \in L(\mathcal{G})$. Let $\overrightarrow{\text{Con}}(N) = (d_1, \ldots, d_m)$ and let

$$y_1 : \beta_1, \ldots, y_m : \beta_m \Rightarrow \alpha$$

be a principal typing of $\widehat{N}[y_1, \ldots, y_m]$. Then

$$\text{database}(N) = \{d_i(\overline{\beta_i}) \mid 1 \leq i \leq m\},$$
$$\text{tuple}(N) = \overline{\alpha}.$$

Since $\mathcal{G}$ is linear, there exists some linear λ-term $P \in \Lambda(\Sigma)$ such that $\vdash_{\mathcal{G}} S(P)$ and $P \twoheadrightarrow_\beta N$. Since the β-reduction from P to N must be non-erasing and non-duplicating, $|\overrightarrow{\text{Con}}(P)| = m$, and $\widehat{P}[y_{h(1)}, \ldots, y_{h(m)}] \twoheadrightarrow_\beta \widehat{N}[y_1, \ldots, y_m]$ for some permutation h on $\{1, \ldots, m\}$. This means

$$\vdash y_1 : \langle \tau(d_1) \rangle (\overline{\beta_1}), \ldots, y_m : \langle \tau(d_m) \rangle (\overline{\beta_m}) \Rightarrow \widehat{P}[y_{h(1)}, \ldots, y_{h(m)}] : \langle f(S) \rangle(\overline{\alpha}).$$

By Lemma 3.31, there is a derivation tree for $\text{program}(\mathscr{G}) \cup \text{database}(N) \vdash S(\overline{\alpha})$ with m extensional nodes. By Lemma 3.2, it follows that there is a derivation tree for $\text{program}(\mathscr{G}) \cup \text{database}(N) \vdash S(\overline{\alpha})$ of size at most $g(m)$. Therefore, $(\text{database}(N), S(\text{tuple}(N)), 1^{|\overrightarrow{\text{Con}(N)}|})$ belongs to the set

$$\{\, (D, q, 1^n) \mid \text{there is a derivation tree for } \text{program}(\mathscr{G}) \cup D \vdash q \text{ of size} \leq g(n) \,\}. \tag{58}$$

Now assume $N \notin L(\mathscr{G})$. Then by Theorem 3.42, it is not the case that $\text{program}(\mathscr{G}) \cup \text{database}(N) \vdash S(\text{tuple}(N))$, and $(\text{database}(N), S(\text{tuple}(N)), 1^{|\overrightarrow{\text{Con}(N)}|})$ does not belong to (58).

By Lemmas 3.1 and 3.43, we conclude that $L(\mathscr{G})$ is log-space reducible to a problem in LOGCFL. Since the class of functions computable in logarithmic space is closed under composition, $L(\mathscr{G})$ itself is in LOGCFL. $\square$

3.6 Almost affine λ-terms

A typed λ-term (M, t), where $M = (\mathcal{T}, f, b) \in \Lambda(\Sigma)$, is *almost affine* if for every $w, w' \in \mathcal{T}$ such that $w \neq w'$, $f(w) = f(w') \in \mathcal{V}$ or $b(w) = b(w')$ implies that $t(w) = t(w')$ is an atomic type. An untyped λ-term M is *almost affine relative to* $\Gamma \Rightarrow \alpha$ if there is a type decoration t for $\Gamma \Rightarrow M : \alpha$ such that (M, t) is almost affine. We say that a typable λ-term is almost affine if it is almost affine relative to some typing, or equivalently, relative to its principal typing.

If a typed λ-term is almost affine, then so is its η-long form. The class of almost affine untyped λ-terms is closed under η-reduction, but not under β-reduction. For example, a pure λ-term $M = (\lambda x.yxx)(zw)$ is almost affine relative to $y : o, z : o \to o, w : o \Rightarrow o$, but $|M|_\beta = y(zw)(zw)$ is not (relative to any typing).

We say that a λ-term $M \in \Lambda(\Sigma)$ is *almost linear* if M is an almost affine λI-term.

A sequent is *negatively non-duplicated* if no atomic type has more than one negative occurrence in it. The following result generalizes the Coherence Theorem (Theorem 3.18):

Theorem 3.45 (Aoto and Ono [3]). *All inhabitants of a negatively non-duplicated sequent are $\beta\eta$-equal.*[34]

The following is a slight generalization of a result by Aoto [2]:

Theorem 3.46. *If $\Gamma \Rightarrow \alpha$ is a principal typing of an almost affine pure λ-term M, then $\Gamma \Rightarrow \alpha$ is negatively non-duplicated.*

---

[34]This theorem can be stated in the same style as the Coherence Theorem: If $\Gamma \cup \Gamma' \Rightarrow \alpha$ is a negatively non-duplicated sequent and both $\vdash \Gamma \Rightarrow M : \alpha$ and $\vdash \Gamma' \Rightarrow M' : \alpha$ hold, then $M =_{\beta\eta} M'$. See [46].

Proof. Let $\Gamma = x_1 : \alpha_1, \ldots, x_n : \alpha_n$ and let t be a principal type decoration for M associated with the typing $\Gamma \Rightarrow \alpha$. Suppose that $(i, v), (i', v')$ are two distinct negative occurrences of the same atomic type p in $\Gamma \Rightarrow \alpha$. By Lemma 3.25, (i, v) is connected to (i', v') in (M, t). This means that $\overline{G}_{(M,t)}$ contains an undirected path from (i, v, d) to (i', v', d') for some $d, d' \in \{\uparrow, \downarrow\}$. Since both (i, v) and (i', v') are negative, by the property of $\overline{G}_{(M,t)}$ mentioned above immediately after its definition, we must have $d = d'$. We may assume $d = d' = \uparrow$. Since there cannot be a directed path from $(i, v, \uparrow)$ to $(i', v', \uparrow)$, this implies that there are three nodes ν_1, ν_2, ν_3 of $\overline{G}_{(M,t)}$ such that

- $\nu_1 \neq \nu_3$,
- there is a directed path from $(i, v, \uparrow)$ to ν_1,
- (ν_1, ν_2) and (ν_3, ν_2) are edges of $\overline{G}_{(M,t)}$, and
- there is an undirected path from $(i', v', \uparrow)$ to ν_3.

The first and third conditions can obtain only in two cases:

- $\nu_2 = (j, u, \downarrow)$ for some $j \in \{1, \ldots, n\}$ and $\nu_1 = (w_1, u, \uparrow), \nu_3 = (w_3, u, \uparrow)$, where $f(w_1) = f(w_3) = x_j$.
- $\nu_2 = (w_2, 1u, \downarrow)$, $\nu_1 = (w_1, u, \uparrow), \nu_3 = (w_3, u, \uparrow)$, and $b(w_1) = b(w_3) = w_2$.

In both cases, since (M, t) is almost affine, it must be the case that $t(w_1) = t(w_3) = p$ and $u = \epsilon$. However, since $\mathrm{pol}(i, v) = -1$ and $\mathrm{pol}(\epsilon) = 1$, there cannot be a directed path from $(i, v, \uparrow)$ to $\nu_1 = (w_1, \epsilon, \uparrow)$, a contradiction. □

Theorems 3.45 and 3.46 show that a principal typing of an almost affine pure λ-term uniquely characterizes it up to $\beta\eta$-equality.

Although we do not need it in establishing the results to follow, we note that the converse of Theorem 3.46 also holds [46]:

Theorem 3.47. *Suppose $\Gamma \Rightarrow \alpha$ is a negatively non-duplicated sequent. For every pure λ-term M such that $\vdash \Gamma \Rightarrow M{:}\alpha$, there exists a λ-term M' such that $M' =_{\beta\eta} M$ and M' is almost affine relative to $\Gamma \Rightarrow \alpha$.*

Let $M \in \Lambda(\Sigma)$ be a typable λ-term, and let t be a principal typing for M. A β-reduction step $M \xrightarrow{w}_\beta M'$ is *almost non-duplicating* if either it is non-duplicating or $\mathrm{subtype}(t(w0), 1) = t(w1)$ is atomic. A β-reduction $M \twoheadrightarrow_\beta M'$ is almost non-duplicating if it consists entirely of almost non-duplicating β-reduction steps.

Example 3.48. Let $M = (\lambda x.(\lambda z.yzz)(xz))(\lambda z.uz)$. Then

$$M \xrightarrow{\epsilon}_\beta (\lambda z.yzz)((\lambda z.uz)z)$$
$$\xrightarrow{\epsilon}_\beta y((\lambda z.uz)z)((\lambda z.uz)z)$$

is almost non-duplicating, whereas

$$M \xrightarrow{00}_\beta (\lambda x.y(xz)(xz))(\lambda z.uz)$$
$$\xrightarrow{\epsilon}_\beta y((\lambda z.uz)z)((\lambda z.uz)z)$$

is not, because the second step duplicates the subterm $(\lambda z.uz)$, whose type must be non-atomic.

A β-reduction $M_0 \xrightarrow{w_1}_\beta M_1 \xrightarrow{w_2}_\beta \cdots \xrightarrow{w_n}_\beta M_n$ is called *leftmost* if for $i = 1, \ldots, n$, w_i is the leftmost β-redex of M_{i-1}, i.e., w_i is the first β-redex of M_{i-1} under the lexicographic ordering $\prec$ of the nodes of M_{i-1}.

Lemma 3.49. *If $M \in \Lambda(\Sigma)$ is almost affine, then the leftmost β-reduction from M to $|M|_\beta$ is almost non-duplicating.*

Proof. Let $M = M_0 \xrightarrow{w_1}_\beta M_1 \xrightarrow{w_2}_\beta \cdots \xrightarrow{w_n}_\beta M_n = |M|_\beta$ by leftmost β-reduction, and let $M_i = (\mathcal{T}_i, f_i, b_i)$. We show that each step of this reduction is almost non-duplicating. Let t be a principal type decoration of M, and for $i = 0, \ldots, n$, let t_i be the type decoration for M_i such that

$$(M, t) = (M_0, t_0) \xrightarrow{w_1}_\beta (M_1, t_1) \xrightarrow{w_2}_\beta \cdots \xrightarrow{w_n}_\beta (M_n, t_n).$$

To prove the lemma, it suffices to show that for every i and every unary node $w \in \mathcal{T}_i^{(1)}$, either

(i) w is to the left of any β-redex in M_i,

(ii) $\mathrm{subtype}(t_i(w), 1)$ is an atomic type, or

(iii) there is at most one $w' \in \mathcal{T}_i$ such that $b_i(w') = w$.

The condition holds of (M_0, t_0) by the assumption that M is an almost affine λ-term. Assume that (M_i, t_i) satisfies the condition, and let v be a unary node of $\mathcal{T}_{i+1}$. Then v is a descendant of a unary node w of $\mathcal{T}_i$ distinct from $w_i 0$.

Suppose that (i) holds of w. Then w is to the left of w_i. Clearly, v must be to the left of any β-redex in M_{i+1}, satisfying (i).

Suppose that (ii) holds of w. Since $t_{i+1}(v) = t_i(w)$, it holds that subtype$(t_{i+1}(v), 1)$ is an atomic type. So v satisfies (ii).

Suppose that (iii) holds of w. Assume that v does not satisfy (iii), i.e., there are $v', v'' \in \mathcal{T}_{i+1}$ such that $v' \neq v''$ and $b_{i+1}(v') = b_{i+1}(v'') = v$. Then v' and v'' must be descendants of the unique node w' of $\mathcal{T}_i$ such that $b_i(w') = w$. For this to hold, it must be the case that $w < w_i$ and $w_i 1 \leq w'$. Then $v = w$. Since w_i is the leftmost β-redex of M_i, there is no β-redex in M_i to the left of w, and it follows that there is no β-redex in M_{i+1} to the left of v. □

We now define a certain equivalence relation between nodes in a λ-term. Let $M = (\mathcal{T}, f, b)$, and let $w, w' \in \mathcal{T}$. We say that w and w' are *congruent* in M and write $w \cong_M w'$, if the following conditions hold:

- $\{ v \mid wv \in \mathcal{T} \} = \{ v \mid w'v \in \mathcal{T} \}$,

- for all v such that $wv \in \mathcal{T}^{(0)}$, either

  - $wv, w'v \in \text{dom}(f)$ and $f(wv) = f(w'v)$,
  - $wv, w'v \in \text{dom}(b)$ and $b(wv) = b(w'v)$, or
  - $wv, w'v \in \text{dom}(b)$ and $b(wv) = wu$ and $b(w'v) = w'u$ for some $u < v$.

It is clear that if $w \cong_M w'$, then for every writing ℓ of M, the λ-expressions $\text{sub}_{M,\ell}(w)$ and $\text{sub}_{M,\ell}(w')$ represent the same λ-term.

The following is clear from the definition of the ancestor-descendant relation for one-step β-reduction.

Lemma 3.50. *Let* $M \xrightarrow{w}_\beta M'$ *be a duplicating β-reduction step. If* $(M, w1) \overset{w}{\blacktriangleright} (M', v_1)$ *and* $(M, w1) \overset{w}{\blacktriangleright} (M', v_2)$, *then* $v_1 \cong_{M'} v_2$.

Let $M = (\mathcal{T}_M, f_M, b_M)$ be a λ-term. Suppose that $v_1, \ldots, v_k$ are nodes in $\mathcal{T}_M$ such that $v_1 \cong_M \ldots \cong_M v_k$. Let w be a node in $\mathcal{T}_M$ such that for all i, $w < v_i$, and $b_M(v_i u) < w$ holds whenever $b_M(v_i u) < v_i$. It is clear that there must be such a w. Define $\text{expand}(M, w, \{v_1, \ldots, v_k\}) = (\mathcal{T}, f, b)$ as follows:

$$\mathcal{T} = \{ v \in \mathcal{T}_M \mid w \not< v \} \cup \{w0\} \cup \{ w00v \mid wv \in \mathcal{T}_M, \neg\exists i (v_i < wv) \} \cup$$
$$\{ w1u \mid v_1 u \in \mathcal{T}_M \},$$
$$f = \{ (v, f_M(v)) \mid w \not< v, v \in \text{dom}(f_M) \} \cup$$
$$\{ (w00v, f_M(wv)) \mid wv \in \text{dom}(f_M), \neg\exists i (v_i < wv) \} \cup$$
$$\{ (w1u, f_M(v_1 u)) \mid v_1 u \in \text{dom}(f_M) \}$$

$$\begin{aligned}
b = &\{ (v, b_M(v)) \mid w \not< v, v \in \operatorname{dom}(b_M) \} \cup \\
&\{ (w00v, b_M(wv)) \mid b_M(wv) < w, \neg \exists i(v_i \leq wv) \} \cup \\
&\{ (w00v, w00u) \mid b_M(wv) = wu, \neg \exists i(v_i \leq wv) \} \cup \\
&\{ (w00v_i, w0) \mid 1 \leq i \leq k \} \cup \\
&\{ (w1u, b_M(v_1 u)) \mid b_M(v_1 u) < w \} \cup \{ (w1u, w1s) \mid b_M(v_1 u) = v_1 s \}.
\end{aligned}$$

It is clear that $\operatorname{expand}(M, w, \{v_1, \ldots, v_k\})$ is a λ-term, and we have $\operatorname{expand}(M, w, \{v_1, \ldots, v_k\}) \overset{w}{\twoheadrightarrow}_\beta M$ and $(\operatorname{expand}(M, w, \{v_1, \ldots, v_k\}), w1) \overset{w}{\blacktriangleright} (M, v_i)$ for $i = 1, \ldots, k$.

Lemma 3.51. *Let $M \in \Lambda(\Sigma)$, where $\Sigma = (A, C, \tau)$, and let t be a type decoration of M. Suppose that w, w' are two nodes of M such that $w \cong_M w'$. If $t(w)$ is an atomic type, then $t(w) = t(w')$.*

Proof. Let ℓ be a writing of M and let $N = \operatorname{sub}_{M,\ell}(w) = \operatorname{sub}_{M,\ell}(w')$. Let

$$\Gamma_w = \{ (x, t(wv)) \mid x = f(wv) \in \mathcal{V} \} \cup \{ (\ell(b(wv)), t(wv)) \mid b(wv) \leq w \}$$
$$\Gamma_{w'} = \{ (x, t(w'v)) \mid x = f(w'v) \in \mathcal{V} \} \cup \{ (\ell(b(w'v)), t(w'v)) \mid b(w'v) \leq w' \}.$$

Then we have

$$\vdash_\Sigma \Gamma_w \Rightarrow N : t(w),$$
$$\vdash_\Sigma \Gamma_{w'} \Rightarrow N : t(w').$$

Since $w \cong_M w'$, we must have $\Gamma_w = \Gamma_{w'}$. By the Subject Reduction Theorem (Theorem 3.14),

$$\vdash_\Sigma \Gamma' \Rightarrow |N|_\beta : t(w),$$
$$\vdash_\Sigma \Gamma' \Rightarrow |N|_\beta : t(w')$$

where $\Gamma' = \Gamma_w \upharpoonright \operatorname{FV}(|N|_\beta)$. By assumption, $t(w)$ is some atomic p, so $|N|_\beta$ must be of the form

$$y P_1 \ldots P_l$$

for some variable y, or else of the form

$$c P_1 \ldots P_l$$

for some constant c. In the former case, $y : \gamma_1 \to \cdots \to \gamma_l \to p$ is in Γ', and in the latter case, $\tau(c) = \gamma_1 \to \cdots \to \gamma_l \to p$ for some types $\gamma_1, \ldots, \gamma_l$. In either case, we must have $t(w') = p$. $\square$

The following lemma generalizes the Subject Expansion Theorem (Theorem 3.15):

Lemma 3.52. *Let $M, M' \in \Lambda(\Sigma)$ be typable λ-terms. Suppose $M \twoheadrightarrow_\beta M'$ by non-erasing, almost non-duplicating β-reduction. If $\vdash_\Sigma \Gamma \Rightarrow M' : \alpha$, then $\vdash_\Sigma \Gamma \Rightarrow M : \alpha$.*

Proof. Clearly, it suffices to consider the case where the β-reduction consists of just one step and $\Gamma \Rightarrow \alpha$ is a principal typing of M'. Let $M = (\mathcal{T}, f, b), M' = (\mathcal{T}', f', b')$, and $M \xrightarrow{w}_\beta M'$. Let t be a principal type decoration of M, t' be the type decoration of M' induced by t (i.e., $(M, t) \xrightarrow{w}_\beta (M', t')$), and $\tilde{t}$ be a principal type decoration of M' (with the associated typing $\Gamma \Rightarrow \alpha$). If the β-reduction step $M \xrightarrow{w}_\beta M'$ is non-erasing and non-duplicating, then $\vdash_\Sigma \Gamma \Rightarrow M : \alpha$ by the Subject Expansion Theorem. So suppose that this β-reduction step is duplicating. Let

$$\{\, v \mid b(w00v) = w0 \,\} = \{v_1, \ldots, v_k\},$$

where $k \geq 2$. Since the β-reduction step is almost non-duplicating, we have $t(w1) = p$ for some atomic type p. For each $i \in \{1, \ldots, k\}$, we have

$$(M, w1) \xrightarrow{w}_\blacktriangleright (M', wv_i)$$

and $t'(wv_i) = p$. Since $\tilde{t}$ is a principal type decoration of M', there is a type substitution σ such that $t' = \sigma \circ \tilde{t}$. It follows that for each $i = 1, \ldots, k$, there is an atomic type q_i such that $\tilde{t}(wv_i) = q_i$. By Lemma 3.50, we have

$$wv_1 \cong_{M'} \ldots \cong_{M'} wv_k,$$

and by Lemma 3.51, it follows that

$$q_1 = \cdots = q_k.$$

Define a function $\tilde{t}_1 \colon \mathcal{T} \to \mathscr{T}(A)$ as follows:

$$\tilde{t}_1(v) = \begin{cases} \tilde{t}(v) & \text{if } w \not\leq v, \\ \tilde{t}(w) & \text{if } v = w, \\ q_1 \to \tilde{t}(w) & \text{if } v = w0, \\ \tilde{t}(wu) & \text{if } v = w00u, \\ \tilde{t}(wv_1 u) & \text{if } v = w1u. \end{cases}$$

It is clear that $\tilde{t}_1$ is a type decoration of M. Although $(M, \tilde{t}_1) \xrightarrow{w}_\beta (M', \tilde{t})$ does not necessarily hold, it is easy to see that $\tilde{t}_1$ is a type decoration for $\Gamma \Rightarrow M : \alpha$. $\square$

Lemma 3.53. *If M is an almost linear λ-term, M and $|M|_\beta$ have the same principal typing.*

Proof. Since M is almost affine, by Lemma 3.49, the leftmost β-reduction from M to $|M|_\beta$ is almost non-duplicating. Since M is a λI-term, this β-reduction must also be non-erasing. By the Subject Reduction Theorem (Theorem 3.14) and Lemma 3.52, any typing of M is a typing of $|M|_\beta$ and vice versa. □

3.7 From CFLGs to Datalog: The case of almost linear CFLGs

Given Aoto and Ono's [3] generalization of the Coherence Theorem (Theorem 3.45), we easily obtain a generalization of Lemma 3.41 to almost affine λ-terms.

Lemma 3.54. *Let M be an almost affine λ-term in strict η-long form relative to γ. Then $|M|_\beta$ is the only λ-term in $\Lambda(\text{database}(M), \langle\gamma\rangle(\text{tuple}(M)))$ that is in η-long β-normal form relative to γ.*

Proof. The proof parallels that of Lemma 3.41. Let $\overrightarrow{\text{Con}}(M) = (d_1, \ldots, d_m)$, and let

$$y_1 : \beta_1, \ldots, y_m : \beta_m \Rightarrow \alpha \tag{59}$$

be a principal typing of $\widehat{M}[y_1, \ldots, y_m]$. Then $\text{database}(M) = \{\, d_i(\overline{\beta_i}) \mid 1 \leq i \leq m \,\}$. Note that $\widehat{M}[y_1, \ldots, y_m]$ is a pure almost affine λ-term in strict η-long form. By Theorem 3.46, (59) is negatively non-duplicated.

We know from Lemmas 3.17, 3.34, and 3.37 that $|M|_\beta$ is in η-long β-normal form relative to γ and that $|M|_\beta \in \Lambda(\text{database}(M), \langle\gamma\rangle(\text{tuple}(M)))$. Suppose $N \in \Lambda(\text{database}(M), \langle\gamma\rangle(\text{tuple}(M)))$ and $|\overrightarrow{\text{Con}}(N)| = n$. Then there is a function $g : \{1, \ldots, n\} \to \{1, \ldots, m\}$ such that $\overrightarrow{\text{Con}}(N) = (d_{g(1)}, \ldots, d_{g(n)})$ and

$$\vdash z_1 : \beta_{g(1)}, \ldots, z_n : \beta_{g(n)} \Rightarrow \widehat{N}[z_1, \ldots, z_n] : \alpha.$$

Substituting $y_{g(i)}$ for z_i, we get

$$\vdash \{\, y_{g(i)} : \beta_{g(i)} \mid 1 \leq i \leq n \,\} \Rightarrow \widehat{N}[y_{g(1)}, \ldots, y_{g(n)}] : \alpha.$$

By Theorem 3.45, $\widehat{N}[y_{g(1)}, \ldots, y_{g(n)}] =_{\beta\eta} \widehat{M}[y_1, \ldots, y_n]$, and so $N =_{\beta\eta} M$. It follows that if N is in η-long β-normal form relative to γ, then $N = |M|_\beta$. □

A CFLG $\mathcal{G} = (\mathcal{N}, \Sigma, f, \mathcal{P}, S)$ is *almost linear* if for every $\pi \in \mathcal{P}$, the λ-term on the left-hand side of π is almost linear. An example of an almost linear CFLG is the grammar in Example 3.28. Almost linear CFLGs can encode IO context-free tree grammars [21] in a straightforward way, similarly to de Groote and Pogodalla's [19] encoding of linear context-free tree grammars.[35]

Lemma 3.39 and Theorem 3.40 do not generalize to the almost linear case, and there is no simple analogue of Theorem 3.42 for almost linear CFLGs. The reason is that $\Lambda(D,\alpha)$ need not be closed under the converse of non-erasing almost non-duplicating β-reduction (in contrast to part (ii) of Lemma 3.34), despite the fact that the Subject Expansion Theorem generalizes to such β-reduction (Lemma 3.52). This is so even when $D = \text{database}(N)$ and $\alpha = \langle \gamma \rangle(\text{tuple}(N))$ for an almost linear λ-term $N \in \Lambda(\Sigma)$ in η-long β-normal form relative to γ.

Example 3.55. Consider the λ-term (19) from Section 2.2:

$$N = \exists(\lambda y.\wedge(\textbf{unicorn}\ y)(\wedge(\textbf{find}\ y\ \textbf{John})(\textbf{catch}\ y\ \textbf{John}))),$$

where the types of the constants $\exists, \wedge, \textbf{unicorn}, \textbf{find}, \textbf{John}, \textbf{catch}$ are as follows:

$$\exists : (e \to t) \to t,$$
$$\wedge : t \to t \to t,$$
$$\textbf{unicorn} : e \to t,$$
$$\textbf{find} : e \to e \to t,$$
$$\textbf{John} : e,$$
$$\textbf{catch} : e \to e \to t.$$

We have $\overrightarrow{\text{Con}}(N) = (\exists, \wedge, \textbf{unicorn}, \wedge, \textbf{find}, \textbf{John}, \textbf{catch}, \textbf{John})$. A principal typing of

$$\widehat{N}[z_1, z_2, z_3, z_4, z_5, z_6, z_7, z_8] = z_1(\lambda y.z_2(z_3 y)(z_4(z_5 y z_6)(z_7 y z_8)))$$

is

$$z_1:(4\to 2)\to 1,\ z_2:3\to 5\to 2,\ z_3:4\to 3,\ z_4:6\to 8\to 5,\ z_5:4\to 7\to 6,\ z_6:7,\ z_7:4\to 9\to 8,\ z_8:9 \Rightarrow 1,$$

which gives rise to

$$\text{database}(N) = \{\exists(1,2,4), \wedge(2,5,3), \textbf{unicorn}(3,4), \wedge(5,8,6), \textbf{find}(6,7,4), \textbf{John}(7),$$
$$\textbf{catch}(8,9,4), \textbf{John}(9)\},$$

---

[35]With respect to string languages, it is easy to see that almost linear CFLGs generating λ-terms representing strings are no more powerful than linear CFLGs (see footnote 41). What this means is that encodings of "non-linear" grammars like IO macro grammars [24] and parallel multiple context-free grammars [66] in terms of CFLGs cannot be almost linear. However, our reduction to Datalog applies to these cases indirectly if we take almost linear CFLGs encoding tree analogues of these grammars (i.e., IO context-free tree grammars and what one might call "parallel multiple regular tree grammars") and use regular sets of trees as input. See Section 4.2 below.

$\mathrm{tuple}(N) = (1)$.

Now consider the λ-term (25):

$$N^\circ = \exists(\lambda y.\wedge(\mathbf{unicorn}\ y)((\lambda x.\wedge(\mathbf{find}\ y\ x)(\mathbf{catch}\ y\ x))\ \mathbf{John})).$$

Although $N^\circ \twoheadrightarrow_\beta N$ by non-erasing almost non-duplicating β-reduction, it is easy to see that $N^\circ \notin \Lambda(\mathrm{database}(N), 1)$. This does not contradict Lemma 3.52, because

$$\widehat{N^\circ}[y_1, y_2, y_3, y_4, y_5, y_6, y_7] \not\twoheadrightarrow_\beta \widehat{N}[z_1, z_2, z_3, z_4, z_5, z_6, z_7, z_8],$$

no matter how one picks the variables $y_1, y_2, y_3, y_4, y_5, y_6, y_7$.

Let $\mathscr{G} = (\mathscr{N}, \Sigma, f, \mathscr{P}, S)$ be an almost linear CFLG, and suppose that $N \in \Lambda(\Sigma)$ is in η-long β-normal form relative to $f(S)$. In order to find a database D and a tuple $\vec{s}$ such that $N \in L(\mathscr{G})$ if and only if $\mathrm{program}(\mathscr{G}) \cup D \vdash S(\vec{s})$, what we do is to β-expand N to a 'most compact' almost linear λ-term N° such that $N^\circ \twoheadrightarrow_\beta N$ in the sense that for any almost linear $P \in \Lambda(\Sigma)$, if $P \twoheadrightarrow_\beta N$ and two occurrences of the same constant in N have a common ancestor in P, then they have a common ancestor in N°. Thus, for every constant $c \in C$, N° contains the fewest occurrences of c among all almost linear λ-terms that β-reduce to N. We have $P \in \Lambda(\mathrm{database}(N^\circ), \mathrm{tuple}(N^\circ))$ if and only if $P \twoheadrightarrow_\beta N$ for all almost linear $P \in \Lambda(\Sigma)$ in η-long form,[36] and the desired equivalence of $N \in L(\mathscr{G})$ and $\mathrm{program}(\mathscr{G}) \cup \mathrm{database}(N^\circ) \vdash S(\mathrm{tuple}(N^\circ))$ follows. We show that such N° can be computed efficiently.

Let us call a node v of a λ-term $M = (\mathcal{T}, f, b)$ a *pivot* if (i) $v \in \mathcal{T}^{(0)} \cup \mathcal{T}^{(2)}$, and (ii) $v = v'0$ implies $v' \in \mathcal{T}^{(1)}$. A pivot v is *duplicated* if $v \notin \mathrm{dom}(b)$ and there is another pivot $v' \in \mathcal{T}$ such that $v \cong_M v'$. If some type decoration of M assigns a node v an atomic type, then v must be a pivot. If M is in η-long form and v is a pivot of M, then the principal type decoration of M assigns v an atomic type.

Algorithm 1.
1: **procedure** Collapse(M)
2:     $M^\circ \leftarrow M$
3:     **while** there is a duplicated pivot in M° **do**
4:         Let V be the set of duplicated pivots of maximal height in M°
5:         Let v_1 be the leftmost (i.e., lexicographically first) node in V
6:         Let $\{v_2, \ldots, v_k\} = \{v \mid v \text{ is a pivot, } v \neq v_1, \text{ and } v_1 \cong_{M^\circ} v\}$
7:         Let w be the pivot of minimal height such that $w < v_i$ for $i = 1, \ldots, k$

---

[36] This corresponds to the special property (28) mentioned in Section 2.3.

8:     $M^\circ \leftarrow \mathrm{expand}(M^\circ, w, \{v_1, \ldots, v_k\})$
9:   **end while**
10:   **return** M°
11: **end procedure**

It is clear that the node w picked in line 7 is such that $\mathrm{expand}(M^\circ, w, \{v_1, \ldots, v_k\})$ in line 8 is defined.

Lemma 3.56. *Let M be a typable λ-term in η-long form, and consider the execution of Algorithm 1 on input M. If w_i is the node w that is picked in line 7 during the ith iteration of the while loop, then the following conditions hold after the ith iteration of the while loop.*

(i) M° *is a typable λ-term in η-long form.*

(ii) $M^\circ \twoheadrightarrow_\beta M$ *by non-erasing, almost non-duplicating β-reduction.*

(iii) *If u_1 and u_2 are pivots of M such that $u_1 \cong_M u_2$ and $(M^\circ, u'_1) \stackrel{w_i,\ldots,w_1}{\blacktriangleright} (M, u_1)$ and $(M^\circ, u'_2) \stackrel{w_i,\ldots,w_1}{\blacktriangleright} (M, u_2)$, then u'_1 and u'_2 are also pivots and $u'_1 \cong_{M_i} u'_2$.*

Proof. Let $M_0 = M$, and let M_i be the value of M° after the ith iteration of the while loop. We show that the conditions (i), (ii), and (iii) hold by induction on i, on the understanding that $u'_1 = u_1$ and $u'_2 = u_2$ when $i = 0$.

The three conditions clearly hold when $i = 0$. Assume that the three conditions hold of M_i and let $v_1, v_2, \ldots, v_k$ be the nodes that the algorithm picks in lines 5–6 during the $(i+1)$st iteration of the while loop. Since M_i is typable and in η-long form, the principal type decoration t_i of M_i assigns v_1 an atomic type p, and by Lemma 3.51, $t_i(v_2) = \cdots = t_i(v_k) = p$. As in the proof of Lemma 3.52, this allows us to define a type decoration for $M_{i+1} = \mathrm{expand}(M_i, w_{i+1}, \{v_1, \ldots, v_k\})$ that assigns p to the node $w_{i+1}1$. It follows that the β-reduction step $M_{i+1} \stackrel{w_{i+1}}{\to}_\beta M_i$ is almost non-duplicating. It is also easy to see that M_{i+1} is in η-long form. So (i) and (ii) hold of M_{i+1}. To see that (iii) holds of M_{i+1}, let s_1 and s_2 be pivots of M_i such that $s_1 \cong_{M_i} s_2$, and let s'_1 and s'_2 be the nodes such that $(M_{i+1}, s'_1) \stackrel{w_{i+1}}{\blacktriangleright} (M_i, s_1)$ and $(M_{i+1}, s'_2) \stackrel{w_{i+1}}{\blacktriangleright} (M_i, s_2)$. It is easy to see that s'_1 and s'_2 are also pivots, so it suffices to show $s'_1 \cong_{M_{i+1}} s'_2$. If $s_1 = s_2$, then clearly $s'_1 = s'_2$. If $s_1 \neq s_2$, since v_1 is a duplicated pivot of maximal height in M_i and $v_1, v_2, \ldots, v_k$ are all of the same height, it cannot be the case that $s_1 < v_i$ or $s_2 < v_i$ for any i. This ensures that $\{\, s \mid s_1 s \text{ is a node of } M_i \,\} = \{\, s \mid s'_1 s \text{ is a node of } M_{i+1} \,\}$, $(M_{i+1}, s'_1 s) \stackrel{w_{i+1}}{\blacktriangleright} (M_i, s_1 s)$, and likewise for s_2 and s'_2. By Lemma 3.4, it is easy to check that the remaining conditions for $s'_1 \cong_{M_{i+1}} s'_2$ are satisfied. □

Lemma 3.57. *Algorithm 1 always terminates.*

Proof. One can prove by induction that the following condition holds at each stage of the algorithm: every pivot v in $M° = (\mathcal{T}_{M°}, f_{M°}, b_{M°})$ that is not in $\operatorname{dom}(b_{M°})$ is either an ancestor of a pivot in M or else a β-redex such that $v00$ and $v1$ are pivots not in $\operatorname{dom}(b_{M°})$. It follows that every pivot in $M°$ that is not in $\operatorname{dom}(b_{M°})$ contains an ancestor of a node in M, and the number of nodes in $M°$ that are ancestors of nodes in M strictly decreases at each iteration of the while loop. □

We now prove two lemmas (Lemmas 3.60 and 3.61) needed to show that COLLAPSE(M) is more compact than any almost affine λ-term that β-reduces to M. These lemmas require us to introduce two new binary relations on nodes. The following lemma is needed to prove the first of these lemmas:

Lemma 3.58. *Let $M = (\mathcal{T}, f, b) \in \Lambda(\Sigma)$, and $v, v' \in \mathcal{T}$ be two nodes such that $v \cong_M v'$. Suppose $M = M_0 \xrightarrow{w_1}_\beta M_1 \xrightarrow{w_2}_\beta \cdots \xrightarrow{w_n}_\beta M_n = |M|_\beta$. Let $v_0 = v, v'_0 = v'$, and for $1 \leq i \leq n$, let v_i and v'_i be nodes of M_i that satisfy one of the following conditions:*

(i) $(M_{i-1}, v_{i-1}) \blacktriangleright^{w_i}_k (M_i, v_i)$ and $(M_{i-1}, v'_{i-1}) \blacktriangleright^{w_i}_{k'} (M_i, v'_i)$, where $k = k'$ if both $w_i 1 \leq v_{i-1}$ and $w_i 1 \leq v'_{i-1}$ hold.

(ii) $v_{i-1} = v_i = w_i$ and $(M_{i-1}, v'_{i-1}) \blacktriangleright^{w_i} (M_i, v'_i)$.

(iii) $(M_{i-1}, v_{i-1}) \blacktriangleright^{w_i} (M_i, v_i)$ and $v'_{i-1} = v'_i = w_i$.

Then $v_n \cong_{|M|_\beta} v'_n$. Moreover, $(M, vu) \blacktriangleright^{w_1,\ldots,w_n} (|M|_\beta, v_n s)$ implies $(M, v'u) \blacktriangleright^{w_1,\ldots,w_n} (|M|_\beta, v'_n s)$.

Proof. Let $M_i = (\mathcal{T}_i, f_i, b_i)$ for $i = 0, 1, \ldots, n$. For $i = 1, \ldots, n$, define $\hat{w}_i$ by:

$$\hat{w}_i = \begin{cases} v'_i r & \text{if } w_i = v_i r, \\ v_i r & \text{if } w_i = v'_i r, \\ w_i & \text{if } v_i \not\leq w_i \text{ and } v'_i \not\leq w_i. \end{cases}$$

Then it is not hard to see that there are λ-terms $\hat{M}_i = (\hat{\mathcal{T}}_i, \hat{f}_i, \hat{b}_i)$ for $i = 0, 1, \ldots, n$ such that

$$v_i, v'_i \in \hat{\mathcal{T}}_i,$$
$$\{u \mid u \in \mathcal{T}_i, v_i \not\leq u, v'_i \not\leq u\} = \{u \mid u \in \hat{\mathcal{T}}_i, v_i \not\leq u, v'_i \not\leq u\},$$

$$\{\, u \mid v_i u \in \mathcal{T}_i \,\} = \{\, u \mid v'_i u \in \hat{\mathcal{T}}_i \,\},$$
$$\{\, u \mid v'_i u \in \mathcal{T}_i \,\} = \{\, u \mid v_i u \in \hat{\mathcal{T}}_i \,\},$$

and

$$M = \hat{M}_0 \xrightarrow{\hat{w}_1}_\beta \hat{M}_1 \xrightarrow{\hat{w}_2}_\beta \cdots \xrightarrow{\hat{w}_n}_\beta \hat{M}_n = |M|_\beta.$$

Moreover, we can check that the following conditions hold for $i = 0, 1, \ldots, n$ by induction:

- $v_i u \in \mathrm{dom}(f_i)$ if and only if $v'_i u \in \mathrm{dom}(\hat{f}_i)$,

- if $v_i u \in \mathrm{dom}(f_i)$, then $f_i(v_i u) = \hat{f}_i(v'_i u)$,

- $v_i u \in \mathrm{dom}(b_i)$ if and only if $v'_i u \in \mathrm{dom}(\hat{b}_i)$,

- if $v_i u \in \mathrm{dom}(b_i)$, then $v_i \leq b_i(v_i u)$ if and only if $v'_i \leq \hat{b}_i(v'_i u)$,

- if $v_i u \in \mathrm{dom}(b_i)$ and $v_i \leq b_i(v_i u)$, then for some s, $b_i(v_i u) = v_i s$ and $\hat{b}_i(v'_i u) = v'_i s$,

- if $v_i u \in \mathrm{dom}(b_i)$ and $b_i(v_i u) < v_i$, then $b_i(v_i u) = \hat{b}_i(v'_i u)$,

- $i \geq 1$ and $(M_{i-1}, v_{i-1} u) \overset{w_i}{\blacktriangleright} (M_i, v_i s)$ imply $(\hat{M}_{i-1}, v'_{i-1} u) \overset{\hat{w}_i}{\blacktriangleright} (\hat{M}_i, v'_i s)$.

From these conditions, we can see that $v_n \cong_{|M|_\beta} v'_n$ and that $(M, vu) \overset{w_1,\ldots,w_n}{\blacktriangleright} (|M|_\beta, v_n s)$ implies $(M, v'u) \overset{\hat{w}_1,\ldots,\hat{w}_n}{\blacktriangleright} (|M|_\beta, v'_n s)$. By Theorem 3.5, we can conclude that $(M, vu) \overset{w_1,\ldots,w_n}{\blacktriangleright} (|M|_\beta, v_n s)$ implies $(M, v'u) \overset{w_1,\ldots,w_n}{\blacktriangleright} (|M|_\beta, v'_n s)$. □

Let $M = (\mathcal{T}, f, b) \in \Lambda(\Sigma)$. We call two nodes w, w' of M *homologous* and write $w \approx_M w'$ if there are v, v', u satisfying the following conditions:

- $w = vu, w' = v'u$,

- $v \cong_M v'$, and

- v and v' are pivots.

The relation $\approx_M$ is symmetric, but not transitive. We call two nodes w, w' of M *similar* if $w \approx_M^* w'$ (i.e., if they stand in the reflexive transitive closure of the relation of being homologous).

Example 3.59. Let

$$M = \lambda yzvw.w(v(\lambda x.z(yx)(yx)))(v(\lambda x.z(yx)(yx))).$$

This is a λ-term in η-long β-normal form, and the following is a natural deduction representation of (M, t), where t is a principal type decoration of M:

$$\cfrac{\cfrac{[2 \to 2 \to 1]^{000} \quad \cfrac{[(5 \to 3) \to 2]^{00} \quad \cfrac{[5 \to 3]}{5 \to 3}\,0000011 \cfrac{[4 \to 4 \to 3]^0 \quad \cfrac{[5 \to 4]^{\epsilon} \quad [5]^{0000011}}{4}}{4 \to 3}^3}{2} \cfrac{[(5 \to 3) \to 2]^{00} \quad \cfrac{[5 \to 3]}{5 \to 3}\,000011 \cfrac{[4 \to 4 \to 3]^0 \quad \cfrac{[5 \to 4]^{\epsilon} \quad [5]^{000011}}{4}}{4 \to 3}^3}{2}}{2 \to 1}^1}{\cfrac{\cfrac{\cfrac{(2 \to 2 \to 1) \to 1}{((5 \to 3) \to 2) \to (2 \to 2 \to 1) \to 1}^{00}}{(4 \to 4 \to 3) \to ((5 \to 3) \to 2) \to (2 \to 2 \to 1) \to 1}^0}{(5 \to 4) \to (4 \to 4 \to 3) \to ((5 \to 3) \to 2) \to (2 \to 2 \to 1) \to 1}^{\epsilon}}^{000}$$

Note that a node v of M is a pivot if and only if $t(v)$ is an atomic type. We have

- $000001 \cong_M 00001$ (the two occurrences of $v(\lambda x.z(yx)(yx))$ (with type 2) are congruent)

- $0000011001 \cong_M 000001101$ (the first and second occurrences of yx (with type 4) are congruent)

- $000011001 \cong_M 00001101$ (the third and fourth occurrences of yx (with type 4) are congruent)

Consequently, we have

- $00000110010 \approx_M 0000110010$ (the first and third occurrences of y are homologous),

- $0000011010 \approx_M 000011010$ (the second and fourth occurrences of y are homologous),

- $00000110010 \approx_M 0000011010$ (the first and second occurrences of y are homologous),

- $0000110010 \approx_M 000011010$ (the third and fourth occurrences of y are homologous), and all four occurrences of y in (the above λ-expression for) M are similar. Note that M β-expands to an almost linear λ-term

$$M' = \lambda yzvw.(\lambda x_1.wx_1x_1)(v(\lambda x.(\lambda x_2.zx_2x_2)(yx))))),$$

in which all four occurrences of y in M have a common ancestor.

Lemma 3.60. *Let $M \in \Lambda(\Sigma)$ be a typable λ-term and suppose $M \twoheadrightarrow_\beta |M|_\beta$ by almost non-duplicating β-reduction. If two distinct nodes of $|M|_\beta$ share a common ancestor in M, then they are similar.*

Proof. Consider two distinct nodes v, v' of $|M|_\beta$ that share a common ancestor in M. Let $M = M_0 \xrightarrow{w_1}_\beta M_1 \xrightarrow{w_2}_\beta \cdots \xrightarrow{w_n}_\beta M_n = |M|_\beta$ be an almost non-duplicating β-reduction, and let t_i be a principal type decoration of M_i. Let $v_n = v$, $v'_n = v'$, and for $i = 1, \ldots, n$, let v_{i-1} and v'_{i-1} be the nodes of M_{i-1} such that $(M_{i-1}, v_{i-1}) \blacktriangleright^{w_i}_{k_i} (M_i, v_i)$ and $(M_{i-1}, v'_{i-1}) \blacktriangleright^{w_i}_{k'_i} (M_i, v'_i)$. By assumption, $v_0 = v'_0$. We first prove the following:

Claim. For some distinct nodes u, u' of $|M|_\beta$, it holds that $u \leq v$, $u' \leq v'$, $u \cong_{|M|_\beta} u'$, and u and u' are pivots.

Since v and v' are distinct, there is an $i \geq 1$ such that $v_{i-1} = v'_{i-1}$ and $v_i \neq v'_i$. We must have $w_i 1 \leq v_{i-1}$, $w_i 1 \leq v'_{i-1}$, and $k_i \neq k'_i$. Let $m = \max\{i \mid 1 \leq i \leq n, w_i 1 \leq v_{i-1}, w_i 1 \leq v'_{i-1}, k_i \neq k'_i\}$. There must be nodes u_m, u'_m of M_m such that $(M_{m-1}, w_m 1) \blacktriangleright^{w_m}_{k_m} (M_m, u_m)$, $(M_{m-1}, w_m 1) \blacktriangleright^{w_m}_{k'_m} (M_m, u'_m)$, $u_m \leq v_m$, and $u'_m \leq v'_m$. By Lemma 3.50, we have $u_m \cong_{M_m} u'_m$. Since by assumption the β-reduction step $M_{m-1} \xrightarrow{w_m}_\beta M_m$ is almost non-duplicating, $t_{m-1}(w_m 1)$ must be an atomic type. It follows that u_m and u'_m are pivots; in particular, neither u_m nor u'_m is a unary node.

For $i = m+1, \ldots, n$, define u_i and u'_i as follows:

$$u_i = \begin{cases} w_i & \text{if } u_{i-1} = w_i, \\ \text{the node such that } (M_{i-1}, u_{i-1}) \blacktriangleright^{w_i}_{k_i} (M_i, u_i) & \text{if } w_i 1 \leq u_{i-1}, \\ \text{the node such that } (M_{i-1}, u_{i-1}) \blacktriangleright^{w_i}_1 (M_i, u_i) & \text{otherwise.} \end{cases}$$

$$u'_i = \begin{cases} w_i & \text{if } u'_{i-1} = w_i, \\ \text{the node such that } (M_{i-1}, u'_{i-1}) \blacktriangleright^{w_i}_{k'_i} (M_i, u'_i) & \text{if } w_i 1 \leq u'_{i-1}, \\ \text{the node such that } (M_{i-1}, u'_{i-1}) \blacktriangleright^{w_i}_1 (M_i, u'_i) & \text{otherwise.} \end{cases}$$

It is easy to see by induction that such u_i and u'_i always exist, and it holds that $u_i \leq v_i$ and $u'_i \leq v'_i$. By the assumption about m, we have that for $i \in \{m+1,\ldots,n\}$, if $w_i 1 \leq u_{i-1}$ and $w_i 1 \leq u'_{i-1}$, then $k_i = k'_i$. By Lemma 3.58, it follows that $u_n \cong_{|M|_\beta} u'_n$.

For $i = m-1,\ldots,n$, define a type decoration t'_i for M_i by

$$t'_{m-1} = t_{m-1},$$
$$(M_{i-1}, t'_{i-1}) \xrightarrow{w_i}_\beta (M_i, t'_i) \quad \text{for } i = m,\ldots,n.$$

Then it is easy to see $t_{m-1}(w_m 1) = t'_i(u_i) = t'_i(u'_i)$ for all $i = m,\ldots,n$. Since $t_{m-1}(w_m 1)$ is an atomic type, it follows that u_n and u'_n are pivots. So we have proved the above claim, with $u = u_n$ and $u' = u'_n$.

Now we show that v and v' are similar by induction on $|v| - |u| + |v'| - |u'|$. Let s, s' be such that $v = us$ and $v' = u's'$. If $s = s'$, then v and v' are homologous and hence similar. Suppose $s \neq s'$. By Lemma 3.58, the nodes $v = us$ and us' of $|M|_\beta$ share a common ancestor in M. By the above claim applied to us, us' in place of v, v', we must have $s = s_1 s_2$, $s' = s'_1 s'_2$, $s_1 \neq s'_1$, $us_1 \cong_{|M|_\beta} us'_1$, and us_1 and us'_1 are pivots. Since $|us| - |us_1| + |us'| - |us'_1| = |s_2| + |s'_2| < |s| + |s'| = |v| - |u| + |v'| - |u'|$, the induction hypothesis applies; hence us and us' are similar. Since us' and $u's'$ are homologous, it follows that us and $u's'$ are similar. □

Lemma 3.61. *Let $M = (\mathcal{T}, f, b) \in \Lambda(\Sigma)$ be a closed typable λ-term in η-long form, and let $M° = \text{COLLAPSE}(M)$. Suppose that u_1 and u_2 are distinct nodes of M such that $u_1 \approx^*_M u_2$. Unless u_1 is a pivot and $u_1 \in \text{dom}(b)$, u_1 and u_2 share a common ancestor in $M°$.*

Proof. Clearly, it suffices to consider the case where $u_1 \approx_M u_2$. There must be a pair of pivots s_1, s_2 such that $s_1 \cong_M s_2$ and for some u, $u_1 = s_1 u$ and $u_2 = s_2 u$. By the assumption about u_1, u_2, we have $s_1, s_2 \notin \text{dom}(b)$. At each stage of the execution of Algorithm 1, let u'_1, u'_2, s'_1, s'_2 be the ancestors of u_1, u_2, s_1, s_2, respectively, in $M°$. By Lemma 3.56, s'_1 and s'_2 are pivots and $s'_1 \cong_{M°} s'_2$. We must have $s'_1 = s'_2$ at the end of the execution of Algorithm 1. Since the nodes $v_1, v_2, \ldots, v_k$ picked in lines 5–6 of the algorithm are duplicated pivots of maximal height in $M°$, we cannot have $s'_1 < v_i$ or $s'_2 < v_i$ for some $i \in \{1, 2, \ldots, k\}$ until $s'_1 = s'_2$. Hence, until $s'_1 = s'_2$, we have $u'_1 = s'_1 u$ and $u'_2 = s'_2 u$. This means that at the first stage where $s'_1 = s'_2$ holds, we have $u'_1 = u'_2$. Therefore, u_1 and u_2 have the same ancestor. □

Lemma 3.62. *Let $M \in \Lambda(\Sigma)$ be a closed λ-term in η-long β-normal form and $M° = \text{COLLAPSE}(M)$. Suppose $M' \twoheadrightarrow_\beta M$ by almost non-duplicating β-reduction. Let $m = |\overrightarrow{\text{Con}}(M°)|$ and $n = |\overrightarrow{\text{Con}}(M')|$. The following hold:*

(i) $|\widehat{M^\circ}[y_1,\ldots,y_m]|_\beta = |\widehat{M'}[y_{g(1)},\ldots,y_{g(n)}]|_\beta$ *for some* $g\colon \{1,\ldots,n\} \to \{1,\ldots,m\}$.

(ii) *If M' is almost affine, then so is M°.*

Proof. (i) Consider two occurrences v_1, v_2 of a free variable z_i in $|\widehat{M'}[z_1,\ldots,z_n]|_\beta$. Since each free variable occurs just once in $\widehat{M'}[z_1,\ldots,z_n]$, the unique occurrence u of z_i in $\widehat{M'}[z_1,\ldots,z_n]$ is the common ancestor of the nodes v_1 and v_2 of $|\widehat{M'}[z_1,\ldots,z_n]|_\beta$. This means that the node u of M' is the common ancestor of the nodes v_1 and v_2 of M. By Lemma 3.60, we have $v_1 \approx_M^* v_2$. Since some constant occurs at v_1 and v_2 in M, Lemma 3.61 implies that v_1 and v_2 have the same ancestor in M°. This means that the same free variable occurs at v_1 and v_2 in $|\widehat{M^\circ}[y_1,\ldots,y_m]|_\beta$. Therefore, there is a function $g\colon \{\, i \mid z_i \in \mathrm{FV}(|\widehat{M'}[z_1,\ldots,z_n]|_\beta) \,\} \to \{1,\ldots,m\}$ such that if v is an occurrence of z_i in $|\widehat{M'}[z_1,\ldots,z_n]|_\beta$, then v is an occurrence of $y_{g(i)}$ in $|\widehat{M^\circ}[y_1,\ldots,y_m]|_\beta$. Some z_i may not occur in $|\widehat{M'}[z_1,\ldots,z_n]|_\beta$, but by extending g to a function from $\{1,\ldots,n\} \to \{1,\ldots,m\}$ in an arbitrary way, we have $|\widehat{M^\circ}[y_1,\ldots,y_m]|_\beta = |\widehat{M'}[y_{g(1)},\ldots,y_{g(n)}]|_\beta$.

(ii) Let t_{M° be a principal type decoration of $M^\circ = (\mathcal{T}_{M^\circ}, f_{M^\circ}, b_{M^\circ})$. Suppose that M° is not almost affine. Then there are two distinct leaves $v_1, v_2 \in \mathcal{T}_{M^\circ}^{(0)}$ such that $b_{M^\circ}(v_1) = b_{M^\circ}(v_2)$ and $t_{M^\circ}(v_1) = t_{M^\circ}(v_2)$ is a non-atomic type. Since M° is in η-long form, we have $v_1 = u_1 0$ and $v_2 = u_2 0$ for some $u_1, u_2 \in \mathcal{T}_{M^\circ}^{(2)}$. Since the β-reduction from M° to $M = (\mathcal{T}_M, f_M, b_M)$ is non-erasing and almost non-duplicating, it is easy to see that by taking the leftmost (i.e., lexicographically first) descendants at each step, we can arrive at $u_1', u_2' \in \mathcal{T}_M^{(2)}$ such that $u_1' 0, u_2' 0$ are descendants of v_1 and v_2, respectively, and $b_M(u_1' 0) = b_M(u_2' 0)$. Now let v_1' and v_2' be the ancestors of $u_1' 0$ and $u_2' 0$, respectively, in $M' = (\mathcal{T}_{M'}, f_{M'}, b_{M'})$. By Lemma 3.4, part (iv), we see that $b_{M'}(v_1') = b_{M'}(v_2')$. Let $t_{M'}$ be a principal type decoration of M'. Since M' is almost affine, either $v_1' = v_2'$ or $t_{M'}(v_1') = t_{M'}(v_2')$ is an atomic type q. Let t_M be a type decoration of M such that $(M', t_{M'}) \twoheadrightarrow_\beta (M, t_M)$. If $t_{M'}(v_1') = t_{M'}(v_2') = q$, then $t_M(u_1' 0) = t_M(u_2' 0) = q$, contradicting $u_1', u_2' \in \mathcal{T}_M^{(2)}$. Hence $v_1' = v_2'$. By Lemma 3.60, $u_1' 0 \approx_M^* u_2' 0$. By Lemma 3.61, $v_1 = v_2$, a contradiction. This contradiction shows that M° is almost affine. □

Lemma 3.63. *Let $M \in \Lambda(\Sigma)$ be a closed λI-term in η-long β-normal form relative to γ, and let $M^\circ = \mathrm{COLLAPSE}(M)$. The following hold:*

(i) *M° is a λI-term in η-long form.*

(ii) *If $M' \twoheadrightarrow_\beta M$ by non-erasing almost non-duplicating β-reduction, then $M' \in \Lambda(\mathrm{database}(M^\circ), \langle \gamma \rangle(\mathrm{tuple}(M^\circ)))$.*

Proof. Part (i) is by Lemma 3.56, parts (i) and (ii).

For part (ii), let $\overrightarrow{\mathrm{Con}}(M^\circ) = (d_1, \ldots, d_m)$ and let

$$y_1 : \beta_1, \ldots, y_m : \beta_m \Rightarrow \alpha$$

be a principal typing of $\widehat{M^\circ}[y_1, \ldots, y_m]$. Then $d_i(\overline{\beta_i}) \in \mathrm{database}(M^\circ)$ for $i = 1, \ldots, m$, and $\langle \gamma \rangle(\mathrm{tuple}(M^\circ)) = \alpha$. Let $n = |\overrightarrow{\mathrm{Con}}(M')|$. By Lemma 3.62, there is a function $g \colon \{1, \ldots, n\} \to \{1, \ldots, m\}$ such that

$$|\widehat{M^\circ}[y_1, \ldots, y_m]|_\beta = |\widehat{M'}[y_{g(1)}, \ldots, y_{g(n)}]|_\beta$$

Since $M' \twoheadrightarrow_\beta M$ by non-erasing almost non-duplicating β-reduction, we also have $\widehat{M'}[y_{g(1)}, \ldots, y_{g(n)}] \twoheadrightarrow_\beta |\widehat{M'}[y_{g(1)}, \ldots, y_{g(n)}]|_\beta$ by non-erasing almost non-duplicating β-reduction. Then by the Subject Reduction Theorem (Theorem 3.14) and Lemma 3.52, we have

$$\vdash \{ y_{g(i)} : \beta_{g(i)} \mid 1 \leq i \leq n \} \Rightarrow \widehat{M'}[y_{g(1)}, \ldots, y_{g(n)}] : \alpha.$$

This means that $M' \in \Lambda(\mathrm{database}(M^\circ), \langle \gamma \rangle(\mathrm{tuple}(M^\circ)))$. □

Lemma 3.63 does not say that $\Lambda(\mathrm{database}(M^\circ), \langle \gamma \rangle(\mathrm{tuple}(M^\circ)))$ is closed under non-erasing almost non-duplicating β-expansion, but together with Lemma 3.54 implies the following, which corresponds to the special property (28) highlighted in the rough proof sketch given in Section 2.3.

Lemma 3.64. *Let $M \in \Lambda(\Sigma)$ be a closed λ-term in η-long β-normal form relative to γ, and suppose that $M^\circ = \mathrm{COLLAPSE}(M)$ is almost linear. Then for every almost linear closed λ-term $M' \in \Lambda(\Sigma)$ in η-long form relative to γ, $M' \twoheadrightarrow_\beta M$ if and only if $M' \in \Lambda(\mathrm{database}(M^\circ), \langle \gamma \rangle(\mathrm{tuple}(M^\circ)))$.*

Proof. First note that $M \in \Lambda(\mathrm{database}(M^\circ), \langle \gamma \rangle(\mathrm{tuple}(M^\circ)))$ by Lemma 3.37 and part (i) of Lemma 3.34.

Suppose $M' \twoheadrightarrow_\beta M$. By Lemma 3.49, the leftmost β-reduction from M' to $M = |M'|_\beta$ is non-erasing and almost non-duplicating. Lemma 3.63 then implies $M' \in \Lambda(\mathrm{database}(M^\circ), \langle \gamma \rangle(\mathrm{tuple}(M^\circ)))$.

Conversely, suppose $M' \in \Lambda(\mathrm{database}(M^\circ), \langle \gamma \rangle(\mathrm{tuple}(M^\circ)))$. By part (i) of Lemma 3.34 again, $|M'|_\beta \in \Lambda(\mathrm{database}(M^\circ), \langle \gamma \rangle(\mathrm{tuple}(M^\circ)))$. Since by Lemma 3.17 $|M'|_\beta$ must be in η-long form relative to γ, Lemma 3.54 implies $|M'|_\beta = M$. □

The following theorem is the main result of the paper.

Theorem 3.65. *Let $\mathscr{G} = (\mathscr{N}, \Sigma, f, \mathscr{P}, S)$ be an almost linear CFLG. Suppose that $N \in \Lambda(\Sigma)$ is a λ-term in η-long β-normal form relative to $f(S)$. Then the following are equivalent:*

(i) $N \in L(\mathscr{G})$.

(ii) $N° = \text{COLLAPSE}(N)$ *is almost linear and* $\text{program}(\mathscr{G}) \cup \text{database}(N°) \vdash S(\text{tuple}(N°))$.

Proof. (i) $\Rightarrow$ (ii). Suppose $N \in L(\mathscr{G})$. Then there is a closed λ-term $P \in \Lambda(\Sigma)$ in η-long form such that $\vdash_{\mathscr{G}} S(P)$ and $P \twoheadrightarrow_\beta N$. Since $\mathscr{G}$ is almost linear, P is almost linear. By part (ii) of Lemma 3.62 and part (i) of Lemma 3.63, $N° = \text{COLLAPSE}(N)$ is almost linear. By Lemma 3.64, $P \in \Lambda(\text{database}(N°), \langle f(S) \rangle(\text{tuple}(N°)))$. Lemma 3.35 then implies $\text{program}(\mathscr{G}) \cup \text{database}(N°) \vdash S(\text{tuple}(N°))$.

(ii) $\Rightarrow$ (i). By Lemma 3.35, there is an almost linear λ-term $P \in \Lambda(\Sigma)$ in η-long form such that $\vdash_{\mathscr{G}} S(P)$ and $P \in \Lambda(\text{database}(N°), \langle f(S) \rangle(\text{tuple}(N°)))$. Since $N°$ is almost linear, Lemma 3.64 implies that $P \twoheadrightarrow_\beta N$. Therefore, $N \in L(\mathscr{G})$. $\square$

Notice that when $\mathscr{G}$ is a linear CFLG and N is a linear λ-term, Theorems 3.42 and 3.65 both hold of $\mathscr{G}$ and N, even though $N° \neq N$ in general.

Let us turn to the complexity analysis of the reduction. Since it is easy to see that Algorithm 1 runs in polynomial time, an immediate corollary to Theorem 3.65 is that the language of every almost linear CFLG belongs to the complexity class P. As in the linear case, we can obtain a tight complexity upper bound. Recall that $\prec$ is the lexicographic order on $\{0,1\}^*$. We write $\approx_M \cap \prec$ for the intersection of the two relations $\approx_M$ and $\prec$, thought of as sets of ordered pairs.

Lemma 3.66. *Let $M = (\mathcal{T}, f, b)$ be a λ-term, and suppose that $u_1 \ (\approx_M \cap \prec) \ v$ and $u_2 \ (\approx_M \cap \prec) \ v$. Then there exists a v' such that*

(i) *either $v' \ (\approx_M \cap \prec) \ u_1$ or $v' = u_1$, and*

(ii) *either $v' \ (\approx_M \cap \prec) \ u_2$ or $v' = u_2$.*

Proof. There are $\hat{u}_1, \hat{u}_2$ and pivots w_1, w, s, s_2 such that

$$w_1 \cong_M w, \qquad w_1 \prec w, \qquad u_1 = w_1 \hat{u}_1, \qquad v = w \hat{u}_1,$$
$$s_2 \cong_M s, \qquad s_2 \prec s, \qquad u_2 = s_2 \hat{u}_2, \qquad v = s \hat{u}_2.$$

Since w and s are prefixes of v, either $w \leq s$ or $s \leq w$. We may assume $w \leq s$. We have

$$s = w\hat{s}, \qquad \hat{u}_1 = \hat{s}\hat{u}_2$$

for some $\hat{s}$.

Case 1. There is an s' such that $w \leq b(ss') < s$. Since $s_2 \cong_M s$, it must be the case that $w \leq b(ss') = b(s_2 s') < s_2$. So there is an $\hat{s}_2$ such that

$$s_2 = w\hat{s}_2, \quad \hat{s}_2 \prec \hat{s}.$$

Since $w\hat{s}_2 = s_2 \cong_M s = w\hat{s}$ and $w_1 \cong_M w$, the definition of the congruence relation $\cong_M$ implies that $w_1\hat{s}_2$ and $w_1\hat{s}$ are pivots and

$$w_1\hat{s}_2 \cong_M w_1\hat{s}.$$

Therefore,

$$u_1 = w_1\hat{u}_1 = w_1\hat{s}\hat{u}_2 \approx_M w_1\hat{s}_2\hat{u}_2 \approx_M w\hat{s}_2\hat{u}_2 = s_2\hat{u}_2 = u_2.$$

Let $v' = w_1\hat{s}_2\hat{u}_2$. Since $\hat{s}_2 \prec \hat{s}$ and $w_1 \prec w$, we have $v' \prec u_1$ and $v' \prec u_2$.

Case 2. There is no s' such that $w \leq b(ss') < s$. Since $w_1 \cong_M w$, we must have

$$s_2 \cong_M s = w\hat{s} \cong_M w_1\hat{s},$$

and $w_1\hat{s}$ is a pivot. Therefore,

$$u_2 = s_2\hat{u}_2 \approx_M w_1\hat{s}\hat{u}_2 = w_1\hat{u}_1 = u_1,$$

and the conclusion clearly holds with either $v' = u_1$ or $v' = u_2$. □

The next lemma easily follows from Lemma 3.66.

Lemma 3.67. *If $u \approx_M^* v$, then there exists a w such that $w \ (\approx_M \cap \prec)^* \ u$ and $w \ (\approx_M \cap \prec)^* \ v$.*

Lemma 3.68. *Let $\mathscr{G} = (\mathscr{N}, \Sigma, f, \mathscr{P}, S)$ be an almost linear CFLG. There is a log-space-bounded deterministic Turing machine that takes as input a λ-term $N \in \Lambda(\Sigma)$ in η-long β-normal form relative to $f(S)$ and decides whether Algorithm 1 returns an almost linear N°, and if so, computes $(\text{database}(N^\circ), S(\text{tuple}(N^\circ)))$.*

Proof (sketch). Let $N = (\mathcal{T}, f, b)$. We assume that N is given as a λ-expression as before. We must avoid computing the output $N^\circ = \text{COLLAPSE}(N)$ of Algorithm 1 explicitly. By Lemmas 3.60 and 3.61, N° is almost affine if and only if for every pair of nodes $v, v' \in \mathcal{T}^{(0)}$ such that $b(v) = b(v')$, it holds that $v \approx_N^* v'$. By Lemma 3.67, this is so if and only if $w \ (\approx_N \cap \prec)^* \ v'$, where w is the leftmost node such that $w \ (\approx_N \cap \prec)^* \ v$.[37] Checking whether two nodes are homologous clearly requires no

---

[37] In fact, by refining the proof of Lemma 3.60, it is not hard to see that it suffices to take the leftmost w such that $b(w) = b(v) = b(v')$ and check $w \ (\approx_N \cap \prec)^* \ v$ and $w \ (\approx_N \cap \prec)^* \ v'$.

more than logarithmic space, so the relation $(\approx_N \cap \prec)^*$ can be decided in logarithmic space as well. It follows that the similarity of v and v' can also be checked in logarithmic space. It is also easy to see that $N°$ is a λI-term if and only if N is, and clearly this can be checked in logarithmic space.

Now suppose that N is λI and $N°$ is almost linear. Let $\overrightarrow{\mathrm{Con}}(N) = (d_1, \ldots, d_m)$, and let

$$y_1 : \beta_1, \ldots, y_m : \beta_m \Rightarrow \beta_0$$
$$z_1 : \gamma_1, \ldots, z_n : \gamma_n \Rightarrow \gamma_0$$

be principal typings of $\widehat{N}[y_1, \ldots, y_m]$ and $\widehat{N°}[z_1, \ldots, z_n]$, respectively. Let $\{w_1, \ldots, w_m\} = \mathrm{dom}(f)$, where $w_1 \prec \ldots \prec w_m$. (We have $f(w_i) = d_i$.) Let

$$I = \{\, i \in \{1, \ldots, m\} \mid \text{there is no } k < i \text{ such that } w_k \approx_N^* w_i \,\}.$$

Let $g \colon \{1, \ldots, m\} \to I$ be the function such that $w_{g(i)} \approx_N^* w_i$ for $i \in \{1, \ldots, m\}$. By Lemmas 3.60 and 3.61, we must have a bijection $h \colon \{1, \ldots, n\} \to I$ such that $\widehat{N°}[y_{h(1)}, \ldots, y_{h(n)}] \twoheadrightarrow_\beta \widehat{N}[y_{g(1)}, \ldots, y_{g(m)}]$. Let σ be a most general unifier of

$$\{\, (\beta_i, \beta_j) \mid i \in I, w_i \approx_N^* w_j \,\}$$

Then

$$\{\, y_i : \beta_i \sigma \mid i \in I \,\} \Rightarrow \beta_0 \sigma \tag{60}$$

is a principal typing of $\widehat{N}[y_{g(1)}, \ldots, y_{g(m)}]$. By Lemma 3.53, (60) is a principal typing of $\widehat{N°}[y_{h(1)}, \ldots, y_{h(n)}]$ as well, and we have

$$\mathrm{database}(N°) = \{\, d_i(\overline{\beta_i \sigma}) \mid i \in I \,\},$$
$$\mathrm{tuple}(N°) = \overline{\beta_0 \sigma}.$$

By Theorem 3.46, (60) is negatively non-duplicated. For every $i \in \{0, \ldots, m\}$ and $v \in \langle \beta_i \rangle^{(0)}$, let $p_{i,v} = \mathrm{subtype}(\beta_i, v)$. Then if $(i_1, v_1), (i_2, v_2)$ are distinct negative occurrences and $i_1, i_2 \in \{0\} \cup I$, then $p_{i_1,v_1}\sigma \neq p_{i_2,v_2}\sigma$. Now consider a positive occurrence (i, v) of $p_{i,v}$ such that $i \in \{0\} \cup I$. The fact that $\widehat{N}[y_1, \ldots, y_m]$ is a λI-term implies that in $(\widehat{N}[y_1, \ldots, y_m], \hat{t})$, where $\hat{t}$ is a principal type decoration of $\widehat{N}[y_1, \ldots, y_m]$, the occurrence (i, v) is linked to some negative occurrence (i', v') of $p_{i,v} = p_{i',v'}$. If $i' \in \{1, \ldots, m\}$, let $j = g(i')$; otherwise let $j = i' = 0$. Then it must be that $p_{i,v}\sigma = p_{i',v'}\sigma = p_{j,v'}\sigma$. Note that although (i, v) may be linked to more than one (i', v') in $(\widehat{N}[y_1, \ldots, y_m], \hat{t})$, the pair (j, v') is uniquely determined independently of the choice of (i', v') because (60) is negatively non-duplicated.

As in the proof of Lemma 3.43, a deterministic log-space-bounded Turing machine can compute a negative occurrence (i', v') linked to a positive occurrence (i, v) by following edges of $\overrightarrow{G}_{(\widehat{N}[y_1,\ldots,y_m],\hat{t})} = \overrightarrow{G}_{(\widehat{N}[y_1,\ldots,y_m],t)}$, where t is the type decoration for $y_1 : \tau(d_1), \ldots, y_m : \tau(d_m) \Rightarrow \widehat{N}[y_1, \ldots, y_m] : f(S)$. Again, the type decoration t is not explicitly computed. There may be more than one maximal directed path starting from (i, v), but any such path will do, so the machine simply picks the first relevant edge that it can find at each point. Once the machine reaches a configuration representing $(w_{i'}, v', \uparrow)$, it can then find the least j such that $w_j \,(\approx_N \cap \prec)^* \, w_{i'}$ using no more than logarithmic space. $\square$

Theorem 3.69. *For every almost linear CFLG $\mathcal{G}$, $L(\mathcal{G})$ belongs to LOGCFL.*

Proof. The proof is similar to that of Theorem 3.44. Note that if $P \twoheadrightarrow_\beta N$ by non-erasing β-reduction, then $|\overrightarrow{\text{Con}}(P)| \leq |\overrightarrow{\text{Con}}(N)|$. This implies that whenever $\text{program}(\mathcal{G}) \cup \text{database}(N^\circ) \vdash S(\text{tuple}(N^\circ))$ holds, there is a derivation tree for it whose size is bounded by a polynomial in the number of occurrences of constants in N. $\square$

4 Some consequences and extensions

4.1 Further complexity-theoretic consequences

We have seen that the problem of recognition for a fixed almost linear CFLG is in LOGCFL. Since there is a context-free language that is LOGCFL-complete [27], it follows that LOGCFL is a tight upper bound on the computational complexity of fixed almost linear CFLG recognition.

Let us sketch some further complexity-theoretic consequences of this work. These concern three different types of problems: (i) the problem of uniform recognition for subclasses of almost linear CFLGs, (ii) the problem of parsing for a fixed almost linear CFLG, and (iii) the problem of finding one target λ-term from an input λ-term for a fixed almost linear synchronous CFLG.

4.1.1 Uniform recognition

If the grammar is not fixed and is part of the input, the recognition problem (known as *uniform recognition*) is known to be P-complete for general context-free grammars, and PSPACE-complete for non-deleting multiple context-free grammars [38, 39]. Since it is easy to translate non-deleting multiple context-free grammars into linear CFLGs, the latter gives a lower bound on the complexity of uniform recognition for almost linear CFLGs. The EXPTIME-completeness of the program complexity of general Datalog query evaluation [16] provides an upper bound; currently I do not know whether either of these bounds is tight, however.

A lower complexity bound for uniform recognition can be obtained for restricted subclasses of almost linear CFLGs. We call a Datalog program **P** *k-bounded* if k is at least as large as the maximal arity of predicates in **P** and the number of variables in any rule of **P**. For a k-bounded Datalog program **P**, the number of work tapes needed in the "storage area" in the log-space-bounded ATM $\mathcal{M}_\mathbf{P}$ simulating **P** does not exceed k. (The description of $\mathcal{M}_\mathbf{P}$ was given in Section 3.1.1.) With additional work tapes to serve as pointers to rules and predicates in the Datalog program, the program can be moved from the finite control of the ATM to part of the input. The resulting log-space-bounded ATM can decide, given input $(\mathbf{P}, D, q)$ with **P** k-bounded, whether $\mathbf{P} \cup D \vdash q$. Now consider the class of k-bounded almost linear CFLGs, i.e., almost linear CFLGs $\mathscr{G}$ such that program($\mathscr{G}$) is k-bounded. As in the proof of Lemma 3.68, it is clear that the translation from $\mathscr{G}$ to program($\mathscr{G}$) can be done in logarithmic space. This means that there is a log-space reduction from the uniform recognition problem for k-bounded almost linear CFLGs to a problem in ALOGSPACE = P. Since uniform recognition for CFGs whose rules are all of the form $A \to BC$ or $A \to \epsilon$ is already P-complete [36], it follows that the uniform recognition problem for k-bounded almost linear CFLGs is P-complete.

It is folklore [55] that the uniform recognition problem for the class of context-free grammars without ϵ-productions is in LOGCFL. What corresponds to an ϵ-production in the case of CFLGs is a rule of the form

$$B(M)$$

(with an empty right-hand side) where M is a pure λ-term. We can eliminate all such ϵ-*rules* from an almost linear CFLG by the same method that Kanazawa and Yoshinaka [47] used for linear CFLGs, so the uniform recognition problem for the class of almost linear CFLGs without ϵ-rules is of interest. If $\mathscr{G}$ is such a CFLG, then all leaves of Datalog derivation trees for program($\mathscr{G}$) are extensional nodes. By the analysis in the proof of Lemma 3.2, we can show that the uniform recognition problem for the class of ϵ-free k-bounded almost linear CFLGs is in LOGCFL.

4.1.2 Parsing

It is also interesting to ask the computational complexity of parsing, as opposed to recognition. *Functional LOGCFL* (written FL^{LOGCFL}) is the class of solution search problems that can be solved by a deterministic log-space-bounded Turing machine with a LOGCFL oracle [26]. It is a natural functional analogue of LOGCFL. Gottlob et al. [26] show that given an ATM $\mathscr{M}$ with simultaneous log-space and poly-size bounds, the problem of finding a first accepting computation tree of $\mathscr{M}$ on input w (within a given polynomial size bound) is in functional LOGCFL. In the course of proving this result, they also show that the set of all accepting computation trees (within a given polynomial size bound), in the form of a 'shared forest', can be computed by a log-space-bounded Turing machine with a LOGCFL oracle. We can use this result to show that the problem of parsing for a fixed almost linear CFLG is in functional LOGCFL, but here we opt to give the following direct proof, which is straightforward and more informative.

Let $\mathbf{P}$ be a Datalog program, D an extensional database for $\mathbf{P}$, and q a ground fact. The 'shared forest' representation of the set of all derivation trees for $\mathbf{P} \cup D \vdash q$ is just the set F all ground instances

$$p(\vec{s}) :- p_1(\vec{s}_1), \ldots, p_l(\vec{s}_l)$$

of rules $p(\vec{x}) :- p_1(\vec{x}_1), \ldots, p_l(\vec{x}_l) \in \mathbf{P}$ that can appear in some derivation tree for $\mathbf{P} \cup D \vdash q$ which use only constants from $D \cup \{q\}$.[38] Suppose that the number of extensional nodes in any derivation tree for $\mathbf{P} \cup D \vdash q$ is bounded by a number k, depending only on D. In order to see whether $p(\vec{s}) :- p_1(\vec{s}_1), \ldots, p_l(\vec{s}_l)$ is in F, one need only check whether there are derivation trees (with no more than k extensional nodes) for

$$\mathbf{P} \cup D \vdash p_i(\vec{s}_i) \quad (i = 1, \ldots, l)$$

and one for

$$\mathbf{P} \cup \{p(\vec{s})\} \cup D \vdash q$$

in which $p(\vec{s})$ appears on exactly one of its leaves. Let $g(n)$ be the polynomial that Lemma 3.2 associates with $\mathbf{P}$. Then derivation trees for $\mathbf{P} \cup D \vdash p_i(\vec{s}_i)$ $(i = 1, \ldots, l)$ can be found from among those with at most $g(k)$ nodes, if there are any. It is not hard to see that the same reasoning as in the proof of Lemma 3.2 shows that the minimal size of the required kind of derivation tree for $\mathbf{P} \cup \{p(\vec{s})\} \cup D \vdash q$ can be bounded by $g(k + 1)$. Thus, answers to these questions can be obtained through oracle queries to two sets

$$\{(D, q_1, 1^m) \mid \text{there is a derivation tree for } \mathbf{P} \cup D \vdash q_1 \text{ of size } \leq g(m)\},$$
$$\{(\{q_2\} \cup D, q_1, 1^m) \mid \text{there is a derivation tree for } \mathbf{P} \cup \{q_2\} \cup D \vdash q_1 \text{ of size } \leq g(m)$$
$$\text{with } q_2 \text{ on exactly one leaf}\}.$$

---

[38] It would be more appropriate to call the set F the "reduced" shared forest, since a shared parse forest in general may contain useless elements.

The former is in LOGCFL by Lemma 3.1. A slight modification of its proof shows that the latter is in LOGCFL, too, and it is easy to combine the two into a single LOGCFL oracle. Thus, if $(D, q, 1^k)$ is given as input, the set F can be computed in logarithmic space with a LOGCFL oracle by cycling through all ground instances of all rules in **P**.

Let **P** = program($\mathscr{G}$) for some almost linear CFLG, and suppose (D, q) is obtained from a λ-term N as in Theorems 3.65 and 3.69. Then we can take the number k to be $|\overrightarrow{\mathrm{Con}}(N)|$, and the set F can be computed in logarithmic space with a LOGCFL oracle.

With the one-one correspondence between the rules of the Datalog program program($\mathscr{G}$) and the rules of the CFLG $\mathscr{G}$, the set F can also be taken to be a shared forest representation of the set of all derivation trees of $\mathscr{G}$ for the input λ-term N. Thus, given an almost CFLG $\mathscr{G}$, the problem of computing the shared forest of all derivation trees of $\mathscr{G}$ for an input λ-term N is in functional LOGCFL.

4.1.3 Transduction with almost linear synchronous CFLGs

Suppose we are given a synchronous CFLG consisting of a pair of almost linear CFLGs. Given an input λ-term M generated by one of the component CFLGs (call it the "source-side" grammar), the set of all derivation trees of M can be efficiently computed in the form of a shared forest, as we have seen above. In order to find a "target-side" λ-term N that the synchronous grammar pairs with M, we can take one of the derivation trees T, construct a λ-term P that the "target-side" CFLG associates with T, and then compute the β-normal form $N = |P|_\beta$ of P. It is of course impossible to explicitly enumerate all such N, because there may be infinitely many derivation trees of M; nor is there any simple "packed" representation of all such N (because the set of all such N is in general as complex as the language of an arbitrary almost linear CFLG). Let us therefore consider the computational complexity of finding *one* λ-term N that the synchronous grammar pairs with M.

As in [26], a deterministic log-space-bounded Turing machine with a LOGCFL oracle can compute a single derivation tree T of M (whenever there is one). It is easy to see that given a derivation tree T, the λ-term P that the target-side grammar associates with T can be computed in logarithmic space. Although the size of $|P|_\beta$ is in general exponential in the size of P and so it is not feasible to compute $|P|_\beta$ explicitly, the pair (database(P), tuple(P)) can be computed in logarithmic space as in the proof of Lemma 3.68. Since P is almost linear, by Lemma 3.54, $|P|_\beta$ is the only λ-term in η-long β-normal form in $\Lambda(\text{database}(P), \langle\gamma\rangle(\text{tuple}(P)))$, where γ is the type that the target-side grammar assigns to P. So (database(P), tuple(P)) serves as a kind of compact representation of $|P|_\beta$. (In fact, when $|P|_\beta$ is a tree, (database(P), tuple(P)) is a representation of a term graph that unfolds to (the hypergraph representation of) $|P|_\beta$.) All in all, given a fixed almost linear synchronous CFLG, the problem of finding one target-side λ-term N corresponding to an input source-side λ-term M is in functional LOGCFL, if we allow as output a compact representation of N in the form of a pair of a database and a tuple of constants.

Note that in the special case where P is linear and $|P|_\beta$ is an encoding of a string or a tree, (database(P), tuple(P)) is nothing but an explicit hypergraph representation of the latter. Thus, with respect to a fixed synchronous grammar consisting of a linear string grammar (e.g., a CFG or MCFG) and an almost linear Montague semantics, the problem of explicitly computing one surface realization of an input logical form is in functional LOGCFL.

4.2 Regular sets as input

4.2.1 Parsing as intersection for linear CFLGs

In ordinary parsing/recognition of string languages, it is sometimes useful to allow as input a regular set of strings (usually represented as a finite automaton), rather than a single string. The resulting generalization of the problem is a key element of the view of "parsing as intersection", where the "shared parse forest" that is the output of parsing is given in the form of a grammar generating the intersection of the language of the original grammar and the input regular set. Various dynamic parsing techniques can then be regarded as variants of Bar-Hillel et al.'s [5] original proof of the closure of the context-free languages under intersection with regular sets [52].

Many well-known grammar formalisms, including context-free grammars, tree-adjoining grammars [37], (parallel) multiple context-free grammars [66], and IO macro grammars [24], have the property that given a regular set R, any grammar G can be "specialized" into a grammar G' generating the intersection of the language of G and R, in such a way that G is the image of G' under a simple "projection" that maps nonterminals of G' to nonterminals of G. Kanazawa [40] has shown that the same property holds of de Groote's [17] abstract categorial grammars. *Linear* context-free λ-term grammars are nothing but abstract categorial grammars whose abstract vocabulary is second-order. Via encoding in linear CFLGs, Kanazawa's [40] result provides a uniform proof of closure under intersection with regular sets for *linear* formalisms such as context-free grammars, (multi-component) tree-adjoining grammars, and multiple context-free grammars.

Theorem 3.40 shows how the recognition problem for linear CFLGs in a generalized form, where an input is a set of λ-terms represented by a pair of a database and a type, reduces to Datalog query evaluation. It is easy to see that any regular set of strings or trees can be represented in this way. In the string case, a nondeterministic finite automaton with an initial state q_I and just one final state q_F translates into the pair $(D, q_F \to q_I)$, where D is the database consisting of all facts of the form $c(q, r)$ such that the automaton has a transition from state q to state r labeled by c. In the tree case, a nondeterministic bottom-up finite automaton with a unique final state q_F translates into the pair (D, q_F), where D is the database consisting of all facts $f(q, q_n, \ldots, q_1)$ such that the automaton has a transition rule $f(q_1(x_1), \ldots, q_n(x_n)) \to q(f(x_1, \ldots, x_n))$. More generally, any set of λ-terms that can be expressed as the set $\Lambda(D, \alpha)$ with a database D and a type α can be used as an input to recognition with a linear CFLG.

With Lemma 3.31, the problem of parsing in this generalized setting reduces to the problem of computing (a representation of) the set of all derivation trees from a Datalog program $\mathbf{P}$ and a database D. The connection to parsing as intersection is that the specialized grammar generating the intersection language corresponds to the propositional Horn clause program consisting of the database D and an appropriate subset of the set $\bigcup_{\pi \in \mathbf{P}} \mathrm{ground}(\pi, U_D)$ of ground instances of rules in $\mathbf{P}$.

4.2.2 Almost linear CFLGs and deterministic databases

As we noted, Theorem 3.40 does not hold of almost linear CFLGs, because there is no analogue of part (ii) of Lemma 3.34 for non-erasing almost non-duplicating β-reduction: if D is a database over $\mathcal{D}_{\Sigma,U}$ and $\alpha \in \mathscr{T}(A)$, the set $\Lambda(D, \alpha)$ is not always closed under the converse of non-erasing almost non-duplicating β-reduction. One sufficient condition for this closure property to hold is given by the following definition:

- D is said to be *deterministic* if for all types $\gamma_1, \ldots, \gamma_m$ and all atomic types p, q,

$$\Lambda(D, \gamma_1 \to \cdots \to \gamma_m \to p) \cap \Lambda(D, \gamma_1 \to \cdots \to \gamma_m \to q) \neq \varnothing$$

  implies $p = q$.

It is not difficult to show that determinism is a decidable property of databases, but I leave a detailed analysis of this notion for another occasion.[39]

Lemma 4.1. *Let $\Sigma = (A, C, \tau)$ be a higher-order signature, $M, M' \in \Lambda(\Sigma)$ be typable λ-terms, and D be a deterministic database over $\mathcal{D}_{\Sigma,U}$. If $M' \twoheadrightarrow_\beta M$ by non-erasing almost non-duplicating β-reduction, then $M \in \Lambda(D, \alpha)$ implies $M' \in \Lambda(D, \alpha)$.*

---

[39]In particular, I have been unable to settle the question whether $\mathrm{database}(N^\circ)$ is deterministic whenever N° is almost linear.

Proof. Let $M = (\mathcal{T}, f, b)$, $M' = (\mathcal{T}', f', b')$ be typable closed λ-terms, and let t' be a principal type decoration of M'. Assume $M \in \Lambda(D, \alpha)$ and $M' \xrightarrow{w}_\beta M$ by a non-erasing almost non-duplicating one-step β-reduction. Let $\{v_1, \ldots, v_k\} = \{ v \mid b'(w00v) = w0 \}$. By assumption, $k \geq 1$. Since the case $k = 1$ is taken care of by Lemma 3.34, part (ii), assume $k \geq 2$ and $t'(w1) = p$ for some atomic type p. Since $M \in \Lambda(D, \alpha)$, there is a type decoration $\hat{t}$ for

$$z_1 : \delta_1, \ldots, z_m : \delta_m \Rightarrow \widehat{M}[z_1, \ldots, z_m] : \alpha,$$

where $\overrightarrow{\mathrm{Con}}(M) = (c_1, \ldots, c_m)$ and for $i = 1, \ldots, m$, we have $c_i(\overline{\delta_i}) \in D$ and $\langle \tau(c_i) \rangle = \langle \delta_i \rangle$.

To show that $M' \in \Lambda(D, \alpha)$, it suffices to prove that

$$\hat{t}(wv_1) = \hat{t}(wv_i) \quad \text{for all } i \in \{1, \ldots, n\}. \tag{61}$$

For, if (61) holds, then it is easy to see that there are a subset $\{i_1, \ldots, i_{m'}\}$ of $\{1, \ldots, m\}$ and a function $g \colon \{1, \ldots, m\} \to \{i_1, \ldots, i_{m'}\}$ satisfying the following conditions:

$$c_i = c_{g(i)} \quad \text{for all } i \in \{1, \ldots, m\},$$
$$\widehat{M'}[z_{i_1}, \ldots, z_{i_{m'}}] \xrightarrow{w}_\beta \widehat{M}[z_{g(1)}, \ldots, z_{g(m)}],$$
$$\vdash z_{i_1} : \delta_{i_1}, \ldots, z_{i_{m'}} : \delta_{i_{m'}} \Rightarrow \widehat{M}[z_{g(1)}, \ldots, z_{g(m)}] : \alpha,$$
$$\vdash z_{i_1} : \delta_{i_1}, \ldots, z_{i_{m'}} : \delta_{i_{m'}} \Rightarrow \widehat{M'}[z_{i_1}, \ldots, z_{i_{m'}}] : \alpha,$$
$$\widehat{M'}[c_{i_1}, \ldots, c_{i_{m'}}] = M'.$$

The reasoning here is similar to that in the proof of Lemma 3.52.

We prove (61). Let ℓ' be a writing of M'. There exists a writing ℓ of M that agrees with ℓ' on $\{ u \in \mathcal{T}'^{(1)} \mid u < w \}$ such that $\mathrm{sub}_{M', \ell'}(w1) = \mathrm{sub}_{M, \ell}(wv_i)$ for $i = 1, \ldots, k$. Let $N = \mathrm{sub}_{M', \ell'}(w1)$ and let n be the number of occurrences of constants in N. Clearly, we have a function $h \colon \{1, \ldots, k\} \times \{1, \ldots, n\} \to \{1, \ldots, m\}$ such that

$$\mathrm{sub}_{\widehat{M}[z_1, \ldots, z_m], \ell}(wv_i) = \widehat{N}[z_{h(i,1)}, \ldots, z_{h(i,n)}].$$

Let $\mathrm{FV}(N) = \{y_1, \ldots, y_r\}$. Then ℓ and $\hat{t}$ determine types $\gamma_1, \ldots, \gamma_r$ such that

$$\vdash y_1 : \gamma_1, \ldots, y_r : \gamma_r, z_{h(i,1)} : \delta_{h(i,1)}, \ldots, z_{h(i,n)} : \delta_{h(i,n)} \Rightarrow \widehat{N}[z_{h(i,1)}, \ldots, z_{h(i,n)}] : \hat{t}(wv_i).$$

Similarly, ℓ' and t' determine types $\gamma'_1, \ldots, \gamma'_r$ such that

$$\vdash_\Sigma y_1 : \gamma'_1, \ldots, y_r : \gamma'_r \Rightarrow N : p.$$

There are two cases to consider.

Case 1. $|N|_\beta = y_j \vec{Q}$ and $\gamma'_j = \vec{\beta}' \to p$. Then we must have $\gamma_j = \vec{\beta} \to \hat{t}(wv_i)$ for all $i \in \{1, \ldots, k\}$. Hence $\hat{t}(wv_1) = \hat{t}(wv_i)$ for all $i \in \{1, \ldots, k\}$.

Case 2. $|N|_\beta = c_j \vec{Q}$ and $\tau(c_j) = \vec{\beta}' \to p$. Then for all $i \in \{1, \ldots, k\}$, $|\widehat{N}[z_{h(i,1)}, \ldots, z_{h(i,n)}]|_\beta = z_{h(i,1)} \vec{P_i}$ and $\delta_{h(i,1)} = \vec{\beta_i} \to \hat{t}(wv_i)$ for some $\vec{P_i}$ and $\vec{\beta_i}$ such that $\vec{\beta}'$ and $\vec{\beta_i}$ are sequences of types of the same length. Since $c_{h(i,1)} = c_j$, it must be that $\langle \tau(c_j) \rangle = \langle \delta_{h(i,1)} \rangle$, which implies that $\hat{t}(wv_i) = q_i$ for some atomic q_i. Then we have

$$\vdash z_{h(i,1)} : \delta_{h(i,1)}, \ldots, z_{h(i,n)} : \delta_{h(i,n)} \Rightarrow \lambda y_1 \ldots y_r.\widehat{N}[z_{h(i,1)}, \ldots, z_{h(i,n)}] : \gamma_1 \to \cdots \to \gamma_r \to q_i,$$

which implies

$$\lambda y_1 \ldots y_r.N \in \Lambda(D, \gamma_1 \to \cdots \to \gamma_r \to q_i).$$

Since D is deterministic, it follows that $q_1 = q_i$. □

Lemma 4.2. *Let* $\Sigma = (A, C, \tau)$ *be a higher-order signature, U be a set of database constants, D be a deterministic database over $\mathcal{D}_{\Sigma,U}$, and $\alpha \in \mathcal{T}(A)$. For every almost linear closed λ-term $M \in \Lambda(\Sigma)$, $M \in \Lambda(D, \alpha)$ if and only if $|M|_\beta \in \Lambda(D, \alpha)$.*

Proof. The "only if" direction is by Lemma 3.34, part (i). Since the β-reduction $M \twoheadrightarrow_\beta |M|_\beta$ must be non-erasing and almost non-duplicating (Lemma 3.49), the "if" direction follows from Lemma 4.1. □

Theorem 4.3. *Let* $\mathcal{G} = (\mathcal{N}, \Sigma, f, \mathcal{P}, S)$ *be an almost linear CFLG and $B \in \mathcal{N}$. Let U be some set of constants, D be a deterministic database over $\mathcal{D}_{\Sigma,U}$, and $\vec{s}$ be a sequence of constants from U such that $|\vec{s}| = |f(S)|$. The following are equivalent:*

(i) $L(\mathcal{G}) \cap \Lambda(D, \langle f(S) \rangle(\vec{s})) \neq \varnothing$.

(ii) $\text{program}(\mathcal{G}) \cup D \vdash S(\vec{s})$.

Proof. The implication from (ii) to (i) is by Lemma 3.36.

(i) $\Rightarrow$ (ii). Assume (i). Then there is an almost linear λ-term $P \in \Lambda(\Sigma)$ such that $\vdash_\mathcal{G} S(P)$ and $|P|_\beta \in \Lambda(D, \langle f(S) \rangle(\vec{s}))$. Since P is almost linear, Lemma 4.2 implies $P \in \Lambda(D, \langle f(S) \rangle(\vec{s}))$. Then (ii) follows by Lemma 3.35. □

It is easy to see that if D is a database representing a finite automaton $\mathcal{A}$ (on strings), then D is deterministic if and only if $\mathcal{A}$ is. If D is a database representing a bottom-up tree automaton $\mathcal{A}$, then, again, D is deterministic if and only if $\mathcal{A}$ is. So Theorem 4.3 applies when a regular set is given as input in the form of a deterministic finite (string or tree) automaton.[40] The string case of this result is not useful, however, because every almost linear CFLG generating λ-term encodings of strings is equivalent to some linear CFLG.[41]

With respect to tree languages, almost linear CFLGs are more powerful than linear CFLGs, and can encode grammars that allow copying of subtrees, like IO context-free tree grammars. For these grammars, Theorem 4.3 implies that parsing as intersection where input is given in the form of a deterministic bottom-up tree automaton reduces to Datalog query evaluation.

4.2.3 An application to string grammars with copying

This last point can be exploited to show that there is a way of representing recognition/parsing (of ordinary single-string input) with respect to some *string* grammars with copying operations, such as IO macro grammars and parallel multiple context-free grammars, in terms of Datalog query evaluation, even though Theorem 3.65 is powerless for that purpose. Such a string grammar can always be turned into a corresponding tree grammar that generates a tree language whose yield image is the language of the string grammar. Since tree copying can be represented by almost linear λ-terms, these tree grammars can be encoded in almost linear CFLGs. Moreover, we can associate with every string w a regular set of trees that yield w so that the language of the tree grammar has a non-empty intersection with that set of trees if and only if w is in the language of the original string grammar.

For example, consider the following parallel multiple context-free grammar [66]:[42]

$$S(x_1 x_2) :\!- A(x_1, x_2).$$
$$A(1, 0).$$
$$A(x_1 x_2 1, x_2 0) :\!- A(x_1, x_2).$$

---

[40]When the automaton has more than one final state, non-empty intersection is equivalent to a disjunction of queries of the form "?– $S(q_I, q)$" (in the string case) or "?– $S(q)$" (in the tree case), one for each final state q. To reduce this to a single query, one can add the rules of the form "$S' :\!- S(q_I, q)$" or "$S' :\!- S(q)$" for all final states q, and use the query "?– S'".

[41]This can be seen as follows. Suppose that $P \in \Lambda(\Sigma)$ is an almost linear closed λ-term such that $|P|_\beta = /c_1 \ldots c_n/ = \lambda z.c_1(\ldots(c_n z)\ldots)$. Then by Lemma 3.49, $P \twoheadrightarrow_\beta |P|_\beta$ by non-erasing, almost non-duplicating β-reduction. But Lemma 3.60 implies that $\overrightarrow{\mathrm{Con}}(P)$ is some permutation $(c_{j_1}, \ldots, c_{j_n})$ of $(c_1, \ldots, c_n)$, and $\widehat{P}[z_{j_1}, \ldots, z_{j_n}] \twoheadrightarrow_\beta \lambda z.z_1(\ldots(z_n z)\ldots)$ by non-erasing, almost non-duplicating β-reduction. However, it is easy to see that the set of non-affine pure λ-terms is closed under non-erasing almost non-duplicating β-reduction. Since $\lambda z.z_1(\ldots(z_n z)\ldots)$ is linear, it follows that $\widehat{P}[z_{j_1}, \ldots, z_{j_n}]$, and hence P, must be linear.

[42]The notation here follows that of *elementary formal systems* [72, 4, 28], which are logic programs on strings.

This grammar generates the language $\{\, w_n \mid n \geq 1 \,\}$, where $w_n = 1010^2 \ldots 10^n$. The third rule involves copying of the variable x_2. The translation of this grammar into a CFLG looks as follows:

$$S(\lambda z.X(\lambda x_1 x_2.x_1(x_2 z))) :\!- A(X).$$
$$A(\lambda w.w(\lambda z.1z)(\lambda z.0z)).$$
$$A(\lambda w.X(\lambda x_1 x_2.w(\lambda z.x_1(x_2(1z)))(\lambda z.x_2(0z)))) :\!- A(X).$$

Here, $f(S) = o \to o$, and $f(A) = ((o \to o) \to (o \to o) \to o) \to o$. This grammar is not almost linear, since the bound variable x_2 in the λ-term on the left-hand side of the third rule must have a non-atomic type in the principal typing of the λ-term.

Here is a grammar that generates a set of trees whose yield image is the language of the above PMCFG:

$$S(c(x_1, x_2)) :\!- A(x_1, x_2).$$
$$A(1, 0).$$
$$A(c(x_1, c(x_2, 1)), c(x_2, 0)) :\!- A(x_1, x_2).$$

Here, c is a symbol of rank 2, and 1 and 0 are symbols of rank 0. A grammar like this, where a nonterminal denotes a relation on trees and a rule may duplicate trees, may be called a *parallel multiple regular tree grammar*, in analogy with a *multiple regular tree grammar* [60, 23]. For example, the tree

$$c(c(1, c(0, 1)), c(0, 0))$$

is generated by the above tree grammar with the following derivation:

$$S(c(c(1, c(0, 1)), c(0, 0)))$$
$$|$$
$$A(c(1, c(0, 1)), c(0, 0))$$
$$|$$
$$A(1, 0)$$

The yield of this tree is $10100 = w_2$.

It is straightforward to encode the above tree grammar into an almost linear CFLG:

$$S(X(\lambda x_1 x_2.cx_1 x_2)) :\!- A(X).$$
$$A(\lambda w.w10).$$
$$A(\lambda w.X(\lambda x_1 x_2.w(cx_1(cx_2 1))(cx_2 0))) :\!- A(X).$$

Here, $f(S) = o$ and $f(A) = (o \to o \to o) \to o$. This CFLG $\mathscr{G}$ translates into the following Datalog program $\mathbf{P}_{\mathscr{G}}$:

$S(i_1) :\!\!- c(i_2, i_4, i_3), A(i_1, i_2, i_4, i_3).$
$A(i_1, i_1, i_3, i_2) :\!\!- \mathtt{1}(i_2), \mathtt{0}(i_3).$
$A(i_1, i_2, i_8, i_3) :\!\!- c(i_3, i_5, i_4), c(i_5, i_7, i_6), \mathtt{1}(i_7), c(i_8, i_9, i_6), \mathtt{0}(i_9), A(i_1, i_2, i_6, i_4).$

The above Datalog program $\mathbf{P}_{\mathscr{G}}$ can be used to parse input strings with respect to the original PMCFG. For example, if the input string is 10100, we first form a deterministic bottom-up tree automaton $\mathscr{A}$ that recognizes the set of trees over the ranked alphabet $\{\mathtt{1}^{(0)}, \mathtt{0}^{(0)}, c^{(2)}\}$ whose yield is 10100. The states of this automaton are of the form q_w, where w is one of the non-empty substrings of this string:

$$0, 1, 00, 01, 10, 010, 100, 101, 0100, 1010, 10100$$

For each of these strings w and nonempty strings u, v such that $w = uv$, the automaton $\mathscr{A}$ has the rule

$$c(q_u(x_1), q_v(x_2)) \to q_w(c(x_1, x_2)).$$

which gives rise to the extensional fact

$$c(q_w, q_v, q_u).$$

Moreover, for each symbol a occurring in w, the automaton has the rule

$$a \to q_a(a)$$

which translates into the extensional fact

$$a(q_a).$$

The database obtained this way is deterministic. In the present case, we get the database D consisting of the following facts (we write w instead of q_w):[43]

$$\mathtt{0}(0). \quad \mathtt{1}(1).$$
$$c(00, 0, 0). \quad c(01, 0, 1). \quad c(10, 1, 0).$$

---

[43]If the PMCFG rules contain occurrences of the empty string ϵ, then the corresponding PMRTG will have a special rank 0 symbol corresponding to ϵ, and one needs to take *all* substrings of the input string, not just non-empty ones, in the construction of the automaton $\mathscr{A}$. The automaton will then represent the *syntactic monoid* of the singleton set consisting of the input string.

$$c(010, 0, 10). \quad c(010, 01, 0).$$
$$c(100, 1, 00). \quad c(100, 10, 0).$$
$$c(101, 1, 01). \quad c(101, 10, 1).$$
$$c(0100, 0, 100). \quad c(0100, 01, 00). \quad c(0100, 010, 0).$$
$$c(1010, 1, 010). \quad c(1010, 10, 10). \quad c(1010, 101, 0).$$
$$c(10100, 1, 0100). \quad c(10100, 10, 100). \quad c(10100, 101, 00). \quad c(10100, 1010, 0).$$

By Theorem 4.3,
$$\mathbf{P}_{\mathscr{G}} \cup D \vdash S(10100) \tag{62}$$

if and only if $\mathscr{G}$ generates (the λ-term representation of) a tree whose yield is 10100. This is so if and only if the original PMCFG generates this string. Since the rules of the PMCFG are in one-one correspondence with the rules of $\mathscr{G}$, parsing the string with this PMCFG reduces to the problem of computing all derivation trees for (62), in the form of a shared forest.

This reduction generally applies to the yield images of the tree languages that can be generated by almost linear CFLGs. It is shown in unpublished work [44] that the class of tree languages generated by almost linear CFLGs coincides with the class of output languages of *tree-valued attribute grammars* or *attributed tree transducers* (see [11]). As a consequence, the class of yield images of these tree languages is simply the class of output languages of *string-valued attribute grammars*, studied by Engelfriet [22].

Clearly, the deterministic bottom-up tree automaton $\mathscr{A}$ (and the corresponding database) associated with the input string can be constructed in logarithmic space. Note that all trees accepted by $\mathscr{A}$ have the same number of constants, namely $2n - 1$ for input string of length n.[44] This implies that recognition and parsing with these grammars are in (functional) LOGCFL, matching the result of Engelfriet [22].[45]

4.2.4 An application to generation from underspecified semantics

Koller et al. [49] have proposed to use a regular tree grammar as an underspecified representation of various readings of sentences with multiple scope-taking operators. However, when the operators include variable-binders, a tree is not ideally suited to represent the scope relation because one needs to associate a variable name to each

---

[44]This number assumes that $\mathscr{A}$ does not have a special symbol representing the empty string.

[45]Note that parsing as intersection with these grammars, where the input is a regular set of strings, can also be represented as Datalog query evaluation. The deterministic bottom-up tree automaton that determines the database and query can be obtained from the syntactic monoid of the input regular set.

```
S(λz.Y(λy₁y₂.y₁(y₂z)), X) :− NP_VP(Y, X).
NP_VP(λw.Y(λy₁y₂.w(λz.y₁z)(λz.didn't(y₂z))), ¬X) :− NP_VP(Y, X).
NP_VP(λw.w(λz.Y₁z)(λz.Y₂z), X₁(λx.X₂x)) :− NP(Y₁, X₁), VP(Y₂, X₂).
NP_VP(λw.Y₁(λy₁y₂.w(λz.y₁z)(λz.y₂(Y₂z))), X₂(λx.X₁x)) :− NP_V(Y₁, X₁), NP(Y₂, X₂).
NP_V(λw.Y(λy₁y₂.w(λz.y₁z)(λz.didn't(y₂z))), λx.¬(Xx)) :− NP_V(Y, X).
NP_V(λw.w(λz.Y₁z)(λz.Y₂z), λy.X₁(λx.X₂yx)) :− NP(Y₁, X₁), V(Y₂, X₂).
VP(λz.didn't(Yz), λx.¬(Xx)) :− VP(Y, X).
VP(λz.Y₁(Y₂z), λx.X₂(λy.X₁yx)) :− V(Y₁, X₁), NP(Y₂, X₂).
NP(λz.Y₁(Y₂z), λv.X₁(λx.X₂x)(λx.vx)) :− Det(Y₁, X₁), N(Y₂, X₂).
Det(/a/, λuv.∃(λx.∧(ux)(vx))).
Det(/every/, λuv.∀(λx.→(ux)(vx))).
Det(/no/, λuv.∀(λx.→(ux)(¬(vx)))).
Det(/not every/, λuv.¬(∀(λx.→(ux)(vx)))).
N(/book/, λx.**book** x).
N(/student/, λx.**student** x).
V(/read/, λyx.**read** y x).
```

Figure 13: A synchronous CFLG.

occurrence of a binder to represent the binding relation. These variable names must be chosen in such a way as to avoid clashes of variables, and some mechanism is needed to identify α-equivalent representations (i.e., representations that differ only in renaming of bound variables).

A compact representation of a set of λ-terms, rather than trees, will improve upon Koller et al.'s [49] approach. We can use a deterministic database D over a database schema $\mathcal{D}_{\Sigma,U}$ associated with a higher-order signature Σ as a representation of a set of λ-terms over Σ. If the syntax-semantics is given as a "synchronous" CFLG whose semantics side is an almost linear CFLG $\mathscr{G}$, then Theorem 4.3 tells us that D can serve as an "underspecified" input to surface realization.

For example, the synchronous CFLG in Figure 13 generates **every student didn't read a book** with six possible readings:

$$\forall(\lambda x.\to(\mathbf{student}\ x)(\neg(\exists(\lambda y.\wedge(\mathbf{book}\ y)(\mathbf{read}\ y\ x)))))$$
$$\forall(\lambda x.\to(\mathbf{student}\ x)(\exists(\lambda y.\wedge(\mathbf{book}\ y)(\neg(\mathbf{read}\ y\ x)))))$$
$$\neg(\forall(\lambda x.\to(\mathbf{student}\ x)(\exists(\lambda y.\wedge(\mathbf{book}\ y)(\mathbf{read}\ y\ x)))))$$
$$\neg(\exists(\lambda y.\wedge(\mathbf{book}\ y)(\forall(\lambda x.\to(\mathbf{student}\ x)(\mathbf{read}\ y\ x)))))$$
$$\exists(\lambda y.\wedge(\mathbf{book}\ y)(\forall(\lambda x.\to(\mathbf{student}\ x)(\neg(\mathbf{read}\ y\ x)))))$$

$$\exists(\lambda y. \wedge (\mathbf{book}\ y)(\neg(\forall(\lambda x. \rightarrow (\mathbf{student}\ x)(\mathbf{read}\ y\ x)))))$$

The set of these λ-terms can be represented by the following database:

$$\mathbf{student(s}, x). \quad \mathbf{book(b}, y). \quad \mathbf{read(r}, x, y).$$
$$\neg(\neg, \mathbf{r}). \quad \neg(\exists\neg, \exists). \quad \neg(\forall\neg, \forall). \quad \neg(\forall\exists\neg, \forall\exists).$$
$$\wedge(\wedge, \mathbf{r}, \mathbf{b}). \quad \wedge(\wedge\neg, \neg, \mathbf{b}). \quad \wedge(\wedge\forall, \forall, \mathbf{b}). \quad \wedge(\wedge\forall\neg, \forall\neg, \mathbf{b}).$$
$$\exists(\exists, \wedge, y). \quad \exists(\exists\neg, \wedge\neg, y). \quad \exists(\forall\exists, \wedge\forall, y). \quad \exists(\forall\exists\neg, \wedge\exists\neg, y).$$
$$\rightarrow(\rightarrow, \mathbf{r}, \mathbf{s}). \quad \rightarrow(\rightarrow\neg, \neg, \mathbf{s}). \quad \rightarrow(\rightarrow\exists, \exists, \mathbf{s}). \quad \rightarrow(\rightarrow\exists\neg, \exists\neg, \mathbf{s}).$$
$$\forall(\forall, \rightarrow, x). \quad \forall(\forall\neg, \rightarrow\neg, x). \quad \forall(\forall\exists, \rightarrow\exists, x). \quad \forall(\forall\exists\neg, \rightarrow\exists\neg, x).$$

In this database (call it D), we use mnemonic names like $\forall\exists\neg$, instead of integers, as database constants. For instance, λ-terms in $\Lambda(D, \forall\exists)$ contain $\forall$ and $\exists$, but not $\neg$. A database like this can be thought of as a hypergraph that can be obtained from the disjoint union of the hypergraphs corresponding to the above six almost linear λ-terms by identifying certain nodes and hyperedges. It is easy to check that this database is deterministic; it can then be used together with the Datalog program associated with the semantic side of the synchronous grammar in Figure 13 to obtain a shared parse forest of all derivation trees of sentences that have at least one reading in common with the sentence every student didn't read a book—namely, no student read a book, not every student read a book, and the same sentence itself. This procedure is more efficient than the brute-force method, where each reading of the sentence is input to a surface realization routine in turn.[46]

4.3 Magic sets and Earley-style algorithms

The *magic-sets* rewriting of a Datalog program allows bottom-up evaluation to avoid deriving useless facts by mimicking top-down evaluation of the original program. The result of the *generalized supplementary magic-sets* rewriting of Beeri and Ramakrishnan [8] applied to the Datalog program representing a CFG essentially coincides with the *deduction system* [69] or *uninstantiated parsing system* [70] for Earley parsing [20]. By applying the same rewriting method to Datalog programs representing almost linear CFLGs, we can obtain efficient parsing and generation algorithms for various grammar formalisms with context-free derivations.

---

[46]There is the question of how a deterministic database representing the range of possible readings of a sentence can be found, if one exists. In the case at hand, there is a way of constructing the desired database from the shared parse forest of the sentence by duplicating certain nodes (namely, the NP nodes and the Det nodes). However, it is easy to see that no such deterministic database may exist in general. It is an open question when and how a desired database can be constructed efficiently.

We illustrate this approach with the program in (6), repeated below, following the presentation of Ullman [77, 78]. We assume the query to take the form "$?- S(0, x).$", so that the input database can be processed incrementally.

$$S(i_1, i_3) :- A(i_1, i_3, i_2, i_2). \qquad (6)$$
$$A(i_1, i_8, i_4, i_5) :- \mathsf{a}(i_1, i_2), \mathsf{b}(i_3, i_4), \mathsf{c}(i_5, i_6), \mathsf{d}(i_7, i_8), A(i_2, i_7, i_3, i_6).$$
$$A(i_1, i_2, i_1, i_2).$$

The program is first made *safe* by eliminating the rule with empty right-hand side:

$$S(i_1, i_3) :- A(i_1, i_3, i_2, i_2).$$
$$A(i_1, i_8, i_4, i_5) :- \mathsf{a}(i_1, i_2), \mathsf{b}(i_3, i_4), \mathsf{c}(i_5, i_6), \mathsf{d}(i_7, i_8), A(i_2, i_7, i_3, i_6).$$
$$A(i_1, i_8, i_4, i_5) :- \mathsf{a}(i_1, i_2), \mathsf{b}(i_2, i_4), \mathsf{c}(i_5, i_6), \mathsf{d}(i_6, i_8).$$

The *subgoal rectification* removes duplicate arguments from subgoals, creating new predicates as needed:

$$S(i_1, i_3) :- B(i_1, i_3, i_2).$$
$$A(i_1, i_8, i_4, i_5) :-, \mathsf{a}(i_1, i_2), \mathsf{b}(i_3, i_4), \mathsf{c}(i_5, i_6), \mathsf{d}(i_7, i_8), A(i_2, i_7, i_3, i_6).$$
$$A(i_1, i_8, i_4, i_5) :- \mathsf{a}(i_1, i_2), \mathsf{b}(i_2, i_4), \mathsf{c}(i_5, i_6), \mathsf{d}(i_6, i_8).$$
$$B(i_1, i_8, i_4) :-, \mathsf{a}(i_1, i_2), \mathsf{b}(i_3, i_4), \mathsf{c}(i_4, i_6), \mathsf{d}(i_7, i_8), A(i_2, i_7, i_3, i_6).$$
$$B(i_1, i_8, i_4) :- \mathsf{a}(i_1, i_2), \mathsf{b}(i_2, i_4), \mathsf{c}(i_4, i_6), \mathsf{d}(i_6, i_8).$$

We then attach to predicates *adornments* indicating the free/bound status of arguments in top-down evaluation, reordering subgoals so that as many arguments as possible are marked as bound:

$$S^{bf}(i_1, i_3) :- B^{bff}(i_1, i_3, i_2).$$
$$B^{bff}(i_1, i_8, i_4) :- \mathsf{a}^{bf}(i_1, i_2), A^{bfff}(i_2, i_7, i_3, i_6), \mathsf{b}^{bf}(i_3, i_4), \mathsf{c}^{bb}(i_4, i_6),$$
$$\mathsf{d}^{bf}(i_7, i_8).$$
$$B^{bff}(i_1, i_8, i_4) :- \mathsf{a}^{bf}(i_1, i_2), \mathsf{b}^{bf}(i_2, i_4), \mathsf{c}^{bf}(i_4, i_6), \mathsf{d}^{bf}(i_6, i_8).$$
$$A^{bfff}(i_1, i_8, i_4, i_5) :- \mathsf{a}^{bf}(i_1, i_2), A^{bfff}(i_2, i_7, i_3, i_6), \mathsf{b}^{bf}(i_3, i_4), \mathsf{c}^{bb}(i_5, i_6),$$
$$\mathsf{d}^{bf}(i_7, i_8).$$
$$A^{bfff}(i_1, i_8, i_4, i_5) :- \mathsf{a}^{bf}(i_1, i_2), \mathsf{b}^{bf}(i_2, i_4), \mathsf{c}^{ff}(i_5, i_6), \mathsf{d}^{bf}(i_6, i_8).$$

The generalized supplementary magic-sets rewriting finally gives the following rule set:

$$r_1 : m\_B(i_1) :- m\_S(i_1).$$
$$r_2 : S(i_1, i_3) :- m\_B(i_1), B(i_1, i_3, i_2).$$
$$r_3 : sup_{2.1}(i_1, i_2) :- m\_B(i_1), \mathsf{a}(i_1, i_2).$$

r_4: $sup_{2.2}(i_1, i_7, i_3, i_6) \mathrel{:\!-} sup_{2.1}(i_1, i_2), A(i_2, i_7, i_3, i_6)$.
r_5: $sup_{2.3}(i_1, i_7, i_6, i_4) \mathrel{:\!-} sup_{2.2}(i_1, i_7, i_3, i_6), \mathsf{b}(i_3, i_4)$.
r_6: $sup_{2.4}(i_1, i_7, i_4) \mathrel{:\!-} sup_{2.3}(i_1, i_7, i_6, i_4), \mathsf{c}(i_4, i_6)$.
r_7: $B(i_1, i_8, i_4) \mathrel{:\!-} sup_{2.4}(i_1, i_7, i_4), \mathsf{d}(i_7, i_8)$.
r_8: $sup_{3.1}(i_1, i_2) \mathrel{:\!-} m\_B(i_1), \mathsf{a}(i_1, i_2)$.
r_9: $sup_{3.2}(i_1, i_4) \mathrel{:\!-} sup_{3.1}(i_1, i_2), \mathsf{b}(i_2, i_4)$.
r_{10}: $sup_{3.3}(i_1, i_4, i_6) \mathrel{:\!-} sup_{3.2}(i_1, i_4), \mathsf{c}(i_4, i_6)$.
r_{11}: $B(i_1, i_8, i_4) \mathrel{:\!-} sup_{3.3}(i_1, i_4, i_6), \mathsf{d}(i_6, i_8)$.
r_{12}: $m\_A(i_2) \mathrel{:\!-} sup_{2.1}(i_1, i_2)$.
r_{13}: $m\_A(i_2) \mathrel{:\!-} sup_{4.1}(i_1, i_2)$.
r_{14}: $sup_{4.1}(i_1, i_2) \mathrel{:\!-} m\_A(i_1), \mathsf{a}(i_1, i_2)$.
r_{15}: $sup_{4.2}(i_1, i_7, i_3, i_6) \mathrel{:\!-} sup_{4.1}(i_1, i_2), A(i_2, i_7, i_3, i_6)$.
r_{16}: $sup_{4.3}(i_1, i_7, i_6, i_4) \mathrel{:\!-} sup_{4.2}(i_1, i_7, i_3, i_6), \mathsf{b}(i_3, i_4)$.
r_{17}: $sup_{4.4}(i_1, i_7, i_4, i_5) \mathrel{:\!-} sup_{4.3}(i_1, i_7, i_6, i_4), \mathsf{c}(i_5, i_6)$.
r_{18}: $A(i_1, i_8, i_4, i_5) \mathrel{:\!-} sup_{4.4}(i_1, i_7, i_4, i_5), \mathsf{d}(i_7, i_8)$.
r_{19}: $sup_{5.1}(i_1, i_2) \mathrel{:\!-} m\_A(i_1), \mathsf{a}(i_1, i_2)$.
r_{20}: $sup_{5.2}(i_1, i_4) \mathrel{:\!-} sup_{5.1}(i_1, i_2), \mathsf{b}(i_2, i_4)$.
r_{21}: $sup_{5.3}(i_1, i_4, i_5, i_6) \mathrel{:\!-} sup_{5.2}(i_1, i_4), \mathsf{c}(i_5, i_6)$.
r_{22}: $A(i_1, i_8, i_4, i_5) \mathrel{:\!-} sup_{5.3}(i_1, i_4, i_5, i_6), \mathsf{d}(i_6, i_8)$.

The following is a version of the seminaive bottom-up evaluation algorithm expressed in the form of chart parsing:

1. (INIT) Initialize the chart to the empty set, the agenda to the singleton $\{m\_S(0)\}$, and n to 0.

2. Repeat the following steps:

    (a) Repeat the following steps until the agenda is exhausted:

       i. Remove a fact from the agenda, called the *trigger*.
       ii. Add the trigger to the chart.
       iii. Generate all facts that are immediate consequences of the trigger together with all facts in the chart, and add to the agenda those generated facts that are neither already in the chart nor in the agenda.

(b) (SCAN) Remove the next fact from the input database and add it to the agenda, incrementing n. If there is no more fact in the input database, go to step 3.

3. If $S(0, n)$ is in the chart, accept; otherwise reject.

The following is the trace of the algorithm on input string aabbccdd; the derived facts are recorded in the order they enter the agenda:

1. $m\_S(0)$ INIT
2. $m\_B(0)$ $r_1, 1$
3. $\mathsf{a}(0, 1)$ SCAN
4. $sup_{2.1}(0, 1)$ $r_3, 2, 3$
5. $sup_{3.1}(0, 1)$ $r_8, 2, 3$
6. $m\_A(1)$ $r_{12}, 4$
7. $\mathsf{a}(1, 2)$ SCAN
8. $sup_{4.1}(1, 2)$ $r_{14}, 6, 7$
9. $sup_{5.1}(1, 2)$ $r_{19}, 6, 7$
10. $m\_A(2)$ $r_{13}, 8$
11. $\mathsf{b}(2, 3)$ SCAN
12. $sup_{5.2}(1, 3)$ $r_{20}, 9, 11$
13. $\mathsf{b}(3, 4)$ SCAN
14. $\mathsf{c}(4, 5)$ SCAN
15. $sup_{5.3}(1, 3, 4, 5)$ $r_{21}, 12, 14$
16. $\mathsf{c}(5, 6)$ SCAN
17. $sup_{5.3}(1, 3, 5, 6)$ $r_{21}, 12, 16$
18. $\mathsf{d}(6, 7)$ SCAN
19. $A(1, 7, 3, 5)$ $r_{22}, 17, 18$
20. $sup_{2.2}(0, 7, 3, 5)$ $r_4, 4, 19$
21. $sup_{2.3}(0, 7, 5, 4)$ $r_5, 20, 13$
22. $sup_{2.4}(0, 7, 4)$ $r_6, 21, 14$
23. $\mathsf{d}(7, 8)$ SCAN
24. $B(0, 8, 4)$ $r_7, 22, 23$
25. $S(0, 8)$ $r_2, 2, 24$

Note that unlike previous Earley-style parsing algorithms for TAGs, the present algorithm is an instantiation of a general schema that applies to parsing with more powerful grammar formalisms as well as to generation with Montague semantics.[47]

5 Conclusion

This paper has shown that recognition and parsing for a wide range of grammars with "context-free" derivations, as well as surface realization (tactical generation) for those grammars coupled with a certain restricted kind of Montague semantics, all reduce to Datalog query evaluation and hence allow highly efficient algorithms. The method of reduction is uniform for both recognition/parsing and surface realization, and the complexity upper bound that has been established, namely, LOGCFL, is

---

[47] The above Earley-style recognition algorithm for tree-adjoining languages does not have the *correct prefix property*, a desirable feature for Earley-style algorithms for string grammars. See [43] for how to supplement magic-sets rewriting with another simple rewriting to achieve the correct prefix property.

tight. By regarding the problem of surface realization as the problem of recognition/parsing of languages of λ-terms, this paper has demonstrated that it is possible to study surface realization abstractly in the style of formal language theory, just like parsing. I hope that the methods employed here help pave the way for eliminating much of the ad hoc methodology that is so common in computational linguistics.

References

[1] Serge Abiteboul, Richard Hull, and Victor Vianu. *Foundations of Databases*. Addison-Wesley, Reading, MA, 1995.

[2] Takahito Aoto. Uniqueness of normal proofs in implicational intuitionistic logic. *Journal of Logic, Language and Information*, 8:217–242, 1999.

[3] Takahito Aoto and Hiroakira Ono. Uniqueness of normal proofs in {→,∧}-fragment of NJ. Research Report IS-RR-94-0024F, School of Information Science, Japan Advanced Institute of Science and Technology, 1994.

[4] Setsuo Arikawa, Takeshi Shinohara, and Akihiro Yamamoto. Learning elementary formal systems. *Theoretical Computer Science*, 95(1):97–113, 1992.

[5] Y. Bar-Hillel, M. Perles, and E. Shamir. On formal properties of simple phrase structure grammars. *Zeitschrift für Phonetik, Sprachwissenschaft und Kommunikationsforschung*, 14(2):143–172, 1961.

[6] Hendrik Pieter Barendregt. *The Lambda Calculus*. North-Holland, Amsterdam, 1984. Revised Edition.

[7] Jon Barwise and Robin Cooper. Generalized quantifiers and natural language. *Linguistics and Philosophy*, 4:159–219, 1981.

[8] Catriel Beeri and Raghu Ramakrishnan. On the power of magic. *Journal of Logic Programming*, 10:255–299, 1991.

[9] N.D. Belnap. The two-property. *Relevance Logic Newsletter*, 1:173–180, 1976.

[10] Inge Bethke, Jan Willem Klop, and Roel de Vrijer. Descendants and origins in term rewriting. *Information and Computation*, 159:59–124, 2000.

[11] Roderick Bloem and Joost Engelfriet. A comparison of tree transductions defined by monadic second order logic and by attribute grammars. *Journal of Computer and System Sciences*, 61:1–50, 2000.

[12] Samuel R. Buss. The undecidability of k-provability. *Annals of Pure and Applied Logic*, 53(1):75–102, 1991.

[13] Ashok K. Chandra, Dexter C. Kozen, and Larry J. Stockmeyer. Alternation. *Journal of the Association for Computing Machinery*, 28(1):114–133, 1981.

[14] Alonzo Church. A formulation of the simple theory of types. *The Journal of Symbolic Logic*, 5:56–68, 1940.

[15] Ann Copestake, Dan Flickinger, Carl Pollard, and Ivan A. Sag. Minimal recursion semantics: An introduction. *Research on Language and Computation*, 3:281–332, 2005.

[16] Evgeny Dantsin, Thomas Eiter, Georg Gottlob, and Andrei Voronkov. Complexity and expressive power of logic programming. *ACM Computing Surveys*, 33:374–425, 2001.

[17] Philippe de Groote. Towards abstract categorial grammars. In *Association for Computational Linguistics, 39th Annual Meeting and 10th Conference of the European Chapter, Proceedings of the Conference*, pages 148–155, 2001.

[18] Philippe de Groote. Tree-adjoining grammars as abstract categorial grammars. In *Proceedings of the Sixth International Workshop on Tree Adjoining Grammar and Related Frameworks (TAG+6)*, pages 145–150. Universitá di Venezia, 2002.

[19] Philippe de Groote and Sylvain Pogodalla. On the expressive power of abstract categorial grammars: Representing context-free formalisms. *Journal of Logic, Language and Information*, 13:421–438, 2004.

[20] Jay Earley. An efficient context-free parsing algorithm. *Communications of the ACM*, 13:94–102, February 1970.

[21] J. Engelfriet and E. M. Schmidt. IO and OI, part I. *The Journal of Computer and System Sciences*, 15:328–353, 1977.

[22] Joost Engelfriet. The complexity of languages generated by attribute grammars. *SIAM Journal on Computing*, 15:70–86, 1986.

[23] Joost Engelfriet. Context-free graph grammars. In G. Rozenberg and A. Salomaa, editors, *Handbook of Formal Langauges, Volume 3: Beyond Words*, pages 125–213. Springer, Berlin, 1997.

[24] Michael J. Fischer. *Grammars with Macro-Like Productions*. PhD thesis, Harvard University, 1968.

[25] Gerald Gazdar, Ewan Klein, Geoffrey K. Pullum, and Ivan A. Sag. *Generalized Phrase Structure Grammar*. Harvard University Press, Cambridge, Mass., 1985.

[26] Georg Gottlob, Nicola Lenoe, and Francesco Scarcello. Computing LOGCFL certificates. *Theoretical Computer Science*, 270:761–777, 2002.

[27] Sheila A. Greibach. The hardest context-free language. *SIAM Journal on Computing*, 2:304–310, 1973.

[28] Annius Groenink. *Surface without Structures*. PhD thesis, Utrech University, 1997.

[29] Leon Henkin. A theory of propositional types. *Fundamenta Mathematicae*, 52:323–344, 1963.

[30] Gerd G. Hillebrand, Paris C. Kanellakis, and Harry G. Mairson. Database query languages embedded in the typed lambda calculus. *Information and Computation*, 127:117–144, 1996.

[31] J. Roger Hindley. *Basic Simple Type Theory*. Cambridge University Press, Cambridge, 1997.

[32] J. Roger Hindley and Jonathan P. Seldin. *Lambda-Calculus and Combinators: An Introduction*. Cambridge University Press, Cambridge, 2008.

[33] Sachio Hirokawa. Balanced formulas, BCK-minimal formulas and their proofs. In Anil Nerode and Mikhail Taitslin, editors, *Logical Foundations of Computer Science — Tver '92*, pages 198–208, Berlin, 1992. Springer.

[34] Gérard Huet. *Résolution d'équation dans des langages d'ordre* $1, 2, \ldots, \omega$. PhD thesis, Université Paris VII, 1976.

[35] David S. Johnson. A catalog of complexity classes. In Jan van Leeuwen, editor, *Handbook of Theoretical Computer Science*, volume A: Algorithms and Complexity, pages 67–161. Elsevier, Amsterdam, 1990.

[36] Neil D. Jones and William T. Laaser. Complete problems for deterministic polynomial time. *Theoretical Computer Science*, 3:105–117, 1977.

[37] Aravind K. Joshi and Yves Schabes. Tree-adjoining grammars. In Grzegoz Rozenberg and Arto Salomaa, editors, *Handbook of Formal Languages*, volume 3, pages 69–123. Springer, Berlin, 1997.

[38] Yuichi Kaji, Ryuichi Nakanishi, Hiroyuki Seki, and Tadao Kasami. The universal recognition problems for multiple context-free grammars and for linear context-free rewriting systems. *IEICE Transactions on Information and Systems*, E 75–D(1):78–88, 1992.

[39] Yuichi Kaji, Ryuichi Nakanishi, Hiroyuki Seki, and Tadao Kasami. The universal recognition problems for parallel multiple context-free grammars and their subclasses. *IEICE Transactions on Informaiton and Systems*, E 75–D(4):499–508, 1992.

[40] Makoto Kanazawa. Abstract families of abstract categorial languages. *Electronic Notes in Theoretical Computer Science*, 165:65–80, 2006. Proceedings of the 13th Workshop on Logic, Language, Information and Computation (WoLLIC 2006), Logic, Language, Information and Computation 2006.

[41] Makoto Kanazawa. Computing interpolants in implicational logics. *Annals of Pure and Applied Logic*, 142(1–3):125–201, 2006.

[42] Makoto Kanazawa. Parsing and generation as Datalog queries. In *Proceedings of the 45th Annual Meeting of the Association for Computational Linguistics*, Prague, Czech Republic, 2007.

[43] Makoto Kanazawa. A prefix-correct Earley recognizer for multiple context-free grammars. In *Proceedings of the Ninth International Workshop on Tree Adjoining Grammars and Related Formalisms (TAG+ 9)*, pages 49–56, 2008.

[44] Makoto Kanazawa. A lambda calculus characterization of MSO definable tree transductions (abstract). *Bulletin of Symbolic Logic*, 15(2):250–251, 2009.

[45] Makoto Kanazawa. Second-order abstract categorial grammars as hyperedge replacement grammars. *Journal of Logic, Language and Information*, 19(2):37–161, 2010.

[46] Makoto Kanazawa. Almost affine lambda terms. In Andrzej Indrzejczak, Janusz Kaczmarek, and Michaał Zawidzki, editors, *Trends in Logic XIII*, pages 131–148, Łódź, 2014. Łódź University Press.

[47] Makoto Kanazawa and Ryo Yoshinaka. Lexicalization of second-order ACGs. NII Technical Report NII-2005-012E, National Institute of Informatics, Tokyo, 2005.

[48] Paris C. Kanellakis. Logic programming and parallel complexity. In Jack Minker, editor, *Foundations of Deductive Databases and Logic Programming*, pages 547–585. Morgan Kaufmann, Los Altos, CA, 1988.

[49] Alexander Koller, Michaela Regneri, and Stafan Thater. Regular tree grammars as a formalism for scope underspecification. In *46th Annual Meeting of the Association for Computational Linguistics: Human Language Technologies: Proccedings of the Conference*, pages 218–226, Columbus, Ohio, 2008.

[50] Alexander Koller and Kristina Striegnitz. Generation as dependency parsing. In *Proceedings of the 40th Annual Meeting of the Association for Computational Linguistics*, pages 17–24, Philadelphia, 2002.

[51] Eric Kow. *Surface Realization: Ambiguity and Determinism*. PhD thesis, l'université Henri Poincaré, 2007.

[52] Bernard Lang. Recognition can be harder than parsing. *Computational Intelligence*, 10(4):486–494, 1994.

[53] J. W. Lloyd. *Logic for Learning: Learning Comprehensible Theories from Structured Data*. Springer, Berlin, 2003.

[54] Ralph Loader. Notes on simply typed lambda calculus. Technical Report ECS-LFCS-98-381, Laboratory for Foundations of Computer Science, School of Informatics, The University of Edinburgh, Edinburgh, 1998.

[55] Markus Lohrey. On the parallel complexity of tree automata. In A. Middeldorp, editor, *RTA 2001*, volume 2051 of *Lecture Notes in Computer Science*, pages 201–215, Berlin, 2001. Springer.

[56] Grigori Mints. *A Short Introduction to Intuitionistic Logic*. Kluwer Academic/Plenum Publishers, New York, 2000.

[57] Richard Montague. The proper treatment of quantification in ordinary English. In J. Hintikka, J. Moravcsik, and P. Suppes, editors, *Approaches to Natural Language: Proceedings of the 1970 Stanford Workshop on Grammar and Semantics*. Reidel, Dordrecht, 1973.

[58] M. H. A. Newman. On theories with a combinatorial definition of "equivalence". *Annals of Mathematics*, 43(2):223–243, 1942.

[59] Detlef Plump. Term graph rewriting. In Hartmut Ehrig, G. Engels, H.-J. Kreowski, and Grzegorz Rozenberg, editors, *Handbook of Graph Grammars and Computing by Graph Transformations*, volume 2, pages 3–61. World Scientific, Singapore, 1999.

[60] Jean-Calude Raoult. Rational tree relations. *Bulletin of the Belgian Mathematical Society*, 4:149–176, 1997.

[61] Walter L. Ruzzo. Tree-size bounded alternation. *Journal of Computer and System Sciences*, 21:218–235, 1980.

[62] Sylvain Salvati. *Problèmes de filtrage et problèmes d'analyse pour les grammaires catégorielles abstraites*. PhD thesis, l'Institut National Polytechnique de Lorraine, 2005.

[63] Sylvain Salvati. Encoding second order string ACG with deterministic tree walking transducers. In Shuly Wintner, editor, *Proceedings of FG 2006: The 11th conference on Formal Grammar*, FG Online Proceedings, pages 143–156, Stanford, CA, 2007. CSLI Publications.

[64] Sylvain Salvati. Recognizability in the simply typed lambda-calculus. In Hiroakira

Ono, Makoto Kanazawa, and Ruy de Queiroz, editors, *WoLLIC 2009*, volume 5514 of *Lecture Notes in Artificial Intelligence*, pages 48–60, Berlin, 2009. Springer.

[65] Sylvain Salvati. On the membership problem for non-linear abstract categorial grammars. *Journal of Logic, Language and Information*, 19:163–183, 2010.

[66] Hiroyuki Seki, Takashi Matsumura, Mamoru Fujii, and Tadao Kasami. On multiple context-free grammars. *Theoretical Computer Science*, 88:191–229, 1991.

[67] Ehud Y. Shapiro. Alternation and the computational complexity of logic programs. *Journal of Logic Programming*, 1:19–33, 1984.

[68] Stuart Shieber. The problem of logical-form equivalence. *Computational Linguistics*, 19:179–190, 1993.

[69] Stuart M. Shieber, Yves Schabes, and Fernando C. N. Pereira. Principles and implementations of deductive parsing. *Journal of Logic Programming*, 24:3–36, 1995.

[70] Klaas Sikkel. *Parsing Schemata*. Springer, Berlin, 1997.

[71] Klaas Sikkel. Parsing schemata and correctness of parsing algorithms. *Theoretical Computer Science*, 199(1–2):87–103, 1998.

[72] Raymond M. Smullyan. *Theory of Formal Systems*. Princeton University Press, Princeton, N.J., 1961.

[73] Morten Heine Sørensen and Paweł Urzyczyn. *Lectures on the Curry-Howard Isomorphism*. Elsevier, Amsterdam, 2006.

[74] Rick Statman. On the complexity of alpha conversion. *Journal of Symbolic Logic*, 72(4):1197–2003, 2007.

[75] A.S. Troelstra and H. Schwichtenberg. *Basic Proof Theory*. Cambridge University Press, Cambridge, second edition edition, 2000.

[76] Jeffrey D. Ullman. *Principles of Database and Knowledge-Base Systems*, volume I. Computer Science Press, Rockville, MD, 1988.

[77] Jeffrey D. Ullman. Bottom-up beats top-down for Datalog. In *Proceedings of the Eighth ACM SIGACT-SIGMOD-SIGART Symposium on Principles of Database Systems*, pages 140–149, Philadelphia, 1989.

[78] Jeffrey D. Ullman. *Principles of Database and Knowledge-Base Systems*, volume II: The New Technologies. Computer Science Press, Rockville, MD, 1989.

[79] Jeffrey D. Ullman and Allen Van Gelder. Parallel complexity of logical query programs. *Algorithmica*, 3:5–42, 1988.

[80] David J. Weir. *Characterizing Mildly Context-Sensitive Grammar Formalisms*. PhD thesis, University of Pennsylvania, 1988.

Foundations as Superstructure
(*Reflections of a practicing mathematician*)

Yuri I. Manin
Max–Planck–Institut für Mathematik,
Bonn, Germany
manin@mpim-bonn.mpg.de

Abstract

This talk presents foundations of mathematics as a historically variable set of principles appealing to various modes of human intuition and devoid of any prescriptive/prohibitive power. At each turn of history, foundations crystallize the accepted norms of interpersonal and intergenerational transfer and justification of mathematical knowledge.

Introduction

Foundations vs Metamathematics. In this talk, I will interpret the idea of Foundations in the wide sense. For me, Foundations at each turn of history embody currently recognized, but historically variable, principles of organization of mathematical knowledge and of the interpersonal/transgenerational transferral of this knowledge. When these principles are studied using the tools of mathematics itself, we get a new chapter of mathematics, *metamathematics*.

Modern philosophy of mathematics is often preoccupied with *informal interpretations of theorems, proved in metamathematics of the XX-th century,* of which the most influential was probably Gödel's incompleteness theorem that aroused considerable existential anxiety.

In metamathematics, Gödel's theorem is a discovery that a certain class of finitely generated structures (statements in a formal language) contains substructures that are not finitely generated (those statements that are true in a standard interpretation).

Talk at the conference "Philosophy, Mathematics, Linguistics: Aspects of Interaction 2012" (PhML-2012), held on May 22–25, 2012 at the Euler International Mathematical Institute.

It is no big deal for an algebraist, but certainly interesting thanks to a new context.

Existential anxiety can be alleviated if one strips "Foundations" from their rigid *prescriptive/prohibitive*, or *normative* functions and considers various foundational matters simply from the viewpoint of their mathematical content and on the background of whatever historical period.

Then, say, the structures/categories controversy is seen in a much more realistic light: contemporary studies fuse (Bourbaki type) structures and categories freely, naturally and unavoidably.

For example, in the definition of *abelian categories* one starts with *structurizing sets of morphisms:* they become abelian groups. In the definition of 2–categories, the sets of morphisms are even *categorified:* they become objects of categories, whose morphisms become then the *morphisms of the second level* of initial category. Since in this way one often obtains vast mental images of complex combinatorial structure, one applies to them principles of *homotopy topology* (structural study of topological structures up to homotopy equivalence) in order to squeeze it down to size etc.

I want to add two more remarks to this personal credo.

First, the recognition of quite restrictive and historically changing normative function of Foundations makes this word somewhat too expressive for its content. In a figure of speech such as "Crisis of Foundations" it suggests a looming crash of the whole building (cf. similar concerns expressed by R. Hersh, [8]).

But, second, the first "Crisis of Foundations" occurred in a very interesting historical moment, when the images of formal mathematical reasoning and algorithmic computation became so precise and detailed that they could be, and were, described as *new mathematical structures*: formal languages and their interpretations, partial recursive functions. They could easily fit Bourbaki's universe, even if Bourbaki himself was too slow and awkward to really appreciate the new development.

At this juncture, contemporary *"foundations"* morphed into a *superstructure*, high level floor of mathematics building itself. This is the reason why I keep using the suggestive word "metamathematics" for it.

This event generated a stream of philosophical thought striving to recover the lost normative function. One of the reasons of my private mutiny against it (see e.g. [11]) was my incapability to find any of the philosophical arguments more convincing than even the simplest mathematical reasonings, whatever "forbidden" notions they might involve.

In particular, whatever doubts one might have about the scale of Cantorial cardinal and ordinal infinities, the basic idea of set embodied in Cantor's famous "definition", as a collection of definite, distinct objects of our thought, is as alive as ever. Thinking about a topological space, a category, a homotopy type, a language or a model, we start with imagining such a collection, or several ones, and continue by adding new types "of distinct objects of our thought", derivable from the previous ones or embodying fresh insights.

To summarize: good metamathematics is good mathematics rather than shackles on good mathematics.

Plan of the article. Whatever one's attitude to mathematical Platonism might be, it is indisputable that human minds constitute an important part of habitat of mathematics. In the first section, I will postulate three basic types of mathematical intuition and argue that one can recognize them at each scale of study: personal, interpersonal and historical ones.

The second section is concerned with historical development of the dichotomy *continuous/discrete* and evolving interrelations between its terms.

Finally, in the third section I briefly recall the discrete structures of linear languages studied in *classical metamathematics,* and then sketch the growing array of language–like non–discrete structures that gradually become the subject–matter of *contemporary metamathematics.*

1 Modes of mathematical intuition

1.1 Three modes. I will adopt here the viewpoint according to which *at the individual level* mathematical intuition, both primary and trained one, has three basic sources, that I will describe as *spatial, linguistic,* and *operational* ones.

The neurobiological correlates of the *spatial/linguistic* dichotomy were elaborated in the classical studies of *lateral asymmetry of brain.* When its mathematical content is objectivized, one often speaks about the opposition *continuous/discrete.*

The *linguistic/operational* dichotomy is observed in many experiments studying proto–mathematical abilities of animals. Animals, when they solve and communicate solutions of elementary problems related to counting, use not words but *actions*: cf. some expressive descriptions by Stanislas Dehaene in [6], Chapter 1: "Talented and gifted animals". Operational mode, when it is externalized and codified, becomes a powerful tool for social expansion of mathematics. Learning by rote of "multiplication table" became almost a symbol of democratic education.

The sweeping subdivision of mathematics into Geometry and Algebra, to which at the beginning of modern era was added Analysis (or Calculus) can be considered as a correlate on the scale of whole (Western) civilization of the trichotomy that we postulated above on the scale of an individual (cf. [2]).

It is less widely recognized that *even at the civilization scale*, at various historical periods, each of the spatial, linguistic and operational modes of mathematical intuition can dominate and govern the way that basic mathematical abstractions are perceived and treated.

I will consider as an example "natural" numbers. Most of us nowadays immediately associate them with their *names*: decimal notation $1, 2, 3, \ldots, 1984, \ldots$, perhaps completed by less systemic signs such as 10^6 or XIX.

This was decidedly not always so as the following examples stretching over centuries and millennia show.

1.2 Euclid and his "Elements": spatial and operational *vs* linguistic.
For Euclid, a number was a "magnitude", a potential result of measurement. Measurement of a geometric figure A by a "unit", another geometric figure U, was conceived as a "physical activity in mental space": moving a segment of line inside another segment, step by step; paving a square by smaller squares etc. Inequality $A < B$ roughly speaking, meant that a figure A could be moved to fit inside B (eventually, after cutting A into several pieces and rearranging them in the interior of B).

In this sense, Euclidean geometry might be conceived as "physics of solid bodies in the dimensions one, two and three" (or more precisely, after Einstein, physics *in gravitational vacuum* of respective dimension). This pervasive identification of Euclidean space with our physical space probably influenced the history of Euclid's "fifth postulate". This history includes repeating attempts *to prove it*, that is, to deduce properties of space "at infinity" from observable ones at a finite distance, and then only reluctant acceptance of the Bólyai and Lobachevsky non–Euclidean spaces as "non–physical" ones.

As opposed to addition and subtraction, *the multiplication of numbers* naturally required passage into *a higher dimension*: multiplying two lengths, we get a surface. This was a great obstacle, but, I think, also opened for trained imagination the door to higher dimensions. At least, when Euclid has to speak about the product of an arbitrary large finite set of primes (as in his proof involving $p_1 \ldots p_n + 1$), he is careful to explain his general reasoning by the case of *three* factors, but without doubt, he had some mental images overcoming this restriction.

In fact, the strength of spatial and operational imagination required and achieved by modern mathematics can be glimpsed on a series of examples, starting, say with Morse theory and reaching Perelman's proof of Poincaré conjecture. Moreover, physicists could produce such wonders as Feynman's path integral and Witten's topological invariants, which mathematicians include in their more rigidly organized world only with considerable efforts.

At first sight, it might seem strange that the notion of *a prime number*, theorem about (potential) *infinity of primes*, and theorem about *unique decomposition* could have been stated and proved by Euclid in his geometric world, when no systematic notation for integers was accepted as yet, and no computational rules *dealing with such a notation* rather than numbers themselves were available.

But trying to rationalize this historical fact, one comes to a somewhat paradoxical realization that an efficient notation, such as Hindu–Arabic numerals, actually *does not help*, and even *hinders* the understanding of properties related to divisibility, primality etc. that is, all properties that refer to numbers themselves rather than their *names*.

In fact, the whole number theory could come into being only unencumbered by any efficient *notation* for numbers.

1.3 "Algorist and Abacist": linguistic *vs* operational.

The dissemination of a positional number system in Europe after the appearance of Leonardo Fibonacci's *Liber Abaci* (1202) was, in essence, the beginning of the expansion of a universal, truly global language. Its final victory took quite some time.

The book by Gregorio Reisch, *Margarita Philosophica*, was published in Strasbourg in 1504. One engraving in this book shows a female figure symbolizing Arithmetics. She contemplates two men, sitting at two different tables, an *abacist* and an *algorist*.

The abacist is bent over his *abacus*. This primitive calculating device survived until the days of my youth: every cashier in any shop in Russia, having accepted a payment, would start calculating change clicking movable balls of her abacus.

The algorist is computing something, writing Hindu–Arabic numerals on his desk. The words "algorist" and modern "algorithm" are derived from the name of the great Al Khwarezmi (born in Khorezm c. 780).

In the context of this subsection, the abacus illustrates the operational mode whereas computations with numerals do the same for linguistic one (although in other contexts the operational side of such computations might dominate).

This engraving in the reception of contemporary readers was more politicised. It symbolized coming of a new epoch of democratic learning.

The Catholic Church supported the Roman tradition, usage of Roman numerals. They were fairly useless for practical commercial bookkeeping, calender computations such as dates of Easter and other moveable feasts etc. Here the abacus was of great help.

The competing tribe of algorists was able to compute things by writing strange signs on paper or sand, and their art was associated with dangerous, magical, secret Muslim knowledge. Al Khwarezmi teaching became their (and our) legacy.

Arithmetics blesses both practitioners.

1.4 John Napier and Alan Turing: operational. The nascent programming languages for centuries existed only as informal subdialects of a natural language. They had a very limited (but crucially important) sphere of applicability, and were addressed to human calculators, not electronic or mechanical ones. Even Alan Turing in the 20th century, when speaking of his universal formalization of computability, later called Turing machine, used the word "computer" to refer to *a person* who mechanically follows a finite list of instructions lying before him/her.

The ninety–page table of natural logarithms that John Napier published in his book *Mirifici Logarithmorum Canonis Descriptio* in 1614 was a paradoxical example of this type of activity that became a cultural and historical monument on a global scale. Napier, who computed the logarithms manually, digit by digit, combined in one person the role of creator of new mathematics and that of computer–clerk who followed his own instructions. His assistant Henry Briggs later performed this function.

Napier's tables were tables of (approximate values of) *natural logarithms*, with the base $e = 2,718281828...$. However, it seems that he neither referred to e explicitly, nor even recognized its existence. Roughly speaking, after having chosen the precision which he wanted to calculate logarithms, say with error $< 10^{-7}$, he dealt with integer powers of the number $1 + 10^{-8}$, whose 10^8 power was close to e.

This is one more example of the seemingly paradoxical fact, that an efficient and unified notation for objects of mathematical world can *hinder a theoretical understanding* of this world.

All the more amazing was the philosophical insight of Leibniz, who in his famous exhortation *Calculemus!* postulated that not only numerical manipulations, but any rigorous, logical sequence of thoughts that derives conclusions from initial axioms can be reduced to computation. It was the highest achievement of the great logicians of the 20th century (Hilbert, Church, Gödel, Tarski, Turing, Markov, Kolmogorov,...) to draw a precise map of the boundaries of the Leibnizian ideal world, in which

- *reasoning is equivalent to computation;*

- *truth can be formalized, but cannot always be verified formally;*

- *the "whole truth" even about the smallest infinite mathematical universe, natural numbers, exceeds potential of any finitely generated language to generate true theorems.*

The central concept of this program, *formal languages*, inherited the basic features of both natural languages (written form fixed by an alphabet) and the positional number systems of arithmetic. In particular, any classical formal language is one–dimensional (linear) and consists of discrete symbols that explicitly express the basic notions of logic.

Euclid found the remedy for the deficiencies of this linearity by strictly restricting role of natural language to the expression of *logic* of his proofs. The *content* of his mathematical imagination was transmitted by pictures.

2 Continuous or discrete? From Euclid to Cantor to homotopy theory

2.1 From continuous to discrete in "Elements". As we have seen, integers (and a restricted amount of other real numbers) for Euclid were results of (mental) measurement: *discrete came from continuous.* This was one–way road: continuous could not be produced from discrete. The idea that a line "consists" of points, so familiar to us today, does not seem to belong to Euclid's mental world and, in fact, to mental worlds of many subsequent generations of mathematicians until Georg Cantor. For Euclid, a point can be (a part of) the boundary of a (segment) of line, but such a segment cannot be scattered to a heap of points.

Geometric images are the source and embodiment not only of numbers, but of logical reasoning as well: in "Elements" at least a comparable part of its logic is encoded in figures rather than in words.

This was made very clear in the London publication of 1847, entitled

THE FIRST SIX BOOKS OF

THE ELEMENTS OF EUCLID

IN WHICH COLOURED DIAGRAMS AND SYMBOLS
ARE USED INSTEAD OF LETTERS FOR THE
GREATER EASE OF LEARNERS

whose author was Oliver Byrne, *"Surveyor of her Majesty's settlements in the Falklands Islands"*, (see a recent republication [5]).

Byrne literally writes algebraic formulas whose main components are triangles, colored sectors of circle, segments of line etc. connected by more or less conventional algebraic signs.

2.2 From discrete to continuous: Cantor, Dedekind, Hausdorff, Bourbaki ... This way is so familiar to my contemporaries that I do not have to spend much time to its description. The description of a mathematical structure, such as a group, or a topological space, according to Bourbaki starts with one or several unstructured sets, to which one adds elements of a these sets or derived sets satisfying restrictions formulated in terms of set theory.

Thus the twentieth century idea of "continuous" is based upon two parallel notions: that of *topological space* X (a set with the system of "open" subsets) and that of a "continuous map" $f\colon X \to Y$ between topological spaces. Further elaboration involving sheaves, topoi etc does not part with this basic intuition.

However, the set–theoretic point of departure helped enrich the geometric intuition by images that were totally out of reach earlier. The discovery of difference between *continuous* and *measurable* (from Lebesgue integral to Brownian motion to Feynman integral) was a radical departure from Euclidean universe.

In a finite–dimensional context, one could now think about Cantor sets, Hausdorff dimension and fractals, curves filling a square, Banach–Tarski theorem. In infinite–dimensional contexts wide new horizons opened, starting with topologies of Hilbert and Banach linear spaces and widening in an immense universe of topology and measure theory of non–linear function spaces.

2.3 From continuous to discrete: homotopy theory. One of the most important development of topology was the discovery of main definitions and results of homotopy theory. Roughly speaking, a homotopy between two topological spaces X, Y is a continuous deformation producing Y from X, and similarly a homotopy between two continuous maps $f, g\colon X \to Y$ is a continuous deformation producing g from f. A *homotopy type* is the class of spaces that are homotopically equivalent pairwise. To see how drastically the homotopy can change a space, one can note that a ball, or a cube, of any dimension is *contractible*, that is, can be homotopically deformed to a point, so that dimension ceases to be invariant.

The basic discrete invariant of the homotopy type of X is the set of its connected components $\pi_0(X)$. To see, how this invariant gives rise to one of the basic structures of mathematics, ring of integers $\mathbf{Z}$, consider a real plane P with a fixed orientation, a point x_0 on it, different from $(0,0)$, and the set of homotopy classes of loops (closed paths) in P, starting and ending at x_0 and avoiding $(0,0)$. This latter set can be canonically identified with $\mathbf{Z}$: just count the number of times the loop goes around $(0,0)$. Each loop going in the direction of orientation counts as $+1$, where as the "counter–clockwise" loops count as -1.

On a very primitive level, this identification shows how the ideas of homotopy naturally introduce negative numbers. In the historically earlier periods when integers were measuring geometric figures (or counting real/mental objects) even idea of zero was very difficult and slowly gained ground in the symbolic framework of positional notation. Introduction of negative numbers required appellation to an extra–mathematical reality, such as *debt* in economics.

More generally, Voevodsky in his research project [14] introduces the following hierarchy of homotopy types graded by their h–levels. Zero level homotopy type consists of one point representing contractible spaces. If types of level n are already defined, types of level $n+1$ consist of spaces such that the space of paths between any two points belongs to type of level n.

He further interprets types of level 1, represented by one point and empty sets, as *truth values*, and types of level 2 as sets. All sets in this universe are thus of the form $\pi_0(X)$.

Higher levels are connected with theory of categories, poly–categories etc, and we will return to them in the next section. At this point, we mention only that Voevodsky hierarchy does not *replace* sets but rather systematically embeds set–theoretical and categorical constructions and intuitions into a vaster universe where continuous and discrete are treated on an equal footing.

3 Language–like mathematical structures and metamathematics

3.1 Metamathematics: mathematical studies of formalized languages of mathematics. Philosophy of mathematics in the XX–th century had to deal with lessons of *metamathematics,* especially of Gödel's incompleteness theorem.

As I have already said, I will consider metamathematics as *a special chapter of mathematics itself,* whose subject is the study of *formal languages and their interpretations.* On the foreground here were the first order formal languages, a formalization of Euclid's and Aristotle's legacy. Roughly speaking, to Euclid we owe the mathematics of spatial imagination (and/or kinematics of solid bodies), whereas Aristotle founded *the mathematics of logical deduction,* expressed in "Elements" by natural language *and* creative usage of drawings.

An important parallel development of formal languages involved languages formalizing *programs for and processes of computation,* of which chronologically first in the XX–th century was *Church's lambda calculus* [9].

An important feature of lambda calculus is the absence of formal distinctions between the language of programs and the language of input/output data (unlike Turing's machines, where a machine "is" the program, whereas input/output are represented by binary words). When, due to von Neumann's insights, this feature became implemented in hardware, lambda calculus was rediscovered and became in the 1960's the basis of development of programming languages.

These languages are *linear*, in the following sense: the set of all syntactically correct expressions in a formal language L could be described as a Bourbaki structure consisting of a certain *words*, – finite sequences of letters in a given *alphabet*, and finite sequences of such words, *expressions*. Words and expressions must be *syntactically correct* (precise description of this is a part of definition of each concrete language). Letters of alphabet are subdivided into *types*: variables, connectives and quantifiers, symbols for operations, relations ... Syntactically correct expressions can be terms, formulas, ...

Such Bourbaki structures can be sufficiently rich to produce formal versions of real mathematical texts, existing and potential ones, and make them an object of study.

I will explain how the advent of category theory (and, to a lesser degree, theory of computability) required enriched languages, that after formalization become at first *non–linear,* and then *multidimensional.* Such languages require for their study *homotopy theory* and suggest a respective enrichment of the universe in which interpretations/models are supposed to live, from Sets to Homotopy Types, as in the Voevodsky's project (cf. above).

3.2 One–dimensional languages of diagrams and graphs.

With the development of homological algebra and category theory in the second half of the XX–th century, the language of commutative diagrams began to penetrate ever wider realms of mathematics. It took some time for mathematicians to get used to "diagram-chasing." A simply looking algebraic identity $kg = hf$, when it expresses a property of four morphisms in a category, means that we are contemplating a simple *commutative diagram*, in which, besides morphisms f, g, k, h, also the objects A, B, C, D invisible in the formula $kg = hf$ play key roles:

$$\begin{array}{ccc} A & \xrightarrow{g} & B \\ f\downarrow & & \downarrow k \\ C & \xrightarrow{h} & D \end{array}$$

Although this square is not a "linear expression", one may argue that it, and its various generalizations of growing size (even the whole relevant category), are still "one–dimensional". This means that they can be encoded in a graph, whose vertices are labeled by (names of) objects of our category, whereas edges are labeled by pairs consisting of an orientation and a morphism between the relevant objects.

Similarly, a program written in a linear programming language can be encoded in a graph whose vertices are labeled by (names of) elementary operations that can be performed over the relevant data. To understand labeling of (oriented) edges, one must imagine that they encode channels, forwarding output data calculated by the operation at the start (input) of the edge to the its endpoint where they become input of the next operation (or the final output, if the relevant vertex is labeled respectively). Labels of edges might then include *types* of the relevant data.

3.3 From graphs to higher dimensions. Generally, a square of morphisms as above need not be commutative (i. e. it is possible that $kg \neq hf$). In order to distinguish these two cases graphically, we may decide to associate with *a commutative square the two–dimensional picture*, by glueing the interior part of the square to the relevant graph.

A well known generalization of this class of spaces are *cell complexes*, or, in more combinatorial and therefore more language–like version, *simplicial complexes*. Of course, we must allow labels of cells as additional structures.

In this way, we can get, for example, a geometric encoding of the category $\mathcal{C}$ by a simplicial complex, in which labeled $(n+1)$–complexes are sequences of morphisms

$$X_0 \xrightarrow{f_0} X_1 \xrightarrow{f_1} \ldots \xrightarrow{f_{n-1}} X_n$$

whereas the face map ∂^i omits one of the objects X_i and, if $1 \leq i \leq n-1$, replaces the pair of arrows around X_i by one arrow labeled by the composition of the relevant morphisms. The resulting simplicial space encodes the whole category in a simplicial complex that is called *the nerve* of the category. Clearly, not only objects and morphisms, but also all compositions of morphisms and relations between them can be read off it.

Thus the language of commutative diagrams becomes a chapter of algebraic topology, and when the study of functors is required, the chapter of homotopical topology.

3.4 Quillen's homotopical algebra and univalent foundations project. In his influential book [13], Quillen developed the idea that the natural language for homotopy theory should appeal *not* to the initial intuition of continuous deformation itself, but rather to a codified list of properties of category of topological spaces stressing those that are relevant for studying homotopy.

Quillen defined *a closed model category* as a category endowed with three special classes of morphisms: *fibrations, cofibrations*, and *weak equivalences*. The list of axioms which these three classes of morphisms must satisfy is not long but structurally quite sophisticated. They can be easily defined in the category of topological spaces using homotopy intuition but remarkably admit translation into many other situations. An interesting new preprint [7] even suggests the definition of these classes in appropriate categories of discrete sets, contributing new insights to old Cantorian problems of the scale of infinities.

Closed model categories become in particular a language of preference for many contexts in which objects of study are quotients of "large" objects by "large" equivalence relations, such as homotopy.

It is thus only natural that the most recent Foundation/Superstructure, Voevodsky's Univalent Foundations Project (cf. [14] and [3]) is based on direct axiomatization of the world of homotopy types.

As a final touch of modernism, the metalanguage of this project is a version of typed lambda calculus, because its goal is to develop a tool for the computer assisted verification of programs and proofs. Thus computers become more and more involved in the interpersonal habitat of "theoretical" mathematics.

It remains to hope that humans will not be finally excluded from this habitat, as some aggressive proponents of databases replacing science suggest (cf. [1]).

Post Scriptum: Truth and Proof in Mathematics

As I have written in [12], the notion of "truth" in most philosophical contexts is a reification of a certain relationship between humans and *texts/utterances/statements*, the relationship that is called "belief", "conviction" or "faith".

Professor Blackburn in [4] in his keynote speech to the Balzan Symposium on "Truth" (where [12] was delivered) extensively discussed other relationships of humans to texts, such as *scepticism, conservatism, relativism, deflationism*. However, in the long range all of them are secondary in the practice of a researcher in mathematics.

I will only sketch here what must be said about texts, sources of conviction, and methods of conviction peculiar to mathematics.

Texts. Alfred North Whitehead said that all of Western philosophy was but a footnote to Plato.

The underlying metaphor of such a statement is: "Philosophy is a text", the sum total of all philosophic utterances.

Mathematics decidedly is *not* a text, at least not in the same sense as philosophy. There are no authoritative books or articles to which subsequent generations turn again and again for wisdom. Already in the XX-th century, researchers did not read Euclid, Newton, Leibniz or Hilbert in order to study geometry, calculus or mathematical logic. The life span of any contemporary mathematical paper or book can be years, in the best (and exceptional) case decades. Mathematical wisdom, if not forgotten, lives as an invariant of all its (re)presentations in a permanently self–renewing discourse.

Sources and methods of conviction. Mathematical truth is not revealed, and its acceptance is not imposed by any authority.

Ideally, the truth of a mathematical statement is ensured by *a proof*, and the ideal picture of a proof is a sequence of elementary arguments whose rules of formation are explicitly laid down before the proof even begins, and ideally are common for all proofs that have been devised and can be devised in future. The admissible starting points of proofs, "axioms", and terms in which they are formulated, should also be discussed and made explicit.

This ideal picture is so rigid that it became the subject of mathematical study in metamathematics.

But in the creative mathematics, the role of proof is in no way restricted to its function of carrier of conviction. Otherwise, there would be no need for Carl Friedrich Gauss to consider eight (!) different proofs the quadratic reciprocity law (cf. [10] for an extended bibliography; I am grateful to Prof. Yuri Tschinkel for this reference).

One metaphor of proof is a route, which might be a desert track boring and unimpressive until one finally reaches the oasis of one's destination, or a foot path in green hills, exciting and energizing, opening great vistas of unexplored lands and seductive offshoots, leading far away even after the initial destination point has been reached.

References

[1] Ch. Anderson. The End of Theory: The Data Deluge Makes the Scientific Method Obsolete. Retrieved Mar. 23, 2016, from http://www.wired.com/2008/06/pb-theory/, Jun. 23, 2008.

[2] M. Atiyah. Geometry and Physics of the 20th Century. In J. Kouneiher, D. Flament, Ph. Nabonnand, and J.-J. Szczeciniarz, editors, *Géométrie au XXe siècle. Histoire et horizons*, pages 4–9. Hermann, Paris, 2005.

[3] S. Awodey. Type theory and homotopy. Retrieved Mar. 23, 2016, from http://arxiv.org/abs/1010.1810, 2010.

[4] S. Blackburn. Truth and Ourselves: the Elusive Last Word. In N. Mout and W. Stauffacher, editors, *Truth in Science, the Humanities, and Religion. Balzan Symposium 2008*, pages 5–14. Springer, 2010.

[5] O. Byrne. *The first six books of the Elements of Euclid.* Taschen, 2013. Facsimile of the first edition published by Pickering in 1847.

[6] S. Dehaene. *The Number Sense: How the Mind Creates Mathematics.* Oxford University Press, 1st edition, 1999.

[7] M. Gavrilovich and A. Hasson. Exercices de style: a homotopy theory for set theory, I. Retrieved Mar. 23, 2016, from http://arxiv.org/abs/1102.5562, 2012.

[8] R. Hersh. Wings, not Foundations! In G. Sica, editor, *Essays on the Foundations of Mathematics and Logic*, volume 1, pages 155–164. Polimetrica Int. Sci. Publisher, Monza, Italy, 2005.

[9] A. Jung. A short introduction to the Lambda Calculus. Retrieved Mar. 23, 2016, from http://www.cs.bham.ac.uk/~axj/pub/papers/lambda-calculus.pdf, Mar. 18, 2004.

[10] F. Lemmermeyer. Proofs of the quadratic reciprocity law. Retrieved Mar. 23, 2016, from http://www.rzuser.uni-heidelberg.de/~hb3/rchrono.html.

[11] Yu. Manin. *A Course in Mathematical Logic for Mathematicians.* Graduate Texts in Mathematics 53. Springer, 2nd edition, 2010. With new Chapters written by Yu. Manin and B. Zilber.

[12] Yu. Manin. Truth as a Value and Duty: Lessons of Mathematics. In N. Mout and W. Stauffacher, editors, *Truth in Science, the Humanities, and Religion. Balzan Symposium 2008*, pages 37–45. Springer, 2010. Preprint arXiv:0805.4057v1 [math.GM].

[13] D. G. Quillen. *Homotopical Algebra*, volume 43 of *Lecture Notes in Mathematics.* Springer-Verlag, Berlin, 1st edition, 1967.

[14] V. Voevodsky. Univalent Foundations Project. Retrieved Mar. 23, 2016, from http://www.math.ias.edu/~vladimir/Site3/Univalent_Foundations_files/univalent_foundations_project.pdf, Oct. 1, 2010.

Classical and Intuitionistic Geometric Logic

Grigori Mints
Department of Philosophy, Stanford University
Stanford, Ca, 94305, USA

Abstract

Geometric sequents "A implies C" where all axioms A and conclusion C are universal closures of implications of positive formulas play distinguished role in several areas including category theory and (recently) logical analysis of Kant's theory of cognition. They are known to form a Glivenko class: existence of a classical proof implies existence of an intuitionistic proof. Existing effective proofs of this fact involve superexponential blow-up, but it is not known whether such increase in size is necessary. We show that any classical proof of such a sequent can be polynomially transformed into an intuitionistic geometric proof of (classically equivalent but intuitionistically) weaker geometric sequent.

Keywords: Geometric Formulas, Glivenko Classes, Intuitionistic Logic.

Introduction

Geometric sequents (see definition below) play distinguished role in several areas including category theory [3]. This fragment of first order logic attracted new attention in the light of recent work by Theodora Achourioti and Michiel van Lambalgen [1] who propose a translation of the philosophical language of Kant's theory of judgements into the language of elementary logic and provide a convincing justification of their view.

Geometric sequents are known to form a Glivenko class: existence of a classical proof of a geometric sequent S implies existence of an intuitionistic proof. Existing proofs of this fact involve superexponential blow-up, but we do not know whether

---

Talk at the conference "Philosophy, Mathematics, Linguistics: Aspects of Interaction 2012" (PhML-2012), held on May 22–25, 2012 at the Euler International Mathematical Institute.

such increase in size is necessary. We show that any classical proof of S can be polynomially transformed into an intuitionistic geometric proof of (classically equivalent but intuitionistically) slightly weaker geometric sequent.

We consider formulas of first order logic.

Definition 1. *Positive formulas* are constructed from atomic formulas and the constant $\bot$ by $\&, \vee, \exists$.

Geometric implications are positive formulas, implications of positive formulas and results of prefixing universal quantifiers to such implications.

Geometric sequents are expressions of the form

$$I_1, \ldots I_n \Rightarrow I$$

where $I_1, \ldots I_n, I$ are geometric implications.

A *geometric derivation* is a derivation consisting of geometric sequents.

The second proof of Theorem 1 given below is non-effective, but the first one allows one to derive some complexity bound. The proof begins with construction of a cut-free derivation, therefore the only obvious bound is the same as for cut-elimination, that is hyperexponential one. This contrasts with the most prominent Glivenko class, namely that of negative formulas. When a classical derivation of a negative formula is given, its intuitionistic derivation is constructed by "negativizing" all formulas in the derivation plus local changes to reinstate the inferences that were destroyed by this transformation. These transformations are polynomial.

We show here a weaker result for geometric sequents. Any classical proof (with cut) of a geometric sequent $\Gamma \Rightarrow I$ can be polynomially transformed into an intuitionistic geometric proof of a geometric sequent sequent $D, \Gamma \Rightarrow I$ where D is obtained by introducing abbreviations for some formulas. In fact $D, \Gamma \Rightarrow I$ is intuitionistically derivable iff $\Gamma \Rightarrow I$ is intuitionistically derivable, but on the surface the definitions in D are only classical.

In section 1 we give two proofs of the Glivenko property of geometric sequents.

Section 2 describes depth-reducing transformations we need for our proofs. As far as I know, this use of formulas (17-19) especially to achieve that the whole proof is new. It is inspired by similar use of (18) by V. Orevkov [5] in a different situation.

Section 3 contains the proof of the main result.

We use $\equiv$ for literal coincidence of syntactic objects and $\leftrightarrow$ for a logical equivalence connective.

LK, LJ are Gentzen's systems for classical and intuitionistic logic, both with cut. $\vdash^c, \vdash^i$ denote derivability in classical or intuitionistic logic, that is in LK, LJ with cut.

A *formula translation* of a sequent $S \equiv A_1, \ldots A_n \Rightarrow B_1, \ldots B_m$ is a formula $S^f \equiv (A_1 \& \ldots \& A_n \to B_1 \vee \ldots \vee B_m)$. Many notions defined for formulas are generalized to sequents via the formula translation. For example $S \leftrightarrow T$ for sequents S, T means $S^f \leftrightarrow T^f$.

c-models are ordinary models for the classical predicate logic, i-models are Kripke models.

1 Geometric sequents have Glivenko property

The next theorem is well-known. The deductive proof given here is due to V. Orevkov [5] and can be traced back to the work of H. Curry [2].

Theorem 1. *A geometric sequent is derivable classically iff it is derivable intuitionistically.*

1. *A deductive proof.* Consider a cut-free proof of a geometric sequent

$$\Gamma \to I$$

in LK. Since the succedent rules for $\to, \forall$ are invertible in LK, we can analyze away initial universal quantifiers and implication in I, then assume that I is a positive formula. After that *the sequent $\Gamma \Rightarrow I$ contains only connective occurrences that give rise to rules*

$$\Rightarrow \&, \Rightarrow \vee, \Rightarrow \exists, \& \Rightarrow, \vee \Rightarrow, \exists \Rightarrow, \to \Rightarrow .$$

These rules are common for LK and LJm, hence our LK-derivation is already LJm-derivation, as required. ⊢

2. *A model-theoretic proof.* The idea here is rather similar, but I have not seen this proof in literature. Suppose a geometric sequent $\Gamma \Rightarrow I$ with positive formula I is underivable in LJm. Consider its proof search tree in LJm (see for example Mints [4]). This tree is not a derivation, and hence has a non-closed branch generating a Kripke countermodel for $\Gamma \Rightarrow I$. The rules for analysis of the connectives $\forall, \to$ in succedent are not applied in this tree. But these are exactly the rules that add new worlds to a model. Therefore the resulting model has just one world, and hence it is a classical model refuting our sequent. ⊢

2 Reducing formula depth

Familiar depth-reducing transformations by introduction of new predicate variables are modified here to preserve geometric sequents. There are subtle points noted below. Let's first recall well-known facts.

Let's define a relation between formulas (widely used in literature without a special name) which is weaker than provable equivalence but in some respects similar to it.

Write $F \trianglerighteq^s G$ where $s \in \{c, i\}$ if

$$G \equiv F' \to F \text{ and } \vdash^s F'[P_1/F_1, \ldots P_n/F_n]$$

where P_i/F_i are substitutions (performed in this order) for predicate variables $P_1, \ldots P_n$ not occurring in F.

Lemma 1. *Assume $F \trianglerighteq^s G$. Then*

1. $\vdash^s F$ *iff* $\vdash^s G$,

2. *s-models for G are expansions (with respect to $P_1, \ldots P_n$) of s-models for F.*

Proof. 1. $\vdash^s F \to G$ is obvious. If $\vdash^s G$ then since $G \equiv (F' \to F)$ the substitutions $P_1/F_1, \ldots P_n/F_n$ and modus ponens yield $\vdash^s F$.

2. Similarly to 1.

$\dashv$

Notation $\mathbf{x}$ below stands for $x_1, \ldots x_n$ with distinct variables $x_1, \ldots x_n$.

Lemma 2. *If $\mathbf{x}$ contains all free variables of formulas $A(\mathbf{x}), B(\mathbf{x})$ then*

$$LJ \vdash \forall \mathbf{x}(A(\mathbf{x}) \leftrightarrow B(\mathbf{x})) \to (F(A) \leftrightarrow F(B)).$$

Proof. Induction on F.

$\dashv$

Lemma 3. *If P is a fresh n-ary predicate symbol, $\mathbf{x}$ contains all free variables of the formula $A(\mathbf{x})$ then for $L \in \{LJ, LK\}$*

$$L \vdash \Rightarrow F(A) \text{ iff } \quad L \vdash \forall \mathbf{x}(A(\mathbf{x}) \leftrightarrow P(\mathbf{x})) \Rightarrow F(P)$$

Proof. If $L \vdash F(A)$, apply the previous Lemma.

If $L \vdash \forall \mathbf{x}(A(\mathbf{x}) \leftrightarrow P(\mathbf{x})) \Rightarrow F(P)$, substitute A for P. The antecedent of the sequent becomes $\forall \mathbf{x}(A(\mathbf{x}) \leftrightarrow A(\mathbf{x}))$.

$\dashv$

For a given formula F assume that for every non-atomic subformula G of F a fresh predicate symbol P_G is chosen with the same arity as the number of free variables of G. In particular P_F has free variables of F as arguments. Atomic subformula $P(t_1, \ldots t_n)$ is not changed.

Symbols P_G can be treated as pointers to subformulas of F. This informal observation can be formalized by assigning equivalences E_G to subformulas G in the following way:

If $G(\mathbf{x}) \equiv H(\mathbf{y}) \odot K(\mathbf{z})$ for $\odot \in \{\&, \vee, \to\}$ then

$$E_G \equiv \forall \mathbf{x}(P_G(\mathbf{x}) \leftrightarrow (P_H(\mathbf{y}) \odot P_K(\mathbf{z}))) \qquad (1)$$

where $\mathbf{y}, \mathbf{z} \subseteq \mathbf{x}$.

If $G(\mathbf{x}) \equiv QyH(\mathbf{x}, y)$ for $Q \in \{\forall, \exists\}$ then

$$E_G \equiv \forall \mathbf{x}(P_G(\mathbf{x}) \leftrightarrow QyP_H(\mathbf{x}, y)). \qquad (2)$$

Lemma 4. *Let $G, H \ldots F$ be all non-atomic subformulas of F. Then for $L \in \{LJ, LK\}$*

$$L \vdash F \leftrightarrow L \vdash E_G, E_H, \ldots E_F \Rightarrow P_F.$$

Proof. Apply previous Lemma successively to subformulas, beginning with the innermost ones. ⊢

Let's rewrite equivalences (1),(2) as pairs or triples of implications, transforming these implications in LJ-equivalent way.

$$\forall \mathbf{x}(P_{G\&H}(\mathbf{x}) \;\to\; P_G(\mathbf{y})), \qquad (3)$$
$$\forall \mathbf{x}(P_{G\&H}(\mathbf{x}) \;\to\; P_H(\mathbf{z})), \qquad (4)$$
$$\forall \mathbf{x}(P_G(\mathbf{y}) \& P_H(\mathbf{z}) \;\to\; P_{G\&H}(\mathbf{x}); \qquad (5)$$
$$\forall \mathbf{x}(P_G(\mathbf{y}) \;\to\; P_{G\vee H}(\mathbf{x})), \qquad (6)$$
$$\forall \mathbf{x}(P_H(\mathbf{z}) \;\to\; P_{G\vee H}(\mathbf{x})), \qquad (7)$$
$$\forall \mathbf{x}(P_{G\vee H}(\mathbf{x}) \;\to\; (P_G(\mathbf{y}) \vee P_H(\mathbf{z})); \qquad (8)$$
$$\forall \mathbf{x}(P_{\exists y P_G}(\mathbf{x}) \;\to\; \exists y P_G(\mathbf{x}, y)) \qquad (9)$$
$$\forall \mathbf{x} \forall y(P_G(\mathbf{x}, y) \;\to\; P_{\exists y P_G}(\mathbf{x})) \qquad (10)$$
$$\forall \mathbf{x} \forall y(P_{\forall y P_G}(\mathbf{x}) \;\to\; P_G(\mathbf{x}, y)); \qquad (11)$$
$$* \; \forall \mathbf{x}(\forall y P_G(\mathbf{x}, y) \;\to\; P_{\forall y P_G}(\mathbf{x})) \qquad (12)$$
$$* \; \forall \mathbf{x}(\neg P_G(\mathbf{x}) \;\to\; P_{\neg G}(\mathbf{x})) \qquad (13)$$
$$\forall \mathbf{x}(P_G(\mathbf{x}) \& P_{\neg G}(\mathbf{x}) \;\to\; \bot) \qquad (14)$$
$$\forall \mathbf{x}(P_{G\to H}(\mathbf{x}) \& P_G(\mathbf{y}) \;\to\; P_H(\mathbf{z})) \qquad (15)$$
$$* \; \forall \mathbf{x}((P_G(\mathbf{y}) \to P_H(\mathbf{z})) \;\to\; P_{G\to H}(\mathbf{x})) \qquad (16)$$

All these universally quantified implications are geometric except the three marked by a *. Let's replace them by classically equivalent geometric implications.

$$\forall \mathbf{x} \exists y (P_G(\mathbf{x}, y) \to P_{\forall y P_G}(\mathbf{x})) \tag{17}$$

$$\forall \mathbf{y}(P_G(\mathbf{y}) \lor P_{\neg G}(\mathbf{y})) \tag{18}$$

$$\forall \mathbf{x}((P_H(\mathbf{z}) \to P_{G \to H}(\mathbf{x})) \;\&\; (P_G(\mathbf{y}) \lor P_{G \to H}(\mathbf{x}))) \tag{19}$$

Denote the resulting set of geometric implications (3-11), (14,15) and (17,18,19) for subformulas of a set $\mathbf{F}$ of formulas by $\mathrm{DEF}_\mathbf{F}$.

3 Transformation of classical derivations

In this section we mean by intuitionistic predicate calculus a multiple-succedent formulation LJm (cf. Mints [4]) which differs from LK only in the requirement that the list Δ is empty in the succedent rules for $\to, \neg, \forall$:

$$\frac{A, \Gamma \Rightarrow \Delta}{\Gamma \Rightarrow \Delta, \neg A} \qquad \frac{A, \Gamma \Rightarrow \Delta, B}{\Gamma \Rightarrow \Delta, A \to B} \qquad \frac{\Gamma \Rightarrow \Delta, A(b)}{\Gamma \Rightarrow \Delta, \forall x A(x)}$$

Definition 2. Formulas $\neg A, A \to B, \forall x A$ introduced by these rules in an LK-derivation are called below *special* formulas when Δ is non-empty.

Let d be a derivation of a geometric sequent S in LK. Then $\mathbf{f}(d)$ denotes the set of all cut formulas in d and DEF_d denotes $\mathrm{DEF}_{\mathbf{f}(d)}$.

Theorem 2.

1. Let d be a derivation of a geometric sequent $\Pi \Rightarrow \Phi$ in LK. Then it can be polynomially *transformed into a geometric derivation in LJm of the sequent*

$$\mathrm{DEF}_d, \Pi \to \Phi$$

   consisting of geometric sequents.

2. $\mathrm{DEF}_d, \Pi \Rightarrow \Phi \;\unrhd^c\; \Pi \Rightarrow \Phi$.

3. $\vdash^c \mathrm{DEF}_d, \Pi \Rightarrow \Phi$ *iff* $\quad \vdash^i \mathrm{DEF}_d, \Pi \Rightarrow \Phi$ *iff* $\quad \vdash^i \Pi \Rightarrow \Phi$

Proof. We assume that all axioms $A, \Gamma \to \Delta, A$ have atomic A. Using if needed inversion transformations we assume that Φ consists of positive formulas. Then every special formula F is traceable to a cut formula. More precisely, $F \equiv F'(\mathbf{t})$ where $F'(\mathbf{x})$ is a subformula of some cut formula. Formula F' has a "representative"

$P_{F'}(\mathbf{x})$ in DEF_d where $\mathbf{x}$ are free variables of F'. In this sense any occurrence of a formula F traceable to a cut formula has a representative which we write as $P_F(\mathbf{t})$.

Denote by d^+ the result of replacing every such occurrence of $F(t)$ as a separate formula in a sequent in d by $P_F(\mathbf{t})$.

This replacement destroys inferences having such $F(t)$ as principal formulas. Consider these inferences in turn to show they can be repaired using DEF_d.

Axioms are assumed to be atomic, therefore they are preserved. The cut inferences become cuts on atomic formulas.

Antecedent inferences are repaired using geometric implications in Def_d. For example $\rightarrow$-antecedent inference

$$\frac{\Gamma \Rightarrow \Delta, G(\mathbf{t}_1) \quad H(\mathbf{t}_2), \Gamma \Rightarrow \Delta}{G(\mathbf{t}_1) \rightarrow H(\mathbf{t}_2), \Gamma \Rightarrow \Delta}$$

goes into the figure

$$\frac{\Gamma \Rightarrow \Delta, P_G(\mathbf{t}_1) \quad P_H(\mathbf{t}_2), \Gamma \Rightarrow \Delta}{P_{G\rightarrow H}(\mathbf{t}), \Gamma \Rightarrow \Delta}$$

which is transformed using the formula $P_{G\rightarrow H}(\mathbf{t}) \& P_G(\mathbf{t}_1) \rightarrow P_H(\mathbf{t}_2)$ denoted below by I which is an instance of a formula (15) in DEF_d.

$$\frac{\dfrac{\text{axiom}}{P_{G\rightarrow H}(\mathbf{t}) \Rightarrow P_{G\rightarrow H}(\mathbf{t}) \quad \Gamma \Rightarrow \Delta, P_G(\mathbf{t}_1)}}{\dfrac{P_{G\rightarrow H}(\mathbf{t}), \Gamma \Rightarrow \Delta, P_{G\rightarrow H}(\mathbf{t}) \& P_G(\mathbf{t}_1) \quad P_H(\mathbf{t}_2), \Gamma \Rightarrow \Delta}{\dfrac{I, P_{G\rightarrow H}(\mathbf{t}), \Gamma \Rightarrow \Delta}{\text{DEF}_d, P_{G\rightarrow H}(\mathbf{t}), \Gamma \Rightarrow \Delta}}} \forall \Rightarrow$$

Other antecedent rules and succedent rules common to LK and LJm are treated similarly. Of the remaining rules consider $\neg, \rightarrow$ and $\forall$ in succedent. Given derivations are transformed as follows. The derivation

$$\frac{G, \Gamma \Rightarrow \Delta}{\Gamma \Rightarrow \Delta, \neg G}$$

goes to

$$\frac{\dfrac{P_G(\mathbf{t}), \Gamma \Rightarrow \Delta \quad P_{\neg G}(\mathbf{t}) \Rightarrow P_{\neg G}(\mathbf{t})}{P_G(\mathbf{t}) \vee P_{\neg G}(\mathbf{t}), \Gamma \Rightarrow \Delta, P_{\neg G}(\mathbf{t})}}{\text{DEF}_d, \Gamma \Rightarrow \Delta, P_{\neg G}(\mathbf{t})} \vee \Rightarrow$$

The derivation

$$\frac{G, \Gamma \Rightarrow \Delta, H}{\Gamma \Rightarrow \Delta, G \rightarrow H}$$

goes to

$$\frac{\dfrac{P_G(\mathbf{t}_1), \Gamma \Rightarrow \Delta, P_H(\mathbf{t}_2) \quad \text{axioms}}{P_H(\mathbf{t}_2) \to P_{G \to H}(\mathbf{t}), P_G(\mathbf{t}_1) \lor P_{G \to H}(\mathbf{t}), \Gamma \Rightarrow \Delta, P_{G \to H}(\mathbf{t})}}{\text{DEF}_d, \Gamma \Rightarrow \Delta, P_{G \to H}(\mathbf{t})} \lor \Rightarrow, \to \Rightarrow$$

The derivation
$$\frac{\Gamma \to \Delta, G(b)}{\Gamma \to \Delta, \forall y G(y)}$$

goes to

$$\frac{\dfrac{\dfrac{\Gamma \to \Delta, P_G(\mathbf{t}, b) \quad P_{\forall y G(y)}(\mathbf{t}) \to P_{\forall y G(y)}(\mathbf{t})}{P_G(\mathbf{t}, b) \to P_{\forall y G(y)}(\mathbf{t}), \Gamma \to \Delta, P_{\forall y G(y)}(\mathbf{t})}}{\exists y (P_G(y, \mathbf{t}) \to P_{\forall y G(y)}(\mathbf{t})), \Gamma \to \Delta, P_{\forall y G(y)}(b, \mathbf{t})}}{\text{DEF}_d, \Gamma \to \Delta, P_{\forall y G(y)}(\mathbf{t})} \begin{array}{c} \to \Rightarrow \\ \exists \to \end{array}$$

This completes the proof of the first part of the theorem.

The second part follows from *classical* derivability of the results of substitution P_G/G into formulas in DEF_d.

For the third part, if $\vdash^c \text{DEF}_d, \Pi \to \Phi$ then substitution P_G/G for $G \in \mathbf{f}(d)$ yields $\vdash^c \Pi \Rightarrow \Phi$, then (by Theorem 1) $\vdash^i \Pi \Rightarrow \Phi$ and hence $\vdash^c \text{DEF}_d, \Pi \Rightarrow \Phi$ completing the chain of equivalences. As pointed out in the Introduction, the transformation in Theorem 1 is not polynomial. ⊢

References

[1] T. Achourioti and M. van Lambalgen. A Formalization of Kant's Transcendental Logic. *The Review of Symbolic Logic*, 4:254–289, 2011.

[2] H. B. Curry. *Foundations of Mathematical Logic*. Dover Publications, New York, 1977.

[3] R. Goldblatt. *Topoi. The Categorial Analysis of Logic*. Elsevier Science Publishers, Amsterdam, 1984.

[4] G. Mints. *A Short Introduction to Intuitionistic Logic*. Kluwer Academic Publishers, New York, 2000.

[5] V. P. Orevkov. Glivenko's sequence classes. In V. P. Orevkov, editor, *Trudy Mat. Inst. Steklov: Vol. 98. Logical and logical-mathematical calculus. Part 1*, pages 131–154. Nauka, Leningrad, 1968.

Commentary on Grigori Mints' "Classical and Intuitionistic Geometric Logic"

Roy Dyckhoff
University of St Andrews, Scotland
roy.dyckhoff@gmail.com

Sara Negri
University of Helsinki, Finland
sara.negri@helsinki.fi

This paper studies a *Glivenko sequent class*, i.e. a class of sequents where classical derivability entails intuitionistic derivability; more specifically, the paper is about "geometric sequents". The main old result in this topic is a direct consequence [11] of Barr's theorem[1]. As background, Mints sketches an old deductive proof (from [7]) and an old model-theoretic proof, as in Exercise 2.6.14 of [10]; but, his interest being in complexity of proof transformations, he gives a third proof, of a result both more and less general.

A modern reconstruction [6] of Orevkov's proof [7, Theorem 4.1, part (1)] relies on what we would now call the "cut-free **G3c** calculus" [9], in which *Cut* and other structural rules are admissible and all the logical rules are invertible (indeed, height-preserving invertible). His result is that the list (or "σ-class") $[\to^+, \neg^+, \forall^+]$ is a "completely Glivenko class"; in other words, he shows that if a sequent with a single succedent has no positive occurrences of $\to$, $\neg$ or $\forall$ then its classical derivability implies its intuitionistic derivability. In modern terminology, this means just that if a sequent $\Gamma \Rightarrow A$ (where Γ consists of geometric implications and A is a positive formula) is derivable in cut-free **G3c**, then it is already derivable in the intuitionistic calculus **m-G3i** (also from [9]). The proof method actually shows the stronger result, that the cut-free **G3c** derivation is already a **m-G3i** derivation. The weaker result extends to the case where A is a geometric implication by using the invertibility in cut-free **G3c** of the succedent rules for the three mentioned connectives. Other work,

---

[1] "Let $\mathcal{E}$ be a Grothendieck topos. Then there is a complete Boolean algebra $\mathcal{B}$ and an exact cotripleable functor $\mathcal{E} \to \mathcal{FB}$", $\mathcal{FB}$ being the topos of sheaves over $\mathcal{B}$ [1].

such as [4], related to the deductive proof of this result, is cited in the bibliographies of [5] and [6]. The usefulness of cut-free **G3** calculi in the study of Glivenko classes has been further demonstrated in [6], with direct proofs of generalisations of results in [7].

Mints' interest, however, in this paper is in derivations in **G3c** with *Cut*. One can apply standard cut-elimination transformations, and then those corresponding to the inversions; but this leads to a "super-exponential blow-up", as can be seen in a similar context in [9, Section 5.2]. How can this be avoided? One solution is just to start with a cut-free derivation. One can go even further, using the cut-free calculi introduced in [5], where the axioms Γ are replaced by inference rules: this avoids proof transformations entirely (since, in such calculi, classical proofs of a geometric implication A are already intuitionistic proofs). But, Mints would insist that **G3c** with *Cut* is a traditional (i.e. respectable) starting point.

The question then arises: can the transformation be changed so that there is an at most polynomial expansion of the derivation? Clearly it should not begin with cut elimination, so a trick is needed to handle instances of the *Cut* rule rather than eliminating them. The trick is attributed to Orevkov [7]; one might also attribute it to Skolem, who pioneered in [8] the use of what [2] should have called "relational Skolemisation", i.e. the replacement, by introduction of new relation symbols, of complex formulae by atomic formulae. When this is sufficiently thorough to ensure that every formula is equivalent to an atomic formula, it is called "atomisation" or "Morleyisation"; this paper doesn't go so far.

The novel result of this paper is now the result (both weaker and stronger) that, if d is a classical proof of a geometric sequent, then it can be polynomially transformed into an intuitionistic proof of the sequent conservatively extended by extra antecedent formulae that are geometric implications. These extra implications are generated by relational Skolemisation of the subformulae of the cut formulae in d. The result is weaker by virtue of having these extra implications; it is stronger by virtue of the complexity reduction.

There are the following points at which the paper is incorrect:

1. Mints' (9) should be $\forall \mathbf{x}(P_{\exists y G}(\mathbf{x}) \to \exists y P_G(\mathbf{x}, y))$ rather than $\forall \mathbf{x}(P_{\exists y P_G}(\mathbf{x}) \to \exists y P_G(\mathbf{x}, y))$;

2. His (10) should be $\forall \mathbf{x} \forall y (P_G(\mathbf{x}, y) \to P_{\exists y G}(\mathbf{x}))$ rather than $\forall \mathbf{x} \forall y (P_G(\mathbf{x}, y) \to P_{\exists y P_G}(\mathbf{x}))$;

3. His (11) should be $\forall \mathbf{x} \forall y (P_{\forall y G}(\mathbf{x}) \to P_G(\mathbf{x}, y))$ rather than $\forall \mathbf{x} \forall y (P_{\forall y P_G}(\mathbf{x}) \to P_G(\mathbf{x}, y))$;

4. His (12) should be $\forall \mathbf{x}(\forall y P_G(\mathbf{x}, y) \to P_{\forall y G}(\mathbf{x}))$ rather than $\forall \mathbf{x}(\forall y P_G(\mathbf{x}, y) \to P_{\forall y P_G}(\mathbf{x}))$;

5. His (19) (replacing (16)) is not a geometric implication;

6. His (17) (replacing (12)) is not a geometric implication.

The first four of these problems are minor: note that in Mints' (9) the suffix $\exists y P_G$ is not a subformula of one of the cut-formulae, and similarly for (10), (11) and (12). The penultimate problem can be fixed by distributing $\forall \mathbf{x}$ across the conjunction, thus obtaining two geometric implications: $\forall \mathbf{x}(P_H(\mathbf{z}) \to P_{G \to H}(\mathbf{x}))$ and $\forall \mathbf{x}(P_G(\mathbf{y}) \lor P_{G \to H}(\mathbf{x}))$. [It has already been made clear that $\mathbf{y}$ and $\mathbf{z}$ are subsets of the set $\mathbf{x}$ of variables.]

The final problem is not so easily fixed: the paper wrongly claims that the formula $\forall \mathbf{x} \exists y (P_G(\mathbf{x}, y) \to P_{\forall y P_G}(\mathbf{x}))$ is a geometric implication. This is not fixed by changing (12) (as proposed above) to $\forall \mathbf{x}(\forall y P_G(\mathbf{x}, y) \to P_{\forall y G}(\mathbf{x}))$ and then obtaining $\forall \mathbf{x} \exists y (P_G(\mathbf{x}, y) \to P_{\forall y G}(\mathbf{x}))$; this is still not geometric, because of the implication within the scope of the existential quantifier.

A partial solution may be had by changing this formula to the geometric implication

$$\forall \mathbf{x}(\exists y P_{\neg G}(\mathbf{x}, y) \lor P_{\forall y G}(\mathbf{x})) \tag{17}$$

but this introduces a new relation symbol $P_{\neg G}$, where $\neg G$ may not be a subformula of one of the cut formulae. To fix this problem, the relational Skolemisation needs to be applied not just to all such subformulae but also to **all** their negations.

With these changes, the application of the extra formulae (i.e. members of DEF_d) to deal with the special formulae of the derivation is unchanged for implication. We show (for example) the effects of improving (9) on the treatment of an antecedent $\exists$-inference and of correcting the treatment of universal quantification.

The improved version of (9) is $\forall \mathbf{x}(P_{\exists y G}(\mathbf{x}) \to \exists y P_G(\mathbf{x}, y))$. The step

$$\frac{G(b), \Gamma \Rightarrow \Delta}{\exists y G(y), \Gamma \Rightarrow \Delta}$$

is transformed to

$$\frac{\dfrac{P_{G(y)}(\mathbf{t}, b), \text{DEF}_d, \Gamma \Rightarrow \Delta}{\exists y P_{G(y)}(\mathbf{t}, y)), \text{DEF}_d, \Gamma \Rightarrow \Delta}}{\text{DEF}_d, P_{\exists y G(y)}(\mathbf{t}), \Gamma \Rightarrow \Delta} \exists \Rightarrow.$$

Using the improved version of (17), the step

$$\frac{\Gamma \Rightarrow \Delta, G(\mathbf{t}, b)}{\Gamma \Rightarrow \Delta, \forall y G(\mathbf{t}, y)}$$

1237

is transformed (with some implicit weakenings to save space and aid readability) to

$$
\cfrac{
  \cfrac{
    \cfrac{
      \cfrac{
        \cfrac{
          \cfrac{\text{DEF}_d, \Gamma \Rightarrow \Delta, P_{G(\mathbf{x},y)}(\mathbf{t},b)}{P_{\neg G(\mathbf{x},y)}(\mathbf{t},b), \text{DEF}_d, \Gamma \Rightarrow \Delta, P_{G(\mathbf{x},y)}(\mathbf{t},b)}\, Wkn
          \quad \Rightarrow\wedge, axiom
        }{P_{\neg G(\mathbf{x},y)}(\mathbf{t},b), \text{DEF}_d, \Gamma \Rightarrow \Delta, P_{\neg G(\mathbf{x},y)}(\mathbf{t},b) \wedge P_{G(\mathbf{x},y)}(\mathbf{t},b)}
        }{P_{\neg G(\mathbf{x},y)}(\mathbf{t},b), \neg(P_{\neg G(\mathbf{x},y)}(\mathbf{t},b) \wedge P_{G(\mathbf{x},y)}(\mathbf{t},b)), \text{DEF}_d, \Gamma \Rightarrow \Delta}\, \neg\Rightarrow
      }{P_{\neg G(\mathbf{x},y)}(\mathbf{t},b), \text{DEF}_d, \Gamma \Rightarrow \Delta}
    }{\exists y P_{\neg G(\mathbf{x},y)}(\mathbf{t},y), \text{DEF}_d, \Gamma \Rightarrow \Delta}\, \exists\Rightarrow
    \qquad
    \cfrac{}{P_{\forall y G(\mathbf{x},y)}(\mathbf{t}), \Gamma \Rightarrow \Delta, P_{\forall y G(\mathbf{x},y)}(\mathbf{t})}\, axiom
  }{\exists y P_{\neg G(\mathbf{x},y)}(\mathbf{t},y) \vee P_{\forall y G(\mathbf{x},y)}(\mathbf{t}), \text{DEF}_d, \Gamma \Rightarrow \Delta, P_{\forall y G(\mathbf{x},y)}(\mathbf{t})}\, \vee\Rightarrow
}{\text{DEF}_d, \Gamma \Rightarrow \Delta, P_{\forall y G(\mathbf{x},y)}(\mathbf{t})}
$$

Note the importance of having $P_{\forall y G(\mathbf{x},y)}(\mathbf{t})$ (rather than, from the succedent of the old (17), Mints' $P_{\forall y P_G}(\mathbf{t})$) in the antecedent of the lowest axiom step. It is not the case that $\forall y P_{G(\mathbf{x},y)}$ (i.e. Mints' $\forall y P_G$) is a subformula of one of the cut formulae; the presence of the fresh predicate symbol $P_{G(\mathbf{x},y)}$ forbids this.

Note also the use of the Weakening rule Wkn; either this rule should be included in the **m-G3i** calculus or the admissibility of the rule exploited once the derivation has been fully transformed.

References

[1] M. Barr. Toposes without points. *J. Pure and Applied Algebra*, 5:265–280, 1974.

[2] R. Dyckhoff and S. Negri. Geometrisation of first-order logic. *Bulletin of Symbolic Logic*, 21:123–163, 2015.

[3] J. E. Fenstad, editor. *T. Skolem, Selected Works in Logic*. Universitetsforlaget, Oslo, 1970.

[4] G. Nadathur. Correspondence between classical, intuitionistic and uniform provability. *Theoretical Computer Science*, 232:273–298, 2000.

[5] S. Negri. Contraction-free sequent calculi for geometric theories, with an application to Barr's theorem. *Archive for Mathematical Logic*, 42:389–401, 2003.

[6] S. Negri. Glivenko sequent classes in the light of structural proof theory. *Archive for Mathematical Logic*, 55:461–473, 2016.

[7] V. P. Orevkov. Glivenko's sequence classes. In V. P. Orevkov, editor, *Trudy Mat. Inst. Steklov: Vol. 98. Logical and logical-mathematical calculus. Part 1*, pages 131–154. Nauka, Leningrad, 1968.

[8] T. Skolem. Logisch-kombinatorische Untersuchungen über die Erfüllbarkeit oder Beweisbarkeit mathematischer Sätze nebst einem Theoreme über dichte Mengen. In *Skrifter utgit av Videnskapsselskapet i Kristiania. I, Matematisk-naturvidenskabelig Klasse*, volume 1920 bd. 1, No 4. Kristiania : I Kommission hos J. Dybwad, 1920. http://www.biodiversitylibrary.org/item/52015#page/111/mode/1up. Also reprinted in [3, pages 103–136].

[9] A. S. Troelstra and H. Schwichtenberg. *Basic Proof Theory*. CUP, 2nd edition, 2000.

[10] A. S. Troelstra and D. van Dalen. *Constructivism in Mathematics (vol 1)*. North Holland, 1988.

[11] G. Wraith. Intuitionistic algebra: some recent developments in topos theory. In *Proceedings of International Congress of Mathematics*, pages 331–337, Helsinki, 1978.

Received 6 October 2016

Proof Theory for Non-normal Modal Logics: The Neighbourhood Formalism and Basic Results

Sara Negri
University of Helsinki, Finland
sara.negri@helsinki.fi

1 Introduction

The advent of Kripke semantics marked a decisive turning point for philosophical logic: earlier axiomatic studies of modal concepts were replaced by a solid semantic method that displayed the connections between modal axioms and conditions on the accessibility relation between possible worlds. However, the success of the semantic method was not followed by equally powerful syntactic theories of modal and conditional concepts and reasoning: Concerning the former, the situation was depicted by Melvin Fitting in his survey in the *Handbook of Modal Logic* [7] as: "No proof procedure suffices for every normal modal logic determined by a class of frames"; In the chapter on tableau systems for conditional logics, Graham Priest stated that "there are presently no known tableau systems of the kind used in this book for *S*" (Lewis' logic for counterfactuals) ([40], p. 93).

The insufficiency of traditional Gentzen systems to meet the challenge of the development of a proof theory for modal and non-classical logic has led to alternative formalisms which, in one way or another, extend the syntax of sequent calculus.

Parts of the results of this paper were presented in workshops and conferences whose respective organisers are gratefully acknowledged: Proof theory of modal and non-classical logics, Helsinki, August 2015 (Giovanna Corsi), Workshop Trends in Proof Theory, Hamburg, September 2015 (Stefania Centrone), Estonian-Finnish Logic Meeting, Rakvere, November 2015 (Tarmo Uustalu). The paper was completed during a stay at the University of Verona, within the "Programma di Internazionalizzazione di Ateneo, Anno 2015, Azione 3, Cooperint." Discussions with my host Peter Schuster have been very useful. Finally, detailed and insightful comments by two referees have contributed to improving the paper.

There have been two main lines of development, one that enriches the structure of sequents (display calculi, hypersequents, nested sequents, tree-hypersequents, deep inference), another that maintains their simple structure but adds *labels*, thus internalizing the possible worlds semantics within the proof system. In particular, for the proof theory of conditional logics there have been several contributions in the literature from both approaches [1, 10, 12, 20, 30, 34–36, 44].[1].

In his work of 1997, Grisha Mints has been among the forerunners[2] of the latter approach to the sequent calculus proof theory of modal logic.[3] In [23], he showed how one can obtain sequent calculi for normal modal logics with any combination of *reflexivity*, *transitivity*, and *symmetry* in their Kripke frames. Possible worlds were represented as prefixes, in fact, finite sequences of natural numbers, with the properties of the accessibility relations of a Kripke frame implicit in the management of prefixes in the logical rules. By this approach, it was possible to give a proof of cut elimination that can be considered as a formalization of Kripke's original completeness proof.

By making explicit the accessibility relation and by using variables, rather than sequences for possible worlds, it is possible to capture a much wider range of modal logics, in particular those characterised by *geometric frame conditions*, with properties such as seriality or directness of the accessibility relation; by using the conversion of geometric implications into rules that extend sequent calculus in a way that maintains the admissibility of structural rules [24], it has been possible to obtain a uniform presentation of a large family of modal logics, including provability logic, with modular proofs of their structural properties [25] and direct semantic completeness proofs [26].

Later, this labelled sequent calculus approach to the proof theory of modal logic has been extended to wider frame classes [27], and in further work it has been shown how the method can capture *any* logic characterized by first-order frame conditions in its relational semantics [5]; the reason is that arbitrary first-order theories can be given an analytic treatment through the extension of G3-style sequent calculi with geometric rules. Notably, in these calculi, all the rules are invertible and a strong form of completeness holds, with a simultaneous construction of formal proofs, for derivable sequents, or countermodels, for underivable ones, as shown in [28].

Despite their wide range of applications, the powerful methods of Kripke semantics are not a universal tool in the analysis of philosophical logics: they impose the straitjacket of *normality*, i.e., validity of the rule of necessitation, from $\vdash A$ to infer

---

[1] See the conclusion of [34] for a discussion and comparison of these different formalisms.

[2] See also the extensive studies of labelled systems for modal logics and non-classical logics in A. Simpson's PhD dissertation from 1994 [42] and L. Viganò's monograph from 2000 [45].

[3] Labelled tableaux, on the other hand, were developed since the early 1970's, cf. [6].

⊢ $\Box A$, and of the K axiom, $\Box(A \supset B) \supset (\Box A \supset \Box B)$. The limitative character of these imposed validities becomes clear in epistemic logic: with the epistemic reading of the modality, an agent knows A if A holds in all the epistemic states available to her, and then the normality properties yield that (1) *whatever has been proved is known* and that (2) *an agent knows all the logical consequences of what she knows*. This leads to *logical omniscience*, clearly inadequate for cognitive agents with human capabilities, and thus to the rejection of both requirements. The same limitation is clear in the interpretation of the modality as a likelihood operator where one sees that the normal modal logic validity $\Box A \& \Box B \supset \Box(A \& B)$ should be avoided.

Another limitation in systems based on a Kripke-style semantics is that the propositional base is classical or intuitionistic logic. In both cases, one is forced to an implication which has been shown since the analysis of C.I. Lewis to be an inadequate form of conditional if a logical analysis is to be pursued in other venues than mathematics: the classical propositional base of modal logic is insufficient to treat conditionals beyond material or strict implication, as shown by David Lewis' path-breaking book *Counterfactuals* [22], and intuitionistic implication shares many of the undesired properties of (classical) material implication.

The early literature on the semantics of conditional logic started with an attempt, in the work of Stalnaker, to reduce the reading of the conditional to a standard possible worlds semantics through the notion of *limit* and *selection functions* [43]. This approach has been criticized as inadequate in many cases: first, the aforementioned limit might not exist (as shown by Lewis in [22]), second, it can be too difficult to achieve and so cannot be taken as a standard basis for a formalization (as in the perfectly moral life of deontic systems), third, it could be impossible to define as in situations with more than one ordering, or, more concretely, conflicting obligations. The inadequacy of a normal modal base as a general framework for modal logic has also been shown in the case of the modal formalization of deontic notions by a series of paradoxes, such as the paradox of the gentle murder [8], that were used in a revisionist way to motivate non-normal modal logics.

The more general *neighbourhood semantics* was introduced in the 1970's to provide a uniform semantic framework for philosophical logics that cannot be accommodated within the setting of normal modal logic. Instead of an accessibility relation on a set of possible worlds, one has for every possible world a family of neighbourhoods, i.e., a collection of some special subsets of the set of possible worlds.

As is usual when a new semantics is introduced, its relationship with the earlier one is investigated. In this case, one can prove that there is a precise link between neighbourhood and relational semantics, in the sense that there is a way to define a neighbourhood frame from a given relational frame, and conversely, a relational frame from a neighbourhood frame. Given a relational frame (W, R), one can define a neighbourhood frame by taking as neighbourhoods of a world w the supersets of the set of worlds accessible from w. Conversely, given a neighbourhood frame (W, I) one can define a relational frame by identfying the worlds accessible from w as the intersection of all the neighbourhoods of w. Neighbourhood frames are more general than relational frames, and in fact the correspondence is a bijection over a certain class of neighbourhood frames called *augmented* ones, those that contains the intersection of all their members and are closed under supersets (cf. [38] for details). The correspondence between relational and neighbourhood frames can be seen also as a way to transfer an intuitive explanation from one semantics to the other: roughly, worlds in a neighbourhood of w replace worlds accessible from w, and correspondingly the intuition on what it means to be an element of a neighbourhood or of the intersection of all the neighbourhoods of a worlds will depend on the kind of modality or conditional that is being modelled; the intuition thus varies from the properties of indistinguishability of worlds as epistemic states to that of plausibility of worlds as factual scenarios.

Among non-normal modal logics, *classical* modal logics are those obtained by requiring that the modality respects logical equivalence, that is, closure under the rule $\frac{A \supset B}{\Box A \supset \Box B}$. One can then obtain other systems below the normal modal logic **K** by removing the normality axiom and the necessitation rule and adding the axiom schemas M, C, N and their combinations. A lattice of eight different logics is obtained (cf. the diagram on p. 237 of [4]). On the logical side, it has been shown by Gasquet and Herzig [9] and Kracht and Wolter [18] that non-normal modal logics can be simulated through an appropriate translation by a normal modal logic with three modalities. This translation has been used by Gilbert and Maffezioli [11] to define modular labelled sequent calculi for the basic classical modal logics. Since the frame conditions considered go beyond the geometric class, *systems of rules* (in the sense of [27]) have been used.

Our goal is to set the grounds for a proof theory of non-normal modal systems based on neighbourhood semantics, to achieve this directly, i.e., without the use of translations, with local rules, and in a modular way, open to extensions in various interweaving directions[4].

The goal will be accomplished through the guidelines of *inferentialism*, that is, by starting from the meaning explanation of logical constants and by converting it into well-behaved rules of a calculus, as detailed in [29].

The paper is organized as follows: In Section 2, after having recalled the basic definitions of neighbourhood semantics, we show how it naturally gives rise to the

---

[4]Such flexibility is already witnessed by developments of the labelled proof theory based on neighbourhood semantics for preferential conditional logic and conditional doxastic logic in [13, 30].

four distinct modalities [ ], ⟨], [⟩, and ⟨ ⟩; the nesting of quantifiers in their semantic explanation is factorized with the help of *local* forcing relations, i.e., relations between (formal) neighbourhoods and formulas. Correspondingly, we have sequent calculus rules for such relations and for the modalities defined upon them. We then show how the basic calculus so obtained can be used to find the rules that correspond to additional properties of the neighbourhoods and the relations between such properties and the normality conditions for the ⟨] modality. It is also shown how the rules obtained validate ⟨](A&B) ⊃ ⟨]A & ⟨]B, and how a modified forcing condition for the modality gives a more general explanation. The link between the two is given by the operation of supplementation in minimal models. In the determination of the rules of the systems we use a sort of 'bootstrapping' procedure, as we use the basic rules of the calculus to find other rules. We also assume at this early stage of the construction of the proof system that the structural properties are available, even if such properties are necessarily proved further on, when all the rules have been determined. In Section 3 we apply the methodology to classical modal logics and generate labelled G3-style sequent calculi for them. The structural properties, height-preserving invertibility of all the rules, height-preserving admissibility of weakening and contraction and admissibility of cut are proved in Section 4. In Section 5, we give a direct proof of completeness for these systems with respect to neighbourhood models as well as an indirect completeness proof via the axiomatic systems. Finally, in the conclusion, our approach to the proof theory of non-normal modal logic is related to other approaches in the literature.

2 The general framework

A *neighbourhood frame* is a pair $\mathcal{F} \equiv (W, I)$, where W is a set of worlds (states) and I is a neighbourhood function

$$I : W \longrightarrow \mathcal{P}(\mathcal{P}(W))$$

that assigns a collection of sets of worlds to each world in W. A *neighbourhood model* is then a pair $\mathcal{M} \equiv (\mathcal{F}, \mathcal{V})$, where $\mathcal{F}$ is a neighbourhood frame and $\mathcal{V}$ a propositional valuation, i.e., a map $\mathcal{V} : \mathtt{Atm} \longrightarrow \mathcal{P}(W)$ from atomic formulas to sets of possible worlds.

Worlds in a neighbourhood are the substitute, in this more general semantics, of accessible worlds. The inductive clauses for truth of a formula in a model are the usual ones for the propositional clauses; for the modal operator we have

$$\mathcal{M}, w \Vdash \Box A \equiv ext(A) \text{ is in } I(w),$$

where $ext(A) \equiv \{u \in W | \mathcal{M}, u \Vdash A\}$.[5]

Starting from the standard forcing relation between possible worlds and formulas, we extend the standard labelled language to a multi-sorted labelled language, with labels for worlds and neighbourhoods, and define two *local*, rather than pointwise, forcing relations, $\Vdash^{\exists}$ and $\Vdash^{\forall}$. These forcing relations are local because unlike the usual forcing of a formula A at a world x, they are relations between elements a of a system of neighbourhoods, that is, sets of subsets of possible worlds, and formulas. The subset a thus ranges in a family of neighbourhoods $I(x)$, which is supposed to be given for every world x. The first relation corresponds to the existence, in the neighbourhood, of a world that forces the formula, the second to the forcing for every world in the neighbourhood; here A is a formula of the propositional modal language (as we shall see below, we shall actually consider an extension of the standard propositional language with four modalities naturally arising from the semantics):

$a \Vdash^{\exists} A$ *is true* iff *there is some world x in a such that $x \Vdash A$*

$a \Vdash^{\forall} A$ *is true* iff *for any world x in a, $x \Vdash A$.*

The standard forcing relation can be then obtained as a special case of both existential and universal forcing through singleton sets (under the condition that they belong to the family of neighbourhoods):

$\{x\} \Vdash^{\exists} A$ iff $\{x\} \Vdash^{\forall} A$ iff $x \Vdash A$

Through the standard method of conversion of forcing clauses into sequent calculus rules [25, 32], we obtain the following rules for the local forcing relations; observe that the language of standard labelled systems is extended by the local forcing relations and has, in place of relational atoms, atoms of the form $x \in a, a \in I(x)$:

$$\frac{x \in a, \Gamma \Rightarrow \Delta, x : A}{\Gamma \Rightarrow \Delta, a \Vdash^{\forall} A} \; R \Vdash^{\forall}, x \, fresh \qquad \frac{x \in a, x : A, a \Vdash^{\forall} A, \Gamma \Rightarrow \Delta}{x \in a, a \Vdash^{\forall} A, \Gamma \Rightarrow \Delta} \; L \Vdash^{\forall}$$

$$\frac{x \in a, \Gamma \Rightarrow \Delta, x : A, a \Vdash^{\exists} A}{x \in a, \Gamma \Rightarrow \Delta, a \Vdash^{\exists} A} \; R \Vdash^{\exists} \qquad \frac{x \in a, x : A, \Gamma \Rightarrow \Delta}{a \Vdash^{\exists} A, \Gamma \Rightarrow \Delta} \; L \Vdash^{\exists}, x \, fresh$$

Table 1: Rules for local forcing

The use of neighbourhood semantics in place of the relational semantics gives a splitting of the standard alethic modalities into four modalities, [ ], ⟨ ⟩, [⟩, ⟨] [38], corresponding to the four different combinations of quantifiers in the semantic explanation:

---
[5]We observe that $ext(A)$ is also denoted by $[A]$.

$x \Vdash [\,]A$ iff *for every neighbourhood a of x, $a \Vdash^\forall A$*

$x \Vdash \langle\,]A$ iff *there is some neighbourhood a of x such that $a \Vdash^\forall A$*

$x \Vdash [\,\rangle A$ iff *for every neighbourhood a of x, $a \Vdash^\exists A$*

$x \Vdash \langle\,\rangle A$ iff *there is some neighbourhood a of x such that $a \Vdash^\exists A$*

The semantic clauses are translated into the following rules:

$$\dfrac{a \in I(x), \Gamma \Rightarrow \Delta, a \Vdash^\forall A}{\Gamma \Rightarrow \Delta, x : [\,]A}\ R[\,],\ a\ fresh \qquad \dfrac{a \in I(x), x : [\,]A, a \Vdash^\forall A, \Gamma \Rightarrow \Delta}{a \in I(x), x : [\,]A, \Gamma \Rightarrow \Delta}\ L[\,]$$

$$\dfrac{a \in I(x), \Gamma \Rightarrow \Delta, x : \langle\,]A, a \Vdash^\forall A}{a \in I(x), \Gamma \Rightarrow \Delta, x : \langle\,]A}\ R\langle\,] \qquad \dfrac{a \in I(x), a \Vdash^\forall A, \Gamma \Rightarrow \Delta}{x : \langle\,]A, \Gamma \Rightarrow \Delta}\ L\langle\,],\ a\ fresh$$

$$\dfrac{a \in I(x), \Gamma \Rightarrow \Delta, a \Vdash^\exists A}{\Gamma \Rightarrow \Delta, x : [\,\rangle A}\ R[\,\rangle,\ a\ fresh \qquad \dfrac{a \in I(x), x : [\,\rangle A, a \Vdash^\exists A, \Gamma \Rightarrow \Delta}{a \in I(x), x : [\,\rangle A, \Gamma \Rightarrow \Delta}\ L[\,\rangle$$

$$\dfrac{a \in I(x), \Gamma \Rightarrow \Delta, x : \langle\,\rangle A, a \Vdash^\exists A}{a \in I(x), \Gamma \Rightarrow \Delta, x : \langle\,\rangle A}\ R\langle\,\rangle \qquad \dfrac{a \in I(x), a \Vdash^\exists A, \Gamma \Rightarrow \Delta}{x : \langle\,\rangle A, \Gamma \Rightarrow \Delta}\ L\langle\,\rangle,\ a\ fresh$$

Table 2: Rules for alethic modalities

Finally, in a G3-style labelled calculus there are two types of initial sequents, those with labelled atomic formulas, from the basic propositional base, of the form $x : P$, $\Gamma \Rightarrow \Delta, x : P$ and those with relational atoms. For labelled calculi based on possible worlds semantics, the latter have the form $xRy, \Gamma \Rightarrow \Delta, xRy$. As observed in [25], such sequents are not needed because none of the rules of the calculus has active relational atoms on the right-hand side, so such initial sequents cannot have an active role in derivations and can thus be dispensed with. Here we have a similar situation, with the two potential types of relational initial sequents being $x \in a, \Gamma \Rightarrow \Delta, x \in a$ and $a \in I(x), \Gamma \Rightarrow \Delta, a \in I(x)$: none of the rules introduced so far has active formulas of the form $x \in a$ or $a \in I(x)$ in the right-hand side, so such initial sequents are not needed. But there is a *caveat*, and, as we shall see, once the assumption of monotonicity which is behind the determination of the above rules is relaxed, we'll have to include rules that have relational atoms of the form $x \in a$ on the right-hand side, and consequently, initial sequents for them.

The *basic calculus for neighbourhood semantics*, **G3n**, is obtained by adding the above rules of local forcing together with the rules for alethic modalities together with the needed relational initial sequents to the standard labelled **G3c** sequent calculus (the propositional part of the calculus **G3K** [25]). For ease of the reader, we give such rules in the table below, with the added relational initial sequents; they are in parentheses since they are needed in extensions but not for the basic system with the rules presented so far.

Initial sequents:

$$x:P, \Gamma \Rightarrow \Delta, x:P \qquad\qquad (x \in a, \Gamma \Rightarrow \Delta, x \in a)$$

Propositional rules:

$$\frac{x:A, x:B, \Gamma \Rightarrow \Delta}{x:A\&B, \Gamma \Rightarrow \Delta} L\& \qquad \frac{\Gamma \Rightarrow \Delta, x:A \quad \Gamma \Rightarrow \Delta, x:B}{\Gamma \Rightarrow \Delta, x:A\&B} R\&$$

$$\frac{x:A, \Gamma \Rightarrow \Delta \quad x:B, \Gamma \Rightarrow \Delta}{x:A \vee B, \Gamma \Rightarrow \Delta} L\vee \qquad \frac{\Gamma \Rightarrow \Delta, x:A, x:B}{\Gamma \Rightarrow \Delta, x:A \vee B} R\vee$$

$$\frac{\Gamma \Rightarrow \Delta, x:A \quad x:B, \Gamma \Rightarrow \Delta}{x:A \supset B, \Gamma \Rightarrow \Delta} L\supset \qquad \frac{x:A, \Gamma \Rightarrow \Delta, x:B}{\Gamma \Rightarrow \Delta, x:A \supset B} R\supset$$

$$\overline{x:\bot, \Gamma \Rightarrow \Delta} L\bot$$

Table 3: The propositional part of system G3n

As an example of the use of the system obtained, we show how to obtain a formal derivation of one of the sequents which gives the known dualities between the compound alethic modalities, namely $x : \langle\,]A \Rightarrow x : \neg[\,\rangle\neg A$ (here and elsewhere in the paper negation is not primitive, but defined through implication); by root-first application of the rules we find the following partial derivation:

$$\frac{\frac{\frac{\frac{\frac{\frac{a \in I(x), y \in A, a \Vdash^\forall A, y:A, x:[\,\rangle\neg A \Rightarrow x:\bot, y:A \quad x:\bot,\ldots \Rightarrow \ldots}{a \in I(x), y \in A, a \Vdash^\forall A, y:A, y:\neg A, x:[\,\rangle\neg A \Rightarrow x:\bot} L\supset}{a \in I(x), y \in A, a \Vdash^\forall A, y:\neg A, x:[\,\rangle\neg A \Rightarrow x:\bot} L\Vdash\forall}{a \in I(x), a \Vdash^\forall A, a \Vdash^\exists \neg A, x:[\,\rangle\neg A \Rightarrow x:\bot} L\Vdash\exists}{a \in I(x), a \Vdash^\forall A, x:[\,\rangle\neg A \Rightarrow x:\bot} L[\,\rangle}{x:\langle\,]A, x:[\,\rangle\neg A \Rightarrow x:\bot} L\langle\,]}{x:\langle\,]A \Rightarrow x:\neg[\,\rangle\neg A} R\supset$$

Derivability of the left topsequent follows from Lemma 4.2 below. In a similar way we obtain the other parts of the dualities between $\langle\,]$ and $[\,\rangle$ and between $\langle\,\rangle$ and $[\,]$, namely we have:

Proposition 2.1. *The following sequents are derivable in* **G3n**:

1. $x : \langle\,]A \Rightarrow x : \neg[\,\rangle\neg A$

2. $x : \neg[\,\rangle\neg A \Rightarrow x : \langle\,]A$

3. $x : \langle\,\rangle A \Rightarrow x : \neg[\,]\neg A$

4. $x : \neg[\,]\neg A \Rightarrow x : \langle\,\rangle A$

We proceed with finding the properties required of the family of neighbourhoods $I(x)$ to obtain a modality that satisfies the K axiom and necessitation[6] through application of the invertible rules of the calculus. Before doing so, we need a formal definition of inclusion between neighbourhoods. Unsurprisingly, inclusion between two neighbourhoods a, b is defined by

$$a \subseteq b \equiv \forall x(x \in a \supset x \in b)$$

with the sequent calculus rules

$$\frac{x \in a, \Gamma \Rightarrow \Delta, x \in b}{\Gamma \Rightarrow \Delta, a \subseteq b}\ R\subseteq,\ x\ fresh \qquad \frac{x \in b, x \in a, a \subseteq b, \Gamma \Rightarrow \Delta}{x \in a, a \subseteq b, \Gamma \Rightarrow \Delta}\ L\subseteq$$

Observe that to keep the notation simpler we use the same symbols ($\in, \subseteq$) both at the semantic and at the syntactic level.

Definition 2.2. *A family of neighbourhoods $I(x)$ is prebasic if for all $a, b \in I(x)$, there exists $c \in I(x)$ such that $c \subseteq a$ and $c \subseteq b$.*

If the definition of a prebasic family of neighbourhoods is not fully unfolded to the level of worlds but left at the level of inclusion between neighbourhoods, the property of being prebasic can be translated into sequent calculus rules that follow the geometric rule scheme:

$$\frac{a \in I(x), b \in I(x), c \in I(x), c \subseteq a, c \subseteq b, \Gamma \Rightarrow \Delta}{a \in I(x), b \in I(x), \Gamma \Rightarrow \Delta}\ Prebasic,\ c\ fresh$$

We have:

Lemma 2.3. *Suppose that for all x the family of neighbourhoods $I(x)$ is prebasic. Then $\langle\,](A \supset B) \supset (\langle\,]A \supset \langle\,]B)$ is valid with respect to the neighbourhood semantics.*

---

[6] Among the four modalities introduced above, the modality now in question is $\langle\,]$.

Proof. Validity is guaranteed by the following derivation in the labelled calculus

$$
\cfrac{\cfrac{\cfrac{\cfrac{\cfrac{\cfrac{\cfrac{...y:A\Rightarrow y:A... \quad ...y:B\Rightarrow ...y:B}{...y:A, y:A\supset B\Rightarrow ...y:B}\,L\supset}{y\in a, y\in b, y\in c, a\in I(x), b\in I(x), c\in I(x), c\subseteq a, c\subseteq b, a\Vdash^{\forall}A, b\Vdash^{\forall}A\supset B\Rightarrow x:\langle\,]B, y:B}\,\substack{L\Vdash^{\forall}\ (\text{twice}) \\ L\subseteq\ (\text{twice})}}{y\in c, a\in I(x), b\in I(x), c\in I(x), c\subseteq a, c\subseteq b, a\Vdash^{\forall}A, b\Vdash^{\forall}A\supset B\Rightarrow x:\langle\,]B, y:B}}{a\in I(x), b\in I(x), c\in I(x), c\subseteq a, c\subseteq b, a\Vdash^{\forall}A, b\Vdash^{\forall}A\supset B\Rightarrow x:\langle\,]B, c\Vdash^{\forall}B}\,R\Vdash^{\forall}}{a\in I(x), b\in I(x), c\in I(x), c\subseteq a, c\subseteq b, a\Vdash^{\forall}A, b\Vdash^{\forall}A\supset B\Rightarrow x:\langle\,]B}\,R\langle\,]}{a\in I(x), b\in I(x), a\Vdash^{\forall}A, b\Vdash^{\forall}A\supset B\Rightarrow x:\langle\,]B}\,Prebasic}{x:\langle\,]A, x:\langle\,](A\supset B)\Rightarrow x:\langle\,]B}\,L\langle\,]\ (\text{twice})}{x:\langle\,](A\supset B)\Rightarrow x:\langle\,]A\supset\langle\,]B}\,R\supset}{\Rightarrow x:\langle\,](A\supset B)\supset(\langle\,]A\supset\langle\,]B)}\,R\supset
$$

where the topsequents are derivable by Lemma 4.2. To conclude the proof one needs to show that the calculus **G3n** is sound with respect to neighbourhood semantics. This will be established as a general result in Theorem 5.3 below. QED

The condition of being prebasic is not only sufficient but also necessary to validate the normality axiom, in fact if rule *prebasic* is not available proof search is limited to the rules of **G3n** and we have:

Lemma 2.4. *Proof search for the K-axiom in the calculus* **G3n** *fails and from the failed proof search it is possible to construct a countermodel in the class of neighbourhood frames.*

Proof. We apply all the rules of **G3n** with conclusion that matches the sequent; we start from the sequent $a\in I(x), b\in I(x), a\Vdash^{\forall} A, b\Vdash^{\forall} A\supset B\Rightarrow x:\langle\,]B$ of the above proof search,[7] and obtain, through two applications resp. of $R\langle\,]$, $R\Vdash^{\forall}$ and $L\Vdash^{\exists}$, the sequent $a\in I(x), b\in I(x), y\in a, z\in b, y:A, z:A\supset B, a\Vdash^{\forall}A, b\Vdash^{\forall}A\supset B\Rightarrow x:\langle\,]B, y:B, z:B$; next a step of $L\supset$ gives a right derivable premiss (that contains both in the left-hand side and in the right-hand side the labelled formula $z:B$) and a left premiss $a\in I(x), b\in I(x), y\in a, z\in b, y:A, a\Vdash^{\forall}A, b\Vdash^{\forall}A\supset B\Rightarrow x:\langle\,]B, z:A, y:B, z:B$. This is not derivable and a countermodel is obtained by taking $I(x)$ to consist of the neighbourhoods a and b inhabited by (only) the worlds in the antecedent, i.e., $a\in I(x), b\in I(x), y\in a, z\in b$ with the forcing relations $y\Vdash A$, $z\nVdash A$, $y\nVdash B$, $z\nVdash B$. Clearly, $a\Vdash^{\forall}A, b\Vdash^{\forall}A\supset B$, but there is no neighbourhood of x that forces universally B. QED

Observe that the above doesn't exclude the possibility that the normality axiom would be derivable in other extensions of **G3n** since proof search in these calculi

---

[7]Since all the rules applied are invertible, this is not restrictive.

would be different, and the countermodel constructed here might not be in the class of frames for the stronger logic.

Next, we look for the property of $I(x)$ that characterizes validity of the rule of necessitation, i.e., the rule

$$\frac{\vdash x : A}{\vdash x : \langle\,]A}\ \text{Nec}$$

If we want to apply root-first the rules of **G3n** from the sequent $\Rightarrow x : \langle\,]A$, the only way to start is to enable the application of rule $R\langle\,]$ by assuming the existence of $a \in I(x)$, i.e., to assume the availability of the geometric rule (with the condition that a is a fresh neighbourhood label)

$$\frac{a \in I(x), \Gamma \Rightarrow \Delta}{\Gamma \Rightarrow \Delta}\ \textit{Nondeg}$$

This justifies the following definition:

Definition 2.5. *A family of neighbourhoods $I(x)$ is nondegenerate if $I(x)$ contains at least a neighbourhood.*

Lemma 2.6. *The rule of necessitation is admissible in the calculus* **G3n** *extended with rule* Nondeg.

Proof. We have the following

$$\cfrac{\cfrac{\cfrac{\cfrac{\cfrac{\Rightarrow x : A}{\Rightarrow y : A}\ \textit{hp-subst}}{a \in I(x), y \in a \Rightarrow x : \langle\,]A, y : A}\ LW, RW}{a \in I(x), \Rightarrow x : \langle\,]A, a \Vdash^\forall A}\ R\Vdash^\forall}{a \in I(x), \Rightarrow x : \langle\,]A}\ R\langle\,]}{\Rightarrow x : \langle\,]A}\ \textit{Nondeg}$$

Here we have used the admissible rules of height-preserving substitution and weakening (to be proved in Propositions 4.3, 4.4 below), hence the statement on admissibility rather than derivability. QED

Relation with minimal models: Neighbourhood models are also called *minimal* models in the literature.[8] Observe that the definition of forcing that we have given for the modality $\langle\,]$ validates

$$\langle\,](A\&B) \supset \langle\,]A \,\&\, \langle\,]B$$

---

[8]See Chapter 7 of Chellas (1980), in particular 7.1, for the definition of minimal models.

and therefore is not minimal in the sense that it automatically imposes some validities. This is avoided if the forcing is modified by requiring that the neighbourhood a not only is included, but *coincides* with the extension $ext(A)$ of A, i.e.,

$$x \Vdash^+ \langle\,]A \text{ iff there is } a \text{ in } I(x) \text{ such that } a \Vdash^\forall A \text{ and } ext(A) \subseteq a$$

Then the rules for $\langle\,]$ justified by the semantics of minimal models are as follows:

$$\frac{a \in I(x), \Gamma \Rightarrow \Delta, x : \langle\,]A, a \Vdash^\forall A \quad a \in I(x), \Gamma \Rightarrow \Delta, x : \langle\,]A, ext(A) \subseteq a}{a \in I(x), \Gamma \Rightarrow \Delta, x : \langle\,]A} \; R\langle\,]'$$

$$\frac{a \in I(x), a \Vdash^\forall A, ext(A) \subseteq a, \Gamma \Rightarrow \Delta}{x : \langle\,]A, \Gamma \Rightarrow \Delta} \; L\langle\,]', \; a \text{ fresh}$$

together with the rules for inclusion and the obvious rules for $ext(A)$, namely[9]

$$\frac{y : A, \Gamma \Rightarrow \Delta}{y \in ext(A), \Gamma \Rightarrow \Delta} \qquad \frac{\Gamma \Rightarrow \Delta, y : A}{\Gamma \Rightarrow \Delta, y \in ext(A)}$$

It is easy to show that with these rules the sequent $\Rightarrow x : \langle\,](A\&B) \supset \langle\,]A \& \langle\,]B$ is not derivable, the reason being that from $ext(A\&B) \subseteq a$ we cannot infer $ext(A) \subseteq a$ and $ext(B) \subseteq a$.

There is however a precise link between the two forcing conditions. We first recall a definition:[10]

Definition 2.7. *The* supplementation *of a neighbourhood model* $\mathcal{M} \equiv (W, I, \Vdash)$ *is the neighbourhood model* $\mathcal{M}^+ \equiv (W, I^+, \Vdash)$ *obtained by taking the superset closure of $I(x)$ for each x in W, i.e., $a \in I(x)^+$ if and only if $a \supseteq b$ for some $b \in I(x)$.*

We also recall the following:[11]

Proposition 2.8. *For all formula A we have*

$$\mathcal{M}, x \Vdash \langle\,]A \text{ if and only if } \mathcal{M}^+, x \Vdash^+ \langle\,]A$$

Proof. In one direction, if there is a in $I(x)$ such that $a \Vdash^\forall A$, i.e., $a \subseteq ext(A)$, then $ext(A)$ is in $I(x)^+$ (and $ext(A) = ext(A)$). For the converse, if there is a in $I(x)^+$ such that $a = ext(A)$, then $b \subseteq ext(A)$ for some b in $I(x)$. QED

---

[9]The rules in terms of ext are intuitively semantically motivated. We shall give below an alternative, more concise, version of the rules in which the inclusion $ext(A) \subseteq a$ is replaced by a binary predicate $A \triangleleft a$ with its own rules which do not require separate rules for inclusion.

[10]This is Definition 7.6 in Chellas (1980).

[11]This is essentially exercise 7.25 (b) in Chellas's book.

3 Classical and other non-normal modal logics

Classical modal logics[12] are non-normal modal logics obtained as extensions of classical propositional logic (**CL**) that contain the schema

$$\Diamond A \supset\subset \neg \Box \neg A$$

and the rule of inference

$$\frac{A \supset\subset B}{\Box A \supset\subset \Box B} \; RE$$

System **E** is the smallest classical system thus obtained. Other classical modal logics are obtained as extensions of **E**. Extensions containing the rule

$$\frac{A \supset B}{\Box A \supset \Box B} \; RM$$

are called *monotonic* logics and the smallest such system is denoted by **M**; extensions containing the rule

$$\frac{A \& B \supset C}{\Box A \& \Box B \supset \Box C} \; RR$$

are called *regular*, and the smallest such system is denoted by **C**.

It is well know (and easily provable) that every normal system is regular, every regular system is monotonic, and every monotonic system is classical.

It can be convenient to give a characterization of extensions **E** through axiom schemas. Among such extension, of particular interest are those obtained by the addition of any combination of the following:

(M) $\Box(A\&B) \supset \Box A \& \Box B$

(C) $\Box A \& \Box B \supset \Box(A\&B)$

(N) $\Box\top$

We recall from Chellas (1980, ch. 8):

Proposition 3.1. *Let Σ be an extension of* **E**. *Then*

1. *Σ is monotonic iff it contains the axiom schema* M.

---

[12] See ch. 8 of Chellas (1980) for a thorough treatment of classical, monotonic and regular modal logics in an axiomatic setting.

2. Σ is regular iff it contains the axiom schema C and is closed under RM.

3. Σ is regular iff it contains the axiom schemas C and M.

4. Σ is normal iff it is regular and contains the axiom schema N.

These logics are denoted with $\mathbf{ES}_1\ldots\mathbf{S}_n$ or simply $\mathbf{S}_1\ldots\mathbf{S}_n$, where $\mathbf{S}_1,\ldots,\mathbf{S}_n$ are the axiom/rule schemas added to system $\mathbf{E}$. With this notation we have $\mathbf{K} = \mathbf{RN} = \mathbf{MCN} = \mathbf{EMCN}$.

We recall that the forcing clause for the alethic modality in neighbourhood semantics is as follows:

$$x \Vdash \Box A \equiv \exists a \in I(x)(a \Vdash^\forall A \,\&\, \forall y(y \Vdash A \supset y \in a))$$

or equivalently

$$x \Vdash \Box A \equiv \exists a \in I(x)(a \Vdash^\forall A \,\&\, ext(A) \subseteq a)$$

The semantic clause is not one of the form that can be directly translated into geometric rules, but we proceed in a way similar to Skolem's definitional extension ([41], see also Section 2 of [5]) and add a new predicate $A \triangleleft a$ for $\forall y(y \Vdash A \supset y \in a)$ together with its definition. The definition is in turn formulated in terms of rules to be added to the calculus. When all the requirement to obtain a calculus with the desired properties are taken care of, the rules are as follows:[13]

$$\frac{y \in a, A \triangleleft a, y : A, \Gamma \Rightarrow \Delta}{A \triangleleft a, y : A, \Gamma \Rightarrow \Delta}\, L\triangleleft \qquad \frac{y : A, \Gamma \Rightarrow \Delta, y \in a}{\Gamma \Rightarrow \Delta, A \triangleleft a}\, R\triangleleft,\, y\, fresh$$

$$\frac{a \in I(x), a \Vdash^\forall A, A \triangleleft a, \Gamma \Rightarrow \Delta}{x : \Box A, \Gamma \Rightarrow \Delta}\, L\Box,\, a\, fresh$$

$$\frac{a \in I(x), \Gamma \Rightarrow \Delta, x : \Box A, a \Vdash^\forall A \quad a \in I(x), \Gamma \Rightarrow \Delta, x : \Box A, A \triangleleft a}{a \in I(x), \Gamma \Rightarrow \Delta, x : \Box A}\, R\Box$$

Table 4: **Modal rules of system E**

---

[13] See [29] for details making explicit a procedure used to obtain such sequent rules starting from the meaning explanation in terms of neighbourhood semantics.

The complete G3-system for **E** is obtained by adding the above rules to the rules for $\Vdash^\forall$ of table 1 and the rules for the propositional part of **G3n** of table 3, including the initial sequents of the form $x \in a, \Gamma \Rightarrow \Delta, x \in a$.[14] We shall denote with **G3E** the resulting system.

In the proofs that follow we use admissibility of the structural rules, that will be proved in Section 4.

Lemma 3.2. *The rule*

$$\frac{x : A, \Gamma \Rightarrow \Delta, x : B}{a \Vdash^\forall A, \Gamma \Rightarrow \Delta, a \Vdash^\forall B} \ (x \notin \Gamma, \Delta)$$

is admissible in **G3E**.

Proof. By admissibility of weakening and steps of $L \Vdash^\forall$ and $R \Vdash^\forall$. QED

Neither rule RE nor a labelled version of the rule has to be added as a rule of **G3E**. The situation is similar to what happens with **G3K**, the sequent calculus for basic normal modal logic, where the rule of necessitation doesn't have to be added as an explicit rule because it is admissible, i.e., whenever it premiss is derivable, also its conclusion is. With the proviso of completeness (proved in Section 5), this amounts to proving that whenever $\Rightarrow x : A \supset\subset B$ is derivable for an arbitrary label x then also $\Rightarrow x : \Box A \supset\subset \Box B$ is derivable for an arbitrary label x:

Lemma 3.3. *Rule RE is admissible in* **G3E**.

Proof. By the following derivation (where we use admissible cut and weakening steps):

$$\frac{\dfrac{\dfrac{x : A \Rightarrow x : B}{a \Vdash^\forall A \Rightarrow a \Vdash^\forall B} \ 3.2}{a \in I(x), a \Vdash^\forall A, A \triangleleft a \Rightarrow x : \Box B, a \Vdash^\forall B} \quad \dfrac{\dfrac{\dfrac{y \in a, \ldots \Rightarrow \ldots, y \in a}{y : B \Rightarrow y : A \quad y : A, a \in I(x), a \Vdash^\forall A, A \triangleleft a \Rightarrow x : \Box B, y \in a} L\triangleleft}{y : B, a \in I(x), a \Vdash^\forall A, A \triangleleft a \Rightarrow x : \Box B, y \in a}\ cut}{\dfrac{a \in I(x), a \Vdash^\forall A, A \triangleleft a \Rightarrow x : \Box B, B \triangleleft a}{a \in I(x), a \Vdash^\forall A, A \triangleleft a \Rightarrow x : \Box B}\ R\Box}\ R\triangleleft}}{\dfrac{a \in I(x), a \Vdash^\forall A, A \triangleleft a \Rightarrow x : \Box B}{x : \Box A \Rightarrow x : \Box B}\ L\Box}$$

Observe that the topsequents in the derivations correspond to both assumptions of rule RE and that it is also required that sequents of the form $x \in a, \Gamma \Rightarrow \Delta, x \in a$ are taken as initial. QED

Next we show how to use this basic calculus to find the extra rules that have to be added to obtain a G3 proof system for each of the above classical modal logics.

---

[14] The reason for the addition will be clear in the proof of Lemma 3.3 below.

Again, as we are "bootstrapping" to find the rules of the calculus, we assume that the desired invertibility and structural properties (to be proved in Section 4 below) are available.

We proceed by root-first proof search in the invertible sequent calculus **G3E**. By abduction we find a sufficient rule for deriving the labelled form of each axiom. Further on, we shall give all the formal definitions and prove that this heuristic method really does yield a complete sequent system for the logic in question.

First, observe that by invertibility of the rules $R\supset$ and $R\&$ the derivability of the sequent $\Rightarrow x : \Box(A\&B) \supset \Box A \& \Box B$ is equivalent to the derivability of both $\Rightarrow x : \Box(A\&B) \supset \Box A$ and $\Rightarrow x : \Box(A\&B) \supset \Box B$. Let us see how the former can be obtained with the following derivation, where we use derivability of initial sequents with arbitrary formulas, a result proved in the next section; the latter sequent is derivable *mutatis mutandis*:

$$\cfrac{\cfrac{\cfrac{\cfrac{\cfrac{a \in I(x), y \in a, y : A, y : B, a \Vdash^\forall A\&B, \ldots \Rightarrow x : \Box A, y : A}{a \in I(x), y \in a, y : A\&B, a \Vdash^\forall A\&B, \ldots \Rightarrow x : \Box A, y : A} L\Vdash^\forall}{a \in I(x), y \in a, a \Vdash^\forall A\&B, \ldots \Rightarrow x : \Box A, y : A} R\Vdash^\forall}{a \in I(x), a \Vdash^\forall A\&B, \ldots \Rightarrow x : \Box A, a \Vdash^\forall A} \quad \cfrac{\vdots}{\cfrac{a \in I(x), a \Vdash^\forall A\&B, A\&B \triangleleft a \Rightarrow x : \Box A, A \triangleleft a}{} R\Box}}{a \in I(x), a \Vdash^\forall A\&B, A\&B \triangleleft a \Rightarrow x : \Box A} R}{x : \Box(A\&B) \Rightarrow x : \Box A} L\Box$$

with the dotted part as follows:

$$\cfrac{\cfrac{\cfrac{\text{Lemma 4.2}}{b \Vdash^\forall A, \ldots \Rightarrow \ldots, b \Vdash^\forall A} \quad \cfrac{\text{Lemma 4.2}}{A \triangleleft b, \ldots \Rightarrow \ldots, A \triangleleft b}}{b \in I(x), b \Vdash^\forall A, A \triangleleft b, a \Vdash^\forall A\&B, a \in I(x), a \Vdash^\forall A\&B, A\&B \triangleleft a \Rightarrow x : \Box A, A \triangleleft a} R\Box}{a \in I(x), a \Vdash^\forall A\&B, A\&B \triangleleft a \Rightarrow x : \Box A, A \triangleleft a} R$$

The extra rule applied (R) amounts to requiring that $ext(A\&B) \in I(x)$ implies $ext(A) \in I(x)$. Since $ext(A\&B) \subseteq ext(A)$ holds by definition, this follows from the property of *monotonicity* of $I(x)$:

$$a \in I(x) \ \& \ a \subseteq b \supset b \in I(x) \quad Mon$$

As a rule, the property is expressed as

$$\cfrac{a \in I(x), a \subseteq b, b \in I(x), \Gamma \Rightarrow \Delta}{a \in I(x), a \subseteq b, \Gamma \Rightarrow \Delta} M$$

Lemma 3.4. *In the presence of monotonicity (Mon), the following forcing conditions give the same class of valid formulas:*

1. $x \Vdash_1 \Box A \equiv \exists a \in I(x)(a \Vdash^\forall A \ \& \ \forall y(y \Vdash A \supset y \in a))$

2. $x \Vdash_2 \Box A \equiv \exists a \in I(x). a \Vdash^\forall A$

Proof. Let $V(1)$ (resp. $V(2)$) be the class of valid formulas according to 1 (resp. 2). We show that $V(1) = V(2)$. We show by induction on formulas that A is in $V(1)$ if and only if A is in $V(2)$. The only non-trivial case is the one for boxed formulas, so suppose that $\models_1 \Box A$, that is, for all models $(W, I, \mathcal{V})$ and for all x we have $\exists a \in I(x)(a \Vdash^\forall A \& A \triangleleft a)$. It is then clear by first-order logic that $\exists a \in I(x). a \Vdash^\forall A$. Therefore $\models_2 \Box A$.

Conversely, if $\models_2 \Box A$, then for an arbitrary x we have $\exists a \in I(x). a \Vdash^\forall A$. Let b be $ext(A)$. By monotonicity, we have that $b \in I(x)$ and b clearly satisfies $A \triangleleft b$, so $x \Vdash_1 \Box A$. Since x was arbitrary, $\models_1 \Box A$. QED

It follows that in the case of logical systems closed under monotonicity the rules for the necessity operator can be simplified to the following form:

$$\frac{a \in I(x), a \Vdash^\forall A, \Gamma \Rightarrow \Delta}{x : \Box A, \Gamma \Rightarrow \Delta} \; L\Box', \; a \; fresh$$

$$\frac{a \in I(x), \Gamma \Rightarrow \Delta, x : \Box A, a \Vdash^\forall A}{a \in I(x), \Gamma \Rightarrow \Delta, x : \Box A} \; R\Box'$$

Table 5: Modal rules of system G3M

Remark 3.5. *Whenever monotonicity is present, we shall consider the above, simplified rules for $\Box$ rather than the original ones with the addition of rule M; this is not just a choice to streamline the sequent calculus, but it follows also from the fact that rule M together with the right rule for inclusion gives a problematic case in the cut elimination procedure.*

Next, we proceed to the determination of the rule for system **C**. We have the following derivation:

$$\frac{\frac{\frac{\frac{\frac{\frac{\frac{\ldots, a \cap b \Vdash^\forall A\&B, \ldots \Rightarrow x : \Box(A\&B), a \cap b \Vdash^\forall A\&B \quad \ldots, A\&B \triangleleft a \cap b, \ldots \Rightarrow x : \Box(A\&B), A\&B \triangleleft a \cap b}{a \cap b \in I(x), a \in I(x), b \in I(x), a \cap b \Vdash^\forall A\&B, a \Vdash^\forall A, b \Vdash^\forall B, A\&B \triangleleft a \cap b, A \triangleleft a, B \triangleleft b \Rightarrow x : \Box(A\&B)} R\Box}{a \cap b \in I(x), a \in I(x), b \in I(x), a \cap b \Vdash^\forall A\&B, a \Vdash^\forall A, b \Vdash^\forall B, A \triangleleft a, B \triangleleft b \Rightarrow x : \Box(A\&B)} Adm_2}{a \cap b \in I(x), a \in I(x), b \in I(x), a \Vdash^\forall A, b \Vdash^\forall B, A \triangleleft a, B \triangleleft b \Rightarrow x : \Box(A\&B)} Adm_1}{a \in I(x), b \in I(x), a \Vdash^\forall A, b \Vdash^\forall B, A \triangleleft a, B \triangleleft b \Rightarrow x : \Box(A\&B)} Rule}{a \in I(x), a \Vdash^\forall A, A \triangleleft a, x : \Box B \Rightarrow x : \Box(A\&B)} L\Box}{x : \Box A, x : \Box B \Rightarrow x : \Box(A\&B)} L\Box}{x : \Box A\&\Box B \Rightarrow x : \Box(A\&B)} L\&$$

Here we have used two steps whose admissibility follows from admissibility of cut and contraction (to be proved below) and the derivability in **G3E** of the sequents

1. $a \Vdash^\forall A, b \Vdash^\forall B \Rightarrow a \cap b \Vdash^\forall A \& B$

2. $A \triangleleft a, B \triangleleft b \Rightarrow A\&B \triangleleft a\cap b$

in a system extended with the following rules for formal intersection:

$$\frac{x \in a, x \in b, x \in a\cap b, \Gamma \Rightarrow \Delta}{x \in a \cap b, \Gamma \Rightarrow \Delta} L\cap \qquad \frac{\Gamma \Rightarrow \Delta, x \in a\cap b, x \in a \quad \Gamma \Rightarrow \Delta, x \in a\cap b, x \in b}{\Gamma \Rightarrow \Delta, x \in a\cap b} R\cap$$

So the extra condition that should be required on the neighbourhoods is just

$$a \in I(x) \ \& \ b \in I(x) \to a\cap b \in I(x)$$

that is, closure of $I(x)$ under intersection. It corresponds to the rule

$$\frac{a\cap b \in I(x), a \in I(x), b \in I(x), \Gamma \Rightarrow \Delta}{a \in I(x), b \in I(x), \Gamma \Rightarrow \Delta} C$$

Observe that if $I(x)$ is closed under supersets, then the above condition can be equivalently replaced by the weaker

$$a \in I(x) \ \& \ b \in I(x) \to \exists c \in I(x).c \subseteq a \ \& \ c \subseteq b$$

which can be translated into the geometric rule *Prebasic* seen already in Section 2:

$$\frac{c \in I(x), a \in I(x), b \in I(x), c \subseteq a, c \subseteq b, \Gamma \Rightarrow \Delta}{a \in I(x), b \in I(x), \Gamma \Rightarrow \Delta} C'$$

where c is a fresh neighbourhood label.

Finally, we determine the rule needed to prove the validity of $\Box\top$. As a preliminary remark, we observe that in the calculus **G3K** which shares the propositional base with **G3E**, the constant $\top$ (for *true*) is not primitive but defined as $\bot \supset \bot$ (or $A \supset A$ for any formula A). The rule to be added for a labelled calculus with $\top$ as a primitive is the dual of the rule $L\bot$, that is, the zero-premiss rule[15]

$$\overline{\Gamma \Rightarrow \Delta, x : \top} R\top$$

We have the search tree[16]

$$\frac{\dfrac{a \in I(x), a \Vdash^\forall \top, \top \triangleleft a \Rightarrow x : \Box\top, a \Vdash^\forall \top \quad a \in I(x), a \Vdash^\forall \top, \top \triangleleft a \Rightarrow x : \Box\top, \top \triangleleft a}{a \in I(x), a \Vdash^\forall \top, \top \triangleleft a \Rightarrow x : \Box\top} R\Box}{\Rightarrow x : \Box\top} rule$$

---

[15] Observe that the rule is actually derivable with $\top$ defined as $\bot \supset \bot$.

[16] Since the topsequents are derivable, the proof search is a derivation once the step indicated by *rule* is taken to be a rule of the system; here, as elsewhere, proof search in the basic calculus is used to determine which additional rules have to be included in the system to make certain sequents derivable.

The extra rule correponds to the following property of neighbourhoods

$$(1.) \; \exists a \in I(x). a \Vdash^\forall \top \; \& \; \top \lhd a$$

which is clearly equivalent to

$$(1.') \; \exists a \in I(x). \top \lhd a$$

and corresponds to the rule

$$\frac{a \in I(x), \top \lhd a, \Gamma \Rightarrow \Delta}{\Gamma \Rightarrow \Delta} N$$

with a fresh. In the presence of monotonicity rule *Nondeg* (Definition 2.4) suffices, because we have

$$\frac{\dfrac{a \in I(x) \Rightarrow x : \Box\top, a \Vdash^\forall \top}{a \in I(x) \Rightarrow x : \Box\top} R\Box'}{\Rightarrow x : \Box\top} Nondeg$$

with topsequent clearly derivable.

3.1 Adding $\Diamond$

The possibility modality is defined in classical modal logic, as in normal modal logic, as the dual of necessity (cf. [4])

$$\Diamond A \equiv \neg \Box \neg A$$

and therefore it is not usually considered as a modality with its own rules. It is however convenient, for the same reasons why it is convenient to have classical logic with all the connectives, not just two (or even one) of them, to have primitive rules for possibility. The rules are found by imposing the above duality and using the rules of $\Box$ and the duality between $\Vdash^\forall$ and $\Vdash^\exists$. In practice, to find the left and right rules for $\Diamond$ we start with the sequents $x : \Diamond A, \Gamma \Rightarrow \Delta$ and $\Gamma \Rightarrow \Delta, x : \Diamond A$, replace them with $x : \neg\Box\neg A, \Gamma \Rightarrow \Delta$ and $\Gamma \Rightarrow \Delta, x : \neg\Box\neg A$, respectively, and apply the rules for $\neg$ and $\Box$. It becomes clear that the former sequent needs also $a \in I(x)$ in the antecedent, else $R\Box$ cannot be applied. The decomposition then gives $a \Vdash^\forall \neg A$ in the succedent (resp. antecedent) which is replaced by the equivalent $a \Vdash^\exists A$ in the antecedent (resp. succedent). The formula $\neg A \lhd a$ instead cannot be moved to the other side with negation removed because the scope of the negation is A, not $A \lhd a$. In the end, the rules for the possibility modality are as follows:

$$\frac{a \in I(x), x : \Diamond A, a \Vdash^\exists A, \Gamma \Rightarrow \Delta \quad a \in I(x), x : \Diamond A, \Gamma \Rightarrow \Delta, \neg A \lhd a}{a \in I(x), x : \Diamond A, \Gamma \Rightarrow \Delta} L\Diamond$$

$$\frac{a \in I(x), \neg A \triangleleft a, \Gamma \Rightarrow \Delta, a \Vdash^{\exists} A}{\Gamma \Rightarrow \Delta, x : \Diamond A} \, R\Diamond$$

In $R\Diamond$, a is a fresh neighbourhood label.

To see the rules at work, we can use them to verify the duality between the two alethic modalities, where both topsequents are derivable[17]:

$$\frac{\dfrac{\dfrac{\dfrac{a \in I(x), x : \Diamond A, y \in a, y : A, y : \neg A, a \Vdash^{\forall} \neg A, \neg A \triangleleft a \Rightarrow x : \bot}{a \in I(x), x : \Diamond A, y \in a, y : A, a \Vdash^{\forall} \neg A, \neg A \triangleleft a \Rightarrow x : \bot} \, L\Vdash^{\forall}}{a \in I(x), x : \Diamond A, a \Vdash^{\exists} A, a \Vdash^{\forall} \neg A, \neg A \triangleleft a \Rightarrow x : \bot} \, L\Vdash^{\exists} \quad a \in I(x), x : \Diamond A, a \Vdash^{\forall} \neg A, \neg A \triangleleft a \Rightarrow x : \bot, \neg A \triangleleft a}{\dfrac{\dfrac{\dfrac{a \in I(x), x : \Diamond A, a \Vdash^{\forall} \neg A, \neg A \triangleleft a \Rightarrow x : \bot}{x : \Diamond A, x : \Box \neg A \Rightarrow x : \bot} \, L\Box}{\dfrac{x : \Diamond A \Rightarrow x : \neg \Box \neg A}{\Rightarrow x : \Diamond A \supset \neg \Box \neg A} \, R\supset} \, R\neg}{}} \, L\Diamond}$$

The derivation of the other direction of the duality, namely $\Rightarrow x : \neg\Box\neg A \supset \Diamond A$, is found in a similar way using the rules for negation, the alethic modalities and the local forcing relations.

If monotonicity is absorbed into the modal rules, also the rules for $\Diamond$ get modified (and simplified). The monotonic version of the rules for $\Diamond$ is as follows:

$$\frac{a \in I(x), x : \Diamond A, a \Vdash^{\exists} A, \Gamma \Rightarrow \Delta}{a \in I(x), x : \Diamond A, \Gamma \Rightarrow \Delta} \, L\Diamond'$$

$$\frac{a \in I(x), \Gamma \Rightarrow \Delta, a \Vdash^{\exists} A}{\Gamma \Rightarrow \Delta, x : \Diamond A} \, R\Diamond'$$

We remark here that all the results below continue to hold when $\Diamond$ is added as an explicit modality, rather than a defined one, in the calculus.

4 Structural properties

In this section we shall give detailed proofs of the structural properties for the systems based on neighbourhood semantics that we have considered. Rather than giving specific proofs for specific systems, we shall indicate how the structural properties can be established by following a generalization of the guidelines presented in [25] and [32], section 11.4. There are some important non-trivial extra considerations caused by the layering of rules for the modalities defined in terms of neighbourhood semantics, which gives a quantifier alternation more complex than in the Kripke-style semantics. Likewise, some preliminary results are needed, namely height-preserving

---

[17]This is a consequence of Lemma 4.2 proved in the following section.

admissibility of substitution (in short, hp-substitution) and height-preserving invertibility (in short, hp-invertibility) of the rules. We recall that the *height* of a derivation is its height as a tree, i.e., the length of its longest branch, and that $\vdash_n$ denotes derivability with derivation height bounded by n in a given system.

In the following we shall denote with **G3n*** any extension of the basic system **G3n** with rules for the modalities $[\,]$, $\langle\,]$, $[\,\rangle$, and $\langle\,\rangle$, $\Box$[18] and with extra (mathematical) rules. This extension is intended to follow the standard *closure condition* for extensions of contraction-free labelled sequent calculi (cf. [25]) to guarantee admissibility of contraction in the resulting system.

As observed above, in the light of Remark 3.5, we can obtain system **G3nM** by modifying the rules $L\Box$, $R\Box$ to the form $L\Box'$ and $R\Box'$; for extensions, we can take in place of C and N the rules C' and *Nondeg*.

In many proofs we shall use an induction on formula weight. In order to find a definition of weight that makes the induction work we have to take into account several constraints that emerge from the proofs of the structural results; the choice for this particular definition will thus become clear from the proofs to follow.

Observe that the definition extends the usual definition of weight from (pure) formulas to labelled formulas and local forcing relations, namely, to all formulas of the form $x : A$, $a \Vdash^\forall A$, $a \Vdash^\exists A$, $A \triangleleft a$, as well as the relational formulas $x \in a$, $a \in I(x)$, $a \subseteq b$.

Definition 4.1. *The label of formulas of the form $x : A$ is x. The label of formulas of the form $a \Vdash^\forall A$, $a \Vdash^\exists A$, $A \triangleleft a$ is a. The label of a formula $\mathcal{F}$ will be denoted by $l(\mathcal{F})$. The pure part of a labelled formula $\mathcal{F}$ is the part without the label and without the forcing relation, either local ($\Vdash^\exists$, $\Vdash^\forall$) or worldwise (:) and will be denoted by $p(\mathcal{F})$.*

The weight of a labelled formula $\mathcal{F}$ is given by the pair $(\mathtt{w}(p(\mathcal{F})), \mathtt{w}(l(\mathcal{F})))$ where

- *For all worlds labels x and all neighbourhood labels a, $\mathtt{w}(x) = 0$ and $\mathtt{w}(a) = 1 + n(\cap)$, where $n(\cap)$ is the number of formal intersections in a.*

- – $\mathtt{w}(P) = \mathtt{w}(\bot) = 1$,
  – $\mathtt{w}(A \circ B) = \mathtt{w}(A) + \mathtt{w}(B) + 1$ *for $\circ$ conjunction, disjunction, or implication,*
  – $\mathtt{w}(\Box A) = \mathtt{w}([\,]A) = \mathtt{w}(\langle\,]A) = \mathtt{w}([\,\rangle A) = \mathtt{w}(\langle\,\rangle A) = \mathtt{w}(A) + 1$

For formulas of the form $a \in I(x)$, $x \in a$, we stipulate $\mathtt{w}(a \in I(x)) = \mathtt{w}(x \in a) = (0, w(a))$ and for formulas of the form $a \subseteq b$, $\mathtt{w}(a \subseteq b) = (w(a), w(b))$. Weights of labelled formulas are ordered lexicographically.

---

[18] We assume that for each modality, the extension has to contain both the right and left rule.

From the definition of weight it is clear that the weight gets decreased if we move from a formula labelled by a neighbourhood label to the same formula labelled by a world label, or if we move (regardless the label) to a formula with a pure part of strictly smaller weight.

Lemma 4.2. *Sequents of the following form are derivable in* **G3n*** *for arbitrary formulas A and B in the propositional modal language of* **G3n***:

1. $a \subseteq b, \Gamma \Rightarrow \Delta, a \subseteq b$

2. $A \triangleleft a, \Gamma \Rightarrow \Delta, A \triangleleft a$

3. $a \Vdash^\forall A, \Gamma \Rightarrow \Delta, a \Vdash^\forall A$

4. $a \Vdash^\exists A, \Gamma \Rightarrow \Delta, a \Vdash^\exists A$

5. $x : A, \Gamma \Rightarrow \Delta, x : A$

Proof. 1. By the following derivation

$$\frac{\dfrac{x \in b, x \in a, a \subseteq b, \Gamma \Rightarrow \Delta, x \in b}{x \in a, a \subseteq b, \Gamma \Rightarrow \Delta, x \in b} L \subseteq}{a \subseteq b, \Gamma \Rightarrow \Delta, a \subseteq b} R \subseteq$$

where the topsequent is initial.

2. By the following derivation

$$\frac{\dfrac{x \in a, x : A, A \triangleleft a, \Gamma \Rightarrow \Delta, x \in a}{x : A, A \triangleleft a, \Gamma \Rightarrow \Delta, x \in a} L \triangleleft}{A \triangleleft a, \Gamma \Rightarrow \Delta, A \triangleleft a} R \triangleleft$$

where the topsequent is initial.

3–5 are proved by simultaneous induction on formula weight.

3. We have the following inference

$$\frac{\dfrac{x : A, x \in a, a \Vdash^\forall A, \Gamma \Rightarrow \Delta, x : A}{x \in a, a \Vdash^\forall A, \Gamma \Rightarrow \Delta, x : A} L \Vdash^\forall}{a \Vdash^\forall A, \Gamma \Rightarrow \Delta, a \Vdash^\forall A} R \Vdash^\forall$$

The topsequent is derivable by induction hypothesis because $\mathtt{w}(x : A) < \mathtt{w}(a \Vdash^\forall A)$.

4. Similar, with $L \Vdash^\exists$ and $R \Vdash^\exists$ in place of $R \Vdash^\forall$ and $L \Vdash^\forall$, respectively, using $\mathtt{w}(x : A) < \mathtt{w}(a \Vdash^\exists A)$.

5. We distinguish subcases according to the structure of A. If it is atomic or $\bot$, the sequent is initial or conclusion of $L\bot$. If the outermost connective of A is a conjunction or a disjunction, or an implication, the sequent is derivable by application of the respective rules and the induction hypothesis. If it is a modality, we have the following further subcases:

5.1. $A \equiv [\,]B$. We have the following inference

$$\dfrac{\dfrac{a \Vdash^\forall B, a \in I(x), x : [\,]B, \Gamma \Rightarrow \Delta, a \Vdash^\forall B}{a \in I(x), x : [\,]B, \Gamma \Rightarrow \Delta, a \Vdash^\forall B} L[\,]}{x : [\,]B, \Gamma \Rightarrow \Delta, x : [\,]B} R[\,]$$

where the topsequent is derivable by induction hypothesis because $\mathtt{w}(a \Vdash^\forall B) < \mathtt{w}(x : [\,]B)$.

5.2. $A \equiv \langle\,]B$. Similar with the rules $L\langle\,]$, $R\langle\,]$, and the inductive hypothesis on $a \Vdash^\forall B$, using $\mathtt{w}(a \Vdash^\forall B) < \mathtt{w}(x : \langle\,]B)$.

5.3. $A \equiv [\,\rangle B$. Similar with the rules $R[\,\rangle$, $L[\,\rangle$, and the inductive hypothesis on $a \Vdash^\exists B$, using $\mathtt{w}(a \Vdash^\exists B) < \mathtt{w}(x : [\,\rangle B)$.

5.4 $A \equiv \langle\,\rangle B$. Similar with the rules $L\langle\,\rangle$, $R\langle\,\rangle$, and the inductive hypothesis on $a \Vdash^\forall B$.

5.5 $A \equiv \Box B$. We have the following inference

$$\dfrac{\dfrac{a \in I(x), a \Vdash^\forall B, B \triangleleft a, \Gamma \Rightarrow \Delta, x : \Box B, a \Vdash^\forall B \quad a \in I(x), a \Vdash^\forall B, B \triangleleft a, \Gamma \Rightarrow \Delta, x : \Box B, B \triangleleft a}{a \in I(x), a \Vdash^\forall B, B \triangleleft a, \Gamma \Rightarrow \Delta, x : \Box B} R\Box}{x : \Box B, \Gamma \Rightarrow \Delta, x : \Box B} L\Box$$

where the left topsequent is derivable by induction hypothesis because $\mathtt{w}(a \Vdash^\forall B) < \mathtt{w}(x : \Box B)$ and the right one by clause 2 above.

For extensions of **G3nM** we have the following inference:

$$\dfrac{\dfrac{a \in I(x), a \Vdash^\forall B, \Gamma \Rightarrow \Delta, x : \Box B, a \Vdash^\forall B}{a \in I(x), a \Vdash^\forall B, \Gamma \Rightarrow \Delta, x : \Box B} R\Box'}{x : \Box B, \Gamma \Rightarrow \Delta, x : \Box B} L\Box'$$

and we can treat this as a sub-case of the above. QED

In our system, in addition to world labels, we have neighbourhood labels. The latter are subject to similar conditions, such as the conditions of being fresh in certain rules, as the world labels. Consequently, we shall need properties of hp-substitution in our analysis. Before stating and proving the property, we observe that the definition of substitution of labels given in [25] can be extended in an obvious way – that need not to be pedantically detailed here – to all the formulas of our language and to neighbourhood labels. We'll have, for example, $x : \langle \, \rangle A(y/x) \equiv y : \langle \, \rangle A$, $a \Vdash^\exists A(b/a) \equiv b \Vdash^\exists A$, and $A \lhd a(b/a) \equiv A \lhd b$. Next, we prove that the calculus enjoys the property of hp-substitution both of world and neighbourhood labels:[19]

Proposition 4.3. *1. If $\vdash_n \Gamma \Rightarrow \Delta$, then $\vdash_n \Gamma(y/x) \Rightarrow \Delta(y/x)$;*

2. If $\vdash_n \Gamma \Rightarrow \Delta$, then $\vdash_n \Gamma(b/a) \Rightarrow \Delta(b/a)$.

Proof. Both statements are proved by induction on the height of the derivation.

If it is 0, then $\Gamma \Rightarrow \Delta$ is an initial sequent or a conclusion of $L\bot$. The same then holds for $\Gamma(y/x) \Rightarrow \Delta(y/x)$ and for $\Gamma(b/a) \Rightarrow \Delta(b/a)$.

If the derivation has height $n > 0$, we consider the last rule applied. If $\Gamma \Rightarrow \Delta$ has been derived by a rule without variable conditions, we apply the induction hypothesis and then the rule. Rules with variable conditions require that we avoid a clash of the substituted variable with the fresh variable in the premiss. This is the case for the logical rules $R \Vdash^\forall$, $L \Vdash^\exists$, $R[\,]$, $L\langle\,]$, $R[\,\rangle$, $L\langle\,\rangle$, $L\square$, $L\square'$ and for the neighbourhood rules $R \subseteq$, $Prebasic/C'$, $Nondeg$. So, if $\Gamma \Rightarrow \Delta$ has been derived by any of these rules, we apply the inductive hypothesis twice to the premiss, first to replace the fresh variable with another fresh variable different, if necessary, from the one we want to substitute, then to make the substitution, and then apply the rule. QED

The rules of weakening for the language of a labelled system with internalized neighbourhood semantics such as **G3n*** have the following form, where ϕ is either a "relational" atom of the form $a \in I(x)$[20] or $x \in a$ or a labelled formula of the form $x : A$, $a \Vdash^\forall A$, $a \Vdash^\exists A$ or a formula of the form $A \lhd a$:

$$\frac{\Gamma \Rightarrow \Delta}{\phi, \Gamma \Rightarrow \Delta} \, L\text{-}Wkn \qquad \frac{\Gamma \Rightarrow \Delta}{\Gamma \Rightarrow \Delta, \phi} \, R\text{-}Wkn$$

Proposition 4.4. *The rules of left and right weakening are hp-admissible in* **G3n***.

Proof. Straightforward induction, with a similar proviso as in the above proof for rules with variable conditions. QED

Next, we prove *hp-invertibility* of the rules of **G3n***, i.e., for every rule of the form $\frac{\Gamma' \Rightarrow \Delta'}{\Gamma \Rightarrow \Delta}$, if $\vdash_n \Gamma \Rightarrow \Delta$ then $\vdash_n \Gamma' \Rightarrow \Delta'$, and for every rule of the form $\frac{\Gamma' \Rightarrow \Delta' \quad \Gamma'' \Rightarrow \Delta''}{\Gamma \Rightarrow \Delta}$ if $\vdash_n \Gamma \Rightarrow \Delta$ then $\vdash_n \Gamma' \Rightarrow \Delta'$ and $\vdash_n \Gamma'' \Rightarrow \Delta''$. Items 7' and 8' are the invertibility

---

[19] We remind that of the two possible notations for substitution we use the one in which $A(y/x)$ indicates the result of substituting y for x in A.

[20] Indeed, such formulas are not needed for right weakenening because they are never active on the right.

for the non-monotonic rules for $\langle\,]$, $R\langle\,]'$ and $L\langle\,]'$, and 15' for the monotonic version of $L\Box$:

Lemma 4.5. *The following hold in* **G3n***:

1. *If* $\vdash_n \Gamma \Rightarrow \Delta, a \Vdash^\forall A$ *then* $\vdash_n x \in a, \Gamma \Rightarrow \Delta, x : A$.

2. *If* $\vdash_n x \in a, a \Vdash^\forall A, \Gamma \Rightarrow \Delta$ *then* $\vdash_n x \in a, x : A, a \Vdash^\forall A, \Gamma \Rightarrow \Delta$.

3. *If* $\vdash_n x \in a, \Gamma \Rightarrow \Delta, a \Vdash^\exists A$ *then* $\vdash_n x \in a, \Gamma \Rightarrow \Delta, x : A, a \Vdash^\exists A$.

4. *If* $\vdash_n a \Vdash^\exists A, \Gamma \Rightarrow \Delta$ *then* $\vdash_n x \in a, x : A, \Gamma \Rightarrow \Delta$.

5. *If* $\vdash_n \Gamma \Rightarrow \Delta, x : [\,]A$ *then* $\vdash_n a \in I(x), \Gamma \Rightarrow \Delta, a \Vdash^\forall A$.

6. *If* $\vdash_n a \in I(x), x : [\,]A, \Gamma \Rightarrow \Delta$ *then* $\vdash_n a \in I(x), x : [\,]A, a \Vdash^\forall A, \Gamma \Rightarrow \Delta$.

7. *If* $\vdash_n a \in I(x), \Gamma \Rightarrow \Delta, x : \langle\,]A$ *then* $\vdash_n a \in I(x), \Gamma \Rightarrow \Delta, x : \langle\,]A, a \Vdash^\forall A$.

7'. *If* $\vdash_n a \in I(x), \Gamma \Rightarrow \Delta, x : \langle\,]A$ *then* $\vdash_n a \in I(x), \Gamma \Rightarrow \Delta, x : \langle\,]A, a \Vdash^\forall A$ *and* $\vdash_n a \in I(x), \Gamma \Rightarrow \Delta, x : \langle\,]A, A \triangleleft a$.

8. *If* $\vdash_n x : \langle\,]A, \Gamma \Rightarrow \Delta$ *then* $\vdash_n a \in I(x), a \Vdash^\forall A, \Gamma \Rightarrow \Delta$.

8'. *If* $\vdash_n x : \langle\,]A, \Gamma \Rightarrow \Delta$ *then* $a \in I(x), a \Vdash^\forall A, A \triangleleft a, \Gamma \Rightarrow \Delta$.

9. *If* $\vdash_n \Gamma \Rightarrow \Delta, x : [\,\rangle A$ *then* $\vdash_n a \in I(x), \Gamma \Rightarrow \Delta, a \Vdash^\exists A$.

10. *If* $\vdash_n a \in I(x), x : [\,\rangle A, \Gamma \Rightarrow \Delta$ *then* $\vdash_n a \in I(x), x : [\,\rangle A, a \Vdash^\exists A, \Gamma \Rightarrow \Delta$.

11. *If* $\vdash_n a \in I(x), \Gamma \Rightarrow \Delta, x : \langle\,\rangle A$ *then* $\vdash_n a \in I(x), \Gamma \Rightarrow \Delta, x : \langle\,\rangle A, a \Vdash^\exists A$.

12. *If* $\vdash_n x : \langle\,\rangle A, \Gamma \Rightarrow \Delta$ *then* $\vdash_n a \in I(x), a \Vdash^\exists A, \Gamma \Rightarrow \Delta$.

13. *If* $\vdash_n A \triangleleft a, y : A, \Gamma \Rightarrow \Delta$ *then* $\vdash_n y \in a, A \triangleleft a, y : A, \Gamma \Rightarrow \Delta$.

14. *If* $\vdash_n \Gamma \Rightarrow \Delta, A \triangleleft a$ *then* $\vdash_n y : A, \Gamma \Rightarrow \Delta, y \in a$.

15. *If* $\vdash_n x : \Box A, \Gamma \Rightarrow \Delta$ *then* $\vdash_n a \in I(x), a \Vdash^\forall A, A \triangleleft a, \Gamma \Rightarrow \Delta$.

15'. *If* $\vdash_n x : \Box A, \Gamma \Rightarrow \Delta$ *then* $\vdash_n a \in I(x), a \Vdash^\forall A, \Gamma \Rightarrow \Delta$.

16. *If* $\vdash_n a \in I(x), \Gamma \Rightarrow \Delta, x : \Box A$ *then* $\vdash_n a \in I(x), \Gamma \Rightarrow \Delta, x : \Box A, a \Vdash^\forall A$ *and* $\vdash_n a \in I(x), \Gamma \Rightarrow \Delta, x : \Box A, A \triangleleft a$.

17. *If* $\vdash_n \Gamma \Rightarrow \Delta, a \subseteq b$ *then* $\vdash_n x \in a, \Gamma \Rightarrow \Delta, x \in b$.

18. If $\vdash_n x \in a, a \subseteq b, \Gamma \Rightarrow \Delta$ then $\vdash_n x \in a, a \subseteq b, x \in b, \Gamma \Rightarrow \Delta$.

Proof. Observe first that all the cases (2, 3, 6, 7, 7′, 10, 11, 13, 16, 18) that are instances of hp-admissibility of weakening follow from Proposition 4.4 above. For the rest, the proof is by induction on n and we show in detail, by way of example item 5., the other cases being shown in a similar way.

Base case: Suppose that $\Gamma \Rightarrow \Delta, x : [\ ]A$ is an initial sequent or conclusion of $L\bot$. Then, in the former case, $x : [\ ]A$ not being atomic or of the form $x \in a$, $a \in I(x), \Gamma \Rightarrow \Delta, a \Vdash^\forall A$ is an initial sequent, in the latter it is a conclusion of $L\bot$.

Inductive step: Assume hp-invertibility up to n, and let $\vdash_{n+1} \Gamma \Rightarrow \Delta, x : [\ ]A$. If $x : [\ ]A$ is principal, then the premiss $a \in I(x), \Gamma \Rightarrow \Delta, a \Vdash^\forall A$ (possibly obtained through hp-substitution) has a derivation of height n. If $x : [\ ]A$ is not principal in the last rule, we distinguish the case in which the last rule is not a rule with eigenvariable from the case in which it is. In the former case, the last rule has one or two premisses of the form $\Gamma' \Rightarrow \Delta', x : [\ ]A$ of derivation height $\leq n$. By induction hypothesis we have $a \in I(x), \Gamma' \Rightarrow \Delta', a \Vdash^\forall A$ for each premiss, with derivation height at most n. Thus, $\vdash_{n+1} a \in I(x), \Gamma \Rightarrow \Delta, a \Vdash^\forall A$. In the latter case, we proceed as in the previous case if the last rule has the eigenvariable for world labels, the critical case being (here) the one with the eigenvariable for neighbourhood labels. So, if the last rule is, say, $L\langle\ ]$, then $\Gamma = \langle\ ]B, \Gamma'$ and we have a premiss that we can assume to be of the form $b \in I(x), b \Vdash^\forall B, \Gamma' \Rightarrow \Delta, x : [\ ]A$ with b different from a (this can be assumed without loss of generality because of hp-substitution). By inductive hypothesis we obtain a derivation of height n of $a \in I(x), b \in I(x), b \Vdash^\forall B, \Gamma' \Rightarrow \Delta, a \Vdash^\forall A$ and by a step of $L\langle\ ]$ we conclude derivability of $a \in I(x), \Gamma \Rightarrow \Delta, a \Vdash^\forall A$ with height $n+1$. Cases 8, 8′, 9, 12, 15, 15′, are proved with a similar analysis. There is a final group of cases (items 1, 4, 14, 17), those of rules with an eigenvariable condition for world labels. The treatment is similar to the case detailed above, with a similar distinction of cases as for the last rule applied in the derivation. A special proviso is needed for the case in which the last rule is not the rule with the principal formula in question and it is a rule with eigenvariable of the same type, namely a world label. The claim is obtained by inductive hypothesis after use, if needed, of hp-substitution on the premisses of such rules to avoid a clash of variables so that the last rule can be applied after the inductive step to restore the original contexts. QED

Lemma 4.6. *All the propositional rules of* **G3n*** *are hp-invertible.*

Proof. Similar to the proof for **G3c** (Theorem 3.1.1 in [31]). QED

Therefore, as a general result, we have:

Corollary 4.7. *All the rules of* **G3n*** *are hp-invertible.*

Proof. By Lemmas 4.5, 4.6, and 4.4 (the latter gives hp-invertibility of the neighbourhood rules). QED

The rules of contraction for the language of a labelled system with internalized neighbourhood semantics such as **G3n*** have the following form, where ϕ is either a "relational" atom of the form $a \in I(x)$ or $x \in a$ or a labelled formula of the form $x : A$, $a \Vdash^\forall A$, $a \Vdash^\exists A$ or a formula of the form $A \triangleleft a$:

$$\frac{\phi, \phi, \Gamma \Rightarrow \Delta}{\phi, \Gamma \Rightarrow \Delta} \; L\text{-}Ctr \qquad \frac{\Gamma \Rightarrow \Delta, \phi, \phi}{\Gamma \Rightarrow \Delta, \phi} \; R\text{-}Ctr$$

Theorem 4.8. *The rules of left and right contraction are hp-admissible in* **G3n***.

Proof. By simultaneous induction on the height of derivation for left and right contraction.

If $n = 0$ the premiss is either an initial sequent or a conclusion of a zero-premiss rule. In each case, the contracted sequent is also an initial sequent or a conclusion of the same zero-premiss rule.

If $n > 0$, consider the last rule used to derive the premiss of contraction. There are two cases, depending on whether the contraction formula is principal or not in the rule.[21]

1. If the contraction formula is not principal in it, both occurrences are found in the premisses of the rule and they have a smaller derivation height. By the induction hypothesis, they can be contracted and the conclusion is obtained by applying the rule to the contracted premisses.

2. If the contraction formula is principal in it, we distinguish two sub-cases:

2.1. The last rule is one in which the principal formulas appear also in the premiss (such as $L \Vdash^\forall$, $R \Vdash^\exists$, $L[\ ]$, $R\langle\ ]$, $L[\ \rangle$, $R\langle\ \rangle$, $L \triangleleft$, $R\square$, $R\square'$, $L \subseteq$, and the neighbourhood rules). In all these cases we apply the induction hypothesis to the premiss(es) and then the rule. For example, if the last rule used to derive the premiss of contraction is $R\square$ we have:

$$\frac{a \in I(x), \Gamma \Rightarrow \Delta, x : \square A, x : \square A, a \Vdash^\forall A \quad a \in I(x), \Gamma \Rightarrow \Delta, x : \square A, x : \square A, A \triangleleft a}{a \in I(x), \Gamma \Rightarrow \Delta, x : \square A, x : \square A} \; R\square$$

---

[21] We recall that the *principal formula* of a logical rule is the formula containing the constant named by the rule in question, which in this case can be a connective, a modality, or a local forcing relation ($\Vdash^\exists$, $\Vdash^\forall$), or the inclusion operator; the other formulas in the rule are active or side formulas. *Side formulas* are the formulas in the contexts and the other formulas, which are neither side not principal formulas are *active formulas*. In the case of labelled systems there can be active formulas in the conclusion of the rules. For example, the formula $a \in I(x)$ in the conclusion of $R\square$ is regarded as an active formula.

By induction hypothesis applied to the premiss we obtain a one step shorter derivation of $a \in I(x), \Gamma \Rightarrow \Delta, x : \Box A, a \Vdash^\forall A$ and $a \in I(x), \Gamma \Rightarrow \Delta, x : \Box A, A \lhd a$ and thus by a step of $R\Box$ we obtain $a \in I(x), \Gamma \Rightarrow \Delta, x : \Box A$ with the same derivation height of the given premiss of contraction.

For the neighbourhood rules we follow the standard procedure as for added extralogical rules and observe that in case the contraction formulas are both principal in the rule (as in the case of rule C) we apply the closure condition

2.2. The last rule is one in which the principal formula does not appear in the premiss(es) (such as the rules for $\&$, $\vee$, $\supset$, $R \Vdash^\forall$, $L \Vdash^\exists$, $R[\ ]$, $L\langle\ ]$, $R[\ \rangle$, $L\langle\ \rangle$, $L\Box$, $R \subseteq$). In all such cases, we apply hp-invertibility to the premiss(es) of the rule so that we have a duplication of formulas at a smaller derivation height, then apply the induction hypothesis (as many times as needed) then apply the rule in question. For example, if the last rule is $L\Box$, we have:

$$\frac{a \in I(x), a \Vdash^\forall A, A \lhd a, x : \Box A, \Gamma \Rightarrow \Delta}{x : \Box A, x : \Box A, \Gamma \Rightarrow \Delta} \; L\Box,\; a\; fresh$$

Using hp-invertibility of $L\Box$ we obtain from the premiss a derivation of height $n-1$ of

$$a \in I(x), a \in I(x), a \Vdash^\forall A, a \Vdash^\forall A, A \lhd a, A \lhd a, \Gamma \Rightarrow \Delta$$

By the induction hypothesis we get a derivation of the same height of the sequent $a \in I(x), a \Vdash^\forall A, A \lhd a, \Gamma \Rightarrow \Delta$ and application of $L\Box$ gives a derivation of height n of $x : \Box A, \Gamma \Rightarrow \Delta$. QED

Cut is a rule of the form

$$\frac{\Gamma \Rightarrow \Delta, \phi \quad \phi, \Gamma' \Rightarrow \Delta'}{\Gamma, \Gamma' \Rightarrow \Delta, \Delta'} \; Cut$$

where ϕ is any formula of the language of the labelled calculus **G3n***. We have:

Theorem 4.9. *Cut is admissible in* **G3n***.

Proof. By double induction, with primary induction on the weight of the cut formula and subinduction on the cut height, i.e., the sum of the heights of derivations of the premisses of cut. The cases in which the premisses of cut are either initial sequents or obtained through the rules for $\&$, $\vee$, or $\supset$ follow the treatment of Theorem 3.2.3 of [31]. Among such cases, we just consider a significant one here, the case in which the initial sequent is $x \in a, \Gamma \Rightarrow \Delta, x \in a$ and the other premiss is conclusion of a rule for inclusion in which $x \in a$ is an active formula. The cut, with $\Gamma' = a \subseteq b, \Gamma''$, is as follows

$$\cfrac{x \in a, \Gamma \Rightarrow \Delta, x \in a \qquad \cfrac{x \in a, x \in b, a \subseteq b, \Gamma'' \Rightarrow \Delta'}{x \in a, \Gamma' \Rightarrow \Delta'} L\subseteq}{x \in a, \Gamma, \Gamma' \Rightarrow \Delta, \Delta'} Cut$$

and it is converted into a cut of reduced height as follows

$$\cfrac{\cfrac{x \in a, \Gamma \Rightarrow \Delta, x \in a \qquad x \in a, x \in b, a \subseteq b, \Gamma'' \Rightarrow \Delta'}{x \in a, x \in b, a \subseteq b, \Gamma, \Gamma'' \Rightarrow \Delta, \Delta'} Cut}{x \in a, \Gamma, \Gamma' \Rightarrow \Delta, \Delta'} L\subseteq$$

For the cases in which the cut formula is a side formula in at least one rule used to derive the premisses of cut, the cut reduction is dealt with in the usual way by permutation of cut, with possibly an application of hp-substitution to avoid a clash with the fresh variable in rules with variable condition. In all such cases the cut height is reduced. We give one example to give concreteness to this qualitative analysis:

$$\cfrac{\Gamma \Rightarrow \Delta, b \Vdash^\forall B \qquad \cfrac{a \in I(x), a \Vdash^\forall A, A \triangleleft a, b \Vdash^\forall B, \Gamma' \Rightarrow \Delta'}{x : \Box A, b \Vdash^\forall B, \Gamma' \Rightarrow \Delta'} L\Box}{x : \Box A, \Gamma, \Gamma' \Rightarrow \Delta, \Delta'} Cut$$

the neighbourhood label in the premiss of $L\Box$ is fresh, but nothing prevents it from appearing in the left premiss of cut; therefore, prior to the permutation of cut, we need to replace it with a neighbourhood label which is fresh not just with respect to the conclusion of $L\Box$ but also with respect to the left premiss of cut. Let the new fresh variable be c. The transformed derivation, with cut reduced to a cut of smaller height, is as follows:

$$\cfrac{\cfrac{\Gamma \Rightarrow \Delta, b \Vdash^\forall B \qquad c \in I(x), c \Vdash^\forall A, A \triangleleft c, b \Vdash^\forall B, \Gamma' \Rightarrow \Delta'}{c \in I(x), c \Vdash^\forall A, A \triangleleft a, \Gamma, \Gamma' \Rightarrow \Delta, \Delta'} Cut}{x : \Box A, \Gamma, \Gamma' \Rightarrow \Delta, \Delta'} L\Box$$

Next we consider in full detail the cases with cut formula principal in both premisses of cut and of the form $a \Vdash^\forall A$, $a \Vdash^\exists A$, $x : [\,]A$, $x : \langle\,]A$, $x : [\,\rangle A$, $x : \langle\,\rangle A$ or $A \triangleleft a$, $x : \Box A$.

1. The cut formula is $a \Vdash^\forall A$, principal in both premises of cut. We have a derivation of the form

$$\dfrac{\dfrac{\begin{array}{c}\mathcal{D}\\ x \in a, \Gamma \Rightarrow \Delta, x : A\end{array}}{\Gamma \Rightarrow \Delta, a \Vdash^\forall A} R\Vdash^\forall \quad \dfrac{y : A, y \in a, a \Vdash^\forall A, \Gamma' \Rightarrow \Delta'}{y \in a, a \Vdash^\forall A, \Gamma' \Rightarrow \Delta'} L\Vdash^\forall}{y \in a, \Gamma, \Gamma' \Rightarrow \Delta, \Delta'} \; Cut$$

This is converted into the following derivation:

$$\dfrac{\dfrac{\begin{array}{c}\mathcal{D}(y/x)\\ y \in a, \Gamma \Rightarrow \Delta, y : A\end{array} \quad \dfrac{\Gamma \Rightarrow \Delta, a \Vdash^\forall A \quad y : A, y \in a, a \Vdash^\forall A, \Gamma' \Rightarrow \Delta'}{y \in a, y : A, \Gamma, \Gamma' \Rightarrow \Delta, \Delta'} \; Cut_1}{\dfrac{y \in a, y \in a, \Gamma, \Gamma, \Gamma' \Rightarrow \Delta, \Delta, \Delta'}{y \in a, \Gamma, \Gamma' \Rightarrow \Delta, \Delta'} \; Ctr^*} Cut_2$$

Here $\mathcal{D}(y/x)$ denotes the result of application of hp-substitution to $\mathcal{D}$, using the fact that x is a fresh variable; compared to the original cut, Cut_1 is a cut of reduced height, Cut_2 is one of reduced weight of cut formula, because $\mathtt{w}(y : A) < \mathtt{w}(a \Vdash^\forall A)$, and Ctr^* denote repreated applications of (hp-)admissible contraction steps.

2. The cut formula is $a \Vdash^\exists A$, principal in both premises of cut. The cut is reduced in a way similar to the one in the case above and the inequality to be used on formula weight is $\mathtt{w}(y : A) < \mathtt{w}(a \Vdash^\exists A)$.

3. The cut formula is $x : [\,]A$, principal in both premises of cut.
We have a derivation of the form

$$\dfrac{\dfrac{\begin{array}{c}\mathcal{D}\\ a \in I(x), \Gamma \Rightarrow \Delta, a \Vdash^\forall A\end{array}}{\Gamma \Rightarrow \Delta, x : [\,]A} R[\,] \quad \dfrac{b \Vdash^\forall A, b \in I(x), x : [\,]A, \Gamma' \Rightarrow \Delta'}{b \in I(x), x : [\,]A, \Gamma' \Rightarrow \Delta'} L[\,]}{b \in I(x), \Gamma, \Gamma' \Rightarrow \Delta, \Delta'} \; Cut$$

The transformed derivation is obtained as follows:

$$\dfrac{\dfrac{\begin{array}{c}\mathcal{D}(b/a)\\ b \in I(x), \Gamma \Rightarrow \Delta, b \Vdash^\forall A\end{array} \quad \dfrac{\Gamma \Rightarrow \Delta, x : [\,]A \quad b \Vdash^\forall A, b \in I(x), x : [\,]A, \Gamma' \Rightarrow \Delta'}{b \Vdash^\forall A, b \in I(x), \Gamma, \Gamma' \Rightarrow \Delta, \Delta'} \; Cut}{\dfrac{b \in I(x), b \in I(x), \Gamma, \Gamma, \Gamma' \Rightarrow \Delta, \Delta, \Delta'}{b \in I(x), \Gamma, \Gamma' \Rightarrow \Delta, \Delta'} \; Ctr^*} Cut$$

where the upper cut is of reduced height and the lower one of reduced weight because $\mathtt{w}(b \Vdash^\forall A) < \mathtt{w}(x : [\,]A)$.

The cases with cut formula of the form $x : \langle\,]A$, $x : [\,\rangle A$, and $x : \langle\,\rangle A$ are all treated in a similar way, using, respectively, the following inequalities that hold for the weight of the cut formulas, namely, $\mathtt{w}(b \Vdash^\forall A) < \mathtt{w}(x : \langle\,]A)$, $\mathtt{w}(b \Vdash^\exists A) < \mathtt{w}(x : [\,\rangle A)$, and $\mathtt{w}(b \Vdash^\exists A) < \mathtt{w}(x : \langle\,\rangle A)$.

We observe that it is essential here that the rules are in harmony in the sense that for each modality each pair of rules has either $\Vdash^\forall$ or $\Vdash^\exists$ in the premises.

4. The cut formula is $A \triangleleft a$, principal in both premises of cut. We have:

$$\dfrac{\dfrac{\mathcal{D}}{x:A,\Gamma \Rightarrow \Delta, x \in a}}{\Gamma \Rightarrow \Delta, A \triangleleft a}\, R\triangleleft \quad \dfrac{y:A, y \in a, A \triangleleft a, \Gamma' \Rightarrow \Delta'}{y \in a, A \triangleleft a, \Gamma' \Rightarrow \Delta'}\, L\triangleleft$$
$$\overline{y:A,\Gamma,\Gamma' \Rightarrow \Delta,\Delta'}\ Cut$$

The cut is converted as follows:

$$\dfrac{\dfrac{\mathcal{D}(y/x)}{y:A,\Gamma \Rightarrow \Delta, y \in a} \quad \dfrac{\Gamma \Rightarrow \Delta, A \triangleleft a \quad y:A,y \in a, A \triangleleft a, \Gamma' \Rightarrow \Delta'}{y:A,y \in a, \Gamma, \Gamma' \Rightarrow \Delta, \Delta'}\ Cut}{\dfrac{y:A,y:A,\Gamma,\Gamma,\Gamma' \Rightarrow \Delta,\Delta,\Delta'}{y:A,\Gamma,\Gamma' \Rightarrow \Delta,\Delta'}\ Ctr^*}\ Cut$$

where the upper cut is of reduced cut height and the lower one of reduced weight of cut formula because $\mathtt{w}(y \in a) < \mathtt{w}(A \triangleleft a)$.

5. The cut formula is $x:\Box A$, principal in both premises of cut. We have a cut of the form

$$\dfrac{\dfrac{b \in I(x), \Gamma \Rightarrow \Delta, x:\Box A, b \Vdash^\forall A \quad b \in I(x), \Gamma \Rightarrow \Delta, x:\Box A, A \triangleleft b}{b \in I(x), \Gamma \Rightarrow \Delta, x:\Box A}\, R\Box \quad \dfrac{\dfrac{\mathcal{D}}{a \in I(x), a \Vdash^\forall A, A \triangleleft a, \Gamma' \Rightarrow \Delta'}}{x:\Box A, \Gamma' \Rightarrow \Delta'}\, L\Box}{b \in I(x), \Gamma, \Gamma' \Rightarrow \Delta, \Delta'}\ Cut$$

This is transformed into derivation with four smaller cuts as follows. First we have

$$\dfrac{\dfrac{b \in I(x), \Gamma \Rightarrow \Delta, x:\Box A, b \Vdash^\forall A \quad x:\Box A, \Gamma' \Rightarrow \Delta'}{b \in I(x), \Gamma, \Gamma' \Rightarrow \Delta, \Delta', b \Vdash^\forall A}\ Cut \quad \dfrac{\mathcal{D}(b/a)}{b \in I(x), b \Vdash^\forall A, A \triangleleft b, \Gamma' \Rightarrow \Delta'}}{(b \in I(x))^2, A \triangleleft b, \Gamma, (\Gamma')^2 \Rightarrow \Delta, (\Delta')^2}\ Cut$$

with two reduced cuts, the upper one with the original cut formula but smaller derivation height, and the lower one with a cut formula of reduced weight because $\mathtt{w}(b \Vdash^\forall A) < \mathtt{w}(x:\Box A)$.

We then continue with two more cuts as follows:

$$\dfrac{\dfrac{b \in I(x), \Gamma \Rightarrow \Delta, x:\Box A, A \triangleleft b \quad x:\Box A, \Gamma' \Rightarrow \Delta'}{b \in I(x), \Gamma, \Gamma' \Rightarrow \Delta, \Delta', A \triangleleft b}\ Cut \quad (b \in I(x))^2, A \triangleleft b, \Gamma, (\Gamma')^2 \Rightarrow \Delta, (\Delta')^2}{\dfrac{(b \in I(x))^3, (\Gamma)^2, (\Gamma')^3 \Rightarrow (\Delta)^2, (\Delta')^3}{b \in I(x), \Gamma, \Gamma' \Rightarrow \Delta, \Delta'}\ Ctr^*}\ Cut$$

where the upper cut is on the original cut formula, but of reduced height, and the lower one of reduced weight because $\mathtt{w}(A \triangleleft b) < \mathtt{w}(x:\Box A)$.

If instead the monotonic rules $R\square'$, $L\square'$ have been used, the conversion is simpler: We have a cut of the form

$$\dfrac{\dfrac{b \in I(x), \Gamma \Rightarrow \Delta, x : \square A, b \Vdash^\forall A}{b \in I(x), \Gamma \Rightarrow \Delta, x : \square A} R\square' \quad \dfrac{\overset{\mathcal{D}}{a \in I(x), a \Vdash^\forall A, \Gamma' \Rightarrow \Delta'}}{x : \square A, \Gamma' \Rightarrow \Delta'} L\square'}{b \in I(x), \Gamma, \Gamma' \Rightarrow \Delta, \Delta'} Cut$$

This is converted into a derivation with two cuts, the upper one of reduced height and the lower one or reduced weight, followed by contractions, so that the inductive hypothesis applies. The details are easy and left to the reader.

For extensions of the basic system, we need to consider also the cases of cut with cut formula of the form $a \subseteq b$ or $x \in a \cap b$ principal in both premisses of cut. In the first case, we have a derivation of the form

$$\dfrac{\dfrac{x \in a, \Gamma \Rightarrow \Delta, x \in b}{\Gamma \Rightarrow \Delta, a \subseteq b} R\subseteq \quad \dfrac{y \in a, y \in b, a \subseteq b, \Gamma' \Rightarrow \Delta'}{y \in a, a \subseteq b, \Gamma' \Rightarrow \Delta'} L\subseteq}{y \in a, \Gamma, \Gamma' \Rightarrow \Delta, \Delta'} Cut$$

This is converted into a derivation with two cuts, the first of reduced height, the second of reduced weight, as follows:

$$\dfrac{y \in a, \Gamma \Rightarrow \Delta, y \in b \quad \dfrac{\Gamma \Rightarrow \Delta, a \subseteq b \quad y \in a, y \in b, a \subseteq b, \Gamma' \Rightarrow \Delta'}{y \in a, y \in b, \Gamma, \Gamma' \Rightarrow \Delta, \Delta'} Cut_1}{\dfrac{y \in a^2, \Gamma^2, \Gamma' \Rightarrow \Delta^2, \Delta'}{y \in a, \Gamma, \Gamma' \Rightarrow \Delta, \Delta'} Ctr^*} Cut_2$$

Here the left premiss of the second cut is obtained by a hp-substitution on the premiss of $R \subseteq$.

In the second case, we have a derivation of the form

$$\dfrac{\dfrac{\Gamma \Rightarrow \Delta, x \in a \cap b, x \in a \quad \Gamma \Rightarrow \Delta, x \in a \cap b, x \in b}{\Gamma \Rightarrow \Delta, x \in a \cap b} R\cap \quad \dfrac{x \in a, x \in b, x \in a \cap b, \Gamma' \Rightarrow \Delta'}{x \in a \cap b, \Gamma' \Rightarrow \Delta'} L\cap}{\Gamma, \Gamma' \Rightarrow \Delta, \Delta'} Cut$$

This is converted into a derivation with five cuts, Cut_1, Cut_2 and Cut_4 of reduced height, and the remaining two of reduced weight of cut furmula:

$$\dfrac{\dfrac{\Gamma \Rightarrow \Delta, x \in a \cap b, x \in b \quad x \in a \cap b, \Gamma' \Rightarrow \Delta'}{\Gamma, \Gamma' \Rightarrow \Delta, \Delta', x \in b} Cut_4 \quad \dfrac{\vdots}{x \in b, \Gamma^2, \Gamma'^2 \Rightarrow \Delta^2, \Delta'^2}}{\dfrac{\Gamma^3, \Gamma'^3 \Rightarrow \Delta^3, \Delta'^3}{\Gamma, \Gamma' \Rightarrow \Delta, \Delta'} Ctr^*} Cut_5$$

where the dotted part is continued as follows:

$$\cfrac{\Gamma \Rightarrow \Delta, x \in a \cap b, x \in a \quad x \in a \cap b, \Gamma' \Rightarrow \Delta'}{\Gamma, \Gamma' \Rightarrow \Delta, \Delta', x \in a} Cut_1 \quad \cfrac{\Gamma \Rightarrow \Delta, x \in a \cap b \quad x \in a, x \in b, x \in a \cap b, \Gamma' \Rightarrow \Delta'}{x \in a, x \in b, \Gamma, \Gamma' \Rightarrow \Delta, \Delta'} Cut_2$$
$$\cfrac{}{x \in b, \Gamma^2, \Gamma'^2 \Rightarrow \Delta^2, \Delta'^2} Cut_3$$

QED

5 Soundness and completeness

Next, we give a proof of soundness and a direct proof of completeness of our calculus with respect to neighbourhood semantics. Specifically, we show that all the rules are sound, and show that proof search in the calculus either produces a proof, or provides us with a saturated branch which is used to define a countermodel. The countermodel will be defined *directly*, that is, using the syntactic elements (labels) and the forcing conditions in the saturated branch, without any need for additional constructions.

Soundness

We recall a definition from Chellas ([4], p. 215):

Definition 5.1. *Let $\mathcal{F} \equiv (W, I)$ be a neighbourhood frame.*

- *$\mathcal{F}$ is **supplemented** if for all subsets α, β of W and for all $x \in W$, if $\alpha \in I(x)$ and $\alpha \subseteq \beta$, we have $\beta \in I(x)$.*

- *$\mathcal{F}$ is **closed under intersection** if for all $x \in W$ for all α, β in $I(x)$, we have $\alpha \cap \beta \in I(x)$.*

- *$\mathcal{F}$ is **contains the unit** if for all $x \in W$, W is in $I(x)$.*

Definition 5.2. *Given a set S of world labels x and a set NL of neighbourhood labels a, and a neighbourhood model $\mathcal{M} = (W, I, \mathcal{V})$, an SN-realisation (ρ, σ) is a pair of functions mapping each $x \in S$ into $\rho(x) \in W$ and mapping each $a \in NL$ into $\sigma(a) \in I(w)$ for some $w \in W$. As SN-realisation (ρ, σ) has to respect formal intersection of the language, i.e., $\sigma(a \cap b) = \sigma(a) \cap \sigma(b)$[22]. We introduce the notion "$\mathcal{M}$ satisfies a formula F under an SN-realisation (ρ, σ)" and denote it by $\mathcal{M} \models_{\rho, \sigma}$*

---

[22]Observe that the symbol on the left denotes formal intersection, the one on the right set-theoretic intersection.

F, where we assume that the labels in F occur in S, NL. The definition extends the usual clauses for the propositional connectives by cases on the form of F.[23]

- $\mathcal{M} \models_{\rho,\sigma} x \in a$ if $\rho(x) \in \sigma(a)$
- $\mathcal{M} \models_{\rho,\sigma} a \in I(x)$ if $\sigma(a) \in I(\rho(x))$
- $\mathcal{M} \models_{\rho,\sigma} a \subseteq b$ if $\sigma(a) \subseteq \sigma(b)$
- $\mathcal{M} \models_{\rho,\sigma} x : A$ if $\rho(x) \Vdash A$
- $\mathcal{M} \models_{\rho,\sigma} a \Vdash^\exists A$ if there exists w in $\sigma(a)$ such that $w \Vdash A$
- $\mathcal{M} \models_{\rho,\sigma} a \Vdash^\forall A$ if for all w in $\sigma(a)$, $w \Vdash A$
- $\mathcal{M} \models_{\rho,\sigma} A \triangleleft a$ if $[A] \subseteq \sigma(a)$
- $\mathcal{M} \models_{\rho,\sigma} x : \Box A$ if for some a, $\sigma(a) \in I(\rho(x))$ and $\sigma(a) = [A]$

Given a sequent $\Gamma \Rightarrow \Delta$, let S, NL be the sets of world and neighbourhood labels occurring in $\Gamma \cup \Delta$, and let (ρ, σ) be an SN-realisation; we define $\mathcal{M} \models_{\rho,\sigma} \Gamma \Rightarrow \Delta$ to hold if whenever $\mathcal{M} \models_{\rho,\sigma} F$ for all formulas $F \in \Gamma$ then $\mathcal{M} \models_{\rho,\sigma} G$ for some formula $G \in \Delta$. We further define $\mathcal{M}$-validity by

$$\mathcal{M} \models \Gamma \Rightarrow \Delta \text{ iff } \mathcal{M} \models_{\rho,\sigma} \Gamma \Rightarrow \Delta \text{ for every } SN\text{-realisation } (\rho, \sigma)$$

We finally say that a sequent $\Gamma \Rightarrow \Delta$ is valid in a neighbourhood frame $\mathcal{F}$ if $\mathcal{M} \models \Gamma \Rightarrow \Delta$ for every neighbourhood model $\mathcal{M}$ based on $\mathcal{F}$.

Below, we shall use the notation $\mathcal{M} \models_{\rho,\sigma} \Gamma$ for $\mathcal{M} \models_{\rho,\sigma} F$ for all $F \in \Gamma$. We shall denote with **G3nM***, **G3nC***, **G3nN*** the extensions of **G3n** which are monotonic, contain rule C, and rule N, respectively. Since extensions are obtained in a modular way, further extensions with rules that correspond to the frame properties * are indicated by the asterisk.

Theorem 5.3. *If $\Gamma \Rightarrow \Delta$ is derivable in* **G3n*** *(respectively* **G3nM***, **G3nC***, **G3nN***), then it is valid in the class of neighbourhood frames (respectively neighbourhood frames which are supplemented, closed under intersection, containing the unit) with the * properties.*

---

[23]Observe that hereafter we use the more compact notation $[A]$, in place of $ext(A)$, for the extension of A.

Proof. By induction on the height n of the derivation of $\Gamma \Rightarrow \Delta$ in **G3nE*** (resp. **G3nM***, **G3nC***, **G3nN***).

For $n = 0$, observe that initial sequents have the same labelled formula in the antecedent and in the succedent so the claim is obvious. Similarly if the antecedent contains $x : \bot$ because we assume that for no $w \in W$, $w \Vdash \bot$.

For the inductive step, consider the last rule in the derivation of $\Gamma \Rightarrow \Delta$. If it is a propositional rule, the claim is immediate by the definition of the forcing clauses for the propositional connectives.

If the last rule is $R \Vdash^\forall$, assume by induction hypothesis that $\mathcal{M} \models x \in a, \Gamma \Rightarrow \Delta, x : A$. Let (ρ, σ) be an arbitrary SN-realisation for the conclusion and assume that $\mathcal{M} \models_{\rho,\sigma} \Gamma$. Since x is fresh, it can be extended to ρ', an S-realization for the premiss with $\rho'(x) \in \sigma(a)$. Then (using the assumption that $x \notin \Gamma$) we have $\mathcal{M} \models_{\rho',\sigma} x \in a, \Gamma$. By the hypothesis $\mathcal{M} \models x \in a, \Gamma \Rightarrow \Delta, x : A$, we have that either (1) $\mathcal{M} \models_{\rho',\sigma} G$ for some G in Δ or (2) $\mathcal{M} \models_{\rho',\sigma} x : A$. In the former case we are done, so let us assume that $\mathcal{M} \models_{\rho',\sigma} G$ for no G in Δ. Since $x \notin \Delta$, this will be the case uniformly, independently of the choice of $\rho'(x)$, so we'll have $\mathcal{M} \models_{\rho',\sigma} x : A$ for all $\rho'(x) \in \sigma(a)$, and therefore $\mathcal{M} \models_{\rho,\sigma} a \Vdash^\forall A$.

If the last rule is $L \Vdash^\forall$, the claim holds because if $\mathcal{M} \models_{\rho,\sigma} x \in a$ and $\mathcal{M} \models_{\rho,\sigma} a \Vdash^\forall A$, then $\mathcal{M} \models_{\rho,\sigma} x : A$ by simply unfolding the definitions.

If the last rule is $R \Vdash^\exists$, consider an arbitrary SN-realisation (ρ, σ) and assume that (1) $\mathcal{M} \models_{\rho,\sigma} x \in a, \Gamma$. Then, by induction hypothesis, either (2) $\mathcal{M} \models_{\rho,\sigma} G$ for some $G \in \Delta$, or (3) $\mathcal{M} \models_{\rho,\sigma} x : A$, or (4) $\mathcal{M} \models_{\rho,\sigma} a \Vdash^\exists A$. If (2) or (4) hold, then the claim follows. If (3) holds, we have $\rho(x) \Vdash A$. Observe that (1) gives in particular $\rho(x) \in \sigma(a)$, so there is $w \in \sigma(a)$ such that $w \Vdash A$. It follows that the conclusion of the rule is $\mathcal{M}$-valid for the SN-realisation (ρ, σ).

If the last rule is $L \Vdash^\exists$, assume that $\mathcal{M} \models_{\rho,\sigma} a \Vdash^\exists A, \Gamma$ for an arbitrary SN-realisation for the conclusion (ρ, σ). Then there is $w \in \sigma(a)$ such that $w \Vdash A$. Since x is fresh, we can extend ρ to and S-realization for the premiss by choosing $\rho'(x) = w$. Then we have $\mathcal{M} \models_{\rho',\sigma} x \in a, x : A$ by definition, and $\mathcal{M} \models_{\rho',\sigma} \Gamma$ because $x \notin \Gamma$. By induction hypothesis, the premiss of the rule is $\mathcal{M}$-valid, and therefore there is G in Δ such that $\mathcal{M} \models_{\rho',\sigma} G$. Since $x \notin \Delta$, this is the same as $\mathcal{M} \models_{\rho,\sigma} G$.

If the last rule is $R \triangleleft$, with premiss $y : A, \Gamma \Rightarrow \Delta, y \in a$, let (ρ, σ) be an arbitrary SN-realisation for the conclusion and assume that $\mathcal{M} \models_{\rho,\sigma} \Gamma$. The claim is that for some formula B in Δ, $\mathcal{M} \models_{\rho,\sigma} B$ or $\mathcal{M} \models_{\rho,\sigma} A \triangleleft a$. Since y is fresh, we can extend ρ to a S-realization for the premiss ρ' by choosing $\rho'(y) \in [A]$. Since $\mathcal{M} \models y : A, \Gamma \Rightarrow \Delta, y \in a$, we have that there exists $B \in \Delta$ such that $\mathcal{M} \models_{\rho',\sigma} B$ or $\mathcal{M} \models_{\rho',\sigma} y \in a$. In the first case, since y does not occur in B, we have also $\mathcal{M} \models_{\rho,\sigma} B$. In the second case, since $\rho'(y)$ was arbitrary in $[A]$, we have $\mathcal{M} \models_{\rho,\sigma} A \triangleleft a$.

If the last rule is $L \triangleleft$, assume that the premiss $y \in a, A \triangleleft a, y : A, \Gamma \Rightarrow \Delta$ is valid, and let (ρ, σ) be an arbitrary SN-realisation with $\mathcal{M} \models_{\rho,\sigma} A \triangleleft a, y : A, \Gamma$. Then we have $\rho(y) \in [A]$ and $[A] \subseteq \sigma(a)$, so that $\rho(y) \in \sigma(a)$, thus $\mathcal{M} \models_{\rho,\sigma} y \in a, A \triangleleft a, y : A, \Gamma$. By the assumption, there is B in Δ such that $\mathcal{M} \models_{\rho,\sigma} B$ and thus the claim follows.

If the last rule is $L\Box$, assume the premiss valid and let (ρ, σ) be an arbitrary SN-realisation with $\mathcal{M} \models_{\rho,\sigma} x : \Box A, \Gamma$. This means in particular that $\rho(x) \in [A]$, i.e., there is α in $I(\rho(x))$ with $\alpha = [A]$. Since a is fresh, we can extend σ to σ' by having $\sigma'(a) = \alpha$. We have $\mathcal{M} \models_{\rho,\sigma'} a \in I(x), A \triangleleft a, a \Vdash^\forall A$ by the definitions and also $\mathcal{M} \models_{\rho,\sigma'} \Gamma$ because $a \notin \Gamma$ and by hypothesis $\mathcal{M} \models_{\rho,\sigma} \Gamma$. Again by hypothesis, there is B in Δ with $\mathcal{M} \models_{\rho,\sigma'} B$ and thus by freshness of a (not in B) we have $\mathcal{M} \models_{\rho,\sigma} B$.

If the last rule is $R\Box$, assume the premisses valid and assume for an arbitrary SN-realisation (ρ, σ) that $\mathcal{M} \models_{\rho,\sigma} a \in I(x), \Gamma$. From the validity of the premisses we have that one of the following alternatives holds: 1: $\mathcal{M} \models_{\rho,\sigma} B$ for some B in Δ. 2. $\mathcal{M} \models_{\rho,\sigma} x : A$. 3. $\mathcal{M} \models_{\rho,\sigma} a \Vdash^\forall A, A \triangleleft a$. Observe that the latter gives, together with $\mathcal{M} \models_{\rho,\sigma} a \in I(x)$ that $\mathcal{M} \models_{\rho,\sigma} x : \Box A$ so in each of the three cases we have proved the claim.

Next, we consider the rules for inclusion. If the last rule is $R \subseteq$, consider an SN-realisation such that $\mathcal{M} \models_{\rho,\sigma} \Gamma$. Since x is fresh, we can extend ρ to ρ' by choosing $\rho'(x) \in \sigma(a)$. Since the premiss in $\mathcal{M}$-valid, by inductive hypothesis we have that $\mathcal{M} \models_{\rho',\sigma} G$ for some $G \in \Delta$ or $\mathcal{M} \models_{\rho',\sigma} x \in b$. Since x is not in Δ, the former gives $\mathcal{M} \models_{\rho,\sigma} G$ for some $G \in \Delta$, whereas the latter gives, by the choice in the range of $\rho'(x)$, $\mathcal{M} \models_{\rho,\sigma} a \subseteq b$.

The case with $L \subseteq$ as the last rule is immediate.

The preservation of validity in the case of rules $[\,]$, $\langle\,]$, $[\,\rangle$, and $\langle\,\rangle$ follows the same pattern of that for the $\Box$ rules. To conclude, it is immediate that rules M, C, N (and the monotonic variants C', $Nondeg$) are valid in frames frames which are supplemented, closed under intersection, containing the unit (and supplemented for the latter two with the monotonic variants) respectively. QED

Definition 5.4. *We say that a branch in a proof search from the endsequent up to a sequent $\Gamma \Rightarrow \Delta$ is saturated with respect to a rule R if condition (R) below holds, where we indicate with $\downarrow\Gamma$ ($\downarrow\Delta$) the union of the antecedents (succedents) in the branch from the end-sequent up to $\Gamma \Rightarrow \Delta$:*

($Init_0$) There is no $x \in a$ in $\Gamma \cap \Delta$.

($Init$) There is no $x : P$ in $\Gamma \cap \Delta$.

($L\bot$) There is no $x :\bot$ in Γ.

(L&) If $x : A\&B$ is in $\downarrow\Gamma$, then $x : A$ and $x : B$ are in $\downarrow\Gamma$.

(R&) If $x : A\&B$ is in $\downarrow\Delta$, then either $x : A$ or $x : B$ is in $\downarrow\Delta$.

(L$\vee$) If $x : A \vee B$ is in $\downarrow\Gamma$, then either $x : A$ or $x : B$ is in $\downarrow\Gamma$.

(R$\vee$) If $x : A \vee B$ is in $\downarrow\Delta$, then $x : A$ and $x : B$ are in $\downarrow\Delta$.

(L$\supset$) If $x : A \supset B$ is in $\downarrow\Gamma$, then either $x : A$ is in $\downarrow\Delta$ or $x : B$ is in $\downarrow\Gamma$.

(R$\supset$) If $x : A \supset B$ is in $\downarrow\Delta$, then $x : A$ is in $\downarrow\Gamma$ and $x : B$ is in $\downarrow\Delta$

(R $\Vdash^\forall$) If $a \Vdash^\forall A$ is in $\downarrow\Delta$, then for some x there is $x \in a$ in Γ and $x : A$ in $\downarrow\Delta$

(L $\Vdash^\forall$) If $x \in a$ and $a \Vdash^\forall A$ and are in Γ, then $x : A$ is in $\downarrow\Gamma$.

(R $\Vdash^\exists$) If $x \in a$ is in Γ and $a \Vdash^\exists A$ is in Δ, then $x : A$ is in $\downarrow\Delta$.

(L $\Vdash^\exists$) If $a \Vdash^\exists A$ is in $\downarrow\Gamma$, then for some x there is $x \in a$ in Γ and $x : A$ is in $\downarrow\Gamma$

(L $\triangleleft$) If $A \triangleleft a$ and $y : A$ are in $\downarrow\Gamma$, then $y \in a$ is in Γ.

(R $\triangleleft$) If $A \triangleleft a$ is in $\downarrow\Delta$, then for some y, $y : A$ is in $\downarrow\Gamma$ and $y \in a$ is in Δ.

(L$\Box$) If $x : \Box A$ is in $\downarrow\Gamma$, then for some a, $a \in I(x)$, $a \Vdash^\forall A$, $A \triangleleft a$ are in $\downarrow\Gamma$.

(L$\Box'$) If $x : \Box A$ is in $\downarrow\Gamma$, then for some a, $a \in I(x)$, $a \Vdash^\forall A$ are in $\downarrow\Gamma$.

(R$\Box$) If $a \in I(x)$ is in Γ and $x : \Box A$ is in $\downarrow\Delta$, then either $a \Vdash^\forall A$ or $A \triangleleft a$ is in $\downarrow\Delta$.

(R$\Box'$) If $a \in I(x)$ is in Γ and $x : \Box A$ is in $\downarrow\Delta$, then $a \Vdash^\forall A$ is in $\downarrow\Delta$.

(L $\subseteq$) If $x \in a$ and $a \subseteq b$ are in $\downarrow\Gamma$, then $x \in b$ is in Γ.

(R $\subseteq$) If $a \subseteq b$ is in $\downarrow\Delta$, then for some x there is $x \in a$ in Γ and $x \in b$ in Δ.

(L$\cap$) If $x \in a \cap b$ is in Γ, then $x \in a$ and $x \in b$ are in Γ.

(R$\cap$) If $x \in a \cap b$ is in Δ, then either $x \in a$ or $x \in b$ are in Δ.

(M) If $a \in I(x), a \subseteq b$ are in Γ, then $b \in I(x)$ is in Γ.

(C) If $a \in I(x), b \in I(x)$ are in Γ, then $a \cap b$ is in Γ.

(C') If $a \in I(x), b \in I(x)$ are in Γ, then for some c, $c \in I(x)$, $c \subseteq a$, $c \subseteq b$ are in Γ.

(N) *For some a, $a \in I(x)$, $\top \triangleleft a$ are in Γ.*

(Nondeg) *For some a, $a \in I(x)$ is in Γ.*

A branch is saturated relative to a systems $\mathcal{S}$ of rules if it is saturated with respect each rule of $\mathcal{S}$.

The definition of saturation with respect to the rules for the modalities $[\,]$, $\langle\,|$, $[\,\rangle$, and $\langle\,\rangle$ has been left out as it involves eight more clauses and it should be by now clear from the meaning of saturation with respect a rule and the pattern of the other cases. The definition of saturated branch is extended to infinite branches $\mathcal{B} \equiv \{\Gamma_i \Rightarrow \Delta_i\}_{i \geq 0}$ by replacing, in the definition above, Γ (or $\downarrow \Gamma$) by $\boldsymbol{\Gamma}$, the union of the Γ_i, and Δ (or $\downarrow \Delta$) by $\boldsymbol{\Delta}$, the union of the Δ_i. The first and second clause ($Init_0$, $Init$) are modified to requiring that for all i, there is no $x \in a$ in $\Gamma_i \cap \Delta_i$ and for all i, there is no $x : P$ in $\Gamma_i \cap \Delta_i$.

Given a sequent $\Gamma \Rightarrow \Delta$ we apply root-first all the available rules. Observe that by invertibility of the rules, there is no prescribed order in which they need to be applied. We want to avoid the possibility that the search produces an infinite branch which is not saturated, something that would result, e.g., from applying the same rule infinitely many times in consecutive steps. This is achieved as usual in such proofs through a counter: if there are m rules, apply at step 1 rule R_1 to all formulas that match its conclusion, at step 2 rule R_2, and in general for all $n \geq 0$ apply at step $n \times m + j$ rule R_j. In this way we'll obtain a *proof-search tree* that can be a derivation, or a non-derivation; the latter can either be a finite search tree that contains finite saturated branches, or an infinite search that, by König's lemma contains an infinite, saturated branch. We shall now prove that a saturated branch (either finite or infinite) for a sequent $\Gamma \Rightarrow \Delta$ gives a countermodel.

Lemma 5.5. *Let $\mathcal{B} \equiv \{\Gamma_i \Rightarrow \Delta_i\}$ be a saturated branch in a proof-search tree for $\Gamma \Rightarrow \Delta$. Then there exists a countermodel $\mathcal{M}$ to $\Gamma \Rightarrow \Delta$, which makes all the formulas in $\boldsymbol{\Gamma}$ true, and all the formulas in $\boldsymbol{\Delta}$ false.*

Proof. Consider a saturated branch and define the countermodel $\mathcal{M} \equiv (W, I, \mathcal{V})$ as follows:

1. The set W of worlds consists of all the world labels in $\boldsymbol{\Gamma}$;

2. For each neighbourhood label a in $\boldsymbol{\Gamma}$, we associate α_a, the set that consists of all the y in W such that $y \in a$ is in $\boldsymbol{\Gamma}$;

3. For each x in W, the set of neighbourhoods of x consists of all the α_a such that $a \in I(x)$ is $\boldsymbol{\Gamma}$;

4. The valuation is defined by $x \in \mathcal{V}(P)$ if $x : P$ is in Γ.

We then define a realization (ρ, σ) by $\rho(x) \equiv x$ and $\sigma(a) \equiv \alpha_a$. Next we prove the following:

1. If A is in Γ, then $\mathcal{M} \models_{\rho,\sigma} A$.

2. If A is in Δ, then $\mathcal{M} \not\models_{\rho,\sigma} A$.

The two claims are proved simultaneously by cases/induction on the weight of A (cf. Definition 4.1).

(a) If A is a formula of the form $a \in I(x)$, $x \in a$, $a \subseteq b$, claim 1. holds by definition of $\mathcal{M}$; if A is $x \in a \cap b$, by saturation we have that $x \in a$ and $x \in b$ are in Γ. These are lighter formulas, so the inductive hypothesis applies and we have $\rho(x) \in \sigma(a)$ and $\rho(x) \in \sigma(b)$, so $\rho(x) \in \sigma(a) \cap \sigma(b)$. The conclusion $\rho(x) \in \sigma(a \cap b)$ follows from the fact that σ respects intersection. Claim 2. is empty for $a \in I(x)$ because such formulas never occur on the right-hand side of sequents. If $x \in a$ is in Δ, then $x \in a$ is not in Γ and thus $\rho(x) \notin \alpha_a$, so $\mathcal{M} \not\models_{\rho,\sigma} x \in a$. If $a \subseteq b$ is in Δ, then for some x, $x \in a$ is in Γ and $x \in b$ is in Δ, so by inductive hypothesis $\mathcal{M} \models_{\rho,\sigma} x \in a$ and $\mathcal{M} \not\models_{\rho,\sigma} x \in b$, and therefore $\mathcal{M} \not\models_{\rho,\sigma} a \subseteq b$.

(b) If A is a labelled atomic formula $x : P$, the claims hold by definition of $\mathcal{V}$ and by the saturation clause *Init* no inconsistency arises. If A is $\bot$, it holds by definition of the forcing relation that it is never forced, and therefore 2. holds, whereas 1. holds by the saturation clause for $L \bot$. If A is a conjunction, or a disjunction, or an implication, the claim holds by the corresponding saturation clauses and inductive hypothesis on smaller formulas.

(c) If $a \Vdash^{\exists} A$ is in Γ, by the saturation clause $(L \Vdash^{\exists})$, for some x there is $x \in a$ in Γ and $x : A$ is in Γ. Then $\mathcal{M} \models_{\rho,\sigma} x \in a$ by (a) and by induction hypothesis $\mathcal{M} \models_{\rho,\sigma} x : A$, therefore $\mathcal{M} \models_{\rho,\sigma} a \Vdash^{\exists} A$. If $a \Vdash^{\exists} A$ is in Δ, consider an arbitrary world x in α_a. Then by definition of $\mathcal{M}$ we have $x \in a$ in Γ and thus by the saturation clause $(R \Vdash^{\exists})$ we also have $x : A$ is in Δ. By induction hypothesis we have $\mathcal{M} \not\models_{\rho,\sigma} x : A$ and therefore $\mathcal{M} \not\models_{\rho,\sigma} a \Vdash^{\exists} A$. The proof for formulas of the form $a \Vdash^{\forall} A$ is similar.

(d) If $A \triangleleft a$ is in Γ, let y be an arbitrary label such that $\mathcal{M} \models_{\rho,\sigma} y : A$. Then by definition of $\mathcal{M}$ we have $y : A$ in Γ and then by saturation $y \in a$ is in Γ thus by inductive hypothesis and by definition of $\mathcal{M}$ we obtain $\mathcal{M} \models_{\rho,\sigma} A \triangleleft a$.

If $A \triangleleft a$ is in Δ, by the corresponding saturation clause we have that for some y, $y : A$ is in Γ and $y : a$ is in Δ, so by induction hypothesis we have that there is y such that $\mathcal{M} \models_{\rho,\sigma} y : A$ and $\mathcal{M} \not\models_{\rho,\sigma} y \in a$. Overall, this means that $\mathcal{M} \not\models_{\rho,\sigma} A \triangleleft a$.

(e) If $x : \Box A$ is in Γ, then for some a, $a \in I(x)$, $a \Vdash^\forall A$, $A \lhd a$ are in Γ. By induction hypothesis we obtain $\mathcal{M} \models_{\rho,\sigma} a \Vdash^\forall A$ and $\mathcal{M} \models_{\rho,\sigma} A \lhd a$, and therefore $\mathcal{M} \models_{\rho,\sigma} x : \Box A$.

If $x : \Box A$ is in Δ, let α_a be a neighbourhood in $I(x)$ in the model. By the saturation clause, we have that either $a \Vdash^\forall A$ or $A \lhd a$ is in Δ. By induction hypothesis we obtain $\mathcal{M} \not\models_{\rho,\sigma} a \Vdash^\forall A$ or $\mathcal{M} \not\models_{\rho,\sigma} A \lhd a$, and therefore $\mathcal{M} \not\models_{\rho,\sigma} x : \Box A$.

In order to prove completeness for extensions of $\mathbf{E}$ we need to prove that the countermodel $\mathcal{M}$ is in the intended class. For M (1), we shall consider the version of the $\Box$ rules with monotonicity built-in and modify the model to impose monotonicity; for C (2) and N (3) instead we shall extend in a consistent way the saturated branch.

(1) Let $\bar{\mathcal{M}}$ be defined as $\mathcal{M}$ above, but taking for $I(x)$ *supersets* of the α_a. In this way $\bar{\mathcal{M}}$ is supplemented. We need to verify that if $x : \Box A$ is in Γ then $\bar{\mathcal{M}} \models_{\rho,\sigma} \Box A$: by the saturation clause for $L\Box$ we have that for some a such that $a \in I(x)$ is in Γ, $a \Vdash^\forall A$ is in Γ. By inductive hypothesis, $\mathcal{M} \models_{\rho,\sigma} a \Vdash^\forall A$ and therefore, since $\bar{\mathcal{M}}$ is supplemented, $\bar{\mathcal{M}} \models_{\rho,\sigma} \Box A$. If $x : \Box A$ is in Δ, let α_a be a neighbourhood of x in the model. This means that $a \in I(x)$ is in Γ. By the $R\Box$ saturation clause, $a \Vdash^\forall A$ is in Δ, so by inductive hypothesis $\mathcal{M} \not\models_{\rho,\sigma} a \Vdash^\forall A$, and therefore it is not the case that for all w in α_a, $w \Vdash A$. Since α_a was an arbitrary neighbourhood of x, we have $\bar{\mathcal{M}} \not\models_{\rho,\sigma} \Box A$.

(2) The saturated branch is extended as follows: whenever Γ contains $x \in a$ and $x \in b$, we add $x \in a \cap b$ to Γ (observe that this move doesn't collapse the saturated branch into an initial sequent since if $x \in a \cap b$ was in Δ, then by saturation either $x \in a$ or $x \in b$ would be in Δ, against the assumption that we started with a saturated branch). We call the branch thus obtained a C-extended saturated branch. Next we prove that the model $\mathcal{M}$ built on the C-extended saturated branch is closed under intersection. Let α_a and α_b be in $I(x)$. This means that $a \in I(x)$ and $b \in I(x)$ are in Γ. We show that $\alpha_a \cap \alpha_b = \alpha_{a \cap b}$ and therefore conclude that $\alpha_a \cap \alpha_b$ is also in $I(x)$. Clearly, if $y \in \alpha_{a \cap b}$, i.e., $y \in a \cap b$ in Γ, then by saturation $y \in a$ and $y \in b$ are in Γ, therefore $y \in \alpha_a \cap \alpha_b$. The converse inclusion is guaranteed by the C-extension of the saturated branch. Observe that the equality just proved also shows that the added formulas $x \in a \cap b$ are true in the model.

(3) The saturated branch is extended as follows: for every label y in the branch, we add the formula $y : \top$ to Γ. The branch thus obtained is an N-extended saturated branch. By the saturation condition for N we have that for some $a \in I(x)$, $\top \lhd a$ is in Γ. By the N-extension and the saturation with respect to $L\lhd$, we have that α_a coincides with W, and therefore W is a neighbourhood of x, so the model contains the unit. QED

We are ready to prove the completeness of the calculus.

Theorem 5.6. *If A is valid then there is a derivation of $\Rightarrow x : A$, for any label x.*

Proof. For every A we either find a derivation or a saturated branch. By the above lemma a saturated branch gives a countermodel to A. It follows that if A is valid it has to be derivable. QED

The above completeness proof gives a method to construct countermodels for unprovable sequents. It is also possible to give a simple completeness proof as a direct consequence of the structural properties of the calculus and the derivability of the characteristic axiom of each of the non-normal systems considered:

Theorem 5.7. *Let A be a formula in the language of the modal propositional logic, and let $\mathbf{E}^*$ be any extension of $\mathbf{E}$ with axioms M, C, N (and combinations thereof) and $\mathbf{G3n}^*$ the corresponding labelled sequent calculus. Then if $\mathbf{E}^* \vdash A$, we have $\mathbf{G3n}^* \vdash \Rightarrow x : A$ where x is an arbitrary world label.*

Proof. By induction on the derivation in the axiomatic system. Observe that the result holds for classical propositional axioms and has been proved for each specific modal axiom in Section 3, so it is enough to prove the inductive step for the only rule of the axiomatic system, i.e., that if $\mathbf{E}^* \vdash A$ is obtained by *modus ponens*, then $\mathbf{G3n}^* \vdash \Rightarrow x : A$. Consider derivable premisses B and $B \supset A$. By inductive hypothesis we have $\mathbf{G3n}^* \vdash \Rightarrow x : B$ and $\mathbf{G3n}^* \vdash \Rightarrow x : B \supset A$. The latter gives by (hp-)invertibility of $R\supset$, $\mathbf{G3n}^* \vdash x : B \Rightarrow x : A$. An admissible step of cut gives the desired conclusion. QED

Computational issues about the calculi are not in the scope of the present paper, and we shall deal with termination and complexity of our calculi in further work. However, following the line of our [30] and [13], we can outline the recipe to obtain a terminating proof search in the calculi here presented. First of all, it is useful to make the distinction between *static* and *dynamic* rules. The former do not introduce new labels in moving from conclusions to premisses, whereas the latter do.[24] The main difficulty in obtaining termination is that a proof branch may potentially introduce infinitely many world and neighbourhood labels by unconstrained application of the dynamic rules. The termination of proof search requires to adopt a suitable strategy of rule application which on the one hand preserves the completeness and on the other ensures that in any proof branch only a finite number of labels will be introduced. The strategy will be specific to each calculus, but it contains at least the following constraints:

---

[24] For example, in **G3n** the rule $L\square$ is dynamic and $R\square$ static.

1. Do not apply a rule R to a sequent $\Gamma \Rightarrow \Delta$ if $\downarrow \Gamma$ and/or $\downarrow \Delta$ satisfy the saturation condition associated to R.

2. Apply static rules before dynamic rules.

The strategy may specify further constraints on the order of applications of rules (e.g. rule R_1 must always be applied before rule R_2) or on the temporal order in which the labels must be treated (e.g. apply all rules to a label x before applying any rule to y if x is "older" than y, that is, introduced earlier in the branch).

There is also an additional difficulty for systems where intersection of neighbourhood labels is allowed, as neighbourhood labels become complex terms so that infinitely many terms can be generated from a finite number of labels. To handle this case we shall need to identify term labels which are equivalent modulo commutativity and associativity of intersection.

We shall carry on a detailed analysis of all computation issues along the above lines in further work.

6 Concluding remarks

We have presented a systematic development of labelled sequent calculi for logical systems based on neighbourhood semantics, with focus on classical modal systems. Other approaches to the proof theory of classical modal logics besides the ones mentioned in the introduction include the nested sequent calculi of [21].[25] Additionally, in [16] standard sequent systems (most of them cut free) are provided for extensions of the monotonic system M by all combinations of the modal axioms D, T, 4, B, and 5. Similar results are obtained for *congruent modal logics* (another name for extensions of E) in [17]. Standard sequents are also considered in [19] via an approach based on a treatment of "sequents as sets" that makes contraction implicit, rather than admissible as in the G3-calculi. When such approach to sequent calculus proof theory is followed, all the rules become context-dependent and the proof of cut elimination presents some difficulties that one does not have with the usual approach to sequents as lists or multisets (cf. [33]). Standard, contraction-free sequent calculi for non-normal systems of deontic logic are presented in [37]. A different approach to the proof-theoretic study of non-normal modal logics, with focus on conditional logics, is pursued in [39]: here a criterion is developed for guaranteeing absorption of the structural rules into a system of sequent rules. The conditions the system has to satisfy are closure conditions and typically generate a large number of rules.

---

[25] We remark that nested sequent calculi have been developed also in other venues in non-classical logics (e.g. for modal logic [3] and bi-intuitionistic logic [14]); complexity has been studied in [2].

Labelled calculi for monotonic and regular modal logics have already been considered in [15]. As in our work the labelling originates from neighbourhood semantics, but there are important differences: first, the proof system is a tableau with signed formulas, rather than a sequent style proof system. Second, the calculus has labels with a path structures and no relations, whereas in our approach we have two sorts of labels and the explicit relation of formal membership. Correspondingly, in one systems there are rules that operate on the structured labels through an unification algorithm, whereas in our system there are rules for the neighbourhood semantics counterpart of the accessibility relation of Kripke semantics.

Labelled systems, on the whole, have several advantages over other formalisms for modal logic. First, the systems originate from a uniform methodology which has reached a wide range of applications; the transparent semantic motivation behind the rules makes them intuitive and allows a direct completeness proof. As we have seen, we can use a ground basic system to find, through proof search by invertible rules, which additional rules are needed to obtain complete systems for extensions; this can be useful especially in the absence of known correspondence results.

This extension of the labelled approach inherits the flexibility and far reach of neighbourhood semantics. Here we have focused on the most basic classical systems but it is possible extend the approach to systems with further requirements on the neighbourhood frames, as those listed in section 7.4 of [4]. A property such as $a \in I(x) \to x \in a$ is straightforward to handle and corresponds to an added rule of the form $\frac{x \in a, a \in I(x), \Gamma \Rightarrow \Delta}{a \in I(x), \Gamma \Rightarrow \Delta}$. Other properties, such as $a \in I(x) \to a^c \notin I(x)$ can also be treated by the method of conversion into rules, but one also needs rules for the formal complement of a neighbourhood.

Rather than dwelling on abstract generality, we stress that alongside with the completion of this ground work, labelled calculi based on neighbourhood semantics have been developed for other logics that cannot be studied *simpliciter* through possible world semantics, such as *preferential conditional logic* [30] and *conditional doxastic logic* [13]. Classical modal logics are decidable. The finite model property and finitary proof search can be established in parallel for labelled calculi; we expect that no special difficulties should arise in the case of classical logics, but a detailed proof, along the lines sketched at the end of Section 5, is left to further work.

References

[1] Alenda, R., N. Olivetti, and G. L. Pozzato. Nested sequent calculi for normal conditional logics. *Journal of Logic and Computation*, vol. 26, pp. 7–50, 2016.

[2] Bruscoli, P. and A. Guglielmi. On the proof complexity of deep inference. *ACM Transactions on Computational Logic*, vol. 10, no. 2/14, pp. 1–34, 2009.

[3] Brünnler, K. Deep sequent systems for modal logic, *Archive for Mathematical Logic*, vol. 48, pp. 551–577, 2009.

[4] Chellas, B. *Modal Logic: An Introduction*. Cambridge University Press, 1980.

[5] Dyckhoff, R. and S. Negri. Geometrization of first-order logic. *The Bulletin of Symbolic Logic*, vol. 21, pp. 123-163, 2015.

[6] Fitting, M. Tableau methods of proof for modal logics. *Notre Dame Journal of Formal Logic*, vol. 2, pp. 237–247, 1972.

[7] Fitting, M. Modal proof theory. In P. Blackburn et al. *Handbook of Modal Logic*, pp. 85–138, Elsevier, 2007.

[8] Forrester, J. W. Gentle murder, or the adverbial samaritan. *The Journal of Philosophy*, vol. 81, pp. 193–197, 1984.

[9] Gasquet, O., and A. Herzig. From classical to normal modal logics, in H. Wansing, (ed), *Proof Theory of Modal Logic*. Kluwer, pp. 293–311, 1996.

[10] Gent, I. P. A sequent- or tableau-style system for Lewis's counterfactual logic VC. *Notre Dame Journal of Formal Logic*, vol. 33, pp. 369–382, 1992.

[11] Gilbert, D. and P. Maffezioli, Modular sequent calculi for classical modal logics. *Studia Logica*, vol. 103, pp. 175–217, 2015.

[12] Giordano, L., V. Gliozzi, N. Olivetti, and C. Schwind. Tableau calculi for preference-based conditional logics: PCL and its extensions. *ACM Transactions on Computational Logic*, vol. 10, no. 3/21, pp. 1–45, 2008.

[13] Girlando, M, S. Negri, N. Olivetti, and V. Risch. The logic of conditional beliefs: neighbourhood semantics and sequent calculus. Proceedings of *Advances in Modal Logic 2016*, in press.

[14] Goré, L. Postniece and A. Tiu. Cut-elimination and proof-search for bi-intuitionistic logic using nested sequents. In C. Areces and R. Goldblatt (eds) *Advances in Modal Logic*, vol. 7, pp. 43–66, 2008.

[15] Governatori, G. and A. Luppi. Labelled tableaux for non-normal modal logics. In E. Lamma and P. Mello (eds) *AI*IA 99: Advances in Artificial Intelligence*, Lecture Notes in Computer Science, vol. 1792, pp. 119–130, 2000.

[16] Indrzejczak, A., Sequent calculi for monotonic modal logics. *Bulletin of the Section of logic*, vol. 34, pp. 151–164, 2005.

[17] Indrzejczak, A., Admissibility of cut in congruent modal logics. *Logic and Logical Philosophy*, vol. 21, pp. 189–203, 2011.

[18] Kracht, M., and F. Wolter. Normal monomodal logics can simulate all others. *The Journal of Symbolic Logic*, vol. 64, pp. 99–138, 1999.

[19] Lavendhomme, R., and T. Lucas. Sequent calculi and decision procedures for weak modal systems. *Studia Logica*, vol. 66, pp.121–145, 2000.

[20] Lellman, B. and D. Pattinson. Sequent Systems for Lewis' conditional logics. In L. Farinas del Cerro, A. Herzig and J. Mengin (eds.), *Logics in Artificial Intelligence*, pp. 320–332, 2012.

[21] Lellmann, B. and E. Pimentel, Proof search in nested sequent calculi. In M. Davis,

A. Fehnker, A. McIver, A. Voronkov (eds) *LPAR-20*. LNCS, vol. 9450, pp. 558–574. Springer, 2015.

[22] Lewis, D. *Counterfactuals*. Blackwell, 1973.

[23] Mints, G. Indexed systems of sequents and cut-elimination. *Journal of Philosophical Logic*, vol. 26, pp. 671–696, 1997.

[24] Negri, S. Contraction-free sequent calculi for geometric theories, with an application to Barr's theorem. *Archive for Mathematical Logic*, vol. 42, pp. 389–401, 2003.

[25] Negri, S. Proof analysis in modal logic. *Journal of Philosophical Logic*, vol. 34, pp. 507–544, 2005.

[26] Negri, S. Kripke completeness revisited. In G. Primiero and S. Rahman (eds.), *Acts of Knowledge - History, Philosophy and Logic*, pp. 247–282, College Publications, 2009.

[27] Negri, S. Proof analysis beyond geometric theories: from rule systems to systems of rules. *Journal of Logic and Computation*, vol. 27, pp. 513–537, 2016.

[28] Negri, S. Proofs and countermodels in non-classical logics, *Logica Universalis*, vol. 8, pp. 25–60, 2014.

[29] Negri, S. Non-normal modal logics: a challenge to proof theory. Proceedings of *Logica 2016*.

[30] S. Negri and N. Olivetti. A sequent calculus for preferential conditional logic based on neighbourhood semantics. In Hans de Nivelle (ed) *Automated Reasoning with Analytic Tableaux and Related Methods*, Lecture Notes in Computer Science, vol. 9323, pp. 115–134, 2015.

[31] Negri, S. and J. von Plato. *Structural Proof Theory*. Cambridge University Press, 2001.

[32] Negri, S. and J. von Plato. *Proof Analysis*. Cambridge University Press, 2011.

[33] Negri, S. and J. von Plato. Cut elimination in sequent calculi with implicit contraction, with a conjecture on the origin of Gentzen's altitude line construction. In D. Probst and P. Schuster (eds) *Concepts of Proof in Mathematics, Philosophy, and Computer Science*. Ontos Mathematical Logic, Walter de Gruyter, Berlin, pp. 269–290, 2016.

[34] Negri, S. and G. Sbardolini. Proof analysis for Lewis counterfactuals. *The Review of Symbolic Logic*, vol. 9, pp. 44–75, 2016.

[35] Olivetti, N. and G.L. Pozzato. A standard and int ernal calculus for Lewis counterfactual logics. In Hans de Nivelle (ed) *Automated Reasoning with Analytic Tableaux and Related Methods*, Lecture Notes in Computer Science, vol. 9323, pp. 270–286, 2015.

[36] Olivetti, N., G. L. Pozzato and C. Schwind. A sequent calculus and a theorem prover for standard conditional logics. *ACM Transactions on Computational Logic*, vol. 8, no.4/22, 2007.

[37] Orlandelli, E. Proof analysis in deontic logics. In Cariani *et al.* (eds) *Deontic Logic and Normative Systems*, LNCS, vol. 8554, pp. 139–148. Springer, 2014.

[38] Pacuit, E. Neighborhood Semantics for Modal Logic. An Introduction. Lecture Notes of a course held at Esslli 2007. Available at http://ai.stanford.edu/~epacuit/classes/esslli/nbhdesslli.pdf.

[39] Pattinson, D. and L. Schröder, Generic modal cut elimination applied to conditional

logics. In M. Giese and A. Waaler (eds) *Automated Reasoning with Analytic Tableaux and Related Methods*, Lecture Notes in Computer Science, vol. 5607, pp. 280–294, 2009.

[40] Priest, G., *An Introduction to Non-Classical Logics*. Cambridge University Press, 2008.

[41] Skolem, T. *Logisch-kombinatorische Untersuchungen über die Erfüllbarkeit und Beweisbarkeit mathematischen Sätze nebst einem Theoreme über dichte Mengen*, Skrifter I **4**, pp. 1–36, Det Norske Videnskaps-Akademi, 1920. Also in J. E. Fenstad, editor, *Selected Works in Logic by Th. Skolem*, pp. 103–136, Universitetsforlaget, Oslo, 1970. Translated in J. van Heijenoort, *From Frege to Gödel*, pp. 254–263, Harvard Univ. Press, 1967.

[42] Simpson, A. Proof Theory and Semantics of Intuitionistic Modal Logic. Ph.D. thesis, School of Informatics, University of Edinburgh, 1994.

[43] Stalnaker, R. and R. H. Thomason. A semantic analysis of conditional logic. *Theoria*, vol. 36, pp. 23–42, 1970.

[44] de Swart, H. C. M. A Gentzen- or Beth-type system, a practical decision procedure and a constructive completeness proof for the counterfactual logics VC and VCS. *The Journal of Symbolic Logic*, vol. 48, pp. 1–20, 1983.

[45] Viganò, L. *Labelled Non-Classical Logics*, Kluwer, 2000.

Constructive Temporal Logic, Categorically

Valeria de Paiva
AI Laboratory, Nuance Communications
Sunnyvale 94085, USA
valeria.depaiva@gmail.com

Harley Eades III
Computer Science Augusta University
Augusta GA, 30912, USA
harley.eades@gmail.com

Preface

This paper is dedicated to the memory of Professor Grigori Mints, to whom the first author owes a huge debt. Not only an intellectual debt (most people working on proof theory nowadays owe this one), not only a friendship one (there are plenty of us who owe this), but also a mentoring (and a personal help when I needed it) debt. Grisha would not be gratuitously conversational, you could say that his style was 'tough love': work hard, then he would talk to you. But when you needed him, he was there for you. This work might not be, yet, at the stage that he would approve of it, especially given all his work on Dynamic Topological Logic with Kremer and others [17], which might be related to what we describe here. However, this is the best that we can do in the time we have, so it will have to do.

The second author does owe Grigori an intellectual debt. While I never got to meet him in person I do remember fondly reading his "A Short Introduction to Intuitionistic Logic" [20]. His book gave me several realizations about intuitionistic logic that I had previously lacked. It was his book that turned on the light, and I thank him for that.

We would like to thank the anonymous reviewers and Jane Spurr for dealing with a last minute, but heartfelt, submission to this volume.

1 Motivation

Generally speaking, Temporal Logic is any system of rules and symbolism for representing, and reasoning about propositions qualified in terms of time. Temporal logic is one of the most traditional kinds of modal logic, introduced by Arthur Prior in the late 1950s, but it is also one of the most controversial kinds of modal logic, as people have different intuitions about time, how to represent it, and how to reason about it.

There has been a large amount of work in Modal Logic in the last sixty years, mainly in classical modal logic. We are mostly interested in constructive systems, not classical ones. In particular we are interested in a constructive version of temporal logic that satisfies some well-known and desirable proof-theoretical properties, but that is also algebraically and category-theoretically *well-behaved*.

Prior's 'Time and Modality' [24] introduced a propositional modal logic with two temporal connectives (modal operators), F and P, corresponding to "sometime in the Future" and "sometime in the Past". This propositional system has been called **tense logic** to distinguish it from other temporal systems.

Ewald [10] produced a first version of an intuitionistically based temporal logic system with not only operators for "sometime in the Future" and "sometime in the Past", but also operators for "in all future times" and "for all past times". The intuitive reading of these operators is very reasonable:

- P "It has at some past time been the case that"

- F "It will at some future time be the case that"

- H "It has always been the case in the past that"

- G "It will always be the case in the future that"

Ewald and most of the researchers that followed his path of constructivization of tense logic, did so assuming a symmetry between past and future. This symmetry, as well as the symmetry between universal and existential quantifiers, both in the past and in the future, are somewhat at odds with intuitionistic reasoning. In particular while an axiom like $A \to GPA$ "What is, will always have been" makes sense in a constructive way of thinking, the dual one $A \to HFA$ paraphrased in the Stanford Encyclopedia of Philosophy as "What is, has always been going to be" feels very classical.

Constructivizing a classical system is alwys prone to proliferation of systems, as is evident when considering the several versions of intuitionistic set theory, for

$$\begin{array}{rcl}
\Diamond(A \vee B) & \to & \Diamond A \vee \Diamond B \\
\Diamond \bot & \to & \bot
\end{array}$$

Figure 1: Distributivity rules

example. In particular the basic constructive modal logic S4 (using Lewis' original naming convention) has two main variants.

The first version of an intuitionistic S4, originally presented by Dag Prawitz in his Natural Deduction book [23] does not satisfy the distributivity of the possibility operator $\Diamond$ over the logical disjunction. Prawitz's system satisfies neither the binary distribution nor its nullary form, as given in Figure 1. We call this system CS4. This system was investigated from a proof theoretical and categorical perspective in [5].

The second main version of an intuitionistic modal S4 does enforce these distributivities and it was thoroughly investigated in Simpson's doctoral thesis [25]. This system is part of a framework for constructive modal logics, based on incorporating, as part of the syntax, the intended semantics of the modal logic, as possible worlds. We call this system IS4.

Ewald's tense logic system consists of a pair of Simpson-style S4 operators [25], representing past and future over intuituionistic propositional logic. This is historically inaccurate, as Simpson based his systems in Ewald's, but it will serve to make some of our main points clearer below. The system we describe in this note is the tense logic system obtained by joining together two pairs of Prawitz-style S4 operators. So it satisfies some of Ewald's rules, but not all.

Simpson remarks that intuitionistic or constructive modal logic is full of interesting questions. As he says:

> Although much work has been done in the field, there is as yet no consensus on the correct viewpoint for considering intuitionistic modal logic. In particular, there is no single semantic framework rivalling that of possible world semantics for classical modal logic. Indeed, there is not even any general agreement on what the **intuitionistic analogue** of the basic modal logic, K, is.

In an intuitionistic logic we do not expect perfect duality between quantifiers, ($\forall x.P(x)$ is not the same as $\neg \exists x.\neg P(x)$) or even between conjunction and disjunction (full De Morgan laws do not hold for intuitionistic propositional logic). So one should not expect a perfect duality between intuitionistic possibility and necessity either. But considerations from first principles do not seem to indicate clearly whether distributivity rules as the ones in Figure 1 should hold or not. Hence it seems sensible to develop different kinds of systems in parallel, proving equivalences, whenever possible. In this paper we develop the idea of tense logic in Prawitz' style. We recall some deductive systems for this tense logic and provide categorical semantics for it.

Much has been done recently in the proof theory of constructive modal logics using more informative sequent systems (e.g. hypersequents, labelled sequents, nested sequents, tree-style sequents, etc.). In particular nested sequents have been used to produce 'modal cubes' for the two variants of constructive modal logics described above. See the pictures below from [2, 26].

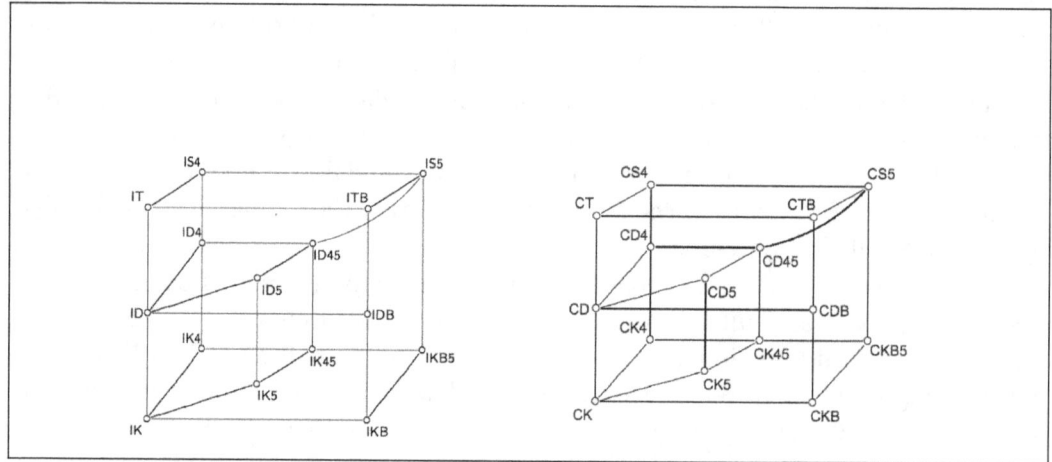

Figure 2: Intuitionistic and constructive modal cubes

Sequent calculi by themselves are not enough to provide us with Curry-Howard correspondences and/or term assignments for these systems. However, using the Prawitz S4 version of these modal systems we can easily produce a Curry-Howard correspondence and a categorical model for the Prawitz-style intuitionistic tense logic, our goal in this paper.

We start by recalling the system using axioms, plain sequent calculus and plain natural deduction. In the next section we describe a term assignment based on the dual calculus described in [12] and show some of its syntactic properties. The next section introduces the categorical model (a cartesian closed category with two intertwined adjunctions) and show the usual soundness and completeness results. Finally we discuss potential applications and limitations of our constructive tense logic.

$$\frac{}{\Delta, A \vdash A} \text{ Id} \qquad \frac{\Gamma \vdash B \quad B, \Delta \vdash C}{\Gamma, \Delta \vdash C} \text{ Cut} \qquad \frac{}{\Gamma, \bot \vdash A} \bot_\mathcal{L}$$

$$\frac{\Gamma, A \vdash C \quad \Gamma, B \vdash C}{\Gamma, A \vee B \vdash C} \vee_\mathcal{L} \qquad \frac{\Gamma \vdash A}{\Gamma \vdash A \vee B} \vee_{\mathcal{R}_1} \qquad \frac{\Gamma \vdash B}{\Gamma \vdash A \vee B} \vee_{\mathcal{R}_2}$$

$$\frac{\Gamma, A \vdash C}{\Gamma, A \wedge B \vdash C} \wedge_{\mathcal{L}_1} \qquad \frac{\Gamma, B \vdash C}{\Gamma, A \wedge B \vdash C} \wedge_{\mathcal{L}_2} \qquad \frac{\Gamma \vdash A \quad \Gamma \vdash B}{\Gamma \vdash A \wedge B} \wedge_\mathcal{R}$$

$$\frac{\Gamma \vdash A \quad \Gamma, B \vdash C}{\Gamma, A \to B \vdash C} \to_\mathcal{L} \qquad \frac{\Gamma, A \vdash B}{\Gamma \vdash A \to B} \to_\mathcal{R}$$

Figure 3: Intuitionistic propositional calculus (LJ)

2 Tense logic CS4-style

We build up to the constructive tense logic we are interested in TCS4 in progressive steps. We start with the intuitionistic basis LJ, add the modalities to get the constructive modal S4 system, CS4, provide the dual context modification (to help with the reuse of libraries, amongst other things), obtaining dual CS4, DCS4 and then finally consider the two adjunctions to obtain the tense constructive system TCS4.

2.1 Intuitionistic sequent calculus

We start by recalling the basic sequent calculus for intuitionistic propositional logic, Gentzen's intuitionistic sequent calculus LJ. The syntax of formulas for LJ is defined by the following grammar:

$$A \quad ::= \quad p \mid \bot \mid A \wedge A \mid A \vee A \mid A \to B$$

The formula p is taken from a set of countably many propositional atoms. The constant $\top$ could be added, but it is the negation of the falsum constant $\bot$. The initial inference rules, which just model propositional intuitionistic logic, are as in Figure 3.

Sequents denoted $\Gamma \vdash C$ consist of a multiset of formulas, (written as either Γ, Δ, or a numbered version of either), and a formula C. The intuitive meaning is that the conjunction of the formulas in Γ entails the formula C. So far this is our intuitionistic basis.

$$\dfrac{\Gamma, A \vdash B}{\Gamma, \Box A \vdash B}\ \Box_{\mathcal{L}} \qquad \dfrac{\Box \Gamma \vdash A}{\Box \Gamma, \Delta \vdash \Box A}\ \Box_{\mathcal{R}}$$

$$\dfrac{\Box \Gamma, A \vdash \Diamond B}{\Delta, \Box \Gamma, \Diamond A \vdash \Diamond B}\ \Diamond_{\mathcal{L}} \qquad \dfrac{\Gamma \vdash A}{\Gamma \vdash \Diamond A}\ \Diamond_{\mathcal{R}}$$

Figure 4: Constructive S4 modal rules (**CS4**)

2.2 Constructive modal S4

Next we recall the sequent calculus formalization of system CS4 as described in [5].

We recap the modality rules in Figure 4. These, in addition to the initial set of inference rules, define the sequent calculus for CS4. In Figure 3, we write $\Box\Gamma$ for the sequence of boxed formulas $\Box G_1, \Box G_2, \ldots, \Box G_k$ where Γ is the set $G_1, G_2, \ldots, G_k$.

Note that we do have right rules and left rules for introducing the new modal operators $\Box$ (necessity) and $\Diamond$ (posssibility), but these rules are not as symmetric as the propositional ones. Most importantly, we have a local restriciton on the rule that introduces the $\Box$ operator: We can only introduce $\Box$ in the conclusion, if all the assumptions are already boxed. Also the rules for the $\Diamond$ operator presuppose that you have already defined $\Box$ operators. This system is indeed constructive, $\Box$ and $\Diamond$ are independent logical operators and $\Box A$ is not logically equivalent to $\neg \Diamond \neg A$, nor is $\Diamond A$ logically equivalent to $\neg \Box \neg A$. Note that the necessity only fragment is well-behaved and closed, while to define the possibility operator you need a necessity operator in place.

This system has a reasonably nice proof theory. Bierman and de Paiva [5] show that it has a Hilbert-style presentation, a Natural Deduction presentation, as well as a sequent calculus presentation and these presentations are provably equivalent, that is, they prove the same theorems. The sequent calculus satisfies cut-elimination, an old result from Ohnishi and Matsumoto [21], as well as a form of the subformula property. The Natural Deduction formulation has a colourful history: one of its distinct features is that it was described in Prawitz' seminal book in Natural Deduction [23], hence it is sometimes called Prawitz' S4 intuitionistic modal logic. Most interestingly the system has both Kripke and categorical semantics, described respectively in [1] and [5] as well as an independent mathematical semantics in terms of simplicial sets, described by Goubault-Larrecq [14].

$$\frac{\Gamma;\emptyset \vdash A}{\Gamma;\Delta \vdash \Box A}\Box_{\mathcal{I}} \qquad \frac{\Gamma;\Delta \vdash \Box A \quad \Gamma,A;\Delta \vdash B}{\Gamma;\Delta \vdash B}\Box_{\mathcal{E}}$$

$$\frac{\Gamma;\Delta \vdash A}{\Gamma;\Delta \vdash \Diamond A}\Diamond_{\mathcal{I}} \qquad \frac{\Gamma;\Delta \vdash \Diamond A \quad \Gamma;A \vdash \Diamond B}{\Gamma;\Delta \vdash \Diamond B}\Diamond_{\mathcal{E}}$$

Figure 5: The dual context modal calculus (DCS4)

2.3 The dual context modal S4 calculus

An equivalent (in terms of provability) but more type-theoretic system can be produced for the modal logic CS4. This is not so well-known, but this system can be given a presentation in terms of a categorical adjunction, between two cartesian closed categories, as we will show in the next section. This categorical presentation has been described both in [5] and in [12], in the former, this is called the **multi-context** formulation of CS4 and the rules are given in Figure 5. (We prefer to call it the dual context sequent calculus.) Note that the rules are Natural Deduction rules, as it should be clear from the fact that they are introduction and elimination rules.

The main difference between the system CS4 and the dual context formulation of CS4 is the fact that the context now has modal formulas and non-necessarily modal ones, separated by a semi-colon as in $\Gamma;\Delta$. The previously difficult rule of $\Box$ introduction now says that to introduce a necessity operator $\Box$ on a conclusion, we need to have an empty context of non-modal assumptions (that is, all the assumptions of this conclusion must be modal). This corresponds to the traditional idea that to prove something is necessarily the case, all its assumptions have to be also necessary (or it must have no assumptions whatsoever).

These rules have been shown by Benton [4] and Barber [3] to correspond to an adjunction of the categories, in the case where the basis is Linear Logic and the modalities correspond to the exponentials. Instead of Linear Logic, we deal with constructive modal logic and the adjunction is between functors corresponding to operators $\Diamond \vdash \Box$.

2.4 The tense CS4 calculus

Finally to get to the tense logic which is the main aim of this note, we need two such adjunctions, but intertwined. This follows the pattern explained by Ewald [10]. Thus $\Diamond$ is left-adjoint to ■ and ♦ is left-adjoint to $\Box$, where we are writing ■ for the operator we called past universal H before and $\Box$ for the future necessity operator

G. The past existential P is ◇ and the future existential is ♦ or F.

A sequent calculus system for this constructive tense logic is given by the rules in Figure 6. This can be transformed into Natural Deduction in the style of [5] as shown in Figure 8. The problem is that the last two adjunction rules in Figure 6 (that relate the two sets of modalities) are extremely badly-behaved proof-theoretically (no cut-elimination and no subformula property even for cut-free proofs), as discussed in page 35 of Benton's full report [4]. In fact they are the reason for moving to a dual context calculus, as explained in that paper and also in Barber's work [3].

The dual context systems, as described in Barber and Benton's work, are proved equivalent to the system with a single modality operator, either '!' or '□'. This is because in Intuitionistic Linear Logic one is not usually interested in either *why not?* '?' or '◇'. (In Classical Linear Logic the possibility modality is defined by negation of the necessity modality, so this extension is easier to make [22].) Given that our main goal is to discuss categorical semantics, which we can do easily for the necessity modalities, in this note we consider only two necessity-like modalities □ and ■.

We would like to have a natural deduction version of the tense calculus in dual context style. A dual context-style presentation of a single necessity modality has been presented in Figure 5. Now we need to add another necessity-like modality and discuss their interaction. A preliminary attempt at such calculus is given in Figure 9.

This corresponds to an intuitionistic tense logic obtained by extending IPL with two pairs of adjoint modalities (♦, □) and (◇, ■), with no explicit relationship between the modalities of the same colour, namely, (♦, ■) and (◇, □).

2.5 Axioms

Axiom sets for the system TCS4 are easier to provide. We need a set for the basic system intuitionistic logic LJ, and any traditional set would do, plus the axioms for modalities, as well as the rules **modus ponens** and **necessitation** for the two necessity operators:

We have similar axioms to Ewald's [10], except that the duality between necessity and possibility is not strict (Ewald's original axioms (7) and (7') in page 171 of [10] are not valid) and that the possibility modalities we deal with, do not distribute over disjunction (Ewald's axioms (4) and (4') are not valid). Also note we do have introspection and reflexivity valid, which correspond to Ewald's extra axioms (12) and (12'), as well as (13) and (13').

We are interested in the term assignment system and its properties, as our aim is to use these as type systems for innovative programming languages. So we needed to provide the systematic work that shows basic properties of the type system TCS4

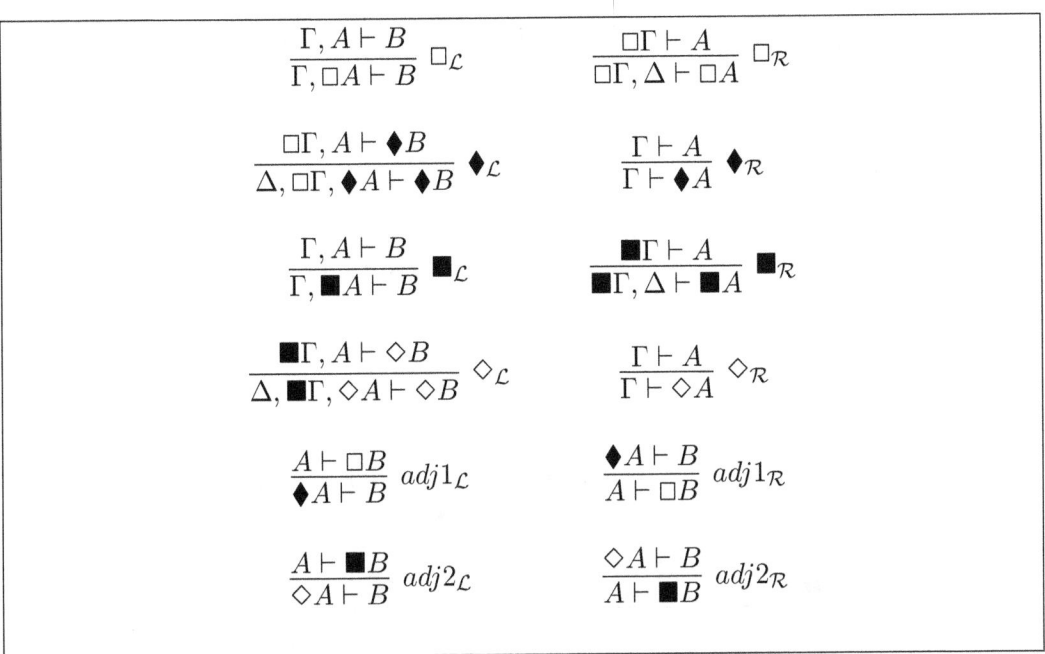

Figure 6: Tense S4 sequent rules (biCS4)

$$\frac{\Theta \vdash_{\mathcal{C}} X \quad Y, \Phi \vdash_{\mathcal{C}} Z}{\Theta, X \to Y, \Phi \vdash_{\mathcal{C}} Z} \to_l^{\mathcal{C}} \qquad \frac{\Theta, X \vdash_{\mathcal{C}} Y}{\Theta \vdash_{\mathcal{C}} X \to Y} \to_r^{\mathcal{C}} \qquad \frac{\Theta; \emptyset \vdash_{\mathcal{L}} A}{\Theta \vdash_{\mathcal{C}} \mathsf{G} A} \mathsf{G}_r$$

$$\frac{\Theta \vdash_{\mathcal{C}} X \quad Y, \Phi; \Gamma \vdash_{\mathcal{L}} A}{\Theta, X \to Y, \Phi; \Gamma \vdash_{\mathcal{L}} A} \to_l^{\mathcal{L}} \qquad \frac{\Theta; \Gamma, A \vdash_{\mathcal{L}} B}{\Theta; \Gamma \vdash_{\mathcal{L}} A \multimap B} \multimap_r^{\mathcal{L}}$$

$$\frac{\Theta; \Gamma \vdash_{\mathcal{L}} A \quad \Phi; \Delta, B \vdash_{\mathcal{L}} C}{\Theta; \Gamma, A \multimap B, \Delta \vdash_{\mathcal{L}} C} \multimap_l^{\mathcal{L}} \qquad \frac{\Theta \vdash_{\mathcal{C}} X}{\Theta; \emptyset \vdash_{\mathcal{L}} \mathsf{F} X} \mathsf{F}_r \qquad \frac{\Theta, X; \Gamma \vdash_{\mathcal{L}} A}{\Theta; \mathsf{F} X, \Gamma \vdash_{\mathcal{L}} A} \mathsf{F}_l$$

$$\frac{\Theta; B, \Gamma \vdash_{\mathcal{L}} A}{\Theta, \mathsf{G} B; \Gamma \vdash_{\mathcal{L}} A} \mathsf{G}_l$$

Figure 7: LNL sequent rules

1295

$$\frac{\Gamma \vdash \Box B}{\Gamma \vdash B} \; \Box_{\mathcal{E}} \qquad \frac{\Gamma \vdash \Box A_1, \ldots, \Gamma \vdash \Box A_k \quad \Box A_1, \ldots \Box A_k \vdash B}{\Gamma \vdash \Box B} \; \Box_{\mathcal{I}}$$

$$\frac{\Gamma \vdash A}{\Gamma \vdash \blacklozenge A} \; \blacklozenge_{\mathcal{I}}$$

$$\frac{\Gamma \vdash \blacksquare A}{\Gamma \vdash A} \; \blacksquare_{\mathcal{E}} \qquad \frac{\Gamma \vdash \blacksquare A_1, \ldots, \Gamma \vdash \blacksquare A_k \quad \blacksquare A_1, \ldots \blacksquare A_k \vdash B}{\Gamma \vdash \blacksquare B} \; \blacksquare_{\mathcal{I}}$$

$$\frac{\Gamma \vdash A}{\Gamma \vdash \Diamond A} \; \Diamond_{\mathcal{I}}$$

$$\frac{\Gamma \vdash \Box A_1 \ldots \Gamma \vdash \Box A_k \quad \Gamma \vdash \blacklozenge B \quad \Box A_1 \ldots \Box A_k, B \vdash \blacklozenge C}{\Gamma \vdash \blacklozenge C} \; \blacklozenge_{\mathcal{E}}$$

$$\frac{\Gamma \vdash \blacksquare A_1 \ldots \Gamma \vdash \blacksquare A_k \quad \Gamma \vdash \Diamond B \quad \blacksquare A_1 \ldots \blacksquare A_k, B \vdash \Diamond C}{\Gamma \vdash \Diamond C} \; \Diamond_{\mathcal{E}}$$

Figure 8: Tense S4 rules ND first version (**NDCS4**)

$$\frac{\Gamma_1; \emptyset \vdash A}{\Gamma_1; \Delta \vdash \Box A} \; \Box_{\mathcal{I}} \qquad \frac{\Gamma_1; \Delta \vdash \Box A \quad \Gamma_1; A; \Delta \vdash B}{\Gamma_1; \Delta \vdash B} \; \Box_{\mathcal{E}}$$

$$\frac{\Gamma \vdash A}{\Gamma \vdash \blacklozenge A} \; \blacklozenge_{\mathcal{I}} \qquad \frac{\Gamma_1; \Delta \vdash \blacklozenge A \quad \Gamma_1; A \vdash \blacklozenge B}{\Gamma_1; \Delta \vdash \blacklozenge B} \; \blacklozenge_{\mathcal{E}}$$

$$\frac{\Gamma; \emptyset \vdash A}{\Gamma; \emptyset \vdash \blacksquare A} \; \blacksquare_{\mathcal{I}} \qquad \frac{\Gamma_1; \Delta \vdash \blacksquare A \quad \Gamma_1; A; \Delta \vdash B}{\Gamma_1; \Delta \vdash B} \; \blacksquare_{\mathcal{E}}$$

$$\frac{\Gamma \vdash A}{\Gamma \vdash \Diamond A} \; \Diamond_{\mathcal{I}} \qquad \frac{\Gamma_1; \Delta \vdash \Diamond A \quad \Gamma_1; A \vdash \Diamond B}{\Gamma_1; \Delta \vdash \Diamond B} \; \Diamond_{\mathcal{E}}$$

Figure 9: biS4 rules, dual context version (**ND2CS4**)

$$\frac{}{\Gamma_1; \Gamma_2, A; \Delta \vdash A} \; \Box\mathrm{Id} \qquad \frac{}{\Gamma_1, A; \Gamma_2; \Delta \vdash A} \; \blacksquare\mathrm{Id}$$

$$\frac{\Gamma_1; \Gamma_2; \emptyset \vdash A}{\Gamma_1; \Gamma_2; \Delta \vdash \Box A} \; \Box_\mathcal{I} \qquad \frac{\Gamma_1; \Gamma_2; \Delta \vdash \Box A \quad \Gamma_1; \Gamma_2, A; \Delta \vdash B}{\Gamma_1; \Gamma_2; \Delta \vdash B} \; \Box_\mathcal{E}$$

$$\frac{\Gamma_1; \Gamma_2; \Delta \vdash A}{\Gamma_1; \Gamma_2; \Delta \vdash \blacklozenge A} \; \blacklozenge_\mathcal{I} \qquad \frac{\Gamma_1; \Gamma_2; \Delta \vdash \blacklozenge A \quad \Gamma_1; \Gamma_2; A \vdash \blacklozenge B}{\Gamma_1; \Gamma_2; \Delta \vdash \blacklozenge B} \; \blacklozenge_\mathcal{E}$$

$$\frac{\Gamma_1; \Gamma_2; \emptyset \vdash A}{\Gamma_1; \Gamma_2; \Delta \vdash \blacksquare A} \; \blacksquare_\mathcal{I} \qquad \frac{\Gamma_1; \Gamma_2; \Delta \vdash \blacksquare A \quad \Gamma_1, A; \Gamma_2; \Delta \vdash B}{\Gamma_1; \Gamma_2; \Delta \vdash B} \; \blacksquare_\mathcal{E}$$

$$\frac{\Gamma_1; \Gamma_2; \Delta \vdash A}{\Gamma_1; \Gamma_2; \Delta \vdash \Diamond A} \; \Diamond_\mathcal{I} \qquad \frac{\Gamma_1; \Gamma_2; \Delta \vdash \blacklozenge A \quad \Gamma_1; \Gamma_2; A \vdash \blacklozenge B}{\Gamma_1; \Gamma_2; \Delta \vdash \blacklozenge B} \; \blacklozenge_\mathcal{E}$$

Figure 10: Dual context 2CS4 calculus (TCS4)

$$\frac{A}{\Box A} \; \Box_{Nec} \qquad \frac{A}{\blacksquare A} \; \blacksquare_{Nec} \qquad \frac{A \to B \quad A}{B} \; \mathcal{MP}$$

Figure 11: Axiomatic rules

we are interested in, this is what we do in the next section.

3 Term assignment

In this section we provide a term assignment to constructive tense logic with only $\Box$ and $\blacksquare$. We leave term assignments to the other varieties of tense logic with $\Diamond$ and $\blacklozenge$ for future work.

The typing rules can be found in Figure 13 with the typed equality rules in Figure 14. Here we can see that types are tense S4 formulas. The sequents have the form $\Gamma \vdash t : A$ and $\Gamma \vdash s = t : A$ where Γ is a multiset of free variables and their types denoted $x : A$, and s and t are terms with the following syntax:

$$t ::= x \mid \lambda x : A.t \mid s\, t \mid \mathsf{let}_\Box x_1 : \Box A_1, ..., x_k : \Box A_k \text{ be } t_1, ..., t_k \text{ in } t \mid$$
$$\mathsf{let}_\blacksquare x_1 : \blacksquare A_1, ..., x_k : \blacksquare A_k \text{ be } t_1, ..., t_k \text{ in } t \mid \mathsf{unbox}_\Box\, t \mid \mathsf{unbox}_\blacksquare\, t$$

Equality is straightforward where it is apparent that the let-expressions model explicit substitutions. These substitutions are triggered when they are applied to an

$$
\begin{array}{c}
\hline
\text{propositional basic intuitionistic axioms} \\
\begin{array}{rcl}
\Box(A \to B) & \to & (\Box A \to \Box B) \\
\Box(A \to B) & \to & (\blacklozenge A \to \blacklozenge B) \\
(\Box A \to A) & \land & (A \to \blacklozenge A) \\
(\Box A \to \Box\Box A) & \land & (\blacklozenge\blacklozenge A \to \blacklozenge A) \\
\blacksquare(A \to B) & \to & (\blacksquare A \to \blacksquare B) \\
\blacksquare(A \to B) & \to & (\Diamond A \to \Diamond B) \\
(\blacksquare A \to A) & \land & (A \to \Diamond A) \\
(\blacksquare A \to \blacksquare\blacksquare A) & \land & (\Diamond\Diamond A \to \Diamond A) \\
\Diamond\blacksquare A \to A & \land & A \to \blacksquare\Diamond A \\
\blacklozenge\Box A \to A & \land & A \to \Box\blacklozenge A \\
\end{array} \\
\hline
\end{array}
$$

Figure 12: Axioms for TCS4

$$\dfrac{}{\Gamma, x:A \vdash x:A}\text{Id} \quad \dfrac{}{\Gamma, x:\bot \vdash \text{contra}:A}\bot\mathcal{E} \quad \dfrac{\Gamma, x:A \vdash t:B}{\Gamma \vdash \lambda x:A.t:A \to B}{\to}\mathcal{I}$$

$$\dfrac{\Gamma \vdash t_1:A \to B \quad \Gamma \vdash t_2:A}{\Gamma \vdash t_1\,t_2:B}{\to}\mathcal{E} \quad \dfrac{\Gamma \vdash t:\Box B}{\Gamma \vdash \text{unbox}_\Box\ t:B}\Box\mathcal{E}$$

$$\dfrac{\Gamma \vdash t_1:\Box A_1,\ldots,\Gamma \vdash t_k:\Box A_k \quad x_1:\Box A_1,\ldots,x_k:\Box A_k \vdash t:B}{\Gamma \vdash \text{let}_\Box\ x_1:\Box A_1,\ldots,x_k:\Box A_k\ \text{be}\ t_1,\ldots,t_k\ \text{in}\ t:\Box B}\Box\mathcal{I}$$

$$\dfrac{\Gamma \vdash t:\blacksquare B}{\Gamma \vdash \text{unbox}_\blacksquare\ t:B}\blacksquare\mathcal{E}$$

$$\dfrac{\Gamma \vdash t_1:\blacksquare A_1,\ldots,\Gamma \vdash t_k:\blacksquare A_k \quad x_1:\blacksquare A_1,\ldots,x_k:\blacksquare A_k \vdash t:B}{\Gamma \vdash \text{let}_\blacksquare\ x_1:\blacksquare A_1,\ldots,x_k:\blacksquare A_k\ \text{be}\ t_1,\ldots,t_k\ \text{in}\ t:\blacksquare B}\blacksquare\mathcal{I}$$

Figure 13: TCS4 typing rules

$$\frac{\Gamma, x : A \vdash t_2 = s_2 : B \quad \Gamma \vdash t_1 = s_1 : A}{\Gamma \vdash (\lambda x : A.t_2)\, t_1 = [s_1/x]s_2 : B}\beta$$

$$\frac{\Gamma \vdash t_1 = s_1 : \Box A_1, ..., \Gamma \vdash t_k = s_k : \Box A_k \quad x_1 : \Box A_1, ..., x_k : \Box A_k \vdash t = s : B}{\Gamma \vdash \mathsf{unbox}_\Box\ (\mathsf{let}_\Box\ x_1 : \Box A_1, ..., x_k : \Box A_k\ \mathsf{be}\ t_1, ..., t_k\ \mathsf{in}\ t) = [s_1/x_1]...[s_k/x_k]s : B}\Box$$

$$\frac{\Gamma \vdash t_1 = s_1 : \blacksquare A_1, ..., \Gamma \vdash t_k = s_k : \blacksquare A_k \quad x_1 : \blacksquare A_1, ..., x_k : \blacksquare A_k \vdash t = s : B}{\Gamma \vdash \mathsf{unbox}_\blacksquare\ (\mathsf{let}_\blacksquare\ x_1 : \blacksquare A_1, ..., x_k : \blacksquare A_k\ \mathsf{be}\ t_1, ..., t_k\ \mathsf{in}\ t) = [s_1/x_1]...[s_k/x_k]s : B}\blacksquare$$

$$\frac{\Gamma \vdash t : A}{\Gamma \vdash t = t : A}\mathsf{refl} \qquad \frac{\Gamma \vdash t_2 = t_1 : A}{\Gamma \vdash t_1 = t_2 : A}\mathsf{sym} \qquad \frac{\Gamma \vdash t_1 = t_2 : A \quad \Gamma \vdash t_2 = t_3 : A}{\Gamma \vdash t_1 = t_3 : A}\mathsf{trans}$$

Figure 14: TCS4 equality rules

unbox-expression.

We have the following basic properties of this term assignment.

Lemma 1 (Substitution for Typing). *If $\Gamma_1 \vdash t_1 : A$, and $\Gamma_1, x : A, \Gamma_2 \vdash t_2 : B$, then $\Gamma_1, \Gamma_2 \vdash [t_1/x]t_2 : B$.*

Proof. This proof holds by straightforward induction on the form of the assumed typing derivation. Please see Appendix A.1.1 for the proof. □

Lemma 2 (Weakening). *If $\Gamma_1, \Gamma_2 \vdash t : B$, then $\Gamma_1, x : A, \Gamma_2 \vdash t : B$.*

Proof. This proof holds by straightforward induction on the form of the assumed typing derivation. Please see Appendix A.1.2 for the proof. □

4 The categorical model

There is not much essentially new in what we discuss here about the tense logic based on CS4. Similar ideas were discussed by Ghilardi and Meloni [13], Makkai and Reyes [18] and more recently in by Dzik et al [7, 9] and Menni and Smith [19].

The upshot of our discussion is that the categorical model we advance is a cartesian closed category endowed with two adjunctions, corresponding to the (limited) universal and existential quantifications relative to the past and to the future that correspond to the two sets of necessity and possibility operators.

This setting is though different enough from the precursors we know about, to justify this note. First, as discussed elsewhere [5], we see no reason for the monads/comonads emerging from this setting to be **idempotent** operators, as they are in [13] or [18] (the idempotency simplification does not seen warranted by the proof theory). Secondly we see no reason to take our models as part of toposes, as we are not interested in the extra structure provided by toposes. However, we also see no reason to confine ourselves to algebraic models such as Heyting algebras with operators, as degenerate posetal categories, as both [7] and [19] do. Different proofs of the same theorem are important to us as they correspond to different morphisms in the category between the same objects. Thus we are interested in **proof relevant** semantics, not simply provability.

We build our main definition in stages. To begin with, a categorical model of propositional intuitionistic logic is a **cartesian closed** category $\mathcal{C}$ with coproducts. Then we recall from [5] that to model a pair of modalities using dual contexts we need a *monoidal* adjunction.

Definition 3 (adjoint model). *An adjoint categorical model of dual context modal logic* DCS4 *consists of the following data:*

1. *A cartesian closed category with coproducts* $(\mathcal{C}, 1, 0, \times, +, \to)$;

2. *A monoidal adjunction* $F \dashv G$, *where* (F, m) *and* $(G, n)\colon \mathcal{C} \to \mathcal{C}$ *are monoidal functors such that their composition* GF *is a monoidal comonad, written as* $\Box$;

3. *The monad* $(\Diamond, \eta, \mu, \mathsf{st}_{A,B})$, *induced by the adjunction* $F \dashv G$, *is* $\Box$-*strong.*

Recall that a *monoidal* comonad $\Box$ implies that there is a natural transformation $m\colon \Box A \times \Box B \to \Box(A \times B)$ (and $m_\top\colon \top \to \Box\top$) satisfying the coherence conditions described in page 23 of [5]. Recall as well that by a monad being $\Box$-strong, we mean that there is a *strength* natural transformation $\mathsf{st}_{A,B}\colon \Box A \times \Diamond B \to \Diamond(\Box A \times B)$ satisfying the four equations in page 27 of [5]. These two natural transformations are required to model the Fisher-Servi axioms, which are a weakening of the duality between $\Box$ and $\Diamond$ that the classical modalities satisfy.

Finally we consider two pairs of modalities (or two adjunctions), intertwined, as in tense logic.

Definition 4 (tense calculus model). *A categorical model of tense calculus dual context modal logic* TCDS4 *is a cartesian closed category* $\mathcal{C}$ *as above, together with two intertwined adjunctions* $(\blacklozenge \dashv \Box, \Diamond \dashv \blacksquare)$. *The adjunctions* $(\blacklozenge \dashv \Box)$ *and* $(\Diamond \dashv \blacksquare)$ *on* $\mathcal{C}$ *are connected by the Fisher-Servi axioms, namely* $\Diamond(A \to B) \to (\blacksquare A \to \Diamond B)$ *and* $(\Diamond A \to \blacksquare B) \to \blacksquare(A \to B)$, *as well as* $\blacklozenge(A \to B) \to (\Box A \to \blacklozenge B)$ *and* $(\blacklozenge A \to \Box B) \to \Box(A \to B)$.

This model is more general than the system, TCS4, given above in that it contains two possibility modalities, which we do not deal with, in the type theory. These possibility operators could be treated as syntactic sugar for $\neg\Box\neg A$ (and respectively $\neg\blacksquare\neg A$), as they usually are in Intuitionistic Linear Logic, for instance. We refrain from doing so explicitly and prefer to consider the necessity-only fragment in the type theory, as this allows us to bypass the discussion of which possibility modality is more appropriate for each setting. More importantly it allows us to dodge the question of how to provide a Curry-Howard categorical interpretation for what we called Simpson-style modal S4. Thus the model should be seen as an over approximation. We give the more general model here to set the stage for future work.

Categorical soundness is proved, as usual, checking the natural deduction rules preserve validity of the constructions used, i.e function spaces, products, coproducts and the two adjunctions.

Define an interpretation $[\![\_]\!] : \mathsf{TCS4} \to \mathcal{C}$ which takes the types and sequents of TCS4 (over a basic set of types) to a model $\mathcal{C}$ as follows:

$$[\![p]\!] = I(p) \text{ for } p \text{ a base type}$$
$$[\![\top]\!] = \top$$
$$[\![A \to B]\!] = [\![A]\!] \to [\![B]\!]$$
$$[\![\Box A]\!] = FG([\![A]\!])$$
$$[\![\blacksquare A]\!] = F'G'([\![A]\!])$$

We extend this interpretation to lists of types by saying that for a list $A_1, ..., A_n$ of types, the interpretation is the product of the interpretations $[\![A_1, ...A_n]\!] = [\![A_1]\!] \times ... \times [\![A_n]\!]$. The interpretation will take a sequent $\Gamma \vdash t : A$ to an arrow $[\![\Gamma \vdash t : A]\!] : [\![\Gamma]\!] \to [\![A]\!]$ in the tense modal category.

Theorem 5. *The type theory* TCS4 *has sound models provided by the structures $\mathcal{C}$ defined above. In other words, given a tense adjoint modal category $\mathcal{C}$, using the above interpretation, the following hold:*

- *Assume $\Gamma \vdash t : A$ in* TCS4. *Then $[\![\Gamma \vdash t : A]\!]$ is a morphism with domain $[\![\Gamma]\!]$ and codomain $[\![A]\!]$;*

- *Assume $\Gamma \vdash t = s : A$. Then $[\![\Gamma \vdash t : A]\!] = [\![\Gamma \vdash s : A]\!]$.*

Proof. The first part holds by induction on $\Gamma \vdash t : A$, and the second by induction on $\Gamma \vdash t = s : A$, but uses the first part. Please see Appendix A.1.3 for the proof. □

We have completeness of the tense modal categories when the model is restricted to box modalities only.

Theorem 6. *The adjoint modal models are complete in the appropriate sense for the type theory* **TCS4**. *This is to say, if we have equality of the interpretations* $[\![\Gamma \vdash t : A]\!] = [\![\Gamma \vdash s : A]\!]$ *(where* $[\![\ ]\!]$ *is the interpretation defined above) in the tense modal category* $\mathcal{C}$ *for any derived sequents* $\Gamma \vdash t : A$ *and* $\Gamma \vdash s : A$ *then we can derive the equation in the type theory* **TCS4** $\Gamma \vdash t = s : A$.

Proof. This result can be shown by constructing a cartesian closed category with coproducts and two comonads, one for $\square$ and one for ■, internal to **TCS4** where the objects are types and the morphisms are α-equivalence classes of terms in context $\Gamma \vdash t : A$. This category is called the syntactic category. Please see Appendix A.1.4 for the remainder of the proof. □

Categorical completeness requires providing an equivalence relation in the Lindenbaum algebra of the formulae, as usual in algebraic semantics. The basic calculations, for traditional algebraic semantics in Heyting algebras were provided, for instance, by Figallo et al in [11] or Dzik et al in [8]. *Mutatis mutantis* these calculations will apply for our version of the system (no distribution of diamonds over disjunctions, no definibility of diamonds in terms of negated boxes).

5 Conclusions

We have described a tense version of constructive temporal logic, conceived as a basic category of propositions, together with two adjunctions, corresponding to two kinds of necessity modalities, in the future and in the past. This system is based on traditional work of Ewald in [10], where we simply do the modifications required to account for the categorical model desired. This work is somewhat inspired by recent work on Functional Reactive Programming (FRP) by Jeltsch [16] and Jeffrey [15], independently. Both of these works consider Curry-Howard correspondents to temporal logic, but they tend to concentrate on the *next* temporal operator, originally considered in LTL (Linear Temporal Logic), as suggested by Davies [6]. The temporal operators we consider are more abstract and one can hope that they may shed some light on the issues of FRP. But this is future work.

References

[1] Natasha Alechina, Michael Mendler, Valeria De Paiva, and Eike Ritter. Categorical and Kripke semantics for constructive S4 modal logic. In *Computer Science Logic*, pages 292–307. Springer, 2001.

[2] Ryuta Arisaka, Anupam Das, and Lutz Straßburger. On Nested Sequents for Constructive Modal Logics. Logic Journal of the IGPL, 22, pp. 992–1018, 2014.

[3] Andrew Barber. *Linear Type Theories, Semantics and Action Calculi*. LFCS, University of Edinburgh, 1997.

[4] Nick Benton. A mixed linear and non-linear logic: Proofs, terms and models. In *Computer Science Logic*, pages 121–135. Springer, 1995.

[5] G.M. Bierman and V.C.V. de Paiva. On an Intuitionistic Modal Logic. *Studia Logica*, 65(3):383–416, 2000.

[6] Rowan Davies. A temporal-logic approach to binding-time analysis. In *Logic in Computer Science, 1996. LICS'96. Proceedings., Eleventh Annual IEEE Symposium on*, pages 184–195. IEEE, 1996.

[7] W. Dzik, J. Järvinen, and M. Kondo. Intuitionistic logic with two Galois connections combined with Fischer Servi axioms. *ArXiv e-prints*, August 2012.

[8] Wojciech Dzik, Jouni Järvinen, and Michiro Kondo. Intuitionistic propositional logic with galois connections. *Logic Journal of IGPL*, 18(6):837–858, 2010.

[9] Wojciech Dzik, Jouni Järvinen, and Michiro Kondo. Characterizing intermediate tense logics in terms of galois connections. *Logic Journal of IGPL*, 22(6):992–1018, 2014.

[10] W. B. Ewald. Intuitionistic tense and modal logic. *The Journal of Symbolic Logic*, 51:166–179, 3, 1986.

[11] Aldo V. Figallo and Gustavo Pelaitay. An algebraic axiomatization of the Ewald's intuitionistic tense logic. *Soft Computing*, 18(10):1873–1883, 2014.

[12] Neil Ghani, Valeria de Paiva, and Eike Ritter. Explicit Substitutions for Constructive Necessity. In *ICALP International Conference on Automata, Languages and Programming*, 1998.

[13] Silvio Ghilardi and GC Meloni. Modal and tense predicate logic: Models in presheaves and categorical conceptualization. In *Categorical algebra and its applications*, pages 130–142. Springer, 1988.

[14] Jean Goubault-Larrecq and Eric Goubault. Order-Theoretic, Geometric and Combinatorial Models of Intuitionistic S4 Proofs. In *In Intuitionistic Modal Logic and Applications (IMLA'99)*, 1999.

[15] Alan Jeffrey. LTL types FRP: Linear-time Temporal Logic Propositions as Types, Proofs as Functional Reactive Programs. In *ACM Workshop Programming Languages meets Program Verification*. ACM, 2012.

[16] Wolfgang Jeltsch. Towards a Common Categorical Semantics for Linear-Time Temporal Logic and Functional Reactive Programming. *Electronic Notes in Theoretical Computer Science, Proceedings of the 28th Conference on the Mathematical Foundations of Programming Semantics (MFPS XXVIII)*, 286:229 – 242, 2012.

[17] Ph. Kremer and G. Mints. Dynamic topological logic. *Annals of Pure and Applied Logic*, 131(1-3):133–158, 2005.

[18] M. Makkai and G.E. Reyes. Completeness results for intuitionistic and modal logic in a categorical setting. *Annals of Pure and Applied Logic*, 72:25–101, 1995.

[19] M. Menni and C. Smith. Modes of Adjointness. *Journal of Philosophical Logic*, 43(2-3):365–391, 2014.

[20] Grigori Mints. *A short introduction to intuitionistic logic*. Kluwer Academic Publishers, Norwell, MA, USA, 2000.

[21] Masao Ohnishi and Kazuo Matsumoto. Gentzen method in modal calculi. *Osaka Math. J.*, 9(2):113–130, 1957.

[22] Jeniffer Paykin and Steve Zdancewic. A linear/producer/consumer model of classical linear logic (extended abstract). In *Third International Workshop on Linearity, LINEARITY*, 2012.

[23] Dag Prawitz. *Natural Deduction, volume 3 of Stockholm Studies in Philosophy*. Almqvist and Wiksell, 1965.

[24] A. N. Prior. *Time and Modality*. Oxford University Press, 1957. Based on his 1956 John Locke lectures.

[25] Alex K Simpson. *The proof theory and semantics of intuitionistic modal logic*. University of Edinburgh. College of Science and Engineering, School of Informatics, 1994.

[26] Lutz Straßburger. Cut elimination in nested sequents for intuitionistic modal logics. In *FOSSACS*, 2013.

A Appendix

A.1 Proofs

A.1.1 Proof of substitution for typing

Lemma (Substitution for Typing). *If $\Gamma_1 \vdash t_1 : A$, and $\Gamma_1, x : A, \Gamma_2 \vdash t_2 : B$, then $\Gamma_1, \Gamma_2 \vdash [t_1/x]t_2 : B$.*

Proof. Suppose $\Gamma_1 \vdash t_1 : A$ and $\Gamma_1, x : A, \Gamma_2 \vdash t_2 : B$. We case split on the structure of the latter, but only show the non-trivial cases. All other cases are similar.

Case Identity.

$$\frac{}{\Gamma_1, x : A, \Gamma_2 \vdash y : C} \text{ Id}$$

In this case $t_2 = y$ and $B = C$. We are not sure if $x = y$, thus, we must consider the case when they are and are not equal.

Suppose $x \neq y$. Then $[t_1/x]t_2 = [t_1/x]y = y$ by the definition of substitution. In addition, it must be the case that either $y : C \in \Gamma_1$ or $y : C \in \Gamma_2$. This implies that $\Gamma_1, \Gamma_2 \vdash y : C$ or $\Gamma_1, \Gamma_2 \vdash [t_1/x]t_2 : B$ hold.

Now suppose $x = y$. Then $A = B$, and $[t_1/x]t_2 = [t_1/x]x = t_1$ by the definition of substitution. Thus, $\Gamma_1, \Gamma_2 \vdash [t_1/x]t_2 : B$ holds, because we know $\Gamma_1, \Gamma_2 \vdash t_1 : A$.

Case Implication Introduction.

$$\frac{\Gamma_1, x : A, \Gamma_2, y : C_1 \vdash t : C_2}{\Gamma_1, x : A, \Gamma_2 \vdash \lambda y : C_1.t : C_1 \to C_2} \to_I$$

In this case $B = C_1 \to C_2$ and $t_2 = \lambda y : C.t$. By the induction hypothesis we know $\Gamma_1, \Gamma_2, y : C_1 \vdash [t_1/x]t : C_2$, and then by reapplying the rule we know $\Gamma_1, \Gamma_2 \vdash \lambda y : C_1.[t_1/x]t : C_2$ holds. However, by the definition of substitution we know $\lambda y : C_1.[t_1/x]t = [t_1/x](\lambda y : C_1.t)$, and thus, we obtain our result.

Case Implication Elimination.

$$\frac{\Gamma_1, x : A, \Gamma_2 \vdash t'_1 : C_1 \to C_2 \quad \Gamma_1, x : A, \Gamma_2 \vdash t'_2 : C_1}{\Gamma_1, x : A, \Gamma_2 \vdash t'_1\, t'_2 : C_2} \to_{\mathcal{E}}$$

We now have that $B = C_2$ and $t_2 = t'_1\, t'_2$. By the induction hypothesis we know that $\Gamma_1, \Gamma_2 \vdash [t_1/x]t'_1 : C_1 \to C_2$ and $\Gamma_1, \Gamma_2 \vdash [t_1/x]t'_2 : C_1$ both hold. Then by reapplying the rule we obtain that $\Gamma_1, \Gamma_2 \vdash ([t_1/x]t'_1)\,([t_1/x]t'_2) : C_2$, and thus, by the definition of substitution $\Gamma_1, \Gamma_2 \vdash [t_1/x](t'_1\, t'_2) : C_2$ holds.

Case $\Box$ Introduction.

$$\frac{\Gamma_1, x : A, \Gamma_2 \vdash t'_1 : \Box C_1,\ \ldots,\ \Gamma_1, x : A, \Gamma_2 \vdash t'_k : \Box C_k \quad x_1 : \Box C_1,\ \ldots,\ x_k : \Box C_k \vdash t : C}{\Gamma_1, x : A, \Gamma_2 \vdash \mathsf{let}_\Box\, x_1 : \Box C_1,\ \ldots,\ x_k : \Box C_k\, \mathsf{be}\, t'_1,\ \ldots,\ t'_k\, \mathsf{in}\, t : \Box C} \Box_{\mathcal{I}}$$

In this case $B = \Box C$ and $t_2 = \mathsf{let}_\Box\, x_1 : \Box C_1,\ \ldots,\ x_k : \Box C_k\, \mathsf{be}\, t'_1,\ \ldots,\ t'_k\, \mathsf{in}\, t$. By the induction hypothesis we know that

$$\Gamma_1, \Gamma_2 \vdash [t_1/x]t'_1 : \Box C_1, \ldots, \Gamma_1, x : A, \Gamma_2 \vdash [t_1/x]t'_k : \Box C_k$$

all hold. Then by reapplying the rule we know that

$$\Gamma_1, \Gamma_2 \vdash \mathsf{let}_\Box\, x_1 : \Box C_1,\ \ldots,\ x_k : \Box C_k\, \mathsf{be}\, [t_1/x]t'_1,\ \ldots,\ [t_1/x]t'_k\, \mathsf{in}\, t : \Box C,$$

but by the definition of substitution and the fact that $[t_1/x]t = t$ because t does not depend on x we know that

$$\Gamma_1, \Gamma_2 \vdash [t_1/x](\mathsf{let}_\Box\, x_1 : \Box C_1,\ \ldots,\ x_k : \Box C_k\, \mathsf{be}\, t'_1,\ \ldots,\ t'_k\, \mathsf{in}\, t) : \Box C$$

holds.

$\Box$

A.1.2 Proof of weakening

Lemma (Weakening). *If* $\Gamma_1, \Gamma_2 \vdash t : B$, *then* $\Gamma_1, x : A, \Gamma_2 \vdash t : B$.

Proof. This proof is by induction on the form of $\Gamma_1, \Gamma_2 \vdash t : B$. We only show a few cases, because the others are similar.

Case Identity.

$$\frac{}{\Gamma_1, \Gamma_2, y : C \vdash y : C}\, \textsc{Id}$$

In this case we have that $B = C$ and $t = y$. We must show that $\Gamma_1, x : A, \Gamma_2, y : C \vdash y : C$ holds, but this clearly holds by reapplying the rule.

Case $\Box$ Introduction.

$$\frac{\Gamma_1, \Gamma_2 \vdash t_1 : \Box C_1, \ldots, \Gamma_1, \Gamma_2 \vdash t_k : \Box C_k \quad x_1 : \Box C_1, \ldots, x_k : \Box C_k \vdash t' : C}{\Gamma_1, \Gamma_2 \vdash \mathsf{let}_\Box \; x_1 : \Box C_1, \ldots, x_k : \Box C_k \; \mathsf{be} \; t_1, \ldots, t_k \; \mathsf{in} \; t' : \Box C} \Box_\mathcal{I}$$

This case is similar to the previous case. First, apply the induction hypothesis to the left-most premise, and then reapply the rule.

$\square$

A.1.3 Proof of soundness of TCS4

Theorem. *The type theory* TCS4 *has sound models provided by the structures $\mathcal{C}$ defined above. In other words, given a tense adjoint modal category $\mathcal{C}$, using the above interpretation, the following hold:*

- *Assume $\Gamma \vdash t : A$ in* TCS4. *Then $[\![\Gamma \vdash t : A]\!]$ is a morphism with domain $[\![\Gamma]\!]$ and codomain $[\![A]\!]$;*

- *Assume $\Gamma \vdash t = s : A$. Then $[\![\Gamma \vdash t : A]\!] = [\![\Gamma \vdash s : A]\!]$.*

Proof. The first part holds by induction on $\Gamma \vdash t : A$, and the second by induction on $\Gamma \vdash t = s : A$. We give a few cases of each part, as the others are similar. Throughout the proof we drop semantic brackets on objects, and we assume, without loss of generality, that the interpretation of contexts are left associated. We begin with the first part.

Case Identity.

$$\frac{}{\Gamma, x : A \vdash x : A}\mathsf{Id}$$

We need to provide a morphism $\Gamma \times A \xrightarrow{f} A$ and we choose $f = \pi_2$ (the 2nd projection), as usual.

Case Implication Introduction.

$$\frac{\Gamma, x : A \vdash t : B}{\Gamma \vdash \lambda x : A.t : A \to B}\to_\mathcal{I}$$

By the induction hypothesis we know that there is a morphism $\Gamma \times A \xrightarrow{f} B$. Then we need to find a morphism $\Gamma \xrightarrow{g} (A \to B)$. Choose $g = \mathsf{curry}(f)$ where $\mathsf{curry} : \mathsf{Hom}_{\mathcal{C}}(A \times B, C) \to \mathsf{Hom}_{\mathcal{C}}(A, B \to C)$ is a natural isomorphism that exists because $\mathcal{C}$ is closed.

Case $\Box$ Introduction.

$$\frac{\Gamma \vdash t_1 : \Box A_1, \ldots, \Gamma \vdash t_k : \Box A_k \quad x_1 : \Box A_1, \ldots, x_k : \Box A_k \vdash t : B}{\Gamma \vdash \mathsf{let}_\Box\; x_1 : \Box A_1, \ldots, x_k : \Box A_k\; \mathsf{be}\; t_1, \ldots, t_k\; \mathsf{in}\; t : \Box B} \Box_{\mathcal{I}}$$

By the induction hypothesis we have the family of morphisms $\Gamma \xrightarrow{f_1} \Box A_1, \ldots, \Gamma \xrightarrow{f_k} \Box A_k$, and a given morphism $\Box A_1 \times \cdots \times \Box A_k \xrightarrow{f} B$. We need to find a morphism $\Gamma \xrightarrow{g} B$. As in previous work, we choose $g = \langle f_1; \delta_{A_1}, \ldots, f_k; \delta_{A_k} \rangle; \mathsf{m}; \Box f$, where $\langle -, - \rangle : \mathsf{Hom}_{\mathcal{C}}(\Gamma, \Box A_1) \times \cdots \times \mathsf{Hom}_{\mathcal{C}}(\Gamma, \Box A_k) \to \mathsf{Hom}_{\mathcal{C}}(\Gamma, \Box A_1 \times \cdots \times \Box A_k)$ exists because $\mathcal{C}$ is cartesian and we make the simplifying assumption that $\Box$ is an endofunctor.

Case $\Box$ Elimination.

$$\frac{\Gamma \vdash t : \Box B}{\Gamma \vdash \mathsf{unbox}_\Box\; t : B} \Box_{\mathcal{E}}$$

By the induction hypothesis there is a morphism $\Gamma \xrightarrow{f} \Box B$. It suffices to find a morphism $\Gamma \xrightarrow{g} B$. Choose $g = f; \eta_B$ where $\eta_B : \Box B \to B$ is the unit of the adjunction.

We now turn to the second part:

Case Unboxing $\Box$.

$$\frac{\Gamma \vdash t_1 = s_1 : \Box A_1, \ldots, \Gamma \vdash t_k = s_k : \Box A_k \quad x_1 : \Box A_1, \ldots, x_k : \Box A_k \vdash t = s : B}{\Gamma \vdash \mathsf{unbox}_\Box\; (\mathsf{let}_\Box\; x_1 : \Box A_1, \ldots, x_k : \Box A_k\; \mathsf{be}\; t_1, \ldots, t_k\; \mathsf{in}\; t) = [s_1/x_1]\ldots[s_k/x_k]s : B} \Box$$

Using the interpretations given above we must show that:

$$\langle f_1; \delta_{A_1}, \ldots, f_k; \delta_{A_k} \rangle; \mathsf{m}; \Box f; \eta_B = \langle f_1, \ldots, f_k \rangle; f : \Gamma \to B.$$

This holds by the following equational reasoning:

$$\begin{aligned}
\langle f_1; \delta_{A_1}, \ldots, f_k; \delta_{A_k} \rangle; \mathsf{m}; \Box f; \eta_B &= \langle f_1; \delta_{A_1}, \ldots, f_k; \delta_{A_k} \rangle; \mathsf{m}; \eta; f \\
&= \langle f_1; \delta_{A_1}, \ldots, f_k; \delta_{A_k} \rangle; (\eta_{A_1} \times \cdots \times \eta_{A_k}); f \\
&= \langle f_1; \delta_{A_1}; \eta_{A_1}, \ldots, f_k; \delta_{A_k}; \eta_{A_k} \rangle; f \\
&= \langle f_1, \ldots, f_k \rangle; f
\end{aligned}$$

$\Box$

A.1.4 Proof of completeness for TCS4

Theorem. *The adjoint modal models are complete in the appropriate sense for the type theory* TCS4. *This is to say, if we have equality of the interpretations* $[\![\Gamma \vdash t : A]\!] = [\![\Gamma \vdash s : A]\!]$ *(where* $[\![\ ]\!]$ *is the interpretation defined above) in the tense modal category* $\mathcal{C}$ *for any derived sequents* $\Gamma \vdash t : A$ *and* $\Gamma \vdash s : A$ *then we can derive the equation in the type theory* TCS4 $\Gamma \vdash t = s : A$.

Proof. This result can be shown by constructing a cartesian closed category with two monoidal comonads, one for $\square$ and one for $\blacksquare$, internal to the type theory TCS4 where the objects are types and the morphisms are α-equivalence classes of terms in context $\Gamma \vdash t : A$. This category is called the syntactic category for the TCS4 type theory.

Showing that this syntactic category is cartesian closed is well known, but we illustrate the proof by describing the case of the $\square$ comonad.

We denote a morphism by the α-equivalence class:

$$[\vec{x}, t]^{\vec{A}, B} = [\vec{x} : \vec{A} \vdash t : B]$$

We then have the following definitions:

- (Identity) $\mathsf{id} = [x, x]^{A, A}$

- (Composition) Given morphisms $[\vec{x}, t]^{\vec{A}, B_i}$ and $[\vec{y}, t']^{\vec{B}, C}$, their composition $[\vec{x}, t]^{\vec{A}, B_i}; [\vec{y}, t']^{\vec{B}, C} = [\vec{x_{i-1}}, \vec{y}, \vec{x_{i+1}}, [t/x_i]t']^{\vec{A}, C}$.

- (Equality) Two parallel morphisms $[\vec{x}, t]^{\vec{A}, B}$ and $[\vec{x}, t']^{\vec{A}, B}$ are equal if and only if $\vec{x} : \vec{A} \vdash t = t' : B$.

Using basic facts about substitution one can show that composition preserves identity and is associative.

We first must show that $\square$ is an endofunctor on the syntactic category. Suppose we have the morphism $[\vec{x}, t]^{\vec{A}, B}$. Then we must construct a morphism $[\vec{y}, t']^{\overrightarrow{\square A}, \square B}$. The latter morphism can be defined in two steps. The first is to change the $\vec{A}$ to $\overrightarrow{\square A}$:

$$[\vec{y}, \wedge \mathsf{unbox}_\square\ y_i]^{\overrightarrow{\square A}, \vec{A}}; [\vec{x}, t]^{\vec{A}, B} = [\vec{y}, [\mathsf{unbox}_\square\ y_i/x_i]t]^{\overrightarrow{\square A}, B}$$

The second step is to change B into $\square B$:

$$[\vec{y}, \mathsf{let}_\square\ \vec{y} : \overrightarrow{\square A}\ \mathsf{be}\ \vec{y}\ \mathsf{in}\ [\mathsf{unbox}_\square\ y_i/x_i]t)]^{\overrightarrow{\square A}, \square B}$$

Straightforward calculations show that this construction preserves identities and composition.

The unit of the comonad is defined as $[y, \mathsf{unbox}_\square\ y]^{\square A, A}$. Next we need to define a morphism between $\square A$ and $\square\ \square\ A$:

$$[y, \mathsf{let}_\square\ y : \square\ A\ \mathsf{be}\ y\ \mathsf{in}\ y]^{\square A, \square\square A}$$

Finally, using these constructions it is possible to show the usual diagrams defining the comonad $\square$ hold. The definition for $\blacksquare$ is similar. $\square$

The Historical Role of Kant's Views on Logic

Andrei Patkul
*Department of Ontology and Epistemology, St. Petersburg State University
7/9, nab. Universitetskaya, St. Petersburg 199034, Russia*
a.patkul@spbu.ru

Abstract

The article represents the outcomes of the reconstruction of Kant's classification of kinds of logic. It demonstrates that it would be impossible to form the traditional understanding of so-called formal logic, as well as of today's symbolic logic, without Kant's notion of pure general logic. It was formed by Kant within the framework of his critique of reason. Critique has changed our understanding of logic from seeing it as *organon* to an understanding of it as a *canon* of finite cognition. In conclusion we pose the question of the status of Kant's transcendental logic with regard to its connection to classification of types of logic given by him and possibility of its formalization in account of Kant's idea of pure general logic.

Keywords: Critique of Reason, General Use of Understanding, Particular Use of Understanding, Pure General Logic, Applied Logic, Transcendental Logic, Epistemology of Logic.

1 Introduction

Currently, at a casual glance, it might appear that logic, of the shape it was created in by Aristotle, is a science of forms of thought. There have been many speculations presuming that logic of such type deals only with the formal thought-structures, independent of any content. Therefore, logic, understood in a such way, could be entitled *formal logic*, i.e. a pure formal science. At the same time, this kind of logic is often treated as wrong, obsolete or, at least, as insufficient one. Such critique generally goes from the point of contemporary logic, which overcame the

---

Talk at the conference "Philosophy, Mathematics, Linguistics: Aspects of Interaction 2012" (PhML-2012), held on May 22–25, 2012 at the Euler International Mathematical Institute.

notion of a thought-form by reducing of object-matter of logic, to the pure symbols and the rules of their combinations. The representatives of so-called symbolic logic understand it as the direct opposite of the traditional formal one. Herein, the question of what the proper *quid juris* of their criticism of traditional logic actually is, arises. In the fact, there is the following reason for such a question to arise. Symbolic logic deals with forms as well, but contemporary logicians understand them quite otherwise. Moreover, being self-evident but not distinct enough the notion of formality is presupposed in both cases. It is obvious that symbols and rules of their combinations are formal as the thought-forms *pari passu*. Namely, both forms of thinking and signs are independent of their possible content. Hence, we have to ask whether symbolic logic is a later descendant of formal logic which has forgotten its own roots.

In other words, is it permissible to consider the definition, given currently both to ontological and epistemological statuses of such symbols, symbolic structures and rules of their combinations, as strict and correct? It is a well-known fact that pure forms of such a kind play a significant role in the process of so-called formalization as one of the basic methods of the contemporary scientific knowledge, but what kind of formalization would allow them to become forms empty of any content? If we tried to formalize the things themselves, which we can accomplish any formalization by, it would lead us into *regressus ad infinitum*. Hence, the formality of logical symbols remains problematic both for ontology and epistemology.

Further, we can inquire into the following matter. Which understanding of the essence of logic and its object domain does play the role of the basis of the differentiation of the mentioned kinds of logic at all? Moreover, it raises another question. Which understanding of the essence of logic and its objects domain does make the basis of the notion of the formality of logical forms in each case?

Indeed, the necessity of the separation of both forms of thought and symbols from their specific content is not self-evident. Moreover, it is perfectly possible that the differentiation existing between them is the extremely later term-division which was being made through the abstractive work of "pure reason" during the history of philosophy. Therefore, this differentiation, as such, has to be justified both ontologically and epistemologically. It is to our regret, that it is impossible to accomplish such a justification in the systematical regard here, but we could undertake a reconstruction of an indicative example of interpretation of logic from the history of philosophy which would emphasize the problematic character of the logical formality. In this way, the validity of the logical formalism could be justified not by a formal deduction of its possibility, which, as it was said, would lead to *regressus ad infinitum*, but by detection of the transcendental genesis of the separation of form and content and, hence, of the notion of logic as of a pure formal discipline in any sense. To be more precise, the matter concerns the transformation of the understanding of the essence of logic made by Immanuel Kant.

It is doubtless that Kant's revolutionary conception of logic is one of the most important points in the historical path of the separation of the notions of form and content which remains, even latently, a live issue nowadays. Hence, it is a matter of great importance for the genuine conception of conditions of possibility and limits of applicability of logic as a formal science. We believe that today it could induce our philosophical disputations on the nature of logical forms and structures.

In this context, on one hand, the main aim of the article is to demonstrate the conditions of shaping of notion of form in its logical sense, by reconstructing Kant's logical views. On the other hand, we should remember that Kant showed of all others that logic in its transcendental shape could have some content. Its content can be only the *a priori* one. Thereby, the correlation between logic as a discipline of the pure universal forms and transcendental logic has to remain controversial. However, we presuppose that the transcendental conditions of Kant's interpretation of logic are historically responsible for the genesis of symbolic logic of nowadays. And therefore, it might be justified theoretically only by an ontological and epistemological justification of Kant's logical views.

We believe that the transformation of the understanding of the essence of logic made by Kant cannot be attacked from the standpoint of today's logic and semantics which finally have been derived from Kant's position. Moreover, there is no need in justifying Kant's logic through them.

In this context, in the course of this consideration, we referred to the following paper by Achourioti and van Lambalgen [1]. It is devoted to a justification of the Kant's idea of logic from the perspective of today's symbolic logic. As it was stated, our thesis is the opposite one. One ought to verify today's logical approaches through the reconstruction of the transcendental and historical genesis of the ontological and epistemological conditions of logic, shaping it as a discipline of formal symbols and structures from the transformation of the essence of logic made by Kant.

2 Traditional logic

Nevertheless, it remains very questionable whether logic could be described as a formal discipline starting right from its origin in Aristotle. It is well-known fact that Aristotle treated form (μορφή) as a shape (εἶδος) or even a prototype (παράδειγμα). For instance, he speaks about form in the sense of *causa formalis*, "The form and template, which is the account of the what-it-was-to-be-that-thing. Also the kinds of form are causes in this way. [...] Also the intrinsic parts of the account."[1] [3, p. 115]. Hence, to be more exact one ought to say that form as εἶδος is connected with *logos* as a meaningful definition. Certainly, εἶδος is one of the possible senses of the term of form. At the same time, it is the preferential one. Anyway, the form, treated in Aristotelian way, is something always rich in content. In this case, no form can be separated from its content. It remains questionable, whether it could be possible to differentiate form from formless content in εἶδος within the framework of Aristotle's views on logic. We believe that no distinction of form and content can exist inside εἶδος at all. Any form is always form of certain content, they correspond in an absolute way to each other in Aristotle's opinion. Any εἶδος, itself, is form (μορφή) as such which holds its own certain content as its "What" in itself. Therefore, one cannot understand logic only as a discipline of pure forms of thinking, independent of any content. Farther, it is doubtful whether logic (the title which Aristotle himself hasn't used) meant ἐπιστήμη, i.e. a science in the strict sense for him. We have to remember that any science, insofar it is a science, should have its own object domain which would have its own ontological status (being in things or only in our mind and so on) and which should be distinct strictly from the object domains of all other sciences in accordance with Aristotle.

Yet we could not find something alike logic in his set of sciences, both theoretical and practical. For instance, logic is absent in Aristotle's set of theoretical sciences, which *philosophia prima*, physics and mathematics belong to. It indicates that logic, from its creator's standpoint, does not have its own object, possessing necessary and invariable principles. Therefore, there is no specific domain of beings (including the domain of mathematical objects which are immovable and dependent upon our mind) which could be the object domain of logical investigations as such. Hence, any specific realm of pure logical forms exists neither *per se* nor *in rebus* nor *in mentis*. Therefore, logic cannot be understood as a true science.

Rather, logic is a kind of τέχνη which deals only with the rules of any correct cognition. Thus, we should learn it before we start to cognize any object domain.

3 The misinterpretation of traditional logic

In opposition, there is an existent conviction, that logic even in its Aristotelian version, is a science in the most rigorous sense. One can compare its status, with regard to exactness, only to the status of mathematics. (Assuming, that the question

---

[1] In his own words Aristotle even says, "... ἄλλον δὲ τὸ εἶδος καὶ τὸ παράδειγμα, τοῦτο δ' ἐστὶν ὁ λόγος τοῦ τί ἦν εἶναι καὶ τὰ τούτου γένη [...] καὶ τὰ μέρη τὰ ἐν τῷ λόγῳ"(*Metaph.*, V, 1013a, 25). Here ἄλλον δὲ means "one can speak about something as cause". See English translation in [2].

of propinquity or heterogeneity of both logic and mathematics remains disputable). Thereby, this conviction has an essential presupposition and, in accordance with it, logic, consistent with the general notion of a science, should have a specific object domain which ought to be cognized by it. This presupposition is generally missed. This domain is considered a region of empty thinking-forms or, in today's version, sign-forms. On the other hand, in this regard it might even appear that today's understanding of logic, in the constructive version, recovers original treating of it as τέχνη in Aristotelian sense. For instance, logic could be understood as τέχνη in a sense of creation of logical formulas by the combination of signs which corresponds to given rules. But what kind of ontological or epistemological status can these rules, as such, have?

Indeed, various types of today's logic do not need to possess any kind of object domain to be considered as an exact and rigorous discipline. But, as it was said, this circumstance does not exclude logic dealing with empty forms in the way which they attribute to Aristotelian *Analytics*. However, such forms mustn't be treated as forms of thinking. As it was stated, the status of such forms is ontologically problematic. In this regard, contemporary logic remains just a specification of an idea of a science of pure forms but technically it is more sophisticated. The truth is that Aristotle himself did not consider his analytic as a science of pure forms.

4 Kant's modification of logic

Now, the following question should be posed. What served as the origin of as well as the reason for transformation, the idea of logic undergone, from its Aristotelian understanding as τέχνη to treating it as a science of pure forms both of thought and signs? Perhaps, such transformation began a great while ago. We could even propose that it began a long while ago before Kant. However, Kant's treating of logic is one of the most notable examples of a reinterpretation of the essence of logical knowledge which could cast the light on the problem of the genesis of the notion of logic as a science which deals only with pure forms apart from their actual content. Nevertheless, Zinkstok emphasizes, "The first thing we should note is that Kant calls logic a *science*. This is, in fact, a break with most of the traditional views..." [13, p. 39]. The possible answer to this question is the following. The origin of transformation of the understanding of the logic's essence could be found in Immanuel Kant's *Critique of Pure Reason*.

4.1 Logic, *noumena* and *phenomena*

It is a well-known fact that the transformation of idea of logic made by Kant is connected with the change in treating of this science as ὄργανον of knowledge (as it was for Aristotle) to treating it as κανών of any possible cognition. From Kant's standpoint, previously it was thought, that logic as ὄργανον is not only necessary condition for any knowledge but also the sufficient one. For instance, before Kant it was considered that logical criteria are absolutely sufficient for rational cognition, i.e. for such kind of cognition which does not refer to any possible experience.

Kant has turned the tide. In accordance with his standpoint, logic can serve as the necessary but not the sufficient condition of cognition. Any knowledge has to correspond to logic, i.e. it must not come into a contradiction with logical laws, in the first instance, to first analytical principle of *tertium non datur*. In other words, any cognition has to be free from contradiction within itself. At the same time, according to Kant, logic (without its connection with a possible experience) is not sufficient for acquiring new knowledge. Hence, it is impossible to obtain new knowledge in the pure rational disciplines with such object domains which cannot be given in any possible experience (soul as simple substance, world as a whole, God). These objects are just ideas of reason in Kant's terminology.

Kant justifies this new conception by the thesis that logic is applicable only to the things as phenomena, but not to the things as such or to entities as entities (*ens qua ens*) in the terms of Aristotle. Thus, logic corresponds to something that does not have objective (in the sense of being *in rebus* themselves) but belongs only to the field of subjectivity. The meaning of subjectivity in this and following expressions does not refer to subject as to a singular person. It is related only to a subject in general or, in other words, to a structures of subjectivity as such. Thereby, these structures should necessarily have a relation to the mode which something what exists in itself can be given to us as subjects in a transcendental sense of subjectivity in. As it was said, it implies that any cognition has to be measured by logic, but logic as such cannot give any new knowledge. Insofar, it is isolated from experience (i.e. from the way which the phenomena could appear in) logic can have only subjective but, at the same time, a general and necessary value. In such a way, logic acquires the meaning of *sine qua non* of any knowledge but not of actual cognition. Hence, it is unacceptable to treat Kant's view of logic without regard for the division made by him between *phenomena* and *noumena*.

At the same time, a science of logic acquires the meaning of a science of pure forms of thought which are originally on the subjective side and, hence, independent of the concrete content of *phenomena* given to us through experience. Such acquiring of the subjective character by logic is a necessary condition for the shaping of the notion of pure form. Nevertheless, such acquiring, insofar it is based on the transcendental character of the subjectivity, cannot relativize logic. By the possible relativization of logic we mean such kind of its treating, which implies that the common value of the logical forms is relative to an empirical subject or, in other words, to a singular person who uses these forms. Kant denied any empirical relativity of logic in this sense. Namely, the matter of experience appears in an accidental mode but forms of thought are general and necessary according to him. They have to be present in us *a priori* to make appearance of a matter through experience possible. However, they justify their objective value only by applying themselves to the things as *phenomena*, i.e. to content of our experience.

On one hand, one can understand the famous division of types of logic made by Kant in his *Critique of Pure Reason* only in the context of the described transformation of the role of logic for human cognition. On the other hand, this division can brilliantly demonstrate the reduction of logic to a science of pure forms of thought made by Kant. In order to do so we have to make an attempt to reconstruct an architectonic of logical disciplines given by Kant, in broad outline. For further information on the matter see, for example, [13].

4.2 The general and the particular use of understanding

Thus, Kant has primarily divided the general notion of logic into (i) logic of the general and (ii) logic of the particular use of our understanding. He stated,

> Now, logic in its turn may be considered as twofold, – namely, as logic of the general [universal], or of the particular use of understanding. (A52, B77) [9, pp. 46–47]

The type of logic, last mentioned, deals with a particular object domain in each case as well as with main rules of its cognition. The logic of the particular use of understanding always refers to a matter of one of object domains. In a manner of speaking, it should follow a content of this domain. Hence, as it depends on concrete content of an object domain, logic of such type cannot be detected as a pure or formal science.

Thus, it is quite noteworthy that Kant considers logic, which would refer to some matter, being possible only as a particular but not as an universal discipline. One can suppose that this circumstance goes back up to Aristotle's fundamental thesis, "That which is is spoken of in many ways" [3, p. 167]², or – in scholastic formula – to *analogia entis*. In fact, Kant does not refer to these formulas. It is unlikely at all that

---

²In Aristotle's own words: (τὸ ὂν λέγεται πολλαχῶς) (*Metaph.*, VII, 1, 1028 a, 10). See [2].

he actually knew this Aristotelian conception. For German-speaking philosophy it was rediscovered later, thanks to the efforts of Brentano.

It is more important that Kant accepted that logic can be non-universal in any case. The possibility of the non-universality of logic stems from the fact that it is connected with its content, i.e. with a matter of a certain object domain. Therefore, following the thesis τὸ ὄν λέγεται πολλαχῶς there is no universal object domain. Moreover, no universal object domain is ontologically possible. Hence, logic can be universal on the assumption of an abstraction from any particular object domain. Any type of universal logic or other formal calculus can, it seems, be only objectless. Here, the objectlessness implies that universal logic does not refer to any object, neither to universal (which is ontologically impossible) nor to particular (which are non-universal) ones. In this sense, it refers only to its own structures which has no objective character but only the (transcendental) subjective one.

On the other hand, as may be supposed, universal logic could not be objectless but it relates to all possible objects by the abstraction from differences of individual entities, of types of objects and so on. Indeed, it is acceptable to interpret Kant's notion of universal logic in this way. In such case, logic would be directed toward an object. But its object would be non-particular. We are opposed to such treating of Kant's view on universal logic. Here, one ought to emphasize two reasons for doing so. (i) We adhere to the above-mentioned Aristotelian thesis which we consider an ontological principle, any understanding of logic has to be founded on. Τὸ ὄν λέγεται πολλαχῶς. Therefore, as it was said, no universal object (even an empty and indifferent one) is ontologically possible. Any object should have its own essence as well as a way-of-being. "Object in general" is a *flatus vocis*. The source of the general validity of logic is quite different from any objectivity. It is subjective in a transcendental sense of subjectivity. (ii) Since Kant made a division between *noumena* and *phenomena* we cannot tell if logical forms could be applied to *noumena* which belong to "object in general". For instance, we do not know whether thinking of God has to be yielded to the principle of *tertium non datur*. After all, the mystics of all time have been telling us that God is being and non-being at the same moment.

Anyway, we assent to an opinion of MacFarlane who has stated the following,

> Kant's claim that logic is purely Formal – that it abstracts entirely from the objective content of thought – is in fact a radical innovation. [11, pp. 44–45]

MacFarlane demonstrated that this "radical innovation" was bounded to Kant's rejection of neo-Leibnizian views on logic, implying that logic is general but not objectless. It has its own most general content. In contraposition to them, Kant started to understand logic as a discipline which deals only with rules of thinking, i.e. which has only subjective sense.

One way or another, Kant himself described logic of the particular use of understanding in the following words,

> The logic of the particular use of the understanding contains the laws of correct thinking upon a particular class of objects. (A52, B77) [9, p. 47]

As opposed to logic of the general use of understanding, logic of its particular use may be *organon* of cognition of a specific object domain in accordance with Kant. He stated,

> The former[3] may be called elemental logic, – the latter, the organon of this or that particular science. The latter is for the most part employed in the schools, as a propaedeutic to the sciences, although, indeed, according to the course of human reason, it is the last thing we arrive at, when the science is already matured, and needs only in finishing touches toward its correction and completion; for our knowledge of the objects of our attempted science must be tolerably extensive and complete before we can indicate the laws by which a science of these objects can be established. (A52, B77) [9, p. 47]

We may presuppose that the logic of the particular use of our understanding could be identified with methodology of a particular science in contemporary word usage. It deals with the rules of cognition of a specified object domain but it can appear only after the maturity of one or another particular science.

However, we think that Kant's idea of the particular logic of understanding remains relevant at the present day. Namely, we believe that an attempt of comparison of Kant's logic of the particular use of understanding and the notion of the regional ontology in the phenomenological branch of today's philosophy could be productive in various methodological perspectives[4]. Though, the notion of the regional ontology is not derived directly from Kant's notion of particular logic, this Kant's term usage could clarify the proper meaning of the term "logic", for instance, in Heidegger's word-combination "productive logic of science" [6, p. 4] which was understood as a regional ontology by him.

Nevertheless, we have to dismiss this analogy between logic of the particular use of understanding and the regional ontology and revert to the above-mentioned division between the logic of the general use of understanding and the logic of its particular use accomplished by Kant. Now we should inquire into first part of this division.

---

[3]"The former" means here the logic of the general use of understanding.
[4]See, for instance, [8].

4.2.1 Back to the division made by Kant

Here, we have to recall that Kant defined logic of the general use of understanding as a discipline which deals with general rules of thought regardless of any matter of applying of this thought. As it was said, in contrast to logic of the particular use of understanding, logic of its general use cannot be general *organon* of our finite cognition. It can be only its *canon*. Then, Kant called it "elemental logic".

Namely, he stated that logic of the general use of understanding

> [...] contains the absolutely necessary laws of thought, no use of understanding at all could be possible without, and therefore gave laws to understanding, regardless of the difference of objects, which it may be applied to. (A52, B77) [9, p. 47]

It is obvious that such distinction was a good step forward in the direction of shaping of the notion of logic as a science of pure forms abstracted from any content. Namely, Kant started to consider a logical form on the base of the notion of the law of thought. The laws of thought are of functional character. So, a form of thought is a function which prescribes the one and only mode which it could act in, regardless of its content. This functional character is based on the spontaneity of thinking as such. The condition, necessary for it, is the universality of a law of thought, i.e. its independence of a concrete matter or content. MacFarlane emphasized the normative character of general logic in this regard,

> The generality of logic, for Frege as for Kant, is a normative generality: logic is general in the sense that it provides constitutive norms for thought as such, regardless of its subject matter. [11, p. 35]

Only logic of such kind can be universal. In other words, it can be used indifferently to the peculiarity of an object domain. In particular, it has to be noted that such understanding of general logic leads to very productive and, yet, very disputable idea of formal ontology which reckons as its object not just subjective "laws of thought" (as it considered by Kant) but also the universal and the only formal definitions of something in general, or of "quasi-region" (Husserl). Nevertheless, here one has to dismiss the reason for a turn from pure formal logic (from the *mathesis universalis* in the widest sense) to formal ontology without prejudice.

In any case, the mentioned step is still insufficient for the ultimate formation of the notion of logic as of a science of pure forms in Kant's interpretation. We just have to point out here, that the distinction between logic of general and logic of the particular use of understanding is based on the quantitative principle of the difference between generality and particularity. Thereby this qualitative principle is aligned with the difference between dependence upon content and its independence stated by Kant. Only the independence of any content of logic ensures its quantitative generality.

4.3 Pure and applied universal logic

As distinct from the notion of logic of the particular use of understanding, the notion of logic of its general use is divisible further from Kant's standpoint. He believed that logic of the general use of understanding can be divided in two parts. Namely, they are, on one hand, pure logic and, on the other hand, applied logic. The common condition for both types of logic is their generality.

The first of the mentioned types of logic, as of a general discipline, deals only with the rules of thought regardless of the concrete conditions of its implementation by an empirical subject. The laws of thought are equally independent of the situation which someone applies them in. Nevertheless, they belong not to objectivity but to the subjective field only, they possess an ideal identity within themselves. The universal and transcendental character of subjectivity in accordance with Kant is the guarantee for such identity. Hence, it is indifferent for the formality of the laws of thought who, where, when and how applies them. In each case, they will remain the same. In this sense, it has to be said, that the laws of logic have "objective value" but it does not mean that the ground of this value lies in objects. It doesn't mean that their source lies in objectivity in its opposition to empirical subjects. This characteristic of the laws of logic refers only to "objectivity" in the sense of the ideal "universal validity".

On the contrary, general but applied logic takes into account such empirical conditions of thinking. This difference could be well-clarified by a few statement made by Kant in his *Critique of Pure Reason*,

> General logic is again either pure or applied. In the former, we abstract all the empirical conditions under which the understanding is exercised; for example, the influence of the senses, the play of the fantasy or imagination, the laws of the memory, the force of habit, of inclination etc, consequently also, the sources of prejudice, – in a word, we abstract all causes from which particular cognitions arise, because these causes regard the understanding under certain circumstances of its application, and, to the knowledge of them experience is required. (A52-3, B77) [9, pp. 47–48]

Hence, we could detect that Kant's term "applied logic" is equal to psychology of logical knowledge in the contemporary usage of terms. The author of *Critique of Pure Reason* has especially emphasized that he used the term "applied" with regard to logic in quite a different sense than it is commonly used. According to Kant's interpretation of applied logic, it doesn't belong to any kind of τέχνη. It is not a technical discipline or a kind of skill which teaches how to apply the logical rules and laws correctly. Applied logic is exactly general logic, not a particular one. Since it is one of subsections of general logic, one cannot use it as ὄργανον of cognition. In any event, it does not consider the logical laws in a strict sense. It treats the domain, which should be already yielded to the rules of the general use of understanding. Hence, it, in fact, remains questionable whether Kant's characteristic of this discipline as of logic is correct. At the same time, it cannot be understood as κανών of knowledge, in contrast to other subsection of logic of the general use of understanding.

The thinker stated the following, concerning general applied logic,

> General logic is called applied, when it is directed to the laws of the use of the understanding, under the subjective empirical conditions which psychology teaches us. It has therefore empirical principles, although, at the same time, it is in so far general, that it applies to the exercise of the understanding, without regard to difference of objects. On this account, moreover, it is neither a canon of the understanding in general, nor an organon of a particular science, but merely a cathartic of the human understanding. (A53, B77-78) [9, p. 48]

Since the domain, which applied general logic inquires into, is already yielded to the general rules of understanding, pure general logic has no need to follow such empirical conditions. It its origin lies not in our actual but in contingent experience,

> Pure general logic has to do, therefore, merely with pure *a priori* principles, and is a canon of understanding and reason, but only in respect of the formal part of their use, be the content what it may, empirical or transcendental. (A53, B77) [9, p. 48]

Therefore, the laws of use of understanding, which pure applied logic discovers, should correspond to the "pure a priori principles". For, in its turn, applied logic as such has to be commensurate to the common κανών of knowledge, i.e. to pure general logic. Hence, general logic is not deducible from pure applied logic. In other words, it is impossible to derive logical forms as such from the modes, we use and apply them in, in our empirical circumstances. As it was said, the logical forms should be already present in a way. They can be applicable in this case only. The usage and the application of logical forms are already yielded to these forms.

In this regard, Kant concluded,

> In general logic, therefore, that part which constitutes pure logic must be carefully distinguished from that which constitutes applied (though still general) logic. The former alone is properly science, although short and dry, as the methodical exposition of an elemental doctrine of the understanding ought to be. (A53-4, B78) [9, p. 48]

It is our belief that in this way Kant derived the notion of logic, very close to "formal logic" in a contemporary sense. In Kant's words, formal logic is the pure general one, i.e. it is logic, independent as well of concrete content given by the experience as of concrete conditions of accomplishing of thinking by an empirical subject.

In summary of this subsection, we would state that principle of the differentiation of pure and applied logic within the framework of logic of the general use of understanding lies in the difference of the notions of the transcendental and the empirical fields. Hence, it is a qualitative or, to be more precise, essential, principle as distinct from the basis of the differentiation of general and particular logics. As it becomes apparent, Kant pointed out two requirements for such kind of logic or conditions it could be formed in:

(i) One ought to differentiate form and content in a thought disregarding to the differences between objects which could be thought by a logical form. Then, one ought to expound form as that what belongs to subjective field. Content has to be considered as that what derives from objects. In that way, we can differentiate form and content finally and, then, get the notion of the form of thought, which would be independent of its content.

(ii) One ought to exclude "empirical principles" of usage of logical forms and, hence, to show that psychological conditions of application of logic have nothing to do with the laws of logic as such. In this way, one can justify why the relativity of the logical forms in their application does not follow from their subjective status. Namely, we show that the universality of these forms does not contradict to their subjectivity because of their subjectivity is not the empirical one. See, for instance, [11, p. 48].

Hence, formality of logical forms is defined in a privative way through the independence (i) of objective content and (ii) circumstances of accomplishing of thinking. As it was said, MacFarlane believed that Kant's understanding of logic as of a formal discipline, hence, the peculiarity of his concept of logical form, became a real "innovation". The path of shaping of the notion of logical form goes through abstraction from content both particular and general.

4.3.1 Pure logic and antipsychologism

One especially ought to emphasize Kant's second requirement for pure general logic. In fact, this thinker formed some conditions for arising of so-called antipsychologism in treating of the essence of logic. He made it by rigorous distinguishing between pure formal principles and subjective empirical conditions of thinking. It is a well-known fact that detailed critique of the reduction of logical laws as well as of mathematical objects and structures to the phenomena of the psychical life was elaborated subsequent to Kant in two very different schools of philosophy, namely in analytical philosophy (Frege) and phenomenology (Husserl). The members of the mentioned schools didn't accept Kant's way of treating nature and status of logic and, especially, of its relation to mathematics and their objects. For instance, Frege has elaborated a program of logical justification of the mathematics, conflicting with Kant's understanding of the essence of the mathematics (see, in particular, [4]). In this regard, we have to remember that logic as a science of pure forms of thought and mathematics as knowledge based on pure forms of sensibility have transcendentally different origin and nature form Kant's standpoint. Hence, he can be acknowledged as the forerunner of the mathematical intuitionism. On the contrary, it would be wrong to speak about the elements of intuitionism in Kant's treating of logic. Thereby, we can find the same situation with regard to a status of the universal validity of logic in all three cases. One can reach the mentioned only by separation of logic forms not only from content delivered from a side of objects but also from private or empirical-subjective conditions of the validity of logical statements.

Nevertheless, all the above-mentioned philosophers have quite different understandings of the logic. It is our belief that one ought to bear in view such kind of difference in presupposing of the possibility in the interpretation of Kant's logical views, for instance, from the point of view of Frege's philosophy. Namely, one should pose a question whether Frege's notion of logic could be justified on the base of the same conception of the (transcendental) subjectivity as Kant's treating of logic. Does Frege's idea of logic require any conception of subjectivity at all?

Conversely, it was demonstrated that Kant's notion of pure general logic is impossible without the admission of a distinction between pure and the applied logic and, thus, without the admission of subjectivity and a subjective character of such kind of logic at all. For instance, MacFarlane has shown that a possible problem in interpretations of the nature of logic made by Kant and Frege lies in the circumstance that both philosophers had very different understanding of a function of logic. MacFarlane stated that Kant's "[...] picture of logic is evidently incompatible with Frege view that logic can supply us with substantive knowledge about objects" [11, p. 29].

On the other hand, Husserl has agreed that one has to consider the logic as well as the *mathesis universalis*, as a whole, and as an effect of the constitutive activity of the transcendental subjectivity. Hence, the idea of transcendental subjectivity is presupposed in Husserl's case of treating of logic as well. However, it is very doubtful that it is the same activity of subjectivity which constitutes logical forms within Kant's and Husserl's understanding.

However, it is very doubtful that this is the same activity of subjectivity which constitutes logical forms in Kant and Husserl.

5 The idea of transcendental logic

Yet, the given reconstruction of Kant's view on logic remains insufficient. It was shown that Kant has created a notion of pure general logic as a science of pure forms of thought. Nonetheless, he laid down demands for creating of a very peculiar type of universal logic which, none the less, would have certain content. Namely, he introduced the notion of transcendental logic beyond his taxonomy of types of logic. It is clear that his notion of pure general logic, as logic in a current sense of formal science, belongs to this taxonomy. But a new type of logic also claims to deal with the universal and necessary knowledge.

In his later *Lectures on Logic*, Kant distinguished between these two types of logic in the following way,

> Now as propaedeutic to all use of the understanding in general, universal logic is distinct also on another side from *transcendental logic*, in which the object itself is represented as an object of the mere understanding; universal logic, on the contrary, deals with all objects in general. [10, p. 530]

Then, we have to ask again. How is it possible for logic to have universal validity and certain content at the same time? Does it not reduce the Kant's breakthrough with regard to the justification of the logic's purity and its universality to absurdity?

There could be an exact following answer to this question. Transcendental logic in Kant's term-use deals with very specific content, namely, with the relations of our cognitions with their objects. These relations are also of very peculiar kind. Namely, the transcendental logic should inquire into the origin of our cognitions of objects, insofar, it cannot be contained directly within these objects.

It deals with the transcendental origin of the cognitions *a priori*, therefore, the nature of transcendental logic is not analytic, but synthetic. Thereby, it differs principally from pure general logic which deals only with analytical forms. In opposition to it, transcendental logic is non-analytic and intends to explicate the grounds of a connection of logical forms with objects. Hence, transcendental logic cannot be explained through any formal analysis. It only could be justified through the transcendental synthesis. Since it has synthetic essence, transcendental logic deals with the transcendental genesis of both knowledge and objectivity.

Kant himself stated with concern to his idea of transcendental logic,

> In this case, there would exist a kind of logic, in which we should not make abstraction of all content of cognition; for that logic which should comprise merely the laws of pure thought (of an object), would of course exclude all those cognitions which were of empirical content. This kind of logic would also examine the origin of our cognitions of objects, so far as that origin cannot be ascribed to the objects themselves; while, on the contrary, general logic has nothing to do with the origin of our cognitions [...] (A55, B81) [9, p. 48]

It might appear at the first sight that Kant, himself, has destroyed his own idea of pure logic with the introduction of the notion of logic which would deal not only with pure forms, as such, but also with their origin *a fortiori*.

Yet, Kant's own opinion was the opposite one. The point is, that transcendental logic is not connected with the origin of cognition of all types of objects but only of the objects, which could be known *a priori*, exclusively. However, in Kant's opinion *a priori* possesses only a formal character. To be more exact, one also ought to limit the notion of a priori in the current context, for it could have a regard to the transcendental use of logic. This logic deals not with all *a priori* cognitions but with the cognitions of such kind, which allow us to know that some concepts are present *a priori* and can be applied only *a priori*. This type of logic clarifies how it can be possible at all. But it excludes from the consideration sensitive *a priori*, i.e. forms of sensibility as well as their relations to objects. Consideration of these forms belongs to transcendental aesthetics. Moreover, transcendental logic does not consider the very notion of the pure understanding, as such, but it treats relations of the mentioned to the objects only. As it was said, its subject is the origin and the limits of their applicability. To be more precise, from Kant's standpoint, transcendental logic deals with the possibility of relation of forms of thought (categories etc.) to objects as phenomena. Therefore, it takes a part in "substantive knowledge about objects".

Hence, we can speak about content of transcendental logic in some peculiar aspect. But if we try to analyze this content we will notice that such kind of content, in its turn, is, in some sense, the formal one. Namely, transcendental logic deals with synthetic formality of thought. For instance, Kant always thought that the categories of the pure understanding are pure forms of understanding. Their relations to objects, in the same measure, are the formal ones. In other words, transcendental logic presupposes abstraction of content too. But this abstraction is not total. One could describe abstraction of such kind as reduction of content to formal relations.

It is very significant that such forms and their formal relations to objects represent the proper content of this discipline. Hence, form, in a way, is content in case of transcendental logic. Thereby, transcendental logic is still pure logic. Since transcendental logical forms have a relation only to objects as *phenomena*, namely, to their realm as whole, insofar, it even is constituted by such forms. Transcendental logic, in a way, is universal. For further details, see [12].

Dealing with the genesis of *a priori* concepts, transcendental logic, in the way it was treated by Kant, has a productive moment within itself, but not absolutely. Namely, Kant searched after the subjective conditions *a priori* of a possible relation of our knowledge to objects. Hence, his understanding of transcendental logic remains subjective (but not empirical). Still, Kant's transcendental logic is not the logic of any, so to say, objective content. Kant's treatment of transcendental logic as of a discipline which has some content rooted in his doctrine of transcendental subjectivity and its structures *a priori*.

Hence, we can conclude that Kant's idea of transcendental logic meets the conditions of pure universal logic mentioned above. On one hand, although it does indeed relate to some specific kind of objects excludes all sensitive objects as well as the rules of the empirical thought from consideration, (i) it can obtain universal validity in a certain sense. Since it inquires only into *a priori* forms of thinking of objects, its content is also only form or, so to say, it is empty of empirical content. On the other hand, since it treats only *a priori* structures of transcendental subjectivity, (ii) transcendental logic does not depend upon the empirical conditions of cognition and therefore upon an empirical accomplishing of our cognition. Therefore, it can have the universal validity in the domain of objects which are also constituted by pure notions of understanding. Yet, from Kant's standpoint, this domain is exclusively inside the realm of *phenomena*.

More to the point, Kant's idea of transcendental logic remains currently relevant. It has been existing in the phenomenological branch of today's philosophy, at least, since Husserl's *Formal and Transcendental Logic*[5]. Phenomenological treating of these two kinds of logic has indicated the following questions. Which correlation between the formal and transcendental kinds of logic is proper? Should formal logic be grounded by transcendental logic? On the contrary, should transcendental logic be understood as widening of formal logic which would underlie to it? Sadly, we have to shelve these questions here. But one ought to emphasize here that the relation of

---

[5]In this context, see [7].

this new type of logic to reconstructed Kant's taxonomy of types of logic is initially ambiguous. Please, find the description of the possible modes of this relation, as well as various reconstructions of Kant's taxonomy of types of logic here: [13, pp. 1–10].

6 Exclusion of "speculative logic"

Now, we have to reject one noteworthy solution which, in our opinion, followed from Kant's dividing form of thought from content as its matter (as well as from division he made between pure general and transcendental logic). We think this solution intends to resolve the fundamental dualism of form and content in logic, which has arisen due to the mentioned divisions made by Kant. The solution we have in mind seems to be very effective and grandiose but, at the same time, very controversial.

Namely, we would like to make an exclusion of so-called "speculative logic", which has originated from Hegel's philosophy of absolute idealism and its materialistic reinterpretation made in the ideology of Marxism (so-called "dialectical logic") from our discussion. Nevertheless, it should be noted that this quasi-logical solution is not absolutely unusual in the history of logic. Moreover, it could shed the light on the problem of the possibility of formalization of logic in some negative way. Namely, it can show possible limits of this formalization.

Thus, this kind of philosophical thinking makes a claim to elaborate the so-called logic of content. It seems to be initially incoherent, being undertaken after a sufficient separation of form from content in logic made by Kant. Yet, we would like just to emphasize the proper way which it was planned to be done in. Sadly, this way often remains undetected by logicians and historians of logic. For instance, Hegel referred not just to the possibility of explication of concrete content of concepts from their implicative condition but also to the possibility of producing and generating of such content through combining of pure logical forms. From his standpoint, the idea of logic, as of a formal discipline, roots in the abstract mode it is considered in. In fact, according to Hegel, logic itself can generate its content and provide a matter of thought to itself. He stated in the Introduction to his *Science of Logic*,

> More to the point is that the emptiness of the logical forms lies rather solely in the manner in which they are considered and dealt with. Scattered in fixed determinations and thus nor held together in organic unity, they are dead forms [...] Therefore they lack proper content [...] But logical reason is itself the substantial or real factor which, within itself, holds together all the abstract determinations and constitutes their proper, absolutely concrete, unity. [5, pp. 27–28]

Nevertheless, it is permissible to notice that the idea of "logic of content", based on Hegel's ontological premise of speculative identity of logic and ontology, should not be identified with Aristotelian understanding of form as of something rich in content. Therefore, "speculative logic" is situated outside of the main path of elaboration of logic as of a science of empty forms, whereas the Aristotelian shape of logic lies in the initial point of this path. Hegel and other "dialectical" logicians tried to get over a chasm between form and content in logic through logical (in the sense which they generally understood the logic in) tools. Hence, they aimed at unifying form and content through the quasi-logical combinations in the situation of historically already-actualized separation between form and content. Therefore, it could not be confused with the initial Aristotelian notion of form as something, rich in content within itself. So, these dialectical ideas don't belong to the mainstream of elaboration of logic starting from the Aristotelian treating of form to an idea of it as of a science of pure forms and form-combinations.

7 Summary and conclusion

In this way, Kant has discovered the possibility of logic which describes the correlation of the pure notions with objects. Thereby he has made room in his architectonic of logic for the kind of logic which would be general and pure, in a sense, yet it could not be independent of content. At the same time, its content doesn't have an empirical source. On this ground, this logic could be titled as properly "philosophical logic" which deals with the origin of our cognitions and their possible relation to objects unlike all the other types of logic which do not have proper philosophical sense. Since this logic considers conditions of our cognitions of objects, we would also call it epistemological logic. Since it discovers condition of relation to objects, we, as well, could define it as ontological logic.

Appositely, one ought to add that our earlier hypothesis implying that the possible conceptual origin of an idea of a regional ontology in phenomenology should lie not just in Kant's idea of logic of the particular use of understanding but also in his concept of transcendental logic. Indeed, logic of the particular use of understanding can, however, play an exclusively methodological role for a particular positive science but the regional ontology should ground one or another particular positive science on the basis of categories and their relation to the subject matter of this science.

Nevertheless, transcendental logic is quite different from pure logic, which has only a formal sense that does not contradict with formal characteristics which are present in both logics, philosophical and non-philosophical (pure formal logical) at the same time. However, the final questions arise. Is it possible to formalize this philosophical logic, which Kant's doctrine of kinds of judgment and categories belongs to? Does logic in interpretation given to it by Kant need any formalization or, at least, allow it? There is the following reason for such kind of questions. As it has been demonstrated already, Kant's transcendental logic is pure and formal (in the sense that its content is the pure formal relations) as well as independent of singular empirical conditions. Hence, we have to ask whether it is possible to formalize this type of logic which has long been logic of forms.

One ought to say that there have been some attempts made recently in order to rehabilitate Kant's transcendental logic with regard to today's semantics through the formalization with the tools of today's symbolic logic. They are of high interest and sophisticated. For instance, we could refer to the paper by Achourioti and van Lambalgen [1] which was mentioned above. In particular, these authors speak about "typical modern dismissal of Kant's formal logic" [1, p. 254]. (They refers to Frege's and Strawson's works in this regard). Regarding the contemporary evaluation of his transcendental logic, they have stated, "Worse, Kant's transcendental logic does not seem to be a logic in the modern sense at all: no syntax, no semantics, inferences" [1, p. 254]. Achourioti and van Lambalgen think magnanimously that they will save Kant's transcendental logic by the demonstrating that "a logical system very much like Kant's formal logic is a distinguished fragment of first-order logic, namely, geometric logic" [1, p. 254]. And we do not think that it is a "hopeless enterprise." [1, p. 254].

Yet, from our standpoint, the following question arises in this regard. Is it still necessary to justify Kant's logics both pure general and transcendental from the point of view of symbolic logic or semantics? It is a problem (i) because, as was shown, the mentioned disciplines are possible in dimension which was cleared away only by the transformation of the understanding of logic with regard to the notion of a pure form made by Kant. Today's logic has just exchanged thought-forms to sign-forms but transcendental-philosophical conditions of the formality which were recognized by Kant remain the same. Moreover, (ii) Kant's logic (even transcendental one) does not need to be, as well as, it and cannot be formalized because, in a way, it has always possessed the formal status, as it is. Therefore, a question posed here should be not of how to justify Kant's views of logic from the perspective of symbolic logic and semantics but of existence of a possibility, as such, for justifying Kant's understanding of logic. It is a fact that both symbolic logic and semantics do not pose the question of their own *quid juris* unlike Kant did with regard to logic. Hence, it is still unclear where an epistemological source of the contemporary fetishism of "syntax, semantics and inferences" is. Maybe we could find it in Kant's philosophy itself. Therefore, it remains disputable whether it is possible or needed to formalize the Kantian conception of logic, even following the semantic character of contemporary logicism which was emphasized by MacFarlane.

References

[1] T. Achourioti and M. van Lambalgen. A Formalization of Kant's Transcendental Logic. *The Review of Symbolic Logic*, 4(2):254–289, 2011.

[2] Aristotle. *Aristotle's Metaphysics*. Clarendon Press, Oxford, 1924.

[3] Aristotle. *Metaphysics*. Penguin Books, London, 1998.

[4] G. Frege. *Foundations of Arithmetic*. Northwestern University Press, Evanston, 1980.

[5] G. W. F. Hegel. *The Science of Logic*. Cambridge University Press, Cambridge, 2010.

[6] M. Heidegger. *History of Concept of Time*. Indiana University Press, Bloomington, 1985.

[7] E. Husserl. *Formal and Transcendental Logic*. Martinus Nijhof, The Hague, 1969.

[8] E. Husserl. *Ideas Pertaining to a Pure Phenomenology and to a Phenomenological Philosophy: First Book: General Introduction to a Pure Phenomenology*. Kluwer Academic Publishing, Dordrecht, 1989.

[9] I. Kant. *Critique of Pure Reason*. Henry G. Bohn, London, 1855.

[10] I. Kant. *Lectures on Logic*. Cambridge University Press, Cambridge, 1992.

[11] J. MacFarlane. Frege, Kant and the Logic in Logicism. *The Philosophical Review*, 11(1):25–65, 2002.

[12] C. Tolley. The Generality of Kant's Transcendental Logic. *Journal of the History of Philosophy*, 50(3):417–446, 2012.

[13] J. H. Zinkstok. *Kant's Anatomy of Logic. Method and Logic in the Critical Philosophy*. Proefschrift, Groningen, 2013.

The Logic for Metaphysical Conceptions of Vagueness

FRANCIS JEFFRY PELLETIER
Department of Philosophy, University of Alberta
Edmonton, Alberta, Canada
Jeff.Pelletier@ualberta.ca

Abstract

Vagueness is a phenomenon whose manifestation occurs most clearly in linguistic contexts. And some scholars believe that the underlying cause of vagueness is to be traced to features of language. Such scholars typically look to formal techniques that are themselves embedded within language, such as supervaluation theory and semantic features of contexts of evaluation. However, when a theorist thinks that the ultimate cause of the linguistic vagueness is due to something other than language – for instance, due to a lack of knowledge or due to the world's being itself vague – then the formal techniques can no longer be restricted to those that look only at within-language phenomena. If, for example a theorist wonders whether the world itself might be vague, it is most natural to think of employing many-valued logics as the appropriate formal representation theory. I investigate whether the ontological presuppositions of metaphysical vagueness can accurately be represented by (finitely) many-valued logics, reaching a mixed bag of results.

Keywords: Vagueness, Many-valued Logic, Evans-argument.

Introduction

Even though people sometimes point to vague memories (e.g., of that very first date you had) or vague objects (like the cloud above me as I write, or the mist that covered St. Petersburg a few nights ago), it is in language where vagueness most clearly manifests itself, and where most theorists focus their attention. The reasons for this are not hard to fathom:

Talk at the conference "Philosophy, Mathematics, Linguistics: Aspects of Interaction 2012" (PhML-2012), held on May 22–25, 2012 at the Euler International Mathematical Institute. This published version has benefited greatly from the comments of a large number of anonymous referees.

- The majority of our linguistic terms admit borderline cases;

- We are unable to resolve the application vs. non-application of many scalar predicates;

- We sometimes may not be able to determine what proposition (if any) is asserted when using certain vague terms.

But even if it is in the realm of language where we find vagueness manifested, there is still the question What is the "ultimate" cause of the vagueness? Is it perhaps a matter of lack of knowledge? Perhaps lack of knowledge of some relevant features of the world? Or perhaps lack of knowledge of the relevant context? Or is it instead that the precise language is correctly representing a vague reality? Or is it merely that language itself does not completely and precisely represent (the non-vague, precise) reality?

It is traditional to divide viewpoints concerning the ultimate cause of vagueness into three sorts: (a) Epistemological Vagueness, where vagueness is claimed to be due to a lack of knowledge – an inability to tell whether some statement is true or false, even though it might correctly represent reality or represent it incorrectly; (b) Linguistic Vagueness, where vagueness is claimed to be due to a shortcoming in the language itself – the language is not adequate to correctly or fully represent the detailed features of the world; and (c) Metaphysical (or Ontological) Vagueness, where vagueness is claimed to be inherent in reality – our language correctly represents reality, but these items are themselves vague. We will look briefly at each in turn, before we focus on the use (and motivation for the use) of many-valued logic.

Most accounts of vagueness, of all these different types, focus on *properties* that can manifest vagueness, particularly properties that characterize a "scale" – such as tallness, or being a heap, or intelligence, or .... Less time has been spent on the possibility of vague *objects*.[1] (Of course, some scholars think that one way to have a vague object is for it to manifest one of the vague properties in a vague manner.) In this paper we investigate the vague objects more closely than vague properties, although of necessity we talk also about vague properties.

1 Epistemological vagueness

The natural way to understand this viewpoint on vagueness is that the "world" is precise, determinate, definite, and so on, but our apprehension of these precise facts is limited in one way or another by our finite epistemological powers. In the "world"

---

[1] Compare the differences in focus and detail of the papers [1] and [20].

there are objects that have precise boundaries, and all properties have sharp cut-off points (or maybe: all objects are in the clear extension or anti-extension of all the properties).

However, because we lack knowledge, vagueness is introduced: For example, old Prof. Worthington, the ancient don at Wembley College, only vaguely/indeterminately/fuzzily remembers who was present at his Doctoral Viva. He can *almost* remember someone with white or maybe blond hair. But he can't recall clearly whether it was the long-dead Coppleston or the equally dead Millingston.

That was a case of "individual vagueness" on Worthington's part. But there can be wider and wider cases of vagueness: *no one* quite remembers what the priest looked like at old Dr. Benoit's baptism. And maybe it is even more pervasive: a feature of the way the world has developed (all relevant people have died, and no one left any unambiguous memoirs) – for instance, did Galileo *actually* drop balls of different weights from on high? And did he also tether together different weighted balls in order to determine how fast the composite object fell? These are events that actually happened or didn't happen – totally and completely – in the actual world. But since there is now no evidence of any sort to decide which way the world actually went, we say it is vague whether Galileo dropped balls of different weights from a height. One might even go so far as to say that this is the category of "verifiable in principle but not actually verifiable".

And it could be more radical than this: For example, the Epistemological Vagueness position holds that in reality there is *in fact* a particular number of grains of sand that would make this pile of sand be a heap (say, m grains). However, we *can't know* that m grains of sand make a heap because all the evidence that we (or anyone) have available is the same for adding one grain of sand to an $(m-2)$-grains pile as it is for adding one grain of sand to an $(m-1)$-grains pile. (Since by hypothesis we can't discern a change when only one grain is added). Yet in the former case we don't know that a pile has become a heap (because by hypothesis it hasn't). So in the latter case we can't know either (even though it *has* become a heap).[2]

What would be an appropriate representational medium for this conception of vagueness? Well, since the view holds that

- *in the world* there is no indeterminacy... every factual sentence either is true or is false, every object is unique, distinct, and separate from all others, and

- vagueness comes from a lack of positive or negative knowledge of these facts, including lack of knowledge as to what proposition is being asserted,

---

[2]The epistemic conception of vagueness is most famously championed by [21, 22, 29].

it seems to follow that some sort of epistemic logic is called for. Thus the epistemic interpretation really involves two logics: classical, two-valued logic for "the world" and the just-mentioned epistemic logic to accommodate the state of knowledge of people. Vagueness seems then just to be identified with conceptual indeterminacy on the part of a speaker. Such an epistemic logic would employ a modal operator that means "is vague", but of course, in this conception, being vague is interpreted as being epistemically indeterminate, and so something can be *non*-vague by being definitely (in the epistemic sense) false, as well as by being definitely true (again, in the epistemic). Thus, if something is vague (epistemically), then so is its negation, under this conception. Using ∇ to represent this indeterminacy (and $\triangle$ for determinacy), typical postulates of such a logic include, among others

$$\text{if } \vDash \varphi, \text{ then } \triangle\varphi$$

$$\triangle\varphi \leftrightarrow \triangle\neg\varphi$$

$$\triangle\varphi \leftrightarrow \neg\nabla\varphi$$

So the required modal logic couldn't be a Kripke-normal modal logic. In [12], I proposed a class of logics of epistemic vagueness (or epistemic indeterminacy): every statement is in fact either true or false (at a world), but when inside the epistemic vagueness operator, we are to evaluate what is going on at a certain class of related worlds. But as I mentioned, this class is not determined in a classical Kripke-manner, but rather in terms of "neighbourhood semantics".

2 Linguistic vagueness

Linguistic Vagueness posits the same ontology as Epistemological Vagueness, namely that the "world" is precise, determinate, definite, and so on. But it differs from the epistemological version by saying that our *description* of these precise facts is limited in one way or another, rather than our knowledge of the precise facts. It holds that in the "world" there are objects that have precise boundaries, and all properties have sharp cut-off points. Vagueness in this conception is a matter of a kind of mismatch between language and "the world" and not a matter of a mismatch between people's knowledge and "the world", as it is in the epistemological conception. (Of course, different versions of Linguistic Vagueness will have differing accounts of what specific parts of language exhibit the mismatch.)

One version of this mismatch might hold, for example, that when Allen says that George is tall, the name 'George' picks out some specific individual in the world (namely, George) who has some specific height such as 180 cm. But it might hold that there is no such *primitive* property in the world as being tall, for only the specific heights count as primitive properties. In this view, either the property TALLNESS doesn't exist, or if it does, then at least it is not a "basic" property[3] but is instead defined, in one way or another, in terms of the more basic, specific properties and (perhaps) "contexts of use" (as in some of the "contextual theories of vagueness", [8, 16, 19]).

It is a shortcoming of our language, according to some (but not all) of the believers in Linguistic Vagueness, that it has developed with these sorts of predicate-terms. Some also hold it to be a shortcoming of our language that *the denotation relation* is not precise: the name 'Mt. Everest' does not unproblematically designate a specific region of the Earth; so when people use this linguistic term they are not accurately identifying what is the case "in the world". When a person says "This rock is a part of Mt. Everest", the imprecision of the denotation relation forces the sentence as a whole to be vague.[4]

Advocates of the explication of vagueness in terms of a linguistic mismatch have formed the largest group of philosophers, at least starting with Frege. Some were dismayed by the fact that natural language had vague predicates, and saw the ideal language as remedying this[5]:

> We have to throw aside concept-words that do not have a *Bedeutung*. These are...such as have vague boundaries. It must be determinate for every object whether it falls under a concept or not; a concept word which does not satisfy this condition on its *Bedeutung* is *bedeutungslos*. [7, p. 178]

Some others who also thought that vagueness was linguistic believed instead that it was a *good* thing in natural language:

---

[3]I use 'basic' and "primitive' in an intuitive manner, allowing that the relevant theories will be obliged to provide a detailed analysis of these notions.

[4]This view of vagueness – although without the feeling that it is a shortcoming – is expressed in [10]:

> The only intelligible account of vagueness locates it in our thought and language. The reason it's vague where the outback begins is not that there's this thing, the outback with imprecise borders; rather, there are many things, with different borders, and nobody's been fool enough to try to enforce a choice of one of them as the official referent of the word 'outback'. (p. 212)

A similar view is expressed in [27].

[5]Actually, it is very difficult to find any theorist of vagueness – of whatever sort – who thinks that vagueness is a shortcoming in language as a whole. What is more problematic, they would say, is the use of some vague term or phrase in a context where more precision, accuracy, or definiteness is desired and is available for use but just not chosen.

> ...a vague belief has a much better chance of being true than a precise one, because there are more possible facts that would verify it. ...Precision diminishes the likelihood of truth." [18, p. 91]

> Vagueness is a natural consequence of the basic mechanism of word learning. The penumbral objects of a vague term are the objects whose similarity to ones for which the verbal response has been rewarded is relatively slight. ...Good purposes are often served by not tampering with vagueness. Vagueness is not incompatible with precision. [15, pp. 113–115]

> There are contexts in which we are much better off using a term that is vague in a certain respect than using terms that lack this kind of vagueness. One such context is diplomacy. [2, pp. 85–86]

(For example, "We will take strenuous measures to block unwanted aggression whenever and wherever it occurs" allows for a wide course of actions, whereas any non-vague statement would not allow such freedom.)

What would be an appropriate representational medium for this conception of vagueness? Well, since the view holds that

- *in the world* there is no indeterminacy... every factual sentence that uses only the *basic* predicates and the correct denotation relation either is true or is false, and

- vagueness comes from the use of non-basic predicates (and "ambiguously denoting" singular terms) where there is no relevantly determined method of stating how they are related to the basic predicates,

it seems to follow that some semantic technique is needed for displaying the various types of results that might hold between the non-basic predicates used in some linguistic expression and the basic predicates that describe "the world".

For example, one might decide that one class of non-basic predicates actually are abbreviations of some (ordered) range of the basic predicates, and that it is "context" that determines which part of this ordered range is relevant to evaluating the truth value of the expression. (Supervaluations and maybe some other semantic techniques, as introduced by [23, 24], and developed by [3, 28], are plausible candidates for this sort of evaluation, as are some of the theories that employ context, like [8, 16, 19]).

Unlike the Epistemic conception of vagueness in which every (declarative) sentence either *is* true or *is* false (but in some cases we may not know which, so that vagueness is a type of epistemic shortfall), in the Linguistic conception only *some* sentences are true and only *some* are false. Among the ones that are true or are false are those composed with basic predicates (and no funny stuff with the denotation relation). Many of the sentences containing non-basic predicates will be given the value 'vague' (i.e., 'neither true nor false'). But not all of these latter type of sentence will be vague, as for example when the specific object of a predication *clearly* satisfies the predicate. For instance, when 200 cm. in height LeBron James is said to be tall, this is true despite the vagueness inherent in 'tall'.

And there can even be true (also false) sentences about the tallness of middle-height people... and similarly for other non-basic terms. For example, supervaluation theory allows that classical logical truths and contradictions are true/false. And perhaps different semantic techniques, such as contextual theories, could generate other examples.

3 Ontological vagueness

Ontological/Metaphysical/Realistic Vagueness locates vagueness "in the world". So, as opposed to being unclear as to whether a situation actually obtains or not (Epistemic Vagueness), and as opposed to being vaguely described by a language that contains non-basic predicates (Linguistic Vagueness), Realistic Vagueness claims that certain objects in the world just plain are vague. (The intent here, which I will in general follow, is to target physical objects with this characterization, although it might also apply to abstracta, events, relations, and so on.) Few writers have explained it, but [18], who is an advocate of Linguistic Vagueness, assures us that it used to be a common view: "...it is a case of the fallacy of verbalism – the fallacy that consists in mistaking the properties of words for the properties of things."

One might also point to fictional entities as neither having nor lacking certain properties: Hamlet neither has nor lacks a 5 mm wart on his left shoulder. Even though this example is from the realm of fiction, Realistic Vagueness might claim that for an vague actual object, there is some property which it neither has nor lacks.

As is well-known, [5] claims that all views advocating ontological vagueness must invoke the claim that, for certain names a and b, the sentence $a = b$ is neither definitely true nor definitely false. (That is, Ontological Vagueness predicts that there are vague objects in the world, and when we have vague objects, then whether they are or are not the same object can also be indeterminate, at least according to some advocates. This gives at least a sufficient condition for metaphysical vagueness.) [25, 26] also proposed that the crucial test would be a situation in which the question 'In talking about x and y, how many things are we talking about?' has the features that 'none' is definitely a wrong answer; 'three', 'four', etc., are all definitely wrong answers; and neither 'one' nor 'two' is either definitely a right or definitely a wrong answer to it. I call this viewpoint about what is the underlying feature of Ontological Vagueness the Evans/van Inwagen criterion, or, when discussing specific argumentation that turns on exactly how this criterion is to be represented formally, the Evans assumption (or the van Inwagen assumption), and when considering the argumentation that makes use of the Evans assumption I call it Evans' argument and sometimes *an* Evans argument (to emphasize that, while it is not exactly Evans' argument, it is an argument inspired by Evans' argument).

What would be an appropriate representational medium for this conception of vagueness? Well, since the view holds that

- *in the world* there is indeterminacy... vague objects actually have the *real* property: being neither red nor not-red, for example, and

- for any object/property pair, either the object has the property (definitely), or the object lacks the property (definitely), or else it neither has nor lacks it – and this last fact is, in its own way, just as definite as the former two,

it seems that the appropriate representation of this conception will employ a many-valued logic. If a has property F, then Fa is true; if a lacks F, then Fa is false; so in the case where a neither has nor lacks F, Fa must take on some other truth value (counting 'neither True nor False' as a truth value).

Employing a modal logic would not accurately capture Realistic Vagueness, for a modal logic presumes that in each world, every sentence either *is* true or *is* false. Employing unusual semantic techniques also does not adequately capture Realistic Vagueness, for the Realist insists that *all* the properties under discussion *are* in fact "real" and "basic". Only a many-valued logic could capture the Realist's attitude toward vagueness.

And it is to many-valued logics that I now turn.

4 A 3-valued logic embodying vagueness

There are three values: intuitively, TRUE, FALSE, INDETERMINATE.[6] These are taken to describe *three different ways the actual world might relate to a sentence describing it*. That is, the portion of the actual world that is under discussion is *actually* one of: definitely the way being described, definitely not the way being described, or correctly described as indeterminate.

---

[6] We turn later to logics with more than three values, when we discuss the possibility of describing "degrees of vagueness" as an account of "higher-order vagueness".

We would like our language to be able to express the facts that sentence φ is TRUE, FALSE or INDETERMINATE (calling these semantic values T, F, I). So let us invent sentence operators ("parametric operators") that do that: $\boldsymbol{D_t}, \boldsymbol{D_f}, \boldsymbol{V}$. They are ordinary, extensional logic operators, having the following truth tables.

φ	$\boldsymbol{D_t}\varphi$	$\boldsymbol{D_f}\varphi$	$\boldsymbol{V}\varphi$
T	T	F	F
F	F	T	F
I	F	F	T

We use standard 3-valued (Łukasiewicz) interpretations of negation, and, or. (And use the convention that the truth values are ordered: $T > I > F$).

φ	$\neg\varphi$
T	F
I	I
F	T

$[\varphi \wedge \psi] = \min([\varphi], [\psi])$
$[\varphi \vee \psi] = \max([\varphi], [\psi])$

I am going to steer clear of the intricacies involved in the interpretation of the conditional and biconditional, other than to advocate on behalf of these principles:

$[\boldsymbol{D_t}\text{-AXIOM}] : \quad \vDash \boldsymbol{D_t}\varphi \to \varphi$

$[\boldsymbol{EQ}\text{-RULES}] : \text{If } \vDash (\varphi \leftrightarrow \psi), \text{then infer } \vDash (\boldsymbol{D_t}\varphi \leftrightarrow \boldsymbol{D_t}\psi)$
$\quad\quad\quad\quad\quad\text{If } \vDash (\varphi \leftrightarrow \psi), \text{then infer } \vDash (\boldsymbol{D_f}\varphi \leftrightarrow \boldsymbol{D_f}\psi)$
$\quad\quad\quad\quad\quad\text{If } \vDash (\varphi \leftrightarrow \psi), \text{then infer } \vDash (\boldsymbol{V}\varphi \leftrightarrow \boldsymbol{V}\psi)$

Although neither $(\varphi \wedge \neg\varphi)$ nor $(\neg\boldsymbol{D_t}\varphi \wedge \neg\boldsymbol{D_f}\varphi)$ is a contradiction in a three-valued logic, contradictions *can* be described by insisting on the Uniqueness of Semantic Value in 3-valued logic[7].

[USV$_3$]: Every sentence takes exactly one of the three values:
$(\boldsymbol{D_t}\varphi \vee \boldsymbol{D_f}\varphi \vee \boldsymbol{V}\varphi) \wedge \neg(\boldsymbol{D_t}\varphi \wedge \boldsymbol{D_f}\varphi) \wedge \neg(\boldsymbol{D_t}\varphi \wedge \boldsymbol{V}\varphi) \wedge \neg(\boldsymbol{D_f}\varphi \wedge \boldsymbol{V}\varphi)$.

Lemma. *If the main operator of formula Φ is $\boldsymbol{D_t}$, $\boldsymbol{D_f}$, or $\boldsymbol{V}$, then $[\![\boldsymbol{V}\Phi]\!] = F$.*

Proof. If the main connective of Φ is one of the three parametric operators, then (as can be seen from their truth tables) the value of Φ is either T or F. But then $\boldsymbol{V}\Phi$ will be F. □

---

[7]Although in Graham Priest's logic *LP* [14], the third value is claimed to be *both T and F simultaneously*, and "not really" a different third value.

Corollary. *If all sentential parts of formula Φ are in the scope of any of D_t, D_f, V, then $[\![V\Phi]\!] = F$, or equivalently, $[\![\neg V\Phi]\!] = T$ or equivalently, $[\![D_f V\Phi]\!] = T$.*

With regards to using a three-valued interpretation in the predicate logic (with identity), I do not give a full characterization, but only three principles:

[**V-$\forall$**]: $\vDash V[(\forall x)F(x)] \to \neg(\exists x)D_f[F(x)]$

(i.e., if it is vague that everything is F, then there cannot be anything of which it is definitely FALSE that it is F)

[**ref$_=$**]: $\vDash D_t[\alpha = \alpha]$

(i.e., self-identities are definitely TRUE).

[**LL**]: $\vDash a = b \leftrightarrow (\forall F)(Fa \leftrightarrow Fb)$

(Leibniz's Law, as this is usually called: Two things are identical if and only if they share all properties.[8])

Now, while some scholars might find some of these principles questionable (I mention Graham Priest in footnote 9 below), the holders of Ontological Vagueness have pretty much uniformly taken them on.

5   An argument against vague objects in this logic

The argument

An Evans argument proceeds by assuming the Evans/van Inwagen criterion of what the believers in Ontological Vagueness hold: that it can be vague whether there is one or two objects before a person; and it continues, using the principles mentioned above about many-valued logic. The following version is given in [13].

a.	$V[a = b]$	the Evans assumption
b.	$a = b \leftrightarrow (\forall F)(Fa \leftrightarrow Fb)$	**LL**
c.	$V[a = b] \leftrightarrow V(\forall F)(Fa \leftrightarrow Fb)$	(b) and [**EQ**-RULE]
d.	$V(\forall F)(Fa \leftrightarrow Fb)$	(a), (c), $\leftrightarrow$-elim
e.	$\neg(\exists F)D_f(Fa \leftrightarrow Fb)$	(d) and [**V-$\forall$**]
f.	$(\forall F)[D_t(Fa \leftrightarrow Fb) \vee V(Fa \leftrightarrow Fb)]$	(e) and [**USV$_3$**]
g.	$D_t[D_t[a = a] \leftrightarrow D_t[a = b]] \vee$ $V[D_t[a = a] \leftrightarrow D_t[a = b]]$	(f), instantiate $(\forall F)$ to $\lambda x D_t[a = x]$ and λ-convert
h.	$D_t[D_t[a = a] \leftrightarrow D_t[a = b]]$	(g), [Lemma], disjunctive syllogism
i.	$D_t[a = a] \leftrightarrow D_t[a = b]$	(h) and [D_t-AXIOM]
j.	$D_t[a = b]$	(i) and [**ref$_=$**], $\leftrightarrow$-elim
k.	$\neg V[a = b]$	(j) and [**USV$_3$**]

---

[8]As both [6, 11] remark, Leibniz himself only took pains to argue for the right-to-left aspect of [LL], and that with a restriction on the types of properties that F designate. Presumably everybody finds the left-to-right direction of [LL] undeniable. As is often noted, there is a peculiarity with this verbalization of the formula, since the formula gives a condition for there being just one thing under consideration, not two, and says that any property this one thing has is a property it has.

Although not every pair of formulas of the form φ and $\neg\varphi$ contradict one another in a three-valued logic, (a) and (k) do really contradict each other. (a) is either T, F, or I (by [**USV**$_3$]); by the truth-table for **V** it cannot be I; so it is either T or F. But this argument shows that if (a) is T then it is F, but if (a) is F then it is T (à la (k) and the truth table for $\neg$). But [**USV**$_3$] claims that no formula can be both T and F.[9]

Comments & defense of the argument

Was there any "cheating" going on in this proof, or with the postulates? Is there something "funny" about the **V**-operator? Might one question the λ-abstract: is it a "real" property? Might one have concerns about one of the principles used: [**D**$_t$-AXIOM], [**EQ**-RULES], [**USV**$_3$], [**LL**], [**V**-$\forall$], [**ref**$_=$]?

The argument I presented proceeds by λ-abstraction, using the predicate: 'being definitely TRUE of x that it is identical to a'. Does that predicate correspond to a *real* property? If not, then this is not a legitimate case of λ-abstraction, by the standards of Ontological Vagueness.

For the advocate of Metaphysical Vagueness, the answer must be 'yes, it is a genuine property'. For, it is a feature of this position that *in the world* there is vagueness, and its contrary, definiteness. These are *real, actual* properties that are designated by these predicates. And unless the advocates of this position want there to be some sort of "ineffability" when it comes to their postulated properties-in-the-world, they have to admit that such expressions *do* designate such properties. The language is entirely extensional – there is no "funny business" going on about 'opaque contexts' or rigid vs. non-rigid names or .... The λ-abstraction picks out what the believer in Ontological Vagueness must acknowledge is a legitimate property.

But let's look again at this presumed notion of vagueness and definiteness. It cannot be the modal notion of the epistemicists, since that characterizes one's epistemic states rather than reality. *That* kind of (in)definiteness does *not* characterize an item "in the world" but rather cognizer's *apprehension* of objects. Any indefiniteness operator of this variety will endorse a principle like $\nabla\varphi \leftrightarrow \nabla\neg\varphi$, as I mentioned above, and such a principle does not characterize the usual ontological vagueness theorists' view.[10]

---

[9]This shows that if we were to interpret the middle value **V** as it is in Priest's [14] – as being *both* **T** *and* **F** – we would have to rephrase the interpretation of [**USV**$_3$]. And in fact, Priest (p.c.) says that he *denies* [**USV**$_3$], believing that there are but *two* truth values, but that some formulas can take both. Therefore, I propose this argument only against those who do not think that vagueness leads to true contradictions.

[10]Well, except perhaps for dialetheic views like that of Priest [14].

The view of Heck in [9] that $\triangledown$ should in fact obey this principle shows that his argumentation is not really directed against nor in favour of metaphysical vagueness, but rather at or in favour of some hybrid view of metaphysical and epistemic vagueness.

I say again: only the many-valued logic viewpoint accurately captures the ontological vagueness theorist's view.

So far as I am aware, no one has faulted the following principles that are used in the proof (well, so long as they are willing to allow a 3-valued logic in the first place, and as long as they see the extra values as being truly distinct from TRUE and FALSE, contrary to Priest's viewpoint expressed in footnote 9):

- **USV$_3$**: $\models (\boldsymbol{D_t}\varphi \vee \boldsymbol{D_f}\varphi \vee \boldsymbol{V}\varphi) \wedge \neg(\boldsymbol{D_t}\varphi \wedge \boldsymbol{D_f}\varphi) \wedge \neg(\boldsymbol{D_t}\varphi \wedge \boldsymbol{V}\varphi) \wedge \neg(\boldsymbol{D_f}\varphi \wedge \boldsymbol{V}\varphi)$;

- **$\boldsymbol{D_t}$-AXIOM**: $\models \boldsymbol{D_t}\varphi \to \varphi$;

- **ref$_=$**: $\models \boldsymbol{D_t}[\alpha = \alpha]$;

- **V-$\forall$**: $\models \boldsymbol{V}[(\forall x)\Phi(x)] \to \neg(\exists x)\boldsymbol{D_f}[\Phi(x)]$.

In the case of those believers in ontological vagueness who hold there to be more "degrees of metaphysical vagueness" than just the three we have been assuming, a strictly analogous proof to the very same conclusion can be crafted, as discussed in [13]. One changes the [**USV**] axiom to accommodate the further truth values, and generalizes the [**V-$\forall$**] axiom for the extra truth values.

We will return to a discussion of the argument after a brief excursion into higher-order vagueness.

6 Higher-order vagueness

The topic of higher-order vagueness concerns the issue of whether it can be vague that something is vague (and even further iterations, such as being definite that it is vague that it is vague). For the believer in Ontological Vagueness, the first iteration amounts to wondering whether it can be vague that some aspect of reality is vague? It is not clear to me that a proponent of Vagueness-in-Reality will wish to accept this as a part of their doctrine concerning *Reality*. They might instead prefer to view it as a mixture of different types of vagueness: "We don't know whether it is true or false that such-and-so is metaphysically vague", and would thereby prefer some mixture of a many-valued logic with an epistemic logic of vagueness (like that of [12]) added on. Certainly, if they *do* wish to have metaphysical higher-order vagueness, they wouldn't represent it by iterating the V-operator (nor with iterated mixtures of any of the V-, D_t, and D_f-operators). The previously-mentioned Lemma precludes this.

Instead they would increase the number of truth values... and with them, the number of truth-operators in the language. For this purpose, it is common in discussions of many-valued logic to take the truth-values to be integers, with **1** being "most true" and (for an n-valued logic) to make **n** be the "most false" value. And then it is common to introduce the so-called J_i-operators [17]. Such an operator is a generalization of the ideas behind our D_t, D_f and V operators – like our operators, the J-operators have a formula as an argument, and are semantically valued as being "completely true" (that is, take the value **1**) if the formula-argument takes the value indicated in the subscript of the J-operator, and "completely false" (that is, take the value **n**) otherwise. Semantically this is to say, for any value i of an n-valued logic ($1 \leq i \leq n$)

$$[\![J_i(\varphi)]\!] = \mathbf{1} \text{ if } [\![\varphi]\!] = \mathbf{i}$$
$$= \mathbf{n} \text{ otherwise}$$

For example, a five-valued logic would have J_1, J_2, J_3, J_4, J_5 as J-operators), with truth tables:

φ	$J_1\varphi$	$J_2\varphi$	$J_3\varphi$	$J_4\varphi$	$J_5\varphi$
1	1	5	5	5	5
2	5	1	5	5	5
3	5	5	1	5	5
4	5	5	5	1	5
5	5	5	5	5	1

and this account might say that $J_3\varphi$ makes the claim that φ is completely vague, while $J_2\varphi$ asserts that it is vague whether φ is completely vague or is true; and that $J_4\varphi$ claims that it is vague whether φ is completely vague or false.

With suitable additions of the number of truth-values, this seems as plausible a way to represent higher-order metaphysical vagueness as it is in modal logic to represent higher order epistemological vagueness by the iteration of a modal operator $\triangledown\triangledown\Phi$. However: a version of the Argument can be made using *any* (finitely-) many valued logic (with suitable emendations to the various principles). I don't rehearse the proof of that fact here; details can be found in [13].

I think the ability to represent higher-order vagueness (of any finite number of iterations) shows that it is not that many-valued logics are incapable of giving some sense to higher-order vagueness, but that the argument in [13] demonstrates that there is some other, perhaps deeper, incoherency with Metaphysical Vagueness.

Although the ploy of increasing the number of truth values shows that many-valued logics are in fact capable of giving a plausible account of higher-order vagueness for any finite number of iterations, there remains still the issue just mentioned: no matter how (finitely) many truth-values our ontological vagueness proponent wishes to invoke as a way of handling higher-order vagueness for a finite number of truth values, there is a generalization of the Argument that can be turned against it. A question naturally arises then concerning the interaction of higher-order vagueness with the number of truth values. We've just seen that to have one iteration of higher-order vague, we would increase the number of truth values from three to five. If we had another iteration, and wanted all possible combinations to be represented, we would need many more. And if we thought it possible to have any level of iterated higher-order vagueness, then the conclusion would be that we need an infinite-valued logic to accommodate this. Infinite-valued logics come with their own share of unusual characteristics, such as that a quantified formula can be assigned true ($[\![J_i(\exists x Fx)]\!]=1$) even without there being any object a in the domain such that $[\![J_i(Fa)]\!]=1$. I think that most vagueness-in-reality theorists either hold that higher-order vagueness of any sort is impossible (as various authors have claimed, even independently of whether they believed in metaphysical vagueness), or else that it is bounded by some finite number of iterations. (It is not clear how this latter possibility might be argued for by our ontological metaphysicians. Most arguments to this conclusion come from the point of view of it being *cognitively impossible* to have infinite iterations... and that's not very relevant to the ontological conception of vagueness.) Anyway, I'm not going to consider it further.

7 Returning to the argument

In [4], Cowles and White object to the statement of Leibniz's Law in the form given in [LL] above, namely

$\vDash a = b \leftrightarrow (\forall F)(Fa \leftrightarrow Fb)$,

and prefer to see it as (what they call "Classical LL"):

$\boldsymbol{D}_t[a=b] \leftrightarrow \boldsymbol{D}_t[\forall F(Fa \leftrightarrow Fb)]$

They also deny the full force of the [**EQ**-RULES]: They claim that just having

$\vDash \varphi \leftrightarrow \psi$

does not justify

$\vDash \boldsymbol{V}[\varphi] \leftrightarrow \boldsymbol{V}[\psi]$,

nor

$\vDash \boldsymbol{D}_f[\varphi] \leftrightarrow \boldsymbol{D}_f[\psi]$.

As they show, their position has the effect of denying both:

- $V[a = b] \leftrightarrow V[\forall F(Fa \leftrightarrow Fb)]$
- $D_f[\forall F(Fa \leftrightarrow Fb)] \to D_f[a = b]$

(although it does validate $D_f[a = b] \to D_f[\forall F(Fa \leftrightarrow Fb)]$).

Just how plausible are these denials? Not very, it seems to me. Is it really plausible to claim that when we have a *logical* truth that two formulas are equivalent, we cannot conclude that one of them is vague just in case the other one is? Nor that one of them is definitely false just in case the other one is? How plausible is it to claim that even when it is *definitely* false that two objects share all properties, it might yet not be definitely false that these are the same object?

On the other hand, I should admit that because of the plausibility of the [**EQ-RULES**], as well as the other rules, I had originally – when I wrote [13] – thought that the Argument showed the complete *im*plausibility of the conception of Metaphysical Vagueness. However, I hadn't internalized these facts (or maybe I hadn't even noticed them):

1. Although the proof given was framed as showing that a contradiction followed from the assumption of $V[a = b]$, *it equally is a proof of*

    $V[a = b] \to \neg V[(\forall F(Fa \leftrightarrow Fb)]$

    i.e., even if it is vague that $a = b$, it can't be vague that they share all the same properties.

2. And of course: $D_t[a = b] \to \neg V[(\forall F(Fa \leftrightarrow Fb)]$

    i.e., if it is definitely TRUE that $a = b$, then it isn't vague that they share all the same properties: intuitively, it is definitely TRUE that they *do* share all the same properties.

3. Furthermore, clearly: $D_f[a = b] \to \neg V[\forall F(Fa \leftrightarrow Fb)]$

    i.e., given that it is definitely FALSE that $a = b$, it can't be vague that they share all the same properties: intuitively, it has to be definitely TRUE that they differ on at least one property.

But [**USV**] asserts that one of the three cases must hold, so we can conclude

$\vDash \neg V[\forall F(Fa \leftrightarrow Fb)]$

That is, it is *never* the case that it is vague that two(?) objects have all properties in common. (Or, that it is *never* vague that an object has all the properties it has).

In conclusion

I used to think that the original argumentation showed:

- The conception of Metaphysical Vagueness is committed to representing its doctrines with a many-valued logic.

- The conception was committed to various logical principles (listed above), as a consequence of its metaphysics.

- Part of Metaphysical Vagueness was a commitment to the Evans/van Inwagen criterion.

- The Argument showed that any many-valued logic which embodied those principles led to a contradiction.

- And I concluded that Metaphysical Vagueness – Vagueness in Reality – was an incoherent notion.

But given that the Argument proves $\vDash \neg \boldsymbol{V}[\forall F(Fa \leftrightarrow Fb)]$, (which by one of the [**EQ**-RULES] shows $\neg \boldsymbol{V}[a=b]$), perhaps we should instead follow a different route:

- Continue to hold to the requirement of a many-valued logic with the specified logical principles to describe the view, *BUT*

- Deny the background assumption given to us by Evans [5] and van Inwagen [25, 26] that Metaphysical Vagueness is committed to instances of $\boldsymbol{V}[a=b]$.

- And similarly deny the [25] version of Evans' assumption to the effect that one cannot count vague objects – because it is *never* TRUE or FALSE that such a thing is one object and it is *never* TRUE or FALSE that it is two objects.

So, by this line of thought the Argument does *not* show Metaphysical Vagueness to be incoherent, it shows instead that *the Evans/van Inwagen criterion of metaphysical vagueness is incorrect*. So believers in Vagueness-in-Reality should turn their attention to finding a different way of stating their basic metaphysical position, and not allow their opponents to define the field for them. Since I am not an advocate of metaphysical vagueness, I cannot therefore offer anything for them.

However, until they do that, Metaphysical Vagueness remains a "deeply dark and dank conception" that one should avoid.

References

[1] A. Abasnezhad and D. Husseini. Vagueness in the World. In K. Akiba and A. Abasnezhad, editors, *Vague Objects and Vague Identity*, pages 239–256. Springer, Dordrecht, 2014.

[2] W. Alston. *Introduction to Philosophy of Language*. Prentice-Hall, Englewood Cliffs, NJ, 1964.

[3] N. Asher, D. Dever, and C. Pappas. Supervaluations Debugged. *Mind*, 118:901–933, 2009.

[4] D. Cowles and M. White. Vague Objects for Those Who Want Them. *Philosophical Studies*, 63:203–216, 1991.

[5] G. Evans. Can There Be Vague Objects? *Analysis*, 38:208, 1978.

[6] P. Forrest. The Identity of Indiscernibles. In E. N. Zalta, editor, *The Stanford Encyclopedia of Philosophy*. Retrieved January 18, 2016, from http://plato.stanford.edu/entries/identity-indiscernible/, Winter 2012 edition, 2012.

[7] G. Frege. Comments on *Sinn* and *Bedeutung* (1892). In M. Beaney, editor, *The Frege Reader*, Wiley Blackwell Readers, pages 172–180. Wiley-Blackwell, Oxford, UK, 1997.

[8] D. Graff Fara. Shifting Sands: An Interest Relative Theory of Vagueness. *Philosophical Topics*, 28:326–335. 2000.

[9] R. Heck. That there Might be Vague Objects (So Far as Concerns Logic). *The Monist*, 81:277–299, 1998.

[10] D. Lewis. *On the Plurality of Worlds*. Blackwell, Oxford, UK, 1986.

[11] B. C. Look. Gottfried Wilhelm Leibniz. In E. N. Zalta, editor, *The Stanford Encyclopedia of Philosophy*. Retrieved January 18, 2016, from http://plato.stanford.edu/entries/leibniz/, Spring 2014 edition, 2014.

[12] F. J. Pelletier. The Not-So-Strange Modal Logic of Indeterminacy. *Logique et Analyse*, 27:415–422, 1984.

[13] F. J. Pelletier. Another Argument Against Vague Objects. *Journal of Philosophy*, 86:481–492, 1989.

[14] G. Priest. *In Contradiction: A Study of the Transconsistent*. Martinus Nijhoff, Amsterdam, 1987.

[15] W. V. O. Quine. *Word and Object*. MIT Press, Cambridge, MA, 1960.

[16] D. Raffman. Vagueness and Context-Relativity. *Philosophical Studies*, 81:175–192. 1996.

[17] J. B. Rosser and A. Turquette. *Many-Valued Logics*. North-Holland Press, Amsterdam, 1952.

[18] B. Russell. Vagueness. *Australasian Journal of Psychology and Philosophy*, 1:84–92, 1923.

[19] S. Shapiro. *Vagueness in Context*. Oxford University Press, Oxford, UK. 2006.

[20] S. Schiffer. Vague Properties. In R. Diaz and S. Monuzzi, editors. *Cuts and Clouds: Vagueness, its Nature and its Logic*, pages 109–130. Oxford UP, Oxford. 2010.

[21] R. Sorensen. *Blindspots*. Clarendon Press, Oxford, 1988.

[22] R. Sorensen. *Vagueness and Contradiction*. Oxford University Press, Oxford, 2001.

[23] B. van Fraassen. Singular Terms, Truth-Value Gaps, and Free Logic. *Journal of Philosophy*, 63:481–495, 1966.

[24] B. van Fraassen. Presuppositions, Supervaluations, and Free Logic. In Karel Lambert, editor, *The Logical Way of Doing Things*, pages 67–91. Yale University Press, New Haven, 1969.

[25] P. van Inwagen. How to Reason About Vague Objects, 1986. Paper read to the American Philosophical Association, Central Division, St. Louis.

[26] P. van Inwagen. *Material Beings*. Cornell Univ. Press, Ithaca, NY, 1990.

[27] A. Varzi. Vagueness in Geography. *Philosophy and Geography*, 4:49–65. 2001.

[28] A. Varzi. Supervaluationism and its Logics. *Mind*, 116:663–676, 2007.

[29] T. Williamson. *Vagueness*. Routledge, Oxford, 1994.

A Note on the Axiom of Countability

Graham Priest
*Departments of Philosophy, the University of Melbourne,
and
the Graduate Center, City University of New York*
`priest.graham@gmail.com`

Abstract

The note discusses some considerations which speak to the plausibility of the axiom that all sets are countable. It then shows that there are contradictory but nontrivial theories of ZF set theory plus this axiom.

In this note, I will make a few comments on a principle concerning sets which I will call the Axiom of Countability. Like the Axiom of Choice, this comes in a weaker and a stronger form (local and global). The weaker form is a principle which says that every set is countable:

WAC $\forall z \exists f (f$ is a function with domain $\omega \wedge \forall x \in z \exists n \in \omega\, f(n) = x)$

(The variables range over pure sets—including natural numbers. ω is the set of all natural numbers.) The stronger form is that the totality of all sets is countable:

SAC $\exists f(f$ is a function with domain $\omega \wedge \forall x \exists n \in \omega\, f(n) = x)$

The stronger form implies the weaker. Any set, a, is a sub-totality of the totality of all sets. Hence, if the latter is countable, so is a. So I focus mainly on this.

Let us start by thinking about the so called Skolem Paradox. Take an axiomatization of set theory, say first-order classical ZF. This proves that some sets, and *a fortiori* the totality of all sets, are uncountable. Standard model theory assures us that there are models of this theory (in which '$\in$' really is the membership relation) where the domain of the model is countable. There *is* a function which enumerates the members of the domain. It is just one which has failed to get into the domain of the interpretation. Why should we not suppose, then, that the universe of sets

---

Talk at the conference "Philosophy, Mathematics, Linguistics: Aspects of Interaction 2012" (PhML-2012), Euler International Mathematical Institute, May 22–25, 2012.

really is countable? From the perspective of the metatheory, ZF$^+$ (ZF + 'There is a model of ZF'), the countable model is not the intended "interpretation". Our metatheory tells us that the domain of all sets is actually uncountable. But ZF$^+$ itself has a countable model, so the situation is exactly the same with this. We might suppose that the countable model of this tells us how things actually are. True, in the metatheory we are now working in, ZF^{++} (ZF$^+$ + 'There is a model of ZF$^+$'), that model will appear not to be the intended model. But we can reply in the exactly same way. Clearly, the situation repeats indefinitely. And at no stage are we *forced* to conclude that the universe of sets is really uncountable. We will always have a countable model at our disposal.

Indeed, it is not just the case that there is nothing that will force us to conclude that the universe of sets is really uncountable. There are certain conceptions of sethood which actually push us to that conclusion. Thus, suppose that one takes the not implausible view that sets are simply the extensions of predicates (or some predicates anyway).[1] Then, given that the language is countable, so it the universe of sets.

Now, imagine that the history of set theory had been slightly different. Suppose that set theory had been investigated for a few years before Cantor, and that those who investigated it took sets to be simply the extensions of predicates. Suppose also that the theory had actually been formalised, say by some mathematician, Zedeff. The (strong) Axiom of Countability, being an *a priori* truth about sets, was one of the axioms. Things were bubbling along nicely, until Cantor came along and showed that within the theory one could prove that some sets are uncountable. The theory was inconsistent. In this history, Cantor was playing Russell to Zedeff's Frege. We can imagine that the community was dismayed by this paradox, and started to try to amend the axiomatization in such a way as to avoid paradox. Perhaps, indeed, the hierarchy ZF, ZF$^+$, ZF^{++}, ... emerged—rather as the hierarchy of Tarski metalanguages emerged in our actual history.

In actual history, set theory was consistentized in response to Russell's paradox and related ones. However, as we now know, there is an alternative: maintain the naive comprehension schema—that is, the schema $\exists x \forall y (y \in x \leftrightarrow \psi)$, where ψ does not contain y—allow the paradoxes, and deploy a paraconsistent logic, which quarantines the paradoxes. The same was an option in our hypothetical history; maintain the Axiom of Countability, the paradoxes it generates, and deploy a paraconsistent logic.

Now back to reality. Is there such a theory? There is. Using the paraconsistent logic LP, we can show the existence of such a theory by applying a result called the

---

[1] See [2, ch. 10]. See also [1].

Collapsing Lemma. Take a first-order language (without function symbols) for LP.[2] Let $M = \langle D, \delta \rangle$ be any interpretation for this. Let $\sim$ be any equivalence relation on D.[3] If $d \in D$, let $[d]$ be its equivalence class under $\sim$. We define a new interpretation (the collapsed interpretation), $M^\sim = \langle D^\sim, \delta^\sim \rangle$, as follows. $D^\sim = \{[d] : d \in D\}$. For any constant, c, $\delta^\sim(c) = [\delta(c)]$. For any n-place predicate, P, $\langle a_1, ..., a_n \rangle$ is in the extension of P in $M^\sim$ iff there are $d_1 \in a_1$, ..., $d_n \in a_n$, such that $\langle d_1, ..., d_n \rangle$ is in the extension of P in M. Similarly for the anti-extension of P. The collapse, in effect, simply identifies all the members of an equivalence class, producing an object with the properties of each of its members. The Collapsing Lemma tells us that any sentence in the language of M (i.e., the language augmented with a name for each member of D) which is true in M is true in $M^\sim$; and any sentence false in M is false in $M^\sim$.[4]

To apply this: let the language be the language of first-order ZF (without set abstracts). Take a (classical) interpretation of this, M, which is a model of ZF. Let k be any countable set in D. (Here, and in what follows, I mean countable—or uncountable—in the sense of M.) Consider the equivalence relation on D which identifies all uncountable sets with k, and otherwise leaves everything alone. That is, $x \sim y$ iff in M:

- x and y are uncountable
- or (x is uncountable and y is k)
- or (y is uncountable and x is k)
- or (x and y are both k)
- or (x and y are countable sets distinct from k, and $x = y$).

Now consider the collapsed model obtained with $\sim$. By the Collapsing Lemma, this is a model of ZF. But in $M^\sim$ every set is countable. For every constant, c, that denotes a countable set in M:

- $\exists f(f$ is a function with domain $\omega \wedge \forall x \in c \exists n \in \omega\, f(n) = x)$

is true in M, and so by the Collapsing Lemma, in $M^\sim$. Since every member of $D^\sim$ has such a name in $M^\sim$, we have the WAC in $M^\sim$:

- $\forall z \exists f(f$ is a function with domain $\omega \wedge \forall x \in z \exists n \in \omega\, f(n) = x)$.

---

[2] For a presentation of the semantics of LP, see [2, sec. 16.3].

[3] If the language were to contain function symbols, $\sim$ would also have to be a congruence on their interpretations.

[4] For full details, including the proof, see [2, sec. 16.8].

A slightly different equivalence relation delivers an interpretation which verifies SAC. Let k now be the object which is V_ω (the sets of rank ω) in M. Consider the equivalence relation which identifies all things of rank greater than ω with V_ω, and leaves everything else alone. That is, $x \sim y$ iff in M:

- $x, y \in V_\omega$ and $x = y$
- or $x, y \notin V_\omega$.

Again, this is a model of ZF. $k \cup \{k\}$ is countable in M. Let i be the name of the function that enumerates it, and let e be the name of any member of $k \cup \{k\}$. Then in M it is true that:

- i is a function with domain $\omega \wedge \exists n \in \omega\, i(n) = e$.

Hence this is true in $M^\sim$. But since every member of $D^\sim$ is named by some e of this kind, we have in $M^\sim$:

- i is a function with domain $\omega \wedge \forall x \exists n \in \omega\, i(n) = x$.

Hence we have the SAC in $M^\sim$:

- $\exists f(f$ is a function with domain $\omega \wedge \forall x \exists n \in \omega\, f(n) = x$.

For good measure, $M^\sim$ is also a model of the naive comprehension schema, $\exists z \forall x(x \in z \equiv A)$, too.[5] If sets just are the extensions of predicates, one would expect this schema to hold. I note also that both of the models we have constructed are non-trivial. Thus, if c and d refer to two distinct objects in D that are not involved in the collapse, $c = d$ is not true in the collapsed model.

What we see, then, is that there are (non-trivial) theories that contain the (strong or weak) Axiom of Countability, plus ZF (plus, in one case, the naive comprehension schema). If T is the set of things true in either of the collapsed models we have constructed, T is one such theory. Within such a theory, every set is countable; but, because of Cantor's Theorem, some sets are uncountable as well. It is Cantor's Theorem that generates the hierarchy of different sizes of infinity. And as seen from the perspective of one of these theories, the Theorem is recognizably paradoxical. The whole hierarchy of infinities is therefore a consequence of the paradox. The transfinite, then, is generated by the transconsistent.[6]

In a nutshell: the Axiom of Countability makes perfectly good paraconsistent sense, even within the context of ZF. And it provides a radically new possible perspective on the universe of sets.

---

[5]For details see [2, sec. 18.4].

[6]It was Zach Weber who first suggested to me that one might see the transfinite in this way. See [3, sec. 8].

Acknowledgements

The author is grateful an anonymous referee for some helpful comments, and to Hamdi Mlika for kind permission to republish this article, which first appeared as *Al-Mukhatabat*, 2012, 1:27–31.

References

[1] J. Myhill. The hypothesis that all Classes are Nameable. *Proceedings of the National Academy of Sciences of the United States of America*, 38(11):979–981, 1952.

[2] G. Priest. *In Contradiction: A Study of the Transconsistent*. Oxford University Press, 2nd edition, 2006.

[3] Z. Weber. Transfinite Cardinals in Paraconsistent Set Theory. *Review of Symbolic Logic*, 5(2):269–293, 2012.

Topologies and Sheaves Appeared as Syntax and Semantics of Natural Language

Oleg Prosorov
St. Petersburg Department of V. A. Steklov Institute of Mathematics RAS
27, nab. r. Fontanki, St. Petersburg 191023, Russia
`prosorov@pdmi.ras.ru`

Abstract

In a sheaf-theoretic framework, we describe the process of interpretation of a text written in some unspecified natural language, say in English. We consider only texts written for human understanding, those we call *admissible*. A *meaning* of a part of a text is accepted as the communicative content grasped in a reading process following the reader's interpretive initiative formalized by the term *sense*. For the meaningfulness correlative with an idealized reader's linguistic competence, the set of all meaningful parts of an admissible text is stable under arbitrary unions and finite intersections, and hence it defines a topology which we call *phonocentric*. We interpret syntactic notions in terms of topology and order; it is a kind of *topological formal syntax*. The connectedness and the T_0-separability of such a phonocentric topology are *linguistic universals*. According to a particular sense of reading, we assign to each meaningful fragment of a given text the set of all its meanings those may be grasped in all possible readings in this sense. This way, to any sense of reading, we assign a sheaf of fragmentary meanings. All such sheaves constitute a category, in terms of which we develop a *sheaf-theoretic formal semantics*. It allows us to generalize Frege's compositionality and contextuality principles related with the *Frege duality* between the category of all sheaves of fragmentary meanings and the category of all bundles of contextual meanings. The acceptance of one of these principles implies the acceptance of the other. This Frege duality gives rise to a representation of fragmentary meanings by continuous functions. Finally, we develop a kind of *dynamic semantics* which describes how the interpretation proceeds as a stepwise extension of a meaning representation function from the initial meaningful fragment to the whole interpreted text.

Keywords: Sense, Meaning, Phonocentric Topology, Linguistic Universals, Sheaf of Fragmentary Meanings, Compositionality Principle, Contextuality Principle, Bundle of Contextual Meanings, Frege Duality, Dynamic Semantics.

Talk at the conference "Philosophy, Mathematics, Linguistics: Aspects of Interaction 2012" (PhML-2012), held on May 22–25, 2012 at the Euler International Mathematical Institute.

1 Introduction and informal outline

In this work, we apply rigorous mathematical methods in studying the process in which the understanding of a written text or an uttered discourse is reached. Our aim is to present a formal model for the understanding of a text or a discourse in a natural language communication process.

Any natural language serves as a means of communication between members of a community that shares this language. The life of a human society, primitive or developed, ancient or contemporary would be impossible without linguistic communication. When we communicate with each other, we are involved in the activity of exchange with two complementary sides: the production and the understanding of language messages in oral or in written form. Any linguistic communication presupposes the emitting activity that produces a message and the receiving activity that produces an understanding. The message is an externalization of thoughts either by utterance or by writing. As a linguistic message unit, a single stand-alone sentence (or phrase) does not suffice to express the variety of thoughts and ideas that people need to communicate. The minimal exchange units that serve as messages in linguistic communications are written texts and uttered discourses. Linguistics is a discipline that studies the use of a language; for empirical objects, it has, therefore, texts and discourses as the units of human interaction, and not stand-alone words or phrases favoured by traditional grammars and the logic in the wake of Aristotelian tradition primarily concerned with questions of reference and truth.

The main parts of traditional grammars are syntax and semantics. A traditional syntax is a study of sentence structures in a given language, specifically in terms of word order. A semantics, of whatever kind, is the study of relationships between the linguistic expressions and their meanings. Traditional approaches are very restrictive or even inadequate to extend grammatical concepts and theories to the level of text or discourse in order to describe linguistic communication in all its forms.

The present work proposes a mathematical framework that generalizes syntax and semantics of a natural language from the traditional level of a stand-alone sentence or phrase to the level of written or spoken discourse. We propose a kind of a discourse analysis that describes the process of a natural language message interpretation in a uniform manner at all semantic levels.

The paper is organized as follows:
- In the next Sect. 2, we discuss in details our acceptance of basic semantic notions *meaning*, *sense*, and *reference*. We study the interpretation of a text in a certain unspecified natural language, say in English, considered as a means of linguistic communication (mostly in written form). We consider the class of minimal communicative units of a language as made up of texts, and so it is broader than the class

of all stand-alone sentences studied in traditional logical and grammatical theories.

From the set-theoretic point of view, any text is a sequence of its constituent sentences[1]. But from the theoretic point of view on linguistic communication, do we need to define somehow what is a genuine text? It seems useless to set some formal criteria of *textuality* which, likewise to formal criteria of *grammaticality*, would decide that a given sequence of sentences is a well-formed text. Although some particular sequence of words or sentences does not appear to be well-formed, nobody can guarantee the contrary for the future, because a natural language is always open for changes. However, the ethics of linguistic communication presupposes that a genuine text is written by its author(s) as a message intended to be understood by a reader. That is why, instead of adopting any criterion of textuality, we restrict the domain of our study to texts that we assume to be written 'with good grace' as messages intended for human understanding; those we call *admissible*. All sequences of words written in order to imitate some human writings are cast aside as irrelevant to the linguistic communication.

A *meaning* of a part of text is accepted as the communicative content grasped in a particular reading of this part following the idealized reader's attitude, presupposition and intention put together in the term *sense*. We adopt this acceptance of terms 'sense' and 'meaning' because it is close to the ordinary usage of these words in everyday English. The advantage of such a choice of terminology is that we can use words 'sense' and 'meaning' sometimes as linguistic terms, sometimes as ordinary words without specifying each time their mode of use. Otherwise, we were to accept in the use their definitions that we reject in the theory. Thus, we may ask e.g., "What does this word (or expression, sentence, text) mean in the literal (or metaphorical, allegorical, moral, Platonic, Fregean, narrow, wide, common, etc.) sense?" So, our acceptance of terms *sense* and *meaning* differs from *Sinn* and *Bedeutung* of Frege's famous paper of 1892. We discuss the difference further.

- In Sect. 3, we discuss topology and order structures underlying an admissible text considered as a means of communication. The linguistic communication may be adequately modelled by a formalism which takes as its object of study texts and discourses in their production and interpretation.

Whatever the human language is, the speaker produces an utterance when putting words one after another in an acoustic string. The listener is forced to interpret such a chain of sounds without the possibility of suspending its course with the purpose to return or to make a leap forward. Everyone knows this prop-

---

[1] It is clear that any such a sequence is made up of so-called 'sentence-tokens', not of so-called 'sentence-types'. Likewise, a sentence is a sequence of word-tokens, and a word is a sequence of morpheme-tokens. Nevertheless, in speaking further about a sequence of certain language units, we shall sometimes omit the word 'token', in order to not overload the terminology.

erty empirically, owing to personal experience of speaker and listener; it should undoubtedly be taken into account by everyone who writes a text intended for a human understanding. We argue that such a fundamental feature of linguistic behaviour enables us to endow an admissible text X with the structure of a finite T_0 topological space where the set of opens $\mathfrak{O}(X)$ is the set of all meaningful parts of a given text X. We call *phonocentric* such a topology defined on the text X.

It is well known[2] that the category **FinTOP**$_0$ of finite T_0 topological spaces with continuous maps is isomorphic to the category **FinORD** of finite partial ordered sets (*posets*, for short) with order preserving maps. We consider two functors L and Q establishing such an isomorphism between these categories. It allows us to define on an admissible text topological and order structures, both of *deep* and *surface* kinds. The writing process consists in endowing the text with the surface structure of so-called linear 'word order' (and corresponding topology). The process of interpretation consists in a backward recovering of the deep structure of the specialization order (and corresponding phonocentric topology) on the text.

Thereafter, we define a phonocentric topology in a similar manner at each semantic level of an admissible text. The mathematical interpretation of different linguistic notions in terms of topology and order is a kind of *topological formal syntax*.

- In Sect. 4, we elaborate in mathematical details the aforesaid topological formal syntax. We argue that the T_0-*separability* and the *connectedness* of a phonocentric topology are two *linguistic universals* of a topological nature.
- In Sect. 5, we study the process of understanding of an admissible text considered as a means of communication. To understand a text or a compound expression is to grasp what it means, i.e. what communicative content it conveys. Thus the understanding of a text during its reading is a dynamic process that develops gradually as the reading progresses over the time.

On the other hand, a speaker (a writer) uses words as a preexisting means to express thoughts, and one combines them to convey thoughts one wants to communicate. So the meaning of a compound expression is determined by the meanings of its (meaningful) constituents, as well as the meaning of the whole text is determined by the meanings of its (meaningful) parts.

In the traditional hermeneutics, the relationship between the understanding of (meaningful) parts and the understanding of the whole text was conceived as a fundamental principle of text interpretation called the *hermeneutic circle*. As its counterpart in linguistic theories, there is a need for some principles those describe how the passage from the meanings of parts to the meaning of the whole and the passage in the reverse direction are proceeding. In logic, linguistics and philosophy

---

[2] See, for instance, [8, 23].

of language, there exist such two complementary principles both traditionally ascribed to Frege: the *compositionality principle* and the *contextuality principle*, both manifest itself in many different formulations.

According to J. F. Pelletier [26, p. 89], R. Carnap was the first to attribute the compositionality principle explicitly to Frege in *Meaning and Necessity* [3], where he stated this principle in terms of a functional dependence. The majority of researchers followed him when formulating their definitions of Frege's compositionality principle in the mathematical paradigm of a function. To illustrate this, we cite a few definitions:

> [...] the meaning (semantical interpretation) of a complex expression is a function of the meanings (semantical interpretations) of its constituent expressions. (J. Hintikka [14, p. 31])

> Like Frege, we seek to do this [...] in such a way that [...] the assignment to a compound will be a function of the entities assigned to its components. (R. Montague [24, p. 217])

> [...] The meaning of a whole is a function of the meanings of the parts. (B. H. Partee [25, p. 313])

In many similar definitions, the meaning of a compound expression is set to be a function of the meanings of its parts, whereas what the meanings are differs substantially. The same reticence concerns with the explicitness of a function in these definitions. In sharpening her definition, B. H. Partee notices that "the Principle of Compositionality requires a notion of partwhole structure that is based on syntactic structure", and then she modifies the latter definition to the following one:

> The meaning of a whole is a function of the meanings of the parts and of the way they are syntactically combined. (B. H. Partee [25, p. 313])

Nevertheless, the modified definition of the compositionality principle remains implicit with regard to the function it refers to. In fact, the pages subsequent to definitions of compositionality principle in [25, p. 313] are devoted to the discussion of how one may explicitly define the input values (arguments) of such a function, and describe how this function acts on its arguments, and what it returns as output values. On this way, B. H. Partee leads the reader to the formal definitions given in the Montague's seminal paper [24].

To sum up our discussion, we have to note that in agreement with the tradition going back to Carnap, almost all generally accepted definitions of the compositionality principle convey the mathematical concept of a function in a set-theoretic paradigm.

In the contemporary mathematics, there are different formalizations of the concept of a function and functional dependence. In a prevailing set-theoretic paradigm, a function (map, mapping) is identified with its *graph*. Formally, a function $f\colon X \to Y$ is a set of ordered pairs $f \subseteq X \times Y$ (a *graph*) which satisfies the following two Claims:

1° For *every* argument's value $x \in X$, *there exists* a function's value $y \in Y$ such that $\langle x, y \rangle \in f$;

2° This function's value y is *unique* as such, that is, whenever $\langle x, y \rangle$ and $\langle x, z \rangle$ are members of f, then $y = z$. Thus all functions are single-valued.

Intuitively, for an ordered pair $\langle x, y \rangle \in f$, a function f is a 'rule' that assigns the element y to the element x. This y is the value of f for the argument x, that is denoted usually as $y = f(x)$.

What is a function in the set-theoretic paradigm is understood in an unambiguous manner by all the scientific community, and the rigorous definition of a function is therefore imposed on any attempt to clarify a vague notion that bears in germ the idea of functional dependence. This is also true for the notion of compositionality in natural language semantics. Any attempt to define explicitly the principle of compositionality as a function $f\colon X \to Y$ in the set-theoretic paradigm meets with serious technical problems to explain what are these sets X, Y, and how is defined the functional graph $f \subseteq X \times Y$. This is a difficult task and even a trap for any attempt to translate literally the set-theoretic notion of a function into the linguistic notion of a compositionality.

The aim of an adequate semantic theory is to conceptualize how the understandings of parts are integrated during the process of reading to produce the understanding of the whole. However, any semantic theory that combines the compositionality defined as the functionality (meant in the 'function as graph' paradigm) with the non-postponed understanding (meant as a dynamic process that develops step by step while the reading progresses over the time) should be obviously inconsistent.

There are two main directions in which the solution of this apparent conflict might be sought:

- either one conserves the compositionality meant as a set-theoretic functionality but refuses to take into account the process of text understanding over the time, and then establishes a kind of *static semantics*;

- or otherwise, one renounces of compositionality meant as a set-theoretic functionality, or somehow redefines it, and then studies the process of text understanding over the time, in order to establish a kind of *dynamic semantics*.

If the semantic compositionality is taken to be the functionality in a set-theoretic paradigm, then it imposes the almost indubitable conclusion that Frege had never explicitly stated (in this way) the principle of semantic compositionality generally ascribed to him, whatever it were, the compositionality of *Sinn* or the compositionality of *Bedeutung*. In several papers, T. M. V. Janssen had carefully analyzed the development of Frege's views on such a semantic compositionality during his long scientific career, and then concluded, as a result, that Frege "would always be against compositionality" [15, p. 19]. Another point of view is expressed by F. J. Pelletier who writes in a solid historical research that "Frege may have believed the principle of semantic compositionality, although there is no straightforward evidence for it and in any case it does not play any central role in any writing of his [...]." [26, p. 111].

However, another theoretical view on the part-whole text structure without prejudice to define the compositionality as a kind of the set-theoretic functionality allows us to interpret Frege's views on the subject in a different way. We notice that in the unpublished work *Logic in Mathematics* of 1914, Frege writes:

> As a sentence is generally a complex sign, so the thought expressed by it is complex too: in fact it is put together in such a way that parts of the thought correspond to parts of the sentence. So as a general rule when a group of signs occurs in a sentence it will have a sense which is part of the thought expressed. (G. Frege [10, pp. 207–208])

In this translation, the expression 'will have a sense' concerning a group of signs should really mean 'will be understandable'. In fact, it is an implicit expression of the hermeneutic circle principle in the particular case of a stand-alone sentence. In a general case, this principle prescribes **'to understand a part in accordance with the understanding of the whole'**. It means that Frege believed the hermeneutic circle principle at the semantic level of a stand-alone sentence. As a logician, Frege was interested primarily in a particular case of sentences, that is, in judgements. It does not really matter whether Frege was familiar with the philological discipline of hermeneutics or not. The principle of hermeneutic circle reveals one of key cognitive operations involved in a natural language text (or discourse) understanding process, and so it is implicitly known by any competent language user. We argue that the hermeneutic circle principle carries in germ the mathematical concept of a *sheaf* which expresses a passage from a local data to the global one, and which is very close to the idea of a functional dependence. From the sheaf-theoretic point of view, one can revise the aforesaid Frege's quotation like this: 'a family of compatible understandings of parts of the sentence are composable into the understanding of the whole sentence'. However, Frege considered words as being elementary units of a sentence, and he believed in the contextuality principle, bearing today his name, in accordance with which words have no meanings in isolation, "but only in the context of a sentence" [9]. We hypothesize that the reluctance to be got involved into the confusion between elements and parts of a whole (between "words [...] in isolation" and "parts of the sentence" in his formulations) prevented Frege from stating explicitly what would be called the compositionality principle. Surely, a meaningful sentence has some meaningful parts, the meanings of which are constitutive to the meaning of this sentence as a whole; but not every of word-tokens may be found among such meaningful parts. This is a kind of the type difference between an element and a subset of a given set.

For an adopted sense $\mathscr{F}$ of reading of a given text X, to each non-empty open (that is to say, meaningful) part $U \subseteq X$ we assign the set $\mathscr{F}(U)$ of all its meanings that may be grasped in all its possible readings in this sense. In fact, it assigns naturally a presheaf $\mathscr{F}$ of fragmentary meanings to the adopted sense of reading. In the beginning of Sect. 5, we argue that such a presheaf $\mathscr{F}$ should satisfy to both Claims **S** and **C** needed for a presheaf to be a sheaf. Thus, the presheaf $\mathscr{F}(U)$ of fragmentary meanings attached to a sense (mode of reading) of an admissible text is really a sheaf. This statement is our generalization of Frege's compositionality principle in the sheaf-theoretic framework. The issuing *sheaf-theoretic formal semantics* takes its departure from another formalization of a functional dependence which is based on the mathematical concept of a sheaf. We use this revised concept of functional dependence in order to define explicitly what is, or rather what should be the compositionality of fragmentary meanings. In this generalized concept of functionality, the arguments and their numbers are not given in advance (one takes for arguments any family of locally compatible *sheaf sections*); but due to the Claim **C**, *for every* such a family of arguments, *there exists* the global sheaf section which is their composition; and due to the Claim **S**, this composition is *unique* as such. In the Subsect. 5.1 we show that these Claims **C** and **S** are analogous to those Claims **1°** and **2°** in the aforesaid formal definition of a function in a set-theoretic paradigm.

- So far, we have considered only the meanings of <u>open sets</u> in the phonocentric topology which we have defined in Sect. 3 at any semantic level. Then, in Sect. 6, we describe how we have to define the meanings of <u>points</u> in the phonocentric topology at any semantic level. For this goal, we recast a famous Frege's contextuality principle in order to define the set of contextual meanings of any point x which belongs to the phonocentric topological space X of some semantic level, whatever this point x may be, a word, a sentence, a paragraph, etc., when considered as an element of a syntactic entity of the higher type. For any semantic level, it is the distinction between the notion of a contextual meaning of a primitive element (a point) at this level and the notion of a fragmentary meaning of a part (a subset) of the whole at this level, that is, of the whole space endowed with a phonocentric topology. The contextual meaning of a point x is defined to be the *inductive limit* of fragmentary meanings s of different open neighbourhoods $U \ni x$ those are got identified on some smaller common open neighbourhood of x. Finally, we generalize Frege's contextuality principle in the categorical terms of *bundles* of contextual meanings.

• In Sect. 7, we show that these generalized Frege's compositionality and contextuality principles are related by a *duality* which we formulate in terms of category theory, and which we name after Frege. This sheaf-theoretic duality sheds new light on the delicate relation between Frege's compositionality and contextuality principles, in revealing that the acceptance of one of them implies the acceptance of the other. It resolves Frege's embarrassing situation with the reconciliation of two principles those bear now his name. As two sides of the same coin, Frege's compositionality and contextuality principles express indeed two complementary parts of the hermeneutic circle principle. That is why they always come together in philosophy, linguistics, and logic. *Grosso modo*, the compositionality principle prescribes to understand a meaningful whole by means of understanding of its meaningful parts, whereas the contextuality principle prescribes to understand the meaning of an entity in accordance with the understanding of its meaningful neighbourhoods.

• Once explicitly stated, Frege duality gives rise to a functional representation of fragmentary meanings. In Sect. 8, this functional representation enables us to develop a kind of *compositional dynamic semantics* which describes how the interpretation proceeds over the time as the step-by-step extension of a meaning representation function, from the initial meaningful fragment to the whole interpreted text. Defined in the proposed sheaf-theoretic framework, such a dynamic semantics conceptualizes the compositionality in a uniform manner at each semantic level: word, clause, sentence, paragraph, section, chapter, text as a whole. Moreover, it treats the polysemy in a realistic manner as one of the essential features of a natural language. This sheaf-theoretic dynamic semantics provides the mathematical model of a text interpretation process, while rejecting attempts to codify interpretative practice as a kind of calculus. We call such a mathematical model of a natural language text interpretation process as *formal hermeneutics* (see, e.g. [29, 31, 32]).

• Then, in Sect. 9, we compare the compositional dynamic semantics proposed in our sheaf-theoretic framework with several algebraic compositional semantics. We notice that an algebraic semantic, of whatever kind, is always static because the meaning of the whole sentence is calculated just after the calculation of meanings of all its syntactic components was done. Algebraic semantic theories are appropriate to study the synonymy, but their irremovable drawback is the inability to describe the polysemy. Any kind of formal grammar which formalizes the compositionality as the functionality in a set-theoretic paradigm shares this fallacy with an algebraic semantics described by T. M. V. Janssen in [15] as "a homomorphism from syntax to semantics".

By contrast, the proposed mathematical framework formalizes the compositionality of fragmentary meanings in a sheaf-theoretic paradigm of functional dependence. In this formal framework, the dynamic semantics describes how the interpretation is incrementally built up as a meaning representation function stepwise extension from the initial meaningful fragment to the whole text. Moreover, in this approach the process of a natural language text interpretation is modelled in a similar manner at all semantic levels.

- The present article culminates in the final Sect. 10 devoted to the statement of a *sheaf-theoretic formal hermeneutics* which describes a natural language in the category of *textual spaces* **Logos**. Appeared as syntax and semantics of a natural language, phonocentric topologies and sheaves of fragmentary meanings constitute together an adequate mathematical framework to formalize different linguistic phenomena in our works, such as linguistic universals of geometric nature in [29], as dynamic semantics in [34], as interpretations of one text by the others, as text summarization and abstracting, as well as many other aspects of intertextuality in [31].

2 Basic semantic concepts

Concerning the linguistic terminology to be used in this work, we have certain difficulties because the sciences of language do not have a unified terminology. According to F. Rastier [37], two traditions seem dominant in the sciences of language: (1) the grammatical tradition centered on the issue of the sign, that confines itself to the word and the sentence; (2) the rhetoric and hermeneutic tradition centered on the communication, that privileges the text and the discourse. Based on different conceptions, these two traditions differ in problematic and in terminology. When using the definition of a technical term proper to one doctrine, we have to privilege this doctrine compared with others, that would not be our goal. The aim of our work is to discern the mathematical structures underlying the process of reading, with the purpose to design a semantic theory that formalizes a natural language understanding process in a uniform manner at all semantic levels (word, sentence, text). We are therefore obliged to accept a terminology based on distinctions that are valid at all semantic levels of an admissible text. In this perspective, we have to study only those spoken or written language segments that are admissible as units of linguistic communication. Therefore, we keep to the hermeneutic tradition in the analysis of a text understanding process. We recognize that there are different scientific trends in discourse analysis; that is why we have to clarify basic semantic terms we use in the present paper. The technical acceptance of terms *meaning*, *sense*, and *reference* as these are used in the present paper may be explained as follows:

Meaning. The term *fragmentary meaning* of some fragment of a given text X is accepted as the communicative content grasped in some particular situation of reading. In this terminological acceptance, a *fragmentary meaning* is immanent not in a given fragment of a text, but in the interpretative process of its reading based on the linguistic competence, which is rooted in the social practice of communication with others through the medium of a language. Any reading is really an interpretative process where the historicity of the reader and the historicity of the text are involved. The understanding of meaning is based not only on the shared language but also on the shared experience as a common life-world, and it deals so with the reality. According to Gadamer, this being-with-each-other is a general building principle both in life and in language. The understanding of a natural language text results from being together in a common world. This understanding as a presumed agreement on 'what this fragment $U \subseteq X$ wants to say' becomes for the reader its fragmentary meaning s. In this acceptance, the meaning of an expression is the communicative content which a competent reader grasps when s/he understands it; and such an understanding can be reached regardless of the ontological status of its *reference*.

The process of coming to some fragmentary meaning s of a fragment $U \subseteq X$ demonstrates a human communicative ability in action. When we qualify some fragment as being meaningful, we state that an idealized competent reader can understand a communicative content which this fragment conveys; the understanding manifests itself as the ability of the reader to express at once this content in other words or in another language (if e.g., the reader is bilingual).

The fact of having such an understanding may be labelled with a certain abstract entity s called *fragmentary meaning* of U. When someone acknowledges the fact that a meaning of U has been understood, this situation may be described by saying that 'this fragment U has the fragmentary meaning s'; it presumes implicitly that the understanding of the meaning s of the fragment U is arrived at through some linguistic communication, direct or mediated. This meaning may be shared in a dialogue with another native speaker, and such a possibility describes the ontological status of the meaning s as being some abstract entity subtracted from the linguistic communication. This situation may be summed up by an external observer as 'the understanding of the fragmentary meaning s of a fragment U', where the 'meaning' may be perceived as a linguistic term in our technical acceptance, and also as an ordinary word of English language. So, our use of the term *fragmentary meaning* corresponds well to the common English usage.

We have noticed above that for any admissible text X, one should distinguish a fragmentary meaning of a meaningful part $U \subseteq X$ and a contextual meaning of an element (point) $x \in X$. It expresses the fact that clauses are parts of a sentence, but idioms and words are its indivisible elements. A fragmentary meaning s is assigned to the part $U \subseteq X$, and this s conveys some part of the communicative content of the whole X in a concrete situation of linguistic communication. This part U is a sequence of primitive elements (tokens) x those have contextual meanings in the context of U.

In the situation of linguistic communication, a unit that is proper to convey a communicative content may be some text or its fragment, some sentence or its clause, some elliptic expression, and yet a word or an exclamation in certain cases of communication. Thus, a meaning is related to the communicative content, regardless of its possible truth value, whatever it may be: true, false or indefinite.

However, the linguistic communication, either spoken or written, consists of the use of words in a conventional way. It is quite difficult to trace the history of how a single word enters the lexicon (vocabulary) of a language. Taken beyond the situation of linguistic communication, a single word is not a discourse nor a part of it, and this word says nothing to nobody. But this word had entered the lexicon in the process of repeated participation in a variety of situations of linguistic communication, with the result that native speakers of the language have a clear idea of the situations in which the use of a particular word is appropriate, and what it then means. These so-called literal meanings of words are recorded in the dictionaries and thesauruses. Generally, by means of examples, these dictionaries allow us to understand what meaning is associated with the use of each word in several standard situations of its use. In this way, dictionaries define those abstract objects, which are called the literal meanings of words. Such definitions carry the entire history of the language and the experience of the numerous uses of this word in the specific situations of communication. The dictionaries thereby demonstrate that the relationship of each word with the set of its possible meanings in specific contexts had gained a normative value. This usage is normative for native speakers of a particular linguistic community, in a particular historic period. These descriptions are aimed to help for a competent reader to adjust better the orientation of his/her efforts to grasp a meaning. In this terminological acceptance, a word, a fragment, a text has a specific meaning only in the situation of linguistic communication, direct or mediated.

However, when using a particular expression in a particular situation of linguistic communication, each interlocutor establishes his/her own connection between this expression and its meaning which is a mental concept (signified) grasped by means of this expression used in this particular situation of communication. This meaning is the mental concept concerning either some physical objects of the world, or some ideas, or some fictional entity, but this meaning is not itself a referred object in the world (in contrast to Frege's *Bedeutung*). As the mental concept, this meaning is apprehended as a being of intersubjective nature because it may be shared with native speakers of the same linguistic community. We equate the 'meaning' with the 'communicative content' because a message (in spoken or written form) is intended by its author as a carrier of a certain communicative content to be grasped by the addressee, that is, as a carrier of a certain meaning to be understood.

Let us take for example the word 'wolf'. A hunter, a scientist zoologist, an adult urban dweller who have never seen of living wolves, or a child who is familiar with them only by fairy tales, they all have different concepts conceived in connection with the word 'wolf'. The ostensive definition of the meaning of this word by pointing out wolves in a zoo, and its definition by dictionaries as a 'wild, flesh-eating animal of the dog family' are conveying different concepts. It implies certainly that an adequate semantic theory should take into account that a lexicon of a competent reader counts not only one but several literal meanings of the word 'wolf'. Every competent native speaker knows also about the use of this word in one of figurative senses, for example, in the moral sense of the proverb: "Who lives with the wolves should howl like a wolf".

It is, therefore, the intention of the reader that controls the choice of meanings during the reading. Which of possible meanings of a particular expression is grasped by the reader depends on the specific situation of reading guided by the reader's intention in the interpretative process, presuppositions and preferences, that we denominate by the term *sense* (or *mode of reading*).

Sense. In our acceptance, the term *sense* (or *mode of reading*) denotes a kind of semantic orientation in the interpretative process that relates to the whole text or its meaningful fragment, to some sentence or its syntagma, and involves the reader's subjective premises that what is to be understood constitutes a meaningful whole. Concerning a word-token of a phrase, one may ask a question "What does this word mean here in a literal sense?", and as we have argued above, an answer consists of the choice of only one meaning from the set of many possible ones. Likewise for a question, "What could it mean in a metaphoric sense?", as for many similar questions in a reading process. In such an acceptance, the term 'sense' is correlative to the intentionality of our interpretative efforts; that is, a sense is not immanent to the text we read, but in some way, it may even precede the reading process. For example, one may intend to read a fable in the moral sense yet in advance of its reading. But when the reading unfolds in time, one still controls own intentions following the current reading situation. These examples illustrate the acceptance of the term 'sense' as the reader's interpretative intention, and the acceptance of the term 'meaning' as the content actualized during the process of communication.

To some extent, our acceptance of the term 'sense' is close to the exegetic conception of four senses of the Holy Scripture. The traditional presentation of this conception of biblical hermeneutics is summarized by the famous distich of Augustine of Dacia: "Littera gesta docet, quid credas allegoria, moralis quid agas, quo tendas anagogia.[3]"

According to the biblical hermeneutics, the readings of the Scripture in literal, allegorical, moral, and anagogical senses are coherent in each of its parts. Suppose we read the whole text of the Scripture by fragments, where each fragment was read in one of four senses: literal, allegorical, moral, or anagogical, but the choice of sense was not the same for all fragments. The composition of these four senses is a method of interpretation which gives rise to a large number of senses of the whole text. Indeed, the overall sense $\mathscr{F}$, as the integral intention in the reading process, is the result of all local intentions taken during these partial readings.

But what guides the subsequent choice of local intentions of an empirical reader? Following Fathers of Church, it is the Holy Spirit which by his very presence guides the soul of the individual believer who reads the text of the Scripture. But for a secular text, how can we characterize in linguistic terms the possibility to join these partial senses? This is the presumed sincerity on the part of the author which makes us suppose that, while being of sound mind and perfect memory, he wrote the text to communicate something to his alleged reader.

However, the local intentions those were taken in the writing process were got integrated into an overall intention of an empirical author; so, these partial writings are consistent to satisfy a certain *gluing condition* of the type that we discuss further in Sect. 5.4. Since the empirical author is almost always inaccessible for a dialogue, how can we understand what does the text mean by virtue of its textual coherence denoted by U. Eco as the *intentio operis*? According to U. Eco [7, p. 65], "it is possible to speak of the text's intention only as the result of a conjecture on the part of the reader. The initiative of the reader basically consists in making a conjecture about the text's intention." He asks further, "How to prove a conjecture about the *intentio operis*?", and he responds: "The only way is to check it upon the text as a coherent whole." He continues then that this idea comes from *De doctrina Christiana* of St. Augustine:

> [...] any interpretation given of a certain portion of a text can be accepted if it is confirmed by, and must be rejected if it is challenged by, another portion of the same text. [7, p. 65]

According to St. Augustine, the presumed textual coherence controls the partial interpretations that are made by an empirical reader. Therefore, in the process of

---

[3] Augustine of Dacia, *Rotulus pugillaris*, I: ed. A. Walz: Angelicum 6 (1929) p. 256. The distich is translated in English as: "The letter tells us what went down, the allegory what faith is sound, the moral how to act well – the anagogy where our course is bound."

reading, all these local intentions to understand a text have also to verify the *gluing condition* of the type that we discuss further in the Sect. 5.4.

In the process of actual communication, a mere consistency of the local interpretations would be insufficient. The inference on the speaker's intention is essential here for the understanding; the contact of interlocutors allows them to get into the coordination between the intention of the sender and the intention of the recipient.

With regard to a text produced not for a single recipient, but for a community of readers, the strategy of a *model author* is to lead his *model reader* to speculate about the text. Among these leading indexes, the central place is held by the *semantic isotopy* that A. J. Greimas defines as "a complex of plural semantic categories which makes possible the uniform reading of a story." [12, p. 188]. Concerning the notion of isotopy, U. Eco notices in [6, pp. 189–190] that "The category would then have the function of textual or transsentential disambiguation, but on various occasions Greimas furnishes examples dealing with sentences and outright noun phrases."

Following B. Pottier, the seme does not exist in isolation but as a part of a sememe, or as the set of coexisting semes.

> Le sémème, l'être de langue (en compétence), s'actualise dans le discours [...].
> Le sémème donne le sens (l'orientation sémantique), et la mise en discours le transforme en signification.[4] [27, pp. 66, 67]

From this definition, we retain the acceptance of the term *sense* as the semantic orientation of the reader's intentions provoked by a sememe, and the fact that a meaning is actualized in the discourse. The reader's conjecture on the subject discussed in a text determines the first interpretive intention which will be clarified in the course of the reading when the recognition of a semantic isotopy becomes possible owing to the context that is more and more revealed. Following U. Eco,

> The first movement toward the recognition of a semantic isotopy is a conjecture about the topic of a given discourse: once this conjecture has been attempted, the recognition of a possible constant semantic isotopy is the textual proof of the 'aboutness' of the discourse in question. [7, p. 63]

In *Two Problems in Textual Interpretation* published in 1980, U. Eco describes the interpretative process as based on the reader's interpretive cooperation:

> Between the theory that the interpretation is wholly determined by the author's intention and the theory that it is wholly determined by the will of the

---

[4] Our translation of this quotation is: "The sememe, the entity of language (in competence), is actualized in the discourse [...]. The sememe gives the sense (the semantic orientation), and the putting into discourse transforms it into meaning."

interpreter there is undoubtedly a third way. Interpretive cooperation is an act in the course of which the reader of a text, through successive abductive inferences, proposes topics, ways of reading, and hypotheses of coherence, on the basis of suitable encyclopedic competence; but this interpretive initiative of his is, in a way, determined by the nature of the text. [2, pp. 43–44]

But later in 1992, in the analysis of so-called *superinterpretation*, U. Eco raises again the problem of a reader's conjectures about the empirical author's intention during the reading. His updated conception of the interpretation of texts "makes the notion of the intention of an empirical author radically unnecessary" [7, p. 60]. He defends this thesis with the support of his own experience as a writer who has discussed with his readers a few different interpretations of his novels.

To summarize now our acceptance of the term *sense* (or *mode of reading*), we have to say that it is close to the latter acceptance described by U. Eco. The term *sense* concerns the reader's initiative in the interpretation of the text; it is wholly determined by the reader's intention to understand possible meanings of the text. In Sect. 5, we identify a particular sense $\mathscr{F}$ (in our acceptance) with the assignment to each meaningful fragment U of a given text X the set of all its meanings $\mathscr{F}(U)$ that may be grasped in all possible readings of U in this sense $\mathscr{F}$. This way, to any *sense* (or *mode of reading*), we assign a *sheaf* of fragmentary meanings.

Remark. It should be noticed that our terminological acceptance of basic semantic notions of *sense* and *meaning* differs from their acceptance in the theories developed within the tradition which goes back to Carnap's semantic theory, sometimes called the theory of "intension and extension". In such theories, expressions of different syntactic kinds refer to entities of different kinds as their extensions, and also refer to entities of different kinds as their intensions. The terms *intension, intensional* are not to be confused with the terms *intention, intentional* we have discussed above. The notions 'intension', 'intensional' primarily concern the domain of logic, whereas 'intention', 'intentional' concern the philosophy of mind. According to A. R. Lacey, "Intuitively extensions can be thought of as the extents which certain kinds of terms range over and intensions as that in virtue of which they do so." [18, p. 164], whereas the intentionality is "that feature of certain mental states by which they are directed at or about objects and states of affairs in the world" [18, p. 50].

Reference. Certainly, the referential function of a language is important in the linguistic communication which concerns the world where the interlocutors live. A natural language has a huge arsenal of denoting expressions to designate real and imaginary objects during communication. The linguistic competence is characterized by the know-how in production and comprehension of natural language expressions realizing the referential relationship called *reference* or *denotation*. In the analytic philosophy of language, the study of denoting expressions plays a considerable role, because the reference to objects with an uncertain ontological status is responsible for some logical paradoxes.

In the present work, we assume a total referential competence of an idealized reader who knows the lexicon of a language and follows the rules of common usage. In short, we assume that the reader has a total language skill, combined with a general knowledge. Such a reader meets no problems to understand the meaning of denotative expressions and the ontological status of objects so defined.

3 Topologies appeared as syntax

The author of an admissible text doesn't suppose that the reader's understanding will be suspended until the end of reading because everybody knows that the words already read trigger intellectual mechanisms of interpretation based on the indissoluble links between the signifier and the signified. To be understood in linguistic communication, one must take it into account and organize one's writing in such a way that the reader's understanding at every moment may be arrived at on the basis of what has been already read. It seems that the primacy of speech over writing is a cause that implies in writing the subordination of graphic expressions to acoustic ones. A spoken utterance is a temporal series of sounds produced by a speaker using a human articulatory apparatus. When written, an acoustic signal is converted into a series of signs whose positions are linearly ordered following an adopted convention; in English, it is from left to right within the lines, and from top to bottom between them. Once a particular sign is taken as the initial, it allows us to specify the position of the following signs by enumeration. From the mathematical point of view, the whole segment may be considered as a finite sequence when the last sign is specified. Thus we ought to consider a text X as a finite sequence $(x_1, x_2, x_3, \ldots, x_n)$ of its constituent sentences x_i, and so it is formally identified with a graph of a function $i \mapsto x_i$ defined on some interval of natural numbers. When reading a particular fragment of the text X, we delete mentally the other sentences but follow the induced order of remaining ones. Important is the induced order of their reading and not the concrete index numbers of their occupied places. Thus, any part of the text is a subsequence whose graph is a subset of the whole sequence graph. Likewise for a sentence considered as a finite sequence of its words.

While reading a text, the understanding is not postponed until the final sentence. So the text should have the meaningful parts, and the meanings of these parts determine the meaning of the whole as it is postulated by the hermeneutic circle principle. For the meaningfulness conveying an idealized reader's linguistic competence, a meaning of a meaningful part is the communicative content grasped in a particular reading of this part guided by the reader's presuppositions and preferences in the interpretative process, that is, guided by the sense (or mode) of reading.

Certainly, there are many meaningful fragments in the text. A simple example of a meaningful fragment is supplied by the interval including all sentences, from the first x_1 till the last x_n. Anybody reads the text as if it would be a written transcription of the story uttered by the author. When telling or writing a story, an author should take into account that the understanding can't be postponed, for "the texts never know the suspense of interpretation. It is compulsive and uncontrollable", as it is noticed by F. Rastier in [36]. If the author don't want to be misunderstood, s/he has to organize the text in such a way that any sentence x is preceded by certain sentences those provide a necessary context for the understanding of x. Thus, any meaningful part contains each sentence together with some its context, and this is characteristic of any part to be meaningful. It is clear that this property fails for a part including, e.g. all sentences x_i whose placehold number i is divisible by 100, and that is why this part is meaningless, and nobody try to read the text in such a manner. In [28, 31–33], we argue that in agreement with our linguistic intuition, the set of all meaningful parts of any admissible text should satisfy two properties:

(t_1) *The union of any set of meaningful parts is a meaningful part.*

(t_2) *The non-empty intersection of two meaningful parts is a meaningful part.*

The first property (t_1) is taken for granted, because it expresses the precept of generally accepted hermeneutic circle principle, which ensures us to understand the union of a given set of meaningful parts through the understanding of all its constitutive members. In the union of any set of meaningful parts, each part contains every its sentence together with some its context, whence the union itself is a part which has such a property. To be more accurate, we have to take into account that the meaning s of a meaningful part U isn't immanent to this part itself, but this meaning is grasped in the reading process following a sense (mode of reading) $\mathscr{F}$ guided by the reader's interpretative intentions. Thus, in the statement (t_1), some sense (or mode of reading) $\mathscr{F}$ is implicitly presumed to be the same for all members of the union. In the following Sect. 5–8, we discuss in details how the resulting meaning of the whole is obtained via the meanings of its constitutive parts.

The second property (t_2) expresses the contextuality of understanding. To understand a meaningful part U of the text X is to understand contextually all sentences $x \in U$, where the context of a particular sentence x is some meaningful part W such that $x \in W \subseteq U$. In the standard process of reading, i.e. from the beginning up to x, this part W should contain a subsequence of sentences those precede x and provide a necessary context for the understanding of x in the sense $\mathscr{F}$. For a particular sense $\mathscr{F}$, there should exist a smaller subsequence $(x_{i_1} \ldots, x_{i_m}) \subseteq W$ whose sentences have been understood during the reading, and then have been taken into account at the moment when the reader understands a meaning of x grasped in the sense $\mathscr{F}$. Let us denote $U_x = (x_{i_1}, \ldots, x_{i_m})$. The tokens x_{i_k} of U_x may be consecutive or dispersed among other tokens of W, it does not matter, but they should be read before the reading of x.

Consider first the case of one session process of reading of X in some sense $\mathscr{F}$. When the part U_x belongs to any meaningful part $W \subseteq X$ such that $x \in W$. Let U, V be two meaningful parts such that $x \in U \cap V$. According to our premises, $x \in U_x \subseteq U$ and $x \in U_x \subseteq V$, hence $x \in U_x \subseteq U \cap V$.

Consider now the case when $x \in U \cap V$, and parts U, V were read in two different sessions of reading, but in the same sense $\mathscr{F}$. This means that the reader is self-identical, and the reading is guided by the same intentionality. It implies that $U_x \subseteq U$ and $U_x \subseteq V$. Hence $x \in U_x \subseteq U \cap V$.

Thus, in both cases, $U \cap V$ is meaningful because $U \cap V = \cup_{x \in U \cap V} U_x$ is the union of meaningful parts, due to (t_1).

Since an admissible text X is supposed to be meaningful as a whole by the very definition, it remains only to define formally the meaning of its empty part (for example, as a singleton) in order to satisfy the third property:

(t_3) *The whole admissible text and the empty part are meaningful.*

This enables us to endow an admissible text X with some *topology* in a strict mathematical sense, where the set $\mathfrak{O}(X)$ of open sets is defined to be the set of all meaningful parts. We call the topology so defined *phonocentric topology* to indicate in its name the subordination of graphic expressions to phonetic ones.

An admissible text X gives rise to a finite space; hence an arbitrary intersection of its open sets is open and so it is an *Alexandrov space*.

In general, a topology on a set X is defined by specifying the set $\mathfrak{O}(X)$ of open subsets of X satisfying axioms similar to ours (t_1), (t_2), and (t_3). But almost always it is impossible to enumerate all the open subsets. Instead, a topology is usually defined by specifying a smaller set of open subsets, called a *basis*, and then generating all the open subsets from this basis.

Likewise, when studying the process of interpretation of an admissible text X, many of linguistic concepts may be well expressed in terms of the phonocentric topology on X which is defined by specifying the set of open subsets $\mathfrak{O}(X)$ to be the set of all meaningful parts satisfying properties (t_1), (t_2), and (t_3). However, it will be more convenient and useful to develop the theory in more concrete, say even constructive, terms of empirically given meaningful parts those constitute a basis for a phonocentric topology.

Fortunately, the set of all meaningful parts $\mathfrak{O}(X)$ of a given text X may be described by specifying a class of fairly simple meaningful parts given as an empirical data related to a reading process. In the reading of a particular text X, the reader is practically concerned with a smaller class of meaningful parts $(U_x)_{x \in X}$, where each part U_x contains a sentence x and provides the smallest context which is necessary for a reader to grasp a particular meaning of x. Because the phonocentric topology $\mathfrak{O}(X)$ is finite, for each x, there exists such a smallest open neighbourhood U_x that is defined as the intersection of all open neighbourhoods of x.

For a given sentence x, the understanding of a whole U_x requires the grasping of meanings of all constitutive sentences of U_x; hence, for any sentence $y \in U_x$, its smallest context U_y should be a part of U_x. Suppose now that we are given two smallest meaningful parts U_x and U_y such that $U_x \cap U_y \neq \varnothing$. Then for each $z \in U_x \cap U_y$, we have $U_z \subseteq U_x$ and $U_z \subseteq U_y$; hence $U_z \subseteq U_x \cap U_y$. Thus, the set $\mathfrak{B}(X) = \{U_x \colon x \in X\}$ is the set of meaningful parts of X satisfying two properties:

(b$_1$) For each $x \in X$, there exists $U_x \in \mathfrak{B}(X)$ such that $x \in U_x$.

(b$_2$) For every two $U_x, U_y \in \mathfrak{B}(X)$ such that $U_x \cap U_y \neq \varnothing$, and for each sentence $z \in U_x \cap U_y$, there exists $U_z \in \mathfrak{B}(X)$ such that $z \in U_z$ and $U_z \subseteq U_x \cap U_y$.

So, the set $\mathfrak{B}(X)$ is a basis for a phonocentric topology on X, because any meaningful part (i.e. open) $V \subseteq X$ is the union $V = \bigcup_{x \in V} U_x$ of the members of some subset of $\mathfrak{B}(X)$. Recall that a set $\mathfrak{B}$ of open sets of a topological space X is called a *basis* for its topology if and only if every open set U of X is the union of the members of a subset of $\mathfrak{B}$. Thus, the class of open sets $\mathfrak{O}(X)$ in a phonocentric topology on X is defined by the subclass $\mathfrak{B}(X)$ of all open sets of the type U_x, that is, a phonocentric topology on X is defined by the empirical data $\mathfrak{B}(X)$.

Any explicitly stated *concept of meaning* or a *criterion of meaningfulness* satisfying conditions (t$_1$), (t$_2$), and (t$_3$) allows us to define some type of *discursive topology* on texts, and then to interpret several problems of discourse analysis in topological terms [31]. In what follows, we consider only admissible texts endowed with a phonocentric topology which is a particular type of discursive topology corresponding to the criterion of meaningfulness conveying the linguistic competence of an idealized reader, meant as the ability to grasp a communicative content.

3.1 Phonocentric topology and partial order

In the ordinary process of reading, any sentence x of a text X should be understood on the basis of the part already read because the interpretation of a natural language text cannot be postponed, although it may be made more precise and corrected in further reading and rereading. In [36], F. Rastier describes this fundamental feature of a competent reader's linguistic behaviour as the following:

> Alors que le régime herméneutique des langages formels est celui du suspens, car leur interprétation peut se déployer après le calcul, les textes ne connaissent jamais le suspens de l'interprétation. Elle est compulsive et incoercible. Par exemple, les mots inconnus, les noms propres, voire les non-mots sont interprétés, validement ou non, peu importe.[5] [36, pp. 165, 166]

Thus for every pair of distinct sentences x, y of X, there exists an open part U containing one of them (to be read first in the natural order $\leq$ of sentences reading) but not the other. This means explicitly that the phonocentric topology satisfies the *separation axiom T_0* of Kolmogorov.

For a sentence $x \in X$, we have defined the open neighbourhood U_x to be the intersection of all the meaningful parts those contain x, that is the smallest open neighbourhood of x. The *specialization relation* $x \preceq y$ (read as 'x is more special than y') on a topological space X is defined by setting $x \preceq y$ if and only if $x \in U_y$ or, equivalently, $U_x \subseteq U_y$. It is clear that $x \in U_y$ if and only if $y \in \mathrm{cl}(\{x\})$, where $\mathrm{cl}(\{x\})$ denotes the topological closure of a one-point set $\{x\}$.

Key properties of these notions are summarized in the Propositions 1, 2 those are linguistic versions of general mathematical results concerning the interplay of topological and order structures defined on a finite set. The proofs may be found in many sources, as for example, in [23].

Proposition 1. *For an admissible text X, the set of all smallest opens $\{U_x \colon x \in X\}$ is a basis for a phonocentric topology on X. Since the phonocentric topology on X satisfies the separation axiom T_0, it defines a partial order $\preceq$ on X by means of the specialization relation. The initial phonocentric topology can be recovered from this partial order $\preceq$ in a unique way as the topology with the basis made up of all sets of the kind $U_x = \{z \colon z \preceq x\}$.*

Proposition 2. *Let X, Y be admissible texts endowed with phonocentric topologies. Then the following statements are equivalent:*
1. *The function $f \colon X \to Y$ is continuous.*
2. *For each $x \in X$, the function f maps a basis set into a basis set: $f(U_x) \subseteq U_{f(x)}$.*
3. *The function f preserves the specialization order: $x \preceq y$ implies $f(x) \preceq f(y)$.*

---

[5]Our translation of this quotation is: "While the hermeneutic regime of formal languages is that of suspense, because their interpretation can be deployed after the calculation, the texts never know the suspense of interpretation. It is compulsive and uncontrollable. For example, unknown words, proper names, even non-words are interpreted, valid or not, whatever."

Example. A continuous function $f_1\colon X_2 \to X_1$ arises in writing process when an author goes from a first plan X_1 of some future text to its more detailed plan X_2, where a sentence x_d of X_1 is substituted by some passage $(x_{d_1}, \ldots, x_{d_m})$. And so on, in going to more and more detailed texts $X_3, \ldots, X_n$, one gets a sequence of continuous functions

$$X_n \xrightarrow{f_{n-1}} X_{n-1} \xrightarrow{f_{n-2}} \cdots \xrightarrow{f_3} X_3 \xrightarrow{f_2} X_2 \xrightarrow{f_1} X_1.$$

3.2 Deep structures and surface structures

Let **FinTOP$_0$** be the category of finite T_0-topological spaces and continuous maps, and let **FinORD** be the category of finite partially ordered sets (posets) and their monotone maps.

Given a finite partially ordered set $(X, \leq)$, one defines a T_0-topology τ on X by means of the basis for τ made up of all *low sets* $\{z\colon z \leq x\}$. Thus one obtains a functor $L\colon \textbf{FinORD} \to \textbf{FinTOP}_0$ acting identically on the maps of underlying set. Conversely, one defines the specialization functor $Q\colon \textbf{FinTOP}_0 \to \textbf{FinORD}$, assigning to each finite T_0-topological space (X, τ) a poset $(X, \preceq)$ with the specialization order $\preceq$, and acting identically on the maps of underlying set. Thus, the functors L and Q establish the isomorphism between the category **FinTOP$_0$** and the category **FinORD**. From the mathematical point of view, the study of one of these two categories is equivalent to the study of the other.

Now we generalize and summarize the considerations of the mathematical structures of topology and order underlying an admissible text:

The considerations in the beginning of Sect. 3 may be slightly modified in order to define a *phonocentric* topology at the semantic level of sentence and even word [31]. Thus, at each semantic level, there exist two topological structures:

(i) *the natural phonocentric topology at a considered semantic level;*

(ii) *the topology defined by applying the functor L to the linear order $x \leq y$ of reading.*

At an arbitrary semantic level (where the whole is a sequence of primitive elements), the difference between topologies can be summed up so that in the phonocentric topology the least neighbourhood U_x of a primitive element x contains only such primitive elements that precede x in the linear order of writing and provide the context necessary to understand the meaning of x in the adopted sense $\mathscr{F}$; whereas in the topology defined by the functor L applied to $(X, \leq)$, the least neighbourhood U_x of a primitive element x contains all primitive elements that precede x in the linear order of writing.

Note that the explicit definition of the phonocentric topology at the semantic level of sentence requires more delicate work in treatment of different grammatical types of sentences due to the lack of space, so to speak. Here there is a certain analogy with the topological classification of varieties which is more difficult in dimensions 3 and 4 compared to that in lower and higher dimensions.

On the other hand, at each semantic level, there exist two order structures:

(i') *the specialization order $x \preceq y$ defined by applying the specialization functor Q to the natural phonocentric topology of a considered semantic level;*

(ii') *the linear order $x \leq y$ of ordinary text reading.*

Similar to a generative grammar, we will qualify the equivalent structures of (i) and (i') as *deep structures* compared to the equivalent structures of (ii) and (ii') qualified as *surface structures*. We notice that this denomination has nothing to do with the acceptance of these terms in a generative grammar.

Remark. The relation $x \preceq y$ implies obviously the relation $x \leq y$, for all the primitive units x, y of the same semantic level. In particular, at the level of text, where the sentences are primitive units, the map id: $L(X, \preceq) \to L(X, \leq)$, which acts as identity $x \mapsto x$ of the underlying set, is a continuous map of topological spaces. Thus the necessary linearization during the writing process, that is the passage from $(X, \preceq)$ to $(X, \leq)$, results in weakening of the phonocentric topology by transition from $L(X, \preceq)$ to $L(X, \leq)$. The process of interpretation consists in a backward recovering of the phonocentric topology (or equally, of the specialization order) on the text.

3.3 Phonocentric topology at the level of text

There is a simple intuitive tool for graphical representation of a finite poset, called Hasse diagram. For a poset $(X, \preceq)$, the *cover relation* $x \prec y$ (read as 'x is covered by y') is defined by setting $x \prec y$ if and only if $x \preceq y$ and there is no other z such that $x \preceq z \preceq y$. For a given poset $(X, \preceq)$, its Hasse diagram is defined as the graph whose vertices are the elements of X and whose edges are those pairs $\langle x, y \rangle$ for which $x \prec y$. In the picture, the vertices of Hasse diagram are labeled by the elements of X and the edge $\langle x, y \rangle$ is drawn by an arrow going from x to y (or sometimes by an indirected line connecting x and y, but in this case the vertex y is displayed lower than the vertex x); moreover, the vertices are displayed in such a way that each line meets only two vertices.

The usage of some kind of Hasse diagram named *Leitfaden* is widely spread in the mathematical textbooks to facilitate the understanding of logical dependence of its chapters or paragraphs. Mostly, the poset is constituted of all chapters of the book. So, in *Local Fields* by J.-P. Serre [39] and in *A Mathematical Logic* by Yu. I. Manin [21], there are such diagrams.

These diagrams may surely be 'split' in order to draw the corresponding ones whose vertices are all the paragraphs, like it is done directly in *Differential Forms in Algebraic Topology* by R. Bott and L. W. Tu [1], where authors suppose indeed the linear reading of paragraphs 1-6, 8-11, 13-16 and 20-22, but it may be drawn explicitly. These three Hasse diagrams are shown in the Figure 1.

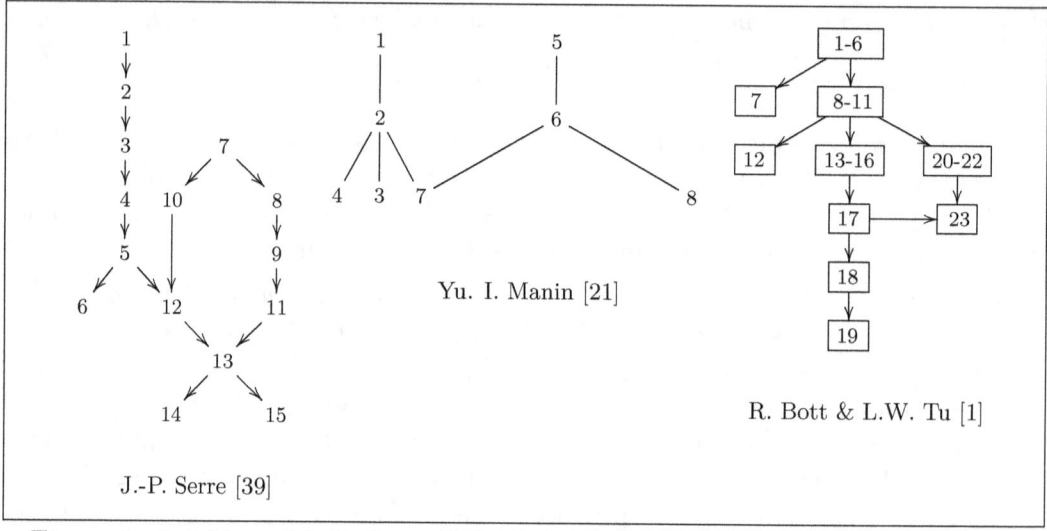

Figure 1: *Leitfiden* of J.-P. Serre [39], Yu. I. Manin [21], R. Bott & L. W. Tu [1].

This way, one may go further and do the next step. For every sentence x of a given admissible text X, one can find a basis open set of the kind U_x in order to define the phonocentric topology at the semantic level of text (where points are sentences), and then to draw the Hasse diagram of the corresponding poset.

In [31], we describe how one may interpret this way the most of diagrams from the *Rhetorical Structure Theory* (RST) conceived in the 1980s the by W. C. Mann and S. A. Thompson [22]. Since then, RST has seen a great development, especially in the computational linguistics, where it is often used for the automatic generation of coherent texts, as well as for the automatic analysis of the structure of texts. The RST aims to describe an arbitrary *coherent text* which is not the random sequence of sentences. The textual coherence demands that for every part of a coherent text there exists a reason for its presence, which is obvious to a competent reader. It seems that RST notion of a coherent text is similar to our notion of an admissible text. In [31] we show that the RST analysis of contextual dependencies between sentences of certain small textual fragments represented as RST diagram may be redrawn as the Hasse diagram for the partial order structure of the corresponding specialization relationship. But the RST diagram may be drawn only for certain small textual fragments such that their sentences are *nucleus* and *satellite* in the sense of the RST. On the other hand, it is not the case when such a fragment is a part of a larger text. Then, according to the RST, there will be no link between a sentence x belonging to such a fragment and any other sentence y that is far enough in the text, because rhetorical relations can only bind adjacent segments. While in our approach, such a link is possible in the specialization relation (of deep order). This link is seen on the corresponding Hasse diagram as a direct edge $\langle x, y \rangle$ or as a sequence of edges that link these two sentences x and y. Thus, our approach is more general than this one of the RST.

3.4 Phonocentric topology at the level of sentence

In order to define a phonocentric topology at the semantic level of sentence, we must distinguish there the meaningful fragments that are similar to meaningful fragments at the level of text. Let x, y be any two word-tokens such that $x \preceq y$ in the specialization order at the level of sentence that is similar to the specialization order coming from the 'logical relations among the different chapters' in a text. This relation $x \preceq y$ means that the word-token x should necessary be an element of the set of word-tokens U_y required to understand the meaning of the word-token y in the interpreted sentence. So we have $x \leq y$ in the order of writing and there should be some syntactic dependence between them. It means that a grammar in which the notion of dependence between pairs of words plays an essential role will be closer to our topological theory than a grammar of Chomsky's type.

There are many formal grammars focused on links between words. The history of this stream of ideas is described by S. Kahane in a detailed review [16]. We think that the theoretical approach of the *special link grammar* of D. Sleator and D. Temperley is most appropriate to define a phonocentric topology at the level of sentence, because in whose formalism "[t]he grammar is distributed among the words" [40, p. 3], and "the links are not allowed to form cycles" [40, p. 13] comparing with *dependency grammars* which draw syntactic structure of sentence as a planar tree with one distinguished root word.

For a given sentence s, the link grammar assigns to it a syntactic structure (called *linkage diagram*) which consists of a set of labeled links connecting pairs of words. We use these diagrams to define all phonocentric topologies on this sentence s.

Example. To explain how to define phonocentric topologies on a particular sentence, let us borrow from [42] the following example of an ambiguous sentence:

(1) John saw the girl with a telescope.

We had yet considered this sentence in [29] by using Chomsky's generative grammar, and also in [31] by using link grammar. The analysis of this sentence by means of the *Link Parser 4.0* of D. Temperley, D. Sleator, and J. Lafferty [41] gives two linkage diagrams shown in the Figure 2.

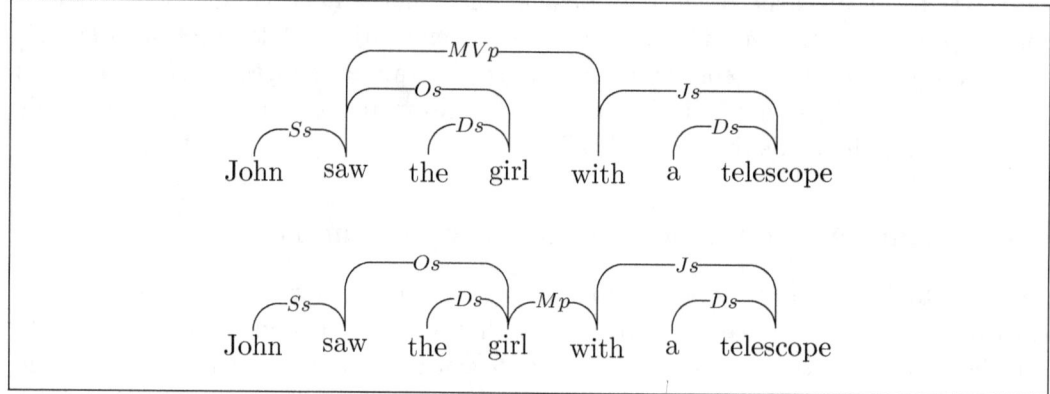

Figure 2: Two linkage diagrams with connector names.

These two diagrams rewritten with arrows that indicate the direction in which the connectors match (instead of connector name) have the appearance shown in the Figure 3.

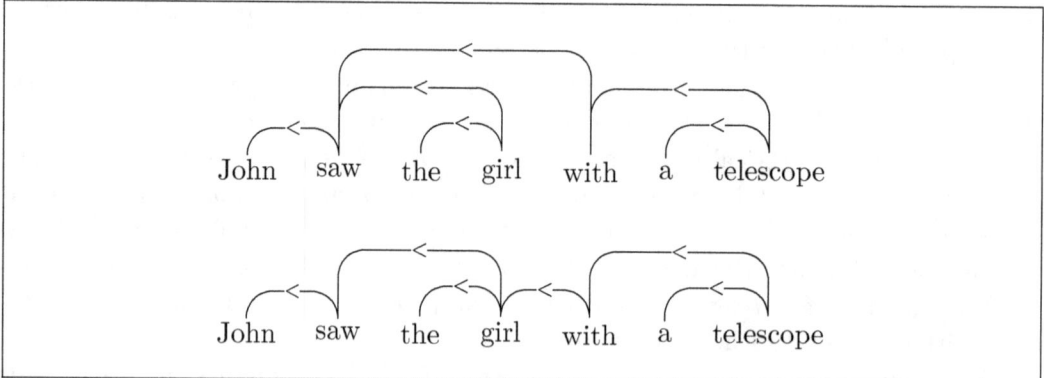

Figure 3: Two linkage diagrams with arrows instead of connector names.

It is clear that the transitive closure $x \preceq y$ of this relation $<$ between pairs of words defines two partial order structures on the sentence (1). By applying the functor L defined in Sect. 3.2, we can endow the sentence (1) with a phonocentric topology in two different ways. The Hasse diagrams of corresponding posets are shown in the Figure 4.

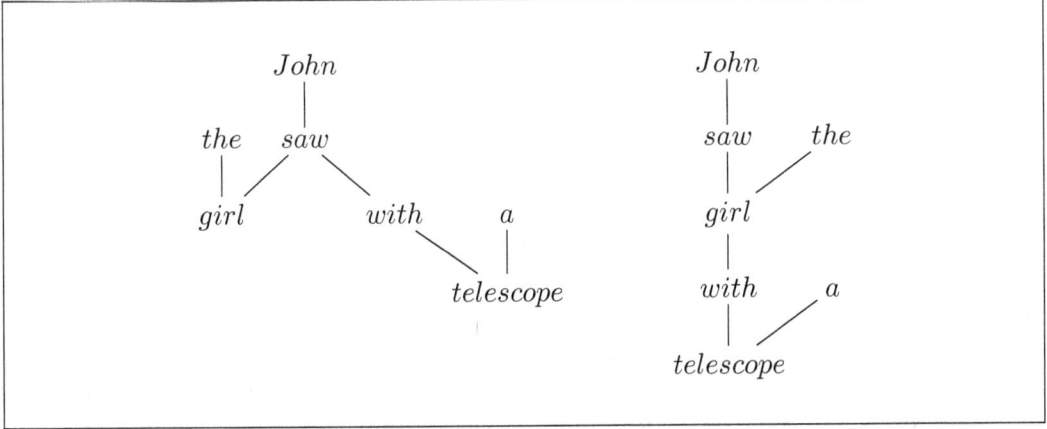

Figure 4: Two Hasse diagrams of the sentence (1) as displayed in [29, 31].

To understand the sentence (1), the reader has to do the *ambiguity resolution* when arriving to the word-token $x =$"with" by choosing only one of two possible basis sets:
$$U_x = \{\langle 1, \text{John}\rangle, \langle 2, \text{saw}\rangle, \langle 5, \text{with}\rangle\};$$
$$U_x = \{\langle 1, \text{John}\rangle, \langle 2, \text{saw}\rangle, \langle 3, \text{the}\rangle, \langle 4, \text{girl}\rangle, \langle 5, \text{with}\rangle\}.$$

In the general case, the step by step choice of an appropriate context $U_x = \{z : z \preceq x\}$ for each word x results in endowing the interpreted sentence with a particular phonocentric topology among many possible.

In [31], we have shown how to define a phonocentric topology at the level of word considered as a sequence of morphemes.

We summarize the results of our analysis presented in Sect. 3 as the following:

Slogan (Phonocentric Topologies as Syntax). Once the phonocentric topology and the corresponding specialization order are determined at a given semantic level, the systematic interpretation of linguistic phenomena in terms of topology and specialization order, and their mathematical study is a formal syntax at this level.

4 Linguistic universals of a topological nature

Throughout the history of scientific study of human languages, researchers are interested in discovering linguistic universals, that is, particular traits common to all languages. Because it is impossible to recognize everything about all languages, it is necessary to first decide where and how to look for linguistic universals. It appears that our sheaf-theoretic approach makes here a small contribution.

By its very origin, a human language is used for linguistic communication; for that reason, written texts and uttered discourses should be considered as communicative units. We must therefore look for linguistic universals, not only in terms of word as it is done by J. H. Greenberg [11] and his successors, but especially in terms of text. A true linguistic universals at the level of text (or discourse) must have a corresponding counterpart at the level of sentence.

By linguistic universals, we understand the characteristic properties of texts those are admissible as messages having communicative purposes, regardless of the language in which they are written. The question is, therefore, reduced to this: What criteria should we accept to be sure that a particular characteristic is truly shared by all admissible texts in any natural language? One can adopt a statistical criterion which ensures to a certain extent that if a certain property is shared by hundreds of natural languages, it is likely that it is shared by all. Such an approach is taken up in the classical works of J. H. Greenberg. But there are no guarantees that a particular trait of the languages already studied is also shared by the language of a lost Indian tribe that escaped the statistical body of research.

To our deep conviction, the way to avoid counter-examples is to adopt a criterion based not only on statistical considerations, but mainly on the analysis of the communicative function of languages. In our talk [30] at the 39th Annual Meeting of *SLE*, we argued that the properties of a phonocentric topology to satisfy the separation axiom T_0 of Kolmogorov and to be connected are linguistic universals. These properties should be required of the underlying phonocentric topology on any text written for the purpose to be understood in the linguistic communication.

A correct translation of an admissible text from one language into another is done by successive translation of each sentence in a manner to conserve their contextual relations. It results in a bijection between the original text and its translation, and also in a homeomorphism between corresponding topological spaces.

It is clear that a phonocentric topology on an admissible text written in one language (as well as the corresponding Hasse diagram) is invariant under translation into another language. Hence, a phonocentric topology on a text X and its properties and geometric invariants (say T_0-separability, connectedness, homology groups, etc.) are stable under translation from one language into another (i.e., under homeomorphism), and so they are formal invariants of the text X.

The properties those are shared by all texts in all natural languages are absolute linguistic universals. In [30–32], we argue that the T_0-*separability*, the *connectedness* of a phonocentric topology, and the *acyclicity* of corresponding Hasse diagram are features shared by the majority of languages.

4.1 Kolmogorov's axiom T_0 as a linguistic universal

One important example of a topological linguistic universal seems to be the *separation axiom T_0* of Kolmogorov. In the Sect. 3, we argued for the relevance of the separation axiom T_0 to all semantic levels of an admissible text on the base of a lucid formulation by F. Rastier [36]. Anyway, there is an essential difference between the hermeneutic regime of formal languages and that one of natural languages; it is important for us that texts written in a natural language "never know the suspense of interpretation." [36, p. 166]. It's still the same idea that Origen expresses in the biblical hermeneutics regarding the non-understanding. According to Origen, yet for an *imbulatum*, there is a meaning as a sign of divine presence in the text.

This empirical truth which everyone knows from his/her own experience of reader still deserves a more nuanced discussion. Firstly, this property of understanding of texts in natural language is obviously taken into account by everyone who writes a text intended for human understanding, whether he/she is a professional writer or not; the rule is accepted as that one of a 'writing game', so to speak.

If we do not want to be misunderstood, we do not propose the reader to suspend understanding until the end of writing because we know that the words already read trigger intellectual interpretation mechanisms based on indissoluble links between signifier and signified. This is well expressed by the colourful Russian saying: *A word is not a sparrow: you can't catch it when it flies away!* In order to be understood, we must organize our writing in such a way that the reader's understanding would always be based on the part of text already read, in total ignorance of its future development.

The second reading (as all subsequent readings) is governed by the same rule, despite the fact that we already know the whole text. The repetitive reading respects the unpredictability of the future; while reading at the time being, we are being in the 'here and now', that leads us to identify the physical real time with the time of the narrative. What lies in the pages that follow makes no context for the understanding of what has been read. In particular, this rule is just applicable to scientific texts.

A question arises: What is the reason for this indisputable empirical phenomenon? It seems to us that it is the primacy of speech over writing, which causes the subordination of graphic expressions to phonetic ones.

Preliterate civilizations existed thousands of years before the advent of writing and even still exist somewhere else. Even today there are thousands of people who cannot read. The cultural history of the human species is repeated in the personal history of each individual because we learn to speak before we learn to read and write. But as a physical phenomenon, a phonetic expression exists in the dimension of time, and here the physiological properties of our speech organs are just involved.

In a conversation, the interlocutors have access only to whatever is already said, because the future remains unpredictable. Once said, the spoken word is flying away and the only chance to get by in such a situation is to understand on the spot all that is said by the others.

For anybody speaking, this attitude quickly becomes a habit and even a conditioned reflex on the situation of linguistic communication. As functional and even physiological in origin, this property of the oral communication is inherited by the written communication. So it becomes a linguistic universal because it is specific to understanding in linguistic communication, regardless of the natural language concerned. In our formalism, this linguistic universal is expressed by the statement that the topological space underlying any semantic level of an admissible text satisfies the separation axiom T_0 of Kolmogorov.

4.2 Topological connectedness as a linguistic universal

In Sect. 3, we have considered some examples of phonocentric topologies at various levels of semantic description of an admissible text. In all these examples, we see that their underlying topological spaces are connected, which shows empirically an important topological property of all genuine natural language texts, namely the connectedness, in the mathematical sense, of their phonocentric topology. The reasons for it aren't accidental, but it reveals a very important topological property of genuine natural language texts. At the conference [30], we presented arguments that the topological connectedness is one of the linguistic universals.

Any literary work has a property to be the communicative unity of meaning. So, for any two novels X and Y yet of the same kind, say historical, detective or biographical, their concatenation Z under one and the same cover doesn't constitute a new one. What does it mean, topologically speaking? We see that for any $x \in X$ there exists an open neighbourhood U of x that doesn't meet Y, and for any $y \in Y$ there exists an open neighbourhood V of y that doesn't meet X. Thus $Z = X \bigsqcup Y$, i.e., Z is a disjoint union of two non-empty open subsets X and Y; hence Z isn't connected. Thus a property of a literary work to be the communicative unity of meaning may be expressed as a connectedness of a topological space related to text.

Recall that a space X is said to be *connected* if it is not the disjoint union of two non-empty open subsets. It is the same to say that X and $\emptyset$ are the only subsets opened and closed at a time. Such a property is called the connectedness of the space X. In any topological space X, a connected set is a subset U of X which is a connected space for the induced topology. It is clear that the union of connected parts having one point in common is also a connected part.

Define on a topological space X the relation $\sim$ by setting $x \sim y$ if and only if

x and y belong to a connected subset of X. It is immediate that this relation is an equivalence; the equivalence class containing a point x is a connected part which is called *connected component* of x. It is clear that a topological space X is the disjoint union of its connected components, and any connected part is contained in exactly one component. If $f\colon X \to Y$ is a continuous mapping of topological spaces where the space X is connected, then $f(X)$ is a connected subset of Y.

Let X be an Alexandrov topological space. It is clear that for all $x \in X$, the smallest open U_x is connected. So, each open set U_x of the basis $\mathfrak{B}(X)$ of a phonocentric topology is connected.

For all $x, y \in X$ such that $x \neq y$, the subspace $\{x, y\}$ is connected if and only if $x \in U_y$ or $y \in U_x$; in terms of the specialization order, this amounts to saying that $x \preceq y$ or $y \preceq x$. The following well-known proposition (see e.g., [23, p. 8]) characterizes connected Alexandrov topological spaces:

Proposition 3. *Let X be a connected Alexandrov topological space. Then for every pair of points x, y of X, there exists a finite sequence $(z_1, \ldots, z_s)$ of points in X such that $z_1 = x$, $z_s = y$ and each $\{z_i, z_{i+1}\}$ is connected (i.e., $z_i \preceq z_{i+1}$ or $z_i \succeq z_{i+1}$) for all $i = 1, \ldots, s-1$.*

Indeed, let Z be a set of points accessible by a finite sequence $(z_1, \ldots, z_s)$ of points in X starting from $x = z_1$, such that each set $\{z_i, z_{i+1}\}$ is connected for $i = 1, \ldots, s-1$. For each $z \in Z$, we have $U_z \subseteq Z$ because any element $y \in U_z$ is itself also accessible by a chain $(z_1, \ldots, z, y)$. We have $Z \subseteq \bigcup_{z \in Z} U_z \subseteq Z$, hence Z is open. For each $z \in Z$, we have also $\mathrm{cl}(\{z\}) \subseteq Z$ because, for all $y \in \mathrm{cl}(\{z\})$, any neighbourhood of y, including U_y, contains z. This implies $z \preceq y$ and $y \in Z$. We have $Z \subseteq \bigcup_{z \in Z} \mathrm{cl}(\{z\}) \subseteq Z$, hence Z is therefore closed because X is an Alexandrov space. Now, the set Z is non-empty because $x \in Z$, opened and closed subset of the connected space X. Hence, $Z = X$.

It should be noticed that the formulation and the proof of the Proposition 3 are valid regardless of the (finite or infinite) number of points in the space X.

Since the relation $x \preceq y$ is transitive, we can, in the assertion of Proposition 3, exclude unnecessary elements of the finite sequence $(z_1, \ldots, z_s)$. Namely, after excluding repetitive elements, we can reduce each subsequence $z_i \prec z_{i+1} \prec z_{i+2}$ to $z_i \prec z_{i+2}$ if any exists, and we can reduce each subsequence $z_j \succ z_{j+1} \succ z_{j+2}$ to subsequence $z_j \succ z_{j+2}$ if any exists.

After a finite number of such steps of reduction, we have a sequence $(z_1, \ldots, z_r)$, such that in this sequence, the relations $\prec$ and $\succ$ follow one after the other, namely:

if $z_i \prec z_{i+1}$, then $z_{i-1} \succ z_i \prec z_{i+1}$ for all i such that $1 < i < s$;
if $z_i \succ z_{i+1}$, then $z_{i-1} \prec z_i \succ z_{i+1}$ for all i such that $1 < i < s$.

Example. In the Hasse diagram of the book [21], one immediately sees such a sequence $(4 \succ 2 \prec 7 \succ 6 \prec 8)$, which connects the Chapter 4 with the Chapter 8, that is shown on the Figure 5.

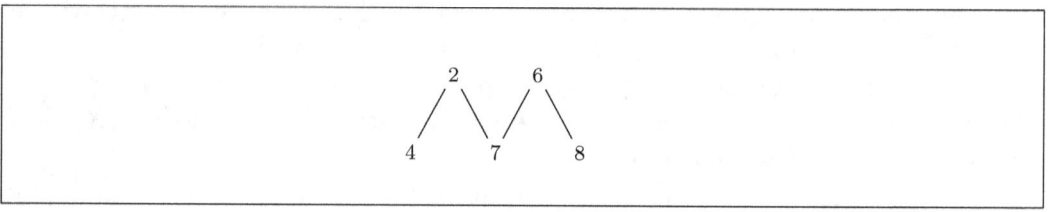

Figure 5: A Khalimsky arc traced in the Leitfaden of [21] shown on the Fig. 1.

The Hasse diagram of the type shown on the Figure 5 is called *Khalimsky arc*.

We define now the *Khalimsky topology* by means of a structure which differs slightly from the original definition of [17]. Let us first define the partition $\mathbb{R} = \bigcup_{m \in \mathbb{Z}} P_m$ of Euclidean line of real numbers $\mathbb{R}$ by setting:

$P_m = [m - \frac{1}{2}, m + \frac{1}{2}]$, closed interval of real numbers $\{t \colon m - \frac{1}{2} \leqslant x \leqslant m + \frac{1}{2}\}$, for each even integer $m \in \mathbb{Z}$;

$P_m =]m - \frac{1}{2}, m + \frac{1}{2}[$, open interval of real numbers $\{t \colon m - \frac{1}{2} < x < m + \frac{1}{2}\}$, for each odd integer $m \in \mathbb{Z}$.

Recall the notion of a *quotient topology*. Let X be a topological space, and let P be an equivalence relation on X. The quotient topology on the quotient set $X/\textsf{P}$ is the finest topology making continuous the canonical projection $X \to X/\textsf{P}$ which associates to each element of X its equivalence class. That is, the set of equivalence classes of $X/\textsf{P}$ is open in the quotient topology if and only if its inverse image is open in X.

Let P be an equivalence relation on $\mathbb{R}$ associated with the partition $\mathbb{R} = \bigcup_{m \in \mathbb{Z}} P_m$. We then define a quotient topology on $X/\textsf{P}$. By identifying $P_m \in X/\textsf{P}$ with $m \in \mathbb{Z}$, we define the *Khalimsky topology* on $\mathbb{Z}$. The set of integers $\mathbb{Z}$ endowed with the Khalimsky topology is called the *Khalimsky line*. Since $\mathbb{R}$ is connected, the Khalimsky line is connected as well.

It is immediate that an even point is closed, and that an odd point is open. Concerning the smallest neighbourhoods, we have $U_m = \{m\}$ if m is odd, and we have $U_m = \{m-1, m, m+1\}$ if m is even. For integers $m \leqslant n$, we define a *Khalimsky interval* to be the interval $[m, n] \cap \mathbb{Z}$ with the topology induced from Khalimsky line, and we denote it by $[m, n]_\mathbb{Z}$. We call a *Khalimsky arc* any topological space which is homeomorphic to a Khalimsky interval $[m, n]_\mathbb{Z}$. We say that the points that are images of m and n are connected by a Khalimsky arc. Now it is clear that the Proposition 3 is equivalent to the following:

Proposition 4. *An Alexandrov topological space X is connected if and only if for every pair of points x, y of X, there exists a Khalimsky arc that connects them.*

In other words, for an Alexandrov space, the connectedness and the connectedness by a Khalimsky arc are equal.

It is obvious that all topological spaces whose Hasse diagrams are shown in the Figure 1 are connected. It is difficult to imagine a book in which there is a single chapter that has no contextual links to other chapters. The same holds not only at the semantic level where primitive elements are chapters, but also at the semantic level where primitive elements are sentences of the text (such a level is called the semantic level of text). If at the end of the reading, we realize that a sentence x has nothing to do with the reminder of the text, we have a feeling that 'a noise crept into the message' because the reading of the text is finished, but the sentence x remains to be its completely strange ingredient.

On the contrary, if during the reading we meet a sentence that does not have direct contextual links with the sentences already read (like the item 7 in the Hasse diagram of the textbook [39] on the Figure 1), we have a feeling to be on a turning point in the narrative, and that the author prepares the reader for the future development, where the suspended sentence will be necessary for the understanding. For an admissible text, these considerations confirm that the connectedness of the underlying topological space expresses mathematically the necessary requirement of a textuality in the sense one understands this concept in the semiotics of text.

This means that a basic unit which is pertinent as a message in the situation of linguistic communication must be an admissible text whose underlying topological space is connected! It is a connected unit because, after communicating of such a message, the transmitter (author, sender) can remain silent to give the floor to its receptor (reader, receiver).

At the level of text, the connectedness of message is also a requirement specific to the kind of linguistic communication qualified as a dialogue, that is, to a bi-directional communication with others. If somebody produces, as a message, a series of phrases that disintegrates into pieces that have no links between, it reveals the disregard for the interlocutor, or the absence of the desire to communicate, or the use of a language for purely expressive purposes without a desire to communicate.

It is the same at the semantic level of sentence with regard to connectedness, although the formal definition of a phonocentric topology at the level of sentence needs more delicate work.

Remark. It should be noticed that for an admissible text, the corresponding Hasse diagram with directed edges is acyclic at any semantic level. It is clear that this property of a phonocentric topology is stable under homeomorphism. This means that the *acyclicity* of the Hasse diagram corresponding to the phonocentric topology is yet another linguistic universal of a topological nature.

5   Sheaves of meanings appeared as semantics

Let X be an admissible text endowed with a phonocentric topology, and let $\mathscr{F}$ be an adopted sense of reading. In a Platonic manner, for each non-empty open (that is meaningful) part $U \subseteq X$, we collect in the set $\mathscr{F}(U)$ all fragmentary meanings of this part U read in the sense $\mathscr{F}$; also we define $\mathscr{F}(\varnothing)$ to be a singleton pt. Thus we are given a map

$$U \mapsto \mathscr{F}(U) \tag{1}$$

defined on the set $\mathfrak{O}(X)$ of all open sets in a phonocentric topology on X.

Following the precept of hermeneutic circle **'to understand a part in accordance with the understanding of the whole'**, for each inclusion $U \subseteq V$ of non-empty opens, the adopted sense of reading $\mathscr{F}$ gives rise to restriction map $\mathrm{res}_{V,U} \colon \mathscr{F}(V) \to \mathscr{F}(U)$. We will consider the inclusion of sets $U \subseteq V$ as being the canonical injection map $U \xhookrightarrow{\mathrm{inj}} V$. Thus we are also given a map

$$\{\, U \xhookrightarrow{\mathrm{inj}} V \,\} \mapsto \{\, \mathscr{F}(V) \xrightarrow{\mathrm{res}_{V,U}} \mathscr{F}(U) \,\} \tag{2}$$

with the properties:

(i) $\mathrm{id}_V \mapsto \mathrm{id}_{\mathscr{F}(V)}$ for all opens V of X;

(ii) $\mathrm{res}_{V,U} \circ \mathrm{res}_{W,V} = \mathrm{res}_{W,U}$ for all nested opens $U \subseteq V \subseteq W$ of X.

The first property means that the restriction $\mathrm{res}_{V,U}$ respects identity inclusions. The second property means that two consecutive restrictions may be done by one step.

As for the empty part $\varnothing$ of X, the restriction maps $\mathrm{res}_{\varnothing,\varnothing}$ and $\mathrm{res}_{V,\varnothing}$ with the same properties are obviously defined.

Let $(X, \mathfrak{O}(X))$ be a topological space. We can consider its topology $\mathfrak{O}(X)$ as the category $\mathbf{Open}_X$ whose objects are open sets of X, and where for two open sets $U, V \in \mathfrak{O}(X)$, the class of morphisms $\mathrm{Mor}(U, V)$ is empty if $U \not\subseteq V$, and $\mathrm{Mor}(U, V)$ is the set reduced to the canonical injection $U \xhookrightarrow{\mathrm{inj}} V$ if $U \subseteq V$. The composition of morphisms is defined as the composition of canonical injections.

From the mathematical point of view, the assignments (1) and (2) give rise to a presheaf $\mathscr{F}$ defined as a contravariant functor from the category $\mathbf{Open}_X$ to the category of sets $\mathbf{Set}$

$$\mathscr{F}: \mathbf{Open}_X \to \mathbf{Set}, \qquad (3)$$

acting on objects as defined by (1), and acting on morphisms as defined by (2).

In sheaf theory, an element $s \in \mathscr{F}(V)$ is called *section* (*over V*); sections over the whole space X are said to be *global*.

We consider the reading process of an open fragment U as its covering by some family of open subfragments $(U_j)_{j \in J}$ already read, that is $U = \bigcup_{j \in J} U_j$.

Following Quine, **"There is no entity without identity"** [35]. We argue that two fragmentary meanings should be equal globally if and only if they are equal locally. It motivates the following *identity criterion*:

Claim S (Separability). *Let X be an admissible text, and let U be an open fragment of X. Suppose that $s, t \in \mathscr{F}(U)$ are two fragmentary meanings of U and there is an open covering $U = \bigcup_{j \in J} U_j$ such that $\mathrm{res}_{U, U_j}(s) = \mathrm{res}_{U, U_j}(t)$ for all fragments U_j. Then $s = t$.*

According to the precept of hermeneutic circle, **'to understand the whole by means of understandings of its parts'**, a presheaf $\mathscr{F}$ of fragmentary meanings satisfies the following:

Claim C (Compositionality). *Let X be an admissible text, and let U be an open fragment of X. Suppose that $U = \bigcup_{j \in J} U_j$ is an open covering of U; suppose we are given a family $(s_j)_{j \in J}$ of fragmentary meanings, $s_j \in \mathscr{F}(U_j)$ for all fragments U_j, such that $\mathrm{res}_{U_i, U_i \cap U_j}(s_i) = \mathrm{res}_{U_j, U_i \cap U_j}(s_j)$. Then there exists some meaning s of the whole fragment U such that $\mathrm{res}_{U, U_j}(s) = s_j$ for all fragments U_j.*

Thus any presheaf of fragmentary meanings defined as above should satisfy both Claims **S** and **C**, and so it is a *sheaf* by the very definition. It motivates the following:

Frege's Generalized Compositionality Principle. *A presheaf of fragmentary meanings naturally attached to any sense (mode of reading) of an admissible text is really a sheaf; its sections over a meaningful fragment of the text are its fragmentary meanings; its global sections are the meanings of the text as a whole.*

Traditionally attributed to Frege, the compositionality principle arises in logic, linguistics and philosophy of language in many different formulations, which all however convey the concept of functionality.

We note that the Claim **S** guarantees the meaning s (whose existence is stated by the Claim **C**) to be unique as such. It is not so hard to see that these two conditions **C** and **S** needed for a presheaf to be a sheaf are analogous to those two conditions $1°$ and $2°$ needed for a binary relation to be a function.

5.1 Sheaf-theoretic conception of a functional dependence

Formally, for a function f of n variables, it is set that: **1°** *for any* family of variables' values $(s_1, \ldots, s_n)$, *there exists* a function's value $f(s_1, \ldots, s_n)$ being dependent on them, and **2°** this function's value is *unique*. Likewise, for a sheaf $\mathscr{F}$, it is set that: (due to **C**) *for any* family of sections $(s_i)_{i \in I}$ which are locally compatible on an open U, *there exists* a section s being their composition dependent on them, and (due to **S**) this composition s is *unique* as such. In this generalized (sheaf-theoretic) conception of a functional dependence, the variables and their number are not fixed in advance (we consider an arbitrary family of pairwise compatible sections as variables), but for any such a family of variables, *there exists* the glued section considered as their composition (analogous to the function's value in a given family of variables) and such a section is *unique*. So the true formulation of Frege's compositionality principle does not demand a set-theoretic functionality, but demands its sheaf-theoretic generalization which states that any presheaf of fragmentary meanings naturally attached to an admissible text ought *de facto* to be a sheaf. The sheaves arise whenever some consistent local data glues into a global one.

5.2 Schleiermacher category of sheaves of fragmentary meanings

The reader should become at home with the senses treated as functors although we call them sometimes as 'modes of readings' instead of 'senses' not only to emphasize the character of intentionality of each actual process of reading but rather to avoid a possible confusion which may be caused by another technical acceptance of the term 'sense'. So one can think, for example, about the historical sense $\mathscr{F}$ and the moral sense $\mathscr{G}$ of some biographical text.

Let us consider now any two senses (modes of reading) $\mathscr{F}$, $\mathscr{G}$ of a given text X, and let $U \subseteq V$ be two arbitrary meaningful fragments of the text X. It seems to be very natural to consider that any meaning s of fragment V understood in the historical sense $\mathscr{F}$ gives a certain well-defined meaning $\phi(V)(s)$ of the same fragment V understood in the moral sense $\mathscr{G}$. Hence, for each $V \subseteq X$, we are given a map $\phi(V) \colon \mathscr{F}(V) \to \mathscr{G}(V)$. To transfer from the meaning s of V in the historical sense to its meaning $\phi(V)(s)$ in the moral sense and then to restrict the latter to a subfragment $U \subseteq V$ is the same operation as to make first the restriction from V to U of the meaning s in the historical sense, and to make then a change of the historical sense to the moral one. This kind of transfer from the understanding in one sense $\mathscr{F}$ to the understanding in another sense $\mathscr{G}$ is a usual matter of linguistic communication. In the Christian theology, the possibility of such a transfer from one of four senses of any biblical verse to some another its sense is considered as the cornerstone method of exegesis.

Formally, this idea is well expressed by the notion of *morphism* of the corresponding sheaves $\phi\colon \mathscr{F} \mapsto \mathscr{F}'$ defined as a family of maps $\phi(V)\colon \mathscr{F}(V) \to \mathscr{F}'(V)$, which commute with restrictions for all opens $U \subseteq V$, that is, $\mathrm{res}'_{V,U} \circ \phi(V) = \phi(U) \circ \mathrm{res}_{V,U}$. This can be expressed in a simple way by saying that the following diagram

$$\begin{array}{ccc} \mathscr{F}(V) & \xrightarrow{\phi(V)} & \mathscr{G}(V) \\ {\scriptstyle \mathrm{res}_{V,U}}\downarrow & & \downarrow{\scriptstyle \mathrm{res}'_{V,U}} \\ \mathscr{F}(U) & \xrightarrow{\phi(U)} & \mathscr{G}(U) \end{array}$$

commutes for all opens $U \subseteq V$ of X.

This notion of morphism is very near to that of *incorporeal transformation* of G. Deleuze and F. Guattari illustrated by several examples, one of which we quote:

> In an airplane hijacking, the threat of a hijacker brandishing a revolver is obviously an action; so is the execution of the hostages, if it occurs. But the transformation of the passengers into hostages, and of the plane-body into a prison-body, is an instantaneous incorporeal transformation, a "mass media act" in the sense in which the English speak of "speech acts." [5, p. 102]

To adapt this example, we need only to transform it into some written story about a hijacking. Hence, the family of maps $(\phi(V))_{V \in \mathfrak{O}(X)}$ defines a change of mode of reading of a given text X, or simply a morphism $\phi\colon \mathscr{F} \mapsto \mathscr{G}$. It is obvious that a family of identical maps $\mathrm{id}_{\mathscr{F}(V)}\colon \mathscr{F}(V) \to \mathscr{F}(V)$ given for each open $V \subseteq X$ defines the identical morphism of the sheaf $\mathscr{F}$ which will be denoted as $\mathrm{id}_{\mathscr{F}}$. The composition of morphisms is defined in an obvious manner: for two arbitrary morphisms $\phi\colon \mathscr{F} \mapsto \mathscr{G}$, $\psi\colon \mathscr{G} \mapsto \mathscr{H}$, we define $(\psi \circ \phi)(V) = \psi(V) \circ \phi(V)$. It is clear that this composition is associative every time it may be defined.

Thus, given an admissible text X, the data of all sheaves $\mathscr{F}$ of fragmentary meanings together with all its morphisms constitutes some *category* in a strict mathematical sense of the term. We name this category of particular sheaves describing the exegesis of the text X as *category of Schleiermacher* and denote it as $\mathbf{Schl}(X)$ because he is generally considered to be the author of the cornerstone principle of a natural language text understanding, called later by Dilthey as the *hermeneutic circle*. The parts are understood in terms of the whole, and the whole is understood in terms of the parts. This part-whole structure in the understanding, he claimed, is principal in the matter of interpretation of any text in natural language.

The theoretical principle of *hermeneutic circle* is a precursor to Frege's principles of *compositionality* and *contextuality* formulated later. The succeeded development of hermeneutics has confirmed the importance of Schleiermacher's concept of circularity in text understanding. From our point of view, the concept of part-whole structure expressed by Schleiermacher in 1829 as the hermeneutic circle principle reveals, in the linguistic form, the fundamental mathematical concept of a sheaf formulated by Leray in 1945, more than a hundred years later. This justifies us to name the particular category of sheaves $\mathbf{Schl}(X)$ after Schleiermacher.

5.3 Building a sheaf of fragmentary meanings from local data

An admissible text X is endowed with a phonocentric topology in such a way that the set $\mathfrak{O}(X)$ of all open sets of this topology is made up of all meaningful parts of X. The Hasse diagram presents a perfect visualization of this topological structure but its construction requires a lot of analytical work. It seems that the author has such a representation about his/her proper text, as well as the structure of text may be rebuild after philological considerations. But for a reader, how this topological structure is obtained during the reading? Obviously, the understanding is manifested in the reader's conscience as an empirical fact of having grasped the meaning of a sentence read in the present moment. Thus, the meaningful parts that are most clearly manifested during the reading process are the opens U_x of the phonocentric topology basis $\mathfrak{B}(X)$ which is defined in the Sect. 3. These meaningful parts U_x provide the set of contexts for the understanding of the whole text.

The Proposition 1 states that the set of all these fragments U_x constitutes the minimal basis for a phonocentric topology. Formally, this means that any arbitrary open set is the union of a family of these basis sets U_x. The liberty in choice of basis sets whose union gives an open set $U \subseteq X$ makes us doubtful whether it would be too strong to impose the satisfaction of Claims **S** and **C** for all opens of the topology $\mathfrak{O}(X)$. Would it be more convenient and more useful to develop the very theory in more concrete terms, say even in constructive terms of opens U_x of the minimal basis $\mathfrak{B}(X)$ of a phonocentric topology on X? The answer is plain and simple: From the psychological point of view, yes, perhaps; but from the mathematical point of view, this approach will be formally equivalent but less technically convenient! Moreover, a general truth is sometimes more understandable that a mass of concrete data. In what follows, we will present formal arguments to justify this point of view.

A topological space $(X, \mathfrak{O}(X))$ may be considered as the category $\mathbf{Open}_X$ with open sets $U \in \mathfrak{O}(X)$ as objects, and injection maps $U \xhookrightarrow{\text{inj}} V$ as morphisms.

Let $\mathfrak{B}(X)$ be a basis for the topology $\mathfrak{O}(X)$ of X. It is obvious that the basis $\mathfrak{B}(X)$ gives rise to a category defined in the same way that we consider the topology $\mathfrak{O}(X)$ as being the category $\mathbf{Open}_X$. By a slight abuse of notations, we will also denote such a category as $\mathfrak{B}(X)$. In the same manner as above, we define a presheaf

$\mathscr{F}$ of sets on the topology basis $\mathfrak{B}$ as a *contravariant functor* on the category $\mathfrak{B}$ with values in the category of sets **Set**.

Namely, for every basis open $U \in \mathfrak{B}(X)$, the presheaf $\mathscr{F}$ attaches a set $\mathscr{F}(U)$, and so we are given a map

$$U \mapsto \mathscr{F}(U) \tag{4}$$

defined on the basis $\mathfrak{B}(X)$ for a topology on X. Also, for every pair of opens $U, V \in \mathfrak{B}(X)$ such that $U \subseteq V$, the presheaf $\mathscr{F}$ attaches a map $\mathrm{res}_{V,U} \colon \mathscr{F}(V) \to \mathscr{F}(U)$, and so we are given a map

$$\{U \subseteq V\} \mapsto \{\mathrm{res}_{V,U} \colon \mathscr{F}(V) \to \mathscr{F}(U)\} \tag{5}$$

with the properties of identity preserving and transitivity:

(i) $\mathrm{id}_V \mapsto \mathrm{id}_{\mathscr{F}(V)}$ for all opens $V \in \mathfrak{B}(X)$;

(ii) $\mathrm{res}_{V,U} \circ \mathrm{res}_{W,V} = \mathrm{res}_{W,U}$ for all nested basis opens $U \subseteq V \subseteq W$ of $\mathfrak{B}(X)$.

Given a basis $\mathfrak{B}(X)$ for a topology on X, the data of $(\mathscr{F}(V), \mathrm{res}_{V,U})_{V,U \in \mathfrak{B}(X)}$ satisfying these properties is called presheaf of sets over the basis $\mathfrak{B}(X)$ for the topology on X. In the case of an admissible text X, the topological basis $\mathfrak{B}(X)$ consists of all fragments of the kind U_x, that may be considered as empirical data.

Let $\mathscr{F}$ be a presheaf of sets over a basis $\mathfrak{B}(X)$ for the topology $\mathfrak{O}(X)$ on X. This presheaf $\mathscr{F}$ is said to be a *sheaf* over the topological basis $\mathfrak{B}(X)$ if the following Claims **Sb** and **Cb** are satisfied:

Claim Sb. *Let U be any open of the basis $\mathfrak{B}(X)$ for the topology on X, and let $s, t \in \mathscr{F}(U)$ be two elements of U. If there exists an open covering $U = \bigcup_{j \in J} U_j$ by basis open sets $U_j \in \mathfrak{B}(X)$ such that for each U_j of this covering, we have $\mathrm{res}_{U, U_j}(s) = \mathrm{res}_{U, U_j}(t)$. Then $s = t$.*

Claim Cb. *Let U be any open of the basis $\mathfrak{B}(X)$ for the topology on X, and let $U = \bigcup_{j \in J} U_j$ be a covering of U by basis open sets $U_j \in \mathfrak{B}(X)$. Suppose we are given a family $(s_j)_{j \in J}$ of elements $s_j \in \mathscr{F}(U_j)$ such that $\mathrm{res}_{U_i, U_i \cap U_j}(s_i) = \mathrm{res}_{U_j, U_i \cap U_j}(s_j)$. Then there exists an element $s \in \mathscr{F}(U)$ such that $\mathrm{res}_{U, U_j}(s) = s_j$ for each open U_j.*

It is obvious that the Claims **Sb** and **Cb** are similar to the Claims **S** and **C** in the definition of a sheaf over a topological space.

Let $\mathscr{F}$ be a presheaf of sets over the basis $\mathfrak{B}(X)$ for a topological space X. For any open $U \in \mathfrak{O}(X)$, the sets $(\mathscr{F}(V))_{\mathfrak{B} \ni V \subseteq U}$ together with maps $\mathrm{res}_{W,V}$ (where $W \in \mathfrak{B}(X)$, $V \in \mathfrak{B}(X)$ such that $V \subseteq W \subseteq U$) form a projective system.

We can associate to $\mathscr{F}$ a presheaf of sets $\mathscr{F}'$ over X in the ordinary sense by assigning to any open $U \in \mathfrak{O}(X)$, the projective limit

$$\mathscr{F}'(U) = \varprojlim_{\mathfrak{B} \ni V \subseteq U} \mathscr{F}(V), \tag{6}$$

where V are running the set (ordered by $\subseteq$) of all opens $V \in \mathfrak{B}(X)$ such that $V \subseteq U$.

In the *EGA* of A. Grothendieck and J. A. Dieudonné [13, p. 75], there is a general proposition which, for a presheaf with values in the category of sets, is interpreted as the following

Proposition 5. *For the presheaf $\mathscr{F}'$ defined over topology $\mathfrak{O}(X)$ by (6) to be a sheaf, that is, to verify the Claims* **S** *and* **C**, *it is necessary and sufficient that the presheaf $\mathscr{F}$ defined over the basis $\mathfrak{B}(X)$ for $\mathfrak{O}(X)$ verifies the Claims* **Sb** *and* **Cb**.

Let $\mathscr{F}$, $\mathscr{G}$ be two presheaves of fragmentary meanings defined over the topological basis $\mathfrak{B}(X)$. We define a morphism $\theta \colon \mathscr{F} \to \mathscr{G}$ as a family $(\theta(V))_{V \in \mathfrak{B}}$ of maps $\theta(V) \colon \mathscr{F}(V) \to \mathscr{G}(V)$ satisfying the conditions of compatibility with the corresponding restriction morphisms. With the notation of Proposition 5, we deduce a morphism $\theta' \colon \mathscr{F}' \to \mathscr{G}'$ of presheaves of fragmentary meanings defined on all opens $U \in \mathfrak{O}(X)$ by taking $\theta'(U)$ to be the projective limit of $\theta(V)$ for $V \in \mathfrak{B}(X)$ and $V \subseteq U \in \mathfrak{O}(X)$.

Let $\mathscr{F}$ be a sheaf of fragmentary meanings over $\mathfrak{O}(X)$, and let $\mathscr{F}_1$ be a sheaf over $\mathfrak{B}(X)$ defined by the restriction of $\mathscr{F}$ to $\mathfrak{B}(X)$. Then, the sheaf $\mathscr{F}_1'$ over $\mathbf{Open}_X$ which is obtained from $\mathscr{F}_1$ according to the Proposition 5 is canonically isomorphic to $\mathscr{F}$, because of the claims **S** and **C**, and by the uniqueness of the projective limit. Usually, we identify $\mathscr{F}$ and $\mathscr{F}_1'$.

For two sheaves $\mathscr{F}$, $\mathscr{G}$ defined over $\mathfrak{O}(X)$ and a morphism $\theta \colon \mathscr{F} \to \mathscr{G}$, one can show that the data of $\theta(V) \colon \mathscr{F}(V) \to \mathscr{G}(V)$ given only for $V \in \mathfrak{B}(X)$ determines completely the morphism θ. For more details, see *EGA* of A. Grothendieck and J. A. Dieudonné [13, p. 76].

Theoretically speaking, this means that we have a good reason to move the considerations from the level of empirical data, where a phonocentric topology is revealed by the minimal basis $(U_x)_{x \in X}$, to the general level, more abstract but more simple, where a phonocentric topology is defined by the set $\mathfrak{O}(X)$ of all opens according to the classical Hausdorff axioms (t_1), (t_2), and (t_3) of a topological space. In mathematics, the axiomatic view on a topology is particularly useful in all sorts of reasonings where topological structures are concerned. Once we have defined a topological space in terms of its basis, we may continue the reasoning in terms of all open sets of this topology.

5.4 Compositionality of locally defined modes of reading

Note that the class of objects in the category $\mathbf{Schl}(X)$ is not limited to a modest list of sheaves corresponding to *literal, allegoric, moral, psychoanalytical* and other senses mentioned above. In the process of text interpretation, the reader's semantic intentionality changes from time to time, with the result that there is some compositionality (or gluing) of these locally defined sheaves of fragmentary meanings, which we consider in details in [31]. There is a standard way to name the result of such a gluing as, for example, this is the case of *Freudo-Marxist* sense.

As the reader's intentionality to interpret an arbitrary text in a certain sense $\mathscr{F}$, this particular sense $\mathscr{F}$ yet precedes a reading; for example, one may intend to read a story in a moral sense. But for a given text X, the *intentional object* 'sense $\mathscr{F}$' is represented by the sheaf of sets $(\mathscr{F}(V), \mathrm{res}_{V,U})_{V,U \in \mathfrak{O}(X)}$ of fragmentary meanings.

To analyze the compositionality of senses (or modes of reading) in our sheaf-theoretic formalism, we recall firstly the notion of *induced sheaf*. Let X be a topological space, let U be an open set of X, and let $i: U \hookrightarrow X$ be the canonical injection of the open U in X. Then, for any sheaf $\mathscr{F}$ of sets over X, one can define a sheaf of sets over U, which is called *sheaf induced by $\mathscr{F}$ on U*, and which is denoted as $\mathscr{F}|_U$, by setting:

$$(\mathscr{F}|_U)(V) = \mathscr{F}(i(V)) \quad \text{for any open } V \subseteq U;$$
$$(\mathrm{res}|_U)_{W,V} = \mathrm{res}_{i(W), i(V)} \quad \text{for all opens } V, W \subseteq U \text{ such that } V \subseteq W.$$

For any morphism $\theta: \mathscr{F} \to \mathscr{G}$ of sheaves of sets over X, we note by $\theta|_U$ the morphism $\mathscr{F}|_U \to \mathscr{G}|_U$ consisting of maps $\theta(i(V))$ for opens $V \subseteq U$.

We have a reason to assume that the reading of the whole text X in a sense $\mathscr{F}$ is represented by an open covering $(U_\lambda)_{\lambda \in L}$ of the text X, where each fragment U_λ is read in a sense $\mathscr{F}_\lambda$ which is defined as $\mathscr{F}_\lambda = \mathscr{F}|_{U_\lambda}$.

The obvious concordance of these senses $\mathscr{F}_\lambda$ means that for all pairs of open fragments $U_\lambda, U_\mu \subseteq X$, we have an isomorphism

$$\theta_{\lambda\mu}: \mathscr{F}_\mu|_{(U_\lambda \cap U_\mu)} \xrightarrow{\sim} \mathscr{F}_\lambda|_{(U_\lambda \cap U_\mu)}. \tag{7}$$

In other words, in the interpretation of the common part $U_\lambda \cap U_\mu$, we can change the sense $\mathscr{F}_\lambda$ to the sense $\mathscr{F}_\mu$ and vice versa.

It is useful to denote $U_{\lambda\mu} = U_\lambda \cap U_\mu$ and $U_{\lambda\mu\nu} = U_\lambda \cap U_\mu \cap U_\nu$. Then, in this notation, the family of isomorphisms

$$\theta_{\lambda\mu}: \mathscr{F}_\mu|_{U_{\lambda\mu}} \xrightarrow{\sim} \mathscr{F}_\lambda|_{U_{\lambda\mu}} \tag{8}$$

satisfies the condition:

$$(\text{for all } U_\lambda, U_\mu, U_\nu) \quad \theta_{\lambda\mu} \circ \theta_{\mu\nu} = \theta_{\lambda\nu} \text{ on } U_{\lambda\mu\nu}. \tag{9}$$

In the theory of sheaves, there is a theorem stating that a family of isomorphisms satisfying the condition (9) allows us to rebuild the sheaf $\mathscr{F}$ uniquely. The following proposition is a linguistic version of this general mathematical result:

Proposition 6. *Let $(U_\lambda)_{\lambda \in L}$ be an open covering of the text X, where each fragment U_λ is read in a sense $\mathscr{F}_\lambda$. Let for each pair of fragments U_λ, U_μ of $(U_\lambda)_{\lambda \in L}$ be given an isomorphism $\theta_{\lambda\mu} \colon \mathscr{F}_\mu|_{U_{\lambda\mu}} \xrightarrow{\sim} \mathscr{F}_\lambda|_{U_{\lambda\mu}}$ of sheaves over $U_{\lambda\mu}$. Assume these isomorphisms are satisfying the condition that for all U_λ, U_μ, U_ν of the covering:*

$$\theta_{\lambda\mu} \circ \theta_{\mu\nu} = \theta_{\lambda\nu} \text{ on } U_{\lambda\mu\nu}. \tag{10}$$

Then, there exists a sheaf $\mathscr{F}$ over X, and for each U_λ of the covering $(U_\lambda)_{\lambda \in L}$ there exists an isomorphism $\theta_\lambda \colon \mathscr{F}|_{U_\lambda} \xrightarrow{\sim} \mathscr{F}_\lambda$ such that $\theta_\mu = \theta_{\mu\lambda} \circ \theta_\lambda$ for U_λ, U_μ of the covering $(U_\lambda)_{\lambda \in L}$. Moreover, $(\mathscr{F}, (\theta_\lambda)_{\lambda \in L})$ is unique up to unique isomorphism.

For the proof, see *EGA* of A. Grothendieck and J. A. Dieudonné [13, p. 77].

The family of isomorphisms $(\theta_{\lambda\mu})$ satisfying the *gluing condition* (10) is called a *1-cocycle*. One says that the sheaf $\mathscr{F}$ is obtained by *gluing* of sheaves $(\mathscr{F}_\lambda)_{\lambda \in L}$ by means of $\theta_{\lambda\mu}$, and usually one identifies $\mathscr{F}_\lambda$ and $\mathscr{F}|_{U_\lambda}$ by means of θ_λ.

For a finite family of sheaves $(\mathscr{F}_\lambda)_{\lambda \in L}$ and their isomorphisms $\theta_{\lambda\mu}$ satisfying the condition of gluing (10), the sheaf $\mathscr{F}$ is called to be their composition obtained by the gluing of sheaves $(\mathscr{F}_\lambda)_{\lambda \in L}$ by means of the $\theta_{\lambda\mu}$; this describes how we define the compositionality of locally defined modes of reading (senses) understood as sheaves of fragmentary meanings.

The gluing of sheaves is a compositionality method which enables us to obtain a large number of globally defined sheaves from a small number of locally defined ones. In fact, the sense $\mathscr{F}$ as a global mode of reading (or an integral intention during the interpretation of the whole text) is composed of all local modes of reading taken during interpretations of parts.

Example. According to the biblical hermeneutics, the readings of the Scripture in the *literal, allegorical, moral,* and *anagogical* senses are consistent over each fragment of the type U_x. Suppose that we have read the whole text of the Scripture by fragments, where each fragment was read in one of these four senses (literal, allegorical, moral, anagogical). These partial readings satisfy the gluing condition (10) above. There exists therefore a sense $\mathscr{F}$ of reading of the whole text of the Scripture such that for each of its sentence, there are a neighbourhood and one of these four senses (literal, allegorical, moral, anagogical) which is consistent with the reading of this neighbourhood in the sense $\mathscr{F}$. The sense $\mathscr{F}$ is a composition of these four senses (literal, allegorical, moral, anagogical), but globally it differs from each of these four senses being applied to the whole text. Hence, for the text E of the Scripture, the class of objects of the category **Schl**(E) of Schleiermacher contains not only these four senses (literal, allegorical, moral, and anagogical) but much more their compound senses, where each compound sense $\mathscr{F}$ is defined by gluing of these four senses according to a particular text covering by fragments, each of which is read in only one sense among the four.

We summarize the results of our analysis presented in Sect. 5 as the following:

Slogan (Sheaves of Fragmentary Meanings as Semantics). The mathematical study of a natural language texts interpretation in terms of the category of sheaves of fragmentary meanings and their morphisms is a sheaf-theoretic formal semantics.

6 Bundles of contextual meanings

So far, we have considered only the meanings of open sets in the phonocentric topology at any semantic level. In this section, we describe how we have to define the meanings of points in the phonocentric topology at a given semantic level. It should be noticed that in general, not every singleton x is open in T_0-topology, and if this is the case, the meaning of such a point x has not yet been defined.

In 1884, Frege wrote in the *Die Grundlagen der Arithmetik* [9, p. X]: "nach der Bedeutung der Wörter muss im Satzzusammenhange, nicht in ihrer Vereinzelung gefragt werden;" This declaration is traditionally named as Frege's principle of contextuality. It was stated eight years before his theoretic distinction between *Sinn* and *Bedeutung*, that is why the word 'Bedeutung' here is usually translated in English as 'meaning': "Never ask for the meaning of a word in isolation, but only in the context of a sentence". As we have yet seen in the Sect. 3.4, the context of a whole sentence is the greatest possible at the semantic level of sentence. We may also ask for the meaning of a word x in the context of a clause to which it belongs, or in the context of some lesser part of this clause as, e.g. of the smallest part U_x. This restatement makes Frege's definition more precise. If we try to recast such a contextuality principle to the level of text, then we would have to say: **"Never ask for the meaning of a sentence in isolation, but only in the context of some meaningful fragment of a text"**. Such a fragment may be chosen in many ways to induce the same contextual meaning of the sentence.

To formalize this definition, let us consider the phonocentric topology at the level of text. Let a sentence x belongs to meaningful fragments (opens) U and V.

Then fragmentary meanings $s \in \mathscr{F}(U)$, $t \in \mathscr{F}(V)$ are said to *induce the same contextual meaning of a sentence* $x \in U \cap V$ if there exists some open neighbourhood W of x, such that $W \subseteq U \cap V$ and $\mathrm{res}_{U,W}(s) = \mathrm{res}_{V,W}(t) \in \mathscr{F}(W)$. The identity of fragmentary meanings is understood here accordingly to the criterion claimed by **S**.

This relation '*fragmentary meanings s, t induce the same contextual meaning of the sentence x*' is clearly an equivalence relation. The equivalence class so defined by a fragmentary meaning s is called a *germ* at x of this s, and is denoted by $\mathrm{germ}_x(s)$. The equivalence class of fragmentary meanings agreeing in some open neighbourhood of a sentence x is natural to define as a *contextual meaning* of x. Let $\mathscr{F}_x$ be the set of all contextual meanings of x. Following S. Mac Lane and I. Moerdijk [20, pp. 83,84], this $\mathscr{F}_x$ is nothing else but the inductive limit $\mathscr{F}_x = \varinjlim(\mathscr{F}(V), \mathrm{res}_{V,U})_{V,U \in \mathfrak{O}(x)}$, where $\mathfrak{O}(x)$ is the set of all open neighbourhoods of x.

In the bundle-theoretic terms, we summarize the aforesaid as being the following:

Frege's Generalized Contextuality Principle. *Let $\mathscr{F}$ be an adopted sense of reading of a fragment U of an admissible text X. For a sentence $x \in U \subseteq X$, its contextual meaning is defined as a $\mathrm{germ}_x(s)$ at x of some fragmentary meaning $s \in \mathscr{F}(U)$. The set $\mathscr{F}_x$ of all contextual meanings of a sentence $x \in X$ is defined as the inductive limit $\mathscr{F}_x = \varinjlim(\mathscr{F}(V), \mathrm{res}_{V,U})_{V,U \in \mathfrak{O}(x)}$, where $\mathfrak{O}(x)$ is the set of all open neighbourhoods of x, that is the set of all meaningful fragments containing x.*

Remark. Note that for an open singleton $\{x\}$, we may canonically identify $\mathscr{F}_x = \mathscr{F}(\{x\})$.

For the coproduct $F = \bigsqcup_{x \in X} \mathscr{F}_x$, we define now a *projection* map $p: F \to X$ by setting $p(\mathrm{germ}_x s) = x$. Every fragmentary meaning $s \in \mathscr{F}(U)$ determines a genuine function $\dot{s}: x \mapsto \mathrm{germ}_x s$ to be well-defined on U.

We define the topology on F by taking as a basis for this topology all the image sets $\dot{s}(U) \subseteq F$. For an open $U \subseteq X$, a continuous function $t: U \to F$ such that $t(x) \in p^{-1}(x)$ for all $x \in U$ is called a *cross-section*. The topology defined on F makes p and every cross-section of the kind of $\dot{s}$ to be continuous.

For a given topological space X, we have so defined a topological spaces F and a continuous surjection $p: F \to X$. In topology, this data (F, p) is called a *bundle over the base space X*. A *morphism* of bundles from $p: F \to X$ to $q: G \to X$ is a continuous map $h: F \to G$ such that the diagram

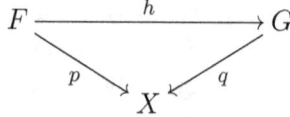

commutes, that is, $q \circ h = p$.

Thus, we have defined a category of bundles over X. A bundle (F,p) over X is called *étale* if $p\colon F \to X$ is a local homeomorphism. It is immediately seen that a bundle of contextual meanings $(\bigsqcup_{x \in X} \mathscr{F}_x, p)$ constructed as above from a given sheaf $\mathscr{F}$ of fragmentary meanings is étale. Thus, for an admissible text X, we have defined the category **Context**(X) of étale bundles (of contextual meanings) over X as a framework for the generalized contextuality principle at the level of text.

The similar definition may be formulated at each semantic level. The definition formulated at the level of sentence returns Frege's classic contextuality principle. Once a semantic level is given, the definition of a contextual meaning for a point x of the corresponding topological space X is stated as $\mathrm{germ}_x s$, where s is some fragmentary meaning defined on some neighbourhood U of x.

7 Frege duality

For a given admissible text X, we have defined two categories formalizing the interpretation process: the Schleiermacher category **Schl**(X) of *sheaves* of fragmentary meanings and the category **Context**(X) of *étale bundles* of contextual meanings. Our intention now is to relate them to each other.

We will firstly define a so-called *germ-functor*

$$\Lambda\colon \mathbf{Schl}(X) \to \mathbf{Context}(X).$$

For each sheaf $\mathscr{F}$, it assigns an étale bundle $\Lambda(\mathscr{F}) = (\bigsqcup_{x \in X} \mathscr{F}_x, p)$, where the projection p is defined as above. For a morphism of sheaves $\phi\colon \mathscr{F} \to \mathscr{F}'$, the induced map of fibers $\phi_x\colon \mathscr{F}_x \to \mathscr{F}'_x$ gives rise to a continuous map $\Lambda(\phi)\colon \bigsqcup_{x \in X} \mathscr{F}_x \to \bigsqcup_{x \in X} \mathscr{F}'_x$ such that $p' \circ \Lambda(\phi) = p$; hence $\Lambda(\phi)$ defines a morphism of bundles. Given another morphism of sheaves ψ, one sees easily that $\Lambda(\psi \circ \phi) = \Lambda(\psi) \circ \Lambda(\phi)$ and $\Lambda(\mathrm{id}_{\mathscr{F}}) = \mathrm{id}_F$. Thus, we have constructed a desired germ-functor $\Lambda\colon \mathbf{Schl}(X) \to \mathbf{Context}(X)$.

We will now define a so-called *section-functor*

$$\Gamma\colon \mathbf{Context}(X) \to \mathbf{Schl}(X).$$

We denote a bundle (F,p) over X simply by F. For a bundle F, we denote the set of all its cross-sections over U by $\Gamma(U,F)$. If $U \subseteq V$ are open sets, one has a restriction map $\mathrm{res}_{V,U}\colon \Gamma(V,F) \to \Gamma(U,F)$ which operates as $s \mapsto s|_U$, where $s|_U(x) = s(x)$ for all $x \in U$. It's clear that $\mathrm{res}_{U,U} = \mathrm{id}_{\Gamma(U,F)}$ for any open U, and that the transitivity $\mathrm{res}_{V,U} \circ \mathrm{res}_{W,V} = \mathrm{res}_{W,U}$ holds for all nested opens $U \subseteq V \subseteq W$. So we have constructed obviously a sheaf $(\Gamma(V,F), \mathrm{res}_{V,U})$.

Then for any given morphism of bundles $h\colon E \longrightarrow F$, we have a map $\Gamma(h)(U)\colon \Gamma(U,E) \to \Gamma(U,F)$ defined as $\Gamma(h)(U)\colon s \mapsto h \circ s$, which is obviously a morphism of sheaves. Thus, we have constructed a desired section-functor $\Gamma\colon \mathbf{Context}(X) \to \mathbf{Schl}(X)$.

The fundamental theorem of topology states that the section-functor Γ and the germ-functor Λ establish a *dual adjunction* between the category of presheaves and the category of bundles (over the same topological space); this dual adjunction restricts to a *dual equivalence* of categories (or *duality*) between corresponding full subcategories of sheaves and of étale bundles (see, e.g. [19, p. 179] or [20, p. 89]). Transferred to linguistics in our [28], it yields at the level of text the following:

Theorem (Frege Duality). *The generalized compositionality and contextuality principles are formulated in terms of categories which are in natural duality*

$$\mathbf{Schl}(X) \xrightleftharpoons[\Gamma]{\Lambda} \mathbf{Context}(X)$$

established by the section-functor Γ and the germ-functor Λ, the pair of adjoint functors.

Each fragmentary meaning $s \in \mathscr{F}(U)$ determines a function $\dot{s}\colon x \mapsto \mathrm{germ}_x s$ to be well-defined on U; for each $x \in U$, its value $\dot{s}(x)$ is taken in the stalk $\mathscr{F}_x$. This gives rise to a functional representation

$$\eta(U)\colon s \mapsto \dot{s} \qquad (11)$$

defined for all fragmentary meanings $s \in \mathscr{F}(U)$. This representation of a fragmentary meaning s as a genuine function $\dot{s}$ provides an insight into the nature of fragmentary meanings. Each fragmentary meaning $s \in \mathscr{F}(U)$, which has been described in Sect. 5 as an abstract entity, may now be thought of as a genuine function $\dot{s}$ defined on the fragment U of a given text. At the argument (sentence) $x \in U$, this function $\dot{s}$ (representing s) takes its value $\dot{s}(x)$ to be the contextual meaning $\mathrm{germ}_x s$ of this sentence x

$$x \mapsto \dot{s}(x) = \mathrm{germ}_x s \qquad (12)$$

Remark. Due to the functional representation (11), the Frege duality is of a great theoretical importance because it allows us to consider any fragmentary meaning s as a genuine function $\dot{s}\colon x_i \mapsto \mathrm{germ}_{x_i} s$ which assigns to each sentence $x_i \in U$ its contextual meaning $\mathrm{germ}_{x_i} s$, and which is continuous on U. It allows us to develop a kind of *dynamic theory of meaning* [28, 31, 34] describing how, during the reading of the text $X = (x_1, x_2, x_3, \ldots, x_n)$, the understanding proceeds through the discrete time $i = 1, 2, 3, \ldots, n$ as a sequence of grasped contextual meanings $(\dot{s}(x_1), \dot{s}(x_2), \dot{s}(x_3), \ldots, \dot{s}(x_n))$. That gives rise to a genuine function $\dot{s}$ on X representing some $s \in \mathscr{F}(X)$ which is one of possible meanings of the whole text X interpreted in the sense $\mathscr{F}$.

Moreover, this duality gives a solution to an old problem concerning delicate relations between Frege's compositionality and contextuality principles, in revealing that the acceptance of one of them implies the acceptance of the other (see, e.g. [31]).

8 Sheaf-theoretic dynamic semantics

We sketch now a formal model of a natural language text understanding which is a kind of dynamic semantics we proposed in [29, 31, 34]. Our approach describes the dynamics of interpretation process which results in the understanding of a certain meaning of the whole text in its integrity. With the notations used above, for a given text $X = (x_1, \ldots, x_n)$ interpreted in a sense $\mathscr{F}$, we have to describe how a reader finally grasps some global section $s \in \mathscr{F}(X)$ of a sheaf $\mathscr{F}$ of fragmentary meanings.

We consider first a particular case of reading from the very beginning of an admissible text $X = (x_1, x_2, x_3, \ldots, x_n)$ whose size is short enough to allow a reading at one sitting. The general case will be reduced to this particular case by means of the generalized Frege's compositionality principle.

The first sentence x_1 in the order $\leq$ of writing must obviously be understood in the context which consists of its own data. This means that a first sentence x_1 constitutes an open one-point set $\{x_1\}$. Thus $U_{x_1} = \{x_1\}$ and therefore sentence x_1 should be a minimal element in the specialization order, and therefore $\mathscr{F}_{x_1} = \mathscr{F}(\{x_1\})$.

This means that the grasping of a contextual meaning of x_1 is equivalent to the grasping of a fragmentary meaning of the fragment $\{x_1\}$ reduced to this sentence x_1. It is obviously equivalent to the grasping of a global meaning of this sentence x_1 at the semantic level of a sentence considered as a sequence of words. We understand first the theme (topic) of this sentence x_1, and then we understand the rheme (comment) as what is being said in the sense $\mathscr{F}$ concerning this theme. Thus we have done a descent from the level of text to the level of sentence. In our reasoning, it is the *basis of induction*.

Let us now do the *induction step*. Let us suppose that we have read and understood the text X in the sense $\mathscr{F}$ from the beginning x_1 up to the sentence x_k, $1 < k < n$. That is, we suppose that we have already endowed $X = (x_1, \ldots, x_k)$ with a phonocentric topology and we have built a suite $(\dot{s}_{x_1}, \ldots, \dot{s}_{x_k})$ of contextual meanings of sentences of the open set $U = (x_1, \ldots, x_k)$ of a given text $X = (x_1, \ldots, x_k, \ldots, x_n)$. The suite $(\dot{s}_{x_1}, \ldots, \dot{s}_{x_k})$ of contextual meanings is a continuous function which represents some fragmentary meaning $s \in \mathscr{F}(U)$.

We consider the interpretation process at its $(k+1)$-th step as the choice of an appropriate context $U_{x_{k+1}}$ for x_{k+1} that endows the initial segment $(x_1, \ldots x_{k+1})$ with a particular phonocentric topology among many possible, and allows us to extend the function s defined on the open $(x_1, \ldots, x_k)$ to a function defined on the open $(x_1, \ldots, x_{k+1})$.

The phrase x_{k+1} is read in the context of the fragment $(x_1, \ldots, x_{k+1})$ of the text X. This neighbourhood is the most large context among possible ones we dispose to understand the contextual meaning of x_{k+1}. To grasp the same contextual meaning of x_{k+1}, it suffices to understand only its minimal neighbourhood $U_{x_{k+1}}$. It may be two cases:

Case 1°: It may happen that the understanding of the sentence x_{k+1} is independent of the understanding of $U = (x_1, \ldots, x_k)$, for it constitutes alone its own context $\{x_{k+1}\} = U_{x_{k+1}}$ because there is here a turning point in the narrative, what may be confirmed by various morphologic markers such as the beginning of a new chapter, etc. The contextual meaning $\dot{s}_{x_{k+1}}$ is defined at a point x_{k+1}, and as such it is a continuous function because $\{x_{k+1}\}$ constitutes an open set.

The process of understanding of x_{k+1} is therefore conducted in the same way as that one of the first sentence x_1 whose case we have considered above as the basis of induction.

Note that the interval $U = (x_1, \ldots, x_k)$ is open. We can therefore extend the suite $(\dot{s}_{x_1}, \ldots, \dot{s}_{x_k})$ which is a continuous function on $U = (x_1, \ldots, x_k)$, to the suite $(\dot{s}_{x_1}, \ldots, \dot{s}_{x_{k+1}})$ which is continuous on $(x_1, \ldots, x_{k+1})$.

Case 2°: The understanding of x_{k+1} is reached with the support of the understanding of the preceding sentences of the interval $U = (x_1, \ldots, x_k)$. Not all the sentences in $U = (x_1, \ldots, x_k)$ are required to determine the understanding of x_{k+1}, but only some subsequence of U. Let V be a subsequence of U, such that V contains only sentences those are required for the understanding of x_{k+1}. We define a phonocentric topology on $(x_1, \ldots, x_{k+1})$ by defining $U_{x_{k+1}} = V \cup \{x_{k+1}\}$.

Now we transform the subsequence V into one sentence in such a way that each sentence of V, except the first in the order $\leq$ of writing, begins with "and then" which assembles it to the preceding sentence in order to get a compound sentence. This single lengthy sentence x is made up of all sentences of V in order to get the thematic context which allows the sentence x_{k+1} to express its communicative content. Finally, we join x_{k+1} to x by means of "and then" inserted at the beginning of the sentence x_{k+1}, that transforms x_{k+1} into another sentence x'_{k+1}.

In the text $(x_1, \ldots, x_k, x'_{k+1})$ so defined, the sentence x'_{k+1} constitutes an open one-point set $\{x'_{k+1}\}$ which is understandable in the context of its own data. A contextual meaning of x'_{k+1} is grasped when we understand the rheme of x_{k+1} as being what is said in the sense $\mathscr{F}$ concerning the theme of x_{k+1} in the context defined by the sentences of V. But obviously the contextual meaning of x'_{k+1} is the same as the contextual meaning of x_{k+1}. So we have extended the sequence of contextual meanings $(\dot{s}_{x_1}, \ldots, \dot{s}_{x_k})$ to the sequence $(\dot{s}_{x_1}, \ldots, \dot{s}_{x_{k+1}})$.

Thus we have done a descent from the level of text to the level of sentence. This trick is inspired by Russell's work *How I write* [38], where he discuss advises he received at the beginning of his career of a writer.

We consider now a general case of reading of an admissible text X whose size does not allow us to finish reading at one sitting. In this case, we consider the reading process of a text X as its covering by some family of meaningful fragments $(U_j)_{j \in J}$ already read, that is $X = \bigcup_{j \in J} U_j$ is an open covering.

Let us suppose given a family $(s_j)_{j \in J}$, where $s_j \in \mathscr{F}(U_j)$ such that all genuine functions $\dot{s}_j : x \mapsto \operatorname{germ}_x s_j$ of the corresponding family $(\dot{s}_j)_{j \in J}$ are pairwise compatible, that is $\dot{s}_i \big|_{U_i \cap U_j}(x) = \dot{s}_j \big|_{U_i \cap U_j}(x)$ for all $x \in U_i \cap U_j$.

Let us define the function t on $X = \bigcup_{j \in J} U_j$ as $t(x) = \dot{s}_j(x)$ if $x \in U_j$ for some j. The Frege duality theorem states that $t = \dot{s}$ where $s \in \mathscr{F}(X)$ is a composition of the family $(s_j)_{j \in J}$, whose existence is ensured by the generalized Frege's compositionality principle.

The formalization of the interpretation process as an extension of a function introduces a *dynamic* view of semantics, and its theory deserves the term *inductive* because the domain of a considered function is naturally endowed with two order structures: the linear order of writing $\leq$ and the specialization order $\preceq$ of context-dependence. We have outlined so a sheaf-theoretic framework for the dynamic semantics of a natural language, where the understanding of a text X in some sense $\mathscr{F}$ is described as a process of step-by-step grasping for each sentence x_i of only one contextual meaning $\dot{s}(x_i)$ from the fiber $\mathscr{F}_{x_i}$ lying over x_i in the étale bundle **Context**(X) of contextual meanings.

9 Algebraic semantics versus sheaf-theoretic semantics

According to T. M. V. Janssen, the compositionality principle is a basis for *Montague grammar*, *Generalized phrase structure grammar*, *Categorial grammar* and *Lexicalized tree adjoining grammar*. These theories propose the different notions of meaning, but follow the compositionality principle in its *standard interpretation*:

> A technical description of the standard interpretation is that syntax and semantics are algebras, and meaning assignment is a homomorphism from syntax to semantics. (T. M. V. Janssen [15, p. 116])

Let us consider this conception of standard interpretation as an algebraic homomorphism $f\colon A \to B$, where the algebra A is representing Syntax, and the algebra B is representing Semantics.

Whatever the algebras A and B would be, the homomorphism f is a function in a set-theoretic paradigm. Given the function f, we define the relation $\mathfrak{q}$ on A so that $\langle x, y \rangle \in \mathfrak{q}$, if and only if $f(x) = f(y)$. Clearly, this $\mathfrak{q}$ is an equivalence relation on A. Any given element $a \in A$ lies in precisely one equivalence class; if $f(a) = b \in B$, then the equivalence class of a is $f^{-1}(b)$. The set of equivalence classes is denoted by $A/\mathfrak{q}$ and called the *quotient set* of A by $\mathfrak{q}$. Let the equivalence classes of a be denoted by $a^{\mathfrak{q}}$. If with each $x \in A$ we associate $x^{\mathfrak{q}}$, we obtain a function $\varepsilon\colon A \to A/\mathfrak{q}$, called the identification associated with $\mathfrak{q}$. Clearly the function ε is surjective, by definition. Following the Theorem 3.1 of [4, p. 15], there is a decomposition of f:

$$\begin{array}{ccc} A & \xrightarrow{f} & B \\ \varepsilon \downarrow & & \uparrow \mu \\ A/\mathfrak{q} & \xrightarrow{f'} & f(A), \end{array}$$

where $\varepsilon\colon A \to A/\mathfrak{q}$ is a surjection, $f'\colon A/\mathfrak{q} \to f(A)$ is a bijection, and $\mu\colon f(A) \to B$ is an injection.

In the category of algebras, an injective homomorphism is called a monomorphism; a surjective homomorphism is called an epimorphism; every homomorphism which is bijective should be an isomorphism (defined as an invertible homomorphism). The above decomposition theorem remains valid in the category of algebras; moreover, $A/\mathfrak{q}$ and $f(A)$ may be endowed with the structures of algebras in such a way that ε, f', μ become homomorphisms.

Linguistically speaking, the Syntax and the Semantics should not be one and the same theory. Thus the meaning assignment homomorphism $f\colon A \to B$ should not be an isomorphism. Nor should this homomorphism f be a monomorphism; otherwise the Syntax A would be isomorph with a proper part of the Semantics B. Hence, f should be an epimorphism with a non-trivial kernel which is defined to be the congruence relation $\mathfrak{q}$ described above. Two different elements of an algebra A representing Syntax are congruent if and only if they are mapped to the same element of an algebra B representing Semantics. Thus, the different syntactical objects will have one and the same meaning as their value under such a homomorphism $f\colon A \to B$. Thus, an algebraic approach is pertinent in the study of synonymy, but the problems of polysemy do resist to algebraic semantic theories. Moreover, an algebraic semantic, of whatever kind, is always static because the meaning $f(x) \in B$ of a syntactic element $x \in A$ under the homomorphism f is calculated in the algebra

B just after the calculation of meanings of all syntactic components of x was done.

However, when studying the process of interpretation of a natural language text, we are confronted with a quite another situation. Any admissible text is really a great universe of meanings to be disclosed or reconstructed in the process of reading and interpretation. But these multiple meanings are offered to a reader as got identified in a single text. Thus, in the process of interpretation of a natural language text, the reader is confronted with a surjection: Semantics $\to$ Syntax. Note that we have turned the arrow round, and this is a **paradigmatic turn**. From a sheaf-theoretic point of view, a discourse interpretation activity proceeds as the following: The text X under interpretation is a given sequence of its sentences $x_1, x_2, x_3, \ldots, x_n$; this is a finite combinatorial object from the universe of Syntax. Over these sentences, there is another sequence of stalks of their contextual meanings $\mathscr{F}_{x_1}, \mathscr{F}_{x_2}, \mathscr{F}_{x_3}, \ldots, \mathscr{F}_{x_n}$; this is a potentially infinite and, in some degree, a virtual object from the universe of Semantics. The total disjoint union of all these stalks, that is, the coproduct $F = \bigsqcup_{x \in X} \mathscr{F}_x$ is projected by a local homeomorphism p on the text X. Thus, we have the surjective projection $p \colon F \to X$ from Semantics to Syntax. The challenge of text interpretation is to create a global cross-section s of the projection p; this s is constructed as a sequence of grasped step-by-step contextual sentences' meanings $(\dot{s}(x_1), \dot{s}(x_2), \dot{s}(x_3), \ldots, \dot{s}(x_n))$; it gives rise to a genuine function $\dot{s}$ on X representing some global cross-section $s \in \mathscr{F}(X)$ which is one of all possible meanings of the whole text X interpreted in the sense $\mathscr{F}$.

The proposed sheaf-theoretic semantics answers to crucial questions about *what* the fragmentary meanings are and *how* they are formally composed. That is, we consider the reading process of a fragment U in a sense $\mathscr{F}$ as its covering by some family of subfragments $(U_j)_{j \in J}$, each read in a unique session. Any family $(s_j)_{j \in J}$ of pairwise compatible fragmentary meanings $s_j \in \mathscr{F}(U_j)$ under a functional representation (11) gives rise to a family $(\dot{s}_j)_{j \in J}$ of genuine functions (where each $\dot{s}_j$ is defined on U_j by (12)), those are pairwise compatible in the sense that $\dot{s}_i \big|_{U_i \cap U_j}(x) = \dot{s}_j \big|_{U_i \cap U_j}(x)$ for all $x \in U_i \cap U_j$. Let a cross-section s be defined on $U = \bigcup_{j \in J} U_j$ as $s(x) = \dot{s}_j(x)$ if $x \in U_j$ for some j. Then this cross-section s over U is clearly a composition of the family $(\dot{s}_j)_{j \in J}$ as it is claimed by the Frege's generalized compositionality principle.

The sheaf-theoretic conception of compositionality serves as the basis for the dynamic semantics we discussed in the Sect. 8. This approach has an advantage because **1°** it extends the area of semantics from the level of sentence or phrase to the level of text or discourse, and it gives a uniform treatment of discourse interpretation at each semantic level (word, sentence, paragraph, text); **2°** it takes into theoretical consideration the polysemy of meanings of words, sentences and texts.

10 Sheaf-theoretic formal hermeneutics

Our approach provides a mathematical model of a text interpretation process while rejecting attempts to codify interpretative practice as a kind of calculus. In a series of previous papers [28, 29, 31–33], we named this text interpretation theory as *formal hermeneutics*. It presents a formal framework for syntax and semantics of texts written in some unspecified natural language, say for us English, French, German, Russian considered as a means of communication. The object of study in this formal hermeneutics are couples $(X, \mathscr{F})$ made up of an admissible text X and a sheaf $\mathscr{F}$ of its fragmentary meanings; we call any such a couple *textual space*. But this representation is possible only in the realm of a language following the famous slogan of Wittgenstein, "to understand a text is to understand a language". Rigorously, this claim may be formulated in the frame of category theory. Likewise, the present sheaf-theoretic formal semantics describes a natural language in the *category of textual spaces* **Logos**. The objects of this category are couples $(X, \mathscr{F})$, where X is a topological space naturally attached to an admissible text and $\mathscr{F}$ is a sheaf of fragmentary meanings defined on X; the morphisms are couples $(f, \theta) \colon (X, \mathscr{F}) \to (Y, \mathscr{G})$ made up of a continuous map $f \colon X \to Y$ and a f-morphism of sheaves θ which respects the concerned sheaves, that is, $\theta \colon \mathscr{G} \to f_*\mathscr{F}$, where f_* is a well-known *direct image* functor.

Given any admissible text E considered to be fixed forever as, for instance, the Scripture, it yields a full subcategory **Schl**(E) in the category **Logos** of all textual spaces. Named after Schleiermacher, the category **Schl**(E) describes the exegesis of this particular text.

The *topological syntax* and the *dynamic sheaf-theoretic semantics* based on Frege duality, as well as different categories and functors related to discourse and text interpretation process are the principal objects of study in the *sheaf-theoretic formal hermeneutics* as we understand it.

References

[1] R. Bott and L. W. Tu. *Differential Forms in Algebraic Topology*. Springer-Verlag, Berlin-Heidelberg-New York, 1982.

[2] R. Capozzi. *Reading Eco: An Anthology*. Advances in Semiotics. Indiana University Press, Bloomington and Indianapolis, 1997.

[3] R. Carnap. *Meaning and Necessity*. The University of Chicago Press, Chicago, 1947.

[4] P. M. Cohn. *Universal Algebra*. D. Reidel Publishing Company, Dordrecht, 1981.

[5] G. Deleuze and F. Guattari. *A Thousand Plateaus: Capitalism and Schizophrenia*, Vol. 2. University of Minnesota Press, Minneapolis, MN, 1987.

[6] U. Eco. *Semiotics and the Philosophy of Language.* Advances in Semiotics. Indiana University Press, Bloomington, 1986.

[7] U. Eco, R. Rorty, J. Culler, and Ch. Brook-Rose. Overinterpreting texts. In S. Collini, editor, *Interpretation and overinterpretation.* Cambridge University Press, UK, 1992.

[8] M. Erné. The ABC of Order and Topology. In H. Herrlich and H.-E. Porst, editors, *Research and Exposition in Mathematics Vol. 18. Category Theory at Work*, pages 57–83. Heldermann, Berlin, 1991.

[9] G. Frege. *Die Grundlagen der Arithmetik. Eine logisch mathematische Untersuchung über den Begriff der Zahl.* Verlag von W. Koebner, Breslau, 1884.

[10] G. Frege. Logic in mathematics. In H. Hermes, F. Kambartel, and F. Kaulbach, editors, *Posthumous Writings*, pages 203–252. University of Chicago Press, Chicago, IL, 1979.

[11] J. H. Greenberg. Some Universals of Grammar with Particular Reference to the Order of Meaningful Elements. In J. H. Greenberg, editor, *Universals of Language*, pages 73–113. MIT Press, Cambridge, MA, 1963.

[12] A. J. Greimas. *Du sens.* Seuil, Paris, 1970.

[13] A. Grothendieck and J. A. Dieudonné. *Eléments de Géométrie Algébrique,* chap. I (2e éd.). Springer-Verlag, Berlin-Heidelberg-New York, 1971.

[14] J. Hintikka. A hundred years later: The rise and fall of Frege's influence in language theory. *Synthese*, 59(1):27–49, April 1984.

[15] T. M. V. Janssen. Frege, contextuality and compositionality. *Journal of Logic, Language, and Information*, 10:115–136, 2001.

[16] S. Kahane. Grammaires de dependance formelles et theorie Sens-Texte, Tutoriel. In *Actes TALN'2001,* volume 2, Tours, France, 2001. Université François-Rabelais.

[17] E. Khalimsky, R. Kopperman, and P. Meyer. Computer graphics and connected topologies on finite ordered sets. *Topology and its Applications*, 36:1–17, 1990.

[18] A. R. Lacey. *A Dictionary of Philosophy.* Routledge, London, 3 edition, 1996.

[19] J. Lambek and P. S. Scott. *Introduction to Higher Order Categorical Logic.* Cambridge University Press, Cambridge, 1986.

[20] S. Mac Lane and I. Moerdijk. *Sheaves in Geometry and Logic. A First Introduction to Topos Theory.* Springer-Verlag, Berlin-Heidelberg-New York, 1992.

[21] Yu. I. Manin. *A Course in Mathematical Logic.* Springer-Verlag, New York Heidelberg Berlin, 1977.

[22] W. C. Mann and S. A. Thompson. Rhetorical Structure Theory: A Theory of Text Organization. URL = <http://www.sfu.ca/rst/pdfs/Mann_Thompson_1987.pdf>, University of Southern California, Information Sciences Institute (ISI), Jun. 1987. ISI/RS-87-190.

[23] J. P. May. Finite Topological Spaces. URL = <http://www.math.uchicago.edu/~may/MISC/FiniteSpaces.pdf>, 2003.

[24] R. Montague. English as a formal language. In B. Visentini et al., editors, *Linguaggi nella Società e nella Tecnica,* page 217. Edizioni di Comunità, Milan, 1970.

[25] B. H. Partee. *An Invitation to Cognitive Science, Vol. 1: Language*, chapter 11, Lexical Semantics and Compositionality, pages 311–360. The MIT Press, Cambridge, Massachusetts, 2nd edition, 1995.

[26] F. J. Pelletier. Did Frege Believe Frege's Principle? *Journal of Logic, Language, and Information*, 10:87–114, 2001.

[27] B. Pottier. *Théorie et analyse en linguistique*. Hachette, Paris, 1992.

[28] O. Prosorov. Compositionality and Contextuality as Adjoint Principles. In M. Werning, E. Machery, and G. Schurz, editors, *The compositionality of meaning and content, (Vol. II : Applications to Linguistics, Psychology and Neuroscience)*, pages 149–174. Ontos-Verlag, Frankfurt, 2005.

[29] O. Prosorov. Sheaf-theoretic formal semantics. *TRAMES Journal of the Humanities and Social Sciences*, 10(1):57–80, 2006a. ISSN 1406-0922.

[30] O. Prosorov. Semantic Topologies as Linguistic Universals. In T. Stolz, editor, *The 39th Annual Meeting of the Societas Linguistica Europaea (SLE). Relativism and Universalism in Linguistcs*, pages 109–110, Bremen, Germany, 2006b. IAAS.

[31] O. Prosorov. Topologies et faisceaux en sémantique des textes. Pour une herméneutique formelle. URL = <http://www.theses.fr/2008PA100145>, Université Paris X, Nanterre, France, Dec. 2008. PhD thesis.

[32] O. Prosorov. Formal hermeneutics as semantics of texts. In Yu. N. Solonin, editor, *Philosophy in the dialogue of cultures*, pages 351–359. St. Petersburg State University, Russia, 2010. In Russian.

[33] O. Prosorov. Topologies and Sheaves in Linguistics. In R. Bhatia, editor, *The International Congress of Mathematicians (ICM 2010). Abstracts of Short Communications and Posters*, pages 623–624, Hyderabad, India, 2010. Hindustan Book Agency.

[34] O. Prosorov. A Sheaf-Theoretic Framework for Dynamic Semantics. In A. Butler, editor, *Proceedings of the Eight International Workshop of Logic and Engineering of Natural Language Semantics (LENLS 8)*, pages 52–67, Takamatsu, Japan, 2011. JSAI.

[35] W. V. Quine. *Theories and Things*. Harvard University Press, Cambridge, 1981.

[36] F. Rastier. Communication ou transmission. *Césure*, 8:151–195, 1995.

[37] F. Rastier. Problématiques du signe et du texte. *Intellectica*, 23:11–52, 1996.

[38] B. Russell. How I Write. In A. G. Bone, editor, *The Collected Papers of Bertrand Russell (Volume 28): Man's Peril, 1954-55*, pages 102–104. Routledge, London, 1983.

[39] J.-P. Serre. *Local Fields*. Springer-Verlag, New York, 1979.

[40] D. Sleator and D. Temperley. Parsing English with a Link Grammar. Computer Science technical report CMU-CS-91-196, Carnegie Mellon University, Pittsburgh, Oct. 1991.

[41] D. Temperley, D. Sleator, and J. Lafferty. Link Grammar. URL = <http://www.link.cs.cmu.edu/link/>, 2008.

[42] M. Werning. The Reasons for Semantic Compositionality. First Düsseldorf Summer Workshop "Philosophy and Cognitive Science", 2003.

Long Sequences of Descending Theories and other Miscellanea on Slow Consistency

Michael Rathjen
*Department of Pure Mathematics, University of Leeds,
Leeds LS2 9JT, United Kingdom*
rathjen@maths.leeds.ac.uk

Abstract

For a provably recursive function $f : \mathbb{N} \to \mathbb{N}$ of **PA** one can consider the notion of f-consistency for **PA**, $\mathrm{Con}_f(\mathbf{PA}) := \forall x\, \mathrm{Con}(\mathbf{PA} \restriction_{f(x)})$, where $\mathbf{PA} \restriction_k$ denotes the fragment of **PA** with induction restricted to Σ_k formulae. It was shown in [8] that for a certain slow growing function f the strength of $\mathbf{PA} + \mathrm{Con}_f(\mathbf{PA})$ lies strictly between **PA** and $\mathbf{PA} + \mathrm{Con}(\mathbf{PA})$. Letting τ_{BH} be the Bachmann-Howard ordinal, this paper exhibits for every $\alpha < \tau_{\mathrm{BH}}$, a hierarchy of ever slower functions $(f_\xi)_{\xi < \alpha}$ of length α such that for $\zeta < \xi < \alpha$ one has $\mathbf{PA} \triangleleft \mathbf{PA} + \mathrm{Con}_{f_\xi}(\mathbf{PA}) \triangleleft \mathbf{PA} + \mathrm{Con}_{f_\zeta}(\mathbf{PA}) \triangleleft \mathbf{PA} + \mathrm{Con}(\mathbf{PA})$, where $T_1 \triangleleft T_2$ conveys that T_2 interprets T_1 but T_1 does not interpret T_2. This confirms a conjecture stated in [8, 3.2.2].

It is also observed that the axioms of Gödel-Löb logic, **GL**, hold for any provability interpretation embodying slow provability. As a result, one obtains the equivalent of Solovay's completeness theorem for **GL** for all of these slow provability notions.

Keywords: Peano Arithmetic, Bachmann-Howard Ordinal, Consistency Strength, Interpretation, Fast Growing Function, Slow Consistency, Slow Provability Interpretations.

2000 MSC: Primary: 03F25, 03F30, Secondary: 03C62, 03F05, 03F15, 03H15.

1 Introduction

The question whether there are "natural" theories lying strictly between Peano arithmetic, **PA**, and **PA** augmented by the standard consistency statement $\mathrm{Con}(\mathbf{PA})$ informed the paper [8] and led to a notion of slow consistency. It was also shown

---

Dedicated to the memory of Grisha Mints.

([8, 4.2]) that iterating slow consistency n-times gives rise to string of n theories of increasing strength lying between **PA** and **PA** + Con(**PA**). In the meantime, it has been shown by A. Freund [6] as well as P. Henk and F. Pakhomov [11] that iterating this operation through all the ordinals $< \varepsilon_0$ yields a hierarchy of theories strictly residing between **PA** and **PA** + Con(**PA**), confirming a conjecture stated in [8, 4.4]. The hierarchy also reaches Con(**PA**) at level ε_0 (see e.g. [6, 3.6]) and thus cannot be extended below **PA** + Con(**PA**). The current paper will show that there are much longer **descending** sequences of theories between **PA** + Con(**PA**) and **PA**.

For a provably recursive function $f : \mathbb{N} \to \mathbb{N}$ of **PA** one can consider the notion of f-consistency for **PA**, $\mathrm{Con}_f(\mathbf{PA}) := \forall x \, \mathrm{Con}(\mathbf{PA} \restriction_{f(x)})$, where $\mathbf{PA} \restriction_k$ denotes the fragment of **PA** with induction restricted to Σ_k formulae. It was shown in [8] that for a certain slow growing function f the strength of $\mathbf{PA} + \mathrm{Con}_f(\mathbf{PA})$ lies strictly between **PA** and **PA** + Con(**PA**). Letting τ_{BH} be the Bachmann-Howard ordinal, this paper exhibits for every $\alpha < \tau_{\mathrm{BH}}$, a hierarchy of ever slower functions $(f_\xi)_{\xi < \alpha}$ of length α such that for $\zeta < \xi < \alpha$ one has $\mathbf{PA} \triangleleft \mathbf{PA} + \mathrm{Con}_{f_\xi}(\mathbf{PA}) \triangleleft \mathbf{PA} + \mathrm{Con}_{f_\zeta}(\mathbf{PA}) \triangleleft \mathbf{PA} + \mathrm{Con}(\mathbf{PA})$, where $T_1 \triangleleft T_2$ conveys that T_2 interprets T_1 but T_1 does not interpret T_2.[1] This confirms ruminations stated in 3.2.2 in [8].

The paper is organized as follows. Section 2 introduces an ordinal representation system for the Bachmann-Howard ordinal together with an assignment of fundamental sequences. This gives rise to the hierarchies of slow and fast growing functions along this ordinal. Some properties about these hierarchies that can be proved in **PA** are considered in section 3. Section 4 presents the heart of the paper as described above. The final section 5 contains the miscellaneous observation that the axioms of Gödel-Löb logic, **GL**, hold for any provability interpretation embodying slow provability. As a consequence, one obtains the equivalent of Solovay's completeness theorem for **GL** for all of these slow provability notions.

2 An ordinal representation system for the Bachmann-Howard ordinal

There are many articles featuring ordinal representation systems for the Bachmann-Howard ordinal. Here we shall use a syntactic approach that suits our purposes.

---

[1] Let S and S' be arbitrary theories. S' is *interpretable in* S or S *interprets* S' (in symbols $S' \trianglelefteq S$) "if roughly speaking, the primitive concepts and the range of the variables of S' are defined in such a way as to turn every theorem of S' into a theorem of S" (quoted from [15, p. 96]; for details see [15, section 6]).

For theories S and S' such that $\mathbf{PA} \subseteq S, S'$ having the same language as **PA**, $S' \trianglelefteq S$ is actually equivalent to saying that every Π_1 statement provable in S is also provable in S'. This is due to Guaspari [10] and Lindström [14] (see also [15, Theorem 6]).

We first define a set of terms $\mathcal{T}$ and a linear ordering $\prec$ on $\mathcal{T}$. The desired ordinal representation system $\mathcal{T}(\Omega)$ will then arise as a proper subset of $\mathcal{T}$. $\mathcal{T}(\Omega)$ will also be equipped with an assignment of fundamental sequences, giving rise to hierarchies of fast and slow growing functions.

Definition 2.1. The set of terms $\mathcal{T}$, its *principal terms*, and the relation $\prec$ are defined inductively by the following clauses. Below $a \preccurlyeq b$ stands for $a \prec b \vee a = b$.

1. $0 \in \mathcal{T}$ and $1 \in \mathcal{T}$. 1 is a principal term.
2. If $a \in \mathcal{T}$, then $\psi a \in \mathcal{T}$ and ψa is a principal term.
3. If $a_0, \ldots, a_n \in \mathcal{T}$ are principal terms, $n \geq 1$, and $a_n \preccurlyeq \ldots \preccurlyeq a_0$, then $(a_0, \ldots, a_n) \in \mathcal{T}$.
4. If $a, b \in \mathcal{T}$, $0 \prec a$ and b is a principal term of either form 1 or ψc, then $\Omega^a b \in \mathcal{T}$ is a principal term.
5. If $a \in \mathcal{T}$ and $a \neq 0$ then $0 \prec a$.
6. $1 \prec a$ for any $a \in \mathcal{T}$ such that $a \neq 0$ and $a \neq 1$.
7. If $b \prec a$ then $\psi b \prec \psi a$.
8. $\psi a \prec \Omega^c b$ whenever $\psi a, \Omega^c b \in \mathcal{T}$.
9. If $\Omega^a b, \Omega^c d \in \mathcal{T}$, then $\Omega^a b \prec \Omega^c d$ whenever $a \prec c$ or $a = c$ and $b \prec d$.
10. If $a = (a_0, \ldots, a_n), b \in \mathcal{T}$, $a_0 \prec b$ and b is a principal term, then $a \prec b$.
11. If $a = (a_0, \ldots, a_n), b \in \mathcal{T}$ and b is a principal term $\preccurlyeq a_0$ then $b \prec a$.
12. If $a = (a_0, \ldots, a_n)$, $b = (b_0, \ldots, b_m)$, $a, b \in \mathcal{T}$, and there exists $i \leq \min(m, n)$ such that $a_i \prec b_i$ and $\forall j < i \, a_j = b_j$, then $a \prec b$.
13. If $a = (a_0, \ldots, a_n)$, $a' = (a_0, \ldots, a_n, a_{n+1}, \ldots, a_m)$ and $a, a' \in \mathcal{T}$, then $a \prec a'$.

>From now on, we will use the shorthand Ω for $\Omega^1 1$.

Corollary 2.2. $\prec$ *furnishes $\mathcal{T}$ with a linear ordering. However, it is not a well-ordering since e.g.* $\psi\Omega \succ \psi(\psi\Omega) \succ \psi(\psi\psi(\Omega)) \succ \psi(\psi(\psi(\psi\Omega))) \succ \ldots$.

Definition 2.3. The terms $0, 1, (1, 1), (1, 1, 1), \ldots$ will be identified with the natural numbers.

We define $\omega := \psi 0$. One then has $a \prec \omega$ iff a is a natural number in this sense.

A term $(a_0, \ldots, a_n)$ of $\mathcal{T}$ can be viewed as a sum of $a_0, \ldots, a_n$ (in the ordinal sense). We extend this to all terms of $\mathcal{T}$ as follows.

1. $a + 0 := 0 + a := a$.

2. Identifying a principal term b with (b), we can write any non-zero term $a \in \mathcal{T}$ as $(a_0, \ldots, a_n)$, where $a_0, \ldots, a_n$ are principal terms and $a_n \preccurlyeq \ldots a_0$, of course allowing the case $n = 0$. We then define

$$(a_0, \ldots, a_n) + (b_0, \ldots, b_m) := (a_0, \ldots, a_{k-1}, b_0, \ldots, b_m)$$

where $k := \max\{l \leq n + 1 \mid \forall i < l\, b_0 \preccurlyeq a_i\}$.

The operation $+$ is obviously associative on $\mathcal{T}$. Clearly for $(a_0, \ldots, a_n) \in \mathcal{T}$, we then have $(a_0, \ldots, a_n) = a_0 + \ldots + a_n$.

For principal terms a put $a \cdot 0 := 0$ and $a \cdot (n+1) := (a \cdot n) + a$, where n is a natural number.

We also define $\Omega^a c$ for arbitrary $a \in \mathcal{T}$ and $c \prec \Omega$. Let $\Omega^0 c := c$ and $\Omega^a 0 := 0$. For $c = (c_0, \ldots, c_n)$ put $\Omega^a c := \Omega^a c_0 + \ldots + \Omega^a c_n$.

Henceforth we shall often write Ω^a for $\Omega^a 1$.

Next we define for each $c \in \mathcal{T}$ a *distinguished fundamental sequence* $(c[x])_{x \prec \mathsf{tp}(c)}$.

Definition 2.4. 1. $\mathsf{tp}(0) := 0$, $\mathsf{tp}(1) := 1$ and $1[0] := 0$.

2. If $(a_0, \ldots, a_n) \in \mathcal{T}$ with $n > 0$, then $\mathsf{tp}(a_0, \ldots, a_n) = \mathsf{tp}(a_n)$ and $(a_0, \ldots, a_n)[x] := a_0 + \ldots + a_{n-1} + (a_n[x])$ for $x \prec \mathsf{tp}(a_n)$.

3. If $\Omega^a b \in \mathcal{T}$ and b is a principal term other than 1, then $\mathsf{tp}(\Omega^a b) := \mathsf{tp}(b)$ and $(\Omega^a b)[x] := \Omega^a b[x]$ for $x \prec \mathsf{tp}(b)$.

4. If $\Omega^a \in \mathcal{T}$, $a \neq 0$ and $\mathsf{tp}(a) \neq 1$, then $\mathsf{tp}(\Omega^a) = \mathsf{tp}(a)$ and $\Omega^a[x] := \Omega^{a[x]}$ for $x \prec \mathsf{tp}(a)$.

5. If $\Omega^a \in \mathcal{T}$ and $\mathsf{tp}(a) = 1$, then $\mathsf{tp}(\Omega^a) = \Omega$ and $\Omega^a[x] := \Omega^{a[0]} x$ for $x \prec \Omega$.

6. Recall that $\omega = \psi 0$. $\mathsf{tp}(\omega) = \omega$ and $\omega[n] := n + 1$ for $n \prec \omega$.

7. If $\psi a \in \mathcal{T}$ and $\mathsf{tp}(a) = \omega$, then $\mathsf{tp}(\psi a) = \omega$ and $(\psi a)[n] := \psi a[n]$ for $n \prec \omega$.

8. If $\psi a \in \mathcal{T}$ and $\mathsf{tp}(a) = \Omega$, then $\mathsf{tp}(\psi a) = \omega$ and $(\psi a)[0] := \psi a[0]$ and $(\psi a)[n+1] := \psi a[(\psi a)[n]]$.

9. If $\psi a \in \mathcal{T}$ and $\mathsf{tp}(a) = 1$, then $\mathsf{tp}(\psi a) := \omega$, $(\psi a)[n] := \psi a[0] \cdot (n+1)$.

Note that clause 5 above entails that $\mathsf{tp}(\Omega) = \Omega$ and $\Omega[x] = x$ for $x \prec \Omega$.

Corollary 2.5. *(i) For $a \in \mathcal{T}$, $\mathsf{tp}(a)$ is either $0, 1, \omega$ or Ω.*

(ii) For $a \in \mathcal{T}$ with $a \prec \Omega$, $\mathsf{tp}(a)$ is either $0, 1$, or ω.

Lemma 2.6. *1. $\mathsf{tp}(a) = 0$ iff $a = 0$.*

2. $\mathsf{tp}(a) = 1$ iff $a = a[0] + 1$.

3. If a is not a term of either form $0, 1, \omega, \Omega$, then $\mathsf{tp}(a) \prec a$.

4. If $x \prec \mathsf{tp}(a)$ then $a[x] \prec a$.

5. If $x \prec y \prec \mathsf{tp}(a)$ then $a[x] \prec a[y]$.

Definition 2.7. The *norm* of a term $a \in \mathcal{T}$, $\mathsf{N}(a)$ measures its syntactic complexity.

1. $\mathsf{N}(0) := 0$, $\mathsf{N}(1) := 1$.

2. $\mathsf{N}(\psi a) := \mathsf{N}(a) + 1$.

3. $\mathsf{N}(a_0, \ldots, a_n) := \mathsf{N}(a_0) + \cdots + \mathsf{N}(a_n)$.

4. $\mathsf{N}(\Omega^a b) := \mathsf{N}(a) + \mathsf{N}(b)$.

As $\mathcal{T}$ is not well-ordered we need to single out a subsystem that is. This will be achieved by collecting the subterms c of a that appear in the shape ψc in a.

Definition 2.8. 1. $\mathsf{K}\, 0 := \mathsf{K}\, 1 := \emptyset$.

2. $\mathsf{K}(a_0, \ldots, a_n) := \mathsf{K}\, a_0 \cup \ldots \cup \mathsf{K}\, a_n$.

3. $\mathsf{K}\, \psi a := \{a\} \cup \mathsf{K}\, a$.

4. $\mathsf{K}\, \Omega^a b := \mathsf{K}\, a \cup \mathsf{K}\, b$.

Definition 2.9. We give an inductive definition of the subset $\mathcal{T}(\Omega)$ of $\mathcal{T}$. We write $\mathsf{K}\, a \prec b$ to convey that for all $x \in \mathsf{K}\, a$, $x \prec b$.

1. $0, 1, \Omega \in \mathcal{T}(\Omega)$.

2. If $(a_0, \ldots, a_n) \in \mathcal{T}$ and $a_0, \ldots, a_n \in \mathcal{T}(\Omega)$, then $(a_0, \ldots, a_n) \in \mathcal{T}(\Omega)$.

3. $a \in \mathcal{T}(\Omega)$ and $\mathsf{K}\, a \prec a$ then $\psi a \in \mathcal{T}(\Omega)$.

4. If $\Omega^a b \in \mathcal{T}$ and $a, b \in \mathcal{T}(\Omega)$ then $\Omega^a b \in \mathcal{T}(\Omega)$.

Lemma 2.10. *1. If $a \in \mathcal{T}(\Omega)$, then $\mathsf{tp}(a) \in \mathcal{T}(\Omega)$.*

2. If $a, x \in \mathcal{T}(\Omega)$ and $x \prec \mathsf{tp}(a)$, then $a[x] \in \mathcal{T}(\Omega)$.

Proof: These results can be proved by induction on $\mathsf{N}(\alpha) + \mathsf{N}(x)$ within a weak fragment of **PA** (e.g. in the fragment with just Σ_1 induction). □

$\mathcal{T}(\Omega)$ is well-ordered by $\prec$, however, transfinite induction on $\prec$ cannot be proved in **PA**.

Convention. For the reminder of this article, lower case Greek letters $\alpha, \beta, \gamma, \ldots$ are always supposed to range over elements of $\mathcal{T}(\Omega)$. In the representation system $\mathcal{T}(\Omega)$, the role of the first fixed point of the ordinal function $\xi \mapsto \omega^\xi$, known as ε_0, is played by $\psi\Omega$.

The Bachmann-Howard ordinal, τ_{BH}, is the first ordinal which is larger than any ordinal in $\mathcal{T}(\Omega) \cap \Omega := \{\alpha \in \mathcal{T}(\Omega) \mid \alpha \prec \Omega\}$. We shall often write $\alpha \prec \tau_{\mathrm{BH}}$ rather than $\alpha \in \mathcal{T}(\Omega) \cap \Omega$.

Definition 2.11. For functions $f : \mathbb{N} \to \mathbb{N}$ we use exponential notation $f^0(x) = x$ and $f^{k+1}(x) = f(f^k(x))$ to denote repeated compositions of f.

We define two hierarchies of functions $G_\alpha, F_\alpha : \mathbb{N} \to \mathbb{N}$ for $\alpha \in \mathcal{T}(\Omega) \cap \Omega$.

$$G_0(n) := 0 \qquad\qquad F_0(n) := n+1$$
$$G_{\alpha+1}(n) := G_\alpha(n) + 1 \qquad\qquad F_{\alpha+1}(n) := F_\alpha^{n+1}(n)$$
$$G_\alpha(n) := G_{\alpha[n]}(n) \qquad\qquad F_\alpha(n) := F_{\alpha[n]}(n)$$

where $\mathsf{tp}(\alpha)$ is assumed to be ω in the last line.

It is well known result (due to Girard [9]) that every function F_α in the fast growing hierarchy with $\alpha \prec \varepsilon_0$ is eventually majorized by a function G_β in the slow growing hierarchy for some $\beta \prec \tau_{\mathrm{BH}}$. Moreover, the "catching-up" doesn't happen earlier than τ_{BH}. Proofs of this hierarchy comparison theorem can be found in [1, 32, 33]. For the particular assignment of fundamental sequences used in the current paper, this is done in [1].

3 Capturing the F_α's and G_α's in PA

The definition of the functions F_α employs transfinite recursion on α. It is therefore not immediately clear how we can speak about these functions in arithmetic. Later on we shall need to refer to a definition of $F_\alpha(x) = y$ which works in an arbitrary model of **PA**. In [12] many facts about the functions F_α for $\alpha \leq \varepsilon_0$, as befits their definition, are proved by transfinite induction on the ordinals $\leq \varepsilon_0$. In [12] there is no attempt to determine whether they are provable in **PA** (let alone in weaker theories). In what follows we will have to assume that some of the properties of the F_α's even for $\alpha \prec \tau_{\mathrm{BH}}$ hold in all models of **PA**. As a consequence, we will have to establish these results in **PA**. As it turns out, this can be done via a formula of low complexity.

Lemma 3.1. *There is a Δ_0-formula expressing $F_\alpha(x) = y$ (as a predicate of α, x, y).*

Proof: This is shown in [29, 5.2] for the hierarchy up to ε_0 but by basically the same proofs it can be extended up to τ_{BH}. The main idea is that the computation of $F_\alpha(x)$ can be described as a rewrite systems, that is, as a sequence of manipulations of expressions of the form

$$F_{\alpha_1}^{n_1}(F_{\alpha_2}^{n_2}(\ldots(F_{\alpha_k}^{n_k}(n))\ldots)),$$

where $n_1, \ldots, n_k \in \omega - \{0\}$ and $\alpha_1 > \ldots > \alpha_k \geq 0$. □

Definition 3.2. The computation of $G_\alpha(x)$ is closely connected with the step-down relations of [12] and [24]. For convenience we define $(\alpha+1)[n] := \alpha$ and $0[n] :=$ for all $n \prec \omega$.

For $\alpha < \beta$ we write $\beta \xrightarrow[n]{} \alpha$ if for some sequence of ordinals $\gamma_0, \ldots, \gamma_r$ we have $\gamma_0 = \beta$, $\gamma_{i+1} = \gamma_i[n]$, for $0 \leq i < r$, and $\gamma_r = \alpha$. If we also want to record the number of steps r, we shall write $\alpha \xrightarrow[n]{r} \beta$.

Note that if r is the smallest number such that $\alpha \xrightarrow[n]{r} 0$ then $r \geq G_\alpha(n)$. With a bit more effort one can also prove that

$$G_\alpha(n+1) \geq r \geq G_\alpha(n).$$

Lemma 3.3. (i) *Let $\alpha \xrightarrow[n]{} \beta$, $\alpha \xrightarrow[n]{} \gamma$, $\beta > \gamma$. Then $\beta \xrightarrow[n]{} \gamma$.*

(ii) *Let $\alpha \xrightarrow[n]{} \beta$, $\beta \xrightarrow[n]{} \gamma$. Then $\alpha \xrightarrow[n]{} \gamma$.*

Proof: This is evident from the definition. □

It is well known that for $\alpha \prec \varepsilon_0$, **PA** $\vdash \forall x \exists y\, F_\alpha(x) = y$. Owing to Gentzen, **PA** proves transfinite induction up to α. As a result, it is easy to prove in **PA** that $\forall \xi \preceq \alpha \forall x \exists y\, F_\xi(x) = y$ by induction on ξ.

The main technical tool for proving properties about the G_α's and F_α's for larger α is the following.

Theorem 3.4. (i) For all $\alpha \in \mathcal{T}(\Omega)$, if $\alpha \prec \Omega$, then $\mathbf{PA} \vdash \forall x \exists y\, \alpha \xrightarrow[x]{y} 0$.

(ii) For $\beta \prec \tau_{BH}$, $\mathbf{PA} \vdash \forall x \exists y\, G_\beta(x) = y$.

It is well-known that G_β is majorized by some F_α with $\alpha \prec \varepsilon_0$, however, most proofs (e.g. [4,9,32]) make use of transfinite induction up to τ_{BH}, a principle that is not available in **PA**. A proof of (i) can be obtained from the results in [33]. But very explicitly (i) is stated in Ulf Schmerl's paper [23, Theorem 18].
(ii) is an immediate consequence of (i). □

Lemma 3.5. Let $\alpha \prec \tau_{BH}$ be a limit. Then **PA** proves the following statements:

1. If $x < y$ then $\alpha[y] \xrightarrow[1]{} \alpha[x]$.

2. For all $\beta \prec \alpha$, if $x < y$ and $\alpha \xrightarrow[x]{} \beta$, then $\alpha \xrightarrow[y]{} \beta$.

3. For all $\beta \prec \alpha$ there exists x such that $\alpha \xrightarrow[x]{} \beta$.

Proof: These results are usually proved by transfinite induction on α (e.g. [12]). However, a careful analysis shows that it is enough to know that for all x and all $\beta \preccurlyeq \alpha$ there exists r such that $\beta \xrightarrow[x]{r} 0$, which is guaranteed by Theorem 3.4. Transfinite induction can then be replaced by ordinary induction on r. □

Lemma 3.6. We use $F_\alpha(x) \downarrow$ to denote $\exists y\, F_\alpha(x) = y$. $F_\alpha \downarrow$ stands for $\forall x\, F_\alpha(x) \downarrow$. Fix $\alpha \prec \tau_{BH}$. The following are provable in **PA**:

(i) For any β and x, if $\alpha \xrightarrow[x]{} \beta$ and $F_\alpha(x) \downarrow$, then $F_\beta(x) \downarrow$ and $F_\alpha(x) \geq F_\beta(x)$.

(ii) For any $\beta \prec \alpha$ and $x > 3$, if $\alpha \xrightarrow[x]{} \beta$ and $F_\alpha(x) \downarrow$, then $F_\beta(x+1) \downarrow$ and $F_\alpha(x) > F_\beta(x+1)$.

(iii) For any $\beta \preccurlyeq \alpha$, if $F_\beta(x) \downarrow$ and $x > y$, then $F_\beta(y) \downarrow$ and $F_\beta(x) \geq F_\beta(y)$.

(iv) If $\beta \preccurlyeq \alpha$ and $F_\alpha \downarrow$, then $F_\beta \downarrow$.

(v) If $i > 0$ and $F_\alpha^i(x) \downarrow$ then $x < F_\alpha^i(x)$.

Proof: Similar properties are stated in [29, Proposition 5.4].
(i): Use induction on r, where $\alpha \xrightarrow[x]{r} \beta$.
(ii): Similar to (i).
(iii) follows from (i) and Lemma 3.5 since $\beta[x] \xrightarrow[1]{} \beta[y]$.

(iv) is a consequence of (ii) and Lemma 3.5.

(v): See [29, Proposition 5.4(i)] for a similar result. □

There is an additional piece of information that is provided by the particular coding and Δ_0 formula denoting $F_\alpha(x) = y$ used in [29, 5.2], namely that there is a fixed polynomial P in one variable such that for all $\alpha \prec \tau_{\mathrm{BH}}$, the number of steps it takes to compute $F_\alpha(x)$ is always bounded by $P(F_\alpha(x))$.

4 The hierarchy

Definition 4.1. For each $\alpha \prec \tau_{\mathrm{BH}}$ we shall define a hierarchy of functions $(F^*_\beta)_{\beta \prec \alpha}$. First let $\alpha^* := \max(\mathsf{K}\alpha \cup \{0\}) + \Omega^\alpha$. note that $\mathsf{K}\alpha^* \cup \mathsf{K}\alpha \prec \alpha^*$, thus $\mathsf{K}(\alpha^* + \Omega \cdot \alpha) \prec \alpha^*$. Moreover, for $0 \prec \beta \prec \alpha$ one shows by induction on the buildup of β that $\max(\mathsf{K}\beta \cup \{0\}) \preccurlyeq \max(\mathsf{K}\alpha \cup \{0\})$ and hence $\mathsf{K}(\alpha^* + \Omega^\beta) \prec \alpha^*$. As a result, for all $0 \prec \beta \prec \alpha$, $\psi(\alpha^* + \Omega^\beta) \in \mathcal{T}(\Omega)$ and whenever $0 \prec \zeta \prec \xi \prec \alpha$, then

$$\psi(\alpha^* + \Omega^\zeta) \prec \psi(\alpha^* + \Omega^\xi).$$

Now put

$$F^*_\beta := F_{\psi(\alpha^* + \Omega^{1+\beta})}$$

for $\beta \prec \alpha$.

We define f_β to be the 'logarithm' of F^*_β, i.e.,

$$f_\beta(n) := \max(\{k < n \mid \exists y \leq n\, F^*_\beta(k) = y\} \cup \{0\}).$$

Note that f_β is a provably recursive function of **PA**. Also note that $\mathrm{Con}_{f_\beta}(\mathbf{PA})$ is equivalent (in **PA**) to the statement

$$\forall x\, (F^*_\beta(x)\!\downarrow \to \mathrm{Con}(\mathbf{PA}\restriction_x)).$$

Let T^α_β be the theory $\mathbf{PA} + \mathrm{Con}_{f_\beta}(\mathbf{PA})$.

A result we shall draw on is that the ordinals $\psi(\alpha^* + \Omega^{1+\beta})$ are ε-numbers, i.e. a fixed point of the enumeration function of the additive principal numbers, $\xi \mapsto \omega^\xi$.

Lemma 4.2. *Let α, α^*, β be as in Definition 4.1. For $\delta \prec \psi(\alpha^* + \Omega^{1+\beta})$ define $h(\delta) := \max(\mathsf{K}\delta \cup \{0\}) + 1$. Now $h(\delta) \prec \alpha^*$ or there exists a unique $\delta' \prec \Omega^{1+\beta}$ such that $h(\delta) = \alpha^* + \delta'$. Set $\delta_0 := 0$ in the former case and $\delta_0 := \delta'$ in the latter case. Now define $\ell(\delta) := \psi(\alpha^* + \delta_0 + \delta)$. Since $\mathsf{K}(\alpha^* + \delta_0 + \delta) \prec \alpha^* + \delta_0 + \delta$ we have $\psi(\alpha^* + \delta_0 + \delta) \in \mathcal{T}(\Omega)$ and $\delta \prec \psi(\alpha^* + \delta_0 + \delta)$. Moreover, if $\eta \prec \delta \prec \psi(\alpha^* + \Omega^{1+\beta})$, then*

$$\ell(\eta) \prec \ell(\delta) \prec \psi(\alpha^* + \Omega^{1+\beta}).$$

Since $\ell(\delta)$ is an additive principal number, this entails that $\psi(\alpha^* + \Omega^{1+\beta})$ is an ε-number.

Definition 4.3. Let E denote the "stack of two's" function, i.e. $E(0) = 0$ and $E(n+1) = 2^{E(n)}$.

Given two elements a and b of a non-standard model $\mathfrak{M}$ of **PA**, we say that 'b **is much larger than** a' if for every standard integer k we have $E^k(a) < b$.

If $\mathfrak{M}$ is a model of **PA** and $\mathfrak{I}$ is a substructure of $\mathfrak{M}$ we say that $\mathfrak{I}$ is an **initial segment** of $\mathfrak{M}$, if for all $a \in |\mathfrak{I}|$ and $x \in |\mathfrak{M}|$, $\mathfrak{M} \models x < a$ implies $x \in |\mathfrak{I}|$. We will write $\mathfrak{I} < b$ to mean $b \in |\mathfrak{M}| \setminus |\mathfrak{I}|$. Sometimes we write $a < \mathfrak{I}$ to indicate $a \in |\mathfrak{I}|$.

Theorem 4.4. *Fix* $\alpha \prec \tau_{BH}$. *For all* $0 \prec \gamma \prec \beta \prec \alpha$,

$$\mathbf{PA} \triangleleft T_\beta^\alpha \triangleleft T_\gamma^\alpha \triangleleft \mathbf{PA} + \mathrm{Con}(\mathbf{PA}).$$

Proof: First we want to show that $T_\beta^\alpha \trianglelefteq T_\gamma^\alpha$. We know that there exists a number n such that $\beta \xrightarrow{n} \gamma$. Provably in **PA** we therefore have that $F_\beta^*(x)\downarrow$ implies $F_\gamma^*(x)\downarrow$, and hence $\mathrm{Con}_{f_\gamma}(\mathbf{PA})$ yields $\mathrm{Con}_{f_\beta}(\mathbf{PA})$.

It remains to find a model of T_γ^α that is not a model of T_β^α.

We shall employ the method of injecting inconsistency from [8, Theorem 4.10].

Let $\mathfrak{M}$ be a countable non-standard model of $\mathbf{PA} + F_\beta^*$ *is total*. Let M be the domain of $\mathfrak{M}$ and $a \in M$ be non-standard. Moreover, let $e = (F_\beta^*)^{\mathfrak{M}}(a)$. As a result of the standing assumption, $\mathfrak{M} \models \mathrm{Con}(\mathbf{PA} \upharpoonright_a)$. Owing to a result of Solovay's [27, Theorem 1.1] (or similar results in [13]), there exists a countable model $\mathfrak{N}$ of **PA** such that:

(i) $\mathfrak{M}$ and $\mathfrak{N}$ agree up to e (in the sense of [8, Definition 3.9]).

(ii) $\mathfrak{N}$ thinks that $\mathbf{PA} \upharpoonright_a$ is consistent.

(iii) $\mathfrak{N}$ thinks that $\mathbf{PA} \upharpoonright_{a+1}$ is inconsistent. In fact there is a proof of $0 = 1$ from $\mathbf{PA} \upharpoonright_{a+1}$ whose Gödel number is less than 2^{2^e} (as computed in $\mathfrak{N}$).

In actuality, to be able to apply [27, Theorem 1.1] we have to ensure that e is much larger than a, i.e., $E^k(a) < e$ for every standard number k (recall that E denotes the It is a standard fact (provable in **PA**) that $E(x) \leq F_3(x)$ holds for all sufficiently large x (cf. [12, p. 269]). In particular this holds for all non-standard elements s of $\mathfrak{M}$ and hence

$$E^k(s) \leq F_3^k(s) \leq F_3^s(s) \leq F_4(s) < F_\beta^*(s),$$

so that $E^k(a) < e$ holds for all standard k, yielding that e is much larger than a.

We will now distinguish two cases.

Case 1: $\mathfrak{N} \models F_\beta^*(a+1) \uparrow$. Then also $\mathfrak{N} \models F_\beta^*(d) \uparrow$ for all $d > a$ by Lemma 3.6, (iii). Hence, in light of (ii), $\mathfrak{N} \models \mathbf{PA} + \mathrm{Con}_{F_\beta}(\mathbf{PA})$.

Now we use the fact (Lemma 3.6,(ii)) that one can show in **PA** that for sufficiently large x,

$$F_\beta^*(x) \downarrow \rightarrow F_\gamma^*(x+1) \downarrow \wedge F_\gamma^*(x+1) < F_\beta^*(x)$$

where sufficiently large means bigger than a (standard) number computable from the (representation) α. Since a is non-standard we certainly have $(F_\gamma^*)^\mathfrak{N}(a+1) \downarrow$ and

$$(F_\gamma^*)^\mathfrak{N}(a+1) \leq (F_\beta^*)^\mathfrak{N}(a).$$

But since $\neg\mathrm{Con}(\mathbf{PA}\restriction_{a+1})$ holds in $\mathfrak{N}$, $\mathfrak{N}$ is not a model of T_γ^α.

Case 2: $\mathfrak{N} \models F_\beta^*(a+1) \downarrow$. We then also have $e = (F_\beta^*)^\mathfrak{N}(a)$, for $\mathfrak{M}$ and $\mathfrak{N}$ agree up to e and the formula '$F_\beta^*(x) = y$' is Δ_0 by Lemma 3.1. Let $c := (F_\beta^*)^\mathfrak{N}(a+1)$. Now an ordinal $\psi(\alpha^* + \Omega^{1+\beta})$ is an epsilon number by Lemma 4.2. Thus by [8, Corollary 3.8], for every standard n there is an initial segment $\mathfrak{J}$ of $\mathfrak{N}$ such $e < \mathfrak{J} < c$ and $\mathfrak{J}$ is a model of Π_{n+1}-induction. Moreover, it follows from the properties of $\mathfrak{N}$ and the fact that $2^{2^e} < \mathfrak{J}$, that

1. $\mathfrak{J}$ thinks that $\mathbf{PA}\restriction_a$ is consistent.

2. $\mathfrak{J}$ thinks that $\mathbf{PA}\restriction_{a+1}$ is inconsistent.

3. $\mathfrak{J}$ thinks that $F_\beta^*(a+1)$ is not defined.

Consequently, $\mathfrak{J} \models \mathrm{Con}_{f_\beta} + \Pi_{n+1}$-induction. Moreover, by the same arguments as in case 1, $\mathfrak{J}$ does not model Con_{f_γ}. Since n was arbitrary, this shows that $\mathbf{PA} + \mathrm{Con}_{f_\beta} + \neg\mathrm{Con}_{f_\gamma}$ is a consistent theory. □

Remark 4.5. Schmerl [22] showed that $\mathbf{PA} + \mathrm{Con}(\mathbf{PA})$ can be reached from **PRA** by a consistency progression $(S_\alpha)_\alpha$ along $\varepsilon_0 \cdot 2$. It is clear from the above that "most" of the theories T_β^α do not correspond to theories in this progression.

4.1 A bit of speculation

One might ponder whether the assumption "α less than the Bachmann-Howard ordinal" could be replaced by "α less than the first non recursive ordinal" in Theorem 4.4. An (anonymous) referee of this paper believes that a more general result than 4.4

could be shown and suggests the following approach.[2] To define a decent hierarchy $(F_\alpha)_{\alpha<\tau}$ of functions, the Bachmann property is usually not needed in full for an assignment of fundamental sequences to ordinals $< \tau$ as long as one defines

$$F_\lambda(n) := \max\{F_{\lambda[y]}(n) \mid y \leq n\}.$$

Lemma 3.5 could presumably be shown for a segment of ordinals τ which exceeds the Bachmann-Howard ordinal by using slowed down fundamental sequences. For epsilon numbers λ one could define

$$\lambda[n] = \max\{\beta < \lambda : N(\beta) \leq n\}$$

where $N : \tau \to \omega$ is some suitable norm function. Such an assignment of fundamental sequences would not be canonical but might (perhaps after some additional fine tuning) be good enough for strengthening Theorem 4.4.

These are interesting ideas for which we thank the referee.

5 Slowness models Gödel-Löb provability logic GL

The language of modal logic has infinitely many propositional variables and the modal operator $\Box$. Formulas are built from propositional variables via the usual propositional connectives (e.g. $\to, \neg, \land, \lor$) and the stipulation that $\Box A$ is a formula if A is. The logic of provability, **GL**, is formulated in this language and has the following axioms and inference rules:

A0. All propositional tautologies are axioms.

A1. $\Box(A \to B) \land \Box A \to \Box B$.

A2. $\Box A \to \Box\Box A$.

A3. $\Box(\Box A \to A) \to \Box A$.

R1. If $\vdash A \to B$ and $\vdash A$, then $\vdash B$.

R2. If $\vdash A$, then $\vdash \Box A$.

A provability interpretation of modal logic in **PA** is determined by an assignment of a sentence p^* of **PA** to each propositional variable p of **GL**. The interpretation A^* of a modal formula A commutes with the propositional connectives in the usual

---

[2]For unexplained notions see [2, 21].

way (e.g., $(B \to C)^*$ is $B^* \to C^*$) and $(\Box B)^*$ is $\mathbf{Pr_{PA}}(\ulcorner B^* \urcorner)$ where $\mathbf{Pr_{PA}}(\ulcorner C \urcorner)$ arithmetizes provability of C in **PA** with $\ulcorner C \urcorner$ denoting the Gödel number of C.

The main result about these interpretations is Solovay's completeness theorem ([26]).

Theorem 5.1. $\mathbf{GL} \vdash A$ *if and only if* $\mathbf{PA} \vdash A^*$ *holds for all provability interpretations* $*$.

Below it is shown that completeness also obtains for a notion of slow provability. Let S be a provably recursive function of **PA** with a fixed Σ_1 definition $\varphi_S(x,y)$ in the language of **PA**, i.e. $\varphi_S(x,y)$ defines the graph of S and $\mathbf{PA} \vdash \forall x \exists! y\, \varphi_S(x,y)$. Below we shall write $S(x) = y$ when we actually mean $\varphi_S(x,y)$.

Moreover, the following further standing assumption will be adopted throughout:

1. **PA** proves that $S(x) \geq 1$ for all x.

2. The range of S is unbounded, i.e., for all k there exists n such that $k < S(n)$.

However, we shall not assume that **PA** can prove the latter fact. Indeed, the whole thing is only interesting when **PA** doesn't 'know' this fact, as in the case of slow growing functions f^α_β.

Definition 5.2. Define

$$\Box_s A := \exists x\, \mathbf{Pr_{PA \upharpoonright S(x)}}(\ulcorner A \urcorner)] \tag{1}$$

where $\mathbf{Pr}_T(\ulcorner A \urcorner)$ arithmetizes provability of A in a theory T and $\ulcorner A \urcorner$ denotes the Gödel number of A.

Lemma 5.3. *(i) If* $\mathbf{PA} \vdash A$ *then* $\mathbf{PA} \vdash \Box_s A$.

(ii) $\mathbf{PA} \vdash \Box_s A \to \Box_s(\Box_s A)$.

(iii) $\mathbf{PA} \vdash \Box_s(A \to B) \wedge \Box_s A \to \Box_s B$.

(iv) $\mathbf{PA} \vdash \Box_s(\Box_s A \to A) \to \Box_s A$.

Proof: (i) $\mathbf{PA} \vdash A$ implies that $\mathbf{PA} \upharpoonright k \vdash A$ for some $k > 0$. There exists n such that $S(n) \geq k$. Thus $\mathbf{Pr_{PA \upharpoonright S(n)}}(\ulcorner A \urcorner)$ is a true Σ_1 statement, and hence $\mathbf{PA} \vdash \Box_s A$.

(ii) We argue in **PA**. Suppose $\Box_s A$. Then $\exists x\, \mathbf{Pr_{PA \upharpoonright S(x)}}(\ulcorner A \urcorner))$. The latter being a Σ_1 statement, formalized Σ_1 completeness yields

$$\mathbf{Pr_{PA \upharpoonright 1}}(\ulcorner \mathbf{Pr_{PA \upharpoonright S(x)}}(\ulcorner A \urcorner) \urcorner)$$

whence $\Box_s(\Box_s A)$ since $S(n) \geq 1$ for some n.

(iii) We argue in **PA**. Suppose $\Box_s(A \to B) \wedge \Box_s A$. Spelling this out there exist x, y such that $\mathbf{Pr}_{\mathbf{PA}\upharpoonright S(x)}(\ulcorner A \to B \urcorner)$ and $\mathbf{Pr}_{\mathbf{PA}\upharpoonright S(y)}(\ulcorner A \urcorner)$. Picking $z \in \{x, y\}$ such that $S(x), S(y) \leq S(z)$ we have $\mathbf{Pr}_{\mathbf{PA}\upharpoonright S(z)}(\ulcorner A \to B \urcorner)$ and $\mathbf{Pr}_{\mathbf{PA}\upharpoonright S(z)}(\ulcorner A \urcorner)$, yielding $\mathbf{Pr}_{\mathbf{PA}\upharpoonright S(z)}(\ulcorner B \urcorner)$, and hence $\Box_s B$.

(iv) We argue in **PA**. Assume that $\Box_s(\Box_s A \to A)$ holds. Then there exists x such that $\mathbf{Pr}_{\mathbf{PA}\upharpoonright S(x)}(\ulcorner \Box_s A \to A \urcorner)$. Spelling the latter out we have

$$\mathbf{Pr}_{\mathbf{PA}\upharpoonright S(x)}(\ulcorner \exists y\, \mathbf{Pr}_{\mathbf{PA}\upharpoonright S(y)}(\ulcorner A \urcorner)) \to A \urcorner). \qquad (2)$$

With $\dot{x}$ denoting the xth numeral, (2) implies

$$\mathbf{Pr}_{\mathbf{PA}\upharpoonright S(x)}(\ulcorner \mathbf{Pr}_{\mathbf{PA}\upharpoonright S(\dot{x})}(\ulcorner A \urcorner) \to A \urcorner). \qquad (3)$$

By the formalized Löb's theorem for $\mathbf{PA}\upharpoonright_{S(x)}$ it follows from (3) that $\mathbf{Pr}_{\mathbf{PA}\upharpoonright S(x)}(\ulcorner A \urcorner)$, whence $\Box_s A$. $\square$

Theorem 5.4. *Let* $*$ *be an assignment of a sentence* p^* *to every propositional variable of* **GL**. $*$ *gives rise to an interpretation* $*_s$ *that commutes with the propositional connectives and satisfies also:*

$$(\Box A)^{*_s} = \Box_s A^{*_s}.$$

Then we have

$$\mathbf{GL} \vdash B \;\Rightarrow\; \mathbf{PA} \vdash B^{*_s}.$$

Proof: Obvious by Lemma 5.3. $\square$

The converse also holds and thus we arrive at the following.

Theorem 5.5. $\mathbf{GL} \vdash B$ *if and only if* $\mathbf{PA} \vdash B^{*_s}$ *holds for all assignments* $*$.

Proof: In view of Theorem 5.4, it remains to prove the direction from right to left. It can be handled by inspection of what happens in §4 of [26]. If one replaces the provability predicate Bew there by slow provability and the consistency notion by slow consistency then the same constructions work as slow provability shares crucial properties (described in Lemma 5.3) with its standard cousin. $\square$

The observations of this section were written down in 2012 (prompted by a question raised by Joost Joosten). In the meantime much more elaborate results have been obtained by Henk and Pakhomov in section 9 of [11]. They study the bimodal provability logic of ordinary and slow provability and show that it coincides with Lindström's logic which was introduced as the bimodal provability logic of ordinary **PA**-provability and **PA**-provability with Parikh's rule.

Acknowledgements. Part of the material is based upon research supported by the EPSRC of the UK through grant No. EP/K023128/1. This research was also supported by a Leverhulme Research Fellowship and a Marie Curie International Research Staff Exchange Scheme Fellowship within the 7th European Community Framework Programme. This publication was made possible through the support of a grant from the John Templeton Foundation.

References

[1] T. Arai. *A slow growing analogue to Buchholz' proof.* Annals of Pure and Applied Logic 54 (1991) 101–120.

[2] W. Buchholz, A. Cichon and A. Weiermann. *A Uniform Approach to Fundamental Sequences and Hierarchies.* Mathematical Logic Quarterly 40 (1994) 273–286.

[3] W. Buchholz, S. S. Wainer. *Provably computable functions and the fast growing hierarchy.* In: S. Simpson (ed.). *Logic and combinatoris.* Contemporary Mathematics 65 (AMS, Providence, 1987) 179–198.

[4] E. A. Cichon, S. S. Wainer. *The slow-growing and the Grzegorczyk hierarchies.* The Journal of Symbolic Logic 48 (1983) 399–408.

[5] C. Dimitracopoulos, J. B. Paris. *Truth definitions for Δ_0 formulae,* in: Logic and Algorithmic, L'Enseignement Mathematique 30 (Univ. Genève, Geneva, 1982) 317–329.

[6] A. Freund. *Slow reflection.* arXiv:1601.08214 (2016) 29 January 2016, 25 pages.

[7] S.-D. Friedman, M. Rathjen, A. Weiermann. *Some results on PA-provably recursive functions,* preprint 2011.

[8] S.-D. Friedman, M. Rathjen, A. Weiermann. *Slow consistency,* Annals of Pure and Applied Logic 164 (2013) 382–393.

[9] J.-Y. Girard. Π_2^1-*logic. I. Dilators.* Annals of Mathematical Logic 21 (1981) 75–219.

[10] D. Guaspari. *Partially conservative extensions of arithmetic,* Trans.Amer.Math.Soc. 254 (1979) 47–68.

[11] P. Henk, F. Pakhomov. *Slow and Ordinary Provability for Peano Arithmetic.* arXiv:1602.01822 (2016) 4 February 2016, 46 pages.

[12] J. Ketonen, R. M. Solovay. *Rapidly growing Ramsey functions,* Annals of Mathematics 113 (1981) 267–314.

[13] J. Krajíček, P. Pudlák. *On the structure of initial segments of models of arithmetic,* Archive for Mathematical Logic 28 (1989) 91–98.

[14] P. Lindström. *Some results on interpretability,* in: Proceedings of the 5th Scandinavian Logic Symposium 1979 (Aalborg University Press, Aalborg, 1979) 329–361.

[15] P. Lindström. *Aspects of incompleteness,* Lecture notes in logic 10, second edition (Association for Symbolic Logic, 2003).

[16] J. Paris, L. Harrington. *A mathematical incompleteness in Peano arithmetic.* In: J. Barwise (ed.). *Handbook of Mathematical Logic* (North Holland, Amsterdam, 1977) 1133-1142.

[17] J. B. Paris. *A hierarchy of cuts in models of arithmetic*, in: Lecture notes in Mathematics, Vol. 834 (Springer, Berlin, 1980) 312–337.

[18] M. Rathjen. *The realm of ordinal analysis.* S. B. Cooper and J. K. Truss (eds.). *Sets and Proofs.* (Cambridge University Press, 1999) 219–279.

[19] M. Rathjen. *Theories and ordinals in proof theory.* Synthese 148 (2006) 719-743.

[20] M. Rathjen. *The Art of Ordinal Analysis.* In: M. Sanz-Solé, J. Soria, J. L. Varona, J. Verdera (eds.). *Proceedings of the International Congress of Mathematicians Madrid 2006*, Volume II (European Mathematical Society, 2006) 45–69.

[21] H. E. Rose. *Subrecursion: Functions and Hierarchies* (Clarendon Press, Oxford, 1984).

[22] U. R. Schmerl. *A fine structure generated by reflection formulas over Primitive Recursive Arithmetic.* In M. Boffa, D. van Dalen, and K. McAloon (eds), Logic Colloquium' 78 (North Holland, Amsterdam, 1979) 335–350.

[23] U. R. Schmerl. *Number theory and the Bachmann-Howard ordinal.* Proceedings of the Herbrand Symposium , editor J. Stern (North-Holland, Amsterdam, 1981) 287–298.

[24] D. Schmidt. *Built-up systems of fundamental sequences and hierarchies of number-theoretic functions*, Arch. Math. Logik 18 (1976) 47–53.

[25] K. Schütte. *Proof Theory* (Springer, Berlin, 1977).

[26] R. M. Solovay. *Provability interpretations of modal logic*, Israel Journal of Mathematics 25 (1976) 287–304.

[27] R. M. Solovay. *Injecting inconsistencies into models of PA*, Annals of Pure and Applied Logic 44 (1989) 101–132.

[28] R. Sommer. *Transfinite induction and hierarchies generated by transfinite recursion within Peano arithmetic.* Ph.D., U.C. Berkeley, 1990.

[29] R. Sommer. *Transfinite induction within Peano arithmetic.* Annals of Pure and Applied Logic 76 (1995) 231–289.

[30] G. Takeuti. *Proof Theory* 2nd edition (North-Holland, Amsterdam, 1987).

[31] O. Veblen. *Continuous increasing functions of finite and transfinite ordinals*, Trans. Amer. Math. Soc. 9 (1908) 280–292.

[32] S. S. Wainer. *Slow Growing Versus Fast Growing.* The Journal of Symbolic Logic 54 (1989) 608–614.

[33] A. Weiermann. *Investigations on slow versus fast growing: How to majorize slow growing functions nontrivially by fast growing ones.* Archive for Mathematical Logic 34 (1995) 313–330.

[34] A. Weiermann. *Sometimes slow growing is fast growing.* Annals of Pure and Applied Logic 90 (1997) 91–99.

Venus Homotopically

Andrei Rodin

Institute of Philosophy RAS - St. Petersburg State University, Russia
andrei@philomatica.org

Abstract

The identity concept developed in the Homotopy Type theory (HoTT) supports an analysis of Frege's famous *Venus* example, which explains how empirical evidences justify judgements about identities. In the context of this analysis we consider the traditional distinction between the extension and the intension of concepts as it appears in HoTT, discuss an ontological significance of this distinction and, finally, provide a homotopical reconstruction of a basic kinematic scheme, which is used in the Classical Mechanics, and discuss its relevance in the Quantum Mechanics.

Keywords: Identity, Homotopy type theory, Intension, Kinematics.

1 Introduction

According to Frege

> Identity is a relation given to us in such a specific form that it is inconceivable that various kinds of it should occur [7, p. 254] [1].

In the second half of the 20th century this view was challenged by Peter Geach [11] who developed a theory of what he called the *relative identity*. Contrary to Frege, Geach holds that the identity concept allows for specifications, which depend on certain associated sortals [2].

Talk at the conference "Philosophy, Mathematics, Linguistics: Aspects of Interaction 2012" (PhML-2012), held on May 22–25, 2012 at the Euler International Mathematical Institute. I thank Danielle Macbeth for very useful comments and discussion.

[1] "Die Identitaet ist eine so bestimmt gegebene Beziehung, dass nicht abzusehen ist, wie bei ihr verschiedene Arten vorkommen können."

[2] Let a, b be parallel lines on Euclidean plane, in symbols $a//b$. Given that $//$ is an equivalence relation, Frege suggests to "take this relation as identity" (in symbols $a = b$) and thus obtain a new abstract object called *direction* [8, p. 74e]; for a more detailed reconstruction of Frege's abstraction see [25]). Geach's analysis of the same example is different: according to Geach $a = b$ reads "a and b are the same *as direction*" even if a and b are different *as lines*.

Geach's unorthodox view on identity has been never developed into an independent formal logical system and remain today rather marginal [2]. However the idea that, contrary to Frege's view, the identity concept can and should be diversified more recently reappeared in a different form in Martin-Löf's Constructive Type theory (MLTT) [15] and in the yet more recent geometrical interpretation of MLTT called Homotopy Type theory (HoTT) [17]. Unlike Geach's original proposal, which has hardly had any influence outside the philosophical logic, HoTT is a piece of new interesting mathematics and mathematical logic closely relevant to Computer Sciences.

The aim of this paper is to analyze some of Frege's ideas about identity in terms of the identity concept as it appears in MLTT and HoTT. In this way I hope to make the technical MLTT-HoTT identity concept more philosophically meaningful and apt to possible applications in science.

The rest of the paper is organized as follows. In the next Section I present Frege's *Venus* example and overview its analysis by the author. In the following three Sections I introduce a basic fragment of MLTT and HoTT and discuss the difference between extensional and intensional versions of these theories. Then I present a reconstruction of Frege's *Venus* with HoTT and discuss in this context an ontological impact of the distinction between extensions and intensions. Finally, I extend my reconstruction of *Venus* to what I call the Basic Kinematic Scheme used in the Classical Mechanics and briefly discuss its relevance in the Quantum Mechanics.

2   How identity statements are known?

Some identity statements are trivial and non-informative while some other are highly informative and in some cases very hard to prove. For example "$2 = 2$" (in words "two is two") is trivial, "2 is the only even prime number" is somewhat more informative but easy (since it follows immediately from the definitions of "even" and "prime"), while "2 is the biggest power n such that the equation $x^n + y^n = z^n$ has a solution in natural numbers" is both informative and highly non-trivial (it is a famous theorem conjectured by Pièrre Fermat in 1637 and proved by Andrew Wiles in 1994).

A non-mathematical example of the same kind is given by Frege in his classical *On Sense and Reference* [5] (English translation [6]). Frege considers three different names - *Venus*, *Morning Star* and *Evening Star* - which all refer to the same planet. Frege wonders how it is possible that while the identity statement

$$\textit{Morning Star} \text{ is } \textit{Morning Star} \qquad (1)$$

and the identity statement,

$$\textit{Morning Star} \text{ is } \textit{Venus} \qquad (2)$$

(which expresses a mere linguistic convention according to which "Venus" is an alternative name of *Morning Star*) are <u>trivial</u> the statement

$$\textit{Evening Star} \text{ is } \textit{Morning Star} \qquad (3)$$

is a non-obvious astronomical fact that needs an accurate justification, which involves both a solid theoretical background and appropriate observational data. [3].

Where does the difference between informative and non-informative identity statements come from? Frege does not provide a full answer to this question but does provide a theoretical framework for answering it. For this end he distinguishes between the *sense* and the *reference* of any given linguistic expression [4].

Whether an identity statement is informative or not depends on its sense (and hence on the sense of its constituents [5]) but not on its reference. Thus there is no mystery in the fact that statements of the form $a = a$ are always trivial (assuming that both the sense and the reference of "a" is fixed), while statements of the form $a = b$ can be either trivial (when terms a, b have the same sense) or non-trivial (when terms a, b have different senses). In expressions (1) and (2) both terms have the same meaning (even if in (2) these terms differ linguistically) but in (3) the senses of two terms are different. This is why (1) and (2) are trivial but (3) is not.

Obviously this is not a complete explanation. Frege's system of symbolic logic aka *Begriffsschrift* [3] does not do full justice to his own distinction between the sense and the reference of a linguistic expression [14]. It provides rules for operating with references of propositions (i.e., with their truth-values) but does not provide rules for operating with their senses. So Frege points to a problem but leaves it largely open. More recently a number of so-called *intensional* logical systems have been

---

[3]Instead of talking about trivial and non-trivial statements Frege uses here a Kantian distinction between synthetic and analytic judgements and talk about the "cognitive value" of the corresponding "thoughts". I shall not use Frege's original way of expressing these ideas in my presentation.

[4]Some writers who want to stress the originality of Frege's logical ideas leave Frege's German terms for *sense* and *reference* (Sinn und Bedeutung) without translation even if they write in English. I use standard English translations instead.

[5]This follows from a general principle known as the *compositionality* of meaning. Frege is sometimes credited for the alleged invention of this principle but the true history is more complicated [16].

developed, some of which have been explicitly motivated by the idea of formalizing certain aspects of Frege's *sense*. The distinction between extensions and intensions of linguistic expressions and logical terms is closely related to Frege's distinction between sense and reference [1]. It has a long history in logic and its philosophy and turns out to be instrumental in MLTT-HoTT, as we shall now see. In the next section I explain the technical meaning of this distinction in MLTT and then discuss its philosophical underpinning.

3 Extension and intension

MLTT [15] comprises two different forms of identity concept [6]. These two forms of identity should look familiar to anyone who has at least a rudimentary experience in programming. It's one thing to assign to a certain symbol or symbolic expression its semantic value (which can be a number, a character, a string of characters and many other things) and it is quite a different thing to state that certain things are equal. (Hereafter I use words "equal" and "identical" interchangeably.) Only in the latter case one forms a *proposition*, which typically has precisely one of the two *Boolean* values: True and False. Outside the context of programming a similar distinction can be made between *naming* or making some more elaborated linguistic convention, on the one hand, and making a *judgement* to the effect that certain things are equal, on the other hand. It is one thing to adopt and use the convention according to which the goddess' name *Venus* is an alias for what is also known as the *Morning Star*, and it is, of course, quite a different thing to judge and state that two apparently different celestial objects known as the *Morning Star* and the *Evening Star* are, in fact, one and the same. In the latter case it is appropriate to ask for a proof. Such a demand is obviously pointless in the former case.

The first kind of identity (one related to conventions) Martin-Löf calls *definitional* or *judgmental*; the second kind of identity he calls *propositional*. Following [17] I shall use sign "$\equiv$" for the definitional identity and the usual sign "$=$" for the propositional identity. Further, we should take *typing* into account. In MLTT both kinds of identity apply only to terms of the same type [7]. Typing is expressed in the notation as follows:

$$s, t : A \qquad (4)$$

---

[6] The original version of this theory involves *four* different kinds of identity [15, p. 59]. I simplify the original account by deliberately confusing some syntactic and semantical aspects. Then we are left with the two forms of identity described below in the main text.

[7] I leave now aside how identity is applied in MLTT to *types* on the formal level. It is sufficient for my present purpose to talk about the "same type" and "different types" in MLTT informally.

is a judgment that states that terms s, t are of type A. Formula

$$s \equiv_A t \tag{5}$$

stands for a judgement, which is tantamount to a convention (aka *definition*) according to which terms s, t of the same type A have the same meaning. Given (5) one says that s, t are *definitionally equal*. The expression

$$s =_A t \tag{6}$$

in its turn, stands for a *proposition* saying that terms s, t of type A are equal. Unlike (5) formula (6) by itself does not express a judgment but only represents a *type*. Under the intended proof-theoretic semantic of MLTT any term p of this type is thought of as a *proof* of the corresponding proposition; in the proof-theoretic jargon proofs are also called *witnesses* and sometimes *evidences*. So the following judgement

$$p : s =_A t \tag{7}$$

states that terms s, t are (propositionally) equal as this is evidenced by proof p.

Let us now see what kind of thing such a proof p can possibly be. In MLTT definitionally equal terms are interchangeable *salva veritate* as usual. Under the intended semantic of this theory this means that definitionally equal terms are interchangeable as proofs. This property of $\equiv$ and the reflexivity of $=$ justify the following rule

$$\frac{s \equiv_A t}{p : s =_A t} \tag{8}$$

where $p \equiv refl_s$ is built canonically [17, p. 46]. In words: the definitional identity (equality) implies the propositional identity (equality).

The converse rule is called the *equality reflection rule* or ER for short:

$$\frac{p : s =_A t}{s \equiv_A t} \tag{ER}$$

In words: the propositional identity implies the definitional identity.

ER does not follow from other principles of MLTT but may be assumed as an independent principle. In this case one obtains a version of MLTT, which is called (definitionally) *extensional*. MLTT without ER is called *intensional*. It can be shown that in the extensional MLTT any (propositional) identity type $s =_A t$ is either empty or has a single term, namely $refl_s$, which is the canonical proof of this identity "by definition".

We see that ER makes the distinction between the definitional and the propositional identity purely formal and epistemologically insignificant. This feature of extensional MLTT can be viewed as a desirable conceptual simplification but it comes with a price. A significant part of this price concerns computational properties in MLTT and is important for applications of this theory in programming: while the intensional MLTT is decidable but the extensional MLTT is not. I shall not discuss this technical feature in this paper. Instead I shall argue that the intensional MLTT has also important epistemic advantages over its extensional cousin.

4 Fixing identities or leaving them evolving?

As we have seen in the extensional MLTT every identity is grounded in a definition. In order to apply this formal theory in reasoning one needs to fix in advance, via appropriate definitions, exact identity conditions for all objects involved in a given reasoning. This logical and epistemic requirement is known in the form of slogan "no entity without identity" due to Quine. It is interesting to notice that Quine himself does not accept this slogan without reservations. In Quine's view the slogan applies only in scientific reasoning and, moreover, only in the contemporary form of scientific reasoning. Bulk terms (aka mass terms) like "water", according to Quine, are remnants of an archaic logical scheme, which does not involve the individuation in its today's form. Quine further speculates that the contemporary "individuative, object-oriented conceptual scheme" can be replaced in a future by a different scheme, that will provide a "yet unimagined pattern beyond individuation" [18, p. 24] [8]. In what follows I argue that the *intensional* MLTT along with HoTT provides such a pattern "beyond individuation" or at least a pattern of individuation beyond its usual extensional mode. But beforehand I would like to stress once again that the standard extensional mode of individuation is not sufficient for certain well-recognized and important scientific purposes. Frege's *Venus* example, if one takes it seriously, demonstrates this clearly. Fixing the identity of *Morning Star* and *Evening Star* and *Venus* via a definition is a prerequisite for applying a standard extensional

---

[8] Here is the full quote:
"[W]e may have in the bulk term a relic, half vestigial and half adapted, of a pre-individuative phase in the evolution of our conceptual scheme. And some day, correspondingly, something of our present individuative talk may in turn end up, half vestigial and half adapted, within a new and as yet unimagined pattern beyond individuation. Transition to some such radically new pattern could occur either through a conscious philosophical enterprise or by slow and unreasoned development along lines of least resistance. A combination of both factors is likeliest [...] ." [18, p. 24]

logical scheme in any reasoning about this celestial object. This condition makes it impossible to support with such a scheme a reasoning, in which the identity of the *Morning Star* and the *Evening Star* is <u>established</u> on the basis of certain sufficient evidences.

Frege's example shows that "half-entities inaccessible to identity" [18, p. 23] may look more familiar than Quine's colorful language suggests. In the *Venus* case we deal with a relatively innocent violation of "no entity without identity" requirement. We start with certain well-defined objects such as the *Morning Star* and the *Evening Star* but do not exclude the possibility that these objects can be eventually proved to be the same - even if we know that this fact does not follow from the corresponding definitions. Following Quine one may think of further deviations from the standard extensional individuating scheme and speculate about a possible conceptual scheme, which does not use the definitional form of identity at all. I do not pursue this further project in this paper. Instead I show how the innocent-looking modification of the extensional individuating scheme, which has been just explained, results into a remarkable diversification of the standard identity concept.

5 Higher identity types

Recall that the intensional version of MLTT has been introduced above via a negative characteristic: it is the core version of MLTT *without* the additional reflexion rule ER.

The absence of ER allows for constructing further identity types as follows. Suppose we have a propositional identity type and a pair of terms of this type:

$$s', t' : s =_A t$$

Terms s', t' witness here the identity of terms s, t. It may now happen that these two witnesses are, in fact, one and the same - as witnessed by two further terms s'', t'':

$$s'', t'' : s' =_{s =_A t} t'$$

Thus we get a tower-like construction, which comprises identity types of two different "levels". It can be further continued indefinitely. In the general case such a construction may have, of course, more than just two elements on each level.

Until the late 1990-ies structural properties of this formal syntactic construction remained opaque. Since the *intentionality* in MLTT is a mere lack of *extensionality*, any model of the extensional MLTT also qualifies as a model of the intensional version of this theory. In 1994-1998 Hofmann and Streicher [12,13] published the first non-extensional model of MLTT where the first-level identity types were modeled by abstract groupoids. This model allows the first-level identity types (i.e., types of the form $s =_A t$ where A is a type other than identity) to have multiple non-trivial terms (proofs) but does not allow the same for higher identity types. In other words, this model verifies the condition called "extensionality one dimension up". A deeper insight into the structure of higher identity types has been obtained around 2006 when Awodey and Voevodsky independently observed that the abstract groupoids of Hofman and Streicher's model can be thought of as *fundamental groupoids* (i.e., groupoids of all continuous *paths*) of topological spaces and be further extended to homotopy- and higher-homotopy groupoids of the same spaces, which model higher-order identity types of MLTT. Thus the Homotopy theory allows for building models of MLTT, which are "intensional all the way up". In such models the identity types of all levels are modeled uniformly. This discovery marked the emergence of a new theory known today under the name of Homotopy Type theory and of a closely related foundational project called the Univalent Foundations of mathematics. For a systematic exposition of HoTT I refer the reader to [17] [9].

Unlike Russell's type theories HoTT does not form its hierarchy of types by considering, first, classes of individuals, second, classes of such classes, and so on. The hierarchy of types in HoTT is of a geometric or, more precisely, homotopic nature. Sets are taken to be types of zero level. Terms of 0-types are points having no non-trivial paths between them. Terms of 1-types are points provided with non-trivial paths between them, but not allowing for non-trivial homotopies between these paths. Terms of 2-types allow for paths and non-trivial homotopies but not for non-trivial higher homotopies. And so on [10].

Notice the cumulative character of the homotopical hierarchy of types described above. Considered in isolation, the identity types $s =_A t$ and $s' =_{s=_A t} t'$ have exactly the same formal properties; correspondingly, there is no intrinsic difference between spaces of points, spaces of paths (aka path spaces) and homotopy spaces of all levels. As usual in the 20-th century geometry one is allowed in HoTT to imagine elements of spaces however one may find it useful - say, as beer mugs after Hilbert's legendary suggestion. However the fact that every path s' is not simply an individual of certain

---

[9] Since this area of research is rapidly developing, the 2013 book [17] does not include certain new results and developments. However it provides an systematic introduction, which is more than sufficient for my present purpose.

[10] Here I follow [17, p. 99–100]. On an alternative count the 0-type is a single point, 1-types are propositional types while sets are 2-types. The count adopted in [17] appears more natural from a logical point view (given the usual understanding of logic) while the latter count used by Voevodsky in his lectures appears more natural from a geometric point of view.

sort but an object with a pair of endpoints s, t, allows for the two-level construction described above. Similarly one obtains n-level constructions by using homotopies and higher homotopies. In order to describe the resulting hierarchy more formally and more precisely we need to complement the bottom-up description used so far but a top-down one. For this end we assume from the outset that every type is a space provided with its infinite-dimensional fundamental groupoid. Then we specify the case of 0-types such that all its paths, homotopies and higher homotopies are trivial; then the case of 1-types such that all its homotopies and higher homotopies (but not paths!) are trivial, and so on.

A given n-type can be transformed into its underlying m-type with $m < n$ by forgetting (or, more precisely, by trivializing) its higher-order structure of all levels $> m$. Such an operation is called in HoTT *truncation*. It will play an important role in what follows.

The logical significance and the possible epistemic function of higher identity types in MLTT are not yet well understood. The present work is an attempt of filling a part of this gap. In what follows I consider only 0- and 1-types and leave a study of higher identity types for a future work.

6 Is Frege's *Venus* example linguistic?

Apparently Frege treats his *Venus* example as purely linguistic on equal footing with his other examples, which involve Alexander the Great, Columbus, Napoleon, Kepler dying in misery, Bucephalus and what not. Accordingly, the main result of his classical paper [5,6], namely the distinction between the sense and the reference of a given linguistic expression, belongs primarily to the philosophy of language. Frege scholarship mostly follows Frege in this respect: a linguistic leaning aka *linguistic turn* became a brand mark of the influential *Analytic* branch of the 20th century and today's philosophy. It is quite remarkable, however, that when Frege first introduces and explains the *Venus* problem he does this not only in linguistic terms:

> The discovery that the rising Sun is not new every morning, but always the same, was one of the most fertile astronomical discoveries. Even today the identification of a small planet [i.e., an asteroid - *A.R.*] or a comet is not always a matter of course. [6, p. 56]

The idea that a logical analysis of ordinary language can be helpful for solving problems of object identification in science in general and in astronomy in particular is based on Frege's strong assumption according to which the identity concept is the same in all these cases, so that "it is inconceivable that various kinds of it should occur" (see the full quote and the reference in the above Introduction). Without trying to challenge this approach on the methodological level I shall provide here an alternative analysis of the same example, which takes its physical content and, even more importantly, its related mathematical form more seriously and applies some basic elements of HoTT introduced in the previous Section. As a matter of course this reconstruction is not intended to be a piece of mathematical physics. Nevertheless it provides a novel formal approach to traditional metaphysical issues concerning the identity through time and motion, which may be possibly helpful for dealing with identity-related problems of modern physics [9, 10].

Frege's remark about the rising Sun quoted above applies both to the *Morning Star* (MS for short) and to the *Evening Star* (ES). These two putative objects are posited as invariants of certain sets of observations made in different places at different times by different people with different astronomical instruments and with the naked eye. However for the sake of the example I leave now this complex underlying structure aside and boldly assume that MS and ES are provided with some appropriate *definitions*, which allow all observers to identify these objects unambiguously. How a *proof* of identity $MS = ES$ may look like in a realistic astronomical context? Classical Celestial Mechanics (CM), or more precisely a very basic fragment of CM that I shall call Basic Kinematic Scheme (BKS) and discuss in more detail in Section 8, provides a definite answer to this question. In order to prove that $MS = ES$ it is necessary and sufficient to present a *continuous path* aka trajectory p, which connects MS and ES and thereby shows that these "two" objects are in fact one and the same. The wanted trajectory p is itself a typical physical object: it is obviously theoretically-laden, it has a canonical mathematical representation, and it is accessible for observations which allow for empirical checks of its theoretically predicted properties. Providing such a proof p amounts to a combination of theoretical work and observation, which is typical in astronomy and any other mature science [11].

Since proof p has empirical contents it can <u>not</u> be called formal. However it has a mathematical *form*, which is expressed within HoTT straightforwardly. As we shall briefly see, this form qualifies both as logical and geometrical. The fact that in HoTT

---

[11] The identity conditions of p depend on those of MS, ES, which are left here without a precise specification. If we assume that MS and ES are enduring spatial objects repeatedly appearing on the sky then we should think of p as a fragment of the planet's orbit. Alternatively (and less realistically), if we think of MS and ES as particular spatio-temporal events which occur in a particular morning and a particular evening, then we should think of p as a continuous process that begins with MS and ends with ES. The HoTT-based reconstruction of Frege's *Venus* example given in this Section does not depend on one's specific assumptions about space, time and motion. The idea of identification of spatial objects or spatio-temporal events via continuous paths, which makes part of BKS, is compatible with many different physical theories and many different ontologies.

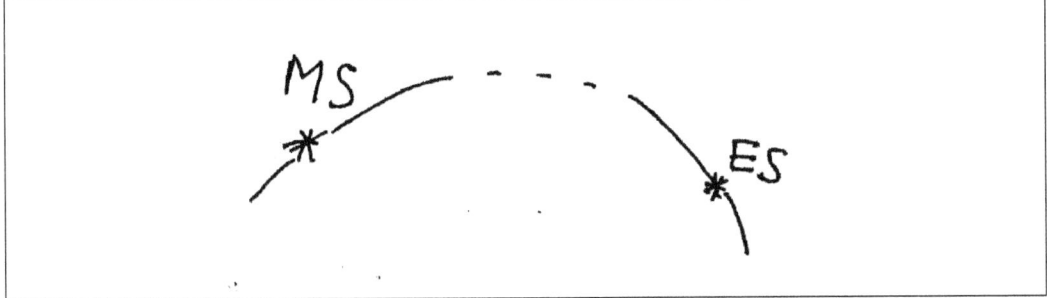

Figure 1: Morning Star and Evening Star are the same

logical and geometrical forms go together, makes HoTT quite unlike other popular formal systems such as the Classical First-Order Logic (FOL), see my [19, ch. 7, 10], for a further discussion on this general issue. Remarkably, the geometrical form of p provided by HoTT (namely, a path) and the standard geometrical representation of the same object provided by CM and BKS (namely, a continuous curve) turn out to be alike [12].

First, we need to specify a type (which under the homotopical interpretation is thought of as a space) where MS and ES belong. Since MS, ES and other celestial bodies are conceived in CM as point-like objects I call the corresponding type/space Pt and think of it as a collection of points:

$$MS, ES : Pt$$

Then we form a new type/space $MS =_{Pt} ES$, which is a space of continuous paths between MS and ES. Finally, we specify a particular path p in this space and form a judgment:

$$p : MS =_{Pt} ES$$

---

[12]In the standard Homotopy theory a *path* is not simply a curve but a parameterized curve. More formally path p with endpoints A, B is a continuous map $[0, 1] \to S$ from the unit interval to space S where points A, B belong, such that $p(0) = A$ and $p(1) = B$. "Paths" about which usually talk HoTT-theorists (as in [17]), cannot be straightforwardly identified with paths of the standard Homotopy theory [22]. But for our purposes the concept of path in the sense of HoTT will suffice: it combines the formalism of HoTT with a mixture of pre-theoretical spatio-temporal intuitions about paths and more elaborated geometrical intuitions (rather than precise concepts) borrowed from the standard Homotopy theory and some other branches of mathematics. By interpreting Frege's *Venus* in terms of HoTT I extend this intuitive part of HoTT with certain additional pre-theoretical intuitions concerning space, time and motion. Conversely, HoTT serves me as a formal tool allowing for putting these pre-theoretical intuitions into an order.

that says that MS and ES are the same as evidenced by p. However little of HoTT's resources we use here, this reconstruction of Frege's example provides some useful lessons as we shall now see [13].

7 Are intensions real?

Recall Frege's question: What is the difference between the sense of proposition (1) ($MS = MS$) and the sense of proposition (3) ($MS = ES$)? It appears to be in accord with Frege to assume that senses of propositions depend functionally on their corresponding proofs (even if proofs and senses are not exactly the same). Then our reconstruction of *Venus* allows for a precise mathematical answer to Frege's question: while the (unique) proof of (1) is trivial loop $refl_{MS}$, the proof of (3) is a non-trivial path p. In both cases a given proposition has a single proof. However these two proofs essentially differ not only in their intuitive "sense" but also in their geometric representation.

Let us now turn to some ontological issues. Albeit the concept of proof is epistemic *par excellence*, the HoTT-based reconstruction of *Venus* makes it clear that proofs in the standard proof-theoretic semantic of MLTT should not be necessary thought of as purely mental constructions. Thinking about such proofs as *truthmakers* opens a way to various forms of *truthmaker realism* [24]. Whether or not one takes *Venus* and/or its trajectory p to be real entities depends, of course, on a particular ontology that one may associate with CM or another theory supporting the relevant astronomical observations. In particular, CM allows for a 4-dimensional ontology where atomic entities are points of Classical aka *Neo-Newtonian* space-

---

[13] The proposed HoTT-based reconstruction of Frege's *Venus* example may not capture some aspects of Frege's volatile notion of sense. This notion may comprise more than HoTT in its existing form is able to detect. For example, arithmetical propositions

$$2 + 2 = 4 \qquad (9)$$

and

$$4 = 4 \qquad (10)$$

arguably have *different* senses. However the standard Peano-style formalization of arithmetic used in HoTT treats both equalities (9) and (10) as definitional and thus doesn't allow for non-trivial proofs of (9), see [17, p. 36 ff]. At the same time, given Frege's specific view on arithmetic as a part of logic developed in his [4], it is not obvious to me that the view that (9) and (10) have one and the *same* sense is indeed untenable in a Fregean conceptual framework. Under this view (9) is a logical truth but $MS = ES$ is a fact of the matter, so the apparent analogy between the two cases should be judged as merely linguistic and superficial. This controversial issue has no bearing on my following argument. I thank an anonymous referee for pointing to this arithmetical example.

time [23, p. 202 ff]. In this ontological framework p, seen as a world-line, qualifies as a full-fledged entity while the moving object *Venus* is its momentary slice. I shall not discuss here details of this and rival ontologies but rely on the fact that p of our example allows for natural realistic interpretations.

According to Frege, senses should not be thought of as psychological entities belonging to individual minds [6, p. 38–39]. However he suggests that senses wholly belong to human collective memories stored in existing natural languages. The only way in which a given sense can be possibly related to the non-human parts of our world, according to Frege's account as I understand it, is via the reference (if any) of the corresponding linguistic expression. For example English word "apple" has a sense, which belongs to this language (and arguably is shared by other natural languages) and a reference, which is a real thing that may exist independently of any linguistic and other human activities. English word "unicorn" equally has a sense but has no reference; so this particular sense is detached from any non-human reality.

The above is a rough interpretation of Frege's view but it points to a common idea about linguistic meaning, which is worth being considered here. Since Frege's concept of sense and the logical concept of *intension* are closely related (see the end of Section 2 above), the standard examples of so-called *intensional contexts* apparently provide a further linguistic support to this idea. Such examples always have to do with intentions, beliefs, knowledge and other human-related issues. So these examples square well with Frege's view according to which propositions (1) and (3) have "different cognitive values" because their senses are different - in spite of the fact that their reference (truth-value) is the same.

Our analysis of *Venus* suggests a revision of this view. Since proofs are constituents of senses (of propositions), and since these proofs admit realistic interpretations, such realistic interpretations may extend to senses. What I have in mind is not a justification of some form of Meinongian existence of unicorns but rather the view that the distinction between the sense and the reference of a given linguistic expression must be freed from all ontological commitments altogether. The idea that the reference is the only linguistic anchor that links human languages and the human cognition to non-human realities is hardly justified. Sense and reference and their logical counterparts such as intensions and extensions of concepts all make part of (various versions of) our conceptual apparatus. How this apparatus connects us, humans, to non-human realities is a question, which cannot be answered only by means of logical and conceptual analysis.

I submit that behind the view on meaning, which I purport now to criticize, is the following strong ontological assumption:

All real entities are individuals. (OE)

For further references I shall call this assumption the *ontic extensionality* or OE for short. The reason why I call this assumption *extensionality* becomes clear from a homotopical reconstruction of Frege's distinction between sense and reference, which generalizes upon the above reconstruction of *Venus* as follows. *References* are point-like individuals belonging to classes of alike individuals, which constitute *extensions* of their corresponding concepts. *Senses* are higher-order homotopical structures, which involve spaces of paths and their homotopies (including higher-order homotopies), and constitute *intensions* of the same concepts. As we have already seen, in the *extensional* version of HoTT the higher-order part of the structure is truncated. Hence the name for OE, which allows the truncated higher-order part of the structure to have an epistemic and cognitive value but includes in the ontology only its basic 0-level part.

From this point view it appears reasonable to claim that talks of apples, of unicorns, of Bucephalus and of Alexander the Great have the same logical form, so the words "apple" and "unicorn" both have a sense and a reference. By the reference of "unicorn" I understand here a fictional individual. Propositions about apples and unicorns may well allow for the same forms of truth-evaluation. The difference between merely fictional, legendary and real entities concerns material (contentful) rather than formal features of truth-evaluation. There is no way to distinguish between a fiction, a legend, and a historical fact on purely formal grounds [14].

I can see no a priori reason for assuming that a part of the homotopic structure is more apt to represent reality than any other. For that reason I don't take OE for granted. Moreover that our reconstruction of *Venus* suggests that terms of 1-types (paths) allow for a realistic interpretation as well as terms of 0-types (points). However in the next Section we shall see that the situation is not so simple, and that BKS is compatible with OE after all.

Concluding this Section I would like to remark that OE goes along the view according to which the Classical first-order logic (FOL) should be seen and used as the basic logical tool for scientific reasoning. In this context the suggestion to drop OE and allow for higher-order entities sounds a part of an argument in favor of a higher-order system of logic with a standard class-based semantics. MLTT and

---

[14] The Bucephalus example demonstrates this particularly clearly. Bucephalus is a legendary horse belonging to Alexander the Great. According to the legend Bucephalus was born the same day as Alexander and, according to a particular version of the same legend, he also died the same day as Alexander. I don't know about a verdict of today's historical science as to how much of this story (if any) is a historical fact and how much of it is a fiction. I don't believe that any advance in formal logic may help for answering this question.

HoTT indeed qualify as higher-order systems in a relevant sense but the homotopical semantic used in HoTT is not standard. In HoTT higher types are formed not by the reiteration of the powerset construction (i.e. not by considering classes of classes of ... of individuals) but in the geometric way, which has been briefly explained in Section 5 above. Our homotopical reconstruction of *Venus* given in Section 6 demonstrates how the geometric semantic of HoTT helps one to use this theory as a tool for mathematical modeling *in* science, not only as a tool for a logical analysis *of* science. I believe that this dummy example points to interesting theoretical possibilities in mathematical physics. For serious attempts to use HoTT and its logical structure in physics see [20, 21].

8 Basic kinematic scheme

Here I supplement the homotopical reconstruction of *Venus* from Section 6 with a similar reconstruction of the Basic Kinematic Scheme (BKS), which captures the usual idea of moving particle. The kinematic space K, in which MS and ES live, allows for multiple paths (trajectories) sharing their ending points. I think about K not as a vehicle of moving particles but rather as a collection Pt of such particles provided with appropriate criteria of identity and an additional structure, which represents their relative motions. The motions are represented by paths between the particles as in the *Venus* example. The additional structure is that of groupoid of paths over Pt. I do not include into K homotopies of paths beyond the trivial ones because such things play no role in BKS. Paths in K are assumed to be reversible and composable by concatenation; the composition is associative [15]. In terms of HoTT K qualifies as a 1-type; Pt is the underlying 0-type of K obtained from K via the (0-)truncation.

Let me now briefly reproduce the above homotopical reconstruction of *Venus* in this slightly extended context. We take two points MS, ES in Pt (and hence in K) and consider the path space $MS =_{Pt} ES$. Then we find in $MS =_{Pt} ES$ a particular path p, which serves us as a proof of identity $MS = ES$. The extended context allows us now to notice an interesting feature of BKS, which so far remained out of the scope of our analysis. Consider the following additional principle, which I'll call the *uniqueness of actual path*:

---

[15] In the usual Homotopy theory the composition of paths in a given space S is defined only up to homotopy; in order to define such an operation one is obliged to provide an appropriate homotopy aka reparameterization by hand. Since in HoTT homotopy types are primitive objects this issue is treated a bit differently. We stipulate an abstract groupoid K without assuming any ambient space S in advance, and then see how much of BKS can be recovered in this way. This approach allows us to describe the composition of paths in K as concatenation without mentioning homotopies.

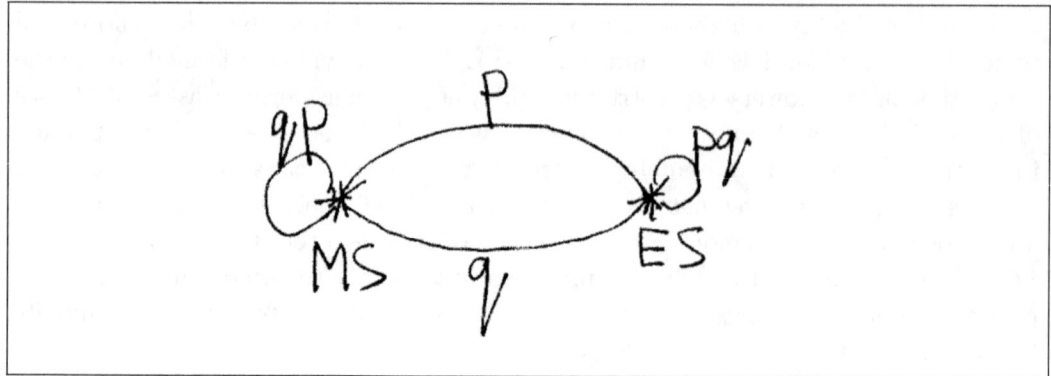

Figure 2: Multiple Paths of Venus

There is at most one path between any two given points. (UAP)

Prima facie *Venus* does not verify this principle. Indeed, *Venus*'s orbit, which is a topological circle, admits two different paths p, q between MS and ES and further, via composition, two non-trivial loops qp and pq for MS and ES correspondingly:

The above picture represents MS and ES as apparently different but in fact the same body, which moves along its circular orbit. But neither this picture nor K construed as above reflects the usual idea that one and the same particle cannot follow two different paths *simultaneously*. This is not particularly surprising since *time* did not feature in our construction of K so far. I am not going now to fill this gap by providing K with an explicit representation of time. Instead, let us consider a model of UAP in the given framework. UAP can be satisfied if we think of MS as "*Venus* at time t_1" and of ES as "*Venus* at (later) time t_2". Then during the time period $\Delta = [t_1, t_2]$ *Venus* follows a unique path p, which can be described as a segment of '*Venus*'s *worldline* in an appropriate spacetime [16]. This shows that we may use UAP for accounting for a time-related feature of BSK without introducing time explicitly. It is quite remarkable because UAP involves only very basic concepts of HoTT and has a purely formal character; it can be itself easily expressed in HoTT.

If we now add a natural assumption that the propositional identity is an equivalence (which excludes "split" or "branching" identities) then UAP reduces possible forms of K to a trivial spaghetti-like form. In this case each particular connected component or "noodle" of K can be called a *worldline* of its corresponding particle (point). Since every noodle is contractible into a point, in this case K and Pt are

---

[16] Since we are talking about the Classical Mechanics but not about the Relativistic Mechanics, the relevant notion of spacetime is that of the *Neo-Newtonian* spacetime, see [23, p. 202 ff].

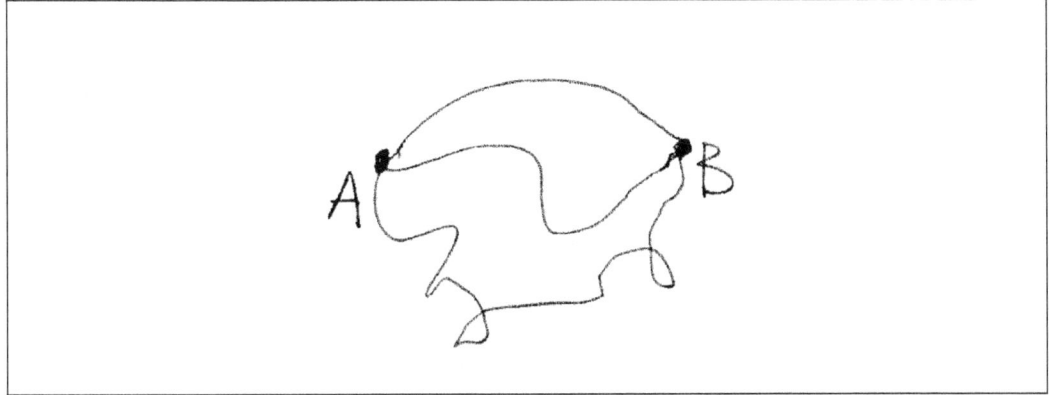

Figure 3: Quantum Paths

homotopically equivalent. They represent the same 0-level homotopy type $K \simeq Pt$ making redundant the very distinction between them. However the distinction between K and Pt becomes useful again when one distinguishes between *actual* and *possible* paths. Indeed, it is plausible to assume that given actual path p with endpoints MS, ES BKS allows for other possible paths with the same endpoints. In other words, BKS allow bodies to follow trajectories, which differ from their actual trajectories. Now we can think of K as groupoid of possible paths where UAP does not hold and distinguish its subgroupoid $A \subset K$ which comprises only actual paths and for which UAP holds. In this case 0-truncation $K \to Pt \simeq A$ becomes non-trivial and represents a *realization* of certain possible paths.

The above analysis of BKS appears to be an appropriate starting point for building a Quantum counterpart of this conceptual scheme. From the homotopical point of view there is nothing impossible or unnatural in the idea that a given particle may follow multiple trajectories simultaneously as this is assumed in the Feynman path integral formulation of Quantum Mechanics:

In the present conceptual framework one may rather inquire into the nature of UAP. What is behind the traditional notion according to which the *actual* trajectory of a given particle during its lifetime is necessary unique?

In order to provide a tentative answer let us return to the issue discussed in the last Section. The above analysis of BKS apparently provides an additional evidence in favor of ontic extensionality (OE). The intensional groupoid structure of K represents *possible* trajectories of particles. But since in the real world each particle has its unique worldline the groupoid K is reduced (truncated) to the extensional set $A \simeq Pt$. Conversely, OE in the given context implies UAP. However OE is compatible with BKS only if one understands the modal property of being *possible* (for paths) in purely epistemic terms - say, as a lack of knowledge about the actual trajectories. Alternatively, one may think about possible paths in K as physically real. This latter view violates OE but it is not wholly unreasonable. Quantum Mechanics where UAP does not apply, provides additional reasons for taking it seriously. I stop here and leave an attempt to develop a HoTT-based theory of identity for Quantum Mechanics for a different occasion.

References

[1] D. J. Chalmers. On Sense and Intension. *(J. Tomberlin, ed.) Philosophical Perspectives 16: Language and Mind: Blackwell*, pages 135–182, 2002.

[2] H. Deutsch. Relative Identity (entry in Stanford Encyclopedia of Philosophy). Retrieved September 29, 2016, from https://plato.stanford.edu/entries/identity-relative/, 2002.

[3] G. Frege. *Begriffsschrift: eine der arithmetischen nachgebildete Formelsprache des reinen Denkens.* Halle, 1879.

[4] G. Frege. *Die Grundlagen der Arithmetik.* Verlag von W. Koebner, Breslau, 1884.

[5] G. Frege. Über Sinn und Bedeutung. *Zeitschrift für Philosophie und philosophische Kritik*, 100:25–50, 1892.

[6] G. Frege. *On Sense and Reference in: Translations from the Philosophical Writings of Gottlob Frege, ed. by Geach and M. Black*, pages 56–78. Oxford: Basil Blackwell, 1952.

[7] G. Frege. *Grundgesetze der Arithmetik.* Hildesheim Olms, 1962.

[8] G. Frege. *The Basic Laws of Arithmetic. Exposition of the System / Translated and edited, with an introduction, by M. Furth.* California Press, 1964.

[9] S. French. Quantum Physics and the Identity of Indiscernibles. *British Journal of the Philosophy of Science*, 39:233–246, 1988.

[10] S. French and D. Krause. *Identity in Physics: A Historical, Philosophical, and Formal Analysis.* Oxford University Press, 2006.

[11] P. T. Geach. *Logic Matters.* Basil Blackwell, 1972.

[12] M. Hofmann and T. Streicher. A Groupoid Model Refutes Uniqueness of Identity Proofs. *Proceedings of the 9th Symposium on Logic in Computer Science (LICS), Paris*, 1994.

[13] M. Hofmann and T. Streicher. The Groupoid Interpretation of Type Theory. *G. Sambin and J. Smith (eds.), Twenty-Five Years of Constructive Type Theory, Oxford University Press*, pages 83–111, 1998.

[14] K. C. Klement. *Frege and the Logic of Sense and Reference.* Routledge, 2002.

[15] P. Martin-Löf. *Intuitionistic Type Theory (Notes by Giovanni Sambin of a series of lectures given in Padua, June 1980).* Napoli: BIBLIOPOLIS, 1984.

[16] F. J. Pelletier. Did Frege Believe Frege's Principle? *Journal of Logic, Language, and Information*, 10:87–114, 2001.

[17] Univalent Foundations Project. *Homotopy Type Theory: Univalent Foundations of Mathematics*. Institute for Advanced Study (Princeton); available at https://homotopytypetheory.org/book/, 2013.

[18] W. V. Quine. *Ontological Relativity and Other Essays*. Columbia University Press, 1969.

[19] A. Rodin. *Axiomatic Method and Category Theory (Synthese Library vol. 364)*. Springer, 2014.

[20] U. Schreiber. *Quantization via Linear homotopy types*. arXiv: 1402.7041, 2014.

[21] U. Schreiber and M. Shulman. *Quantum Gauge Field Theory in Cohesive Homotopy Type Theory*. arXiv:1408.0054, 2014.

[22] M. Shulman. *Brouwer's Fix-Point Theorem in Real-Cohesive Homotopy Type Theory*. arXiv:1509.07584, 2016.

[23] L. Sklar. *Space, Time, and Spacetime*. University of California Press, 1974.

[24] B. Smith. Truthmaker realism. *Australasian Journal of Philosophy*, 77(3):274–291, 1999.

[25] C. Wright. *Frege's Conception of Numbers as Objects*. Aberdeen University Press, 1983.

On a Combination of Truth and Probability: Probabilistic IF Logic

Gabriel Sandu
Department of Philosophy, History, Culture and Arts, University of Helsinki
Unioninkatu 40 A, 00014 Helsinki, Finland
`sandu@mappi.helsinki.fi`

Abstract

I will give a short exposition of Independence-Friendly logic (IF logic), a system of logic which extends ordinary first-order logic with arbitrary patterns of dependent and independent quantifiers. Truth and falsity of IF sentences is defined in terms of the existence of winning strategies in a 2-player win-lose games of imperfect information. One consequence of imperfect information is the existence of indeterminate IF sentences (on finite models). I sketch how indeterminacy may be overcome using von Neumann's Minimax Theorem. My exposition draws on ideas in [6].

Keywords: IF Logic, IF Games.

Introduction

In a seminal paper [2], Goldfarb points out that "The connection between quantifiers and choice functions or, more precisely, between quantifier-dependence and choice functions, is the heart of how classical logicians in the twenties viewed the nature of quantification." [2, p. 357]. For a less historical but more systematic point of view [8], Terence Tao, notices that we know how to render in first-order logic statements like:

1. For every x, there exists a y *depending on* x such that $B(x, y)$ is true

and

2. For every x, there exists a y *independent of* x such that $B(x,y)$ is true.

The first one can be rendered by

$$\forall x \exists y B(x,y)$$

and the second one by

$$\exists y \forall x B(x,y).$$

(Here $B(x,y)$ is a binary relation holding of two objects x, y). Things become more complicated when four quantifiers and a 4-place relation $Q(x,x',y,y')$ are involved. We can express in first-order logic statements like:

3. For every x and x', there exists a y depending only on x and a y' depending on x and x' such that $Q(x,x',y,y')$ is true

and

4. For every x and x', there exists a y depending on x and x' and a y' depending only on x' such that $Q(x,x',y,y')$ is true

by

$$\forall x \exists y \forall x' \exists y' Q(x,x',y,y')$$

and

$$\forall x' \exists y' \forall x \exists y Q(x,x',y,y')$$

respectively. However, Tao continues, one cannot always express the statement

5. For every x and x', there exists a y depending only on x and a y' depending only on x' such that $Q(x,x',y,y')$ is true.

His conclusion is that

> It seems to me that first order logic is limited by the linear (and thus totally ordered) nature of its sentences; every new variable that is introduced must be allowed to depend on all the previous variables introduced to the left of that variable. This does not fully capture all of the dependency trees of variables which one deals with in mathematics. (Idem)

1 Independence-friendly logic

Independence-friendly logic (IF logic) introduced by Hintikka and Sandu in [4], is intended to represent patterns of dependence and independence of quantifiers like those exemplified by 5 which go beyond those expressible in ordinary first-order logic. More exactly, IF logic contains quantifiers of the form

$$(\exists x/W) \; (\forall x/W)$$

where W is a finite set of variables. The intended interpretation of $(\exists x/W)$ is: the existential quantifier $\exists x$ is independent of the quantifiers which bind the variables in W. The notion of independence involved here is a game-theoretical one and corresponds to the mathematical notion of uniformity. The example 5 above will be rendered in the new formalism by:

$$\forall x \forall x' (\exists y/\{x'\})(\exists y'/\{x,y\}) Q(x, x', y, y').$$

The original interpretation of IF formulas is given by semantical games of imperfect information. An alternative, equivalent interpretation is by skolemization. We shall adopt the latter.

2 Truth in IF logic

Let φ be a formula of IF logic in a given vocabulary L and U a finite set of variables which contains the free variables of φ. We expand the vocabulary L of φ to $L^* = L \cup \{f_\psi : \psi \text{ is a subformula of } \varphi\}$. The skolemized form or skolemization of φ with variables in U is defined by the following clauses, as detailed in [6]:

$Sk_U(\psi) = \psi$, for ψ an atomic subformula of φ or its negation

$Sk_U(\psi \circ \theta) = Sk_U(\psi) \circ Sk_U(\theta)$, for $\circ \in \{\vee, \wedge\}$

$Sk_U((\forall x/W)\psi) = \forall x Sk_{U \cup \{x\}}(\psi)$

$Sk_U((\exists x/W)\psi) = Subst(Sk_{U \cup \{x\}}(\psi), x, f_{\exists x}(y_1, ..., y_n))$

where $y_1, ..., y_n$ enumerate all the variables in $U - W$. We notice that if $W = \emptyset$ the last clause becomes

$$Sk_U((\exists x)\psi) = Subst(Sk_{U \cup \{x\}}(\psi), x, f_{\exists x}(y_1, ..., y_n))$$

where $y_1, ..., y_n$ enumerate all the variables in U. That is, we recover the notion of skolemization for the standard quantifiers. We abbreviate $Sk_\emptyset(\varphi)$ by $Sk(\varphi)$. An interpretation of $f_{\exists x}(y_1, ..., y_n)$ is called a Skolem function.

Example. We skolemize the sentence φ

$$\forall x \forall x' (\exists y/\{x'\})(\exists y'/\{x,y\}) Q(x,x',y,y').$$

We denote $(\exists y'/\{x,y\})Q(x,x',y,y')$ by ψ. $Sk(\varphi)$ is obtained through the following steps:

$$\begin{aligned} Sk_{\{x,x',y,y'\}}(Q(x,x',y,y')) &= Q(x,x',y,y') \\ Sk_{\{x,x',y\}}(\psi) &= Q(x,x',y,f_{y'}(x')) \\ Sk_{\{x,x'\}}((\exists y/\{x'\})\psi) &= Q(x,x',f_y(x),f_{y'}(x')) \\ Sk_{\{x\}}(\forall x'(\exists y/\{x'\})\psi) &= \forall x' Q(x,x',f_y(x),f_{y'}(x')) \\ Sk_{\varnothing}(\forall x \forall x'(\exists y/\{x'\})\psi) &= \forall x \forall x' Q(x,x',f_y(x),f_{y'}(x')). \end{aligned}$$

The original vocabulary L receives an interpretation through an L-structure $\mathbb{M}$ in the usual way. We are now ready for the truth-definition.

Definition 1. Let φ be an L-sentence of IF logic and $\mathbb{M}$ an L-structure. We say that φ is true in $\mathbb{M}$, $\mathbb{M} \models^+ \varphi$, if and only if there exist functions $g_1, ..., g_n$ of appropriate arity in M to be the interpretations of the new function symbols $f_{x_1}, ..., f_{x_n}$ in $Sk(\varphi)$ such that

$$\mathbb{M}, g_1, ..., g_n \models Sk(\varphi).$$

3 Falsity in IF logic

In order to deal with falsity, we shall define another translation procedure, $Kr_U(\varphi)$ (we continue to follow [6]):

$Kr_U(\psi) = \neg\psi$, for ψ an atomic subformula or its negation

$Kr_U(\psi \vee \theta) = Kr_U(\psi) \wedge Kr_U(\theta)$,

$Kr_U(\psi \wedge \theta) = Kr_U(\psi) \vee Kr_U(\theta)$,

$Kr_U((\exists x/W)\psi) = \forall x Kr_{U \cup \{x\}}(\psi)$

$Kr_U((\forall x/W)\psi) = Subst(Kr_{U \cup \{x\}}(\psi), x, f_{\forall x}(y_1, ..., y_m)$

where $y_1, ..., y_m$ are all the variables in $U - W$. We call the value of interpretation of $f_{\forall x}(y_1, ..., y_m))$ a Kreisel counter-example.

By analogy with the truth definition, we stipulate that an IF sentence φ is false in a structure $\mathbb{M}$, $\mathbb{M} \models^- \varphi$, if and only if there exist functions $h_1, ..., h_m$ of appropriate arity in M to be the interpretations of the new function symbols $f_{x_1}, ..., f_{x_m}$ in $Kr(\varphi)$ such that

$$\mathbb{M}, h_1, ..., h_m \models Kr(\varphi).$$

4 Indeterminacy and signaling

Here is an example of an IF sentence which is neither true nor false in any structure $\mathbb{M}$ which contains at least two elements:

$$\varphi = \forall x (\exists y / \{x\}) x = y.$$

It may be checked that $Sk(\varphi) = \forall x\, x = c$, where c is a new 0-place function (individual constant); and $Kr(\varphi) = \forall y \neg d = y$. Then by the definitions above, we have:

$\mathbb{M} \models^+ \varphi$ iff there is $a \in \mathbb{M}$ such that $\mathbb{M}, a \models \forall x\, x = c$
$\mathbb{M} \models^- \varphi$ iff there is $b \in \mathbb{M}$ such that $\mathbb{M}, b \models \forall y \neg d = y$.

As the structure $\mathbb{M}$ contains at least two elements, none of the assertions on the right side is true. Thus we have both $\mathbb{M} \not\models^+ \varphi$ and $\mathbb{M} \not\models^- \varphi$.

It is interesting to compare the previous example with ψ

$$\forall x \exists z (\exists y / \{x\}) x = y$$

whose skolemization is

$$\forall x\, x = g(f(x)).$$

It may be checked that this sentence is a logical truth. Unlike in ordinary first-order logic, the example shows that inserting a dummy existential quantifier in an IF sentence changes its semantical value. Hodges has discussed this example in [5], and the phenomenon of signaling in IF logic.

5 Expressive power

Example of sentences of the form

$$\forall x \forall x' (\exists y / \{x'\})(\exists y' / \{x, y\}) Q(x, x', y, y')$$

which are mentioned by Tao and which are not first-order definable are not difficult to find. We prefer to use a different example, which will turn up to be useful for other purposes too. There is an IF sentence which expresses the (Dedekind) infinity of the universe M. M is said to be (Dedekind) infinite iff there is a function $h: M \to M$ which is an injection and in addition there is an element in M which is not the image under h of any element of M. The sentence we look for is φ_{inf}

$$\exists w \forall x (\exists y / \{w\})(\exists z / \{w, x\})(x = z \land w \neq y).$$

The Skolem form of φ_{inf} is

$$\forall x(x = g(f(x)) \land c \neq f(x)).$$

It can be checked that φ_{inf} is true in a model iff the function f is an injection which range is not the entire universe. On the other side if $\mathbb{M}$ is finite, it may be shown that we have both $\mathbb{M} \not\models^+ \varphi_{inf}$ and $\mathbb{M} \not\models^- \varphi_{inf}$. Thus we have produced another example of an indeterminate IF sentence.

6 Strategic games

Consider our earlier IF sentence $\varphi = \forall x(\exists y/\{x\})x = y$ and a finite model $\mathbb{M}$. The set $S_\exists$ of Skolem functions of Eloise in this game reduces to the set of all individuals in M which can be the values of the new function symbols in $Sk(\varphi) = \forall x\, x = c$. In this case $S_\exists = \mathbb{M}$. And the set $S_\forall$ of Kreisel counter-examples of Abelard in this game reduces to the set of all individuals in M which can be the values of the new function symbols in $Kr(\varphi) = \forall y \neg d = y$. Thus $S_\forall = \mathbb{M}$. We can now formulate a two-player strategic game in which we let $S_\exists$ be the set of (pure) strategies of Eloise, and $S_\forall$ the set of (pure) strategies of Abelard. The two players choose simultaneously $s \in S_\exists$ and $t \in S_\forall$, respectively. The payoff of the outcome is determined in a very simple way: if s and t satisfy the equation $x = y$, Eloise wins (1 euro). Otherwise Abelard wins. Here is the complete matrix of the game for the case in which $S_\exists = \{1, 2, 3\} = S_\forall = \mathbb{M}$:

	1	2	3
1	(1,0)	(0,1)	(0,1)
2	(0,1)	(1,0)	(0,1)
3	(0,1)	(0,1)	(1,0)

The rows represent the strategies of Eloise and the columns the strategies of Abelard. In (m, n), $m \in \{0, 1\}$ is the payoff of Eloise, i.e. $u_\exists(m, n) = m$, and n is the payoff for Abelard for the corresponding pair of strategies.

It is interesting to compare this game to the one associated with the IF sentence $\psi = \forall x(\exists y/\{x\})x \neq y$ and $M = \{1, 2, 3\}$:

	1	2	3
1	(0,1)	(1,0)	(1,0)
2	(1,0)	(0,1)	(1,0)
3	(1,0)	(1,0)	(0,1)

We shall call these games strategic IF games, and denote them by $\Gamma(M, \varphi) = (S_\exists, S_\forall, u_\exists, u_\forall)$. Obviously these games are

- win-lose: Every game has exactly two payoffs, 0 and 1.
- 1 sum: For every $s \in S_\exists$ and $t \in S_\forall$ we have: $u_\exists(s,t) + u_\forall(s,t) = 1$.

Let $\Gamma = (S_\exists, S_\forall, u_\exists, u_\forall)$ be a finite strategic IF game. For $s^* \in S_\exists$ and $t^* \in S_\forall$, the pair (s^*, t^*) is an equilibrium in Γ iff the following two conditions are jointly satisfied:

(i) $u_\exists(s^*, t^*) \geq u_\exists(s, \tau t^*)$ for every strategy s in $S_\exists$. In other words

$$u_\exists(s^*, t^*) = max_s u_\exists(s, t^*)$$

(ii) $u_\forall(s^*, t^*) \geq u_\forall(s^*, t)$ for every strategy t in $S_\forall$. In other words

$$u_\forall(s^*, t^*) = max_t u_\forall(s^*, t).$$

We can check that in our earlier strategic IF games $\Gamma(\mathbb{M}, \forall x (\exists y/\{x\}) x = y)$ and $\Gamma(\mathbb{M}, \forall x (\exists y/\{x\}) x \neq y)$ where $\mathbb{M} = \{1, 2, 3\}$, there are no equilibria. Obviously this is a reflection of the fact that these games are undetermined.

6.1 Mixed strategies equilibria in IF games

There is an equilibrium in every IF game if, instead of pure strategies, we switch to mixed strategies. Let $\Gamma = (S_\exists, S_\forall, u_\exists, u_\forall)$ be a finite IF strategic game. A mixed strategy ν for player i in this strategic game is a probability distribution over S_i, that is, a function $\nu : S_i \to [0, 1]$ such that $\sum_{\tau \in S_i} \nu(\tau) = 1$. ν is uniform over $S'_i \subseteq S_i$ if it assigns equal probability to all strategies in S'_i and zero probability to all the strategies in $S_i - S'_i$. Obviously we can simulate a pure strategy s with a mixed strategy ν such that ν assigns s probability 1. Given a mixed strategy μ for player $\exists$ and a mixed strategy ν for player $\forall$, the expected utility for player i is given by:

$$U_i(\mu, \nu) = \sum_{s \in S_\exists} \sum_{t \in S_\forall} \mu(s)\nu(t)u_i(s,t).$$

When $s \in S_\exists$ and ν is a mixed strategy for player $\forall$, we let

$$U_i(s, \nu) = \sum_{t \in S_\forall} \nu(t)u_i(s,t).$$

Similarly if $t \in S_\forall$ and μ is a mixed strategy for player $\exists$, we let

$$U_i(\mu, t) = \sum_{s \in S_\exists} \mu(s)u_i(s,t).$$

Von Neumann's well known Minimax Theorem shows that every finite, constant sum, two player game has an equilibrium in mixed strategies. It is also well known that every two equilibria in such a game returns the same expected utility to the two players. Thus we can talk about the expected utility returned to player $\exists$ by an IF strategic game. This justifies the next definition:

Definition 2. Let φ be an IF sentence and $\mathbb{M}$ a finite model. When $0 \leq \varepsilon \leq 1$ we define: $\mathbb{M} \vDash^{eq}_\varepsilon \varphi$ iff the expected utility returned to player $\exists$ by the strategic game $\Gamma(\mathbb{M}, \varphi)$ is ε.

The above definition gives us the (probabilistic) value of an IF sentence φ on a given finite model $\mathbb{M}$. It can be shown that this interpretation is a conservative interpretation of the earlier interpretation, in the following sense.

Proposition 1. *For every IF sentence φ and finite model $\mathbb{M}$ we have: $\mathbb{M} \vDash^+ \varphi$ iff $\mathbb{M} \vDash^{eq}_1 \varphi$; and $\mathbb{M} \vDash^- \varphi$ iff $\mathbb{M} \vDash^{eq}_0 \varphi$.*

The next proposition is often useful for checking that a pair of mixed strategies is an equilibrium.

Proposition 2. *Let μ^* be a is a mixed strategy for player $\exists$ and ν^* is a mixed strategy for player $\forall$ in the strategic IF game Γ. The pair (μ^*, ν^*) is an equilibrium in Γ if and only if the following conditions hold:*

1. *$U_\exists(\mu^*, \nu^*) = U_\exists(\sigma, \nu^*)$ for every $\sigma \in S_\exists$ in the support of μ^*;*

2. *$U_\forall(\mu^*, \nu^*) = U_\forall(\mu^*, \tau)$ for every $\tau \in S_\forall$ in the support of ν^*;*

3. *$U_\exists(\mu^*, \nu^*) \geq U_\exists(\sigma, \nu^*)$ for every $\sigma \in S_\exists$ outside the support of μ^*;*

4. *$U_\forall(\mu^*, \nu^*) \geq U_\forall(\mu^*, \tau)$ for every $\tau \in S_\forall$ outside the support of ν^*.*

Recall our earlier examples $\Gamma(\mathbb{M}, \forall x(\exists y/\{x\})x = y)$ and $\Gamma(\mathbb{M}, \forall x(\exists y/\{x\})x \neq y)$ where $\mathbb{M} = \{1, 2, 3\}$. In both cases the uniform strategies $\mu^*(1) = \mu^*(2) = \mu^*(3) = \frac{1}{3}$ and $\nu^*(1) = \nu^*(2) = \nu^*(3) = \frac{1}{3}$ form an equilibrium. The value of the first game is $\frac{1}{3}$ and that of the second game is $\frac{2}{3}$. Thus $\mathbb{M} \vDash^{eq}_{\frac{1}{3}} \forall x(\exists y/\{x\})x = y$ and $\mathbb{M} \vDash^{eq}_{\frac{2}{3}} \forall x(\exists y/\{x\})x \neq y$.

A more complex argument shows that for M a finite model with n elements we have $\mathbb{M} \vDash^{eq}_{\frac{n-1}{n}} \varphi_{inf}$. Thus when n grows to infinity the value of φ_{inf} approaches 1, as expected.

Notes

IF logic has been introduced by Hintikka and Sandu in [4]. In [3], Hintikka discusses the foundational role of IF logic in the philosophy of mathematics. The basic model theoretical properties of IF logic from a game-theoretical perspective are described by Mann, Sandu, and Sevenster in [6]. In that work the probababilistic interpretation of IF logic, which is the source of our exposition in section 6, is thoroughly studied. The idea to use von Neumann's Minimax Theorem in the context of partially ordered quantifiers is due to Ajtai as mentioned in [1]. The first systematic investigation of strategic IF games is provided by [7]. Recently, an alternative approach to IF logic has been developed which replaces the independence of quantifiers by the dependence between terms. In this new setting, (5) is rendered by

$$\forall x \forall x' \exists y \exists y' (= (x,y) \wedge = (x',y') \wedge Q(x,x',y,y')).$$

The intended meaning of $' = (x,y)'$ is: y functionally depends on x. The semantical interpretation of this language is based on Hodges' compositional interpretation, introduced in [5]. A self-contained introduction to this logic is [9].

References

[1] A. Blass and Y. Gurevich. Henkin quantifiers and complete problems. *Annals of Pure and Applied Logic*, 32(C):1–16, 1986.

[2] W. Goldfarb. Logic in the twenties: The nature of the quantier. *Journal of Symbolic Logic*, 44:351–368, 1979.

[3] J. Hintikka. *Principles of Mathematics Revisited*. Cambridge University Press, Cambridge, 1996.

[4] J. Hintikka and G. Sandu. Informational independence as a semantic phenomenon. In J. E. Fenstad et al., editors, *Logic, Methodology and Philosophy of Science*, volume 8, pages 571–589. Elsevier, Amsterdam, 1989.

[5] W. Hodges. Compositional semantics for a language of imperfect information. *Logic Journal of the IGPL*, 5:539–563, 1997.

[6] A. Mann, G. Sandu, and M. Sevenster. *Independence-Friendly Logic: A Game-theoretic Approach*. Cambridge University Press, Cambridge, 2011.

[7] M. Sevenster. *Branches of Imperfect Information: Logic, Games and Computation*. Phd thesis, University of Amsterdam, Amsterdam, 2006.

[8] T. Tao. Printer-friendly CSS, and nonfirstorderisability. Retrieved on January 15, 2016, from the website: URL = <https://terrytao.wordpress.com/2007/08/27/printer-friendly-css-and-nonfirstorderizability/>, 2007. What's new: Updates on my research and expository papers, discussion of open problems, and other maths related topics.

[9] J. Väänänen. *Dependence Logic*. Cambridge University Press, Cambridge, 2007.

Towards Analysis of Information Structure of Computations

Anatol Slissenko
Laboratory for Algorithmics, Complexity and Logic (LACL)
University Paris East Créteil (UPEC)
61 av. du Général de Gaulle 94010, Créteil France
and
ITMO, St. Petersburg, Russia
slissenko@u-pec.fr

Abstract

The paper discusses how one can try to analyze computations, and maybe computational problems from the point of view of information evolution. The considerations presented here are very preliminary. The long-standing goal is twofold: on the one hand, to find other vision of computations that may help to design and analyze algorithms, and on the other hand, to understand what is realistic computation and what is real practical problem. The concepts of modern computer science, that came from classical mathematics of pre-computer era, are overgeneralized, and for this reason are often misleading and counter-productive from the point of view of applications. The present text discusses mainly what classical notions of entropy might give for analysis of computations. In order to better understand the problem, a philosophical discussion of the essence and relation of knowledge/information/uncertainty in algorithmic processes might be useful.

Keywords: Computation, Problem, Partition, Entropy, Metric.

1 Introduction

The goal of this paper is to discuss along what lines one can look for ways to describe the quantity of information transformed by computations. This may permit to better understand the computations themselves and, possibly, what is practical

---

Talk at the conference "Philosophy, Mathematics, Linguistics: Aspects of Interaction 2012" (PhML-2012), held on May 22–25, 2012 at the Euler International Mathematical Institute.

computation and what is practical algorithmic problem. The considerations presented here are very preliminary, more of philosophical than of mathematical flavor. We consider rather straightforward geometrical and information ideas that come to mind. Usually they are not sufficient taken directly. Making explicit the obstacles may help to devise more productive approaches[1].

In Introduction we give some arguments that illustrate that the mathematical formulations of computational problems we usually consider, are overgeneralized, and sometimes this hinders the development of practical algorithms or the understanding why certain algorithms for theoretically hard problems work well in practice. In Section 2 we outline some approaches to measuring information in computations, and discuss their weak and strong points. Section 3 is about the structure of problems for which we can presumably develop measures of information along the lines described in the previous section. It contains also a short discussion of the role of linguistic considerations in describing practical problems.

Why traditional mathematical settings look too general for practical computer science? And when it is inevitable and when maybe not?

Most notions used in theoretical computer science either come from mathematics of pre-computer era or are developed along mathematical lines of that epoch. From mathematics of pre-computer era the computational theory borrows logics, logical style algorithms (lambda-calculus, recursive function, Turing machine), general deductive systems (grammars), Boolean functions, graphs. More specific notions like finite automata, Boolean circuits, random access machines etc., though motivated by modeling of computations, are of traditional mathematical flavor. All these concepts played and continue to play fundamental role in theoretical computer science, however other, more adequate concepts are clearly needed.

I can illustrate this thesis by Boolean functions and their realization by circuits. Almost all Boolean functions of n variables have exponential circuit complexity $(2^n/n)$ [9], and there is an algorithmic method to find such an optimal realization for a given 'random' function [6]. But it is clear that even for $n = 64$, that is not so big from practical viewpoint, one cannot construct a circuit with $2^n/n$ gates. So one can state that almost all Boolean functions will never appear in applications. The notion of Boolean function is of evident practical value, but not in its generality. All this does not say that the general notion and the mentioned result on the complexity of realization are useless in theory (moreover, they are known to be useful). But an optimal circuit construction for almost all Boolean functions is not of great value for practical Boolean functions.

Consider another example. We know that the worst-case complexity of the de-

---

[1]Some of them were developed later to become quite mathematical.

cidability of the theory of real addition is exponential [2]. This theory is a set of valid closed formulas that are constructed from linear inequalities with integer coefficients with the help of logical connectives, including quantifiers over real numbers (in fact, only rational numbers are representable by such formulas, as the only admissible constants are integers). In particular, one can express in this theory the existence of a solution of a system of linear inequalities, and various parametric versions of this problem, e.g., whether such a solution exists for any value of some variable in some interval. The complexity of recognition of validity of the formulas grows up with the number of quantifier alternations.

The mentioned exponential lower bound on the computational complexity of the theory of real addition is proven along the following lines. Denote $\mathbb{B} =_{df} \{0, 1\}$ and denote by $\mathbb{B}^*$ the set of all strings over $\mathbb{B}$. Under some technical constraints for any algorithm f from $\mathbb{B}^*$ to $\mathbb{B}$, whose complexity is bounded by some exponential function φ, and for any its input $x \in \mathbb{B}^*$ one can construct a formula $\Phi(f, x)$ of sufficiently small size (polynomial in the size of f and x) that is valid if and only if $f(x) = 0$.

Within a reasonable algorithmic framework (e.g., for some random access machines, like LRAM from [10]) one can construct a predicate $f : \mathbb{B}^* \to \mathbb{B}$ whose upper bound on computational complexity is φ, and any algorithm that computes this predicate has lower bound $\theta \cdot \varphi$, for some $0 < \theta < 1$. This f is a diagonal algorithm, I do not know other kind of algorithms for this context. Such a diagonal algorithms works like follows. Assume that the complexity of computing $\varphi(|x|)$, where $|x|$ is the length of $x \in \mathbb{B}^*$, is bounded by its value $\varphi(|x|)$. The algorithm f computes $\varphi(|x|)$ and makes roughly $\varphi(|x|)$ steps of simulation of algorithm with the code x applied to input x. If the process ends within less that $\varphi(|x|)$ steps then f outputs the value different from the value computed by the algorithm with the code x, otherwise it outputs say, 0 (in the latter case the value is not important).

Thus, the recognition of the validity of formulas $\Phi(f, x)$ has a high complexity. But they are not formulas that appear in practice. Moreover, practical formulas, that may have a good amount of quantifier alternations, are semantically much simpler, they never speak about diagonal algorithms, though may speak about practical algorithms, e.g., about execution and properties of hard real-time controllers.

The just presented argument is valid for all negative complexity results (undecidability, high lower bounds, relative hardness) with the existing proofs. And here one arrives at another 'incoherence' between theory and practice that can be illustrated by the TAUT problem, i.e., by the problem of recognition of the validity of propositional formulas. This problem is considered as relatively hard (more precisely, coNP-complete) in theory, but existing algorithms solve very efficiently practical instances of this problem, and the problem is considered as an easy one by people working in applications. This is not the only example.

There are similar examples of another flavor, like the practical efficiency of linear programming algorithms. Here one finds mathematically interesting results of their average behavior. However, traditional evaluation of the average or Teng-Spielman smooth analysis [13] deal with sets of inputs almost none of which appears in practice. If one accepts Kolmorogov algorithmic vision of randomness, i.e., a string (or other combinatorial construct) is random if its Kolmogorov complexity is close to the maximal value, then one gets another argument that random constructs cannot appear from physical or human activity.

Many people believe that physical processes may produce truly random data. Many years ago, it was somewhere in the 70th, G. M. Adelson-Velsky[2] told me that M. M. Bongard[3] showed, using not very complicated learnability algorithm, that Geiger counter data, that were considered as truly random, can be predicted with a probability definitely higher that $1/2$. Who else analyzed physical 'random data' in this way? Notice that standard statistical tests that are used to prove randomness can be easily fooled by simple deterministic sequences, e.g., Champernowne's sequence. Happily, in practice 'sufficiently random' sequences suffice.

The practical inputs are always described in a natural language whose constructs are numerous but incomparably less numerous than arbitrary constructs, so they are not so random.

One may refer to the ideology of modern mathematics. Modern mathematics does not study arbitrary functions, nor arbitrary continuous functions, nor even arbitrary smooth functions. It studies particular, often rather smooth, manifolds on which often, though not always, acts a group with some properties modeling properties inspired by applications in mind.

It is not so evident how to find a structure to study in algorithmic problems, but it is much simpler to see a structure in computations, namely, in sets of runs (executions). One can try to find geometry in these sets. An intuitive sentiment is that any algorithm transforms information, so we can try to find geometry in computations using this or that concept of information.

It is improbable that one approach will work for all types of algorithms that appear in practice. The frameworks we use to study different types of algorithms are different. For example, reactive real-time systems are studied not as data base queries, computer algebra algorithms are studied not in the same way as combinatorial algorithms etc. In this paper I try to look at off-line 'combinatorial' algorithms without defining this class rigorously. Roughly speaking such an algorithm processes

---

[2] Georgy Maximovich Adelson-Velsky (1922–2014) was a well-known Soviet and Israeli mathematician and computer scientist.

[3] Mikhail Moiseevich Bongard (1924–1971) was a well-known Soviet computer scientist.

a finite 'combinatorial' input accessible from the very beginning, where each bit is 'visible' except maybe some integers that are treated as abstract atoms or 'short' integers with addition and comparison. Examples are string matching, binary convolution, TAUT, shortest path in graphs with integer weights etc.

But algorithms of this vaguely defined class may be very different from the point of view of their analysis. For example, take diagonal algorithms and compare such an algorithm with an algorithm like just mentioned above. One can see that runs of diagonal algorithms are highly diverse, within the same length of inputs we may see a run that corresponds to an execution of a string-matching algorithm, another run that correspond to solving a linear system etc. In the algorithms mentioned above the runs are more or less 'similar'. My first idea was to say that this distinguishes practical algorithms from non-practical ones. However, E. Asarin immediately drew my attention to interpreters that are quite practical and whose sets of runs are of the same nature that the set of runs of diagonal algorithms. It is interesting that compilers (to which N. Dershowitz drew my attention in the context of a discussion on practical and impractical algorithms some time ago) are in the same class that the mentioned combinatorial algorithms because they do not execute the programs that they transform. But interpreters are not in the same class as the combinatorial algorithms that are under study here. We do not demand that an interpreter diminish the computational complexity of the interpreted algorithm. And the interpretation itself slows down the interpreted algorithm by a small multiplicative constant that we can try to diminish. In some way, the output of the interpreter is a trace of the interpreted algorithm, so their diversity is intrinsic, and the length of their outputs is compared with their time complexity. We consider algorithms whose outputs are 'much shorter' than their time complexity.

2 How to evaluate similarity of computations?

Some syntactic precisions on the representation of runs of algorithms are needed. Suppose that F is an algorithm of bounded computational complexity that has as its inputs some structures (strings, graphs etc.) and whose outputs are also some structures.

By the size of an input we mean not necessarily the length of its bit code but some value that is more intuitive and 'not far' from its bit size. E.g., the number of vertices for a weighted graph, the length of vectors in binary convolution etc. In any case the bit size is polynomially bounded by our size. Thus, for a weighted graph we assume that weights are integers whose size is of the order of logarithm of the number of vertices if the weights are treated as binary numbers or whose size is $O(1)$

if they are treated abstractly.

We mention two very simple examples, namely palindrome recognition and sum of elements of a string over $\mathbb{B}$.

Assume that for the structures under consideration a reasonable notion of size is defined, and the set of all inputs of size n, that are in the domain of F, is denoted by $\boldsymbol{dm}_n(F)$ or $\boldsymbol{dm}$ if F and n are clear from the context. The set of corresponding values of F is denoted $\boldsymbol{rn}_n(F)$ or $\boldsymbol{rn}$. We assume that n is a part of inputs. Below n is fixed and often omitted in the notations.

We look at algorithms from the viewpoint of logic. Though in programming, as well as in logic, any program may be seen as an abstract state machine, there is no terminology that is commonly accepted in logic and programming. For example, what is called variable in programming is not variable in logic; from the point of view of logic it is a function without arguments but that may have different values during the execution of the program. In order to avoid such discrepancy we use logical terminology that was developed by Yu. Gurevich for his Abstract State Machines [3], and may be applicable to any kind of programs. Our framework is not that of Yu. Gurevich machines, we deal with executions of low-level programs seen as some kind of abstract state machines.

An algorithm computes the values of outputs using *pre-interpreted* constants like integers, rational numbers, Boolean values, characters of a fixed alphabet, and pre-interpreted functions like addition, order relations over numbers and other values, Boolean operations. These functions are *static*, i.e., they do not change during the executions of F. The other functions are *abstract* and *dynamic*. The inputs are given by the values of functions (that constitute the respective structure) that F can only read; they are *external* (as well as pre-interpreted functions). The functions that can be changed by F are its *internal functions*, they are subdivided into *output functions* and *proper internal* functions. We assume for simplicity that the output functions are updated only once. Dynamic functions may have arguments, like, e.g., arrays, and we limit ourselves to such functions that have one natural argument. When the argument i in such a function f is fixed, this $f(i)$ can be considered as nullary function, i.e. as a function without arguments. All these functions constitute a vocabulary of the algorithm.

We consider computations only for inputs from a finite set $\boldsymbol{dm}_n(F)$. These computations are represented as sets of traces that we describe below. We can treat such sets abstractly without precise notion of algorithm. However, for better intuitive vision, we describe a simple algorithmic language that gives a general notion of algorithm and that suffices for our examples.

Term is defined as usual, and without loss of generality, we consider non nested terms, i.e., terms whose arguments are only variables if any. *Guard* is a literal.

Update (assignment) is an expression $g := \theta$, where g is an internal function, and θ is a term. Constructors of a program (algorithm) are: update, sequential composition (denoted ;), branching **if** *guard* **then** P **else** P' , where P and P' are programs, **goto**, **halt**. As delimiters we use brackets.

A *state* is an interpretation of the vocabulary of the algorithm. A state is changed by updates in the evident way. The initial state is common for all inputs, we assume that the initial value of any internal function f is symbol $\natural$ that represents *undefined*, is never used in updates, and that $f^{-1}(\natural) = \emptyset$. A *run* is usually defined as a sequence of states, but we use an equivalent representation of executions as traces.

Given an input $X \in \boldsymbol{dm}_n(F)$ a trace $\boldsymbol{tr}(X)$ is constructed as follows according to the executed operators: update is written as it is in the program; in the case of conditional branching **if** *guard* **then-else** we put in the trace either *guard* or its negation depending on what is true in this trace. For simplicity the initial state and **halt** are not explicitly mentioned in traces, neither **goto**. Thus, a trace is a sequence of updates and guards that are called *events*. The tth event in a trace $\boldsymbol{tr}(X)$ is denoted $\boldsymbol{tr}(X,t)$. These events are *symbolic*. An execution gives values to the internal functions, and thus, an interpretation of any event.

Denote by $\boldsymbol{t}_F^*(X)$ the time complexity of F for input X, and by $\boldsymbol{t}_F(n)$ the maximum of these values, i.e., the worst-case time complexity of F over $\boldsymbol{dm}_n(F)$.

For an input $X \in \boldsymbol{dm}_n(F)$ and a time instant t, $1 \leq t \leq \boldsymbol{t}^*(X)$, we denote by $f[X,t]$ the value of a internal function f in $\boldsymbol{tr}(X)$ at t, the value is defined recursively together with the recursive definition of trace given just above. If f is not undated at t then $f[X,t] = f[X, t-1]$. If $\boldsymbol{tr}(X,t)$ is of the form $f := g(\eta)$ then $f[X,t] = g(\eta[X, t-1])[X, t-1]$.

Consider two examples.

Palindrome recognition. Inputs are non empty strings of length n over an alphabet $\mathcal{A}$ with $\alpha \geq 2$ characters. For simplicity assume that n is even and set $\nu =_{df} \frac{n}{2}$. We denote the input by w, and the character in the ith position by $w(i)$. We take a straightforward algorithm φ that compares characters starting from the ends and going to the middle of the input. We use % to mark comments, and we omit **halt** that is evident.

Algorithm φ:

% i is a loop counter, r is the output (0 means non palindrome, 1 palindrome)
1:   $i := 0$;
2:   **if** $i < \nu$ **then** $\Big(\, i := i+1;$
           if $w(i) = w(n-i+1)$ **then goto** 2 **else** r:=0$\Big)$
        else r:=1

Algorithm φ has two types of traces (one with output 0 and the other with output 1):

$i := 0$, $i < \nu$, $i := i+1$, $w(i) = w(n-i+1)$, ..., $i < \nu$, $i := i+1$, $w(i) = w(n-i+1)$, $i < \nu$, $i := i+1$, $w(i) \neq w(n-i+1)$, r:=0

$i := 0$, $i < \nu$, $i := i+1$, $w(i) = w(n-i+1)$, ..., $i < \nu$, $i := i+1$, $w(i) = w(n-i+1)$, $i \geq \nu$, r:=1

A trace of the first type may have different lengths starting from 5, but the length of the trace of the second type is always the same.

For a string $aabaaa$ with $a \neq b$, if in the respective trace we replace the internal functions, as well as n, by their values we can write:

$i := 0$, $0 < 3$, $i := 0+1$, $w(1) = w(6)$, $1 < 3$, $i := 1+1$, $w(2) = w(5)$, $2 < 3$, $i := 2+1$, $w(3) \neq w(4)$, r:=0

Sum modulo 2 of bits of a string. Inputs are strings of the set $\mathbb{B}^n$.
Algorithm σ:

% x is input, r is output, i is a loop counter, s is an intermediate value
1:      $i := 0$; $s := 0$;   %Initialization
2:      **if** $i < n$ **then** $i := i+1$; $s := s + x(i)$; **goto** 2
3:            **else** $r := s$ % case $i \geq n$

All traces of σ are 'symbolically' the same (the algorithm is oblivious), for clarity we put in an event the value of i acquired before this event:

$i := 0$, $s := 0$, $0 < n$, $i := 0+1$, $s := s + x(1)$, $1 < n$, $i := 2$, $s := s + x(2)$, ..., $n - 1 < n$, $i := n$, $s := s + x(n)$, $n \geq n$, $r := s$

Remark. For Boolean circuits we can also produce traces that are even simpler, as a Boolean circuit is a non branching oblivious algorithm. Such a trace consists of updates, each one being an application of the Boolean function attributed to a vertex of the circuit, to the values attributed to its predecessors.

Denote by $\boldsymbol{Tr_n}$ the set of all traces for inputs from $\boldsymbol{dm_n}$. The length $|\boldsymbol{tr}(X)|$ of a trace $\boldsymbol{tr}(X)$, $X \in \boldsymbol{dm_n}$, is the number of occurrences of events in it, i.e., the time complexity $\mathbf{t}_F^*(X)$.

2.1 A syntactic similarity of traces

A straightforward way to compare two traces is the following one. We look in $\boldsymbol{tr}(X)$ and $\boldsymbol{tr}(Y)$ for a longest common subsequence (we tacitly assume that some equivalence between events is defined), and take as a measure of similarity the size of the rest. More precisely, if S is the longest common subsequence then we take as measure the value $|\boldsymbol{tr}(X)| + |\boldsymbol{tr}(Y)| - 2|S|$, where $|S|$ is the size (the number of elements) of a sequence S. This measure is something like the size of symmetric difference of two sequences.

We can go further, and to take into account only causal order in what concerns the order of events, and to permit a renaming of proper internal functions and their values. The causal order is defined as follows. If the function updated or used (in the case of guard verification) in an event e depends on a function updated earlier in an event e' then e' *causally precedes* e. Taking a transitive closure of this relation we get *causal order* between events in a given trace. This generalization is too technical (details can be found in [12]), and as I cannot give examples of realistic applications, it is just mentioned as a theoretical possibility.

The measure introduced above gives a pseudo-metric (it is like metric except that two different traces may have zero distance; in our case the zero distance relation is an equivalence) over traces. As the trace space $\boldsymbol{Tr}_n$ is clearly compact, this metric permits to define epsilon-entropy [5] on it. This entropy is defined as follows. For a given ε (in our case it is a natural number) take an ε-net of minimal size such that the ε-balls centered at the points of the net cover all the space. Then $\log s$, where s is the size of this net, is the ε-entropy. It gives the size complexity of the ε-approximation of the space, or to say it differently, how much information one needs to have, in order to describe an element of the space with accuracy ε.

Consider our examples.

Trace space of φ. We define similarity as follows (it is a rather general way to define it). First, in the right-hand side of each update $f := \theta$ replace all proper internal functions of θ by their values. In guards replace all internal functions by their values. We get as transformed events the expressions: $i := m$, where $m \in \mathbb{N}$ and $m = (\ldots((0+1)+1)+\cdots+1)$, $m < \nu$, $m \geq \nu$, $w(m) = w(n-m+1)$, $w(m) \neq w(n-m+1)$, $r = 0$, $r = 1$. As similarity (we refer to it as 'weak similarity') we take the syntactic equality of these transformed events.

With this similarity we have $(\nu+1)$ different (*classes of similar*) traces (ν classes with $r = 0$ at the end, and 1 class with $r = 1$): denote by P_k traces with $(k-1)$ equalities and one inequality in the k comparison, $1 \leq k \leq \nu$, and by P the only trace with $r = 1$. The distance between P_k and P_l is $3|k-l|$, and between P_k and P is $3(\nu-k)+2$. If we take $\varepsilon = 2$ then ε-net should include all the traces but, however, it is of size $(\log n + \mathcal{O}(1))$. If we take $\varepsilon = 3p$, $p \in \mathbb{N}$, then as an ε-net we can take each pth trace ordered according to their lengths; hence, $3p$-entropy is of size $\lceil \frac{n}{2p} \rceil = \lceil \frac{\nu}{p} \rceil$ (maybe plus 1).

The situation changes if we take stronger similarity. We say that $w(m) = w(n-m+1)$ and $w(m') = w(n-m'+1)$ are similar if $m = m'$ (as before) and the respective values of inputs are the same $w(m) = w(m')$ (for $\neq$ we demand also $w(n-m+1) = w(n-m'+1)$). In this case the trace space becomes of exponential size. We illustrate this kind of similarity for algorithm σ.

Trace space of σ. We define similarity of event of the form $i := i+1$ and of the form $i < n$ as in the previous case: values of $(i+1)$ in similar events of the form $i := i+1$ and the value of i in similar events of the form $i < n$ should be equal. Two events of the form $s := s + x(i)$ are similar if the values of i, as well as of the acquired values of s, are equal. Any string from $\mathbb{B}^n$ may be a string of consecutive values s starting from $s := 0 + x(1)$ that equals to $x(1)$. Thus the set of traces of σ with this similarity and our metric divided by 2 is isometric to the Boolean cube $\mathbb{B}^n$ with Hamming metric. This space is studied in the coding theory, and I cannot say more than can be found there.

Unfortunately, the metric spaces in the examples above do not say much about the advancement of the algorithm towards the result. If we take spaces of traces up to some time instant and their dynamics with growing time, it does not help much neither. Moreover, the size of the space $\boldsymbol{Tr_n}$ is bounded by $|\boldsymbol{dm_n}|$, and does not depend on the complexity of F, and this is also a shortcoming of this approach.

2.2 Remark on Kolmogorov complexity approach

Why not to measure distance between traces on the basis of Kolmogorov complexity? This question was put by some of my colleagues.

A direct application of Kolmorogov algorithmic entropy [4] to measure similarity of traces does not give results corresponding to our intuition. Indeed, in [4] Kolmogorov defines entropy as conditional complexity $\boldsymbol{K}(\alpha|\beta)$. Similarity of structures α and β may be measured as $\mathfrak{K}(\alpha, \beta) = \boldsymbol{K}(\alpha|\beta) + \boldsymbol{K}(\beta|\alpha)$. This is not a metric, strictly speaking, however, we call this function $\mathfrak{K}$-*distance* as it has a flavor of intuitive distance-like measure.

Denoting by $|F|$ and $|X|$ binary lengths of respectively F and X we get
$$\boldsymbol{K}(\boldsymbol{tr}(X)/\boldsymbol{tr}(Y)) \leq |F| + \boldsymbol{K}(X/Y) + O(1) \leq |F| + |X| + O(1).$$
This formula follows from an observation that X and F are sufficient to calculate the trace $\boldsymbol{tr}(X)$. Thus, whatever be an algorithm F and whatever be its computational complexity, the $\mathfrak{K}$-distance between traces from $\boldsymbol{Tr_n}$ is not greater than $O(|X|)$ that we assume, for simplicity, to be $O(n)$. On the other hand, given a minimal length program G that computes $\boldsymbol{tr}(X)$ from $\boldsymbol{tr}(Y)$ (thus, $|G| = \boldsymbol{K}(\boldsymbol{tr}(X)/\boldsymbol{tr}(Y))$) one can get X from Y as follows: from Y one computes $\boldsymbol{tr}(Y)$ using F (whose size is a constant), then using G one computes $\boldsymbol{tr}(X)$ and finally extracts X from $\boldsymbol{tr}(X)$ with the help of a simple fixed program, say E, whose length is a constant (without loss of generality, we can assume that the input is reproduced at the beginning of each trace). All this gives (we put 'absolute' constants $|F|, |E|$ in the last $O(1)$)
$$\boldsymbol{K}(X|Y) \leq |F| + |G| + |E| + O(1) \leq \boldsymbol{K}(\boldsymbol{tr}(X)/\boldsymbol{tr}(Y)) + O(1).$$

We assume that the cardinality of binary codes of $dm_n(F)$ is at least 2^n (hence, almost all inputs have Kolmogorov complexity $n - o(n)$), then the chain rule for Kolmogorov complexity (e.g., see [4]) for almost all X, Y gives
$$K(X|Y) = K(X,Y) - K(Y) - O(\log K(X,Y)) \geq n - c\log n$$
for some constant $c > 0$.
Together with the previous formula this gives a lower bound for $K(tr(X)/tr(Y))$ that shows that $\mathfrak{K}$-distance is almost always of order of n that can hardly be seem as satisfactory for evaluation of similarity of traces from Tr_n.

What is said above, does not exclude that other types of Kolmogorov style complexity could work better (e.g., a more general notion of entropy [11] is based on inference complexity.). In particular, resource bounded complexity approaches may prove to be productive if we find a 'good' description of information extracted by algorithm as datum (structure); however, this remains an open question.

2.3 Similarity via entropy of partitions

In this subsection we outline another approach to measure similarity of traces. It refers to the classical entropy of partitions. We use partitions of the inputs. For this reason a probabilistic measure over the inputs is needed. Such a measure is a technical means, so there is no evident way to introduce it. We do it taking into account an intuition related to the evolution of the 'knowledge' of the algorithm. When an algorithm F starts its work it 'knows' nothing about its output. So all values from rn_n are equiprobable.

Let $M = |rn_n(F)|$. As any of these M values is equiprobable (imagine that an input is given by an adversary who plays against F), we set $P_n(F^{-1}(Y)) = \frac{1}{M}$ for all $Y \in rn_n(F)$, and inside $F^{-1}(Y)$ the measure is uniform as the algorithm a priori has no preferences. In particular, if F is a 2-valued function, say $rn(F) = \mathbb{B}$, then its domain is partitioned into two sets $F^{-1}(0)$ and $F^{-1}(1)$ with the same measure $1/2$ of each set. E.g., for palindromes we measure of a palindrome is $\frac{1}{2\alpha^\nu}$ and that of a non palindrome is $\frac{1}{2(\alpha^n - \alpha^\nu)}$. There is nothing random in the situation we consider, we wish only to model the evolution of the knowledge of an algorithm during its work. So this way to introduce a measure may be not the best one.

Suppose that f is updated at t and $f[X,t] = v$. How to describe the knowledge acquired by F via this event at t that gives $v = f[X,t]$? This value v may be acquired by f in different traces, even several times in the same trace, and at different time instants. The traces are not 'synchronized' in time, however, we can compare events, as in subsection 2.1, due to this or that similarity relation, that is determined by our goal and our vision of the situation. Notice that formally speaking similarity is a relation between pairs (X, t), where $X \in \boldsymbol{dm}(F)$ and $1 \leq t \leq \boldsymbol{t}_F^*(X)$. Similarity can be defined not only along the lines described in subsection 2.1. One may think about quite different ways. Just to give an idea, one can, for example, consider as similar events corresponding to the kth execution of the same command of the program with or without demanding equality of these or that values. Or one can permit renaming of internal function as it was mentioned at the beginning of subsection 2.1.

Suppose that some similarity relation $\sim$ is fixed.

To compare traces we attribute to each event of a trace a partition of inputs. Thus, to each trace there will be attributed a sequence of partitions. Taking into account that the set of inputs is a space with probabilistic measure we can define a distance between partitions and furthermore a distance between sequences or sets of partitions.

For any input X and an instant t, $1 \leq t \leq \boldsymbol{t}^*(X)$, denote by $\boldsymbol{sm}(X, t)$ all the inputs X' such that $(X, t) \sim (X', t')$ for some t'. Clearly, $X \in \boldsymbol{sm}(X, t)$. Denote by $\boldsymbol{pt}(X, t)$ the partition of $\boldsymbol{dm}_n$ into $\boldsymbol{sm}(X, t)$ and its complement that we denote $\boldsymbol{sm}(X, t)^c =_{df} \boldsymbol{dm}_n \setminus \boldsymbol{sm}(X, t)$.

Thus, each input X determines a *sequence* $(\boldsymbol{pt}(X, t))_t$ or a *set* $\{\boldsymbol{pt}(X, t)\}_t$ of *partitions* of $\boldsymbol{dm}_n(F)$. These constructions, namely sequence or set, provide different opportunities for further analysis, e.g., we can define distance between metric spaces, e.g., see [1, ch. 7].

For measurable partitions of a probabilistic space $\mathcal{P} = (\Omega, \Sigma, P)$ one can define entropy (no particular technical constraints are needed in our case of finite sets), see [8] or books like [7].

Let $\mathcal{A}$ and $\mathcal{B}$ be measurable partitions of $\mathcal{P}$ (in our situation all the sets are measurable).

Entropy $H(\mathcal{A})$ and conditional entropy $H(\mathcal{A}/\mathcal{B})$ are defined as

$$H(\mathcal{A}) = - \sum_{A \in \mathcal{A}} P(A) \log P(A), \; H(\mathcal{A}/\mathcal{B}) = - \sum_{B \in \mathcal{B}, A \in \mathcal{A}} P(A \cap B) \log \frac{P(A \cap B)}{P(B)}. \quad (1)$$

The conditional entropy permits to introduce Rokhlin metric [8] between partitions:
$\rho(\mathcal{A}, \mathcal{B}) = H(\mathcal{A}/\mathcal{B}) + H(\mathcal{B}/\mathcal{A}) = 2H(\mathcal{A} \vee \mathcal{B}) - H(\mathcal{A}) - H(\mathcal{B})$,
(here $\mathcal{A} \vee \mathcal{B}$ is common refinement of partitions $\mathcal{A}$ and $\mathcal{B}$, that is the partition formed by all pairwise intersection of sets of $\mathcal{A}$ and $\mathcal{B}$).

There are other ways to introduce distance between partitions, e.g., see [7, 4.4], so one can take or invent maybe more productive metrics or entropy-like measures.

Unfortunately, the combinatorial difficulties of estimating such distancies are discouraging, they do not justify what we get form them. We illustrate this for the palindrome recognition algorithm φ.

Denote $w^=(1..k) =_{df} \{w : \bigwedge_{1 \leq i \leq k} w(i) = w(n-i+1)\}$ (the set of words whose prefix of length k permits to extend it to a palindrome), denote by $w^{\neq}(1..k)$ the complement of $w^=(1..k)$; in particular, $w^=(k..k) = \{w : w(k) = w(n-k+1)\}$ and $w^{\neq}(k..k) = \{w : w(k) \neq w(n-k+1)\}$. Probabilities are easy to calculate (we use them in the next subsection), here $1 \leq k < m \leq \nu$:

$$\boldsymbol{P}(w^=(1..k) \cap w^{\neq}(k+1..m)) = \frac{\alpha^{\nu-m}(\alpha^{m-k}-1)}{2(\alpha^\nu-1)}, \qquad (2)$$

$$\boldsymbol{P}(w^=(1..k)) = \frac{1}{2} + \frac{\alpha^{\nu-k}-1}{2(\alpha^\nu-1)}, \quad \boldsymbol{P}(w^{\neq}(1..k)) = \frac{\alpha^{\nu-k}(\alpha^k-1)}{2(\alpha^\nu-1)}. \qquad (3)$$

(We omit technical details, the role of the formulas is illustrative.)

However, when we try to calculate the distance between partitions, take for example $\rho(\pi^=(k), \pi^=(m))$, where $\pi^=(s) = (w^=(1..s), w^{\neq}(1..s))$, we arrive at a formula that is a sum of several expressions like $\left(\frac{1}{2} + \frac{\alpha^{\nu-s}-1}{2(\alpha^\nu-1)}\right) \log \left(\frac{1}{2} + \frac{\alpha^{\nu-s}-1}{2(\alpha^\nu-1)}\right)$, that is hard to evaluate. And what is worse the result is not very instructive, e.g.,

$$\rho(\pi^=(1), \pi^=(\nu)) \approx \begin{cases} 0.9 & \text{if } \alpha = 2 \\ 0.67 & \text{if } \alpha = 3 \\ 0.6 & \text{if } \alpha = 4 \end{cases}$$

Technical combinatorial difficulties do not discard the idea of geometry of spaces of events or traces, the point is to find a geometry and its interpretation that really deepens our understanding of algorithms and problems.

2.4 The question of information convergence

Now we discuss how similarity of events may serve to evaluate the rate of convergence of a given algorithm towards the result.

Among the first ideas that come to mind is the following one. The result $F(X)$ for an input X is represented in terms of a partition of $\boldsymbol{dm}_n(F)$ into $F^{-1}(F(X))$ and its complement $F^{-1}(F(X))^c$. The current knowledge of F at an instant t is in its current event that also defines a partition denoted above $\boldsymbol{pt}(X,t)$.

How this local knowledge represented by $\boldsymbol{pt}(X,t)$, is related to the partition $(F^{-1}(F(X)), F^{-1}(F(X))^c)$ mentioned just above? A possible answer is: compare $\boldsymbol{pt}(X,t)$ (the local knowledge at an instant t in terms of partitions) with the partition $(F^{-1}(F(X)), F^{-1}(F(X))^c)$. This idea can be a priori implemented differently, for example, in terms of conditional probabilities or in terms of conditional entropies.

If we try to apply this idea to any of our examples, we find that the commands that control the loops give trivial partition $(\boldsymbol{dm}, \emptyset)$ because they are in all traces, and these events give nothing useful. So we take only events that process inputs.

Consider φ (the example of palindromes). Using (2), (3) we get, omitting technicalities and taking sufficient approximations:

$$\boldsymbol{P}(r=1|w^=(1..k)) \approx \frac{1}{1+A(k)}, \quad \boldsymbol{P}(r=0|w^=(1..k)) \approx \frac{A(k)}{1+A(k)}, \tag{4}$$

where $A(k) = \alpha^{-k} - \alpha^{-\nu}$. The probabilities (4) do not reflect our information intuition that φ converges to the result when $k \to \nu$ as one goes to 1, and the other to 0. But if we take the respective entropy

$$-\frac{1}{1+A(k)} \log \frac{1}{1+A(k)} - \frac{A(k)}{1+A(k)} \log \frac{A(k)}{1+A(k)}, \tag{5}$$

we see that it goes to 0, thus, to total certainty.

Consider σ (sum modulo 2). The similarity that we used for the trace space of σ in subsection 2.1 we call here *weak similarity*. Denote $\sigma^{-1}(a)$ by $\sigma = a$. Clearly, $|\sigma = 0| = |\sigma = 1| = 2^{n-1}$, $\boldsymbol{P}(\sigma = a) = \frac{1}{2}$, and $\boldsymbol{P}$ is a uniform distribution over $\mathbb{B}^n$.

Denote by $S_k(a)$, where $a \in \mathbb{B}$, the set $\{x : s + x(k) = a\}$; it is a set of type $\boldsymbol{sm}(X,t)$. For $k < n$ and all $a, b \in \mathbb{B}$ we have

$$\boldsymbol{P}(S_k(a)) = \frac{|S_k(a)|}{2^n} = \frac{2^{n-1}}{2^n} = \frac{1}{2}, \quad \boldsymbol{P}(S_n(a) \cap S_k(b)) = \frac{1}{4} \tag{6}$$

$$\boldsymbol{P}(\sigma = a|S_k(b)) = \frac{\boldsymbol{P}(S_n(a) \cap S_k(b))}{\boldsymbol{P}(S_k(b))} = \frac{1}{2} \tag{7}$$

We see that nothing changes with advancing of time, i.e., with $k \to n$. If we apply formula (1) for conditional entropy, it gives a constant. Hence, with this similarity, we do not see any convergence of σ to the result.

Let us try a stronger similarity: we say that a event $s := s + x(k)$ is (strongly) similar to $s := s + x'(k)$ if $x(i) = x'(i)$ for all $1 \le i \le k$. Denote by $Z(\chi)$, where $\chi \in \mathbb{B}^k$, $1 \le k < n$, the set of inputs x such that for event $s := s + x(k)$ there holds $x(i) = \chi(i)$ for $1 \le i \le k$; this set describes the set of inputs of strongly similar events. We have $|Z(\chi)| = 2^{n-k}$, and $|(\sigma = a) \cap Z(\chi)| = 2^{n-k-1}$, thus, $\boldsymbol{P}(Z(\chi)) = 2^{-k}$ and $\boldsymbol{P}((\sigma = a) \cap Z(\chi)) = 2^{-k-1}$. So the measure of the space of continuations of the known part of the input diminishes. The respective term in conditional entropy (1) gives $-2^{-k-1} \log \frac{1}{2} = 2^{-k-1}$ that is encouraging but the term related to $Z(\chi)^c$ (notice that $\boldsymbol{P}(Z(\chi)^c) = 1 - 2^{-k}$ and $\boldsymbol{P}((\sigma = a) \cap Z(\chi)^c) = \frac{1}{2} - 2^{-k-1}$) bring us back to values that practically do not diminish. All this means only that the classical entropy does not work, and we are to seek for entropy-like measures that truly reflect our intuition.

The partition based measures of convergence look promising. However, one can say that the number of partitions is limited by an exponential function of $|\mathbf{dm}|$. So if the complexity is very high, e.g., hyper-exponential, then there is 'not enough' of partitions to represent the variety computations. In fact, we think about certain class of problems that are outlined in the next section for which there seems to be 'enough' of partitions. As for high complexity problems, another interpretation of input data is needed. Some hints are given in the next section 3.

3 On the structure of problems

Here are presented examples of problems together with a reference to their inner structure that may be useful for further study of information structure of computations and that of problems themselves along the lines discussed in the paper. The examples below concern only simple 'combinatorial problems'. The instances of these problems are finite graphs (in particular, strings, lists, trees etc.) whose edges and vertices may be supplied with additional objects that are either abstract atoms with some properties or strings. As examples of problems that are not in this class one can take problems with exponential complexity like theory of real addition or Presburger arithmetics. The problems in the examples below are divided into 'direct' and the respective 'inverse' ones.

Direct Problems

(A1) *Substring verification.* Given two strings U, W over an alphabet with at least two characters and a position k in W, to recognize whether $U = W(k, k+1, \ldots, k + |U| - 1)$, i.e., whether U is a substring of W from position k.

(A2) *Path weight calculation.* Given a weighted (undirected) graph and a path, calculate the weight of the path.

(A3) *Evaluation of a Boolean formula for a given value of variables.* Given a Boolean formula Φ and a list X of values of its variables, calculate the value $\Phi(X)$ for these values of variables.

(A4) *Permutation.* Given a list of elements and a permutation, apply the permutation to the list.

(A5) *Binary convolution (or binary multiplication).* For simplicity we consider binary convolution that represents also the essential difficulties of multiplication. Given 2 binary vectors or strings $x = x(0) \ldots x(n-1)$ and $y = y(0) \ldots y(n-1)$ calculate

$$z(k) = \sum_{i=0}^{i=k} x(i) y(k-i), \quad 0 \leq k \leq (2n-2),$$

assuming that $x(i) = y(i) = 0$ for $n - 1 < i \leq (2n - 2)$.

Inverse Problems

(B1) *String matching.* Given two strings W and U over an alphabet with at least two characters, to recognize whether U is a substring of W.

(B2) *Shortest path.* Given a weighted (undirected) graph G and its vertices u and v, find a shortest path (a path of minimal weight) from u to v.

(B3) *Propositional tautology TAUT* Given a propositional formula Φ, to recognize whether it is valid, i.e., is true for all assignment of values to its variables. A variant that is more interesting in out context is MAX-SAT: given a CNF (conjunctive normal form), to find the longest satisfying assignment of variables, i.e. an assignment that satisfies the maximal number of clauses.

(B4) *Sorting.* Given a list of elements of a linearly ordered set, to find a permutation that transforms it into an ordered list.

(B5) *Factorization.* Given z, to find x and y whose convolution or product (in the case of multiplication) is z.

Examples (A1)–(A4) give algorithmic problems whose solution, based directly on their definitions, is practically and theoretically the most efficient. Each solution consists in a one-directional walk through a simple data structure making, again rather simple, calculations – something that is similar to scalar product calculation.

In (A1) the structure is a list $(k, k+1, \ldots, k+|U|-1)$, and while walking along it, we calculate conjunction of $U(i) = W(i)$ for $k \leq i < (k+|U|)$ until i reaches the last value or *false* appears.

Example (A2) is similar, where the list of vertices constituting the linear structure is explicitly given, and the role of conjunction of (A1) is played by addition.

The structures used in (A3) depend on the representation of Φ and of the distribution of values of its variables. In any case one simple linear structure does not suffice here. Suppose Φ is represented in DNF (Disjunctive Normal Form), i.e., as a disjunction of conjunctions. This can be seen as a list of lists of literals, and a given distribution of values is represented as an array corresponding to a fixed order of variables. So given a variable, its value is immediately available. Thus, the representation of values is a linear structure, and DNF is a linear structure of linear structures. It is more interesting to suppose that Φ is a tree. Then we deal with the representation of values and with a walk, again without return, through a tree with calculating the respective Boolean functions at the vertices of the tree. So we see another simple basic structure, namely a tree.

In example (A4), while walking through two given lists, namely a list of elements and a permutation, a third list (a list of permuted elements) is constructed.

Example (A5) is more complicated, and the definition of problem does not give an algorithm that may be considered as the best; it is known that the direct algorithm for convolution is not the fastest one. Here there is no search, and for this reason this problem is put in the class of direct ones, but there is a non-trivial intermixing of data. One may see the description of the problem as a code of data structures to extract, and then to calculate the resulting values by simple walks through these data structures. The number of the data structures to extract is quadratic. In order to find a faster algorithm, one should ensure the same intermixing but using different data structures and operations.

Examples (B1)–(B5) give algorithmic problems of search among substructures coded in inputs. The number of these substructures, taken directly from the definition, is quadratic for (B1), and exponential for (B2)–(B5). The substructures under search should satisfy conditions that characterize the corresponding direct problem. More complicated problems code substructures not so explicitly as in examples (B1)–(B5). To illustrate this, take e.g., quantifier elimination algorithm for the formulas of the theory of real addition, not necessarily closed formulas. Here it is not evident how to define the substructures to consider. The quantifier elimination by any known algorithm produces a good amount of linear inequalities that are not in the formula. So the formula codes its expansion that is more than exponentially bigger as compared with the initial formula itself.

Whatever be the mentioned difficulties, intuitively the substructures and constraints generated by a problem may be viewed as an extension of the set of inputs. And in this extended set one can introduce not only measure but also metrics that give new opportunities to analyze the information contents and the information evolution. One can see that the cardinality constraints on the number of partitions that was mentioned in subsections 2.3 and 2.4 is relaxed. This track has not been yet studied, though one observation can give some hint to how to proceed. When comparing substructures it seems productive to take into account its context, i.e., how it occurs in the entire structure. For example, we can try to understand the context of an assignment A of values to variables of a propositional formula Φ in the following manner. Pick up a variable x_1 and its value v_1 from A and calculate the result $\Phi(\dot{x}_1, v_1)$ of the standard simplification of Φ where x_1 is replaced by Boolean value v_1. This resulting residue formula gives some context of (x_1, v_1). We can take several variables and look at the respective residue as at a description of context. This or that set of residues may be considered as a context of A. It is just an illustration of what I mean here by 'context'.

A metric over substructures may distinguish 'smooth' inputs from 'non-smooth' ones, and along this line we may try to distinguish practical inputs from non practical ones. Though it is not so evident.

For some 'simple' problems such a distinction is often impossible. It looks hard to do for numerical values. The set of such values often constitutes a variety with specific properties that may represent realistic features but almost all elements of such varieties will never appear in practical computations. An evident example is binary multiplication. Among 2^{128} possible inputs of multiplication of 64-bit numbers most of them will never be met in practice.

A remark on the usage of linguistical frameworks

One more way to narrow the sets of inputs to take into account, is a language based one. Inputs describing human constructions, physical phenomena, and their properties, when they are not intended to be hidden, have descriptions in a natural language. Encrypted data are not of this nature. So for input data with non hidden information, we have a grammar that generates these inputs. Such a grammar dramatically reduces the number of possible inputs and, what is more important, defines a specific structure of inputs. The diminishing of the number of generated inputs is evident. For example, the number of 'lexical atoms' of the English is not more than 250 thousands, i.e., not more than 2^{18}. On the other hand, the number of strings with at most, say, 6 letters is at least $26^2 = 2^{6 \cdot \log 26} > 2^{6 \cdot 4.7} > 2^{28}$ (here 26 is the number of letters in English alphabet). The set of cardinality 2^{18} is tiny with respect to the set of cardinality 2^{28}. If one tries to evaluate the number of phrases, the difference becomes much higher.

But this low density of 'realistic' inputs does not help much without deeper analysis. The particular structure of inputs may help to devise algorithms more efficient over these inputs than the known algorithms over all inputs; there are examples, however not numerous and mainly of more theoretical value. So if one wishes to describe practical inputs in a way that may help to devise efficient algorithms, one should find grammars well aimed at the representation of particular structures of inputs. This point of view does not go along traditional mathematical lines when we look for simple and general descriptions, that are usually too general to be adequate to the computational reality.

The grammar based view of practical inputs may influence theoretical vision of a problem. For example, consider the question of quality of encryption. The main property of any encryption is to be resistant to cryptanalysis. Notice that linguistic arguments play an essential role in practical cryptanalysis. In reality the encryption is not applied to all strings, it mostly deals only with strings produced by this or that natural language, often rather primitive. Thus, there are relations defined over plain texts. E.g., some substrings are related as subject-predicate-direct compliment, etc. A good encryption should not leave traces of these relations in the encrypted text. What does it mean? Different precisions come to mind. A simple example: let P be a predicate of arity 2 defined over plain texts, and its arguments be of small bounded size. Take concrete values A and B of arguments of P. Assume that we introduced a probabilistic measure on all inputs (plain texts), and hence we have a measure of the set S^+ of inputs where $P(A, B)$ holds and of its complement S^-. Now suppose that we have chosen a predicate Q over 'substructures' of encrypted texts (I speak about 'substructures' to underline that the arguments of Q are not necessarily substrings, as for P), again simple to understand. Denote by E^+ the set of encrypted texts for which Q is true for at least one argument and by E^- its complement. The encryption well hides $P(A, B)$ if the measures of all 4 sets $(S^\alpha \cap E^\beta)$, where $\alpha, \beta \in \{+, -\}$, are very 'close'. This example gives only an idea but not a productive definition.

However, in order to find grammars that help to solve efficiently practical problems 'semantical' nature of sets of practical inputs should be studied.

Conclusion

The considerations presented above are very preliminary. The crucial question is to define information convergence of algorithms, not necessarily of general algorithms, but at least of practical ones.

One can imagine also other ways of measuring similarity of traces. We can hardly avoid syntactical considerations when keeping in mind the computational complexity. However semantical issues are crucial, and may be described not only in the terms chosen in this paper.

The analysis of philosophical question of relation of determinism versus uncertainty in algorithmic processes could clarify the methodology to choose. Here algorithmic process is understood at large, not necessarily as executed by a computer. Though the process is often deterministic, and if we adhere to determinism then it is always deterministic, at a given time instant, when it is not yet accomplished, we do not know with certainty the result, though some knowledge has been acquired. The question is: what is or how to formalize the knowledge (information) that the algorithm acquires after each step of its execution?

Acknowledgments

I am thankful to Edward Hirsch and Eugene Asarin for their remarks and questions that stimulated this study. I am also grateful to anonymous referees for their remarks and questions that considerably influenced the final text.

The research was partially supported by French *Agence Nationale de la Recherche* (ANR) under the project EQINOCS (ANR-11-BS02-004) and by the Government of the Russian Federation, Grant 074-U01.

References

[1] D. Burago, Yu. Burago, and S. Ivanov. *A Course in Metric Geometry*, volume 33, Graduate Studies in Mathematics. Americal Mathematical Society, Providence, Rhode Island, 2001.

[2] M. Fischer and M. Rabin. Super-exponential complexity of presburger arithmetic. In *Complexity of Computation, SIAM-ASM Proceedings, vol. 7*, pages 27–41, 1974.

[3] Yu. Gurevich. Evolving algebra 1993: Lipari guide. In E. Börger, editor, *Specification and Validation Methods*, pages 9–93. Oxford University Press, 1995.

[4] A. N. Kolmogorov. On the Logical Foundations of Information Theory and Probability Theory. *Probl. Peredachi Inf.*, 5(3):3–7, 1969. In Russian. English translation in: Problems of Information Transmission, 1969, 5:3, 1–4, or in: Selected Works of A.N. Kolmogorov: Volume III: Information Theory and the Theory of Algorithms (Mathematics and its Applications), Kluwer Academic Publishers, 1992.

[5] A. N. Kolmogorov and V. M. Tikhomirov. ε-entropy and ε-capacity of sets in function spaces. *Uspekhi Mat. Nauk*, 14(2(86)):3–86, 1959. In Russian. English translation in: Selected Works of A.N. Kolmogorov: Volume III: Information Theory and the Theory of Algorithms (Mathematics and its Applications), Kluwer Academic Publishers, 1992.

[6] O. B. Lupanov. The synthesis of contact circuits. *Dokl. Akad.Nauk SSSR*, 119:23–26, 1958.

[7] N. F. G. Martin and J. W. England. *Mathematical Theory of Entropy*. Addison-Wesley, Reading, Massachusetts, 1981.

[8] V. A. Rokhlin. Lectures on the entropy theory of measure-preserving transformations. *Russian Math. Surveys*, 22(5):1–52, 1967.

[9] C. Shannon. The synthesis of two-terminal switching circuits. *Bell System Technical Journal*, 28(1):59–98, 1949.

[10] A. Slissenko. Complexity problems of theory of computation. *Russian Mathematical Surveys*, 36(6):23–125, 1981. Russian original in: *Uspekhi Matem. Nauk*, 36(2):21–103, 1981.

[11] A. Slissenko. On Measures of Information Quality of Knowledge Processing Systems. *Information Sciences: An International Journal*, 57–58:389–402, 1991.

[12] A. Slissenko. On Entropy in Computations. In *Proc. of the 8th Intern. Conf. on Computer Science and Information Technology (CSIT'2011), September 26–30, 2011, Yerevan, Armenia. Organized by National Academy of Science of Armenia*, pages 25–30. National Academy of Science of Armenia, 2011. ISBN 978-5-8080-0797-0.

[13] D. A. Spielman and S.-H. Teng. Smoothed analysis of algorithms: Why the simplex algorithm usually takes polynomial time. *J. ACM*, 51(3):385–463, 2004.

Horizons of Scientific Pluralism: Logics, Ontology, Mathematics

Vladimir L. Vasyukov
Institute of Philosophy, Russian Academy of Science
12/1, ul. Goncharnaya, Moscow 109240, Russia
vasyukov4@gmail.com

Abstract

Discussions on the scientific pluralism typically involve the unity of science thesis, which has been first advanced by Neo-Positivists in the 1930-ies and later widely criticized in the late 1970-ies. In the present paper the problem of scientific pluralism is examined in the context of modern logic, where it became particularly pertinent after the emergence of non-Classical logics. Usual arguments in favor of a unique choice of "the" logical system are of an extralogical nature. The conception of Universal Logic as a theory of mutual translatability and combination of alternative logical systems allows for a more constructive approach to the issue. Logical pluralism gives rise not only to the ontological pluralism but also to non-Classical mathematics based on various non-Classical logics. Our analysis of ontological pluralism rises the following question: is our mathematics globally Classical and locally non-Classical (i.e. having non-Classical parts) or rather, the other way round, is globally non-Classical and only locally Classical? We conclude that in the context of post-non-Classical science the logical pluralism justifies one's freedom to chose logical tools in conformity with one's aims, norms and values.

Keywords: Unity of Science, Scientific Pluralism, Logical Pluralism, Universal Logic, Ontological Pluralism, Non-Classical Mathematics.

1 An issue of the unity of science

Presently many philosophers and scientists are inclined to take a pluralistic position regarding scientific theories or methods. It is a common wisdom that the totality

---

Talk at the conference "Philosophy, Mathematics, Linguistics: Aspects of Interaction 2012" (PhML-2012), held on May 22–25, 2012 at the Euler International Mathematical Institute.

of natural phenomena cannot be possibly explained with a single theory or a single approach. (cf. [14]). Current debates on the scientific pluralism usually involve the 'Unity of Science' thesis first advanced by Neo-Positivists in the 1930-ies. According to this thesis

> Laws and concepts of particular sciences have to belong to the one system and be reciprocally related. They have to form certain unified science with a common system of concepts (common language), separate sciences are just the members of it and their languages are parts of the common language. [15, p. 147–148]

In 1978 Patrick Suppes [27] in his presidential address to the Philosophy of Science Association claimed that the time for defending science against metaphysics (which he took to be the original rationale for the unity of science movement) had passed. Suppes argued that neither the languages of scientific disciplines nor their subject matters were reducible to one language and one subject matter. Nor was there any unity of method beyond the trivially obvious such as use of elementary mathematics.

The majority of philosophers of science were not particularly enthusiastic about Suppes's ideas. A noticeable exception was Nancy Cartwright and her collaborators who stressed the irreducible variety of scientific disciplines involved in solving concrete scientific problems. Later Cartwright [7] elaborated a pluralistic account of a 'dappled world' composed of a number of separate areas. Each particular area of this world is ruled by its own laws, so that this system laws form a loose patchwork, which does not reduce to a single compact system of fundamental laws. A similar view has been put forward by John Dupré [11] who also supports a pluralist metaphysical position called the "promiscuous realism".

One has to distinguish between the pluralism *in* science and the pluralism *about* science. At any stage of their development sciences typically use a variety of different approaches corresponding to different aspects of studied phenomena. They use various representational or classificatory schemes, various explanatory strategies, various models and theories, etc. This is a pluralism *in* science. The pluralism *about* science is a view according to which such a plurality of approaches in science is ineliminable as a matter of principle, and that it does not constitute any deficiency in knowledge. According to this view, an analysis of meta-scientific concepts (such as theory, explanation, evidence) should take into consideration the possibility that in the long run the explanatory and investigative aims of science can be best achieved with a pluralistic science.

Modern scientific monism can be described as follows [14, p. x]:

- the ultimate aim of a science is to establish a single, complete, and comprehensive account of the natural world (or the part of the world investigated by the science) based on a single set of fundamental principles;

- the nature of the world is such that it can, at least in principle, be completely described or explained by such an account;

- there exist, at least in principle, methods of inquiry that if correctly pursued will yield such an account;

- methods of inquiry are to be accepted on the basis of whether they can yield such an account;

- individual theories and models in science are to be evaluated in large part on the basis of whether they provide (or come close to providing) a comprehensive and complete account based on fundamental principles.

Notice that the above description does not imply that the wanted complete theory of everything is necessarily unique. Nevertheless such the uniqueness assumption is often taken for granted.

The Vienna's Circle's thesis of the Unity of Science describes this unity in ontological terms. As Alan Richardson notes, when Rudolf Carnap claims to establish the unity of 'the object domain of science' he

> does this by presenting a language in which all significant scientific discourse can be formulated. Putative metaphysical things such as essences, however, cannot be constructed — that is, they cannot be defined in the language — and this is the fact that Carnap uses to expunge metaphysical talk. Metaphysics does not speak of things in the object domain of science; there is only one such domain, and it contains all the objects that can be referred to, so metaphysics strictly does not speak of anything at all. [23, p. 6]

Carnap adds that

> we can, of course, still differentiate various types of objects if they belong to different levels of the constructional system, or, in case they are on the same level, if their form of construction is different. [6, p. 9]

He gives an example of synthetic geometry where complex constructions are built from basic elements such as points, straight lines, and planes. Such constructions may involve several different layers but all statements about these constructions are ultimately the statements about their basic elements. So we have here different types of objects and yet a unified domain of objects from which they all arise.

The question arises: how big and how independent can be such complexes? It turns out that the "global" monism in the sense of the above definition allows, after all, for a pluralistic picture if one splits it into a number of "local" monisms based on independent complexes. A good example is a situation in today's non-Classical logics to which we now turn.

2 Logical pluralism and logical monism

The Tower of Babel is a cultural pattern, which recurs again and again. The first attempt of its erection, as it is well known, ended up with a catastrophe and produced multiple languages and the lack of understanding between the builders of this monster. However this was not the end of the story. A new Babel Tower dating back to Aristotle and the Stoics was the project of developing a unique and uniform logic supposed to provide rules of correct reasoning for all. This attempt seemed successful throughout the last two thousand years but eventually it failed as a result of the development and proliferation of the so-called non-Classical logics. Some thinkers including Aristotle himself considered certain deviations from the Classical logic earlier but only in the beginning of the 20-th century researches began to explore this new territory systematically. As a result many today's logicians hold a view according to which there exist many alternative systems of logic rather than a single "right" logic. This view is known under the name of *logical formalism*. Although the philosophical analysis of logical pluralism is still in its infancy the soundness of this view is hardly any longer questionable. It is possible that the logical pluralism will point to ways out of some deadend of modern logic and determine a strategy for developing logic in the 21st century. Implications of logical pluralism for the modern also still wait to be studied. In what follows we shall consider some problems of logical and metalogical pluralism and explore their implications for ontology and foundations of mathematics.

It may appear that the logical monism does not need an argumentative defense because it is supported by more then two thousand years of the history of logic. However the situation is not so is simple. Does the Classical logic in some sense imply the logical monism? Or perhaps some non-Classical logic can play the same role of the only "right" logic common for all? The Intuitionistic logic at certain point of history was considered as a candidate for this role. Later were considered some other candidates such as the Relevant logic, which allows one to avoid certain paradoxes appearing in the Classical logic. According to Stephen Read [21], the only purpose of logic is to distinguish between valid and invalid inferences. Hence, the argument goes, there is only one "true" logic, which can be nothing but the Relevant logic.

However if one takes into account how the concept of relevance has been modified in the course of the 20-th century, one can hardly accept this and similar arguments of logical monists. All such arguments are ultimately ethical or aesthetic arguments rather than properly logical. They call for the "lost paradise", from where logics and logicians have been earlier expelled. The existing experience of metalogical researches indicates that there is no logical system satisfying all wanted metalogical properties and free from all paradoxes. As a matter of fact, it is difficult to single out even a short list of universal meta-properties which the ideal logical system of logical monists should necessarily possess.

Earlier R. Carnap [5] put forward the Principle of Tolerance in logic according to which logic should justify conclusions rather than establish some bans. There is no moral in logic and everyone has a liberty of building his or her own system of logic. As a matter of fact Carnap talks about the choice of formal language rather than the choice of logic. As it has been shown by G. Restall [22] one and the same language may admit for different logical consequence relations. So the distinction between language and logic is essential in this context.

Beall and Restall point to the following problem of logical pluralism:

> Which of these many logics governs your reasoning about how many logics there really are? In other words, which logic ought to govern your reasoning about the nature of logic itself? And indeed, which logic ought to govern your reasoning about the nature of logic itself? [1, p. 6]

Indeed, a goal of logical pluralist is to study mutual relationships between the known logical systems. These logical systems can be seen either as a list of candidates for the same role of "the" unique "true" logic or as a friendly "logic community" providing different answers to the same questions. The builders of the Babel Tower eventually lost a common language and a mutual understanding. Does the existing logical community await the same fate?

A basic problem of logical pluralism is the problem of relationships between different systems of logic. How such systems can be compared and evaluated? If we recall that a logical theory is always a theory of some individual domain then the logical pluralism can be understood as the thesis according to which the one and the same domain, generally, admits for several alternative logics. Logical rules do not depend on empirical reliability, they cannot be cancelled because of empirical observations: logic is aprioristic by its very nature. Hartry Field argues [12] that a system of logic accepted a priori can be eventually replaced by an alternative logical system, equally designed a priori, under the pressure of facts. This view qualifies as a sort of *fallibilistic apriorism* (borrowing the term from the philosophy of science). However such a revision of logic can be possibly viewed as a mere recognition of the fact that the old logic simply did not correspond to the studied individual domain. As notices Ottavio Bueno [4] this possibility cannot be ruled out a priori.

3 Logical eclecticism and logical relativism

The logical monism is a dogmatic position. The logical eclecticism, in its turn, is a variety of logical pluralism, which makes a choice of the best logical system from a list of such systems and aims at harmonizing competing approaches. On the other hand it operates like logical monism when it rejects certain moments of known logical systems as "erroneous".

A problem of logic eclecticism, as well as of any other sort of eclecticism, is the arbitrariness of choices: one chooses and uses certain principles without having any general theory justifying the choice. However the choice between logical systems becomes interesting when one translates problems formulated in some given logical framework into a different logical framework. This allows one to look at the given problem from a different viewpoint and sometimes helps to find an unexpected solution.

The same feature belongs to the position called *logical relativism*. Roy Cook describes it as follows: one qualifies as a relativist about a particular phenomenon if and only if one thinks that the correct account of it is a function of some distinct set of facts [9, p. 493]. How many similar correct accounts of the same set of facts can exist in principle? If the answer is that such accounts are multiple then this position reduces to a version of pluralism; of one assumes that there is only one such account then it reduces to monism. In this context Cook distinguishes between the *dependent* and *simple* varieties of pluralism. While former variety of pluralism is based on the relativism the latter is not. It may appear that an obvious example of the dependent pluralism is given by the *Tarskian Relativism* [29] according to which every term in a formal language can be equally treated either as logical or non-logical. But, as Varzi rightly notices, the Tarskian Relativism implies a stronger form of logical relativism according to which different ways of specifying the semantics of terms are equally admissible. It is possible, for example, that you and I agree that identity is a logical constant but you may think that it stands for a transitive relation whereas I may not accept this assumption.

4 Metalogical relativism as the consequence of logical pluralism

Varzi's paper referred to above makes it clear that Tarskian Relativism adds to the logical pluralism a new dimension related to the choice of logical semantics. Each variant of logical semantics comes with its own conception of logical consequence. Indeed, the usual definition of logical consequence – the conclusion follows from the given premises when in every case where the premises are true the consequence is also true – only looks neutral. In fact it involves the concept of truthfulness which depends on the chosen semantics of logical terms. Alternatively one may use in this definition a metaimplication opening thus yet a further dimension of pluralism.

Should be one's metalogic necessarily Classical? Graham Priest, considering Tarski's theory of truth and his T-construction, writes that

> sometimes it is said that Tarskian theory must be based on Classical logics: this logic is required for the construction to be performed. Such a claim is just plain false. It can be carried out in intuitionistic logics, paraconsistent logics, and, in fact, most logics. [20, p. 45]

Thus the Tarskian Relativism turns into the metalogical relativism and the metalogical pluralism. It allows for considering various alternative definitions of logical consequence such as: "the conclusion follows from premises if and only if any case in which each premise is true is also a case in which conclusion is relevantly true" (a case of Relevant metalogic), "the conclusion follows from premises if and only if any case in which each premise is true is also a case in which conclusion is intuitionistically true" (a case of Intuitionistic metalogics), "the conclusion follows from premises if and only if any case in which each premise is true is also a case in which conclusion is paraconsistently true" (a case of Paraconsistent metalogic), "the conclusion follows from premises if and only if any case in which each premise is true is also a case in which conclusion is quantum logically true" (a case of quantum metalogic), etc.

Moreover, apparently nothing prevents one from correlating one's concept of logical consequence with a non-Classical logic. Then the above definition can be modified as follows: "the conclusion intuitionistically follows from premises if and only if any case in which each premise is intuitionistically true is also a case in which conclusion is intuitionistically true" (the case of Intuitionistic logic and metalogic), "the conclusion relevantly follows from premises if and only if any case in which each premise is relevantly true is also a case in which conclusion is relevantly true" (the case of Relevant logic and metalogic), "the conclusion intuitionistically follows from premises if and only if any case in which each premise is relevantly true is also a case in which conclusion is relevantly true" (the case of Relevant metalogic for Intuitionistic logic), "the conclusion relevantly follows from premises if and only if any case in which each premise is intuitionistically true is also a case in which conclusion is intuitionistically true" (the case of Relevant metalogic for Intuitionistic logic), etc. Here the choice may be limited by certain specific properties of these 'cases' [24, p. 396].

Thus we can formulate a "metalogical" definition of logical consequence as follows:

A conclusion is valid in the given logic if in the corresponding metalogic the validity of premises implies the validity of the conclusion.

On this basis it is possible to construe two further different versions of the above metalogical definition:

(i) a conclusion is valid in some logic if in some metalogic the validity of premises implies the validity of the conclusion.

(ii) a conclusion is valid in some logic if in all metalogics the validity of premises implies the validity of the conclusion.

The second version is hardly realistic since all possible metalogics can be hardly taken into account. One may also suspect that the choice of metalogic may depend on the existence of 'translation' from certain logic to the given logic. Indeed, all "mixed" principles arise via a meddling or substituting semantics of one logic to another. These semantic operations may provide grounds for further arguments pro or contra the monistic (when logic always coincides with the metalogic) and (when logic and metalogic may differ conceptually) points of view.

A non-Classical metaimplication gives rise to a meta-metalogical definition of logical consequence as follows:

- A conclusion follows from premises iff the truth of the conclusion follows from the truth of premises iff in all cases the truth of premises implies the truth of the conclusion.

On the one hand, this is a bad infinity. But on the other hand, this situation can be described in terms of S. Kripke's theory of truth [16]:

- A conclusion logically follows from premises if and only if the truth of the conclusion follows from the truth of premises if and only if the truth of the truth of the conclusion follows from the truth of the truth of premises.

- *Mutatis mutandis* in case of the 'mixed' principle. In this case in addition to Kripke's considerations of cases of the truth or falsity at corresponding meta-levels we need also to construe the truth on pluralistic variants of meta-levels.

5 Logical pluralism and universal logics

How statements of the form 'A follows from B iff B is true implies A is true in metalogic $\mathbf{M}$' can be compared in the case of different metalogics? Some authors suggest that this can be done with a theory of *Universal Logic* that would provided criteria for such a comparison (see [32, 33]).The Universal Logic (UL) is a theory of translatability and combination of logical systems. The above statements can be compared with UL as follows. First one constructs a translation F from (meta)logic Y_1 to (meta)logic Y_2. Then

'A follows from B iff B *is true* implies in Y_1 A *is true*'

translates under F into

'A follows from B iff $F(B$ *is true*$)$ implies in Y_2 F (A *is true*)'.

If such translations between different metalogics exist then we can speak about a local metalogical monism: the translatability gives us an invariant kernel preserved through translations.

Instead of linking by means of translation we would consider, using methods of universal logic, the combinations of two formulations, e.g. join of two formulations. In this case join of two logics gives us the uniform logic possessing properties of both initial logics. In particular, in union $Y_1 \oplus Y_2$ of two metalogics Y_1 and Y_2 "joint" consequence relation is defined by means of a condition:

if from A *is true* in one metalogic (Y_1 or Y_2) follows B *is true* in the same metalogic then B *jointly* follows from A (i.e. within the framework of the metalogic $Y_1 \oplus Y_2$).

To put it more precisely "jointly follows" gives us that

- A follows from B iff A *is true* implies in $Y_1 \oplus Y_2$ B *is true*.

Instead of unions of metalogics one can also use their product (taking pairs of metaformulas as new metaformulas), so the definition becomes

if B *multiplicatively* follows from A (i.e. within the framework of the metalogic $Y_1 \otimes Y_2$) then from A *is true* in both metalogics (Y_1 and Y_2) follows B *is true*.

"Multiplicativeness" gives us that

- if A *is true* follows in metalogic $Y_1 \otimes Y_2$ from B *is true* then A follows from B.

Similarly one can consider the *exponential* and *co-exponential* local metalogical monism combining metasystems Y_1, Y_2 into $Y_1 \Rightarrow Y_2$ and $Y_1 \Leftarrow Y_2$ respectively and then use the "implications" of these combined metasystems in the definition of logical consequence of the same form (provided such combinations as allowed in UL).

An obstacle for this project is the omniscience problem: we cannot explicitly describe all possible logics in advance and hence cannot accomplish all possible combinations of logics. The above types of combinations of (meta)logical systems do not exhaust all possible combinations being only the most common ones.

6 From logical to ontological pluralism

According to J. Bocheński, the modern logic is "a most abstract theory of objects whatsoever" or a "physics of the object in general". Thus "logic, as it is now constituted, has a subject matter similar to that of ontology" [3, p. 288].

In effect, ontology is a prolegomenon to logic. While ontology is an informal, intuitive inquiry into the basic properties and basic aspects of entities in general, logic is the systematic, formal, axiomatic elaboration of these ontological intuition. While ontology as it is usually practiced is the most abstract theory of real entities, logic in its present state is the general ontology of both real and ideal entities [3, p. 290].

Thereby logical pluralism is 'dangerous' because it implies the ontological pluralism. Since any logical theory is always a theory of some domain of individuals, the acceptance of this or that logic compels to certain assumptions, hypotheses about the cognizable objects inhabiting this area and described by our theory. It is a good thing if we are in a position to control these assumptions; too often such assumptions remain tacit.

Ontological assumptions are specific to languages – artificial or natural. The term "ontological commitment" that denotes this phenomenon can be understood either as an ontological assumption, or an ontological obligation or as an ontological hypothesis. Scientific artificial languages, which are always designed for a definite purpose, may enforce certain ontological commitments not intended by their designers.

Such troubles are rooted in the fact that formal languages designed for the scientific purposes should cope with two different ontologies, one of which represents the domain of scientific inquiry while the other belongs to the language itself and depends on its formal properties. The history of science of the 20-th century makes it clear that interactions between these two ontological layers cannot be ignored.

How ontological assumptions of a given formal language can be identified? An answer is given by A. Church's criterion: a language carries an ontological commitment associated with every sentence, which is analytic in this language, i.e., of every true sentence whose truth is granted by the semantics of this language. The distinction between analytical and synthetic sentences is made here as follows:

One can single out two types of propositions: propositions, whose truth or falsity should be established on the basis of semantic rules of the system, and propositions, whose truth or falsity cannot not be seen from them. Such division of statements of language in respect to fixed semantic system, division on analytical and synthetic in this sense, in our opinion, is indisputable. The question consists in their exact definition and interpretation. [25, p. 88]

The usual semantics of the first-order Classical logic is given in terms of its Tarskian models. The *universe of all sets* and the related set theory provide in this case the proper ontology for this language. Thus in the case of this particular language the 'theory of objects in general' coincides with some version of set theory (possibly with urelements and empirical predicates, see [8]).

However the set theory is itself an elementary theory, i.e., a set of formal statements deduced from a conservative axiomatic extension of predicate logic with certain non-logical axioms, which describe formal properties of predicate $\in$. By modifying the logical part of this theory one can obtain a new theory based on some non-Classical logic: Paraconsistent, Relevant, Quantum, Fuzzy etc. Thus one obtains a class of non-Classical set-theoretic universes associated with their non-Classical underlying logics.

There is another simple argument supporting the claim that logical pluralism implies the pluralism of universes. Consider usual definitions of operations of join $\cup$, meet $\cap$ and complement / on sets

$x \cup y =_{def} \{a : a \in x \vee a \in y\}$,
$x \cap y =_{def} \{a : a \in x \wedge a \in y\}$,
$x/y =_{def} \{a : a \in x \wedge \neg(a \in y)\}$.

A pluralist may ask: what type of connectives $\vee$ (or), $\wedge$ (and), $\neg$ (it is incorrect, that) are used in these definitions? If these are Classical connectives then the algebra of subsets of a given set is Boolean.

But what happens, if one modifies the operations on sets using non-Classical logic connectives $\vee$, $\wedge$, $\neg$ and then construes an algebra for the obtained new operations? Since in Tarskian models set-theoretic operations are responsible for truth values of formulas this provides us with an interpretation of a non-Classical logic in the Classical universe. In this way one can interpret in the given Classical universe as many non-Classical logics as one wants. One can also use a non-Classical universe and introduce in it Classical set-theoretic operations. So one gets an interpretation of Classical logic (along with non-Classical ones) in a non-Classical universe.

Is there a way to check whether "our" universe is Classical or non-Classical? Logical pluralism gives an answer in negative. One can assume the existence of a global underlying logic for a given universe but this global logic does not determine any set of local logics, which this universe may admit. Of course, we talk about global and local logics in this context only metaphorically as markers fixing a state of affairs.

7 Non-Classical mathematics: as many logics as mathematics

The 20-th century has witnessed how the original intuitionist and constructivist renderings of set theory, arithmetic, analysis, etc. were later accompanied by those based on relevant, paraconsistent, non-contractive, modal, and other non-Classical logical frameworks. This development led to the ongoing scientific program of "Non-Classical Mathematics". At the conference "Non-Classical Mathematics 2009" (June 2009, Hejnice, Czech Republic) the non-Classical Mathematics 2009 has been defined as a study of mathematics which is formalized by means of non-Classical logics. The Program of this conference included the following sections:

- Intuitionistic mathematics: Heyting arithmetic, Intuitionistic set theory, topos-theoretic foundations of mathematics;

- Constructive mathematics: constructive set or type theories, pointless topology;

- Substructural mathematics: Relevant arithmetic, non-contractive naive set theories, axiomatic fuzzy set theories;

- Inconsistent mathematics: calculi of infinitesimals, inconsistent set theories;

- Modal mathematics: arithmetic or set theory with epistemic, alethic, or other modalities, modal comprehension principles, modal treatment of vague objects, modal structuralism.

It is obvious, that there is not one but many true mathematics. But it remains unclear how these different mathematics interact. Are they complementary or mutually exclusive? This situation resembles that with non-Euclidean geometries. This analogy suggests questions like this: is our mathematics globally Classical, and only locally non-Classical or, on the contrary, it is globally non-Classical and locally Classical?

Gaisi Takeuti develops a *quantum set theory*, which involves a quantum-valued universe. It remains however unclear whether the mathematics based on quantum logic has a very rich mathematical content. This is clearly shown by the fact that there are many complete Boolean algebras inside quantum logic. For each complete Boolean algebra $\mathcal{B}$, mathematics based on $\mathcal{B}$ has been shown by our work on Boolean valued analysis to have rich mathematical meaning. Since mathematics based on $\mathcal{B}$ can be considered as a sub-theory of mathematics based on quantum logic, there is no doubt about the fact that mathematics based on quantum logic is very rich. The situation seems to be the following. Mathematics based on quantum logic is too gigantic to see through clearly. [28, p. 303]

Robert Meyer proposes a construction of Relevant arithmetic built along the same 'pluralistic' line on a basis of Relevant logic [18]. Recall that Peano Arithmetic (PA) is based on the first-order Classical logic (FOL) and involves a number of non-logical axioms. Relevant Peano arithmetic R# according to Meyer is obtained from PA via a replacement of FOL by a system of Relevant logic R, leaving the non-logical axioms unchanged.

One more instance of a non-Classical mathematical theory is given by K. Mortensen in his book *Inconsistent Mathematics* [19]. Claiming that "philosophers have hitherto attempted to understand the nature of contradiction, the point however is to change it", Mortensen describes the mathematics based on the Paraconsistent logic.

In a more sophisticated way a non-Classical logical basis is used in theories of formal topology. A topological structure is usually specified via a specification of set of opens closed under the set-theoretical intersection. By modifying the concept of intersection one obtains a family of new topologies. In particular the set-theoretic intersection can be replaced by the operation of monoidal multiplication. Such constructions can be made with a non-Classical set theory interpreted in a Classical universe.

When one accepts logical pluralism and allows for various logical foundations formal topological properties can be equally taken into account. An example of such an account can be found in the Quantum theory (QT). Garrett Birkhoff and John von Neumann demonstrated an equivalence between experimental statements of QT and subspaces of Hilbert spaces. The set-theoretic intersection of two given experimental statements (represented as the closed vector subspaces of Hilbert space) is also an experimental statement (i.e., a closed vector subspace of Hilbert space). Whence one easily defines a topological structure using the standard definition of boundary.

However when one takes into account the fact that the negation of an experimental statement is its orthogonal complementation, one obtains a formal topology, which differs from its Classical counterpart.

Today's mathematics is going through a paradigm shift in its foundations from the set-theoretic paradigm to the category-theoretic one. From a logical point of view Category theory like Set theory is an elementary theory based on the Classical first-order calculus with equality.

Following N. C. A. da Costa, O. Bueno and A. Volkov [10] one can build *Paraconsistent* elementary theory of categories using the paraconsistent logic $C_1^=$. The axioms of the Paraconsistent category theory include all usual axioms with the Classical negation and some new axioms with the paraconsistent negation. One can also construct a Paraconsistent category theory [33] using axioms for category theory proposed by G. Blanc and M.-R. Donnadieu [2].

Recall that *topos* is a category of a special kind in which there exists a special object bearing a structure of Heyting algebra. The above algorithm for developing non-Classical mathematical theories allows one to build various 'quasi-toposes' by replacing Heyting algebra with some other algebras of logics. For example, the replacement of Heyting algebra by the paraconsistent da Costa algebra brings a 'potos' (aka da Costa topos). A potos is a paraconsistent universe in which one can develop paraconsistent mathematical theories just as in the case of the Intuitionistic mathematics. While in the usual topos the paraconsistency features only in special constructions and in this sense remain local artefacts, in a potos the paraconsistency is organic and underlies all further constructions. In the paraconsistent universe the Classical mathematics features as an artefact, i.e. as a local deviation from the paraconsistent regularities.

Similarly one can replace Heyting algebra with the Relevant one and thus obtain a category called 'reltos' which interprets the Relevant logic and allows for developing the Relevant mathematics [34]. This short list does not exhaust all possibilities for developing the non-Classical mathematics.

Toposes, generally, are non-Classical constructions, namely, constructive intuitionistic universes.

> By imposing natural conditions on a topos (extensionality, sections for epics, natural numbers object), we can make it correspond precisely to a model of Classical set theory. Thus, to the extent that set theory provides a foundation for mathematics, so too does topos theory. [13, p. 344]

What a "natural condition" means precisely in this context?

In a topos-theoretic context a Classical universe is a local construction (being a special case of general topos) while the nature of general topos is purely intuitionistic, i.e., essentially non-Classical. Thus the general topos serves as a global non-Classical foundation of mathematics, which can be Classical locally.

Other kinds of non-Classical mathematics can be similarly obtained locally in the same global intuitionistic context. This can be achieved with Lawvere's 'variable sets' aka intensional sets aka "set-theoretical concepts" (R. Goldblatt's terminology). According to Goldblatt, the intension or meaning of a given expression, is an "individual concept expressed by it". For example, if $\varphi(x)$ is the statement 'x is a finite ordinal' then the intension of φ is the *concept* of a finite ordinal. In the categorical language this concept is represented by a functor that assigns to each $p \in P$ a set of things known "at stage p" to be finite ordinals [13, p. 212].

By varying p, one can impose different "natural restrictions" on given sets of individuals and thus obtain set-theoretic concepts, which describe non-Classical sets. In particular, such a variation can be used for interpreting quantum logics in toposes; in this case the obtained set-theoretic concepts characterize quantum sets.

Likewise it is possible to use functor category Set^A from the so-called CN-category (which is a category-theoretic equivalent of da Costa algebra) to category *Set*. This category is a topos. Notice that the completeness of da Costa C^1 paraconsistent system has been proved with respect to a similar topos [30]. A similar approach can be used in the case of Relevant logic R [31].

Presently only a small minority of mathematicians expresses an interest in the non-Classical mathematics (beyond its intuitionistic and constructive varieties, which are related to the theory of computability). There are two reasons for this. First, the non-Classical mathematics so far did not bring anything interesting for the viewpoint of mathematical novelty. Researches in this field still focus on mathematical characteristics of non-Classical logics and their models. This common tendency is evident in spite of some noticeable exceptions (e.g. Kris Mortensen's book 'Inconsistent Mathematics', an attempt by K. Piron to reformulate quantum mechanics on quantum logic foundations). Perhaps the development of interactive non-Classical provers and decision-making systems will be able to make this research filled more vivid. The effectiveness and the convenience of the human-machine interaction may serve as a strong argument in favour of this or that non-Classical mathematics.

Second, there is a danger for non-Classical mathematician to become a 'hero of deserted landscapes'. Polish science-fiction writer Stanisław Lem distinguished between three kinds of genius [17, p. 89]. A genius of the third kind is an ordinary genius who is beyond the intellectual scope of his age. A genius of the second kind is a hard nut, which his contemporaries cannot crack. Such a genius usually gets a postmortem recognition. Geniuses of the first and the highest kind remain wholly unknown – both during their lifetimes and after their deaths. Their intellectual impact is so revolutionary that no one can evaluate it. Lem provides a fictitious historical example of a manuscript by an anonymous Florentian mathematician of 18-th century, which *prima facie* appeared to be a work in Alchemy but at a closer examination turned out to be a project of alternative mathematics, which differed drastically from our mathematics as we know it. Checking whether this alternative mathematics is better or worse than the usual one would require a lifetime work of hundreds of scientists working on the manuscript by the Florence Anonymous in a way similar to which Bolyai, Lobachevsky and Riemann worked on Euclid. In reality most mathematicians simply avoid developing any 'parallel' mathematics.

8 Conclusion

Recent developments in logic support a pluralistic logical picture of the world. Besides, it should not be expected that such situation is true only for logics. The emergence of non-Classical mathematics should not be seen as a supporting evidence for logicism. It should be rather understood as a natural consequence of the internal pluralism of logic which has been made explicit in recent developments. Having in mind D. Hilbert's view according to which logic is a metamathematics one can see that logical pluralism implies the plurality of mathematics, i.e., the plurality of mathematical pictures of the world.

Describing the Classical science Kant famously remarked that "each science is as much a science as much there is mathematics in it". Can one really expect a 'pluralization' of such scientific disciplines as physics and biology along with the pluralization of mathematics? From the Classical point of view the answer should be affirmative. However, we are living in the epoch of post-non-Classical rather than Classical science. For this reason scientific pluralism is limited with a variety of systems of social values and goals, which dictate choices of our research strategies. According to V. S. Stepin

> The post-non-classical type of scientific rationality broadens the field of reflection over activity. It takes into account correlation of obtained knowledge of the object not only with specificity of means and operations of activity, but also with value-goal structures. Here we explicate the connection between intrascience goals and extra-scientific, social values and goals. [26, p. 634]

So the pluralism of the modern logic is rather a precondition of freedom in our choices of logical toolkits, which determines directions of our researches.

The development of logic in the 20-th century made clear that certain metalogical characteristics which were earlier believed to be universal were actually not universal. This concerns, in particular, the completeness and the consistency of logical systems, which make no sense in the case of paraconsistent logical systems (albeit they have such properties as paraconsistency and paracompleteness). Notice that

Relevant logics can be paraconsistent and at the same time consistent and complete. Such facts provide an additional evidence in favour of the post-non-Classical view according to which a logician or a mathematician should select his or her formal toolkit on the basis of certain goals, values and norms.

References

[1] J. C. Beall and G. Restall. Defending Logical Pluralism. In John Woods and Bryson Brown, editors, *Logical Consequence: Rival Approaches. Proceedings of the 1999 Conference of the Society of Exact Philosophy*, pages 1–22. Hermes, Stanmore, 2001.

[2] G. Blanc and M.-R. Donnadieu. Axiomatisation de la catégorie des catégories. *Cah. Topol. Géom. Différent.*, XVII, 2:1–38, 1976.

[3] J. M. Bocheński. Logic and Ontology. *Philosophy East and West* 24, VII(3):275–292, 1974.

[4] O. Bueno. Is Logic A Priori? *The Harvard Review of Philosophy*, XVII:105–117, 2010.

[5] R. Carnap. *The Logical Syntax of Language*. Adams and Co., Littlefield, 1959. Translated by Amethe Smeaton.

[6] R. Carnap. *The Logical Structure of the World and Pseudoproblems in Philosophy*. Open Court Publishing, Chicago and La Salle, Illinois, 2005.

[7] N. Cartwright. *The Dappled World*. Cambridge University Press, Cambridge, 1999.

[8] N. B. Cochiarella. Predication Versus Membership in the Distinction between Logic as Language and Logic as Calculus. *Synthese* 77:37–72, 1988.

[9] R. T. Cook. Let a thousand flowers bloom: A tour of logical pluralism. *Philosophy Compass*, 5(6):492–504, 2010.

[10] N. C. A. da Costa, O. Bueno, and A. Volkov. Outline of a Paraconsistent Category Theory. In P. Weingartner, editor, *Alternative Logics. Do Science Need Them?*, pages 95–114. Springer, Berlin, Heidelberg, New York, 2004.

[11] J. Dupré. *The Disorder of Things: Metaphysical Foundations of the Disunity of Science*. Harvard University Press, Cambridge, Mass., 1993.

[12] H. Field. Epistemological Nonfactualism and the A Prioricity of Logic. *Philosophical Studies* 92:1–24, 1998.

[13] R. Goldblatt. *Topoi. The Categorial Analysis of Logic*. North-Holland, Amsterdam, New York, Oxford, 1984.

[14] S. H. Keller, H. E. Longino, and C. K. Waters. Introduction: The Pluralist Stance. In S. H. Keller, H. E. Longino, and C. K. Waters, editors, *Scientific Pluralism (Minnesota studies in the philosophy of science; 19)*, pages vii–xxix, University of Minnesota Press, Minneapolis, 2006.

[15] V. Kraft and Der Wiener Kreis. *Der Ursprung des Neopositivismus*. Springer-Verlag, Wien, New York, 1968.

[16] S. Kripke. Outline of Truth Theory. *The Journal of Philosophy*, 72(9):695–717, 1975.

[17] S. Lem. *Doskonała próżnia. Wielkośćurojona.* Wydawnictwo Literackie, Krakow, 1974.

[18] R. K. Meyer. Relevant arithmetic. *Bulletin of the Section of Logic*, 5:133–137, 1976.

[19] K. Mortensen. *Inconsistent Mathematics*. Kluwer, Dordrecht, 1995.

[20] G. Priest. *Doubt Truth to be a Liar*. Clarendon Press, Oxford, 2008.

[21] S. Read. Monism: the one true logic. In D. De Vidi and T. Kenyon, editors, *A Logical Approach to Philosophy: Essays in Honour of Graham Solomon*, pages 193–209, Springer, 2006.

[22] G. Restall. Carnap's Tolerance, Meaning and Logical Pluralism. *Journal of Philosophy* 99:426–443, 2002.

[23] A. W. Richardson. The Many Unities of Science: Politics, Semantics, and Ontology. In *Scientific Pluralism (Minnesota studies in the philosophy of science; 19)*, pages 1–25. University of Minnesota Press, Minneapolis, 2006.

[24] R. Routley and R. Meyer. Semantics of Entailment. In H. Leblanc, editor, *Truth, Syntax and Modality*, pages 199–243. Amsterdam London, 1973.

[25] E. D. Smirnova. Analiticheskaya istinnost' (An Analytical Truth). *Metodologicheskiye aspekty kognitivnykh protsessov (Vychislitel'nye sistemy, 172)*, pages 74–134. Novosibirsk, 2002. In Russian.

[26] V. S. Stepin. *Theoretical Knowledge*. Synthese Library, volume 326. Springer, 2005.

[27] P. Suppes. The Plurality of Science. In Peter Asquith and Ian Hacking, editors, *PSA 1978: Proceedings of the 1978 Biennial Meeting of the Philosophy of Science Association*, vol. 2, pages 3–16. Philosophy of Science Association, East Lansing, Mich., 1978.

[28] G. Takeuti. Quantum Set Theory. In S. Beltrametti and B. van Fraassen, editors, *Current Issues on quantum logic*, pages 303–322. Plenum, New York London, 1981.

[29] A. C. Varzi. On Logical Relativity. *Philosophical Issues*, 10:197–219, 2002.

[30] V. L. Vasyukov. Paraconsistency in Categories. In D. Batens, C. Mortensen, G. Priest, and J.-P. van Bendegem, editors, *Frontiers of Paraconsistent Logic*, pages 263–278, Research Studies Press Ltd., Baldock, Hartfordshire, England, 2000.

[31] V. L. Vasyukov. Paraconsistency in Categories: Case of Relevant Logic. *Studia Logica*, vol. 98, 3:429–443, 2011.

[32] V. L. Vasyukov. Structuring the Universe of Universal Logic. *Logica Universalis*, vol. 1, 2:277–294, 2007.

[33] V. L. Vasyukov. Logicheskiy pluralism i neklassicheskaya teoriya kategoriy (Logical Pluralism and Non-Classical Category Theory). *Logicheskiye Issledovaniya*, issue 18, pages 60–76. Tsentr gumanitarnych initsiativ, Moscow St. Petersburg, 2012. In Russian.

[34] V. L. Vasyukov. Reltoses Semantics for Relevant Logic. In *Devyatye Smirnovskiye chteniya po logike. Materialy mezhdunarodnoy nauchnoy konferentsiyi, 17-19 iyunia 2015*, pages 13–15, Sovremennye tetradi, Moscow, 2015.

Received 29 May 2016

Contents

Reminiscences

In Memoriam: Grigori E. Mints, 1939–2014 — 813
Solomon Feferman and Vladimir Lifschitz

Grigori Mints, a Proof Theorist in the USSR: Some Personal Recollections in Scientific Context — 817
Sergei Soloviev

About Grisha Mints — 841
Anatoly Vershik

Articles

Kant's Logic Revisited — 845
Theodora Achourioti and Michiel van Lambalgen

An Ordinal-free Proof of the Complete Cut-elimination Theorem for $\Pi^1_1 - CA + BI$ with the ω-rule — 867
Ryota Akiyoshi

Model Theory of Some Local Rings — 885
Paola D'Aquino and Angus Macintyre

On the Complexity of Translations from Classical to Intuitionistic Proofs — 901
Matthias Baaz and Alexander Leitsch

Unification for Multi-agent Temporal Logics with Universal Modality — 939
S. I. Bashmakov, A. V. Kosheleva and V. V. Rybakov

Implicit Dynamic Function Introduction and Ackermann-like Function Theory — 955
Marcos Cramer

Locology and Localistic Logic: Mathematical and Epistemological Aspects — 967
Michel De Glas

Secure Multiparty Computation without One-way Functions — 993
Dima Grigoriev and Vladimir Shpilrain

IF Logic and Linguistic Theory — 1011
Jaakko Hintikka

Commentary on Jaakko Hintikka's "IF Logic and Linguistic Theory" — 1023
Gabriel Sandu

Medieval Modalities and Modern Method: Avicenna and Buridan — 1029
Wilfrid Hodges and Spencer Johnston

On the Existence of Alternative Skolemization Methods — 1075
Rosalie Iemhoff

The Logical Cone — 1087
Reinhard Kahle

Parsing and Generation as Datalog Query Evaluation — 1103
Makoto Kanazawa

Foundations as Superstructure
(*Reflections of a practicing mathematician*) — 1213
Yuri I. Manin

Classical and Intuitionistic Geometric Logic — 1227
Grigori Mints

Commentary on Grigori Mints' "Classical and Intuitionistic Geometric Logic" — 1235
Roy Dyckhoff and Sara Negri

Proof Theory for Non-normal Modal Logics: The Neighbourhood Formalism and Basic Results — 1241
Sara Negri

Constructive Temporal Logic, Categorically — 1287
Valeria de Paiva and Harley Eades III

The Historical Role of Kant's Views on Logic — 1311
Andrei Patkul

The Logic for Metaphysical Conceptions of Vagueness — 1333
Francis Jeffry Pelletier

A Note on the Axiom of Countability — 1351
Graham Priest

Topologies and Sheaves Appeared as Syntax and Semantics of Natural Language — 1357
Oleg Prosorov

Long Sequences of Descending Theories and other Miscellanea on Slow Consistency — 1411
Michael Rathjen

Venus Homotopically — 1427
Andrei Rodin

On a Combination of Truth and Probability: Probabilistic IF Logic — 1447
Gabriel Sandu

Towards Analysis of Information Structure of Computations — 1457
Anatol Slissenko

Horizons of Scientific Pluralism: Logics, Ontology, Mathematics — 1477
Vladimir Vasyukov

ISBN 978-1-84890-240-4

www.ingramcontent.com/pod-product-compliance
Lightning Source LLC
Chambersburg PA
CBHW081836230426
43669CB00018B/2731